Joanna Parish (Ed.)

Tumour Viruses

MDPI

This book is a reprint of the special issue that appeared in the online open access journal *Viruses* (ISSN 1999-4915) in 2014 (available at: http://www.mdpi.com/journal/viruses/special_issues/tumour_viruses).

Guest Editor
Joanna Parish
University of Birmingham
UK

Editorial Office
MDPI AG
Klybeckstrasse 64
Basel, Switzerland

Publisher
Shu-Kun Lin

Managing Editor
Delia Mihaila

1. Edition 2015

MDPI • Basel • Beijing • Wuhan

ISBN 978-3-03842-151-1 (Hbk)
ISBN 978-3-03842-152-8 (PDF)

Table of Contents

List of Contributors

Ibrahim Abdulsalam: University of Tromsø, Faculty of Health Sciences, Institute of Medical Biology, NO-9037 Tromsø, Norway.

Shahrzad Akbari: MRC-University of Glasgow Centre for Virus Research, Institute of Infection, Immunity and Inflammation, College of Medical, Veterinary and Life Sciences, University of Glasgow, Garscube Estate, Glasgow G61 1QH, Scotland, UK.

Lawrence Banks: International Center for Genetic Engineering and Biotechnology, Area Science Park, Padriciano 99, Trieste 34149, Italy.

Jesus Omar Muñoz Bello: Instituto de Investigaciones Biomédicas, Universidad Nacional Autónoma de México, Av. Universidad 3000, Col. Ciudad Universitaria, Del. Coyoacán, México DF CP. 04510, Mexico.

Jennifer Biryukov: Department of Microbiology and Immunology, The Pennsylvania State University, 500 University Drive, Hershey, PA 17033, USA.

David J. Blackbourn: School of Biosciences and Medicine, University of Surrey, Surrey GU2 7XH, UK.

George Eric Blair: School of Molecular and Cellular Biology, Faculty of Biological Sciences and Astbury Centre for Structural Molecular Biology, University of Leeds, Leeds LS2 9JT, UK.

Melissa J. Blumenthal: Institute of Infectious Disease and Molecular Medicine; Division of Medical Biochemistry; MRC/UCT Receptor Biology Research Unit, Faculty of Health Sciences, University of Cape Town, Observatory 7925, South Africa.

Justyna Broniarczyk: International Center for Genetic Engineering and Biotechnology, Area Science Park, Padriciano 99, Trieste 34149, Italy.

Özlem Cesur: School of Molecular and Cellular Biology, Faculty of Biological Sciences and Astbury Centre for Structural Molecular Biology, University of Leeds, Leeds LS2 9JT, UK.

Lu Dai: Research Center for Translational Medicine and Key Laboratory of Arrhythmias of the Ministry of Education of China, East Hospital, Tongji University School of Medicine, 150 Jimo Road, Shanghai 200120, China.

Tina Dalianis: Department of Oncology-Pathology, Karolinska Institutet, Cancer Center Karolinska R8:01, Karolinska University Hospital, 171 76 Stockholm, Sweden.

Li Dong: MRC-University of Glasgow Centre for Virus Research, Institute of Infection, Immunity and Inflammation, College of Medical, Veterinary and Life Sciences, University of Glasgow, Garscube Estate, Glasgow G61 1QH, Scotland, UK.

John Doorbar: Division of Virology, Department of Pathology, University of Cambridge, Tennis Court Road, Cambridge CB2 1QP, UK; The Francis Crick Institute Mill Hill Laboratory, the Ridgeway, Mill Hill, London NW7 1AA, UK.

Penelope Duerksen-Hughes: Department of Basic Science, Loma Linda University, Loma Linda, CA 92354, USA.

Michael Edward: Dermatology, School of Medicine, College of Medical, Veterinary and Life Sciences, University of Glasgow, Glasgow G12 8TT, Scotland, UK.

Nagayasu Egawa: Division of Virology, Department of Pathology, University of Cambridge, Tennis Court Road, Cambridge CB2 1QP, UK; The Francis Crick Institute Mill Hill Laboratory, the Ridgeway, Mill Hill, London NW7 1AA, UK.

Kiyofumi Egawa: Department of Dermatology, The Jikei University School of Medicine, 3-25-8 Nishi-shinbashi, Minato-ku, Tokyo 105-0003, Japan; Department of Microbiology, Kitasato University School of Allied Health and Sciences, 1-15-1 Kitasato, Sagamihara, Kanagawa 252-0373, Japan.

Maria Filippova: Department of Basic Science, Loma Linda University, Loma Linda, CA 92354, USA.

Ezequiel M. Fuentes-Pananá: Unidad de Investigación en Virología y Cáncer, Hospital Infantil de México Federico Gómez. Dr. Márquez 162, Col. Doctores, C.P. 06720. México D.F., Mexico.

Ketaki Ganti: International Center for Genetic Engineering and Biotechnology, Area Science Park, Padriciano 99, Trieste 34149, Italy.

Mariana Fuentes Gonzalez: Instituto de Investigaciones Biomédicas, Universidad Nacional Autónoma de México, Av. Universidad 3000, Col. Ciudad Universitaria, Del. Coyoacán, México DF CP. 04510, Mexico.

Sheila V. Graham: MRC-University of Glasgow Centre for Virus Research, Institute of Infection, Immunity and Inflammation, College of Medical, Veterinary and Life Sciences, University of Glasgow, Garscube Estate, Glasgow G61 1QH, Scotland, UK.

Roger J. Grand: School of Cancer Sciences, College of Medicine and Dentistry, University of Birmingham, Birmingham B15 2TT, UK.

Heather Griffin: Division of Virology, Department of Pathology, University of Cambridge, Tennis Court Road, Cambridge CB2 1QP, UK; The Francis Crick Institute Mill Hill Laboratory, the Ridgeway, Mill Hill, London NW7 1AA, UK.

Helen Groves: School of Molecular and Cellular Biology, Faculty of Biological Sciences and Astbury Centre for Structural Molecular Biology, University of Leeds, Leeds LS2 9JT, UK.

Nathalie Grün: Department of Oncology-Pathology, Karolinska Institutet, Cancer Center Karolinska R8:01, Karolinska University Hospital, 171 76 Stockholm, Sweden.

Patrick Hearing: Department of Molecular Genetics and Microbiology, Stony Brook University, Stony Brook, NY 11794-5222, USA.

Andrew D. Hislop: School of Cancer Sciences, the College of Medicine and Dentistry, University of Birmingham, Birmingham B15 2TT, UK.

Malcolm B. Hodgins: Dermatology, School of Medicine, College of Medical, Veterinary and Life Sciences, University of Glasgow, Glasgow G12 8TT, Scotland, UK.

Robert Hollingworth: School of Cancer Sciences, College of Medicine and Dentistry, University of Birmingham, Birmingham B15 2TT, UK.

Katherine L. Hussmann: Laboratory of Viral Diseases, National Institute of Allergy and Infectious Diseases, National Institutes of Health, Bethesda, MD 20892, USA.

Scott R. Johnstone: Institute of Cardiovascular and Medical Sciences, College of Medical, Veterinary and Life Sciences, University of Glasgow, Glasgow G12 8TT, Scotland, UK.

Arieh A. Katz: Institute of Infectious Disease and Molecular Medicine; Division of Medical Biochemistry; MRC/UCT Receptor Biology Research Unit, Faculty of Health Sciences, University of Cape Town, Observatory 7925, South Africa.

Tamar Kleinberger: Department of Microbiology, Faculty of Medicine, Technion—Israel Institute of Technology, 1 Efron St., Bat Galim, Haifa 31096, Israel.

Laurie T. Krug: Department of Molecular Genetics and Microbiology, Stony Brook University, Stony Brook, NY 11794-5222, USA.

Marcela Lizano: Instituto de Investigaciones Biomédicas, Universidad Nacional Autónoma de México, Av. Universidad 3000, Col. Ciudad Universitaria, Del. Coyoacán, México DF CP. 04510, Mexico.

Alasdair I. MacDonald: MRC-University of Glasgow Centre for Virus Research, Institute of Infection, Immunity and Inflammation, College of Medical, Veterinary and Life Sciences, University of Glasgow, Garscube Estate, Glasgow G61 1QH, Scotland, UK.

Jamel Mankouri: School of Molecular and Cellular Biology, Faculty of Biological Sciences and Astbury Centre for Structural Molecular Biology, University of Leeds, Leeds LS2 9JT, UK.

Wiem Manoubi: International Center for Genetic Engineering and Biotechnology, Area Science Park, Padriciano 99, Trieste 34149, Italy.

Paola Massimi: International Center for Genetic Engineering and Biotechnology, Area Science Park, Padriciano 99, Trieste 34149, Italy.

Alison A. McBride: Laboratory of Viral Diseases, National Institute of Allergy and Infectious Diseases, National Institutes of Health, Bethesda, MD 20892, USA.

Caleb C. McKinney: Laboratory of Viral Diseases, National Institute of Allergy and Infectious Diseases, National Institutes of Health, Bethesda, MD 20892, USA.

Craig Meyers: Department of Microbiology and Immunology, The Pennsylvania State University, 500 University Drive, Hershey, PA 17033, USA.

Suruchi Mittal: International Center for Genetic Engineering and Biotechnology, Area Science Park, Padriciano 99, Trieste 34149, Italy.

Ugo Moens: University of Tromsø, Faculty of Health Sciences, Institute of Medical Biology, NO-9037 Tromsø, Norway.

Abigail Morales-Sánchez: Unidad de Investigación en Virología y Cáncer, Hospital Infantil de México Federico Gómez. Dr. Márquez 162, Col. Doctores, C.P. 06720. México D.F., Mexico; Programa de Doctorado en Ciencias Biomédicas, Facultad de Medicina, Universidad Nacional Autónoma de Mexico. Av. Universidad 3000, Col. Universidad Nacional Autónoma de México, C.U., C.P. 04510 México D.F., México.

Clare Nicol: School of Molecular and Cellular Biology, Faculty of Biological Sciences and Astbury Centre for Structural Molecular Biology, University of Leeds, Leeds LS2 9JT, UK.

Leslie Olmedo Nieva: Facultad de Química, Universidad Nacional Autónoma de México, Av. Universidad 3000,Col. Ciudad Universitaria, Del. Coyoacán, México DF CP. 04510, Mexico.

Adriana Contreras Paredes: Unidad de Investigación Biomédica en Cáncer/Instituto Nacional de Cancerología-Instituto de Investigaciones Biomédicas, Universidad Nacional Autónoma de México, Av. San Fernando 22, Col. Sección XVI, Del. Tlalpan, México DF CP. 14080, Mexico.

Joanna L. Parish: School of Cancer Sciences, University of Birmingham, Edgbaston, Birmingham, B15 2TT, UK.

Ieisha Pentland: School of Cancer Sciences, University of Birmingham, Edgbaston, Birmingham, B15 2TT, UK.

Francesca Peruzzi: Neurological Cancer Research, Stanley S. Scott Cancer Center, Department of Medicine, Louisiana State University Health Sciences Center, 1700 Tulane Ave., New Orleans, LA 70112, USA.

David Pim: International Center for Genetic Engineering and Biotechnology, Area Science Park, Padriciano 99, Trieste 34149, Italy.

Zhiqiang Qin: Research Center for Translational Medicine and Key Laboratory of Arrhythmias of the Ministry of Education of China, East Hospital, Tongji University School of Medicine, 150 Jimo Road, Shanghai 200120, China; Department of Microbiology, Immunology and Parasitology, Louisiana State University Health Sciences Center, Louisiana Cancer Research Center, 1700 Tulane Ave., New Orleans, LA 70112, USA.

Torbjörn Ramqvist: Department of Oncology-Pathology, Karolinska Institutet, Cancer Center Karolinska R8:01, Karolinska University Hospital, 171 76 Stockholm, Sweden.

Kashif Rasheed: University of Tromsø, Faculty of Health Sciences, Institute of Medical Biology, NO-9037 Tromsø, Norway.

Krzysztof Reiss: Neurological Cancer Research, Stanley S. Scott Cancer Center, Department of Medicine, Louisiana State University Health Sciences Center, 1700 Tulane Ave., New Orleans, LA 70112, USA.

Georgia Schäfer: Institute of Infectious Disease and Molecular Medicine; Division of Medical Biochemistry, Faculty of Health Sciences, University of Cape Town, Observatory 7925, South Africa.

George L. Skalka: School of Cancer Sciences, the College of Medicine and Dentistry, University of Birmingham, Birmingham B15 2TT, UK.

Thomas Stamminger: Institute for Clinical and Molecular Virology, University of Erlangen-Nuremberg, Schlossgarten 4, Erlangen 91054, Germany.

Grant S. Stewart: School of Cancer Sciences, the College of Medicine and Dentistry, University of Birmingham, Birmingham B15 2TT, UK.

Nicola J. Stonehouse: School of Molecular and Cellular Biology, Faculty of Biological Sciences and Astbury Centre for Structural Molecular Biology, University of Leeds, Leeds LS2 9JT, UK.

Peng Sun: Feinberg School of Medicine, North Western University, Chicago, IL 60611, USA.

Baldur Sveinbjørnsson: University of Tromsø, Faculty of Health Sciences, Institute of Medical Biology, NO-9037 Tromsø, Norway.

Anita Szalmas: International Center for Genetic Engineering and Biotechnology, Area Science Park, Padriciano 99, Trieste 34149, Italy.

Jayashree Thatte: International Center for Genetic Engineering and Biotechnology, Area Science Park, Padriciano 99, Trieste 34149, Italy.

Miranda Thomas: International Center for Genetic Engineering and Biotechnology, Area Science Park, Padriciano 99, Trieste 34149, Italy.

Vjekoslav Tomaić: International Center for Genetic Engineering and Biotechnology, Area Science Park, Padriciano 99, Trieste 34149, Italy.

Bart Tummers: Department of Clinical Oncology, Leiden University Medical Center, Albinusdreef 2, 2333 ZA Leiden, The Netherlands.

Sjoerd H. van der Burg: Department of Clinical Oncology, Leiden University Medical Center, Albinusdreef 2, 2333 ZA Leiden, The Netherlands.

Elizabeth I. Vink: Department of Molecular Genetics and Microbiology, Stony Brook University, Stony Brook, NY 11794-5222, USA; Current address: Department of Microbiology, New York University Langone Medical Center, New York, NY 10016, USA.

Sonia N. Whang: Department of Basic Science, Loma Linda University, Loma Linda, CA 92354, USA.

Rukhsana Yeasmin: Department of Computer Science, Stony Brook University, Stony Brook, NY 11794, USA.

Leticia Rocha Zavaleta: Instituto de Investigaciones Biomédicas, Universidad Nacional Autónoma de México, Av. Universidad 3000, Col. Ciudad Universitaria, Del. Coyoacán, México DF CP. 04510, Mexico.

Yueting Zheng: Department of Molecular Genetics and Microbiology, Stony Brook University, Stony Brook, NY 11794-5222, USA.

About the Guest Editor

Dr. Joanna Parish completed her Ph.D. at the University of Bristol in 2002, where she elucidated the mechanism of the induction of apoptosis by the HPV E2 protein under the supervision of Dr. Kevin Gaston. Following her Ph.D. studies, she moved to the University of Massachusetts Medical School, USA, to work with Professor Elliot Androphy. During this time, Joanna continued her study of the papillomavirus E2 protein and discovered several novel virus–host interactions including an interaction with the cellular DNA helicase ChlR1. Her work elegantly demonstrated that ChlR1 is required for the maintenance and persistence of papillomavirus by tethering viral genomes to cellular chromatin during mitosis and was published in Molecular Cell. In 2007, Joanna was awarded a highly competitive Royal Society University Research Fellowship and returned to the UK to establish her independent research group at the University of St Andrews. Joanna was recruited to the University of Birmingham, UK, in 2012 and is Group Leader of the HPV Persistence Group. Joanna is a passionate mentor of early career researchers and, as an elected member of the Microbiology Society Virus Division, continues to promote virology on the national and international stage.

Dr. Parish's research is currently focused on the molecular biology of HPV persistence and the study of virus–host interactions that are important for transcriptional control and persistence of HPV and other small DNA viruses. Her recent research has identified an essential role for the host transcription factor CTCF in the control of differentiation-dependent virus gene expression and she continues to address long-standing questions about HPV gene expression control and replication using novel and state-of-the-art technologies.

Preface

Current worldwide estimates suggest that nearly 11% of all cancers are caused by viral infections. At present, there are eight viruses that have a strong association with cancer development namely, human papillomavirus, Epstein-Barr virus, Kaposi's sarcoma-associated herpes virus, human T-cell lymphotrophic virus type I, Merkel cell polyomavirus, hepatitis B and C viruses and human immunodeficiency virus. Some of these viruses and associated cancers, such as human papillomavirus and cervical cancer, are well studied and the causal link between infection and cancer development established. However, the involvement of these known oncogenic viruses in cancer development at other body sites is not well understood and further study of these viruses continues to highlight novel mechanisms of cellular transformation. Other cancer-associated viruses are only recently discovered, such as Merkel cell polyomavirus, and further work is required to formally prove their role in cancer development.

The articles published in this Special Issue cover many of the known mechanisms of carcinogenesis of these oncogenic viruses and raise important questions about the known unknowns. Review articles cover Merkel cell polyomavirus driven disease, epithelial cancers caused by human papillomavirus, pathogenesis of Kaposi's sarcoma associated herpesvirus and some mechanistic insights into host cell manipulation by adenovirus. Several articles describe how these viruses target the host cell DNA damage response to support viral replication and utilize host cell transcriptional regulators to control virus gene expression. In addition, the ability of human papillomavirus to target cellular tumor suppressors and immune pathways is summarized.

I would like to thank all of the contributors who used their valuable time to write excellent articles and summarize their respective areas of expertise for this Special Issue.

Joanna Parish, Ph.D.
Royal Society University Research Fellow and Deputy Director of the
Centre for Human Virology at the University of Birmingham, UK
Guest Editor

Human Viruses and Cancer

Abigail Morales-Sánchez and Ezequiel M. Fuentes-Panamá

Abstract: The first human tumor virus was discovered in the middle of the last century by Anthony Epstein, Bert Achong and Yvonne Barr in African pediatric patients with Burkitt's lymphoma. To date, seven viruses -EBV, KSHV, high-risk HPV, MCPV, HBV, HCV and HTLV1- have been consistently linked to different types of human cancer, and infections are estimated to account for up to 20% of all cancer cases worldwide. Viral oncogenic mechanisms generally include: generation of genomic instability, increase in the rate of cell proliferation, resistance to apoptosis, alterations in DNA repair mechanisms and cell polarity changes, which often coexist with evasion mechanisms of the antiviral immune response. Viral agents also indirectly contribute to the development of cancer mainly through immunosuppression or chronic inflammation, but also through chronic antigenic stimulation. There is also evidence that viruses can modulate the malignant properties of an established tumor. In the present work, causation criteria for viruses and cancer will be described, as well as the viral agents that comply with these criteria in human tumors, their epidemiological and biological characteristics, the molecular mechanisms by which they induce cellular transformation and their associated cancers.

Reprinted from *Viruses*. Cite as: Morales-Sánchez, A.; Fuentes-Panamá, E.M. Human Viruses and Cancer. *Viruses* **2014**, *6*, 4047-4079.

1. Introduction: Historical and Epidemiological Aspects

The first observations about a possible infectious etiology of cancer arose at the beginning of the past century. Ellermann and Bang in 1908 and Rous in 1911 transmitted avian leukemias and sarcomas, respectively, through cell-free tumor extracts, suggesting a viral etiology [1–3]. About 50 years later, the first human tumor virus was discovered. Sir Anthony Epstein, Bert Achong and Yvonne Barr observed viral particles in cell cultures from equatorial African pediatric patients with Burkitt's lymphoma; this virus was named Epstein Barr virus (EBV) in honor of their discoverers [4]. In the following years, a set of experimental evidence demonstrated that EBV was the causative agent of endemic Burkitt's lymphoma and other neoplasias. Currently, there is clear evidence that several viruses are oncogenic to humans and the first century of tumor virology research has culminated with the Medicine Nobel Price granted to Harald zur Hausen for the discovery of HPV as the causative agent of cervical cancer [5,6]. To date, EBV, Kaposi's sarcoma-associated herpesvirus (KSHV), human high-risk papillomaviruses (HPV), Merkel cell polyomavirus (MCPV), hepatitis B virus (HBV), hepatitis C virus (HCV) and Human T-cell Lymphotropic virus type 1 (HTLV1) have been classified as type 1 carcinogenic agents (the most strongly associated with human cancers) by the International Agency for Research on Cancer (IARC) (reviewed in [7]). It is estimated that infections are responsible for up to 15% of cancer cases worldwide and about 20% in developing countries [8]. With advent of new technologies allowing genetic identification, it is very likely that this numbers will continue to increase.

Virus-mediated oncogenesis results from the cooperation of multiple events, including different mechanisms bound to the viral life cycle. The knowledge derived from the study of tumor viruses has allowed the construction of a conceptual biological framework to understand not only cancers of infectious origin but also of almost any type of cancer. However, to change the traditional scientific thinking to accept the participation of infectious agents in cancer was difficult, mostly because the biological processes involved do not adjust to the causation dogmatic principles postulated by Koch [9] (Table 1). Koch original observations about the transmission of acute infectious agents are difficult to apply to cancer because of the multi-factorial nature of cancer and because tumorigenic viruses are generally present in a large part of the population without causing disease. Sir Austin Bradford Hill's epidemiologic causation criteria, which were originally proposed to establish the causation between smoking and lung cancer, are more suitable as a base to infer a causative relationship between a viral infection and cancer (Table 1) [10].

Table 1. Koch and Bradford Hill's postulates for causative relations.

Henle Koch's Postulates [9]	Bradford Hill's Causative Principles [10]
1. The pathogen agent must be present in sick population and absent in healthy population. 2. The agent must not appear randomly in another disease. 3. The agent can be isolated and cultured from a diseased organism and should cause disease when introduced into a healthy organism. 4. The agent isolated from the new host should be identical to the original causative agent.	1. Strength of the association. The agent must be more common in cases than in healthy controls. 2. Consistency. Different researchers must corroborate the association. 3. Specificity. The disease must coexist with the agent in the same space, preferably over other associations of the same agent and another disease. 4. Temporality. Exposure to the agent must predict the appearance of the disease. 5. Gradient. A higher exposure must correlate with a higher probability to develop the disease. 6. Plausibility. The association must be founded on the known biological aspects of the causative agent. 7. Coherence. The association must be based on the known aspects of the disease. 8. Experimental. Controlled conditions must reproduced and coincide with the disease and the blockage of biological mechanisms involved must reduce or prevent its appearance. 9. Analogy. Agents with similar mechanisms must be associated to similar diseases.

It is also accepted that none of the Bradford Hill's criteria could by itself conclude causation, neither it is necessary to comply with all of them to accept the virus-cancer association. For example, the geographic distribution of endemic Burkitt's lymphoma (equatorial Africa) does not coincide with the world distribution of EBV. However, we know today that malaria, endemic to this region, is a critical co-factor to develop Burkitt's lymphoma (reviewed in [11]). The Bradford Hill criteria applied to virus and cancer associations consider that causation is established if the virus is present in the tumor cells and not in the surrounding healthy tissue and if there exists plausibility and coherence between infection and cancer. For example, EBV resides in B-lymphocytes that reactivate in the epithelium of the upper digestive tract and EBV has been associated to B-cell lymphomas and carcinomas in tongue, nasopharynx and stomach. Also, transgenic animals that express EBV latent proteins develop neoplasias [12]. These combined data grant a minimal context for biologic plausibility and coherence required by the Bradford Hill criteria.

Arguable, the most powerful tool to indicate direct association is the viral monoclonal analysis in the tumor; the presence of a specific viral variant or viral quasispecie in all tumor cells indicates that the event of infection preceded the malignant cell transformation. This strongly supports that the virus was part of the initial genetic lesion that allowed the appearance of the cancerous clone, satisfying the Bradford Hill criteria for temporality.

2. General Principles of Viral Oncogenic Mechanisms

Oncogenic viruses generally maintain chronic infections in which there is not or little production of viral particles, and that last for the whole life of the infected individual. These mechanisms of viral persistency and/or latency are biologically compatible with the carcinogenic process, because they avoid cell death most common in acute lytic infections, while maintaining the infectious agent hidden from the immune system. Viral persistence in the host is achieved by integrating the viral genome into the cell genome or by expressing viral proteins that equally segregate the viral genome into daughter cells during cell partitioning. Both mechanisms ensure that the virus is not lost during cellular replication. Viral persistence is usually characterized by expression of proteins that control cell death and proliferation; in this manner, oncogenic viruses nurture infection of a controlled number of cells establishing a balance between virus and host, preserving the integrity of both. Cell transformation is probably not an evolutionary viral strategy, but rather a biological accident that rarely occurs in the virus-host interaction. Cancer leads to the death of the host, and thus, it also represents the end of the virus. The existence of viral oncogenes is explained as part of the viral persistence mechanisms, which only under altered conditions may lead to cancer. All virus-associated tumors result from the cooperation of various events, involving more than persistent infection and viral transformation mechanisms. Additional oncogenic hits are necessary for full-blown transformation. The occurrence of mutations impairing expression and function of viral and/or cellular oncogenes is necessary in the carcinogenic process, in line with that, an increased mutation rate of infected over normal cells is frequently observed (reviewed in [13,14]). In this scenario, latently infected cells by oncogenic viruses might be more susceptible targets of additional oncogenic hits; e.g., due to smoking, a diet scarce in fruits and vegetables or/and increased exposure to environmental oncogenic agents. All these insults, plus the host genetic component driving inflammatory responses triggered by the infection itself result in cell transformation and cancer development.

2.1. Direct and Indirect Viral Carcinogenesis

Infectious agents can contribute to carcinogenesis by direct and/or indirect mechanisms (Figure 1). The direct-acting carcinogenic agents are generally found in a monoclonal form within the tumor cells. These agents help to keep the tumor phenotype through expression of either viral or cellular oncogenes (reviewed in [7]). Retroviruses, whose replication cycle requires the integration of the viral genome into the host genome, commonly transform because integration deregulates expression of cellular oncogenes or tumor suppressor genes (insertional mutagenesis, see Section 4.2). On the other hand, EBV is an example of a virus that does not need to integrate and transforms through expression of its own oncogenes.

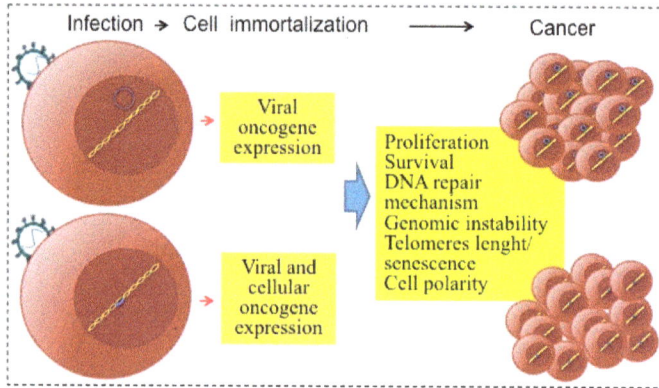

Figure 1. Direct mechanisms of viral carcinogenesis. After infecting target cells, tumor viruses are persistently maintained as genetic elements; viral genomes can form episomes (upper panel example, herpesviruses) or integrate into the host genomic DNA (lower panel example, retroviruses and HBV).

The indirect transforming viruses are not conditioned to exist within the cell that forms the tumor. These agents act through two main mechanisms: (i) triggering chronic inflammation and oxidative stress that persistently damage local tissues; and (ii) by producing immunosuppression that reduces or eliminates anti-tumor immune surveillance mechanisms (Figure 2). Among the most documented viral agents belonging to the first group are HBV and HCV; chronic inflammation produced by persistent infection associated with any of these viruses is a major risk to develop hepatocellular carcinoma (HCC) (reviewed in [15,16]). On the other hand, HIV belongs to the second group; patients with non-controlled infection and low T cell counts frequently develop lymphomas associated with EBV or KSV infection (reviewed in [17]).

Tumorigenic viruses were previously considered either exclusively direct or indirect transforming agents. However, some agents may require both mechanisms to induce carcinogenesis; for instance HBV and HCV [15,16]. *Helicobacter pylori* is the prototype indirect carcinogen through chronic inflammation [18]. Nevertheless, the bacterium also encodes the CagA oncoprotein, which is translocated to epithelial cells though a type IV secretion system (reviewed in [19]). Therefore, direct and indirect mechanisms are not mutually exclusive and some tissues may be equally dependent in both mechanisms for oncogenic transformation, such as the liver and stomach.

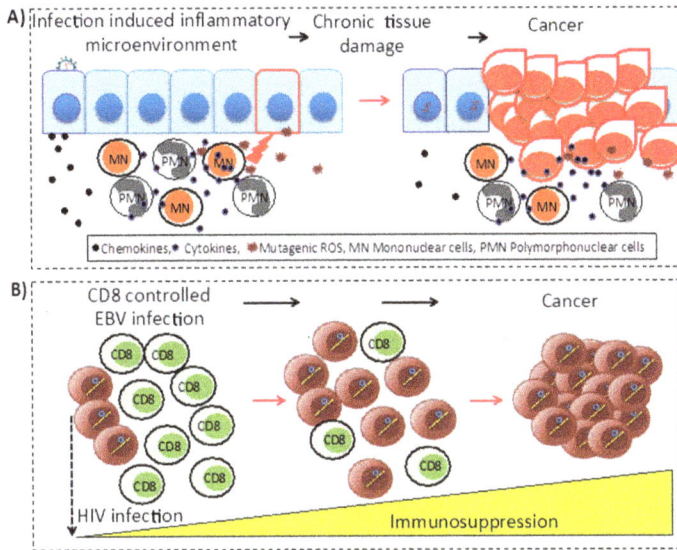

Figure 2. Indirect mechanisms of viral carcinogenesis. (**A**) Chronic inflammation. Infected cells produce chemokines attracting immune cells, which establish a chronic inflammatory microenvironment that persistently damage the local tissue. Cancer evolves within this cycle of infection, induced inflammation and tissue damage. (**B**) Immunosuppression. The prototype agent for immunosuppression is HIV. In immunocompetent individuals EBV infection is efficiently controlled by cytotoxic CD8 T cells; as HIV infection progresses and immune responses collapse, individuals become at increased risk of developing EBV associated lymphomas.

3. Human Oncogenic Viruses and Associated Cancers

Many different viruses have direct transformation characteristics; however, they have not been convincingly associated to human neoplasias based in the Bradford Hill criteria, and the IARC does not include them in the group of human type 1 carcinogens. Thus, adenoviruses, polyomaviruses SV40, JCV and BKV and others, will not be discussed here. The human cancers associated with viral infection are summarized in Table 2.

Table 2. Human oncogenic viruses and their associated tumors.

Human Virus	Associated Tumors	Reference
EBV	Burkitt's lymphoma, Hodgkin's lymphoma, immunosuppression-related lymphoma, T and NK cell lymphomas; nasopharyngeal and stomach carcinomas.	Reviewed in [11]
KSHV	Primary effusion lymphoma and Kaposi sarcoma	[20]
High-risk HPVs	Cervical, head and neck and anogenital tract carcinomas	Reviewed in [21]
MCPV	Merkel cell carcinoma	[22]
HBV	Hepatocellular carcinoma	[23]
HCV	Hepatocellular carcinoma	[24]
HTLV1	Adult T-cell leukemia/lymphoma	[25]

3.1. Herpesviruses: Epstein Barr Virus and Kaposi Sarcoma-Associated Herpesvirus

Herpesviruses are enveloped viruses with double-stranded linear DNA that after infecting the host cell remain in the nucleus as episomes (reviewed in [26]). Both EBV and KSHV show a biphasic life cycle consisting of a latent and a lytic phase. The latent phase seems to be the primary choice in which most of viral gene expression is shut down. This phase allows these viruses to coexist with the host generally asymptomatically and only in unusual situations may cause disease, e.g., during pharmacological or HIV induced immunosuppression. The lytic phase occurs in healthy individuals only in poorly understood sporadic events of reactivation.

EBV, also known as HHV4 (Human *Herpesvirus Type 4*), is found in approximately 95% of the adult population worldwide [27]; its principal routes of transmission are oral and blood [28,29], while intrauterine transmission has been documented too [30,31]. Early acquisition of this agent does not cause disease but when primary infection occurs during adolescence or early adulthood it causes infectious mononucleosis (reviewed in [32]). Interestingly, this condition represents a risk factor for developing Hodgkin's lymphoma (reviewed in [33]).

B cells are the main target of EBV infection (reviewed in [34]); more rarely and less understood, EBV can also infect epithelial cells, mainly in the upper digestive tract, which is thought to occur in viral reactivation events [35]. EBV has mainly been associated with malignancies of B and epithelial cells of the upper digestive tract (Table 2), which provides biological plausibility and coherence to the role of EBV in these neoplasias.

EBV is found in a latent stage in both lymphomas and carcinomas, and within the latent genes, there are several with oncogenic properties. The viral protein best recognized as oncogenic is LMP1, a signaling protein that imitates a constitutively active TNF receptor. LMP1 activates MAP kinases and STAT and NFκB transcription factors in B cells, and also PI3K in epithelial cells [36]. LMP1 increases proliferation and survival of the infected cell. Of note, STAT and NFκB activation potentially stimulates expression of cytokines and chemokines important to establish the inflammatory microenvironment critical to create the niche from which infectious and non-infectious tumors emerge. LMP2A is another constitutively active viral protein with ITAM (*immunoreceptor tyrosine activated motif*) signaling domains [37]. LMP2A expression promotes the activation of PLCγ and PI3K pathways, which correlates with its capacity to transform epithelial cells and to confer a migratory phenotype to the transformed cell [38]. LMP1 and LMP2A provide antigen recognition-like signals to B cells, required for differentiation into long-lived memory cells in which the virus persists hidden from antagonistic immune responses. Although, both proteins can be expressed in EBV-induced carcinomas, their normal function in non-lymphoid tissue is not clear.

EBV-associated tumors are characterized by the expression of a different set of viral transcripts or latencies. In lymphomas arousing in immunosuppressed individuals (latency III) the family of EBNA proteins provides with additional oncogenic insults. For instance, EBNA-LP, -3A and -3C directly interfere with p53 and pRb functions, as well as with other proteins of the G0 to G1 phase transition. EBNA-1 is the common protein expressed in all EBV-associated neoplasias; it is expressed in latency III, latency II (Hodgkin's lymphoma and carcinomas) and it is the sole viral protein expressed in latency I (Burkitt's lymphoma). This absolute requirement for EBNA-1 is probably due to its capacity to equally segregate EBV episomes to both daughter cells during cell division [39].

The prevalence of KSHV infection varies among geographic regions, being 5% in Europe, Asia and some parts of North America and more than 50% in sub-Saharan Africa. KSHV is transmitted from casual contacts as well as through sexual contact, blood transfusion and organ transplant. In non-endemic regions, the main via of transmission is probably through sexual contact and the use of contaminated syringes [40]. KSHV is the etiological agent of both lymphomas and sarcomas [41] (Table 2). Neoplasias associated with KSVH were not frequent before the AIDS pandemic, but currently represent one of the most important signs of this disease [42,43].

Several KSVH genes have potential oncogenic properties, for example, modulation of transduction of signals by K1 and K5; regulation of cell cycle by v-Cyclin and LANA 1; apoptosis inhibition by K1, vFLIP and v-Bcl2 and immune modulation by v-IRF, K3 and K5 (reviewed in [41]). LANA1 cooperates with h-Ras to transform fibroblasts and immortalize endothelial cells [44]. v-FLIP induces the development of lymphomas in transgenic mice primarily through its anti-apoptosis activity, which has been related to the activation of NFkB [45,46]. K1 also contains an ITAM signaling domain similar to the one found in EBV LMP2A and activates PI3K. K1 expressed in mice as a transgene promotes the development of sarcomas and lymphomas [47]. These similarities in the transformation mechanisms of both herpesviruses satisfy the principle of analogy of Bradford Hill's causation criteria.

KSHV is often lytic in a small number of tumor cells [48], and some of its potentially oncogenic genes are products of the lytic cycle. Also similar to HCMV in gliomas, a few K1 lytic genes provide growth and angiogenic functions in a paracrine fashion, favoring tumor growth [49–52]. Of note, EBV defective viruses unable to switch to lytic cycle trigger less aggressive lymphomas in SCID mice than their wild type counterparts [53], and a small fraction of nasopharyngeal carcinomas (NPC) also harbors the virus in lytic phase [54]. Whether analogous mechanisms are in place for EBV, KSHV and HCMV in their paracrine requirement of lytic cycle proteins is not clear.

3.2. High-Risk Papillomaviruses

Human papilloma viruses belong to the *Papillomaviridae* family; they contain a double-strand DNA genome of approximately 8000 bp and are not enveloped viruses. More than 100 members of this family have been described and from them, more than a dozen (types 16, 18, 31, 33, 35, 45, 51, 52, 56, 58, 59, 62, 66 and 68) have been classified as high-risk due to their epidemiological association with cervical and other cancers (Table 2; reviewed in [21]). HPV subtypes 16 and 18 are the most frequently found in tumors; the first is mainly associated with invasive cervical cancer and the second is the most frequent in squamous cell carcinoma [55,56]. Low-risk HPVs generally cause benign lesions, such as warts (reviewed in [57]).

HPV is transmitted by skin contact, including genital contact during sexual intercourse; thus HPV infection in the genital area tents to be common in sexually active persons. Infection is generally controlled by the immune system and only in a low number of people, HPV persists, increasing the risk to develop epithelial lesions (reviewed in [58]). Viral persistence seems to be greatly helped by the inability of infected cells to present antigenic epitopes to adaptive immune cells, which is common in individuals with alterations in the HLA (Human Leucocyte Antigen) antigen presentation pathway (reviewed in [59]).

The neoplastic progression involves a series of histological changes that have been stratified in clinical stages, which correlate with differential expression of viral oncogenes and accumulation of mutations in the host genome. The main oncogenic proteins are E6 and E7, which are required since the first lesions and are necessary for the maintenance of the malignant phenotype. HPV is usually not integrated into the host genomic DNA, and E2 negatively regulates the expression of E6 and E7. An important event in the oncogenic process is the integration of the viral genome, a step usually resulting in loss of E2 and over-expression of E6 and E7 (reviewed in [60]). Increased expression of E6 and E7 correlates with progression to high grade lesions and eventually to carcinoma *in situ* (reviewed in [58]).

3.3. Merkel Cell Polyomavirus

Polyomaviruses are non-enveloped viruses with a circular, double-stranded DNA of approximately 5000 bp. The members of this family are present in all regions of the world infecting several species. Historically, it was considered that only JCV and BKV polyomaviruses infected humans, but next generation sequencing techniques have enabled the identification of at least nine other members in humans, among them MCPV. MCPV was identified in 2008 in an aggressive skin cancer denominated Merkel cell carcinoma (MCC) [22]. Virtually the whole adult population worldwide is infected by MCPV. Evidence supporting the participation of this agent in MCC carcinogenesis includes the presence of MCPV genomes in about 80% of the tumors but not in healthy tissue, and the clonal integration of the viral genome [22,61–63]. MCPV oncogenic transformation may result from loss of immune surveillance, as MCC mainly occurs in immunosuppressed individuals. MCC was a very rare cancer before the AIDS pandemia, and today, there are around 1700 new cases per year in the US [64,65].

The MCPV genome is inserted into the host genome during viral carcinogenesis. Integration is characterized by preserving the viral induced cell proliferation functions while abrogating viral replication; the latter probably due to deletion of some of the viral T antigen gene regions [66,67]. Viral integration also favors host resistance to cell death promoting viral persistence in a latent state [68]. This is a significant difference between the presence of the virus in MCC and in non-tumor tissue.

Due to the recent discovery of MCPV, we still do not understand the function of viral proteins. However, some viral proteins present homology in functional domains with tumorigenic polyomaviruses from non-human species. For example, like SV40 MCPV T antigens are generated by differential splicing to produce large T and small T antigens [69]. The large T antigen presents the structural motif that inactivates pRb (LXCXE) [70], and the T antigen is generally expressed in MCC, and even in its truncated form it maintains intact the pRb-inactivating domain [71]. Inactivation of the T antigen in MCC cell lines results in cell death, further supporting the causative role of MCPV in MCC [72]. Also, the small T antigen conserves the AKT/mTOR activating domain, which is responsible for loss of contact inhibition and promoting independent growth of substrate and serum [73].

3.4. Hepatitis B Virus

The *Hepadnaviridae* family groups a series of viruses that cause liver disease in animals, with Hepatitis B virus (HBV) infecting humans. HBV is an enveloped virus with an approximate 3.2 Kb genome of a partially double stranded DNA chain and a single stranded fragment. HBV replicates through an intermediary RNA via a viral reverse transcriptase. The main target of infection by HBV is the hepatocyte and infection can occur through vertical or horizontal transmission starting in the first years of life or during adulthood (reviewed in [74]).

Chronic infection by HBV is one of the main causes of hepatocellular carcinoma (HCC). The carcinogenesis process triggered by HBV is complex, involving direct and indirect mechanisms with the latter being driven by chronic inflammation (reviewed in [75]). Direct mechanisms such as expression of viral oncogenes and insertional mutagenesis have also been documented [76]. HBV X (HBx) is the main oncogenic viral protein. HBx is a viral replication protein that participates in transcription and DNA repair through which it regulates cell cycle, apoptosis and genomic instability [77]. Furthermore, HBx transgenic mice develop liver carcinomas [78].

3.5. Hepatitis C Virus

Hepatitis C virus (HCV) is a member of the *Flaviviridae* family; there are at least six genotypes that are regionally distributed and divided into subtypes [79]. The HCV genome consists of a single strand RNA of positive polarity of approximately 9600 nucleotides from which a polyprotein is translated from an unique open reading frame and later subdivided into different viral polypeptides by viral proteases (reviewed in [80]). HCV infects hepatocytes causing an acute infection that may turn chronic when the immune system cannot eliminate it. In those cases, the carrier may progress to hepatitis, cirrhosis and eventually to HCC (reviewed in [80]). It is estimated that more than 170 million persons worldwide are infected by HCV from which about 40% will develop some form of liver disease and 1%–4% HCC [81]. Transmission commonly occurs through blood and infected blood products.

Direct and indirect transforming mechanisms have also been described for HCV. The viral oncoprotein Core is the only viral product that in transgenic mice promotes the appearance of HCC [82]. Core is the main trigger of steatosis, an abnormal retention of lipids within the hepatocyte, and oxidative stress leading to chronic liver damage and HCC [83]. Different functions have been attributed to this protein, including altered cellular gene transcription, cell proliferation and cell death. For instance, Core expression correlates with changes in the activity of bona fide cellular tumor suppressors and oncogenes, and also of intermediaries of MAP kinases, NFκB and β-catenin signaling pathways [84]. Core protein regulates ROS production by inducing nitric oxide synthase (iNOS) which activates cyclooxygenase-2 (COX-2), importantly contributing with oxidative stress [85]. iNOS and COX-2 are also important components of the inflammatory pathway leading to cancer (reviewed in [86,87]). Core localizes in the mitochondria where it regulates levels of the mitochondrial chaperone prohibitin; it is proposed that altered binding of prohibitin and cytochrome c oxidase results in increased oxidative stress that favors DNA damage [88]. Taken together all these data has contributed to the formation of a model in which accelerated cell division by the inhibition of p53, pRb and other cell proteins in the presence of DNA damage by oxidative stress and the inflammatory response leads to the development of HCC.

3.6. Human T-Lymphotropic Virus Type 1

The *Retroviridae* family groups several viruses with two copies of a positive sense single stranded RNA genome that is retro-transcribed to DNA and integrated into the host cell genome. Retroviruses are classified as simple and complex. Simple retroviruses encode *gag*, *pol* and *env* genes from which structural proteins are expressed, plus other proteins involved in viral replication and integration. Complex viruses encode additional regulatory genes besides the mentioned above.

HTLV1 is a potent direct carcinogenic agent that has been associated with a spectrum of lymphoproliferative diseases collectively referred as adult T-cells leukemia/lymphoma (ATL) (reviewed in [89]). HTLV1 is endemic of Japan, the Western African coast, Central America and the Caribbean, with 15–25 million people infected worldwide [90]. There are three demonstrated ways of transmission for HTLV1: sexual contact, intravenous and breast feeding. The virus infects T- and B-lymphocytes and dendritic cells *in vivo*.

Although, the main retroviral mechanism of transformation is by insertional mutagenesis (see Section 4.2), HTLV1 is a complex retrovirus whose genome also encodes the Tax oncoprotein. Tax has the ability to immortalize cells *in vitro* and its enforced expression in transgenic mice results in development of leukemia/lymphoma [91–95]. Tax is a transcriptional activator/repressor capable of modulating expression of multiple cellular genes and it also directly interacts with a plethora of cellular proteins. Tax principal mechanism of transformation is related to reprogramming cell cycle and inhibition of DNA repair [96]. Tax induces NFκB activity, which stimulates the expression of cytokines and their receptors, including those of IL-13, IL-15, IL-2, IL-2Rα and co-stimulatory surface receptors (OX40/OX40L) [97–99]. Importantly, this activity mimics the chronic inflammatory process critical in the oncogenic progression of many types of cancers. These molecules trigger T cell proliferation, which may help to amplify the pool of HTLV1 infected cells. Thus, contrary to other cancers in which the inflammatory process is mediated by immune cells in response to the oncogenic insult, in HTLV1 infection this is directly induced by Tax. Besides NFkB promoters, Tax also regulates expression of cellular transcriptional promoters through interaction with cyclic-AMP response element binding protein (CREB) and serum response factor (SRF) (reviewed in [96]).

4. Common Mechanisms of Direct Carcinogenesis

4.1. Viral Oncogenes and Oncoproteins

4.1.1. p53 and pRb Inactivation and Other Targets of Increased Proliferation and Survival

Viral oncogenes often increase the rate of cell proliferation and resistance to apoptosis, which eventually leads to alterations in DNA repair mechanisms and genomic instability. Increased mutation rates then alter cell polarity, with substrate-independent growth, and acquisition of cell migration properties, among other malignancy-associated features. The mechanisms used by viruses to induce these cellular changes are similar and often converge on common signaling pathways and transcription factors. For instance, inactivation of p53 and pRb tumor suppressor genes is an event that occurs in most pathways of viral oncogenesis, both human and animal (reviewed in [100–102]). In conditions of DNA damage, p53 arrests cell cycle until the damage has been repaired. When this does not occur, p53 induces cell apoptosis or cell senescence (reviewed in [103,104]). pRb also arrests cell cycle progression after binding and inactivating members of the E2F family of transcription factors [102]. pRb specifically inhibits the G1-S transition in response to DNA damage. Thus, an accumulation of mutations and chromosomal abnormalities is favored in the absence of p53 and pRb function. Since tumorigenic viruses are not usually associated with the massive production of viral infectious particles that characterize acute-infecting viruses, they relay in triggering cell proliferation mechanisms to increase the pool of infected cells. Furthermore, the termini of the viral genomes could be sensed as nicked DNA by the p53 and pRb machinery, and this would trigger host cell apoptosis immediately after infection, if both proteins were not inactivated.

HPV E6 and E7 induce the degradation of tumor suppressor proteins, p53 and pRb, respectively. E6 catalyzes the degradation of p53 by binding to the E6 associated protein (E6AP), a cellular protein with ubiquitin-ligase activity. The E6/E6AP complex binds to the p53 central region, which is then ubiquitinated and subsequently degraded in the proteasome [105,106]. E6 also blocks the cell cycle inhibitor p16^{INK4}, which increases cell proliferation [107]. E7 directly induces release of E2F1-3 from pRb/E2F1-3 complexes, E2F1-3 in turn activates transcription of genes involved in cell cycle progression such as cyclins E and A [60,108]. HTLV1 Tax induces hyper-phosphorylation of pRb while promoting its degradation in the proteasome [109]. The mechanism by which Tax affects p53 function is less well understood and many mechanisms have been proposed, including hyper-phosphorylation, interfering with p53 function through competitive binding of cellular co-activators and through direct binding mediated by NFκB [110]. Tax also interferes with the activity and/or expression levels of cyclins and cyclin-dependent kinases [111–113]. Another HTLV1 protein, HBZ, induces over-expression of E2F1 target genes stimulating the proliferation of T lymphocytes [114]. KSHV LANA 1 inactivates p53 and it induces pRb phosphorylation and subsequent inactivation through its association with Cdk6 [115–117]. Most EBV latency III proteins target p53 and pRb for inactivation, along with other cell proliferation proteins: HA95, HAX1, cyclin A and D, p27kip1, p16INK4A and c-Myc [34].

Cell cycle progression and cell survival are conjointly regulated mechanisms. Still, tumor viruses often trigger additional survival mechanisms besides p53 and pRb inactivation. EBV LMP1 and LMP2A constitutively activate NFκB and PI3K/Akt signaling pathways, which results in increased activity of anti-apoptotic proteins Bcl-2, Bcl-xl, Mcl1 and A20 [118–120]. EBV also encodes BHRF1, a Bcl-2 homologue that in a subset of Burkitt's lymphoma seems to counteract c-Myc pro-apoptotic activity [121]. HTLV1 Tax is also an important activator of NFκB and PI3K/Akt signaling pathways, and HTLV p12 and p13 proteins regulate Bcl-2 and caspase 3 and 9 activity [122]. HPV E6 and E7 function has been associated with degradation of pro-apoptotic proteins pro-caspase 8, FADD and BAK, and upregulation of expression of anti-apoptotic proteins c-IAP2 and survivin [123–127]. E6 binds to the E6-associated protein ligase (E6AP), an ubiquitin ligase that targets E6-interacting pro-apoptotic proteins to the proteasome. HBV HBx interacts with Damaged DNA Binding Protein 1 (DDB1) inhibiting proteasome activity resulting in resistance to apoptosis [128]. Anti-apoptotic mechanisms have also been described for HCV core and NS5A proteins (reviewed in [129]). Because virally infected cells are at high risk of elimination by apoptosis, anti-apoptotic mechanisms are critical for viral persistence and carcinogenesis.

4.1.2. Genomic Instability

Another common carcinogenic route promoted by infectious agents is genomic instability, which leads to gene amplification and deletion, changes in the number of chromosomes (polyploidy and aneuploidy) and aberrant fusion of non-homologous chromosomes (translocations). For instance, HPV-16 E6 and E7 proteins promote gene amplification, structural chromosomal alterations and centrosome replication errors leading to aneuploidy and polyploidy. Thus, HPV immortalized cell lines are characterized by gain and loss of whole chromosomes [130–133]. In agreement, aneuploidy can be found as early as in HPV-associated noninvasive lesions (reviewed in [134]). HBV HBx also interferes with genomic instability. HBx forms complexes with HBx interacting protein (HBXIP) altering the formation of the mitotic spindle and the centrosome function [77]. EBV EBNA-1 may promote genomic instability through activation of the recombinase-activating genes RAG1 and RAG2 [7], which may be responsible for the Myc chromosomal translocation present in Burkitt's lymphoma [135]. Another enzyme associated with genomic instability is activation-induced cytidine deaminase (AID), whose expression is induced by EBV during the transit through germinal center reaction. Increased rate of mutations are observed in the variable regions of heavy and light chains after EBV infection [136]. Whether other host genomic regions are also targeted by AID is not know, but potentially this would facilitate EBV-induced transformation.

A mutator phenotype has also been attributed to Tax and both small and gross changes in DNA and chromosomes are often found in HTLV1 transformed cells [137,138]. On one hand, Tax multiple targets operating during the G2/M transition impair the DNA-damage-induced response, allowing cells to scape this transition with accumulated mutations [139]. On the other hand, Tax directly induces chromosomal instability by transcriptionally repressing various targets, including the DNA polymerase-β, an enzyme involved in base-excision repair [96]. Tax can also independently suppress the nucleotide excision repair mechanism, which is normally utilized by cells following UV irradiation [140]. Furthermore, ATL cells often contain an abnormal number of chromosomes (aneuploidy), and a role for Tax has been proposed, although the mechanism is not clear. Tax directly binds and inactivates MAD1, a mitotic spindle assembly checkpoint (SAC) kinetochore protein in charge of ensuring proper chromosomal segregation during mitosis [141]. Tax also promotes premature activation of the CDC20-associated anaphase promoting complex [142]. Overall these mechanisms would lead to faulty chromosomal segregation resulting in aneuploidy in HTLV1 infected cells.

4.1.3. Interfering with Telomere Shortening

Telomere shortening and cell senescence are the natural consequence of unlimited cell proliferation, and tumor viruses also display mechanisms of telomere maintenance. Telomere length maintenance is a fine regulated mechanism involving a complex set of proteins and the enzyme telomerase (reviewed in [143]). Expression of telomerase in physiological conditions is restricted to cells with stem properties, e.g., germinal cells or somatic stem/progenitor cells, but telomerase expression is turned off in differentiated cells. How tumor viruses regulate telomere length is not clear, but HPV E6, EBV LMP1, KSHV LANA, HTLV1 Tax and HBV HBx have all been shown to induce expression of telomerase [144–147]. Tumor viruses interference with DNA repair mechanisms and concomitant genomic instability may be in great measure a consequence of bypassing regulatory checkpoints of telomere length and p53- and pRb-dependent senescence (reviewed in [148]). In this scenario, tumor viruses have evolved with these mechanisms in order to achieve replicative immortality and thus persistency.

4.1.4. Interfering with Cell Polarity

Viral oncoproteins may also promote carcinogenesis by inactivating proteins related to cell polarity. Proteins containing PDZ (post synaptic density protein, drosophila disc large tumor suppressor and zonula occludens-1 protein) domains function like scaffolds for both membrane and cytosolic supramolecular complexes, which have an important role in cell-cell contact and cell signaling. PDZ domains interact with target proteins through PBMs (PDZ domain-binding motif). A class I PBM was first described in the E4-ORF1 oncoprotein from adenovirus 9, and subsequently identified in other human virus oncoproteins, such as HPV E6 and HTLV1 Tax [149–152]. The E6 PBM is necessary for both *in vitro* and *in vivo* E6-mediated transformation [153,154]. On the other hand, mutational disruption of the Tax PBM reduces Tax-mediated cellular transformation and the capacity of HTLV1 to induce persistent infections [155,156]. Inactivation of cell polarity-associated proteins likely favors carcinogenesis by impairing morphogenesis, asymmetric division, migration and normal cell proliferation, survival and differentiation programs.

4.1.5. Viral miRNAs

MicroRNAs (miRNAs) have recently being shown to also participate in cell transformation. miRNAs are strongly conserved single stranded RNAs of approximately 22 nucleotide long that regulate expression of most genes. miRNAs inhibit mRNA translation mainly by translational repression based on base pair complementarity (reviewed in [157,158]). Almost all cancers present altered expression of cellular miRNAs (reviewed in [159,160]). However, a new and interesting topic in viral oncology concerns to viruses encoding miRNAs with oncogenic capabilities. The first five viral miRNAs were described in the EBV positive B95 cell line; to date, more than 40 miRNAs produced from the EBV BARTs and BHFR1 transcripts have been identified [161,162]. Those miRNAs are able to inhibit apoptosis, and some target cellular tumor suppressor genes, such as: PUMA, Bin, TOMM22 and WIF1 [163–166]. EBV infection of gastric carcinoma cells (AGS) induced anchorage independence in absence of viral protein synthesis, highlighting the importance of EBV miRNAs in the malignant process [167].

4.2. Insertional Mutagenesis

In the retrovirus life cycle, the integrated viral genome (the provirus) is replicated as a cellular genetic element during the host cell cycle. Expression of the provirus is controlled by viral regulatory elements, the long terminal repeats (LTRs), which are powerful transcriptional activators that often control the expression of cellular genes in the vicinity of the insertion area. When the provirus is close to a cell proto-oncogene, the LTR can upregulate its expression to oncogenic results. Although, all best characterized examples of cell transformation are due to upregulation of proto-oncogenes, retroviruses can potentially also interrupt tumor suppressor genes with similar effects (reviewed in [168]). During viral progeny formation, infective particles sometimes carry cellular oncogenes that were close to the insertion site and which are transduced to new hosts, now under the control of the LTRs. Among those genes frequently transduced by retroviruses are cell receptors such as ErbB and Fms, kinases such as Src and Abl and transcription factors such as Jun, Fos and Myc. These chimerical viruses become non-competent and unable to trigger lytic infections. Thus, these viruses have an augmented transforming capacity, as they are the only infectious agents capable of inducing tumors in just a few days. Due to this characteristic, such retroviruses are known as acute transformers. Preferential insertion sites are known as "hot spots", for example, the murine mammary tumor virus (MMTV) responsible for breast cancer in several mouse species, frequently inserts in regions near to *Wnt* and *Notch* proto-oncogenes [169].

The mutagenesis mechanism by insertion or retro-transduction of cell proto-oncogenes has not been demonstrated in human retroviruses (HTLV1 mainly transforms by Tax expression. See Section 3.6) (reviewed in [170]). Next generation sequencing techniques have shown that HBV is preferably integrated in tumor cells in comparison to non-tumor infected hepatic tissue and its integration correlates with deregulated expression of *TERT*, *MLL4* and *CCNE* cellular oncogenes [76]. The similarity with animal retroviruses has granted analogy to this HBV mechanism of insertional mutagenesis. Although HPV and MCPV require integration to become oncogenic, in these cases the biological consequence is more similar to acute transforming retroviruses. Here, viral regulatory regions are lost and proviruses become defective and unable to produce infection competent progeny. Hence, integration correlates with the establishment of a latent stage, over-expression of viral oncogenes and host cell transformation (see Section 3.2 and 3.3).

5. Common Mechanisms of Indirect Carcinogenesis

The mechanisms of indirect oncogenesis are more difficult to demonstrate since they cannot be measured by *in vitro* assays, nor the expression of viral genes in transgenic animal models recapitulates the oncogenic process. These mechanisms have been proposed from epidemiologic evidence and coherence and plausibility principles are more difficult to fulfill. Besides chronic inflammation and immunosuppression (Figure 2), other proposed indirect mechanisms of transformation are described below (Figure 3).

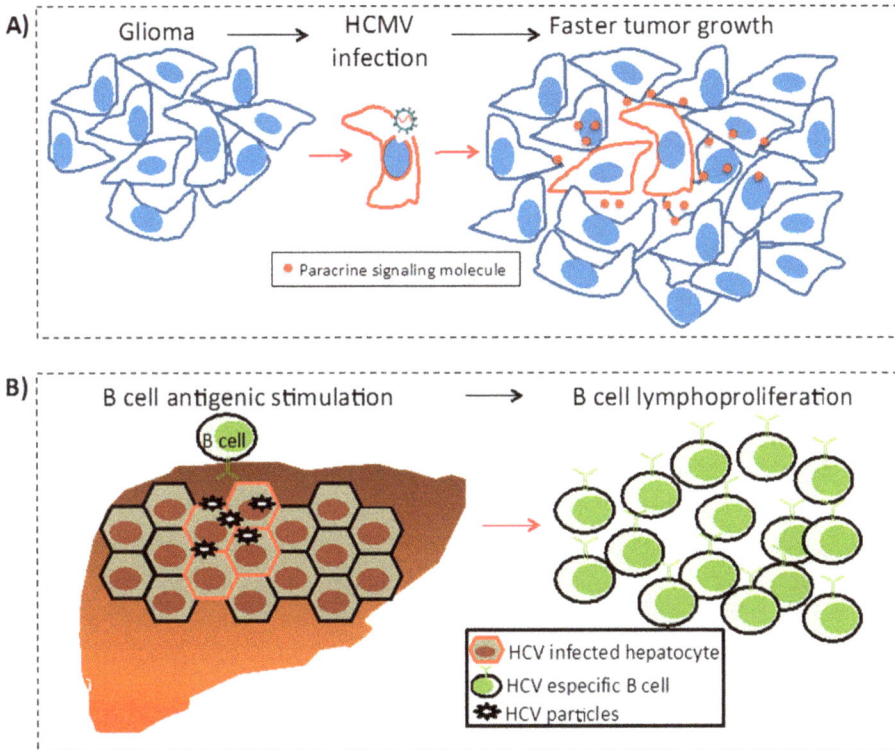

Figure 3. Other indirect mechanisms: Oncomodulation (**A**) and chronic antigen-driven lymphoproliferation (**B**). (**A**) In oncomodulation HCMV does not participate in the initial transformation of the glia; perhaps the virus has an increased tropism for tumor cells once the glioma has formed. Here, the virus only infects a fraction of the tumor cells activating signaling pathways that favor tumor growth; (**B**) B cells with antigen receptors specific for HCV antigens chronically respond to infected hepatocytes and free virus. This chronic stimulation increases the risk of unregulated lymphoproliferation and lymphoma.

5.1. Chronic Inflammation

Persistent infection is generally accompanied by local chronic inflammation, still in the presence of evasion mechanisms of the immune response. It has been proposed that this chronic inflammation induces a constant and progressive local damage, closely associated to regeneration events of the damaged tissue. The inflammatory response is characterized by local expression of pro-inflammatory cytokines, chemokines, adhesion molecules, growth factors and anti-apoptotic genes that regulate the sequential recruitment of leukocytes and stimulates fibroblasts and endothelial cells to divide and produce components of tissue remodeling and neovascularization (reviewed in [171]). A normal inflammatory response is self-limiting; chemoattraction of immune cells is gradually eliminated, pro-inflammatory cells already in the site of infection suffer apoptosis and are phagocytosed, while vascular changes are reversed. In contrast, in chronic inflammation associated with persistent infections, leukocytes remain in the lesion site and their apoptosis is suppressed. Additionally, to eliminate the infectious agent, immune cells produce oxygen and nitrogen free radicals, which are highly mutagenic. In this scenario, chronic inflammation favors the appearance of a cancerous clone, while tissue regeneration functions can also favor tumor growth, invasion and metastasis (for review, see [172], Figure 2A).

Helicobacter pylori is the prototype indirect carcinogen; it generates chronic gastric inflammation leading to gastric cancer through a series of progressive inflammatory lesions: non-atrophic gastritis, atrophic gastritis, intestinal metaplasia, dysplasia and gastric cancer [18]. *Helicobacter pylori* is also an IARC type I carcinogenic pathogen [173]. *Helicobacter pylori* stimulates the gastric epithelium to secrete IL-8 that attracts and activates neutrophils, favoring the establishment of a microenvironment populated by leukocytes and high concentrations of pro-inflammatory cytokines, such as TNF-α, IL-6, IL-1β and IL-12 [174]. Neutrophils release reactive oxygen species (superoxide anions, hydrogen peroxide, hydroxyl radicals and hydroperoxyl) and nitrogen oxides (nitric oxide, peroxynitrite and nitrogen dioxide) that induce mutations in cells of the gastric mucosa (reviewed in [175]). In agreement, individuals with IL-8, IL-1β and TNF-α polymorphisms present and increased risk to develop gastric cancer [176–179]. Of note, *Helicobacter pylori* pharmacological eradication in patients with pre-neoplastic lesions reverses tissue damage and halts the appearance of cancer, providing further support to the role of the bacterium in gastric cancer progression.

Although all oncogenic viruses maintain persistent infections, the role of inflammatory responses in oncogenesis is not clear. HCV and HBV triggered inflammation correlates with necrosis and tissue regeneration that eventually progresses to hepatic lesions such as steatosis, fibrosis and cirrhosis, from which liver cancer emerges. It has been observed that during steatosis and fibrosis, the liver is highly infiltrated by immune cells and there is a microenvironment of inflammatory cytokines and chemokines, among which TGF-β and IL-1β stand out (reviewed in [180]). We have also observed that increased EBV reactivation correlates with severe gastric inflammation and increased tissue damage leading to advanced gastric lesions, arguing for an important role for inflammation in the EBV-associated transformation of the gastric mucosa [181,182].

5.2. Immunosuppression

The role of the immune system in onco-surveillance has been clearly established since the AIDS pandemic. Although, HIV is not capable of inducing tumors in its host cell, 40% of patients with AIDS develop cancers associated to the disease. Thus, severe immunosuppression induced by HIV infection indirectly promotes the development of tumors (reviewed in [17,183]). Individuals with a low $CD8^+$ cytotoxic T lymphocyte count are more susceptible to infectious cancers [17], such as EBV- and KSHV-associated lymphomas (exemplified in Figure 2B), KSHV sarcomas, HPV head and neck and cervical carcinomas, and MCPV Merkel cell carcinomas. Due to these features, HIV is classified as an indirect carcinogenic agent, while, direct transformation mechanisms mediated by EBV, KSHV, HPV and MCPV are still operating.

A similar phenomenon is observed in individuals with pharmacological immunosuppression due to solid organ or bone marrow transplants. The post-transplant lymphoproliferative disorders (PTLD) are often EBV-associated B-cell proliferations. PTLDs begin as polyclonal proliferations with a high risk to evolve into monoclonal aggressive lymphomas [184,185]. Although, most PTLD arise from host lymphocytes, a donor origin is possible too. The use of T-cell depleting agents is a risk factor for EBV-positive PTLD, highlighting the importance of T-cells in the antitumor immunosurveillance mechanisms (reviewed in [186]). Infusion of autologous T-cells specific to the immunodominant EBV EBNA3A/3B/3C proteins has proven a very successful therapy [187,188].

5.3. Oncomodulation

There is evidence that viruses also participate in tumor growth modulating the biological course of an already-established cancer (Figure 3A). The term "oncomodulation" was suggested by Martin Michaelis *et al.* to describe the role of human cytomegalovirus (HCMV) in tumor progression [189]. HCMV is a herpesvirus whose worldwide prevalence is between 50% and 100% in the adult population; infection is normally asymptomatic and only produces disease under immunosuppressive conditions.

Up to today, there is not enough evidence of HCMV being capable of participating in the transformation process. Nevertheless, HCMV may infect tumor cells and through the expression of viral genes affecting signaling pathways important for proliferation, survival, angiogenesis, invasiveness and immune modulation, could increase the aggressiveness of the tumor [190–194]. The best documented example is HCMV participation in high degree gliomas, a brain cancer with extremely bad prognosis. HCMV genome and proteins have been found more frequently in high degree gliomas than in other central nervous system tumors or non-tumorous brain samples (e.g., from epilepsy) [195–197]. However, even when HCMV resides in the tumor, it only infects a fraction of the tumor cells and does not exhibit viral monoclonality; therefore, the HCMV association with high grade gliomas does not fulfill the Bradford Hill's temporality criteria, arguing that infection happened after the event of transformation.

5.4. Chronic Antigen-Driven Lymphoproliferation

Cells of the immune system are expanded in response to infection; particularly, B cells exhibit extensive proliferation during the germinal center (GC) reaction in which they undergo antigen receptor isotype switch and increased target affinity (somatic hypermutation). The activation-induced cytidine deaminase (AID) is in charge of both processes in which the antigen receptor is modified. B cell lymphomas frequently emerge from the GC reaction due to the risky combination of increased proliferation and expression of mutagenic enzymes. EBV LMP1 and LMP2A provide decoy signals inducing infected B cells to go through the GC reaction and emerge as memory cells in which the virus can persist for the life-time of the host. EBV-associated lymphomas may partially result from this EBV-induced GC reaction [136]. Similarly, chronic antigenic stimulation resulting from other persistent infections can potentially increase the risk to develop lymphomas; among the most widely documented are, infection by bacteria *Helicobacter pylori*, *Borrelia burgdorferi*, *Campylobacter jejuni* and *Chlamydia psittaci* and by HCV. Concerning the latter, clonal B cell expansions have been observed in HCV infected patients correlating with longer chronic infections and with receptor specificity against HCV proteins [198–200]. Still, the most reliable evidence for a causal association comes from HCV pharmacological eradication, which is frequently associated with lymphoma remission [201]. Similarly, anti- *Helicobacter pylori* treatment results in regression of associated gastric MALT lymphomas [202].

6. Conclusions

The last 100 years have seen the birth and evolution of tumor virology with seven viral agents already been convincingly associated with the pathogenesis of cancer in humans. Tumor virology has importantly contributed to the understanding of the molecular mechanisms operating during carcinogenesis. However, causation is especially difficult to demonstrate because in most cases the tumor viruses are wide spread in the population without causing disease. It is essential to consider that infection by tumor viruses is never sufficient but always required for development of associated tumors. Cancer cannot be the aim of the virus since it compromises both host and virus survival. However, viral mechanisms of persistence in which cellular processes are impaired such as proliferation, survival, DNA repair, among others, provide a suitable substrate from which cancer can emerge after additional environmental aggressions and permissive host genetics. Today, Sir Austin Bradford Hill's causation criteria are considered the experimental, epidemiological and clinical conceptual base from which to infer a virus-cancer causative relationship.

Two different modes of cellular transformation have been documented based in whether the virus is acting from within (directly) or outside (indirectly) the cell that will form the tumor. Direct mechanisms infer expression of viral oncogenes together with deregulation of cellular oncogenes and/or tumor suppressor genes. Among the most important indirect mechanisms are (i) the establishment of an inflammatory milieu in which chronic production of mutagenic molecules is persistently damaging the surrounding tissue, and (ii) immunosuppression with loss of the cancer immunosurveillance mechanisms. However, all tumor viruses probably present direct and indirect mechanisms and this separation mostly alludes to the main mechanism of cell transformation. The tumor microenvironment is always inflammatory whether infectious or aseptic, and inflammatory molecules importantly contribute with tumor initiation and progression. *Helicobacter pylori* is considered the prototype infectious agent transforming through chronic inflammation, still a bacterial oncogene able to induce gastric tumors in transgenic mice has recently been described. Similarly, EBV importantly cooperates with gastric inflammation and progression though a series of inflammatory lesions of increased severity [181,182] and HBV- and HCV-mediated liver cancer also progresses from a series of precursor inflammatory lesions besides their known capacity to express viral and cellular oncogenes. It is also possible that the contribution of direct and indirect mechanisms is mandated by the transforming tissue, with liver and gastric tissue equally depending on both modes of transformation.

Infection-associated tumors are responsible for 15%–20% of all cancer cases worldwide, representing an important challenge in public health programs. With the advent of new technologies it is highly probably that this frequency will increase. Wide genome sequencing technologies have recently allowed the discovery of MCPV and have helped to establish its causal association with Merkel cell carcinoma. Together with new tumor agents, it is probably that new mechanisms of infection-induced transformation will emerge while others could be better understood. The hit-and-run transforming mechanism proposes that a viral agent takes part in the carcinogenesis but it is later lost as the tumor cell acquires additional oncogenic hits [203]. This "non classical" oncogenic pathway is not compatible with current causality criteria. However, it is very likely that current causation criteria will be modified and extended in the future. For instance, Birdwell *et al.* used a model of transient infection in EBV-infected keratinocytes to analyze the pattern of methylation in CpG islands. They found that the epigenetic changes caused by infection correlated with a tumorigenic phenotype, which was maintained even after of loss of the virus [204]. Although, there is not evidence of transient infection by EBV does not happen naturally, the Birdwell's work highlights the potential role of infections that are not maintained throughout cancer development. Also, facilitated by massive sequencing of tumor samples, a widespread APOBEC3B fingerprint was found among many types of cancers [205]. Because the APOBEC family of cytidine deaminases is part of an innate antiviral response, it is possible that this fingerprint reflects a history of past infections in which the antiviral response also collaterally triggered cellular somatic mutations leading to cancer.

The discovery of cancers with an infectious origin is critical to develop vaccines and preventive and therapeutic pharmacological therapies. This knowledge has already leaded to vaccines against HBV and high risk-HPV and targeted therapies against HIV and HCV.

Acknowledgments

This work constitutes a partial fulfillment of the Graduate Program of Doctor Degree in Biomedical Sciences, Medicine Faculty, National Autonomous University of Mexico, Mexico City, Mexico. A Morales-Sánchez acknowledges to the National Autonomous University of Mexico and the scholarship and financial support provided by the National Council of Science and Technology (CONACyT) and National Autonomous University of Mexico. This study was supported by grant

HIM-2013-051 (to E.M. Fuentes Pananá) from Fondo de Apoyo a la Investigación Hospital Infantil de México Federico Gómez.

Conflicts of Interest

The authors declare no conflict of interest.

References and Notes

1. Rous, P. A transmissible avian neoplasm. (sarcoma of the common fowl.). *J. Exp. Med.* **1910**, *12*, 696–705.
2. Rous, P. A sarcoma of the fowl transmissible by an agent separable from the tumor cells. *J. Exp. Med.* **1911**, *13*, 397–411.
3. Ellerman, V.; Bang, O. Experimentelle leukämie bei hühnern. *Zent. Bakteriol. Parasitenkd. Infectionskr. Hyg. Abt. Orig.* **1908**, *46*, 595–609.
4. Epstein, M.A.; Achong, B.G.; Barr, Y.M. Virus particles in cultured lymphoblasts from burkitt's lymphoma. *Lancet* **1964**, *1*, 702–703.
5. Durst, M.; Gissmann, L.; Ikenberg, H.; zur Hausen, H. A papillomavirus DNA from a cervical carcinoma and its prevalence in cancer biopsy samples from different geographic regions. *Proc. Natl. Acad. Sci. USA* **1983**, *80*, 3812–3815.
6. Boshart, M.; Gissmann, L.; Ikenberg, H.; Kleinheinz, A.; Scheurlen, W.; zur Hausen, H. A new type of papillomavirus DNA, its presence in genital cancer biopsies and in cell lines derived from cervical cancer. *EMBO J.* **1984**, *3*, 1151–1157.
7. Moore, P.S.; Chang, Y. Why do viruses cause cancer? Highlights of the first century of human tumour virology. *Nat. Rev. Cancer* **2010**, *10*, 878–889.
8. Parkin, D.M. The global health burden of infection-associated cancers in the year 2002. *Int. J. Cancer* **2006**, *118*, 3030–3044.
9. Koch, R. Untersuchungen über bakterien: V. Die ätiologie der milzbrand-krankheit, begründet auf die entwicklungsgeschichte des bacillus anthracis [investigations into bacteria: V. The etiology of anthrax, based on the ontogenesis of bacillus anthracis]. *Cohns Beitr. Biol. Pflanz.* **1876**, *2*, 277–310.
10. Hill, A.B. The environment and disease: Association or causation? *Proc. Royal Soc.Med.* **1965**, *58*, 295–300.
11. Thompson, M.P.; Kurzrock, R. Epstein-barr virus and cancer. *Clin. Cancer Res.* **2004**, *10*, 803–821.
12. Kulwichit, W.; Edwards, R.H.; Davenport, E.M.; Baskar, J.F.; Godfrey, V.; Raab-Traub, N. Expression of the epstein-barr virus latent membrane protein 1 induces b cell lymphoma in transgenic mice. *Proc. Natl. Acad. Sci. USA* **1998**, *95*, 11963–11968.
13. Loeb, L.A.; Springgate, C.F.; Battula, N. Errors in DNA replication as a basis of malignant changes. *Cancer Res.* **1974**, *34*, 2311–2321.
14. Prindle, M.J.; Fox, E.J.; Loeb, L.A. The mutator phenotype in cancer: Molecular mechanisms and targeting strategies. *Curr. Drug Targets* **2010**, *11*, 1296–1303.

15. Coleman, W.B. Mechanisms of human hepatocarcinogenesis. *Curr. Mol. Med.* **2003**, *3*, 573–588.

16. Zucman-Rossi, J.; Laurent-Puig, P. Genetic diversity of hepatocellular carcinomas and its potential impact on targeted therapies. *Pharmacogenomics* **2007**, *8*, 997–1003.

17. Chadburn, A.; Abdul-Nabi, A.M.; Teruya, B.S.; Lo, A.A. Lymphoid proliferations associated with human immunodeficiency virus infection. *Arch. Pathol. Lab. Med.* **2013**, *137*, 360–370.

18. Correa, P.; Piazuelo, M.B. The gastric precancerous cascade. *J. Dig. Dis.* **2012**, *13*, 2–9.

19. Hatakeyama, M.; Higashi, H. *Helicobacter pylori* caga: A new paradigm for bacterial carcinogenesis. *Cancer Sci.* **2005**, *96*, 835–843.

20. Chang, Y.; Cesarman, E.; Pessin, M.S.; Lee, F.; Culpepper, J.; Knowles, D.M.; Moore, P.S. Identification of herpesvirus-like DNA sequences in aids-associated kaposi's sarcoma. *Science* **1994**, *266*, 1865–1869.

21. Zur Hausen, H. Papillomaviruses and cancer: From basic studies to clinical application. *Nat. Rev. Cancer* **2002**, *2*, 342–350.

22. Feng, H.; Shuda, M.; Chang, Y.; Moore, P.S. Clonal integration of a polyomavirus in human merkel cell carcinoma. *Science* **2008**, *319*, 1096–1100.

23. Beasley, R.P.; Hwang, L.Y.; Lin, C.C.; Chien, C.S. Hepatocellular carcinoma and hepatitis b virus. A prospective study of 22,707 men in taiwan. *Lancet* **1981**, *2*, 1129–1133.

24. Choo, Q.L.; Kuo, G.; Weiner, A.J.; Overby, L.R.; Bradley, D.W.; Houghton, M. Isolation of a cDNA clone derived from a blood-borne non-a, non-b viral hepatitis genome. *Science* **1989**, *244*, 359–362.

25. Poiesz, B.J.; Ruscetti, F.W.; Gazdar, A.F.; Bunn, P.A.; Minna, J.D.; Gallo, R.C. Detection and isolation of type c retrovirus particles from fresh and cultured lymphocytes of a patient with cutaneous t-cell lymphoma. *Proc. Natl. Acad. Sci. USA* **1980**, *77*, 7415–7419.

26. Rickinson, A.B.; Kieff, E. Epstein-barr virus. In *Fields Virology*, Lippincott-Raven Publishers: Philadelphia, PA, USA, 1996; Volume 2, pp. 2397–2446.

27. Henle, G.; Henle, W.; Clifford, P.; Diehl, V.; Kafuko, G.W.; Kirya, B.G.; Klein, G.; Morrow, R.H.; Munube, G.M.; Pike, P.; *et al.* Antibodies to epstein-barr virus in burkitt's lymphoma and control groups. *J. Natl. Cancer Inst.* **1969**, *43*, 1147–1157.

28. Gerber, P.; Walsh, J.H.; Rosenblum, E.N.; Purcell, R.H. Association of eb-virus infection with the post-perfusion syndrome. *Lancet* **1969**, *1*, 593–595.

29. Hoagland, R.J. The transmission of infectious mononucleosis. *Am. J. Med. Sci.* **1955**, *229*, 262–272.

30. Goldberg, G.N.; Fulginiti, V.A.; Ray, C.G.; Ferry, P.; Jones, J.F.; Cross, H.; Minnich, L. In utero epstein-barr virus (infectious mononucleosis) infection. *JAMA: J. Am. Med. Assoc.* **1981**, *246*, 1579–1581.

31. Meyohas, M.C.; Marechal, V.; Desire, N.; Bouillie, J.; Frottier, J.; Nicolas, J.C. Study of mother-to-child epstein-barr virus transmission by means of nested pcrs. *J. Virol.* **1996**, *70*, 6816–6819.

32. Stock, I. Infectious mononucleosis—A "childhood disease" of great medical concern. *Med. Mon. Pharma.* **2013**, *36*, 364–368.

33. Hjalgrim, H. On the aetiology of hodgkin lymphoma. *Dan. Med. J.* **2012**, *59*, B4485.

34. Klein, G.; Klein, E.; Kashuba, E. Interaction of epstein-barr virus (ebv) with human b-lymphocytes. *Biochem. Biophys. Res. Commun.* **2010**, *396*, 67–73.

35. Morgan, D.G.; Niederman, J.C.; Miller, G.; Smith, H.W.; Dowaliby, J.M. Site of epstein-barr virus replication in the oropharynx. *Lancet* **1979**, *2*, 1154–1157.

36. Shair, K.H.; Schnegg, C.I.; Raab-Traub, N. Ebv latent membrane protein 1 effects on plakoglobin, cell growth, and migration. *Cancer Res.* **2008**, *68*, 6997–7005.

37. Caldwell, R.G.; Wilson, J.B.; Anderson, S.J.; Longnecker, R. Epstein-barr virus lmp2a drives b cell development and survival in the absence of normal b cell receptor signals. *Immunity* **1998**, *9*, 405–411.

38. Scholle, F.; Bendt, K.M.; Raab-Traub, N. Epstein-barr virus lmp2a transforms epithelial cells, inhibits cell differentiation, and activates akt. *J. Virol.* **2000**, *74*, 10681–10689.

39. Yates, J.L.; Warren, N.; Sugden, B. Stable replication of plasmids derived from epstein-barr virus in various mammalian cells. *Nature* **1985**, *313*, 812–815.

40. Dukers, N.H.; Rezza, G. Human herpesvirus 8 epidemiology: What we do and do not know. *Aids* **2003**, *17*, 1717–1730.

41. Cai, Q.; Verma, S.C.; Lu, J.; Robertson, E.S. Molecular biology of kaposi's sarcoma-associated herpesvirus and related oncogenesis. *Adv. Virus Res.* **2010**, *78*, 87–142.

42. Viejo-Borbolla, A.; Ottinger, M.; Schulz, T.F. Human herpesvirus 8: Biology and role in the pathogenesis of kaposi's sarcoma and other aids-related malignancies. *Curr. HIV/AIDS Rep.* **2004**, *1*, 5–11.

43. Beral, V.; Peterman, T.A.; Berkelman, R.L.; Jaffe, H.W. Kaposi's sarcoma among persons with aids: A sexually transmitted infection? *Lancet* **1990**, *335*, 123–128.

44. Radkov, S.A.; Kellam, P.; Boshoff, C. The latent nuclear antigen of kaposi sarcoma-associated herpesvirus targets the retinoblastoma-e2f pathway and with the oncogene hras transforms primary rat cells. *Nat. Med.* **2000**, *6*, 1121–1127.

45. Chugh, P.; Matta, H.; Schamus, S.; Zachariah, S.; Kumar, A.; Richardson, J.A.; Smith, A.L.; Chaudhary, P.M. Constitutive nf-kappab activation, normal fas-induced apoptosis, and increased incidence of lymphoma in human herpes virus 8 k13 transgenic mice. *Proc. Natl. Acad. Sci. USA* **2005**, *102*, 12885–12890.

46. Sun, Q.; Zachariah, S.; Chaudhary, P.M. The human herpes virus 8-encoded viral flice-inhibitory protein induces cellular transformation via nf-kappab activation. *J. Biol. Chem.* **2003**, *278*, 52437–52445.

47. Wang, L.; Dittmer, D.P.; Tomlinson, C.C.; Fakhari, F.D.; Damania, B. Immortalization of primary endothelial cells by the k1 protein of kaposi's sarcoma-associated herpesvirus. *Cancer Res.* **2006**, *66*, 3658–3666.

48. Staskus, K.A.; Zhong, W.; Gebhard, K.; Herndier, B.; Wang, H.; Renne, R.; Beneke, J.; Pudney, J.; Anderson, D.J.; Ganem, D.; *et al.* Kaposi's sarcoma-associated herpesvirus gene expression in endothelial (spindle) tumor cells. *J. Virol.* **1997**, *71*, 715–719.

49. Xie, J.; Pan, H.; Yoo, S.; Gao, S.J. Kaposi's sarcoma-associated herpesvirus induction of ap-1 and interleukin 6 during primary infection mediated by multiple mitogen-activated protein kinase pathways. *J. Virol.* **2005**, *79*, 15027–15037.

50. Naranatt, P.P.; Krishnan, H.H.; Svojanovsky, S.R.; Bloomer, C.; Mathur, S.; Chandran, B. Host gene induction and transcriptional reprogramming in kaposi's sarcoma-associated herpesvirus (kshv/hhv-8)-infected endothelial, fibroblast, and b cells: Insights into modulation events early during infection. *Cancer Res.* **2004**, *64*, 72–84.

51. Masood, R.; Cesarman, E.; Smith, D.L.; Gill, P.S.; Flore, O. Human herpesvirus-8-transformed endothelial cells have functionally activated vascular endothelial growth factor/vascular endothelial growth factor receptor. *Am. J. Pathol.* **2002**, *160*, 23–29.

52. Cerimele, F.; Curreli, F.; Ely, S.; Friedman-Kien, A.E.; Cesarman, E.; Flore, O. Kaposi's sarcoma-associated herpesvirus can productively infect primary human keratinocytes and alter their growth properties. *J. Virol.* **2001**, *75*, 2435–2443.

53. Hong, G.K.; Gulley, M.L.; Feng, W.H.; Delecluse, H.J.; Holley-Guthrie, E.; Kenney, S.C. Epstein-barr virus lytic infection contributes to lymphoproliferative disease in a scid mouse model. *J. Virol.* **2005**, *79*, 13993–14003.

54. Martel-Renoir, D.; Grunewald, V.; Touitou, R.; Schwaab, G.; Joab, I. Qualitative analysis of the expression of epstein-barr virus lytic genes in nasopharyngeal carcinoma biopsies. *J. Gen. Virol.* **1995**, *76*, 1401–1408.

55. Munoz, N.; Bosch, F.X.; de Sanjose, S.; Herrero, R.; Castellsague, X.; Shah, K.V.; Snijders, P.J.; Meijer, C.J. International Agency for Research on Cancer Multicenter Cervical Cancer Study, G. Epidemiologic classification of human papillomavirus types associated with cervical cancer. *N. Engl. J. Med.* **2003**, *348*, 518–527.

56. Clifford, G.M.; Smith, J.S.; Plummer, M.; Munoz, N.; Franceschi, S. Human papillomavirus types in invasive cervical cancer worldwide: A meta-analysis. *Br. J. Cancer* **2003**, *88*, 63–73.

57. Chow, L.T.; Broker, T.R. Human papillomavirus infections: Warts or cancer? *Cold Spring Harbor Perspect. Biol.* **2013**, *5*, 10.1101/cshperspect.a012997.

58. Woodman, C.B.; Collins, S.I.; Young, L.S. The natural history of cervical hpv infection: Unresolved issues. *Nat. Rev. Cancer* **2007**, *7*, 11–22.

59. Zur Hausen, H. Papillomaviruses—To vaccination and beyond. *Biochem. Biokhimiia* **2008**, *73*, 498–503.

60. Ghittoni, R.; Accardi, R.; Hasan, U.; Gheit, T.; Sylla, B.; Tommasino, M. The biological properties of e6 and e7 oncoproteins from human papillomaviruses. *Virus Genes* **2010**, *40*, 1–13.

61. Becker, J.C.; Houben, R.; Ugurel, S.; Trefzer, U.; Pfohler, C.; Schrama, D. Mc polyomavirus is frequently present in merkel cell carcinoma of european patients. *J. Investig. Dermatol.* **2009**, *129*, 248–250.

62. Carter, J.J.; Paulson, K.G.; Wipf, G.C.; Miranda, D.; Madeleine, M.M.; Johnson, L.G.; Lemos, B.D.; Lee, S.; Warcola, A.H.; Iyer, J.G.; *et al.* Association of merkel cell polyomavirus-specific antibodies with merkel cell carcinoma. *J. Natl. Cancer Inst.* **2009**, *101*, 1510–1522.

63. Duncavage, E.J.; Zehnbauer, B.A.; Pfeifer, J.D. Prevalence of merkel cell polyomavirus in merkel cell carcinoma. *Mod. Pathol.* **2009**, *22*, 516–521.

64. Agelli, M.; Clegg, L.X. Epidemiology of primary merkel cell carcinoma in the United States. *J. Am. Acad. Dermatol.* **2003**, *49*, 832–841.

65. Hodgson, N.C. Merkel cell carcinoma: Changing incidence trends. *J. Surg. Oncol.* **2005**, *89*, 1–4.

66. Kassem, A.; Schopflin, A.; Diaz, C.; Weyers, W.; Stickeler, E.; Werner, M.; Zur Hausen, A. Frequent detection of merkel cell polyomavirus in human merkel cell carcinomas and identification of a unique deletion in the vp1 gene. *Cancer Res.* **2008**, *68*, 5009–5013.

67. Kwun, H.J.; Guastafierro, A.; Shuda, M.; Meinke, G.; Bohm, A.; Moore, P.S.; Chang, Y. The minimum replication origin of merkel cell polyomavirus has a unique large t-antigen loading architecture and requires small t-antigen expression for optimal replication. *J. Virol.* **2009**, *83*, 12118–12128.

68. Moens, U.; van Ghelue, M.; Johannessen, M. Oncogenic potentials of the human polyomavirus regulatory proteins. *Cell. Mol. Life Sci.* **2007**, *64*, 1656–1678.

69. Shuda, M.; Arora, R.; Kwun, H.J.; Feng, H.; Sarid, R.; Fernandez-Figueras, M.T.; Tolstov, Y.; Gjoerup, O.; Mansukhani, M.M.; Swerdlow, S.H.; *et al.* Human merkel cell polyomavirus infection i. Mcv t antigen expression in merkel cell carcinoma, lymphoid tissues and lymphoid tumors. *Int. J. Cancer* **2009**, *125*, 1243–1249.

70. Shuda, M.; Feng, H.; Kwun, H.J.; Rosen, S.T.; Gjoerup, O.; Moore, P.S.; Chang, Y. T antigen mutations are a human tumor-specific signature for merkel cell polyomavirus. *Proc. Natl. Acad. Sci. USA* **2008**, *105*, 16272–16277.

71. Houben, R.; Adam, C.; Baeurle, A.; Hesbacher, S.; Grimm, J.; Angermeyer, S.; Henzel, K.; Hauser, S.; Elling, R.; Brocker, E.B.; *et al.* An intact retinoblastoma protein-binding site in merkel cell polyomavirus large t antigen is required for promoting growth of merkel cell carcinoma cells. *Int. J. Cancer* **2012**, *130*, 847–856.

72. Houben, R.; Shuda, M.; Weinkam, R.; Schrama, D.; Feng, H.; Chang, Y.; Moore, P.S.; Becker, J.C. Merkel cell polyomavirus-infected merkel cell carcinoma cells require expression of viral t antigens. *J. Virol.* **2010**, *84*, 7064–7072.

73. Shuda, M.; Kwun, H.J.; Feng, H.; Chang, Y.; Moore, P.S. Human merkel cell polyomavirus small t antigen is an oncoprotein targeting the 4e-bp1 translation regulator. *J. Clin. Investig.* **2011**, *121*, 3623–3634.

74. Kew, M.C. Epidemiology of chronic hepatitis b virus infection, hepatocellular carcinoma, and hepatitis b virus-induced hepatocellular carcinoma. *Pathol.-Biol.* **2010**, *58*, 273–277.

75. Tarocchi, M.; Polvani, S.; Marroncini, G.; Galli, A. Molecular mechanism of hepatitis b virus-induced hepatocarcinogenesis. *World J. Gastroenterol.* **2014**, *20*, 11630–11640.

76. Sung, W.K.; Zheng, H.; Li, S.; Chen, R.; Liu, X.; Li, Y.; Lee, N.P.; Lee, W.H.; Ariyaratne, P.N.; Tennakoon, C.; *et al.* Genome-wide survey of recurrent hbv integration in hepatocellular carcinoma. *Nat. Genet.* **2012**, *44*, 765–769.

77. Wen, Y.; Golubkov, V.S.; Strongin, A.Y.; Jiang, W.; Reed, J.C. Interaction of hepatitis b viral oncoprotein with cellular target hbxip dysregulates centrosome dynamics and mitotic spindle formation. *J. Biol. Chem.* **2008**, *283*, 2793–2803.

78. Kim, C.M.; Koike, K.; Saito, I.; Miyamura, T.; Jay, G. Hbx gene of hepatitis b virus induces liver cancer in transgenic mice. *Nature* **1991**, *351*, 317–320.

79. Simmonds, P.; Bukh, J.; Combet, C.; Deleage, G.; Enomoto, N.; Feinstone, S.; Halfon, P.; Inchauspe, G.; Kuiken, C.; Maertens, G.; *et al.* Consensus proposals for a unified system of nomenclature of hepatitis c virus genotypes. *Hepatology* **2005**, *42*, 962–973.

80. Tang, H.; Grise, H. Cellular and molecular biology of hcv infection and hepatitis. *Clin. Sci.* **2009**, *117*, 49–65.

81. Morgan, R.L.; Baack, B.; Smith, B.D.; Yartel, A.; Pitasi, M.; Falck-Ytter, Y. Eradication of hepatitis c virus infection and the development of hepatocellular carcinoma: A meta-analysis of observational studies. *Ann. Intern. Med.* **2013**, *158*, 329–337.

82. Moriya, K.; Fujie, H.; Shintani, Y.; Yotsuyanagi, H.; Tsutsumi, T.; Ishibashi, K.; Matsuura, Y.; Kimura, S.; Miyamura, T.; Koike, K. The core protein of hepatitis c virus induces hepatocellular carcinoma in transgenic mice. *Nat. Med.* **1998**, *4*, 1065–1067.

83. Moriya, K.; Yotsuyanagi, H.; Shintani, Y.; Fujie, H.; Ishibashi, K.; Matsuura, Y.; Miyamura, T.; Koike, K. Hepatitis c virus core protein induces hepatic steatosis in transgenic mice. *J. Gen. Virol.* **1997**, *78 (Pt 7)*, 1527–1531.

84. Pei, Y.; Zhang, T.; Renault, V.; Zhang, X. An overview of hepatocellular carcinoma study by omics-based methods. *Acta Biochim. Biophys. Sin.* **2009**, *41*, 1–15.

85. Nunez, O.; Fernandez-Martinez, A.; Majano, P.L.; Apolinario, A.; Gomez-Gonzalo, M.; Benedicto, I.; Lopez-Cabrera, M.; Bosca, L.; Clemente, G.; Garcia-Monzon, C.; *et al.* Increased intrahepatic cyclooxygenase 2, matrix metalloproteinase 2, and matrix metalloproteinase 9 expression is associated with progressive liver disease in chronic hepatitis c virus infection: Role of viral core and ns5a proteins. *Gut* **2004**, *53*, 1665–1672.

86. Murakami, A.; Ohigashi, H. Targeting nox, inos and cox-2 in inflammatory cells: Chemoprevention using food phytochemicals. *Int. J. Cancer* **2007**, *121*, 2357–2363.

87. Ohshima, H.; Tazawa, H.; Sylla, B.S.; Sawa, T. Prevention of human cancer by modulation of chronic inflammatory processes. *Mutat. Res.* **2005**, *591*, 110–122.

88. Fujinaga, H.; Tsutsumi, T.; Yotsuyanagi, H.; Moriya, K.; Koike, K. Hepatocarcinogenesis in hepatitis c: Hcv shrewdly exacerbates oxidative stress by modulating both production and scavenging of reactive oxygen species. *Oncology* **2011**, *81* (Suppl. S1), 11–17.

89. Kannian, P.; Green, P.L. Human t lymphotropic virus type 1 (htlv-1): Molecular biology and oncogenesis. *Viruses* **2010**, *2*, 2037–2077.

90. Proietti, F.A.; Carneiro-Proietti, A.B.; Catalan-Soares, B.C.; Murphy, E.L. Global epidemiology of htlv-i infection and associated diseases. *Oncogene* **2005**, *24*, 6058–6068.

91. Grassmann, R.; Dengler, C.; Muller-Fleckenstein, I.; Fleckenstein, B.; McGuire, K.; Dokhelar, M.C.; Sodroski, J.G.; Haseltine, W.A. Transformation to continuous growth of primary human t lymphocytes by human t-cell leukemia virus type i x-region genes transduced by a herpesvirus saimiri vector. *Proc. Natl. Acad. Sci. USA* **1989**, *86*, 3351–3355.

92. Grassmann, R.; Berchtold, S.; Radant, I.; Alt, M.; Fleckenstein, B.; Sodroski, J.G.; Haseltine, W.A.; Ramstedt, U. Role of human t-cell leukemia virus type 1 x region proteins in immortalization of primary human lymphocytes in culture. *J. Virol.* **1992**, *66*, 4570–4575.

93. Akagi, T.; Shimotohno, K. Proliferative response of tax1-transduced primary human t cells to anti-cd3 antibody stimulation by an interleukin-2-independent pathway. *J. Virol.* **1993**, *67*, 1211–1217.

94. Hasegawa, H.; Sawa, H.; Lewis, M.J.; Orba, Y.; Sheehy, N.; Yamamoto, Y.; Ichinohe, T.; Tsunetsugu-Yokota, Y.; Katano, H.; Takahashi, H.; *et al.* Thymus-derived leukemia-lymphoma in mice transgenic for the tax gene of human t-lymphotropic virus type i. *Nat. Med.* **2006**, *12*, 466–472.

95. Benvenisty, N.; Ornitz, D.M.; Bennett, G.L.; Sahagan, B.G.; Kuo, A.; Cardiff, R.D.; Leder, P. Brain tumours and lymphomas in transgenic mice that carry htlv-i ltr/c-myc and ig/tax genes. *Oncogene* **1992**, *7*, 2399–2405.

96. Azran, I.; Schavinsky-Khrapunsky, Y.; Aboud, M. Role of tax protein in human t-cell leukemia virus type-i leukemogenicity. *Retrovirology* **2004**, *1*, 20.

97. Li, X.H.; Gaynor, R.B. Mechanisms of nf-kappab activation by the htlv type 1 tax protein. *AIDS Res. Hum. Retrovir.* **2000**, *16*, 1583–1590.

98. Harhaj, E.W.; Harhaj, N.S.; Grant, C.; Mostoller, K.; Alefantis, T.; Sun, S.C.; Wigdahl, B. Human t cell leukemia virus type i tax activates cd40 gene expression via the nf-kappa b pathway. *Virology* **2005**, *333*, 145–158.

99. Mori, N.; Fujii, M.; Cheng, G.; Ikeda, S.; Yamasaki, Y.; Yamada, Y.; Tomonaga, M.; Yamamoto, N. Human t-cell leukemia virus type i tax protein induces the expression of anti-apoptotic gene bcl-xl in human t-cells through nuclear factor-kappab and c-amp responsive element binding protein pathways. *Virus Genes* **2001**, *22*, 279–287.

100. Hollstein, M.; Sidransky, D.; Vogelstein, B.; Harris, C.C. P53 mutations in human cancers. *Science* **1991**, *253*, 49–53.

101. Levine, A.J.; Momand, J.; Finlay, C.A. The p53 tumour suppressor gene. *Nature* **1991**, *351*, 453–456.

102. Khidr, L.; Chen, P.L. Rb, the conductor that orchestrates life, death and differentiation. *Oncogene* **2006**, *25*, 5210–5219.

103. Lee, J.M.; Bernstein, A. Apoptosis, cancer and the p53 tumour suppressor gene. *Cancer Metastasis Rev.* **1995**, *14*, 149–161.

104. Amundson, S.A.; Myers, T.G.; Fornace, A.J., Jr. Roles for p53 in growth arrest and apoptosis: Putting on the brakes after genotoxic stress. *Oncogene* **1998**, *17*, 3287–3299.

105. Scheffner, M.; Werness, B.A.; Huibregtse, J.M.; Levine, A.J.; Howley, P.M. The e6 oncoprotein encoded by human papillomavirus types 16 and 18 promotes the degradation of p53. *Cell* **1990**, *63*, 1129–1136.

106. Scheffner, M.; Huibregtse, J.M.; Vierstra, R.D.; Howley, P.M. The hpv-16 e6 and e6-ap complex functions as a ubiquitin-protein ligase in the ubiquitination of p53. *Cell* **1993**, *75*, 495–505.

107. Reznikoff, C.A.; Yeager, T.R.; Belair, C.D.; Savelieva, E.; Puthenveettil, J.A.; Stadler, W.M. Elevated p16 at senescence and loss of p16 at immortalization in human papillomavirus 16 e6, but not e7, transformed human uroepithelial cells. *Cancer Res.* **1996**, *56*, 2886–2890.

108. Dyson, N.; Howley, P.M.; Munger, K.; Harlow, E. The human papilloma virus-16 e7 oncoprotein is able to bind to the retinoblastoma gene product. *Science* **1989**, *243*, 934–937.

109. Giam, C.Z.; Jeang, K.T. Htlv-1 tax and adult t-cell leukemia. *Front. Biosci.: J. Virtual Libr.* **2007**, *12*, 1496–1507.

110. Pise-Masison, C.A.; Brady, J.N. Setting the stage for transformation: Htlv-1 tax inhibition of p53 function. *Front. Biosci.* **2005**, *10*, 919–930.

111. Haller, K.; Wu, Y.; Derow, E.; Schmitt, I.; Jeang, K.T.; Grassmann, R. Physical interaction of human t-cell leukemia virus type 1 tax with cyclin-dependent kinase 4 stimulates the phosphorylation of retinoblastoma protein. *Mol. Cell. Biol.* **2002**, *22*, 3327–3338.

112. Suzuki, T.; Narita, T.; Uchida-Toita, M.; Yoshida, M. Down-regulation of the ink4 family of cyclin-dependent kinase inhibitors by tax protein of htlv-1 through two distinct mechanisms. *Virology* **1999**, *259*, 384–391.

113. Suzuki, T.; Kitao, S.; Matsushime, H.; Yoshida, M. Htlv-1 tax protein interacts with cyclin-dependent kinase inhibitor p16ink4a and counteracts its inhibitory activity towards cdk4. *EMBO J.* **1996**, *15*, 1607–1614.

114. Satou, Y.; Yasunaga, J.; Yoshida, M.; Matsuoka, M. Htlv-i basic leucine zipper factor gene mrna supports proliferation of adult t cell leukemia cells. *Proc. Natl. Acad. Sci. USA* **2006**, *103*, 720–725.

115. Cai, Q.L.; Knight, J.S.; Verma, S.C.; Zald, P.; Robertson, E.S. Ec5s ubiquitin complex is recruited by kshv latent antigen lana for degradation of the vhl and p53 tumor suppressors. *PLoS Pathog.* **2006**, *2*, e116.

116. Friborg, J., Jr.; Kong, W.; Hottiger, M.O.; Nabel, G.J. P53 inhibition by the lana protein of kshv protects against cell death. *Nature* **1999**, *402*, 889–894.

117. Si, H.; Robertson, E.S. Kaposi's sarcoma-associated herpesvirus-encoded latency-associated nuclear antigen induces chromosomal instability through inhibition of p53 function. *J. Virol.* **2006**, *80*, 697–709.

118. Spender, L.C.; Cannell, E.J.; Hollyoake, M.; Wensing, B.; Gawn, J.M.; Brimmell, M.; Packham, G.; Farrell, P.J. Control of cell cycle entry and apoptosis in b lymphocytes infected by epstein-barr virus. *J. Virol.* **1999**, *73*, 4678–4688.

119. Mei, Y.P.; Zhou, J.M.; Wang, Y.; Huang, H.; Deng, R.; Feng, G.K.; Zeng, Y.X.; Zhu, X.F. Silencing of lmp1 induces cell cycle arrest and enhances chemosensitivity through inhibition of akt signaling pathway in ebv-positive nasopharyngeal carcinoma cells. *Cell Cycle* **2007**, *6*, 1379–1385.

120. Portis, T.; Longnecker, R. Epstein-barr virus (ebv) lmp2a mediates b-lymphocyte survival through constitutive activation of the ras/pi3k/akt pathway. *Oncogene* **2004**, *23*, 8619–8628.

121. Desbien, A.L.; Kappler, J.W.; Marrack, P. The epstein-barr virus bcl-2 homolog, bhrf1, blocks apoptosis by binding to a limited amount of bim. *Proc. Natl. Acad. Sci. USA* **2009**, *106*, 5663–5668.

122. Saggioro, D.; Silic-Benussi, M.; Biasiotto, R.; D'Agostino, D.M.; Ciminale, V. Control of cell death pathways by htlv-1 proteins. *Front. Biosci.* **2009**, *14*, 3338–3351.

123. Underbrink, M.P.; Howie, H.L.; Bedard, K.M.; Koop, J.I.; Galloway, D.A. E6 proteins from multiple human betapapillomavirus types degrade bak and protect keratinocytes from apoptosis after uvb irradiation. *J. Virol.* **2008**, *82*, 10408–10417.

124. Garnett, T.O.; Filippova, M.; Duerksen-Hughes, P.J. Accelerated degradation of fadd and procaspase 8 in cells expressing human papilloma virus 16 e6 impairs trail-mediated apoptosis. *Cell Death Differ.* **2006**, *13*, 1915–1926.

125. Yuan, H.; Fu, F.; Zhuo, J.; Wang, W.; Nishitani, J.; An, D.S.; Chen, I.S.; Liu, X. Human papillomavirus type 16 e6 and e7 oncoproteins upregulate c-iap2 gene expression and confer resistance to apoptosis. *Oncogene* **2005**, *24*, 5069–5078.

126. Filippova, M.; Parkhurst, L.; Duerksen-Hughes, P.J. The human papillomavirus 16 e6 protein binds to fas-associated death domain and protects cells from fas-triggered apoptosis. *J. Biol. Chem.* **2004**, *279*, 25729–25744.

127. Du, J.; Chen, G.G.; Vlantis, A.C.; Chan, P.K.; Tsang, R.K.; van Hasselt, C.A. Resistance to apoptosis of hpv 16-infected laryngeal cancer cells is associated with decreased bak and increased bcl-2 expression. *Cancer Lett.* **2004**, *205*, 81–88.

128. Becker, S.A.; Lee, T.H.; Butel, J.S.; Slagle, B.L. Hepatitis b virus x protein interferes with cellular DNA repair. *J. Virol.* **1998**, *72*, 266–272.

129. Aweya, J.J.; Tan, Y.J. Modulation of programmed cell death pathways by the hepatitis c virus. *Front. Biosci.* **2011**, *16*, 608–618.

130. Plug-DeMaggio, A.W.; Sundsvold, T.; Wurscher, M.A.; Koop, J.I.; Klingelhutz, A.J.; McDougall, J.K. Telomere erosion and chromosomal instability in cells expressing the hpv oncogene 16e6. *Oncogene* **2004**, *23*, 3561–3571.

131. Duensing, S.; Munger, K. The human papillomavirus type 16 e6 and e7 oncoproteins independently induce numerical and structural chromosome instability. *Cancer Res.* **2002**, *62*, 7075–7082.

132. Duensing, S.; Duensing, A.; Crum, C.P.; Munger, K. Human papillomavirus type 16 e7 oncoprotein-induced abnormal centrosome synthesis is an early event in the evolving malignant phenotype. *Cancer Res.* **2001**, *61*, 2356–2360.

133. Duensing, S.; Lee, L.Y.; Duensing, A.; Basile, J.; Piboonniyom, S.; Gonzalez, S.; Crum, C.P.; Munger, K. The human papillomavirus type 16 e6 and e7 oncoproteins cooperate to induce mitotic defects and genomic instability by uncoupling centrosome duplication from the cell division cycle. *Proc. Natl. Acad. Sci. USA* **2000**, *97*, 10002–10007.

134. Chen, J.J. Genomic instability induced by human papillomavirus oncogenes. *North Am. J. Med. Sci.* **2010**, *3*, 43–47.

135. Bornkamm, G.W. Epstein-barr virus and its role in the pathogenesis of burkitt's lymphoma: An unresolved issue. *Semin. Cancer Biol.* **2009**, *19*, 351–365.

136. Heath, E.; Begue-Pastor, N.; Chaganti, S.; Croom-Carter, D.; Shannon-Lowe, C.; Kube, D.; Feederle, R.; Delecluse, H.J.; Rickinson, A.B.; Bell, A.I. Epstein-barr virus infection of naive b cells *in vitro* frequently selects clones with mutated immunoglobulin genotypes: Implications for virus biology. *PLoS Pathog.* **2012**, *8*, e1002697.

137. Marriott, S.J.; Lemoine, F.J.; Jeang, K.T. Damaged DNA and miscounted chromosomes: Human t cell leukemia virus type i tax oncoprotein and genetic lesions in transformed cells. *J. Biomed. Sci.* **2002**, *9*, 292–298.

138. Lemoine, F.J.; Marriott, S.J. Genomic instability driven by the human t-cell leukemia virus type i (htlv-i) oncoprotein, tax. *Oncogene* **2002**, *21*, 7230–7234.

139. Chandhasin, C.; Ducu, R.I.; Berkovich, E.; Kastan, M.B.; Marriott, S.J. Human t-cell leukemia virus type 1 tax attenuates the atm-mediated cellular DNA damage response. *J. Virol.* **2008**, *82*, 6952–6961.

140. Kao, S.Y.; Lemoine, F.J.; Marriott, S.J. P53-independent induction of apoptosis by the htlv-i tax protein following uv irradiation. *Virology* **2001**, *291*, 292–298.

141. Jin, D.Y.; Spencer, F.; Jeang, K.T. Human t cell leukemia virus type 1 oncoprotein tax targets the human mitotic checkpoint protein mad1. *Cell* **1998**, *93*, 81–91.

142. Liu, B.; Hong, S.; Tang, Z.; Yu, H.; Giam, C.Z. Htlv-i tax directly binds the cdc20-associated anaphase-promoting complex and activates it ahead of schedule. *Proc. Natl. Acad. Sci. USA* **2005**, *102*, 63–68.

143. Cohen, S.B.; Graham, M.E.; Lovrecz, G.O.; Bache, N.; Robinson, P.J.; Reddel, R.R. Protein composition of catalytically active human telomerase from immortal cells. *Science* **2007**, *315*, 1850–1853.

144. Terrin, L.; dal Col, J.; Rampazzo, E.; Zancai, P.; Pedrotti, M.; Ammirabile, G.; Bergamin, S.; Rizzo, S.; Dolcetti, R.; de Rossi, A. Latent membrane protein 1 of epstein-barr virus activates the htert promoter and enhances telomerase activity in b lymphocytes. *J. Virol.* **2008**, *82*, 10175–10187.

145. Zhang, X.; Dong, N.; Zhang, H.; You, J.; Wang, H.; Ye, L. Effects of hepatitis b virus x protein on human telomerase reverse transcriptase expression and activity in hepatoma cells. *J. Lab. Clin. Med.* **2005**, *145*, 98–104.

146. Verma, S.C.; Borah, S.; Robertson, E.S. Latency-associated nuclear antigen of kaposi's sarcoma-associated herpesvirus up-regulates transcription of human telomerase reverse transcriptase promoter through interaction with transcription factor sp1. *J. Virol.* **2004**, *78*, 10348–10359.

147. Gewin, L.; Myers, H.; Kiyono, T.; Galloway, D.A. Identification of a novel telomerase repressor that interacts with the human papillomavirus type-16 e6/e6-ap complex. *Genes Dev.* **2004**, *18*, 2269–2282.

148. Chen, X.; Kamranvar, S.A.; Masucci, M.G. Tumor viruses and replicative immortality—Avoiding the telomere hurdle. *Semin. Cancer Biol.* **2014**, *26*, 43–51.

149. Ohashi, M.; Sakurai, M.; Higuchi, M.; Mori, N.; Fukushi, M.; Oie, M.; Coffey, R.J.; Yoshiura, K.; Tanaka, Y.; Uchiyama, M.; *et al.* Human t-cell leukemia virus type 1 tax oncoprotein induces and interacts with a multi-pdz domain protein, magi-3. *Virology* **2004**, *320*, 52–62.

150. Glaunsinger, B.A.; Lee, S.S.; Thomas, M.; Banks, L.; Javier, R. Interactions of the pdz-protein magi-1 with adenovirus e4-orf1 and high-risk papillomavirus e6 oncoproteins. *Oncogene* **2000**, *19*, 5270–5280.

151. Rousset, R.; Fabre, S.; Desbois, C.; Bantignies, F.; Jalinot, P. The c-terminus of the htlv-1 tax oncoprotein mediates interaction with the pdz domain of cellular proteins. *Oncogene* **1998**, *16*, 643–654.

152. Lee, S.S.; Weiss, R.S.; Javier, R.T. Binding of human virus oncoproteins to hdlg/sap97, a mammalian homolog of the drosophila discs large tumor suppressor protein. *Proc. Natl. Acad. Sci. USA* **1997**, *94*, 6670–6675.

153. Spanos, W.C.; Hoover, A.; Harris, G.F.; Wu, S.; Strand, G.L.; Anderson, M.E.; Klingelhutz, A.J.; Hendriks, W.; Bossler, A.D.; Lee, J.H. The pdz binding motif of human papillomavirus type 16 e6 induces ptpn13 loss, which allows anchorage-independent growth and synergizes with ras for invasive growth. *J. Virol.* **2008**, *82*, 2493–2500.

154. Spanos, W.C.; Geiger, J.; Anderson, M.E.; Harris, G.F.; Bossler, A.D.; Smith, R.B.; Klingelhutz, A.J.; Lee, J.H. Deletion of the pdz motif of hpv16 e6 preventing immortalization and anchorage-independent growth in human tonsil epithelial cells. *Head Neck* **2008**, *30*, 139–147.

155. Xie, L.; Yamamoto, B.; Haoudi, A.; Semmes, O.J.; Green, P.L. Pdz binding motif of htlv-1 tax promotes virus-mediated t-cell proliferation *in vitro* and persistence *in vivo*. *Blood* **2006**, *107*, 1980–1988.

156. Hirata, A.; Higuchi, M.; Niinuma, A.; Ohashi, M.; Fukushi, M.; Oie, M.; Akiyama, T.; Tanaka, Y.; Gejyo, F.; Fujii, M. Pdz domain-binding motif of human t-cell leukemia virus type 1 tax oncoprotein augments the transforming activity in a rat fibroblast cell line. *Virology* **2004**, *318*, 327–336.

157. Bartel, D.P. Micrornas: Genomics, biogenesis, mechanism, and function. *Cell* **2004**, *116*, 281–297.

158. Lewis, B.P.; Burge, C.B.; Bartel, D.P. Conserved seed pairing, often flanked by adenosines, indicates that thousands of human genes are microrna targets. *Cell* **2005**, *120*, 15–20.

159. Calin, G.A.; Croce, C.M. Microrna signatures in human cancers. *Nat. Rev. Cancer* **2006**, *6*, 857–866.

160. Lu, J.; Getz, G.; Miska, E.A.; Alvarez-Saavedra, E.; Lamb, J.; Peck, D.; Sweet-Cordero, A.; Ebert, B.L.; Mak, R.H.; Ferrando, A.A.; *et al*. Microrna expression profiles classify human cancers. *Nature* **2005**, *435*, 834–838.

161. Cai, X.; Schafer, A.; Lu, S.; Bilello, J.P.; Desrosiers, R.C.; Edwards, R.; Raab-Traub, N.; Cullen, B.R. Epstein-barr virus micrornas are evolutionarily conserved and differentially expressed. *PLoS Pathog.* **2006**, *2*, e23.

162. Pfeffer, S.; Zavolan, M.; Grasser, F.A.; Chien, M.; Russo, J.J.; Ju, J.; John, B.; Enright, A.J.; Marks, D.; Sander, C.; *et al*. Identification of virus-encoded micrornas. *Science* **2004**, *304*, 734–736.

163. Choy, E.Y.; Siu, K.L.; Kok, K.H.; Lung, R.W.; Tsang, C.M.; To, K.F.; Kwong, D.L.; Tsao, S.W.; Jin, D.Y. An epstein-barr virus-encoded microrna targets puma to promote host cell survival. *J. Exp. Med.* **2008**, *205*, 2551–2560.

164. Dolken, L.; Malterer, G.; Erhard, F.; Kothe, S.; Friedel, C.C.; Suffert, G.; Marcinowski, L.; Motsch, N.; Barth, S.; Beitzinger, M.; *et al.* Systematic analysis of viral and cellular microrna targets in cells latently infected with human gamma-herpesviruses by risc immunoprecipitation assay. *Cell Host Microbe* **2010**, *7*, 324–334.

165. Marquitz, A.R.; Mathur, A.; Nam, C.S.; Raab-Traub, N. The epstein-barr virus bart micrornas target the pro-apoptotic protein bim. *Virology* **2011**, *412*, 392–400.

166. Marquitz, A.R.; Raab-Traub, N. The role of mirnas and ebv barts in npc. *Semin. Cancer Biol.* **2012**, *22*, 166–172.

167. Marquitz, A.R.; Mathur, A.; Shair, K.H.; Raab-Traub, N. Infection of epstein-barr virus in a gastric carcinoma cell line induces anchorage independence and global changes in gene expression. *Proc. Natl. Acad. Sci. USA* **2012**, *109*, 9593–9598.

168. Blattner, W.A. Human retroviruses: Their role in cancer. *Proc. Assoc. Am. Phys.* **1999**, *111*, 563–572.

169. Kim, H.H.; van den Heuvel, A.P.; Schmidt, J.W.; Ross, S.R. Novel common integration sites targeted by mouse mammary tumor virus insertion in mammary tumors have oncogenic activity. *PLoS One* **2011**, *6*, e27425.

170. Sourvinos, G.; Tsatsanis, C.; Spandidos, D.A. Mechanisms of retrovirus-induced oncogenesis. *Folia Biol.* **2000**, *46*, 226–232.

171. Chimal-Ramirez, G.K.; Espinoza-Sanchez, N.A.; Fuentes-Panana, E.M. Protumor activities of the immune response: Insights in the mechanisms of immunological shift, oncotraining, and oncopromotion. *J. Oncol.* **2013**, *2013*, 835956.

172. Elinav, E.; Nowarski, R.; Thaiss, C.A.; Hu, B.; Jin, C.; Flavell, R.A. Inflammation-induced cancer: Crosstalk between tumours, immune cells and microorganisms. *Nat. Rev. Cancer* **2013**, *13*, 759–771.

173. Schistosomes, Liver Flukes and Helicobacter Pylori. Iarc working group on the evaluation of carcinogenic risks to humans. Lyon, 7–14 June 1994. *IARC Monogr. Eval. Carcinog. Risks Hum.* **1994**, *61*, 1–241.

174. Fuentes-Panana, E.; Camorlinga-Ponce, M.; Maldonado-Bernal, C. Infection, inflammation and gastric cancer. *Salud Publica Mex.* **2009**, *51*, 427–433.

175. Kusters, J.G.; van Vliet, A.H.; Kuipers, E.J. Pathogenesis of *Helicobacter pylori* infection. *Clin. Microbiol. Rev.* **2006**, *19*, 449–490.

176. Xu, J.; Yin, Z.; Cao, S.; Gao, W.; Liu, L.; Yin, Y.; Liu, P.; Shu, Y. Systematic review and meta-analysis on the association between il-1b polymorphisms and cancer risk. *PLoS One* **2013**, *8*, e63654.

177. Xue, H.; Lin, B.; Ni, P.; Xu, H.; Huang, G. Interleukin-1b and interleukin-1 rn polymorphisms and gastric carcinoma risk: A meta-analysis. *J. Gastroenterol. Hepatol.* **2010**, *25*, 1604–1617.

178. Crusius, J.B.; Canzian, F.; Capella, G.; Pena, A.S.; Pera, G.; Sala, N.; Agudo, A.; Rico, F.; Del Giudice, G.; Palli, D.; *et al.* Cytokine gene polymorphisms and the risk of adenocarcinoma of the stomach in the european prospective investigation into cancer and nutrition (epic-eurgast). *Ann. Oncol.* **2008**, *19*, 1894–1902.

179. Cheng, D.; Hao, Y.; Zhou, W.; Ma, Y. Positive association between interleukin-8 -251a>t polymorphism and susceptibility to gastric carcinogenesis: A meta-analysis. *Cancer Cell Int.* **2013**, *13*, 100.

180. Poli, G. Pathogenesis of liver fibrosis: Role of oxidative stress. *Mol. Asp. Med.* **2000**, *21*, 49–98.

181. Cardenas-Mondragon, M.G.; Carreon-Talavera, R.; Camorlinga-Ponce, M.; Gomez-Delgado, A.; Torres, J.; Fuentes-Panana, E.M. Epstein barr virus and helicobacter pylori co-infection are positively associated with severe gastritis in pediatric patients. *PLoS One* **2013**, *8*, e62850.

182. Cárdenas-Mondragón, M.; Carreón-Talavera, R.; Flores-Luna, L.; Camorlinga-Ponce, M.; Gómez-Delgado, A.; Torres, J.; Fuentes-Pananá, E. Epstein barr virus reactivation is an important trigger of gastric inflammation and progression to intestinal type gastric cancer. In Proceedings of the Epstein Barr Virus 50th Anniversary Conference, Keble College, Oxford, UK, 23–25 March, 2014.

183. Bernstein, W.B.; Little, R.F.; Wilson, W.H.; Yarchoan, R. Acquired immunodeficiency syndrome-related malignancies in the era of highly active antiretroviral therapy. *Int. J. Hematol.* **2006**, *84*, 3–11.

184. Young, L.S.; Murray, P.G. Epstein-barr virus and oncogenesis: From latent genes to tumours. *Oncogene* **2003**, *22*, 5108–5121.

185. Young, L.; Alfieri, C.; Hennessy, K.; Evans, H.; O'Hara, C.; Anderson, K.C.; Ritz, J.; Shapiro, R.S.; Rickinson, A.; Kieff, E.; *et al.* Expression of epstein-barr virus transformation-associated genes in tissues of patients with ebv lymphoproliferative disease. *N. Engl. J. Med.* **1989**, *321*, 1080–1085.

186. Nourse, J.P.; Jones, K.; Gandhi, M.K. Epstein-barr virus-related post-transplant lymphoproliferative disorders: Pathogenetic insights for targeted therapy. *Am. J. Transplant.* **2011**, *11*, 888–895.

187. Gandhi, M.K.; Wilkie, G.M.; Dua, U.; Mollee, P.N.; Grimmett, K.; Williams, T.; Whitaker, N.; Gill, D.; Crawford, D.H. Immunity, homing and efficacy of allogeneic adoptive immunotherapy for posttransplant lymphoproliferative disorders. *Am. J. Transplant.* **2007**, *7*, 1293–1299.

188. Haque, T.; Wilkie, G.M.; Jones, M.M.; Higgins, C.D.; Urquhart, G.; Wingate, P.; Burns, D.; McAulay, K.; Turner, M.; Bellamy, C.; *et al.* Allogeneic cytotoxic t-cell therapy for ebv-positive posttransplantation lymphoproliferative disease: Results of a phase 2 multicenter clinical trial. *Blood* **2007**, *110*, 1123–1131.

189. Michaelis, M.; Doerr, H.W.; Cinatl, J. The story of human cytomegalovirus and cancer: Increasing evidence and open questions. *Neoplasia* **2009**, *11*, 1–9.

190. Boldogh, I.; Huang, E.S.; Rady, P.; Arany, I.; Tyring, S.; Albrecht, T. Alteration in the coding potential and expression of h-ras in human cytomegalovirus-transformed cells. *Intervirology* **1994**, *37*, 321–329.

191. Geder, K.M.; Lausch, R.; O'Neill, F.; Rapp, F. Oncogenic transformation of human embryo lung cells by human cytomegalovirus. *Science* **1976**, *192*, 1134–1137.

192. Geder, L.; Kreider, J.; Rapp, F. Human cells transformed *in vitro* by human cytomegalovirus: Tumorigenicity in athymic nude mice. *J. Natl. Cancer Inst.* **1977**, *58*, 1003–1009.

193. Geder, L.; Laychock, A.M.; Gorodecki, J.; Rapp, F. Alterations in biological properties of different lines of cytomegalorivus-transformed human embryo lung cells following *in vitro* cultivation. *IARC Sci. Publ.* **1978**, *24*, 591–601.

194. Cinatl, J.; Scholz, M.; Kotchetkov, R.; Vogel, J.U.; Doerr, H.W. Molecular mechanisms of the modulatory effects of hcmv infection in tumor cell biology. *Trends Mol. Med.* **2004**, *10*, 19–23.

195. Scheurer, M.E.; Bondy, M.L.; Aldape, K.D.; Albrecht, T.; El-Zein, R. Detection of human cytomegalovirus in different histological types of gliomas. *Acta Neuropathol.* **2008**, *116*, 79–86.

196. Cobbs, C.S.; Soroceanu, L.; Denham, S.; Zhang, W.; Kraus, M.H. Modulation of oncogenic phenotype in human glioma cells by cytomegalovirus ie1-mediated mitogenicity. *Cancer Res.* **2008**, *68*, 724–730.

197. Maussang, D.; Verzijl, D.; van Walsum, M.; Leurs, R.; Holl, J.; Pleskoff, O.; Michel, D.; van Dongen, G.A.; Smit, M.J. Human cytomegalovirus-encoded chemokine receptor us28 promotes tumorigenesis. *Proc. Natl. Acad. Sci. USA* **2006**, *103*, 13068–13073.

198. Vallat, L.; Benhamou, Y.; Gutierrez, M.; Ghillani, P.; Hercher, C.; Thibault, V.; Charlotte, F.; Piette, J.C.; Poynard, T.; Merle-Beral, H.; *et al.* Clonal b cell populations in the blood and liver of patients with chronic hepatitis c virus infection. *Arthritis Rheum.* **2004**, *50*, 3668–3678.

199. Quinn, E.R.; Chan, C.H.; Hadlock, K.G.; Foung, S.K.; Flint, M.; Levy, S. The b-cell receptor of a hepatitis c virus (hcv)-associated non-hodgkin lymphoma binds the viral e2 envelope protein, implicating hcv in lymphomagenesis. *Blood* **2001**, *98*, 3745–3749.

200. Chan, C.H.; Hadlock, K.G.; Foung, S.K.; Levy, S. V(h)1–69 gene is preferentially used by hepatitis c virus-associated b cell lymphomas and by normal b cells responding to the e2 viral antigen. *Blood* **2001**, *97*, 1023–1026.

201. Hermine, O.; Lefrere, F.; Bronowicki, J.P.; Mariette, X.; Jondeau, K.; Eclache-Saudreau, V.; Delmas, B.; Valensi, F.; Cacoub, P.; Brechot, C.; *et al.* Regression of splenic lymphoma with villous lymphocytes after treatment of hepatitis c virus infection. *N. Engl. J. Med.* **2002**, *347*, 89–94.

202. Zullo, A.; Hassan, C.; Cristofari, F.; Andriani, A.; de Francesco, V.; Ierardi, E.; Tomao, S.; Stolte, M.; Morini, S.; Vaira, D. Effects of helicobacter pylori eradication on early stage gastric mucosa-associated lymphoid tissue lymphoma. *Clin. Gastroenterol. Hepatol.* **2010**, *8*, 105–110.

203. Skinner, G.R. Transformation of primary hamster embryo fibroblasts by type 2 simplex virus: Evidence for a "hit and run" mechanism. *Br. J. Exp. Pathol.* **1976**, *57*, 361–376.

204. Birdwell, C.E.; Queen, K.J.; Kilgore, P.C.; Rollyson, P.; Trutschl, M.; Cvek, U.; Scott, R.S. Genome-wide DNA methylation as an epigenetic consequence of epstein-barr virus infection of immortalized keratinocytes. *J. Virol.* **2014**, *88*, 11442–11458.

205. Burns, M.B.; Temiz, N.A.; Harris, R.S. Evidence for apobec3b mutagenesis in multiple human cancers. *Nat. Genet.* **2013**, *45*, 977–983.

The Role of Merkel Cell Polyomavirus and Other Human Polyomaviruses in Emerging Hallmarks of Cancer

Ugo Moens, Kashif Rasheed, Ibrahim Abdulsalam and Baldur Sveinbjørnsson

Abstract: Polyomaviruses are non-enveloped, dsDNA viruses that are common in mammals, including humans. All polyomaviruses encode the large T-antigen and small t-antigen proteins that share conserved functional domains, comprising binding motifs for the tumor suppressors pRb and p53, and for protein phosphatase 2A, respectively. At present, 13 different human polyomaviruses are known, and for some of them their large T-antigen and small t-antigen have been shown to possess oncogenic properties in cell culture and animal models, while similar functions are assumed for the large T- and small t-antigen of other human polyomaviruses. However, so far the Merkel cell polyomavirus seems to be the only human polyomavirus associated with cancer. The large T- and small t-antigen exert their tumorigenic effects through classical hallmarks of cancer: inhibiting tumor suppressors, activating tumor promoters, preventing apoptosis, inducing angiogenesis and stimulating metastasis. This review elaborates on the putative roles of human polyomaviruses in some of the emerging hallmarks of cancer. The reciprocal interactions between human polyomaviruses and the immune system response are discussed, a plausible role of polyomavirus-encoded and polyomavirus-induced microRNA in cancer is described, and the effect of polyomaviruses on energy homeostasis and exosomes is explored. Therapeutic strategies against these emerging hallmarks of cancer are also suggested.

Reprinted from *Viruses*. Cite as: Moens, U.; Rasheed, K.; Abdulsalam, I.; Sveinbjørnsson, B. The Role of Merkel Cell Polyomavirus and Other Human Polyomaviruses in Emerging Hallmarks of Cancer. *Viruses* **2015**, *7*, 1871-1901.

1. Introduction

Polyomaviruses are naked, circular double-stranded DNA viruses that infect birds and mammals, and recently the first fish-associated polyomavirus was described [1,2]. The genome of most polyomaviruses is approximately 5000 base-pairs and encodes regulatory proteins and structural proteins. The major regulatory proteins are the large tumor antigen (LT-ag) and the small tumor antigen (st-ag), while at least two structural proteins (VP1 and VP2) form the capsid. The regulatory proteins are expressed early during infection and participate in viral replication and viral transcription, while the structural proteins are expressed later in the infection cycle [3]. Many polyomaviruses encode additional regulatory and structural proteins (e.g., ALTO, VP3, VP4, agnoprotein) [4–6].

Studies with mice in the 1950s initiated by Ludwik Gross, and extended by Sarah Stewart and Bernice Eddy led to the identification of the first polyomavirus. They showed that a filtrate from a mouse leukaemia could cause multiple tumors in new-born mice and later it was demonstrated that these multiple tumors were virus-indeed. Hence the virus was referred to as polyomavirus from the Greek πολύσ for many and ωμα for tumors (reviewed in [7]). The first primate polyomavirus was isolated in 1960 [8]. This virus, Simian virus 40 (SV40), was shown to transform cells, including human cells, to induce tumors in animal models, and to be present in human cancers. The oncogenic potential of SV40 primarily depends on its LT-ag, which can bind the tumor suppressor proteins p53 and pRb, interfere with DNA repair, apoptosis, cellular transcription, protein degradation, telomerase activity, immune- and inflammatory responses, and stimulate angiogenesis and cell migration. SV40 st-ag can contribute to transformation by inactivating protein phosphatase 2A [9,10]. Besides SV40 and murine polyomavirus, other non-human polyomavirus such as hamster polyomavirus, lymphotropic polyomavirus, and simian agent 12 were shown to possess oncogenic properties in cell cultures or animal models [11–13]. However, the oncogenic role of these viruses in their natural host is unclear. In fact, only one mammalian polyomavirus seems to be firmly associated with cancer in its genuine host. Raccoon polyomavirus (RacPyV) was first identified in tumors of frontal lobes and olfactory tracts from raccoons. Ten out of 52 (19%) raccoons had brain tumors within the cranial portion of their frontal lobe(s), and all tumors contained RacPyV DNA, though not tissues from 20 unaffected animals. RacPyV genome was episomal in all tumors tested [14]. One case of hamster polyomavirus-induced lymphoma in a hamster outside of the laboratory environment has been described [15], while two novel mammalian polyomaviruses have been isolated from benign tumors. A polyomavirus was isolated from fibropapilloma on the tongue of a sea lion, and the complete genome of another polyomavirus was amplified in a biopsy from a fibroma on the trunk of an African elephant [16,17]. Further studies are required to assess whether these mammals are the genuine host, and whether these polyomaviruses are the causal infectious agent of such hyperplastic fibrous tissue in their natural host.

In contrast to mammalian polyomaviruses, bird polyomaviruses do not seem to induce tumors. Despite a similar genetic organization to that of mammalian polyomavirus, their LT-ag lacks homologies to the p53 binding sequences of mammalian polyomavirus and not all avian polyomavirus LT-ag possess the consensus sequence LXCXE required for pRb binding [18].

2. Human Polyomaviruses and Cancer

The first two human polyomavirus viruses were isolated in 1971, and were named after the initials of the patient in which the virus was found: the BK virus (BKPyV) and the JC virus (JCPyV) [19,20]. Both BKPyV and JCPyV possess a genomic organization that resembles SV40 more than the murine polyomavirus. The former three viruses lack the middle T-antigen that is encoded by the murine polyomavirus, but have an additional late gene referred to as the agnogene [3]. Because the genomic organization of SV40 displays a higher functional and sequence similarity with the BKPyV and JCPyV, SV40 became the polyomavirus model system for unveiling the oncogenic mechanisms of this family [21,22]. Since 2007, 11 novel human polyomaviruses have been described: KIPyV, WUPyV, Merkel cell PyV (MCPyV), HPyV6, HPyV7, *Trichodysplasia spinulosa*-associated PyV (TSPyV), HPyV9, HPyV10 (and the isolates MW and MX), STLPyV, HPyV12, and NJPyV-2013 [23–35]. The seroprevalence of the different human polyomavirus ranges from ~25% to ~100% depending on the virus. The high seropositivity therefore demonstrates that these viruses are common in the adult human population [36–38].

Whereas the oncogenic properties of BKPyV, JCPyV and MCPyV in cell culture and animal models are well-documented [39–42], only MCPyV seems to be associated with cancer in its natural host. Approximately 80% of Merkel cell carcinoma tumors are positive for the MCPyV genome, which is typically integrated and encodes a truncated form of LT-ag [43]. BKPyV and JCPyV DNA, RNA and proteins have been detected in several tumor tissues, but are also often present in control non-malignant tissues [44–46]. Hence, a causal role for these viruses in human cancers remains controversial, although the presence of BKPyV may increase the risk of the development of renal and prostate cancer, while JCPyV may be associated with colorectal cancer and CNS tumors [47–50]. Polyomavirus-associated colorectal cancer may be due to other polyomaviruses present in meat as suggested by Harald zur Hausen [51]. Recent analyses of beef samples have identified several bovine polyomaviruses related to the human polyomaviruses MCPyV, HPyV 6, HPyV7 or other animal polyomaviruses including fruit bat polyomavirus, RacPyV and chimpanzee polyomavirus [52,53]. It remains to be established whether these viruses can be detected in human colorectal biopsies. The possible association of the other human polyomaviruses with cancer has been scarcely examined, and in only few cases was viral DNA or protein detected in tumor tissue (Table 1). Based on our present knowledge, convincing proof of their role in these cancers is lacking.

Table 1. Prevalence of the novel human polyomaviruses in human cancers. BK virus (BKPyV), JC virus (JCPyV), Merkel cell PyV (MCPyV) are not included.

	Number of samples	Method	Number of positive samples	Comments	Reference
Melanoma (st-age IV)	18	PCR and IHC (HPyV6 VP1moAb)	HPyV6: 18 HPyV7: 17 TSPyV: 4 HPyV9: 1 HPyV10: 12	Low viral DNA loads, but higher for HPyV6	[54]
Mucosal melanoma	37	PCR	KIPyV: 0 WUPyV: 0 HPyV6:0 HPyV7:0 TSPyV: 0 HPyV9:0 MWPyV: 0		[55]
Squamous cell carcinoma	63	PCR	HPyV6: 2 HPyV7: 1	Low viral DNA loads	[56]
Basal cell carcinoma	50	PCR	HPyV6: 1 HPyV7: 2	Low viral DNA loads	[56]
Melanoma	47	PCR	HPyV6: 2 HPyV7: 2	Low viral DNA loads	[56]
Basal cell carcinoma	41	PCR	HPyV6:3 HPyV7:0 TSPyV: 0 HPyV9:0		[57]

Table 1. *Cont.*

	Number of samples	Method	Number of positive samples	Comments	Reference
Squamous cell carcinoma	52	PCR	HPyV6:2 HPyV7:0 TSPyV: 0 HPyV9:0		[57]
SCC *in situ*	8	PCR	HPyV6:1 HPyV7:0 TSPyV: 0 HPyV9:0		[57]
Keratoacanthoma	42	PCR	HPyV6:2 HPyV7:0 TSPyV: 0 HPyV9:0		[57]
Microcystic adnexal carcinoma	5	PCR	HPyV6:0 HPyV7:0 TSPyV: 0 HPyV9:0		[57]
Atypical fibroxanthoma	14	PCR	HPyV6:0 HPyV7:0 TSPyV: 0 HPyV9:0		[57]
Actinic keratosis	31	PCR	HPyV6:1 HPyV7:0 TSPyV: 0 HPyV9:0		[57]
Breast cancer	54	PCR	HPyV6: 1 HPyV7:1		[58]
Merkel cell carcinoma		deep sequencing	HPyV6: 1 HPyV7:1 HPyV9:1		[59]
Extracutaneous melanoma	38	PCR	KIPyV: 0 WUPyV: 0		[60]
SCC+AK	142	deep sequencing	HPyV6: 1		[61]
Chronic lymphocytic leukaemia	27	PCR	HPyV9: 0		[62]
Primary cutaneous B-cell lymphomas (CBCLs) or cutaneous T-cell lymphomas (CTCLs)	130	PCR	HPyV6: 6 HPyV7: 1 TSPyV: 0		[63]

Table 1. *Cont.*

	Number of samples	Method	Number of positive samples	Comments	Reference
MCC	28	PCR	HPyV6: 0 HPyV7:0		[64]
Pilomatricomas (benign skin tumor associated with hair follicles	?	?	TSPyV: 0		[65]
Lung cancer	20	PCR	KIPyV:9		[66]
CNS tumors	25	PCR	KIPyV: 0 WUPyV: 0		[67]
Neuroblastoma	31	PCR	KIPyV: 0 WUPyV: 0		[67]
Acute lymphoblastic leukaemia	50	PCR	KIPyV: 0 WUPyV: 0		[68]
Lung cancer	30	PCR	KIPyV: 0 WUPyV: 0		[69]
	32	PCR	KIPyV: 0 WUPyV: 0		[70]
Neuroendocrine tumors	50	PCR	KIPyV: 0 WUPyV: 0 HPyV6:0 HPyV7:0 TSPyV: 0		[71]
Skin lesions from CTCL patients	39	PCR	HPyV6:11 HPyV7:5 TSPyV: 0 HPyV9:0		[72]
Blood from CTCL patients	39	PCR	HPyV6:0 HPyV7:0 TSPyV: 0 HPyV9:0		[72]
Glioblastoma multiforme	39	PCR	HPyV6:0 HPyV7:0 HPyV9:0		[73]
Thymic epithelial tumors	37	PCR, FISH, IHC	PCR FISH IHC HPyV7: 20 23 17 HPyV6: 0		[74]
Thymic hyperplasias	20		HPyV7: 8 14 6		
Foetal thymus tissue	20		HPyV7: 0		

The cancer biology of BKPyV, JCPyV and MCPyV has been extensively reviewed by others [39,43,45,46,75–77] and is also discussed by others in this special issue on Tumor Viruses.

This review will focus on novel strategies that human polyomaviruses may use to transform cells. Figure 1 summarizes the novel mechanisms by which HPyV may contribute to cancer.

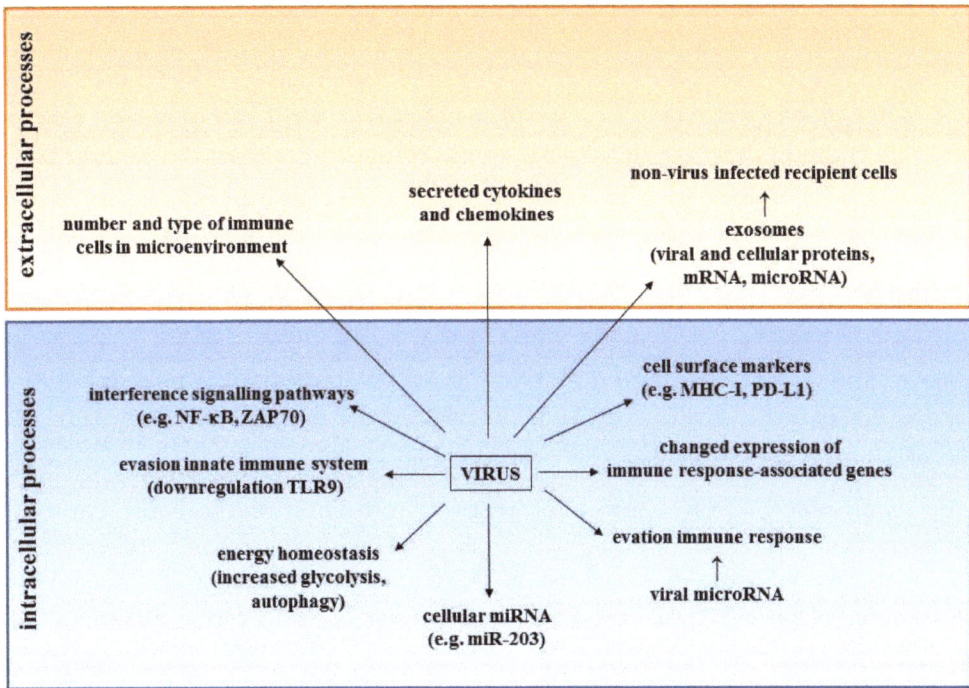

Figure 1. Novel mechanisms by which HPyV may contribute to cancer. See text and Table 2 for details.

3. HPyV and Emerging Hallmarks of Cancer

3.1. The Immune System and HPyV in Cancer

Individuals with a dysfunctional immune system are more disposed to diseases, infections and (viral-induced) cancers. Moreover, oncoviruses can induce inflammation, which may predispose host cells to acquire carcinogenic mutations [78]. In accordance with the cancer immunoediting hypothesis, tumor cells need to proficiently traverse separate phases in a sequential order to attain cancer manifestation and progression. These phases constitute interactions between the immune system and the cancer cell, and include the elimination of newly transformed cells, an equilibrium in which the immune system restrains the outgrowth of tumors, and an escape in which the tumor cells are able to circumvent the host immune response phases [79–81]. For a virus to induce tumors, they need to circumvent elimination by the immune system and to induce alternations in the tumor microenvironment, including in the infected cell allowing the virus-transformed cell to progress [82,83]. Because MCPyV is the only HPyV associated with cancer, the main focus will be on MCPyV's interaction with the immune system.

Epidemiologic data show that patients with T cell dysfunction are at a 5- to 50-fold increased risk of developing MCC, thereby indicating the importance of the immune system (reviewed in [83]). However, immunocompetent individuals may also develop MCPyV-positive MCC, suggesting that the virus and virus-infected cells can avoid elimination by the immune system.

3.1.1. HPyV and Evasion of the Innate Immune System

One mechanism by which MCPyV circumvents the immune system is to abate the innate defence mechanism. MCPyV LT-ag and st-ag downregulate the Toll-like receptor 9 (TLR9), an important receptor of the host innate immune system that senses viral dsDNA in epithelial and MCC cells [84]. LT-ag inhibits TLR9 expression by decreasing the mRNA levels of the transcription factor C/EBPβ. LT-ag of BKPyV, but not JCPyV, KIPyV and WUPyV, is also able to repress TLR9 expression. Interestingly, C/EBPβ has a vital role in regulating IL-6, IL-8, and TNF-α cytokine transcription [85]. Moreover, it is also suggested that C/EBPβ has a tumor-suppressive activity by down-regulating CDK2, CDK4, and E2F complex activity [86,87]. Thus MCPyV LT-ag mediated suppression of C/EBPβ expression may perturb immune responses and provoke cell proliferation.

3.1.2. Immune Cells in the Microenvironment of MCC

To investigate inflammatory modulators in MCC required for escaping of the tumor from immune surveillance, and to deduce a possible contribution of MCPyV in oncogenesis, several groups have examined immune cells and inflammatory mediators virus-positive and virus-negative MCC. Differences in immune and inflammatory cells, markers, and gene expression in MCPyV-positive and MCPyV-negative MCC tumors are summarized in Table 2. Compared to virus-negative tumors, a higher number of infiltrating CD8+ T-cells in MCPyV-positive MCC has been observed [89–91], while others group have not detected a relationship with virus status and the number of intratumoral CD8+ T-cells [92,93]. Other differences in the microenvironment of virus-positive and virus-negative MCC include a higher number of CD3+ T-cells, CD20+ B cells, CD16+ natural killer cells, and CD68+, CD69+, CD163+ macrophages [88–90,93–95]. FoxP3+ regulatory T-cells were present in 4/4 LT-ag positive MCC, whereas 3/6 LT-ag negative tumors did not contain FoxP3+ regulatory T-cells [93].

Afanasieve *et al.* proposed that MCC tumors may prevent the invasion of lymphocytes by a reduction of E-selectin-positive vessels within the tumors because the downregulation of E-selectin in human squamous cell carcinomas was associated with a restricted entry of T-cells into tumors [100,101]. Of 56 tested MCC biopsies, approximately half displayed a reduction of E-selectin-positive vessels within the tumors compared with vessels in peritumoral areas [102]. However, the association between the presence of virus and E-selectin levels was not investigated.

Table 2. Immune cells and inflammatory mediators in MCPyV-positive and MCPyV-negative Merkel cell carcinoma (MCC).

Component	MCPyV-positive *versus* MCPyV-negative MCC	Reference
Cells in tumor microenvironment		
-CD3+ T-cells		
	higher number in MCPyV-positive MCC	[88–90]
-CD4+ T-cells		
	high number associated with high LT-ag expression	[90]
-CD8+ T-cells	higher number in MCPyV-positive MCC	
		[89,91,92]
- CD16+ natural killer cell	higher number in MCPyV-positive MCC	
		[88,90]
-CD20+ B cells	more common in MCPyV-positive MCC;	
	no significant difference between MCPyV-positive and –	[93]
	negative MCC	[89]
-CD68+ macrophages	higher number in MCPyV-positive MCC	
		[88,90,94,95]
-CD69+ macrophages	higher number in MCPyV-positive MCC	
		[90,94,95]
-FoxP3+ regulatory T-cells	more common in MCPyV-positive MCC	
		[93]
Cell surface markers:		
-CD3D	enrichment of transcripts in MCPyV-positive MCC	[89]
	enrichment of transcripts in MCPyV-positive MCC	
-CD3G	lacking in CD8+ T-cells	[89]
-CXCR3	lower levels in MCPyV-positive MCC	[93]
-MHC-I	higher in MCPyV-positive MCC	[96]
-PD1	higher in MCPyV-positive MCC	[95,97,98]
-Tim-3		[97]
Signal transduction proteins		
-NFκB levels	lower in MCPyV-positive MCC	[99]
-IκB levels	lower in MCPyV-positive MCC	[99]
-TANK		
	reduction in MCPyV st-ag expressing cells MCC13 cells	[99]
- ZAP70	compared to virus-negative cells	
	enrichment of transcripts in MCPyV-positive MCC	[89]

Table 2. *Cont.*

Component	MCPyV-positive *versus* MCPyV-negative MCC	Reference
Cytokines/chemokines		
-CCL20	reduction in MCPyV st-ag expressing cells MCC13 cells compared to virus-negative cells	[99]
-CXCL-9	reduction in MCPyV st-ag expressing cells MCC13 cells compared to virus-negative cells	[99]
-IL-2	reduction in MCPyV st-ag expressing cells MCC13 cells compared to virus-negative cells	[99]
-IL-8	reduction in MCPyV st-ag expressing cells MCC13 cells compared to virus-negative cells	[99]
-Prokineticin 1 mRNA	higher in MCPyV-negative MCC	[90]
-Prokineticin 2 mRNA	higher in MCPyV-positive MCC	[90]
Other differentially expressed proteins		
-granzyme B (role in apoptosis)	Expression was rare in CD8+ cells	[93]

3.1.3. Changes in Expression of Cell Surface Markers on MCC Cells

Expression of cell surface markers was performed to determine the functionality of the immune cells. These analyses revealed that the expression of MHC-I in MCPyV-positive MCC was significantly lower than in virus-negative MCC [96]. Cell-surface MHC-I expression was down-regulated in 84% (n = 114) of MCC, and approximately half of the tumors had poor or undetectable MHC-I levels. The downregulation of MHC-I expression has been identified as a vital immune evasion strategy used by several viruses, including oncoviruses [103–108]. An identical mechanism can be employed by MCPyV, but it remains to be determined as to whether viral proteins are implicated in MHC-I down-regulation. Tumors that undergo a significant downregulation of MHC-I should become a target of natural killer cells. MCC can avoid this by e.g., reducing the expression of NK-activating receptors such as natural killer group 2, member D (NKG2D) [109]. Interestingly, BKPyV and JCPyV microRNA target ULBP3, which is the ligand of NKG2D (see further), though it is not known whether MCPyV microRNA targets ULBP3 or NKG2D. Another surface marker that was differentially expressed on MCPyV-positive and negative tumors is the immune-inhibitory ligand programmed death ligand-1 (PD-L1) [95,97,98]. The major receptor for PD-L1, PD-1 is expressed by activated T lymphocytes, and when this receptor is engaged by its ligands PD-L1 it serves to inhibit the T-cell response. PD-L1 may be aberrantly expressed by tumor cells and protect against immune attack [110]. The number of intra-tumor T-cells is commonly higher in virus-positive MCC than virus-negative MCC, and PD-1 was expressed on a high percentage of MCPyV-positive tumors [95,97,98]. Moreover, approximately 50% of MCPyV-positive MCC express PD-L1 on tumor cells, while no expression was detected in MCPyV-negative MCC. Hence, the association between PD-1-positive cells and PD-L1 expression in the tumor microenvironment seems to create immune resistance by the tumor, thereby allowing the tumor to progress [97,98]. The mechanism by which MCPyV provokes the expression of PD-L1 remains to be determined, but Lipson and co-workers anticipated that IFN-γ may drive PD-L1 expression, but other interleukins such as IL-6, IL-10, IL-17 and IL-21 cannot be excluded. A role for PD-1 positive cells in protecting PD-L1-expressing MCC cells is buttressed by observations in a complete or partial regression of MCC. The exact mechanism for spontaneous regression is not known, although T-cell-mediated response and apoptosis by T-cells has been suggested [111]. The rate of regression of MCPyV-positive *versus* MCPyV-negative MCC has not been evaluated, but complete regression has been reported in a 76-year old Japanese man with virus-positive MCC [112]. In this patient, only ~3% of the tumor-infiltrating T-cells were PD-1 positive, while in three other patients (females, mean age 81.3 years) with MCPyV-positive MCC who did not show any regression of the tumor, 18.2%–23.0% of the T-cells were of PD-1 positive. This suggests that a reduction of PD-1-positive T-cells may be associated with spontaneous tumor regression [112]. Another surface protein that was aberrantly expressed on immune cells in the tumor microenvironment was CXCR3 [97]. All CD8+ cells lacked CXCR3, thus indicating that these T-cells were functionally compromised. CXCL12 or stromal cell-derived factor 1, a chemokine with pleiotropic functions, including the attraction of inflammatory cells [113], was expressed outside malignant nodules, but its receptor CXCR4 was expressed by tumor cells, though not on infiltrating CD8+ cells. Finally, the cell-surface protein T-cell immunoglobulin and mucin domain-3 (Tim-3), which also functions to inhibit T-cell responses, was also upregulated on infiltrating T-cells in MCPyV-positive MCC [97].

3.1.4. Expression Profile of Genes Associated with the Immune Response in MCC

Gene expression profile analysis has been applied to identify differentially expressed genes in MCPyV-positive and MCPyV-negative MCC. Microarray technology, using >54,000 probes, identified 1593 genes that were differently (≥2-fold) expressed comparing virus-positive and virus-negative MCCs [89]. An enrichment of genes associated with the immune response included genes encoding the δ and γ chains of CD3, the tyrosine kinase ZAP70, which plays an important role in the T-cell response, and the C-region of the μ heavy chain. Another approach compared the transcriptome from cells with an inducible expression of MCPyV st-ag with that of control cells, and revealed that the induction of st-ag expression resulted in ≥2-fold reduced transcript levyels of genes associated with the immune response such as CCL20, CXCL-9, IL-2, IL-8 and TANK, a negative regulator of TLR signaling. Less CCL20 and IL-8 were secreted by MCC13 cells expressing MCPyV st-ag compared with virus-negative MCC13 cells after TNFα stimulation [99]. mRNA profiling of 35 MCC tumors (both MCPyV-positive and negative) with favorable prognoses overexpressed genes such as components of cytotoxic granules (granzymes A, B, H and K), chemokine CCL19 and chemokine receptor 2, MHC-II and NKG2D [92]. The contribution of MCPyV on the expression of these genes cannot be appreciated because the data originate from both MCPyV-positive and MCPyV-negative tumors. Gene expression profiling of MCC tumor cells showed the lack of expression of IL-2 and IFN-γ, whereas IL-12 was expressed [113]. However, this study was performed before MCPyV was identified, so therefore a role of the virus in altered gene expression cannot be deduced. Another study monitored the transcript levels of the chemokine-like proteins prokineticin-1 and prokineticin-2, which are involved in angiogenesis, inflammation and cancer. MCPyV-positive MCCs had a higher than median prokineticin-2 mRNA levels, while virus-negative tumors had a higher than median prokineticin-1 transcript levels [90]. A high tumor prokineticin-2 mRNA content was associated with the expression of MCPyV LT-ag. The biological relevance of this observation for virus-induced MCC remains to be established. Wheat and co-workers observed that the expression of granzyme B, a mediator of apoptosis [114], was rare in MCC infiltrating CD8+ cells, hence suggesting that these cytotoxic T cells were functionally compromised [93].

3.1.5. Effect of st-ag on the NF-κB Pathway

The molecular mechanism by which MCPyV may perturb gene expression in virus-positive MCC tumor cells is not known, but several of the genes listed in Table 2 (e.g., CXCL9, IL-2, IL-8, MHC-I, IκB) are known to be a target for NF-κB [115,116]. Interestingly, MCPyV st-ag was shown to downregulate NF-κB-mediated transcription [99]. St-ag-mediated inhibition of the NF-κB pathway seems to require an interaction of st-ag with NF-κB essential modulator (NEMO) adaptor protein and protein phosphatases 2A and 4C. This will prevent IKKα/IIKβ-mediated phosphorylation of IκB, thus leading to a reduced nuclear translocation of NF-κB. MCPyV interference with the NF-κB pathway is further sustained by the observations that IκB levels were 60% lower in the MCPyV-positive MCC cell line MKL-1 compared with MCPyV-negative MCC13 cells, and by a declined expression of NF-κB and NF-κB-associated genes in virus-positive MCC compared to virus-negative MCC [99,117]. All these findings indicate that MCPyV interferes with the NF-κB pathway, and that MCPyV st-ag may help the virus to evade the host antiviral defence and to persist in the infected cell [99]. It is not known whether the st-ag of other HPyV has the same property, but residues 95 to 111, which are crucial for the interaction between MCPyV st-ag, NEMO and PP2A and PP4C are not conserved [118]. Interestingly, ultraviolet (UV) exposure, a risk factor for MCC [119], was shown to stimulate mutations in LT-ag and increase the expression of st-ag in the tumor cells [120]. Hence, UV exposure may be a virus-dependent mechanism that promotes MCPyV-induced MCC through the aforementioned st-ag:NF-κB interaction.

3.1.6. Viral Microrna and Evation of the Immune Response

Another mechanism by which HPyV may affect gene expression is by microRNA. MicroRNAs (miRNAs) are small RNAs that can down-regulate protein production by either degrading transcripts or inhibiting the translation of mRNA. SV40 miRNA, the first PyV miRNA to be described, was shown to reduce cytotoxic T lymphocyte-mediated lysis and IFN-γ release [121], whereas other HPyV seem to apply different strategies to escape the immune system. BKPyV, JCPyV and MCPyV miRNA were unable to inhibit IFN-induced transcription of the luciferase reporter gene [122], but BKPyV and JCPyV miRNAs inhibited the translation of UL16-binding protein 3 (ULBP3) mRNA [123]. ULBP3 is a ligand recognized by natural killer group 2, member D (NKG2D) receptor. NKG2D is expressed by NK and CD8+ T-cells and binding to ULBP3 triggers killing of the target cell [124]. Consequently, BKPyV- and JCPyV-infected cells may escape from NKG2D-mediated killing and circumvent the immune system. The proteins PSME3 and PIK3CD/p110δ, which are implicated in immune functions, were predicted to be putative targets for MCPyV miRNA [125]. PSME3 is a subunit of a proteasome responsible for the generation of peptides loaded onto MHC I, and PI3KCD plays a unique role in antigen receptor signaling by activating T-cells and B-cell proliferation [126–128]. The depletion of these proteins may prevent MCPyV infection to be cleared by the immune system, thereby allowing the viral infection to sustain. One of the SV40 strain RI257 miRNA targets is α-actinin 4 (ACTN4), a protein that activates the NFκB pathway [129]. Stable knockdown of ACTN4 reduces TNFα-mediated induction of NFκB and expression of e.g., IL-1β [130]. SV40-RI257I miRNA may therefore interfere with inflammatory responses. The 3p, and the 5p miRNAs of BKPyV and JCPyV share sequence identity (16 out of 22 nucleotides) with SV40-RI257I miRNA [6], but it is not known whether they also target ACTN4. SV40 strain 776 microRNA was shown to diminish the expression of the Serine/Threonine kinase MST4 in the African green monkey kidney epithelial cell line BSC-40, though not in human embryonal kidney 293T cells [129]. Interestingly, knockdown of MST4 in mice resulted in an exacerbated inflammation upon septic shock [131]. It is not known whether any of the HPyV encodes a miRNA that targets MST4, but if so, the following scenario can be imagined: A persistent HPyV infection may result in the depletion of MST4, thus causing the aggravation of inflammatory responses and a contribution to malignancy.

3.2. The Role of HPyV microRNA and HPyV-induced microRNA in Cancer

Some polyomaviruses have been shown to express viral miRNA, while others may encode a putative miRNA [6,132–134]. Although several viral miRNAs have been suggested to play a role in cancer [135], a direct implication of HPyV miRNA in cancer is lacking. Because RacPyV and MCPyV are the only PyV to so far be associated with cancer in their natural host, the expression of their miRNAs was examined in tumors. RacPyV miRNA was among the most abundant miRNAs detectable in RacPyV-associated tumors, but was not observed in RacPyV-negative non-tumor raccoon tissue [134]. This stands in contrast to MCPyV-positive MCC tumors, in which viral miRNA is only detectable in less than half of the tumors tested, and when present, MCPyV miRNA levels were <0.025% of total miRNAs in MCPyV-positive MCC [125,136]. This observation suggests that MCPyV miRNA is not involved in MCC.

PyV miRNA can modulate biological activities that can contribute to malignancy such as evading the immune system, apoptosis and perturbing cellular gene expression [137,138]. The role of HPyV-encoded miRNA in immune evasion was discussed above. PyV miRNAs may also prevent apoptosis. MCPyV miRNA targets the host cell protein AMBRA1, which is involved in autophagy and apoptosis [139], while mouse PyV miRNA downregulates the pro-apoptotic factor Smad2, resulting in a suppression of apoptosis *in vivo* [140]. Aberrant cellular gene expression may promote neoplastic progression, and PyV miRNAs may perturb cellular gene expression by interfering with splicing, thereby targeting transcription factors or proteins controlling the activity of transcription factors, or by inducing the expression of cellular miRNAs. SV40 miRNA is predicted to target the dual-specificity protein phosphatase DUSP8, a negative regulator of the JNK and p38 mitogen-activated protein kinases, whereas MCPyV may downregulate the expression of transcription factor RUNX1, the splicing factor RBM9/FOX2, as well as the repressor MECP2 [125,129]. Viral infection can induce a unique signature of host cell miRNAs, which may contribute to viral pathogenic processes [141]. MiRNAs are initially transcribed by RNA polymerase II, and SV40 LT-ag has been shown to interfere with RNA polymerase II-dependent transcription [142,143]. Hence, PyV infection may alter the pattern of cellular miRNA expression. However, a common feature shared by all known PyV miRNA is the silencing expression of LT-ag so that no effect on cellular miRNA expression is expected [121,122,132,134,144,145]. On the other hand, interference with cellular miRNA expression is plausible in MCPyV-positive tumors because MCC do express LT-ag [146], and RacPyV-positive tumors also express LT-ag [147]. The effect of LT-ag on cellular miRNA expression has not been investigated, but the proteins of the oncovirus HBV, EBV, KSHV and HCV help regulate the levels of cellular miRNAs, including oncogenic miRNAs [148–154].

Xie and co-workers compared miRNA profile in MCPyV-positive and negative MCC. One miRNA that was significantly lower expressed in MCPyV-positive MCC compared to MCPyV-negative MCC was miR-203. The overexpression of miR-203 in MCPyV-negative MCC inhibited cell growth and induced cell cycle arrest [155]. This finding suggests that MCPyV may cause cell proliferation by repressing the expression of miR-203, but the exact mechanism by which MCPyV may regulate this miRNA remains to be elucidated.

3.3. Effect of HPyV on Energy Homeostasis

In healthy cells, glycolysis initiates in the cytoplasm where glucose is metabolized into pyruvate, which then enters the mitochondria where it is converted into acetyl-CoA and enters the Krebs' cycle to generate ATP. In cancer cells, a metabolic switch occurs: the suppression of mitochondrial glucose oxidation and the upregulation of aerobic breakdown of glucose. This phenomenon was first described by Otto Heinrich Warburg, and is known as the Warburg effect [156]. While mitochondrial glucose oxidation generates 36 molecules of ATP per molecule of glucose, only two molecules of ATP are produced per molecule of glucose by aerobic breakdown. Cancer cells compensate for this by increasing the uptake of glucose and by stimulating the transcription of almost all the glycolytic enzymes in the cytoplasm [79,157,158]. Metabolism in Merkel cells and (MCPyV-positive) MCC has not been studied, but a Positron Emission Tomography (PET) scan of MCC with glucose analogues suggests a high rate of glycolysis in these tumors [159]. However, a possible role for MCPyV in enhanced glucose metabolism in MCC remains to be determined. Several studies with polyomavirus-transformed cells indicate that these viruses may affect glucose metabolism. Additionally, a redistribution of membrane glucose transporters, increased aerobic glycolysis and an increased activity of glycolytic enzymes were observed in SV40-transformed cells compared to non-transformed cells [160–162]. A role for JCPyV LT-ag in regulating the metabolic utilization of glucose in brain tumors has been recently suggested [163]. Another study showed that JCPyV LT-ag expressing medulloblastoma cells had a significantly lower mitochondrial respiration and glycolysis, but a three-fold higher consumption of glutamine compared to non-LT-ag expressing cells [164]. Oxygen consumption and glucose uptake were compared in fibroblast transduced with the telomerase catalytic subunit, or in combination with SV40 LT-ag or LT-ag plus st-ag. A progressive increase in both metabolic markers was measured, as cell lines expressed more oncogenes. This observation underscores a role for LT-ag and st-ag in decreasing the cell's dependence on mitochondrial energy production [165]. SV40 st-ag can activate Akt, while Akt can stimulate the expression of glycolytic enzymes and aerobic glycolysis [166–168]. SV40 st-ag can also activate the MEK/ERK mitogen-activated protein kinase pathway, which can then increase glucose transport [169,170]. These findings suggest that st-ag may stimulate glucose uptake and aerobic glycolysis. The tumor suppressor p53 acts as an anti-Warburg molecule because it acts as a potent inhibitor of glycolysis [171]. The LT-ag of BKPyV and JCPyV has been shown to bind and inactivate p53 [172,173]. Although the LT-ag-mediated inactivation of p53 may promote glucose uptake and stimulate the glycolytic pathway, it may not be operational in MCPyV-positive MCC because the truncated form of LT-ag expressed in Merkel cell carcinomas does not bind p53 [40]. The interaction between p53 and LT-ag of other HPyV has not been investigated, but they all encompass a putative p53-binding motif [6].

Autophagy is another mechanism that allows cancer cells to maintain the levels of nutrients and energy in nutrient-limited environments, which helps to facilitate the survival of tumor cells. Moreover, autophagy regulates cellular invasion and metastasis [173]. The human tumor viruses EBV, KSHV, HBV, and HCV can modulate the autophagy pathway to favor viral infection by enhancing viral replication, prevent apoptosis or maintain a persistent and life-long infection [174]. Hence, viral interference with the autophagy pathway may contribute to tumorigenesis by these viruses, though less is known about the effect of HPyV on autophagy and the biological consequences. Bouley et al. could establish a supporting role for autophagy in BKPyV infection [175]. Using transformed human foreskin fibroblasts and HEK cells expressing or lacking the SV40 st-ag, it was demonstrated that st-ag helps to maintain energy homeostasis in glucose-deprivation cancer cells by activating AMP-activated protein kinase (AMPK), thereby inhibiting the mammalian target of rapamycin (mTOR) to shut down protein translation, and inducing autophagy as an alternate energy source. This protective role of st-ag under conditions of glucose deprivation depends on its ability to interact with protein phosphatase 2A [176]. It is not known whether the st-ag of other HPyV may exert similar functions, but BKPyV, JCPyV, MCPyV and MWPyV (HPyV10) st-ag have also been shown to interact with PP2A [99,177–181]. Other effects of HPyV on autophagy came from a study by Khalili et al., who found that JCPyV LT-ag suppressed the expression of Bcl-2-associated athanogene Bag3, a protein implicated in apoptosis and autophagy [182]. On the other hand, overexpression of Bag3 induces autophagy-mediated degradation of JCPyV LT-ag. Bag3 interacts with the C-terminal half region of LT-ag which encompasses a zinc finger structure and partially overlaps with the p53 binding domain [183]. Consequently, LT-ag may repress the expression of Bag3 and protect itself from being degraded by Bag3-mediated autophagy. MCPyV-positive MCCs express C-terminal truncated LT-ag, which may impede interaction with Bag3. Under stress conditions, primary neuroglial cells immortalized with SV40 LT-ag had increased levels of the autophagy marker LC3B compared to non-LT-ag expressing cells [184]. However, the biological implication was not investigated.

3.4. HPyV and Exosomes

Exosomes are endosome-derived membrane vesicles of approximately 50 nm–100 nm in diameter that are shed by cells and act as a communication tool between cells. Exosomes contain cellular proteins, carbohydrates, lipids, DNA, rRNA, mRNA, siRNA and other non-coding RNAs (for recent reviews, see [185–187]). Exosomes secreted by tumor cells participate in the modulation of angiogenesis, cell proliferation, cell invasion, gene regulation and immune evasion, thereby creating advantages for malignant growth [186]. Exosomes released by virus-infected cells can also contain viral-derived components and are implicated in the pathogenesis of viruses. Indeed, the human oncoviruses Epstein-Barr virus, Kaposi sarcoma-associated herpes virus, hepatitis B virus, hepatitis C virus and human T-lymphotropic virus type 1 all utilize exosomes to transfer viral (onco)proteins, mRNA and miRNAs to non-infected cells [188–194]. Exosomes captured by target cells may facilitate the spread of (onco)viral proteins and nucleic acids, thereby promoting malignancy in the recipient cells in the absence of an infection by virions. Exosomes can provoke immune alterations that may play a role to create an immunotolerogenic microenvironment during the carcinogenesis process. They can promote host immune and inflammatory responses by activating T- and B-cells, and by releasing exosome-trapped inflammatory molecules such as TNFα and IL1β in the recipient cells. Even so, exosomes have also been shown to inhibit immune responses by preventing CD4$^+$ T-cell proliferation, CD8$^+$ CTL response or transporting anti-inflammatory molecules (reviewed in [187]).

The generation of exosomes by HPyV-infected cells has scarcely been investigated. Studies with mouse primitive glioblastoma-like brain tumor cell lines harbouring integrated SV40 large T-antigen DNA revealed the presence of SV40 large T-antigen sequences in exosomes produced by these cells [195]. Recently, JCPyV microRNA was detected in exosomes derived from human plasma and urine [196], although studies on the possible roles of exosomes released by HPyV-infected hosts cells are lacking.

4. Therapeutic Strategies against Emerging Hallmarks of Cancer

Specific inhibitors against HPyV are lacking, and the development of vaccines and vaccination are still in a very preliminary phase [197–201]. Therapeutic strategies directed against emerging features of cancer such as inflammation, immune evasion, exosomes, microRNA and energy homeostasis may offer alternatives to help combat HPyV-positive tumors. *In vitro* studies and MCC xenograft mouse models suggested a beneficial effect of IFN. Intratumoral administering of a mixture of different IFNα subtypes and IFNβ resulted in a regression of MCPyV-positive, but not MCPyV-negative xenografts of MCC cells, while IFNα-2b, IFNβ-1b, and IFNγ-1b challenge resulted in an increased cell-surface expression of MHC-I on MCC cell lines [96,202]. However, studies in patients with MCPyV-positive MCC have been proven to exhibit variable effects. Subcutaneous administration of IFN-β resulted in a complete regression of MCPyV-positive MCC tumors in a Japanese patient, but IFN-α-2b treatment of an 84-year-old man and an 81-year-old woman had no effect [203,204]. IFNβ stimulated MHC-I expression on tumor cells of 3/3 MCPyV-positive MCC patients [96]. Intralesional treatment of a 67-year old MCPyV-positive MCC patient with INFβ-Ib, followed by re-infusion of expanded MCPyV LT-ag-specific CD8+ T-cells resulted in a complete response in two of three metastatic lesions and a delayed appearance of new metastasis compared to controls [205]. Tumor necrosis factor (TNF) can be considered as an alternative for treating MCC because this cytokine displayed a high efficacy in three patients [206,207], and in one patient out of three treated with IFNγ plus TNFα a complete response was noticed, while a partial and no response was observed in two others [208]. Because the three aforementioned TNFα studies were performed before the discovery of MCPyV, the presence of virus in the tumors was not known. Interestingly, patients who have been treated with TNFα inhibitors show an increased risk of developing MCC [209,210]. Another immunotherapy approach for the treatment of MCC could be PD-1 ligand and Tim-3, with these receptors highly expressed on MCPyV-specific CD+ T-cells. Drugs targeting the PD-1/PDL-1 pathway such as nivolumab (a blocking antibody against PD-1), pembrolizumab (anti-PD-1 antibody) and BMS-936559 (anti-PDL-1 antibody) have been used in other cancers and may be used to treat MCC [211].

Therapeutic strategies aimed at other emerging hallmarks of cancer have been little explored. Intratumoral delivery of anti-microRNA may help in silencing viral microRNA or viral-induced cellular microRNA, while RNA interference may turn off the expression of viral oncoproteins. RNA interference targeting LT-ag has been shown to abrogate HPyV replication *in vitro* and suppress tumor growth *in vitro* and in an animal model [180,212–217]. The use of exosomes as vaccines against cancer and infectious diseases has been suggested and exosomes pulsed with MCPyV LT-ag could be considered to treat MCPyV-positive MCC patients [187,218]. Lastly, therapeutic strategies directed against the metabolic changes in tumor cells (anti-glycolysis therapy) may be considered. The hexokinase inhibitor 2-deoxyglucose inhibited growth of fibroblasts transformed by the telomerase catalytic subunit plus SV40 LT-ag, and by the telomerase catalytic subunit plus SV40 LT-ag plus st-ag [165]. To our best knowledge, the effect of 2-deoxyglucose on the growth of MCPyV-positive MCC cells has not been investigated.

5. Conclusion and Future Perspectives

Seroprevalence studies demonstrate that HPyV viruses are common in the human population [36]. Although HPyV LT-ag and st-ag possess proven or putative transforming properties, only MCPyV seems to be associated with human cancer. This virus encodes additional early proteins (ALTO protein and 57 kD protein), whose functions are not completely understood [4,219]. Proper immune surveillance may explain why HPyVs establish a harmless life-long infection in most individuals, while immune deficiencies may lead to viral-associated pathologies, including malignancy. The role of HPyVs in the emerging hallmarks of cancer has been little investigated and further investigations are required to elucidate the mechanisms by which HPyV-positive tumors can evade the antiviral responses of the host and affect energy homeostasis. A better understanding of the tumor microenvironment is required to comprehend the development of MCC. A possible involvement of exosomes in HPyV-induced cancer and modulation of the immune system have not been addressed, and the role of viral microRNA or HPyV-induced microRNA in tumorigenesis is incompletely understood. Unveiling the mechanisms by which these viruses participate in emerging hallmarks of cancer may therefore enable the development of novel therapeutic strategies.

Author Contributions

U.M, K.R., I.A. and B.S. were all involved in gathering information, reading articles and writing and critically revising the manuscript.

Conflicts of Interest

The authors declare no conflict of interest.

References

1. Johne, R.; Buck, C.B.; Allander, T.; Atwood, W.J.; Garcea, R.L.; Imperiale, M.J.; Major, E.O.; Ramqvist, T.; Norkin, L.C. Taxonomical developments in the family Polyomaviridae. *Arch. Virol.* **2011**, *156*, 1627–1634.
2. Peretti, A.; FitzGerald, P.C.; Bliskovsky, V.; Pastrana, D.V.; Buck, C.B. Genome Sequence of a Fish-Associated Polyomavirus, Black Sea Bass (*Centropristis striata*) Polyomavirus 1. *Genome Announc.* **2015**, *3*, e01476–14.
3. DeCaprio, J.A.; Imperiale, M.J.; Major, E.O. Polyomaviruses. In *Fields Virology*, 6th ed.; Knipe, D.M., Howley, P.M., Eds.; Lippincott Williams & Wilkins, and Wolters Kluwer: Philadelphia, PA, USA, 2013; Volume 2, pp. 1633–1661.
4. Carter, J.J.; Daugherty, M.D.; Qi, X.; Bheda-Malge, A.; Wipf, G.C.; Robinson, K.; Roman, A.; Malik, H.S.; Galloway, D.A. Identification of an overprinting gene in Merkel cell polyomavirus provides evolutionary insight into the birth of viral genes. *Proc. Natl. Acad. Sci. USA* **2013**, *110*, 12744–12749.
5. Schowalter, R.W.; Buck, C.B. The merkel cell polyomavirus minor capsid protein. *PLOS Pathog.* **2013**, *9*, e1003558.
6. Ehlers, B.; Moens, U. Genome analysis of non-human primate polyomaviruses. *Infect. Genet. Evol.* **2014**, *26*, 283–294.
7. Fulghieri, C.; Bloom, S. SarahElizabeth Stewart. *Emerg. Infect. Dis.* **2014**, *20*, 893–895.
8. Sweet, B.H.; Hilleman, M.R. The vacuolating virus, SV40. *Proc. Soc. Exp. Biol. Med.* **1960**, *105*, 420–427.
9. Cheng, J.; DeCaprio, J.A.; Fluck, M.M.; Schaffhausen, B.S. Cellular transformation by Simian Virus 40 and Murine Polyomavirus T antigens. *Semin. Cancer Biol.* **2009**, *19*, 218–228.
10. An, P.; Sáenz Robles, M.T.; Pipas, J.M. Large T antigens of polyomaviruses: Amazing molecular machines. *Annu. Rev. Microbiol.* **2012**, *66*, 213–236.
11. Valis, J.D.; Strandberg, J.D.; Shah, K.V. Transformation of hamster kidney cells by simian papovavirus SA12. *Proc. Soc. Exp. Biol. Med.* **1979**, *160*, 208–212.
12. Bastien, C.; Feunteun, J. The hamster polyomavirus transforming properties. *Oncogene* **1988**, *2*, 129–135.
13. Chen, J.D.; Neilson, K.; van Dyke, T. Lymphotropic papovavirus early region is specifically regulated in transgenic mice and efficiently induces neoplasia. *J. Virol.* **1989**, *63*, 2204–2214.
14. Dela Cruz, F.N., Jr.; Giannitti, F.; Li, L.; Woods, L.W.; del Valle, L.; Delwart, E.; Pesavento, P.A. Novel polyomavirus associated with brain tumors in free-ranging raccoons, western United States. *Emerg. Infect Dis.* **2013**, *19*, 77–84.
15. Simmons, J.H.; Riley, L.K.; Franklin, C.L.; Besch-Williford, C.L. Hamster polyomavirus infection in a pet Syrian hamster (*Mesocricetus auratus*). *Vet. Pathol.* **2001**, *38*, 441–446.

16. Colegrove, K.M.; Wellehan, J.F., Jr.; Rivera, R.; Moore, P.F.; Gulland, F.M.; Lowenstine, L.J.; Nordhausen, R.W.; Nollens, H.H. Polyomavirus infection in a free-ranging California sea lion (*Zalophus californianus*) with intestinal T-cell lymphoma. *J. Vet. Diagn. Invest.* **2010**, *22*, 628–632.

17. Stevens, H.; Bertelsen, M.F.; Sijmons, S.; van Ranst, M.; Maes, P. Characterization of a novel polyomavirus isolated from a fibroma on the trunk of an African elephant (*Loxodonta africana*). *PLOS ONE* **2013**, *8*, e77884.

18. Johne, R.; Müller, H. Polyomaviruses of birds: Etiological agents of inflammatory diseases in a tumor virus family. *J. Virol.* **2007**, *81*, 11554–11559.

19. Gardner, S.D.; Field, A.M.; Coleman D.V.; Humle, B. New human papovavirus (B.K.) isolated from urine after renal transplantation. *Lancet* **1971**, *1*, 1253–1257.

20. Padgett, B.L.; Walker, D.L.; ZuRhein, G.M.; Eckroade, R.J.; Dessel, B.H. Cultivation of papova-like virus from human brain with progressive multifocal leukoencephalopathy. *Lancet* **1971**, *1*, 1257–1260.

21. Atkin, S.J.; Griffin, B.E.; Dilworth, S.M. Polyoma virus and simian virus 40 as cancer models: History and perspectives. *Semin. Cancer Biol.* **2009**, *19*, 211–217.

22. Pipas, J.M. SV40: Cell transformation and tumorigenesis. *Virology* **2009**, *384*, 294–303.

23. Allander, T.; Andreasson, K.; Gupta, S.; Bjerkner, A.; Bogdanovic, G.; Persson, M.A.; Dalialis, T.; Ramqvist, T.; Andersson, B. Identification of a third human polyomavirus. *J. Virol.* **2007**, *81*, 4130–4136.

24. Gaynor, A.; Nissen, M.D.; Whiley, D.M.; Mackay, I.M.; Lambert, S.B.; Wu, G.; Brennan, D.C.; Storch, G.A.; Wang, D. Identification of a novel polyomavirus from patients with acute respiratory tract infections. *PLOS Pathog.* **2007**, *3*, 595–604.

25. Feng, H.; Shuda, M.; Chang, Y.; Moore, P.S. Clonal integration of a polyomavirus in human Merkel cell carcinoma. *Science* **2008**, *319*, 1096–1100.

26. Schowalter, R.M.; Pastrana, D.V.; Pumphrey, K.A.; Moyer, A.L.; Buck, C.B. Merkel cell polyomavirus and two previously unknown polyomaviruses are chronically shed from human skin. *Cell Host Microbe* **2010**, *7*, 509–515.

27. van der Meijden, E.; Janssens, R.W.; Lauber, C.; Bouwens Bavinck, J.N.; Gorbalenya, A.E.; Feltkamp, M.C. Discovery of a new human polyomavirus associated with trichodysplasia spinulosa in an immunocompromized patient. *PLOS Pathog.* **2010**, *6*, e1001024.

28. Scuda, N.; Hofmann, J.; Calvignac-Spencer, S.; Ruprecht, K.; Liman, P.; Kühn, J.; Hengel, H.; Ehlers, B. A novel human polyomavirus closely related to the African green monkey-derived lymphotropic polyomavirus. *J. Virol.* **2011**, *85*, 4586–4590.

29. Sauvage, V.; Foulongne, V.; Cheval, J.; Pariente, K.; Le Gouil, M.; Burguiere, A.M.; Manuguerra, J.C.; Caro, V.; Eloit, M. Human polyomavirus related to the green monkey lymphotropic polyomavirus. *Emerg. Infect. Dis.* **2011**, *17*, 1364–1370.

30. Buck, C.B.; Phan, G.Q.; Raiji, M.T.; Murphy, P.M.; McDermott, D.H.; McBride, A.A. Complete genome sequence of a tenth human polyomavirus. *J. Virol.* **2012**, *86*, 10887.

31. Siebrasse, E.A.; Reyes, A.; Lim, E.S.; Zhao, G.; Mkakosya, R.S.; Manary, M.J.; Gordon, J.I.; Wang, D. Identification of MW polyomavirus, a novel polyomavirus in human stool. *J. Virol.* **2012**, *86*, 10321–10326.

32. Yu, G.; Greninger, A.L.; Isa, P.; Phan, T.G.; Martinez, M.A.; de la Luz Sanchez, M.; Contreras, J.F.; Santos-Preciado, J.I.; Parsonnet, J.; Miller, S.; *et al.* Discovery of a novel polyomavirus in acute diarrheal samples from children. *PLOS ONE* **2012**, *7*, e49449.

33. Lim, E.S.; Reyes, A.; Antonio, M.; Saha, D.; Ikumapayi, U.N.; Adeyemi, M.; Stine, O.C.; Skelton, R.; Brennan, D.C.; Mkakosya, R.S.; *et al.* Discovery of STL polyomavirus, a polyomavirus of ancestral recombinant origin that encodes a unique T antigen by alternative splicing. *Virology* **2013**, *436*, 295–303. Erratum in: *Virology* **2013**, *439*, 163–164.

34. Korup, S.; Rietscher, J.; Calvignac-Spencer, S.; Trusch, F.; Hofmann, J.; Moens, U.; Sauer, I.; Voight, S.; Schmuck, R.; Ehlers, B. Identification of a novel human polyomavirus in organs of the gastrointestinal tract. *PLOS ONE* **2013**, *8*, e58021.

35. Mishra, N.; Pereira, M.; Rhodes, R.H.; An, P.; Pipas, J.M.; Jain, K.; Kapoor, A.; Briese, T.; Faust, P.L.; Lipkin, W.I. Identification of a novel polyomavirus in a pancreatic transplant recipient with retinal blindness and vascular myopathy. *J. Infect. Dis.* **2014**, *210*, 1595–1599.

36. Moens, U.; van Ghelue, M.; Song, X.; Ehlers, B. Serological cross-reactivity between human polyomaviruses. *Rev. Med. Virol.* **2013**, *23*, 250–264.

37. Lim, E.S.; Meinerz, N.M.; Primi, B.; Wang, D.; Garcea, R.L. Common exposure to STL polyomavirus during childhood. *Emerg. Infect. Dis.* **2014**, *20*, 1559–1561.

38. Nicol, J.T.; Leblond, V.; Arnold, F.; Guerra, G.; Mazzoni, E.; Tognon, M.; Coursaget, P.; Touzé, A. Seroprevalence of human Malawi polyomavirus. *J. Clin. Microbiol.* **2014**, *52*, 321–323.

39. Moens, U.; van Ghelue, M.; Johannessen, M. Oncogenic potentials of the human polyomavirus regulatory proteins. *Cell Mol. Life Sci.* **2007**, *64*, 1656–1678.

40. Borchert, S.; Czech-Sioli, M.; Neumann, F.; Schmidt, C.; Wimmer, P.; Dobner, T.; Grundhoff, A.; Fischer, N. High-affinity Rb binding, p53 inhibition, subcellular localization, and transformation by wild-type or tumor-derived shortened Merkel cell polyomavirus large T antigens. *J. Virol.* **2014**, *88*, 3144–3160.

41. Verhaegen, M.E.; Mangelberger, D.; Harms, P.W.; Vozheiko, T.D.; Weick, J.W.; Wilbert, D.M.; Saunders, T.L.; Ermilov, A.N.; Bichakjian, C.K.; Johnson, T.M.; Imperiale, M.J.; Dlugosz, A.A. Merkel cell polyomavirus small t antigen is oncogenic in transgenic mice. *J. Invest. Dermatol.* **2014**, doi:10.1038/jid.2014.446.

42. Spurgeon, M.E.; Cheng, J.; Bronson, R.T.; Lambert, P.F.; DeCaprio, J.A. Tumorigenic activity of Merkel cell polyomavirus T antigens expressed in the stratified epithelium of mice. *Cancer Res.* **2015**, doi:10.1158/0008-5472.CAN-14-2425.

43. Chang, Y.; Moore, P.S. Merkel cell carcinoma: A virus-induced human cancer. *Annu. Rev. Pathol.* **2012**, *7*, 123–144.

44. Moens, U.; Johannessen, M. Human polyomaviruses and cancer: Expanding repertoire. *J. Dtsch. Dermatol. Ges.* **2008**, *6*, 704–708.

45. Abend, J.R.; Jiang, M.; Imperiale, M.J. BK virus and cancer: Innocent until proven guilty. *Semin. Cancer Biol.* **2009**, *19*, 252–260.

46. Maginnis, M.S.; Atwood, W.J. JC virus: An oncogenic virus in animals and humans? *Semin. Cancer Biol.* **2009**, *19*, 261–269.

47. Enam, S.; Del Valle, L.; Lara, C.; Gan, D.D.; Ortiz-Hidalgo, C.; Palazzo, J.P.; Khalili, K. Association of human polyomavirus JCV with colon cancer: Evidence for interaction of viral T-antigen and beta-catenin. *Cancer Res.* **2002**, *62*, 7093–7101.

48. Bulut, Y.; Ozdemir, E.; Ozercan, H.I.; Etem, E.O.; Aker, F.; Toraman, Z.A.; Seyrek, A.; Firdolas, F. Potential relationship between BK virus and renal cell carcinoma. *J. Med. Virol.* **2013**, *85*, 1085–1089.

49. Delbue, S.; Ferrante, P.; Provenzano, M. Polyomavirus BK and prostate cancer: An unworthy scientific effort? *Oncoscience* **2014**, *1*, 296–303.

50. Carluccio, S.; Signorini, L.; Elia, F.; Villani, S.; Delbue, S.; Ferrante, P. A potential linkage between the JC and BK polyomaviruses and brain and urinary tract tumors: A review of the literature. *Adv. Tumor. Virol.* **2014**, *4*, 17–24.

51. zur Hausen, H. Red meat consumption and cancer: Reasons to suspect involvement of bovine infectious factors in colorectal cancer. *Int. J. Cancer* **2012**, *130*, 2475–2483.

52. Zhang, W.; Li, L.; Deng, X.; Kapusinszky, B.; Delwart, E. What is for dinner? Viral metagenomics of US store bought beef, pork and chicken. *Virology* **2014**, *468–470*, 303–310.

53. Peretti, A.; FitzGerald, P.C.; Bliskovsky, V.; Buck, C.B.; Pastrana, D.V. Hamburger polyomaviruses. *J. Gen. Virol.* **2015**, doi:10.1099/vir.0.000033

54. Schrama, D.; Groesser, L.; Ugurel, S.; Hafner, C.; Pastrana, D.V.; Buck, C.B.; Cerroni, L.; Theiler, A.; Becker, J.C. Presence of human polyomavirus 6 in mutation-specific BRAF inhibitor-induced epithelial proliferation. *JAMA Dermatol.* **2014**, *150*, 1180–1186.

55. Ramqvist, T.; Nordfors, C.; Dalianis, T.; Ragnarsson-Olding, B. DNA from human polyomaviruses, TSPyV, MWPyV, HPyV6, 7, and 9 was not detected in primary mucosal melanomas. *Anticancer Res.* **2014**, *34*, 639–643.

56. Imajoh, M.; Hashida, Y.; Nakajima, H.; Sano, S.; Daibata, M. Prevalence and viral DNA loads of three novel human polyomaviruses in skin cancers from Japanese patients. *J. Dermatol.* **2013**, *40*, 657–660.

57. Scola, N.; Wieland, U.; Silling, S.; Altmeyer, P.; Stücker, M.; Kreuter, A. Prevalence of human polyomaviruses in common and rare types of non-Merkel cell carcinoma skin cancer. *Br. J. Dermatol.* **2012**, *167*, 1315–1320.

58. Antonsson, A.; Bialasiewicz, S.; Rockett, R.J.; Jacob, K.; Bennett, I.C.; SLoots, T.P. Exploring the Prevalence of Ten Polyomaviruses and Two Herpes Viruses in Breast Cancer. *PLOS ONE* **2012**, *7*, e39842.

59. Foulongne, V.; Sauvage, V.; Hebert, C.; Dereure, O.; Cheval, J.; Gouilh, M.A.; Pariente, K.; Segondy, M.; Burguière, A.; Manuguerra, J.C.; Caro, V.; Eloit, M. Human skin microbiota: High diversity of DNA viruses identified on the human skin by high throughput sequencing. *PLOS ONE* **2012**, *7*, e38499.

60. Giraud, G.; Ramqvist, T.; Ragnarsson-Olding, B.; Dialialis, T. DNA from BK virus and JC virus and from KI, WU, and MC polyomaviruses as well as from simian virus 40 is not detected in non-UV-light-associated primary malignant melanomas of mucous membranes. *J. Clin. Microbiol.* **2008**, *46*, 3595–3598.

61. Bzhalava, D.; Johansson, H.; Ekström, J.; Faust, H.; Möller, B.; Eklund, C.; Nordin, P.; Stenquist, B.; Paoli, J.; Persson, B.; Forslund, O.; Dillner, J. Unbiased approach for virus detection in skin lesions. *PLOS ONE* **2013**, *8*, e655953.

62. Imajoh, M.; Hashida, Y.; Taniguchi, A.; Kamioka, M.; Daibata, M. Novel human polyomaviruses, Merkel cell polyomavirus and human polyomavirus 9, in Japanese chronic lymphocytic leukemia cases. *J. Hematol. Oncol.* **2012**, *5*, e25.

63. Kreuter, A.; Silling, S.; Dewan, M.; Stücker, M.; Wieland, U. Evaluation of 4 recently discovered human polyomaviruses in primary cutaneous B-cell and T-cell lymphoma. *Arch. Dermatol.* **2011**, *147*, 1449–1451.

64. Duncavage, E.J.; Pfeifer, J.D. Human polyomaviruses 6 and 7 are not detectable in Merkel cell polyomavirus-negative Merkel cell carcinoma. *J. Cutan. Pathol.* **2011**, *38*, 790–796.

65. Kanitakis, J.; Kazem, S.; van Der Meijden, E.; Feltkamp, M. Absence of the trichodysplasia spinulosa-associated polyomavirus in human pilomatricomas. *Eur. J. Dermatol.* **2011**, *21*, 453–454.

66. Babakir-Mina, M.; Ciccozzi, M.; Campitelli, L.; Aquaro, S.; Lo Coco, A.; Perno, C.F.; Ciotti, M. Identification of the novel KI Polyomavirus in paranasal and lung tissues. *J. Med. Virol.* **2009**, *81*, 558–561.

67. Giraud, G.; Ramqvist, T.; Pastrana, D.V.; Pavot, V.; Lindau, C.; Kogner, P.; Orrego, A.; Buck, C.B.; Allander, T.; Holm, S.; Gustavsson, B.; Dalianis, T. DNA from KI, WU and Merkel cell polyomavirus is not detected in childhood central nervous system tumors or neuroblastomas. *PLOS ONE* **2009**, *4*, e8239.

68. Gustafsson, B.; Honkaniemi, E.; Goh, S.; Giraud, G.; Forestier, E.; von Döbeln, U.; Allander, T.; Dalianis, T.; Bogdanovic, G. KI, WU, and Merkel cell polyomavirus DNA was not detected in guthrie cards of children who later developed acute lymphoblastic leukemia. *J. Pediatr. Hematol. Oncol.* **2012**, *34*, 364–367.

69. Teramoto, S.; Kaiho, M.; Takano, Y.; Endo, R.; Kikuta, H.; Sawa, H.; Ariga, T.; Ishiguro, N. Detection of KI polyomavirus and WU polyomavirus DNA by real-time polymerase chain reaction_in nasopharyngeal swabs and in normal lung and lung adenocarcinoma tissues. *Microbiol. Immunol.* **2011**, *55*, 525–530.

70. Duncavage, E.J.; Le, B.M.; Wang, D.; Pfeifer, J.D. Merkel cell polyomavirus: A specific marker for Merkel cell carcinoma in histologically similar tumors. *Am. J. Surg. Pathol.* **2009**, *33*, 1771–1777.

71. Schmitt, M.; Höfler, D.; Koleganova, N.; Pawlita, M. Human polyomaviruses and other human viruses in neuroendocrine tumors. *Cancer Epidemiol. Biomarkers Prev.* **2011**, *20*, 1558–1561.

72. Du-Thanh, A.; Foulongne, V.; Guillot, B.; Dereure, O. Recently discovered human polyomaviruses in lesional and non-lesional skin of patients with primary cutaneous T-cell lymphomas. *J. Dermatol. Sci.* **2013**, *71*, 140–142.

73. Hashida, Y.; Taniguchi, A.; Yawata, T.; Hosokawa, S.; Murakami, M.; Hiroi, M.; Ueba, T.; Daibata, M. Prevalence of human cytomegalovirus, polyomaviruses, and oncogenic viruses in glioblastoma among Japanese subjects. *Infect Agent Cancer* **2015**, *10*, e3.

74. Rennspiess, D.; Pujari, S.; Keijzers, M.; Abdul-Hamid, M.A.; Hochstenbag, M.; Dingemans, A.; Kurz, A.K.; Speel, E.; Haugg, A.; Pastrana, D.V.; *et al.* Detection of Human Polyomavirus 7 in human thymic epithelial tumors. *J. Thorac. Oncol.* **2015**, *10*, 360–366.

75. Gjoerup, O.; Chang, Y. Update on human polyomaviruses and cancer. *Adv. Cancer Res.* **2010**, *106*, 1–51.

76. DeCaprio, J.A.; Garcea, R.L. A cornucopia of human polyomaviruses. *Nat. Rev. Microbiol.* **2013**, *11*, 264–276.

77. Spurgeon, M.E.; Lambert, P.F. Merkel cell polyomavirus: A newly discovered human virus with oncogenic potential. *Virology* **2013**, *435*, 118–130.

78. Moore, P.S.; Chang, Y. Why do viruses cause cancer? Highlights of the first century of human tumor virology. *Nat. Rev. Cancer* **2010**, *10*, 878–889.

79. Dunn, G.P.; Bruce, A.T.; Ikeda, H.; Old, L.J.; Schreiber, R.D. Cancer immunoediting: From immunosurveillance to tumor escape. *Nat. Immunol.* **2002**, *3*, 991–998.

80. Hanahan, D.; Weinberg, R.A. Hallmarks of cancer: The next generation. *Cell* **2011**, *144*, 646–674.

81. Banerjee, A.; Vasanthakumar, A.; Grigoriadis, G. Modulating T regulatory cells in cancer: How close are we? *Immunol. Cell Biol.* **2013**, *91*, 340–349.

82. Vossen, M.T.; Westerhout, E.M.; Söderberg N.C.; Wiertz, E.J. Viral immune evasion: A masterpiece of evolution. *Immunogenetics* **2002**, *54*, 527–542.

83. Bhatia, S.; Afanasiev, O.; Nghiem, P. Immunobiology of Merkel cell carcinoma: Implications for immunotherapy of a polyomavirus-associated cancer. *Curr. Oncol. Rep.* **2011**, *13*, 488–497.

84. Shahzad, N.; Shuda, M.; Gheit, T.; Kwun, H.J.; Cornet, I.; Saidj, D.; Zannetti, C.; Hasan, U.; Chang, Y.; Moore, P.S.; *et al.* The T Antigen Locus of Merkel Cell Polyomavirus downregulates Human Toll-Like Receptor 9 Expression. *J. Virol.* **2013**, *87*, 13009–13019.

85. Li, H.; Gade, P.; Xiao, W.; Kalvakolanu, D.V. The interferon signaling network and transcription factor C/EBP-beta. *Cell Mol. Immunol.* **2007**, *4*, 407–418.

86. Johnson, P.F. Molecular stop signs: Regulation of cell-cycle arrest by C/EBP transcription factors. *J. Cell Sci.* **2005**, *118*, 2545–2555.

87. Nerlov, C. The C/EBP family of transcription factors: A paradigm for interaction between gene expression and proliferation control. *Trends Cell Biol.* **2007**, *17*, 318–324.

88. Sihto, H.; Joensuu, H. Tumor-infiltrating lymphocytes and outcome in Merkel cell carcinoma, a virus-associated cancer. *Oncoimmunology* **2012**, *1*, 1420–1421.

89. Harms, P.W.; Patel, R.M.; Verhaegen, M.E.; Giordano, T.J.; Nash, K.T.; Johnson, C.N.; Daignault, S.; Thomas, D.G.; Gudjonsson, J.E.; Elder, J.T.; *et al.* Distinct gene expression profiles of viral- and nonviral-associated merkel cell carcinoma revealed by transcriptome analysis. *J. Invest. Dermatol.* **2013**, *133*, 936–945.

90. Lauttia, S.; Sihto, H.; Kavola, H.; Koljonen, V.; Böhling, T.; Joensuu, H. Prokineticins and Merkel cell polyomavirus infection in Merkel cell carcinoma. *Br. J. Cancer* **2014**, *110*, 1446–1455.

91. Andea, A.A.; Coit, D.G.; Amin, B.; Busam, K.J. Merkel cell carcinoma: Histologic features and prognosis. *Cancer* **2008**, *113*, 2549–2558.

92. Paulson, K.G.; Iyer, J.G.; Tegeder, A.R.; Thibodeau, R.; Schelter, J.; Koba, S.; Schrama, D.; Simonson, W.T.; Lemos, B.D.; Byrd, D.R.; *et al.* Transcriptome-wide studies of Merkel cell carcinoma and validation of intratumoral CD8+ lymphocyte invasion as an independent predictor of survival. *J. Clin. Oncol.* **2011**, *29*, 1539–1546.

93. Wheat, R.; Roberts, C.; Waterboer, T.; Steele, J.; Marsden, J.; Steven, N.M.; Blackbourn, D.J. Inflammatory cell distribution in primary merkel cell carcinoma. *Cancers* **2014**, *6*, 1047–1064.

94. Sihto, H.; Böhling, T.; Kavola, H.; Koljonen, V.; Salmi, M.; Jalkanen, S.; Joensuu, H. Tumor infiltrating immune cells and outcome of Merkel cell carcinoma: A population-based study. *Clin. Cancer Res.* **2012**, *18*, 2872–2881.

95. Triozzi, P.L.; Fernandez, A.P. The role of the immune response in merkel cell carcinoma. *Cancers (Basel)* **2013**, *5*, 234–254.

96. Paulson, K.G.; Tegeder, A.; Willmes, C.; Iyer, J.G.; Afanasiev, O.K.; Schrama, D.; Koba, S.; Thibodeau, R.; Nagase, K.; Simonson, W.T.; *et al.* Downregulation of MHC-I expression is prevalent but reversible in Merkel cell carcinoma. *Cancer Immunol. Res.* **2014**, *2*, 1071–1079.

97. Afanasiev, O.K.; Yelistratova, L.; Miller, N.; Nagase, K.; Paulson, K.; Iyer, J.G.; Ibrani, D.; Koelle, D.M.; Nghiem, P. Merkel polyomavirus-specific T cells fluctuate with merkel cell carcinoma burden and express therapeutically targetable PD-1 and Tim-3 exhaustion markers. *Clin. Cancer Res.* **2013**, *19*, 5351–5360.

98. Lipson, E.J.; Vincent, J.G.; Loyo, M.; Kagohara, L.T.; Luber, B.S.; Wang, H.; Xu, H.; Nayar, S.K.; Wang, T.S.; Sidransky, D.; *et al.* PD-L1 expression in the Merkel cell carcinoma microenvironment: Association with inflammation, Merkel cell polyomavirus and overall survival. *Cancer Immunol. Res.* **2013**, *1*, 54–63.

99. Griffiths, D.A.; Abdul-Sada, H.; Knight, L.M.; Jackson, B.R.; Richards, K.; Prescott, E.L.; Peach, A.H.; Blair, G.E.; Macdonald, A.; Whitehouse, A. Merkel cell polyomavirus small T antigen targets the NEMO adaptor protein to disrupt inflammatory signaling. *J. Virol.* **2013**, *87*, 13853–13867.

100. Clark, R.A.; Huang, S.J.; Murphy, G.F.; Mollet, I.G.; Hijnen, D.; Muthukuru, M.; Schanbacher, C.F.; Edwards, V.; Miller, D.M.; Kim, J.E.; *et al.* Human squamous cell carcinomas evade the immune response by down-regulation of vascular E-selectin and recruitment of regulatory T cells. *J. Exp. Med.* **2008**, *205*, 2221–2234.

101. Gehad, A.E.; Lichtman, M.K.; Schmults, C.D.; Teaque, J.E.; Calarese, A.W.; Jiang, Y.; Watanabe, R.; Clark, R.A. Nitric oxide-producing myeloid-derived suppressor cells inhibit vascular E-selectin expression in human squamous cell carcinomas. *J. Invest Dermatol.* **2012**, *132*, 2642–2651.

102. Afanasiev, O.K.; Nagase, K.; Simonson, W.; Vandeven, N.; Blom. A.; Koelle, D.M.; Clark, R.; Nghiem, P. Vascular E-selectin expression correlates with CD8 lymphocyte infiltration and improved outcome in Merkel cell carcinoma. *J. Invest. Dermatol.* **2013**, *133*, 2065–2073.

103. Hayashi, H.; Tanaka, K.; Jay, F.; Khoury, G.; Jay, G. Modulation of the tumorigenicity of human adenovirus-12-transformed cells by interferon. *Cell* **1985**, *43*, 263–267.

104. Cromme, F.V.; van Bommel, P.F.; Walboomers, J.M.; Gallee, M.P.; Stern, P.L.; Kenemans, P.; Helmerhorst, T.J.; Stukart, M.J.; Meijer, C.J. Differences in MHC and TAP-1 expression in cervical cancer lymph node metastases as compared with the primary tumors. *Br. J. Cancer* **1994**, *69*, 1176–1181.

105. Hill, A.; Jugovic, P.; York, I.; Russ, G.; Bennink, J.; Yewdell, J.; Ploeggh, H.; Johnson, D. Herpes simplex virus turns off the TAP to evade host immunity. *Nature* **1995**, *375*, 411–415.

106. Koopman, L.A.; van Der Slik, A.R.; Giphart, M.J.; Fleuren, G.J. Human leukocyte antigen class I gene mutations in cervical cancer. *J. Natl. Cancer Inst.* **1999**, *91*, 1669–1677.

107. Haque, M.; Ueda, K.; Nakano, K.; Hirata, Y.; Parravicini, C.; Corbellino, M.; Yamanishi, K. Major histocompatibility complex class I molecules are down-regulated at the cell surface by the K5 protein encoded by Kaposi's sarcoma-associated herpesvirus/human herpesvirus-8. *J. Gen. Virol.* **2001**, *82*, 1175–1180.

108. Hansen, T.H.; Bouvier, M. MHC class I antigen presentation: Learning from viral evasion strategies. *Nat. Rev. Immunol.* **2009**, *9*, 503–513.

109. Chretien, A.S.; Le Roy, A.; Vey, N.; Prebet, T.; Blaise, D.; Fauriat, C.; Olive, D. Cancer-Induced Alterations of NK-Mediated Target Recognition: Current and Investigational Pharmacological Strategies Aiming at Restoring NK-Mediated Anti-Tumor Activity. *Front. Immunol.* **2014**, *5*, e122.

110. Riella, L.V.; Paterson, A.M.; Sharpe, A.H.; Chandraker, A. Role of the PD-1 pathway in the immune response. *Am. J. Transplant.* **2012**, *12*, 2575–2587.

111. Takenaka, H.; Kishimoto, S.; Shibagaki, T.; Nagata, M.; Yasuno, H. Merkel cell carcinoma with spontaneous regression: An immunohistochemical, ultrastructural, and TUNEL labeling study. *Am. J. Dermatopathol.* **1997**, *19*, 614–618.

112. Fujimoto, N.; Nakanishi, G.; Kabuto, M.; Nakano, T.; Eto, H.; Nakajima, H.;, Sano, S.; Tanaka, T. Merkel cell carcinoma showing regression after biopsy: Evaluation of programmed cell death 1-positive cells. *J. Dermatol.* **2015**, doi:10.1111/1346-8138.12805.

113. Arany, I.; Tyring, S.K. Status of cytokine and antigen presentation genes in Merkel cell carcinoma of the skin. *J. Cutan. Med. Surg.* **1998**, *2*, 138–141.

114. Lord, S.J.; Rajotte, R.V.; Korbutt, G.S.; Bleackley, R.C. Granzyme B: A natural born killer. *Immunol. Rev.* **2003**, *193*, 31–38.

115. Pahl, H.L. Activators and target genes of Rel/NF-kappaB transcription factors. *Oncogene* **1999**, *18*, 6853–6866.

116. Hiroi, M.; Ohmori, Y. Constitutive nuclear factor kappaB activity is required to elicit interferon-gamma-induced expression of chemokine CXC ligand 9 (CXCL9) and CXCL10 in human tumor cell lines. *Biochem. J.* **2003**, *376*, 393–402.

117. Arora, R.; Shuda, M.; Guastafierro, A.; Feng, H.; Toptan, T.; Tolstov, Y.; Normolle, D.; Vollmer, L.L.; Vogt, A.; Dömling, A.; *et al.* Survivin is a therapeutic target in Merkel cell carcinoma. *Sci. Transl. Med.* **2012**, *4*, 133ra56.

118. Van Ghelue, M.; Khan, M.T.; Ehlers, B.; Moens, U. Genome analysis of the new human polyomaviruses. *Rev. Med. Virol.* **2012**, *22*, 354–377.

119. Youlden, D.R.; Youl, P.H.; Peter Soyer, H.; Fritschi, L.; Baade, P.D. Multiple primary cancers associated with Merkel cell carcinoma in Queensland, Australia, 1982–2011. *J. Invest. Dermatol.* **2014**, *134*, 2883–2889.

120. Mogha, A.; Fautrel, A.; Mouchet, N.; Guo, N.; Corre, S.; Adamski, H.; Watier, E.; Misery, L.; Galibert, M.D. Merkel cell polyomavirus small T-antigen mRNA level is increased following *in vivo* UV-radiation. *PLOS ONE* **2010**, *5*, e11423.

121. Sullivan, C.S.; Grundhoff, A.T.; Tevethia, S.; Pipas, J.M.; Ganem, D. SV40-encoded microRNAs regulate viral gene expression and reduce susceptibility to cytotoxic T cells. *Nature* **2005**, *435*, 682–686.

122. Cox, J.E.; McClure, L.V.; Goga, A.; Sullivan, C.S. Pan-viral-microRNA screening identifies interferon inhibition as a common function of diverse viruses. *Proc. Natl. Acad. Sci. USA* **2015**, *112*, 1856–1861.

123. Bauman, Y.; Mandelboim, O. MicroRNA based immunoevasion mechanism of human polyomaviruses. *RNA Biol.* **2011**, *8*, 591–594.

124. Bauman, Y.; Nachmani, D.; Vitenshtein, A.; Tsukerman, P.; Drayman, N.; Stern-Ginossar, N.; Lankry, D.; Gruda, R.; Mandelboim, O. An identical miRNA of the human JC and BK polyoma viruses targets the stress-induced ligand ULBP3 to escape immune elimination. *Cell Host Microbe* **2011**, *9*, 93–102.

125. Lee, S.; Paulson, K.G.; Murchison, E.P.; Afanasiev, O.K.; Alkan, C.; Leonard, J.H.; Byrd, D.R.; Hannon, G.J.; Nghiem, P. Identification and validation of a novel mature microRNA encoded by the Merkel cell polyomavirus in human Merkel cell carcinomas. *J. Clin. Virol.* **2011**, *52*, 272–275.

126. Groettrup, M.; Soza, A.; Eggers, M.; Kuehn, L.; Dick, T.P.; Schild, H. A role for the proteasome regulator PA28alpha in antigen presentation. *Nature* **1996**, *381*, 166–168.

127. Okkenhaug, K.; Bilancio, A.; Farjot, G.; Priddle, H.; Sancho, S.; Peskett, E.; Pearce, W.; Meek, S.E.; Salpekar, A.; Waterfield, M.D.; Smith, A.J.; Vanhaesevroeck, B. Impaired B and T cell antigen receptor signaling in p110delta PI 3-kinase mutant mice. *Science* **2002**, *297*, 1031–1034.

128. Yan, Q.; Sharma-Kuinkel, B.K.; Deshmukh, H.; Tsalik, E.L.; Cyr, D.D.; Lucas, J.; Woods, C.W.; Scott, W.K.; Sempowski, G.D.; Thaden, J.T.; *et al.* Dusp3 and Psme3 Are Associated with Murine Susceptibility to Staphylococcus aureus Infection and Human Sepsis. *PLOS Pathog.* **2014**, *10*, e1004149.

129. Chen, C.J.; Cox, J.E.; Kincaid, R.P.; Martinez, A.; Sullivan, C.S. Divergent MicroRNA targetomes of closely related circulating strains of a polyomavirus. *J. Virol.* **2013**, *87*, 11135–11147.

130. Zhao, X.; Hsu, K.S.; Lim, J.H.; Bruggeman, L.A.; Kao, H.Y. α-Actinin 4 Potentiates Nuclear Factor κ-Light-chain-enhancer of Activated B-cell (NF-κB) Activity in Podocytes Independent of Its Cytoplasmic Actin Binding Function. *J. Biol. Chem.* **2015**, *290*, 338–349.

131. Jiao, S.; Zhang, Z.; Li, C.; Huang, M.; Shi, Z.; Wang, Y.; Song, X.; Liu, H.; Li, C.; Chen, M.; *et al.* The kinase MST4 limits inflammatory responses through direct phosphorylation of the adaptor TRAF6. *Nat. Immunol.* **2015**, *16*, 246–257.

132. Kincaid, R.P.; Sullivan, C.S. Virus-Encoded microRNAs: An Overview and a Look to the Future. *PLOS Pathog.* **2012**, *8*, e1003018.

133. Lagatie, O.; Tritsmans, L.; Stuyver, L.J. The miRNA world of polyomaviruses. *Virol. J.* **2013**, *10*, e268.

134. Chen, C.J.; Cox, J.E.; Azarm, K.D.; Wylie, K.N.; Woolard, K.D.; Pesavento, P.A.; Sullivan, C.S. Identification of a polyomavirus microRNA highly expressed in tumors. *Virology* **2015**, *476*, 43–53.

135. Zhu, Y.; Haecker, I.; Yang, Y.; Gao, S.J.; Renne, R. γ-Herpesvirus-encoded miRNAs and their roles in viral biology and pathogenesis. *Curr. Opin. Virol.* **2013**, *3*, 266–275.

136. Renwick, N.; Cekan, P.; Masry, P.A.; McGeary, S.E.; Miller, J.B.; Hafner, M.; Li, Z.; Mihailovic, A.; Morozov, P.; Brown, M.; *et al.* Multicolor microRNA FISH effectively differentiates tumor types. *J. Clin. Invest.* **2013**, *123*, 2694–2702.

137. Grundhoff, A.; Sullivan, C.S. Virus-encoded microRNAs. *Virology* **2011**, *411*, 325–343.

138. Cullen, B.R. MicroRNAs as mediators of viral evasion of the immune system. *Nat. Immunol.* **2013**, *14*, 205–210.

139. Fimia, G.M.; Corazzari, M.; Antonioli, M.; Piacentini, M. Ambra1 at the crossroad between autophagy and cell death. *Oncogene* **2013**, *32*, 3311–3318.

140. Sung, C.K.; Yim, H.; Andrews, E.; Benjamin, T.L. A mouse polyomavirus-encoded microRNA targets the cellular apoptosis pathway through Smad2 inhibition. *Virology* **2014**, *468–470*, 57–62.

141. Skalsky, R.L.; Cullen, B.R. Viruses, microRNAs, and host interactions. *Annu. Rev. Microbiol.* **2010**, *64*, 123–141.

142. Moens, U.; Seternes, O.M.; Johansen, B.; Rekvig, O.P. Mechanisms of transcriptional regulation of cellular genes by SV40 large T- and small t-antigens. *Virus Genes* **1997**, *15*, 135–154.

143. Cullen, B.R. Transcription and processing of human microRNA precursors. *Mol. Cell* **2004**, *16*, 861–865.

144. Seo, G.J.; Fink, L.H.; O'Hara, B.; Atwood, W.J.; Sullivan, C.S. Evolutionary conserved function of a viral microRNA. *J. Virol.* **2008**, *82*, 9823–9828.

145. Broekema, N.M.; Imperiale, M.J. miRNA regulation of BK polyomavirus replication during early infection. *Proc. Natl. Acad. Sci. USA* **2013**, *110*, 8200–8205.

146. Shuda, M.; Arora, R.; Kwun, H.J.; Feng, H.; Sarid, R.; Fernández-Figueras, M.T.; Tolstov, Y.; Gjoerup, O.; Mansukhani, M.M.; Swerdlow, S.H.; *et al.* Human Merkel cell polyomavirus infection I. MCV T antigen expression in Merkel cell carcinoma, lymphoid tissues and lymphoid tumors. *Int. J. Cancer* **2009**, *125*, 1243–1249.

147. Brostoff, T.; Dela Cruz, F.N., Jr.; Church, M.E.; Woolard, K.D.; Pesavento, P.A. The raccoon polyomavirus genome and tumor antigen transcription are stable and abundant in neuroglial tumors. *J. Virol.* **2014**, *88*, 12816–12824.

148. Braconi, C.; Valeri, N.; Gasparini, P.; Huang, N.; Tacciolo, C.; Nuovo, G.; Suzuki, T.; Croce, C.M.; Patel, T. Hepatitis C virus proteins modulate microRNA expression and chemosensitivity in malignant hepatocytes. *Clin. Cancer Res.* **2010**, *16*, 957–966.

149. Wu, Y.H.; Hu, T.F.; Chen, Y.C.; Tsai, Y.N.; Tsai, Y.H.; Cheng, C.C.; Wang, H.W. The manipulation of miRNA-gene regulatory networks by KSHV induces endothelial cell motility. *Blood* **2011**, *118*, 2896–2905.

150. Bala, S.; Tilahun, Y.; Taha, O.; Alao, H.; Kodys, K.; Catalano, D.; Szabo, G. Increased microRNA-155 expression in the serum and peripheral monocytes in chronic HCV infection. *J. Transl. Med.* **2012**, *10*, e151.

151. Lin, L.; Yin, X.; Hu, X.; Wang, Q.; Zheng, L. The impact of hepatitis B virus x protein and microRNAs in hepatocellular carcinoma: A comprehensive analysis. *Tumor. Biol.* **2014**, *35*, 11695–11700.

152. Ma, L.; Deng, X.; Wu, M.; Zhang, G.; Huang, J. Down-regulationof miRNA204 by LMP-1 enhances CDC2 activity and facilitates invasion of EBV-associated nasopharyngeal carcinoma cells. *FEBS Lett.* **2014**, *588*, 1562–1570.

153. Mansouri, S.; Pan, Q.; Blencowe, B.J.; Claycomb, J.M.; Frappier, L. Epstein-Barr virus EBNA1 protein regulates viral latency through effects on let-7 microRNA and dicer. *J. Virol.* **2014**, *88*, 11166–11177.

154. Vernin, C.; Thenoz, M.; Pinatel, C.; Gessain, A.; Gout, O.; Delfau-Larue, M.H.; Nazaret, N.; Legras-Lachuer, C.; Wattel, E.; Mortreux, F. HTLV-1 bZIP factor HBZ promotes cell proliferation and genetic instability by activating OncomiRs. *Cancer Res.* **2014**, *74*, 6082–6093.

155. Xie, H.; Lee, L.; Caramuta, S.; Höög, A.; Browaldh, N.; Björnhagen, V.; Larsson, C.; Lui, W.O. MicroRNA expression patterns related to merkel cell polyomavirus infection in human merkel cell carcinoma. *J. Invest. Dermatol.* **2014**, *134*, 507–517.

156. Warburg, O. On respiratory impairment in cancer cells. *Science* **1956**, *124*, 269–270.

157. Zhao, Y.; Liu, H.; Riker, A.I.; Fodstad, O.; Ledoux, S.P.; Wilson, G.L.; Tan, M. Emerging metabolic targets in cancer therapy. *Front. Biosci.* **2011**, *16*, 1844–1860.

158. Kinnaird, A.; Michelakis, E.D. Metabolic modulation of cancer: A new frontier with great translational potential. *J. Mol. Med.* **2015**, *93*, 127–142.

159. George, A.; Girault, S.; Testard, A.; Delva, R.; Soulié, P.; Couturier, O.F.; Morel, O. The impact of (18)F-FDG-PET/CT on Merkel cell carcinoma management: A retrospective study of 66 scans from a single institution. *Nucl. Med. Commun.* **2014**, *35*, 282–290.

160. Kitagawa, K.; Nishino, H.; Iwashima, A. Analysis of hexose transport in untransformed and sarcoma virus-transformed mouse 3T3 cells by photoaffinity binding of cytochalasin B. *Biochim. Biophys Acta* **1985**, *821*, 63–66.

161. Bravard, A.; Beaumatin, J.; Luccioni, C.; Fritsch, P.; Lefrançois, D.; Thenet, S.; Adolphe, M.; Dutrillaux, B. Chromosomal, mitochondrial and metabolic alterations in SV40-transformed rabbit chondrocytes. *Carcinogenesis* **1992**, *13*, 767 772.

162. Lachaise, F.; Martin, G.; Drougard, C.; Perl, A.; Vuillaume, M.; Wegnez, M.; Sarasin, A.; Daya-Grosjean, L. Relationship between posttranslational modification of transaldolase and catalase deficiency in UV-sensitive repair-deficient xeroderma pigmentosum fibroblasts and SV40-transformed human cells. *Free Radic. Biol. Med.* **2001**, *30*, 1365–1373.

163. Noch, E.; Sariyer, I.K.; Gordon, J.; Khalili, K. JC virus T-antigen regulates glucose metabolic pathways in brain tumor cells. *PLOS ONE* **2012**, *7*, e35054. doi:10.1371/journal.pone.0035054.

164. Duracher, D.; Wyczechowska, D.; Wilk, A.; Lassak, A.; Zea, A.; Estrada, J.; Reis, K. The Effects of JCV T-antigen on Tumor Cell Metabolism. Available online: http://www.medschool.lsuhsc.edu/cancer_center/docs/Dustin%20Duracher%20Poster%2036x60.pdf (accessed on 10 March 2015).

165. Ramanathan, A.; Wang, C.; Schreiber, S.L. Perturbational profiling of a cell-line model of tumorigenesis by using metabolic measurements. *Proc. Natl. Acad. Sci. USA* **2005**, *102*, 5992–5997.

166. Yuan, H.; Veldman, T.; Rundell, K.; Schlegel, R. Simian virus 40 small tumor antigen activates AKT and telomerase and induces anchorage-independent growth of human epithelial cells. *J. Virol.* **2002**, *76*, 10685–10691.

167. Elstrom, R.L.; Bauer, D.E.; Buzzai, M.; Karnauskas, R.; Harris, M.H.; Plas, D.R.; Zhuang, H.; Cinalli, R.M.; Alavi, A.; Rudin, C.M.; Thompson, C.B. Akt stimulates aerobic glycolysis in cancer cells. *Cancer Res.* **2004**, *64*, 3892–3899.

168. Rodriguez-Viciana, P.; Collins, C.; Fried, M. Polyoma and SV40 proteins differentially regulate PP2A to activate distinct cellular signaling pathways involved in growth control. *Proc. Natl. Acad. Sci. USA* **2006**, *103*, 19290–19295.

169. Sontag, E.; Fedorov, S.; Kamibayashi, C.; Robbins, D.; Cobb, M.; Mumby, M. The interaction of SV40 small tumor antigen with protein phosphatase 2A stimulates the map kinase pathway and induces cell proliferation. *Cell* **1993**, *75*, 887–897.

170. Agnetti, G.; Maraldi, T.; Giorentini, D.; Giordano, E.; Prata, C.; Hakim, G.; Muscari, C.; Guarnieri, C.; Caldarera, C.M. Activation of glucose transport during simulated ischemia in H9c2 cardiac myoblasts is mediated by protein kinase C isoforms. *Life Sci.* **2005**, *78*, 264–270.

171. Madan, E.; Gogna, R.; Bhatt, M.; Pati, U.; Kuppusamy, P.; Mahdi, A.A. Regulation of glucose metabolism by p53: Emerging new roles for the tumor suppressor. *Oncotarget* **2011**, *2*, 948–957.

172. Shivakumar, C.V.; Das, G.C. Interaction of human polyomavirus BK with the tumor-suppressor protein p53. *Oncogene* **1996**, *13*, 323–332.

173. Staib, C.; Pesch, J.; Gerwig, R.; Gerber, J.K.; Brehm, U.; Stangl, A.; Grummt, F. p53 inhibits JC virus DNA replication *in vivo* and interacts with JC virus large T-antigen. *Virology* **1996**, *219*, 237–246.

174. Kenific, C.M.; Debnath, J. Cellular and metabolic functions for autophagy in cancer cells. *Trends Cell Biol.* **2015**, *25*, 37–45.

175. Silva, L.M.; Jung, J.U. Modulation of the Autophagy Pathway by Human Tumor Viruses. *Semin. Cancer Biol.* **2013**, *23*, 323–328.

176. Bouley, S.J.; Maginnis, M.S.; Derdowski, A.; Gee, G.V.; O'Hara, B.A.; Nelson, C.D.; Bara, A.M.; Atwood, W.J.; Dugan, A.S. Host cell autophagy promotes BK virus infection. *Virology* **2014**, *456–457*, 87–95.

177. Kumar, S.H.; Rangarajan, A. Simian virus 40 small T antigen activates AMPK and triggers autophagy to protect cancer cells from nutrient deprivation. *J. Virol.* **2009**, *83*, 8565–8574.

178. Rundell, K.; Major, E.O.; Lampert, M. Association of cellular 56,000- and 32,000-molecular-weight protein with BK virus and polyoma virus t-antigens. *J. Virol.* **1981**, *37*, 1090–1093.

179. Bollag, B.; Hofstetter, C.A.; Reviriego-Mendoza, M.M.; Frisque, R.J. JC virus small T antigen binds phosphatase PP2A and Rb family proteins and is required for efficient viral DNA replication activity. *PLOS ONE* **2010**, *5*, e10606.

180. Shuda, M.; Kwun, H.J.; Feng, H.; Chang, Y.; Moore, P.S. Human Merkel cell polyomavirus small T antigen is an oncoprotein targeting the 4E-BP1 translation regulator. *J. Clin. Invest.* **2011**, *121*, 3623–3634.

181. Berrios, C.; Jung, J.; Primi, B.; Wang, M.; Pedamallu, C.; Duke, F.; Marcelus, C.; Cheng, J.; Garcea, R.L.; Meyerson, M.; DeCaprio, J.A. Malawi polyomavirus is a prevalent human virus that interacts with known tumor suppressors. *J. Virol.* **2015**, *89*, 857–862.

182. Kwun, H.J.; Shuda, M.; Camacho, C.J.; Gamper, A.M.; Thant, M.; Chang, Y.; Moore, P.S. Restricted Protein Phosphatase 2A Targeting by Merkel Cell Polyomavirus Small T Antigen. *J. Virol.* **2015**, *89*, 4191–4200.

183. Basile, A.; Darbinian, N.; Kaminski, R.; White, M.K.; Gentilella, A.; Turco, M.C.; Khalili, K. Evidence for modulation of BAG3 by polyomavirus JC early protein. *J. Gen. Virol.* **1999**, *90*, 1629–1640.

184. Sariyer, I.K.; Merabova, N.; Patel, P.K.; Knezevic, T.; Rosati, A.; Turco, M.C.; Khalili, K. Bag3-induced autophagy is associated with degradation of JCV oncoprotein, T-Ag. *PLOS ONE* **2012**, *7*, e45000.

185. Ribbens, J.J.; Moser, A.B.; Hubbard, W.C.; Bongarzone, E.R.; Maegawa, G.H. Characterization and application of a disease-cell model for a neurodegenerative lysosomal disease. *Mol. Genet. Metab.* **2014**, *111*, 172–183.

186. Meckes, D.G., Jr.; Raab-Traub, N. Microvesicles and viral infection. *J. Virol.* **2011**, *85*, 12844–12854.

187. Brinton, L.T.; Sloane, H.S.; Kester, M.; Kelly, K.A. Formation and role of exosomes in cancer. *Cell Mol. Life Sci.* **2015**, *72*, 659–671.

188. Schorey, J.S.; Cheng, Y.; Singh, P.P.; Smith, V.L. Exosomes and other extracellular vesicles in host-pathogen interactions. *EMBO Rep.* **2015**, *16*, 24–43.

189. Ramakrishnaiah, V.; Thumann, C.; Fofana, I.; Habersetzer, F.; Pan, Q.; de Ruiter, P.E.; Willemsen, R.; Demmers, J.A.; Stalin Raj, V.; Jenster, G.; *et al.* Exosome-mediated transmission of hepatitis C virus between hepatoma Huh7.5 cells. *Proc. Natl. Acad. Sci. USA* **2013**, *110*, 13109–13113.

190. Jaworski, E.; Narayanan, A.; van Duyne, R.; Shabbeer-Meyering, S.; Iordanskiy, S.; Saifuddin, M.; Das, R.; Afonso, P.V.; Sampey, G.C.; Chung, M.; *et al.* Human T-lymphotropic virus type 1-infected cells secrete exosomes that contain Tax protein. *J. Biol. Chem.* **2014**, *289*, 22284–22305.

191. Zhao, X.; Wu, Y.; Duan, J.; Ma, Y.; Shen, Z.; Wei, L.; Cui, X.; Zhang, J.; Xie, Y.; Liu, J. Quantitative proteomic analysis of exosome protein content changes induced by hepatitis B virus in Huh-7 cells using SILAC labeling and LC-MS/MS. *J. Proteome Res.* **2014**, *13*, 5391–5402.

192. Lee, T.H.; Chennakrishnaiah, S.; Audemard, E.; Montermini, L.; Meehan, B.; Rak, J. Oncogenic ras-driven cancer cell vesiculation leads to emission of double-stranded DNA capable of interacting with target cells. *Biochem. Biophys. Res. Commun.* **2014**, *451*, 295–301.

193. Giovannelli, I.; Martelli, F.; Repice, A.; Massacesi, L.; Azzi, A.; Giannecchini, S. Detection of JCPyV microRNA in blood and urine samples of multiple sclerosis patients under natalizumab therapy. *J. Neurovirol.* **2015**, doi:10.1007/s13365-015-0325-3.

194. Meckes, D.G., Jr.; Shair, K.H.; Marquitz, A.R.; Kung, C.P.; Edwards, R.H.; Raab-Traub, N. Human tumor virus utilizes exosomes for intercellular communication. *Proc. Natl. Acad. Sci. USA* **2010**, *107*, 20370–20375.

195. Dreux, M.; Garaigorta, U.; Boyd, B.; Décembre, E.; Chung, J.; Whitten-Bauer, C.; Wieland, S.; Chisari, F.V. Short-range exosomal transfer of viral RNA from infected cells to plasmacytoid dendritic cells triggers innate immunity. *Cell Host Microbe* **2012**, *12*, 558–570.

196. Chugh, P.E.; Sin, S.H.; Ozgur, S.; Henry, D.H.; Menezes, P.; Griffith, J.; Eron, J.J.; Damania, B.; Dittmer, D.P. Systemically circulating viral and tumor-derived microRNAs in KSHV-associated malignancies. *PLOS Pathog.* **2013**, *9*, e1003484.

197. Meckes, D.G., Jr.; Gunawardena, H.P.; Dekroon, R.M.; Heaton, P.R.; Edwards, R.H.; Ozgur, S.; Griffith, J.D.; Damania, B.; Raab-Traub, N. Modulation of B-cell exosome proteins by gamma herpesvirus infection. *Proc. Natl. Acad. Sci. USA* **2013**, *110*, E2925–E2933.

198. Velders, M.P.; Macedo, M.F.; Provenzano, M.; Elmishad, A.G.; Holzhütter, H.G.; Carbone, M.; Kwast, W.M. Human T cell responses to endogenously presented HLA-A*0201 restricted peptides of Simian virus 40 large T antigen. *J. Cell Biochem.* **2001**, *82*, 155–162.

199. Vlastos, A.; Andreasson, K.; Tegerstedt, K.; Holländerová, D.; Heidari, S.; Forstová, J.; Ramqvist, T.; Dalianis, T. VP1 pseudocapsids, but not a glutathione-S-transferase VP1 fusion protein, prevent polyomavirus infection in a T-cell immune deficient experimental mouse model. *J. Med. Virol.* **2003**, *70*, 293–300.

200. Lowe, D.B.; Shearer, M.H.; Tarbox, J.A.; Kang, H.S.; Jumper, C.A.; Bright, R.K.; Kennedy, R.C. *In vitro* simian virus 40 large tumor antigen expression correlates with differential immune responses following DNA immunization. *Virology* **2005**, *332*, 28–37.

201. Zeng, Q.; Gomez, B.P.; Viscidi, R.P.; Peng, S.; He, L.; Ma, B.; Wu, T.C.; Hung, C.F. Development of a DNA vaccine targeting Merkel cell polyomavirus. *Vaccine* **2012**, *30*, 1322–1329.

202. Sospedra, M.; Schippling, S.; Yousef, S.; Jelcic, I.; Bofill-Mas, S.; Planas, R.; Stellmann, J.P.; Demina, V.; Cinque, P.; Garcea, R.; *et al.* Treating progressive multifocal leukoencephalopathy with interleukin 7 and vaccination with JC virus capsid protein VP1.*Clin. Infect Dis.* **2014**, *59*, 1588–1592.

203. Willmes, C.; Adam, C.; Alb, M.; Völkert, L.; Houben, R.; Becker, J.C.; Schrama, D. Type I and II IFNs inhibit Merkel cell carcinoma via modulation of the Merkel cell polyomavirus T antigens. *Cancer Res.* **2012**, *72*, 2120–2128.

204. Nakajima. H.; Takaishi, M.; Yamamoto, M.; Kamijima, R.; Kodama, H.; Tarutani, M.; Sano, S. Screening of the specific polyoma virus as diagnostic and prognostic tools for Merkel cell carcinoma. *J. Dermatol. Sci.* **2009**, *56*, 211–213.

205. Biver-Dalle, C.; Nguyen, T.; Touzé, A.; Saccomani, C.; Penz, S.; Cunat-Peultier, S.; Riou-Gotta, M.O.; Humbert, P.; Coursaget, P.; Aubin, F. Use of interferon-alpha in two patients with Merkel cell carcinoma positive for Merkel cell polyomavirus. *Acta Oncol.* **2011**, *50*, 479–480.

206. Chapuis, A.G.; Afanasiev, O.K.; Iyer, J.G.; Paulson, K.G.; Parvathaneni, U.; Hwang, J.H.; Lai, I.; Roberts, I.M.; Sloan, H.L.; Bhatia, S.; *et al.* Regression of metastatic Merkel cell carcinoma following transfer of polyomavirus-specific T cells and therapies capable of re-inducing HLA class-I. *Cancer Immunol. Res.* **2014**, *2*, 27–36.

207. Hata, Y.; Matsuka, K.; Ito, O.; Matsuda, H.; Furuichi, H.; Konstantinos, A.; Nuri, B. Two cases of Merkel cell carcinoma cured by intratumor injection of natural human tumor necrosis factor. *Plast Reconstr. Surg.* **1997**, *99*, 547–553.

208. Lampreave, J.L.; Bénard, F.; Alavi, A.; Jimeneze-Hoyuela, J.; Fraker, D. PET evaluation of therapeutic limb perfusion in Merkel's cell carcinoma. *J. Nucl. Med.* **1998**, *39*, 2087–2090.

209. Olieman, A.F.; Liénard, D.; Eggermont, A.M.; Kroon, B.B.; Lejeune, F.J.; Hoekstra, H.J.; Koops, H.S. Hyperthermic isolated limb perfusion with tumor necrosis factor alpha, interferon gamma, and melphalan for locally advanced nonmelanoma skin tumors of the extremities: A multicenter study. *Arch. Surg.* **1999**, *134*, 303–307.

210. Krishna, S.M.; Kim, C.N. Merkel cell carcinoma in a patient treated with adalimumab: Case report. *Cutis* **2011**, *87*, 81–84. Erratum in *Cutis* **2011**, *87*, 185.

211. Linn-Rasker, S.P.; van Albada-Kuipers, G.A.; Dubois, S.V.; Janssen, K.; Zweers, P.G. Merkel cell carcinoma during treatment with TNF-alpha inhibitors: Coincidence or warning? *Ned. Tijdschr. Geneeskd.* **2012**, *156*, A44–A64.

212. Reiss, K.A.; Forde, P.M.; Brahmer, J.R. Harnessing the power of the immune system via blockade of PD-1 and PD-L1: A promising new anticancer strategy. *Immunotherapy* **2014**, *6*, 459–475.

213. Orba, Y.; Sawa, H.; Iwata, H.; Tanaka, S.; Nagashima, K. Inhibition of virus production in JC virus-infected cells by postinfection RNA interference. *J. Virol.* **2004**, *78*, 7270–7273.

214. Sabbioni, S.; Callegari, E.; Spizzo, R.; Veronese, A.; Altavilla, G.; Corallini, A.; Negrini, M. Anticancer activity of an adenoviral vector expressing short hairpin RNA against BK virus T-ag. *Cancer Gene Ther.* **2007**, *14*, 297–305.

215. Houben, R.; Shuda, M.; Weinkam, R.; Schrama, D.; Feng, H.; Chang, Y.; Moore, P.S.; Becker, J.C. Merkel cell polyoma virus-infected Merkel cell carcinoma cells require expression of viral T antigens. *J. Virol.* **2010**, *84*, 7064–7072.

216. Angermeyer, S.; Hesbacher, S.; Becker, J.C.; Schrama, D.; Houben, R. Merkel cell polyomavirus-positive Merkel cell carcinoma cells do not require expression of the viral small T antigen. *J. Invest Dermatol.* **2013**, *133*, 2059–2064.

217. Lin, M.C.; Wang, M.; Fang, C.Y.; Chen, P.L.; Shen, C.H.; Chang, D. Inhibition of BK virus replication in human kidney cells by BK virus large tumor antigen-specific shRNA delivered by JC virus-like particles. *Antiviral Res.* **2014**, *103*, 25–31.

218. Hartman, Z.C.; Wei, J.; Glass, O.K.; Guo, H.; Lei, G.; Yang, X.Y.; Osada, T.; Hobeika, A.; Delcayre, A.; Le Pecq, J.B.; Morse, M.A.; Clay, T.M.; Lyerly, H.K. Increasing vaccine potency through exosome antigen targeting. *Vaccine* **2011**, *29*, 9361–9367.

219. Shuda, M.; Feng, H.; Kwun, H.J.; Rosen, S.T.; Gjoerup, O.; Moore, P.S.; Chang, Y. T antigen mutations are a human tumor-specific signature for Merkel cell polyomavirus. *Proc. Natl. Acad. Sci. USA* **2008**, *105*, 16272–16277.

Human Papillomaviruses; Epithelial Tropisms, and the Development of Neoplasia

Nagayasu Egawa, Kiyofumi Egawa, Heather Griffin and John Doorbar

Abstract: Papillomaviruses have evolved over many millions of years to propagate themselves at specific epithelial niches in a range of different host species. This has led to the great diversity of papillomaviruses that now exist, and to the appearance of distinct strategies for epithelial persistence. Many papillomaviruses minimise the risk of immune clearance by causing chronic asymptomatic infections, accompanied by long-term virion-production with only limited viral gene expression. Such lesions are typical of those caused by Beta HPV types in the general population, with viral activity being suppressed by host immunity. A second strategy requires the evolution of sophisticated immune evasion mechanisms, and allows some HPV types to cause prominent and persistent papillomas, even in immune competent individuals. Some Alphapapillomavirus types have evolved this strategy, including those that cause genital warts in young adults or common warts in children. These strategies reflect broad differences in virus protein function as well as differences in patterns of viral gene expression, with genotype-specific associations underlying the recent introduction of DNA testing, and also the introduction of vaccines to protect against cervical cancer. Interestingly, it appears that cellular environment and the site of infection affect viral pathogenicity by modulating viral gene expression. With the high-risk HPV gene products, changes in E6 and E7 expression are thought to account for the development of neoplasias at the endocervix, the anal and cervical transformation zones, and the tonsilar crypts and other oropharyngeal sites. A detailed analysis of site-specific patterns of gene expression and gene function is now prompted.

Reprinted from *Viruses*. Cite as: Egawa, N.; Egawa, K.; Griffin, H.; Doorbar, J. Human Papillomaviruses; Epithelial Tropisms, and the Development of Neoplasia. *Viruses* **2015**, *7*, 3863–3890.

1. Introduction

Papillomaviruses have been discovered in a wide array of vertebrates. More than 300 papillomaviruses have been identified and completely sequenced, including over 200 human papillomaviruses (PaVE: Papillomavirus Episteme [1]). One of the most distinctive characteristics of the papillomavirus group is their genotype-specific host-restriction, and the preference of particular papillomavirus types for distinct anatomical sites, where they cause lesions with distinctive clinical pathologies [2]. These include benign hyper-proliferative lesions such as warts, as well as unapparent or asymptomatic precursor lesions, that can in some instances progress to high-grade neoplasia and invasive malignant cancer. The association of "high-risk" human papillomavirus (HPV) types (see legend to Figure 1) with cervical cancer is now well established, and provides a rationale for the introduction of HPV DNA testing in cervical screening, as well as the development of prophylactic vaccines against HPV16 and 18 which are the major papillomavirus types responsible for cervical cancer. Such typing studies have also revealed a plethora of HPV types, both high and low-risk, in oral mouthwash samples despite the absence of apparent clinical disease [3], as well as in skin swabs and plucked hairs taken from immunocompetent individuals [4–9]. The predominant asymptomatic skin HPV types come primarily from the genera Betapapillomavirus and Gammapapillomavirus, and at a population level, members of these genera are very successful, infecting children at a young age to produce persistent subclinical infections [4]. Changes in the epithelial micro-environment, as can occur following immunosuppression or in individuals suffering from epidermodysplasia verruciformis (EV), can allow these HPV types to produce visible papillomas, and in some situations can facilitate the development of cancers [10]. Certain Beta HPV types are a significant cause of non-Melanoma skin cancer in susceptible individuals, although the molecular mechanism by which they facilitate cancer progression appears to be somewhat different from what has been worked out for the Alphapapillomavirus types [11]. When considered together, it appears that different papillomavirus types have evolved distinct life-cycle strategies, which allow them to thrive and produce viral progeny at different epithelial sites. At least part of this variation reflects the different ability of papillomaviruses to interact with the immune system, and to produce productive infections that are visible at the macroscopic level. The complex immune evasion strategies that underlie the ability to produce such lesions are a particular characteristic of the Genus Alphapapillomavirus, with low-risk Alphapapillomavirus types in particular (Figure 1), often being responsible for recalcitrant warts even in immunocompetent hosts [12]. By contrast, most members of the genera β- and Gammapapillomavirus only cause such visible lesions when the normal immune response of the host is compromised [13–16]. The clinical importance of papillomaviruses, and their relatively small genomic size, has led to many full genomic sequences becoming available in recent years. The availability of such extensive sequence information, combined with a developing clinical and biochemical understanding of disease-biology, means that papillomaviruses are an ideal model system to understand how evolution can influence viral tropisms, pathogenicity, and the underlying molecular processes that govern disease outcome [17–20]. As discussed below, this "natural experiment" in infection is also providing us with insight into epithelial cell biology and immunology.

2. Papillomavirus Diversity at the Level of Genotype, Epithelial Tropism and Pathogenicity

Papillomaviruses comprise a diverse group of viruses that infect both humans and animals. Their origin is linked to changes in the epithelium of their ancestral host that occurred at least 350 million years ago. Since then, they have co-evolved as their different host species have evolved, with little cross-transfer between species [17,18]. They are now found in birds, reptiles, marsupials and other mammals, pointing to an earlier evolutionary appearance than was initially suspected. Recent phylogenetic analysis suggests, however, that the "generalist" ancestral papillomavirus may not have followed an identical evolutionary path to that of their hosts, but paralleled the evolution of host resources or attributes, such as the presence or absence of fur or the evolution of sweat glands. These host-adaptations are thought to have created new ecological niches for papillomavirus to colonise, which in turn drove viral diversity followed by co-speciation with their hosts [17]. Through this route, papillomaviruses have developed their remarkable species specificity as well as a great diversity of epithelial tropisms. With over 240 distinct papillomavirus types classified into 37 genera, papillomavirus may perhaps be considered as one of the most successful families of vertebrate viruses [17,21,22].

The classification of papillomaviruses is based on nucleotide sequence comparison rather than on serology, with individual HPVs being referred to as genotypes [22]. Data from vaccine trials has shown that there is limited antibody cross-reactivity, except between closely related genotypes [23], with the major coat protein of the virus (L1) containing hypervariable loops that are exposed on the virion surface [24,25]. The L1 gene was chosen early on as the standard for PV classification, and for some papillomavirus types, there is little sequence information outside of this region. To be classified as distinct types, individual papillomaviruses must be at least 10% divergent from each other in their L1 nucleotide sequence. These papillomavirus "types" are grouped into larger phylogenetic groupings or genera, which are categorised with a Greek letter followed by a number that indicates the species (Figure 1) [22].

Thus the species Alphapapillomavirus 9 includes HPV types 16, 31, 35, 33, 52, 58 and 67. Ideally, papillomavirus (PV) classification should integrate phylogeny, genome organization, biology and pathogenicity as a single property, rather than being based simply on genomic sequence analysis, or the analysis of genomic fragments [21]. This phylogenetic species concept, although potentially useful, is complex to implement, however, and for many papillomavirus types, detailed information regarding their biology is only poorly defined.

In general, sequence-based phylogeny does provide some useful insight into disease association, although closely related types can in some instances show distinct pathologies. HPV4, 65 and 95 are encompassed in the Gammapapillomavirus 1 species, for instance, but while HPV4 and 65 induce indistinguishable pigmented wart-like lesions mainly on the palmoplantar or lateral surface of hands and feet [27–29], HPV95 induces less obvious unpigmented papules primarily on plantar epithelial surfaces [30]. HPV6 and 11 are similarly contained within a common species-grouping (Alphapapillomavirus 10), and although both cause papillomas of similar appearance, HPV6 shows a marked predilection for genital sites, when compared to HPV11, which is the most prominent type at oral sites [31–34]. A third example comes from the Alphapapillomavirus 8 species which includes HPV40, which causes mucosal lesions, and HPV7, which is the cause of "butchers" warts that develop at cutaneous sites, particularly the hands [35]. Explaining such subtle tropism differences is beyond our current understanding of virus biology, and is not of obvious medical importance. It is however likely that such tropism differences reflect differences in viral gene function, patterns of gene expression and epithelial regulation at different body sites. Perhaps of greater importance are the differences in cancer risk associated with the high-risk types. HPV16, 31 and 35 are all contained within the Alphapapillomavirus 9 species group and are classified as "carcinogenic to humans" by The International Agency for the Research on Cancer (IARC) [36]. The association between HPV16 and cervical cancer is more than 10 times stronger than that between either HPV31 or HPV35 however [37], and of these three types, HPV16 is uniquely associated with tumours of the oropharyngeal region [38]. To explain this will require a dissection of virus-specific gene expression at this particular epithelial site, and an understanding of virus protein function and the extent to which proteins encoded by different HPV types affect common molecular pathways.

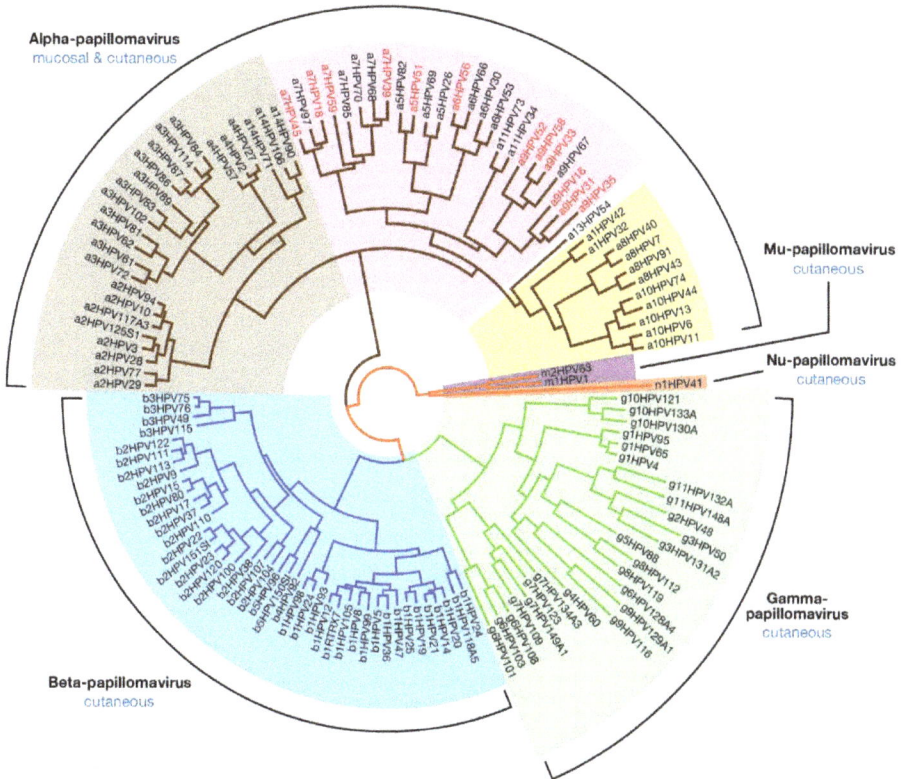

Figure 1. Evolutionary Relationship between Human Papillomaviruses. The human papillomaviruses types found in humans fall into five genera, with the Alpha-, Beta- (**blue**) and Gammapapillomavirus (**green**) representing the largest groups; Human papillomaviruses types from the Alphapapillomavirus genus are often classified as low-risk cutaneous (**light brown**); low-risk mucosal (**yellow**); or high-risk (**pink**) according to their association with the development of cancer. The high-risk types highlighted with red text are confirmed as "human carcinogens" on the basis of epidemiological data. The remaining high-risk types are "probable" or "possible" carcinogens. Although the predominant tissue associations of each genus are listed as either cutaneous or mucosal, these designations do not necessarily hold true for every member of the genus. The evolutionary tree is based on alignment of the *E1*, *E2*, *L1*, and *L2* genes [26]. HPV sequence data was be obtained from PaVE [1].

From the above, it is apparent that viral pathogenicity depends on multiple factors, including the virus genotype, the nature of the cell infected (tropism) and the status of host immunity. Our current thinking suggest that tropism is controlled primarily at the level of viral gene expression rather than at the level of viral entry into the cell [39], and that regulatory elements within the long control region (LCR) are of key importance in determining the tissue range of different HPV types [40,41]. Differences in infectivity may however be a contributing factor, and it has been suggested that this may correlate with differences in surface charge distribution between cutaneous and mucosal virions [42]. Although the diseases caused by specific HPV types sometimes occur at non-typical sites, this is uncommon, with such lesions often exhibiting non-typical morphology and pathology [43]. The idea that papillomaviruses have epithelial sites where they have evolved to complete their productive life-cycle, and epithelial sites where they do this less-well or not at all, makes good sense when we consider the small number of instances where cross-species infection has been documented. Bovine Papillomavirus type 1 infection of cattle leads to the development of benign cutaneous lesions that may regress, whereas infection of horses typically results in the development of locally aggressive, non-regressing equine sarcoids that are non-permissive for virus production [44–47]. In this case, it appears that the BPV E5 gene, which is important for the production of fibropapillomas in cattle, can stimulate aberrant cell proliferation in the dermal layers in the horse [47]. Although cottontail rabbit papillomavirus (CRPV) is not a fibropapillomavirus like BPV, a similar concept explains its ability to generate non-productive lesions in domestic rabbits, that have a tendency to develop over time into neoplasia and cancer [48,49]. In this case, it is the E6 and E7 proteins, along with the viral transcription factor E2, that are ultimately important for the development of the cancer phenotype [50,51]. Historically, these ideas are not new to the field of tumour virus biology, with a large body of experimental work using animal models to understand how members of the polyoma and adenoviruses families can cause tumours [52,53]. In the case of adenoviruses, it is the E1A and E1B proteins, and with polyomaviruses it is the T antigens, that when deregulated can cause tumours in experimental systems [54,55], and it is now well known that these proteins share functional similarity with the E6 and E7 proteins of the high-risk papillomaviruses [56,57]. In fact, our current knowledge suggests that it is the deregulated expression of the high-risk E6 and E7 proteins that leads to the development of neoplasia at specific epithelial sites, and that these infections should be regarded as "non-productive or abortive infections" rather than "ordered" productive infections, where the level and pattern of viral gene expression is properly controlled [58,59]. It has been reported that several factors such as the type of cell, hormone as well as inflammatory cytokines control viral gene expression [41,60–62]. With this in mind, it is clearly important to understand how epithelial environment, cell type and host immunity act together to drive neoplasia, and we suspect that this will be an important topic of future research. Although this line of thinking fits well with our understanding of how high-risk HPV types cause neoplasia and cancer at specific sites, including the cervix, anus and oropharynx, the same broad principles also explain the elevated risk of non-melanoma skin cancer posed by Betapapillomaviruses, even though the molecular detail of how these viruses cause cancer is different [11]. Betapapillomavirus types are prevalent as asymptomatic infections in normal skin and mucosa in the general population [3,63,64], with infection occurring soon after birth [9]. In immunocompromised individuals, including those suffering from epidermodysplasia verruciformis, it appears that normal patterns of viral gene expression can become deregulated, allowing a level of viral gene expression that would not be tolerated in immune competent individuals (Figure 2) [11].

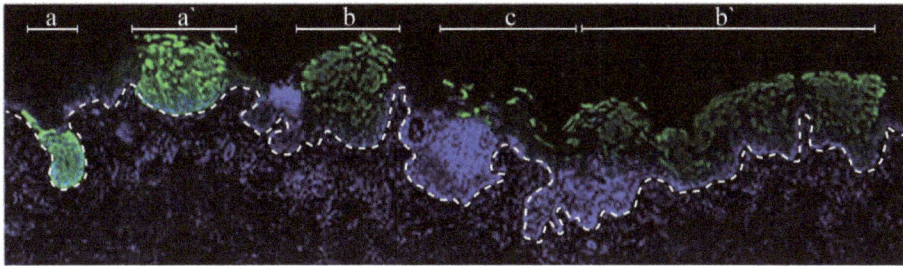

E4 DAPI

Figure 2. Viral gene-expression in adjacent Betapapillomavirus lesions. Immunostaining for the Betapapillomavirus E4 protein (**green**) reveals distinct patterns of expression in different lesions in an immunosuppressed individual. Lesion (**a**) and (**a′**) are atypical, and show marked basal and suprabasal staining; This pattern of E4 expression is distinct from typical E4 pattern of expression, which is usually restricted to the mid/upper epithelial layers as seen in region (**c**); Lesions (**b**) and (**b′**) show an intermediate pattern of E4 staining with expression close to the basal layer, as well as in the suprabasal cell layers (see also review [11]).

Interestingly, we have also seen that certain papillomaviruses, which are usually associated with symptomatic lesions following infection, can persist at the site of infection following immune regression, with reactivation occurring as the immune environment changes [65,66]. For both α- and Betapapillomavirus types, this work suggests that the immune system provides an important check on the extent of viral gene expression, and that overexpression of viral gene products can predispose to cancer progression following infection by certain β HPV types as well as for the high-risk Alphapapillomaviruses. The existence of latent papillomavirus infections remains to be conclusively validated in humans, but experiments in animal models suggest that this is likely [67,68].

3. Genome Structure and the Classification of Viral Gene Products

As outlined above, papillomaviruses can complete their life-cycle at many different anatomical sites, including the skin, anogenital tract and the oral cavity, but do not always produce clinically apparent lesions, sometimes persisting in the epithelium as non-productive latent infections. To progress beyond these basic observations requires an understanding of viral protein function and the expression patterns of the viral gene products in the epithelium. The viral genome is a circular double-stranded DNA episome, which ranges in length from 6953 bp for Cheloniamydas papillomavirus type 1 (CmPV1) to 8607 bp for Canine papillomavirus type 1 (CPV1). Viral genomes generally contain one regulatory region designated as the URR (upstream regulatory region) or LCR (long control region), which contains transcription factor-binding sites and the replication origin, and two groups of ORFs, which are designated early (E) or late (L) [69]. Some papillomavirus types, such as CPV, contain a second regulatory region between the end of the early region and the start of the late regions [70,71]. Despite variation in the size and number of ORFs, all PV members include a

URR, along with a group of proteins with a high level of conservation that are found in all sequenced papillomavirus types (Figure 3A).

These "core" genes include the E1 and E2 proteins necessary for viral replication [72,73], and the viral late proteins L1 and L2 [24,74]. The remaining viral genes (E6, E7, E5) may be considered as "accessory" genes that have evolved to facilitate replication in stratified epithelium (Figure 3B) [56,57,75]. These early gene products are generally more divergent than E1 and E2, both functionally and at the primary amino acid sequence level, and are not universally present in all papillomavirus types. Interestingly, the viral E4 protein has some characteristics of a core protein, in that it is present in most if not all papillomaviruses, but is also variable at the primary amino acid sequence level, which at first site could suggest a divergent function [76]. It is thought however that E4 plays a key role in virus escape from the cornified epithelial layers, and that evolution has driven E4 divergence to allow it to carry out its common function at different anatomical sites where epithelial structure differs. In general, the viral core genes are thought to have existed early during papillomavirus evolution, and carry out essential functions during the virus life cycle in the epithelium. L1 encodes primary structural protein in the virus capsid [77], with the minor capsid protein L2 binding to the circular viral DNA to facilitate optimal genome encapsidation [78]. E1 encodes a virus-specific DNA helicase [72], while E2 functions in viral transcription, replication and genome partitioning [73]. The remaining genes (E6, E7 and E5) encode proteins that modify the cellular environment, and in many cases perform similar but not necessarily identical functions during the life cycle of different papillomaviruses to support the production of progeny virions and to affect virulence. As such, it is thought that a major determinant of virus tropism may lie in the function of these accessory genes as well as in their regulation.

4. Productive Papillomavirus Life Cycle and Non-Productive Infection

Because the study of papillomaviruses has been driven by the association of the high-risk Alphapapillomavirus types with anogenital and oropharyngeal cancers, our knowledge of these types is perhaps more complete. However, there are some general principles that extend across the different genera that are worth overviewing before commenting further on the specific characteristics of different papillomavirus groups [79]. It is widely accepted that papillomavirus infection occurs once virus particles gain access to the epithelial basal cells or stem cells, which in many cases involves some level of epithelial trauma. In all cases, it is thought that virions bind initially to the glycosaminoglycan (GAG) chains of heparan sulfate proteoglycan (HSPG) in a charge-based manner, and that this leads to conformational changes in the virion which expose a furin/proprotein convertase cleavage site at the amino terminus of L2. It is generally thought that virus entry requires interaction with a secondary receptor, which still remains to be properly characterised [39]. Importantly, these early HSPG-dependent events have been shown to occur on the extracellular basement membrane (BM) in the murine cervico-vaginal model of HPV16 infection [80], close to where the target basal cells reside.

After PV infection, an initial phase of genome amplification takes place prior to maintenance of the viral episome at low copy number in the infected basal cells (Figure 4A) [81,82].

A

B

LCR

p97

7904/1

7000

1000

p670

6000

Alpha

2000

5000

3000

4000

Core Proteins

E1	ATP-dependent helicase. Role in papillomavirus genome replication.
E2	Coactivator of viral genome replication through the recruitment of E1 to the viral replication origin. Transcription factor of E6 and E7, also important for viral genome segregation.
L1	Major capsid protein. Assembles into pentomeric capsomeres, which are the primary components of the icosahedral virion shell.
L2	Minor capsid protein. Involved in encapsidation of viral DNA. Facilitate virus entry and trafficikng to the nucleus.
E4	E4 gene is embedded within the E2 gene, and is expressed abundantly as an E1^E4 fusion protein during the late stages of the viral life cycle. Binds to cytokeratin filaments and disrupts their structure. E4 is thought also to contribute to virus release and transmission.

Accessory Proteins

E5	Small transmembrane protein. In Alpha-PVs, E5 interacts with the EGF receptor and activates mitogenic signaling pathways. Has a role in evading the immune response and apoptosis. Beta-, Gamma- and Mu-PVs lack E5 gene.
E6	Drives cell cycle entry to allow genome amplification in the upper epithelial layers. E6 of high-risk Alpha-PV types binds and degrades p53 and can also activate telomerase and contribute to transformation. Also involved in immune evasion. Some Gamma-PVs lack E6 gene.
E7	Drives cell cycle entry to allow genome amplification in the upper epithelial layers. E7 of high-risk Alpha-PV types binds and degrades pRb and can induces chromosomal instability. E7 is necessary for cell transformation.

Both E6 and E7 have a large number of cellular substrates, with the identity of these substrates differing between HPV types [78].

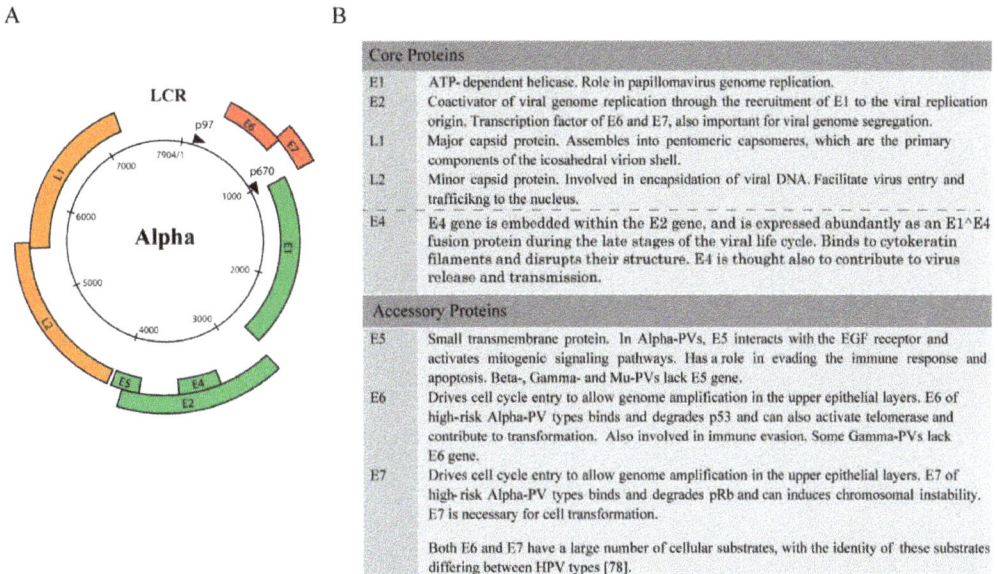

Figure 3. Alphapapillomavirus genome organization and the function of HPV proteins. (**A**) Genome organization typical of the high-risk Alphapapillomavirus types is illustrated by the genome of HPV16. The early (p97) and late (p670) promoters are marked by arrows. The six early ORFs (E1, E2, E4 and E5 (**in green**) and E6 and E7 (**in red**)) are expressed from the different promoters at different stages during epithelial cell differentiation. The late ORFs (L1 and L2 (**in yellow**)) are expressed from the p670 promoter in the upper epithelial layers as result of changes in splicing. The LCR/URR also contains the replication origin as well as post-transcriptional control sequences that contribute to viral gene expression. (**B**) The function of viral proteins. All known papillomavirus encodes a group of "core" proteins that were present early on during papillomavirus evolution, and which are conserved in sequence and in function between PV types. These include E1, E2, L2 and L1. The E4 protein may also be a core protein that has evolved to meet papillomaviruses epithelial specialization. The accessory proteins have evolved in each papillomavirus type during adaptation to different epithelial niches. The sequence and function of these genes are divergent between types. In general, these proteins are involved in modifying the cellular environment to facilitate virus life cycle completion, contributing to the virulence and pathogenicity. Knowledge of accessory protein function comes primarily from the study of Alphapapillomavirus types.

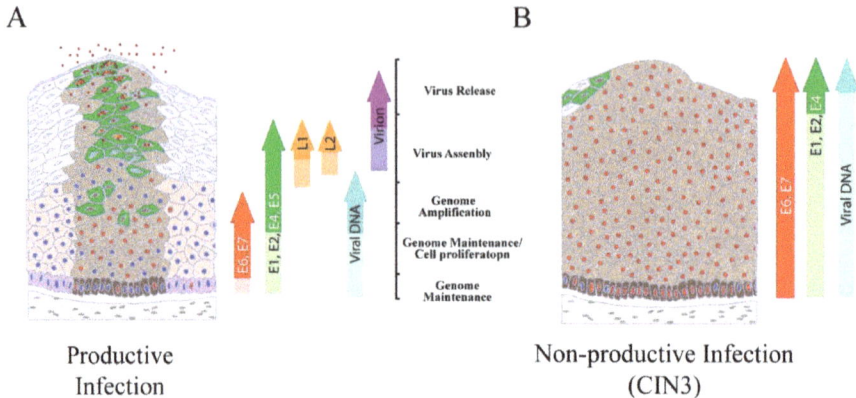

Figure 4. Regulation and deregulation of the high-risk Alphapapillomavirus life cycle. (**A**) The papillomavirus life cycle is regulated during epithelial cell differentiation and is shown diagrammatically. Cells that are driven through the cell cycle as a result of E6 and E7 expression are marked with red nuclei. The up-regulation of viral proteins necessary for genome amplification (*i.e.*, E1 and E2) requires activation of the viral late promoter in the upper epithelial layers (cells shown in green with red nuclei), with virus particles subsequently being released from the epithelial surface; (**B**) In HPV-associated neoplasia, late gene expression is retarded, and although the order of events remains the same, the production of infectious virions is restricted to smaller and smaller areas close to the epithelial surface. This situation is thought to be accompanied elevated E6/E7 expression, and represents a non-productive or poorly productive abortive infection. Integration of HPV DNA into the host cell genome is facilitated by deregulated E6/E7 expression. If integration disrupts the E1/E2 region this can allow the persistent high-level expression of E6 and E7 and the accumulation of genetic errors in the host genome. Eventually, the productive virus life cycle is no longer supported and viral episomes are lost (reviewed in [83]).

The viral replication proteins E1 and E2 are thought to be essential for this initial amplification, but E1 may be dispensable for maintenance-replication once copy number has stabilised at 50–100 [65]. E2 has an established role in genome partitioning, replication and viral gene expression. Experimental systems have shown at least a two-log increase in viral copy number during genome amplification [65], with E1, E2 and cellular DNA replication proteins being crucial for viral genome amplification *in vitro* [72,73,84]. The extent of genome amplification may be higher (four-log increase) during infection *in vivo*, as suggested from laser capture analysis of productive infections in rabbits [65]. The essential role of E6 and E7 in the viral lifecycle is primarily to modify the cellular environment to allow viral genome amplification, mainly by driving S-phase re-entry in the upper epithelial layers. Genome amplification occurs as the infected cell moves from an S to a G2-like phase before committing to full differentiation. In the case of high-risk types, E6 and E7 appear also to drive cell proliferation in the basal and parabasal layers [83]. By contrast, in the case of low-risk types (HPV6, HPV11 or other HPV types that have a tendency to cause benign lesions), the precise role of these proteins in the infected basal cells is unclear. In fact, functional differences in E6 and E7 represent a major determinant of HPV disease pathogenicity between HPV types [85]. During the virus life cycle, E5 also contributes to viral genome amplification as a result of its ability to stabilize epidermal growth factor (EGF) receptor and to enhance mitogen-activated protein (MAP) kinase activity. E5 also modulates both extracellular-signal-regulated kinase 1/2 (ERK 1/2) and p38 independently of EGF receptor [75]. Interestingly, the nuclear localisation and export signals in E1 are phosphorylated by these MAP kinases, which in turn enhances the nuclear localisation of E1 protein which is necessary for genome amplification [86,87]. In the upper layers of the epithelium, amplified viral genomes are packaged into virus particles produced from the major (L1) and minor (L2) virus coat proteins. The E4 protein, which accumulates at very high levels in cells supporting virus synthesis [76], appears to have a primary function in virus release and/or transmission, but acts also to optimise the success of virus genome amplification. In high-risk HPV types, the E4 protein assembles into amyloid fibrils that can disrupt keratin structure and compromise the normal assembly of the cornified envelope [88,89]. Although not yet precisely defined, E4 amyloid fibres may contribute to virion release from the upper epithelial layers, and therefore infectivity and transmission.

Although high-risk HPV infection is common, cervical cancer arises rarely as a result of infection, with most infections being cleared by the host without clinical symptoms. Regression of anogenital warts is accompanied by a CD4+ T cell-dominated Th1 response, which is also seen in animal models [90–95]. In general, HPV infections evade both the adaptive and innate immune response, with the life cycle being totally intra-epithelial, without viraemia, cell lysis or inflammation. During persistent infection, pro-inflammatory cytokines are not released, and the signals for Langerhans cell and dendritic cell activation and recruitment are largely absent [96]. In fact, cells supporting viral late gene expression, and which may contain high levels of viral proteins, are shed from the surface of the epithelium away from immune surveillance. In general, a failure to develop an effective host immune response correlates with persistent infection and an increased probability of progression toward invasive cancer.

In HPV-associated neoplasias, the ordered expression of viral gene products necessary for virus synthesis does not occur, leading to a non-productive or abortive infection rather than a productive infection (Figure 4B). In this situation, the deregulated expression of the high-risk E7 protein can stimulate host genome instability through deregulation of the centrosome cycle, with the deregulated expression of the high-risk E6 protein contributing to the accumulation of cellular mutations as a result of inhibition or loss of E6 function. This deregulation and the subsequent effects on the cell can eventually lead to the development of cancer (reviewed in [2]).

5. Epithelial Stem Cells and Their Postulated Role in Papillomavirus-Associated Disease

Current thinking suggests that the development of papillomavirus-associated disease requires the infection not just of an epithelial basal cell, but more specifically an epithelial tissue stem cell at a pluristratified cutaneous or mucosal site [97–99]. In general, papillomavirus-associated lesions often persist for many years. Epithelial tissue renewal is achieved by proliferation of stem or stem-like cells in the basal layer, to produce transit amplifying cells that have a limited proliferative capacity, and which subsequently go on to terminally differentiate [100]. Once committed to differentiate, these cells are pushed towards the cell surface by the production of new cells beneath them, and are eventually lost from the epithelial surface as terminally differentiated squames. This movement comprises the "epidermal flow", which occurs over a shorter time period than the life span of a papillomavirus infection [101]. Thus, we assume that papillomaviruses target basal cells with stem cell-like properties [97], or in the case of the high-risk papillomaviruses, they confer such properties on the cells that they infect. The idea that papillomaviruses generally reside in an epithelial stem cell following infection, is compatible with our understanding of latency and reactivation from latency [68], and the clonal origin of papillomas which was established several decades ago [102]. Indeed, studies on animal models have shown that CRPV primarily targets the bulge region of the hair follicle, where epithelial stem cells reside [103], and in cervical neoplasia, the epithelial reserve cells have long been considered the target cell from which cervical disease arises [99,104]. However, the cellular targets of infection have not yet been mapped more precisely, and this still remains a hypothesis, albeit a plausible one.

6. Local Epithelial Structure and Sites of Papillomavirus Infection

Epithelial tissue is composed of three components; epidermis, dermis and subepidermal tissue (subcutaneous tissue at skin sites). The epidermis is composed of multiple keratinocyte layers, and is the component that papillomaviruses target. In addition, the skin also consists of various functional appendages, including hair follicles, sebaceous glands, eccrine and apocrine sweat glands, finger and toe nails, as well as the inter-appendageal epidermis. Specialised epithelial sites contain additional appendages, such as the salivary glands of the oral cavity and the tonsillar crypts of the oropharynx. For papillomaviruses, these "specialist" structures represent particularly vulnerable sites, as they lack the highly structured barrier function usually associated with the epithelium. An additional level of complexity resides at transformation zone regions, where stratified epithelium abuts columnar epithelium, and it is generally accepted that high-risk HPV-associated neoplasias develop primarily at these sites [105]. High-risk HPV types are associated with the vast majority of cervical cancers (>99%), most of which arise at the cervical transformation zone, as well as the more than 90% of cancers that arise at the anal transformation zone. It is reasonable to suspect that transformation zones at other anatomical sites, such as at the boundary between the oesophagus and the stomach, may also be susceptible to HPV-associated neoplasia, but that these sites only rarely become infected because of their location. At the cervix, the transformation zone is maintained by a specialised type of cell known as the reserve cell [104], and possibly also by a cluster of cuboidal cells localised more precisely at the squamo-columnar junction [98]. Our current thinking suggests that these cell types represent the stem cells that maintain either the columnar epithelium of the endocervix, or the stratified epithelium of the transformation zone depending on their extracellular environment. These cells are thought to respond differently to signals from neighbouring epithelial cells and from the dermis, when compared to the more conventional epithelial stem cells that populate the stratified layers of the ectocervix. Indeed, our current model suggests that cervical neoplasia develops primarily at the squamo-columnar junction because these cells fail to properly regulate viral gene expression, leading to a non-productive or abortive infection rather than a productive infection (Figure 5) [59,106].

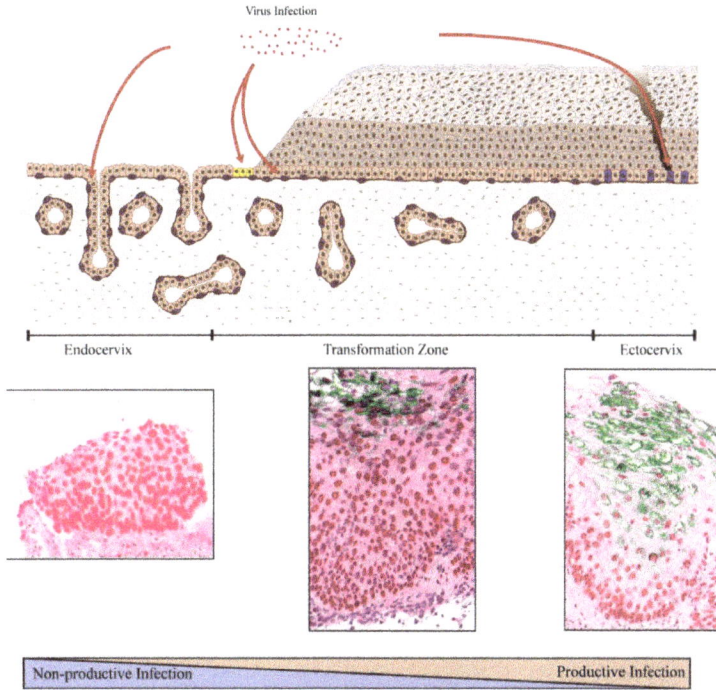

Figure 5. The difference of host cells (the site of infection) affects viral pathogenicity. Most cervical cancers arise at the cervical transformation zone. The transformation zone is maintained by a specialized type of tissue stem cell known as the reserve cell (shown in purple in the transformation zone), and possibly also by a cluster of cuboidal cells (**yellow**) localized more precisely at the squamo-columnar junction. These cells can maintain either the columnar epithelium of the endocervix or the stratified epithelium of transformation zone depending on their extracellular environment. In the ectocervix, the epithelium is populated by conventional epithelial tissue stem cells (purple in the ectocervix). The different characteristics of the various tissue stem cells that HPV infects are thought to influence the pattern of viral gene expression differently [41]. Current thinking suggests that productive infection is favoured at the ectocervix, while a non-productive or abortive infection is more likely at the endocervix. In the immunostains, MCM expression (**red**) indicates the expression of viral E7 protein. E4 expression (**green**) indicates the productive infection [59].

In this situation, deregulated viral gene expression facilitates the accumulation of genetic errors in the infected cell that no longer supports a productive infection (as described previously), leading eventually to the development of cancer.

By comparison with the cervical reserve cells, the stem cells that populate cutaneous epithelial sites are better characterised, and are known to reside in the bulge region of the hair follicle (Figure 6A) [107,108].

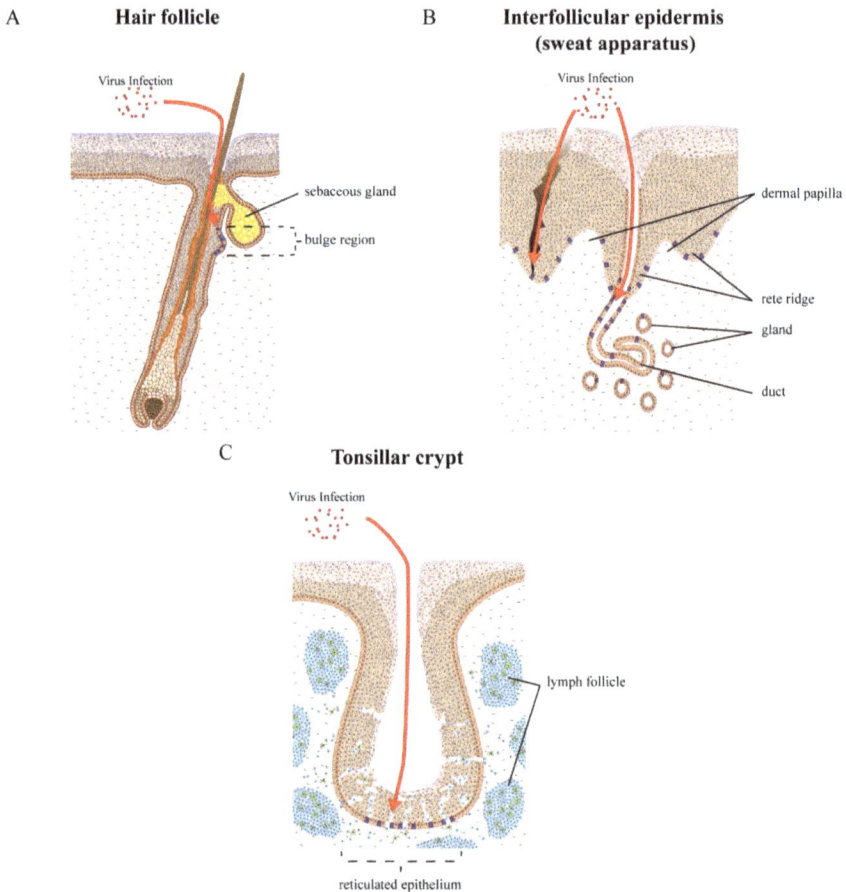

Figure 6. Epithelial tissue sites and tissue stem cells as targets for HPV infection. HPVs infect a variety of epithelial tissue sites, and can cause lesions in the vicinity of hair follicles, eccrine and apocrine sweat apparatus), nails, and also the inter-appendageal epidermis. Specialized epithelial sites contain other appendages, such as the salivary glands of the oral cavity and the tonsillar crypts of the oropharynx, where oropharyngeal cancers arise. The transformation zone regions, where stratified epithelium abuts columnar epithelium such as the cervical (Figure 5) or anal transformation zones, are other target sites where infection is thought to be facilitated. (**A**) The bulge region of the

hair follicle is a well characterized region where the stem cells that populate cutaneous epithelial sites reside. HPV virions are thought to gain access to the epithelial stem cells (coloured purple), either through a wound or possibly through the hair follicle; (**B**) Between the hair follicles, the tissue stem cells are thought to reside in both rete ridge and over the dermal papilla, and are not thought to be clustered at any specific location in the basal component [109]. With the sweat apparatus, at least two distinct stem cell populations have been identified that may be accessible for infection, either in the gland or the duct. These are able to repair damaged epidermis. HPV virions are thought to gain access to these stem cells (coloured purple), either through a wound or maybe through the eccrine duct; (**C**) The tonsillar crypts are a highly specialized lymphoepithelial tissue. A dense lymphocyte infiltrate generally obscures the junction between the lymphoid and epithelial components and splinters the epithelial sheath into irregular nests and cords. This reticulated epithelium may facilitate viral access to tissue stem cells at an immune-privileged site, which can inhibit virus-specific T cell activity and thereby facilitate immune evasion during initial HPV infection and subsequent virus-induced malignant transformation.

Because these cells can mediate the repair of damaged interfollicular epidermis, the interfollicular stem cells were for some time, thought to be supplied by follicular stem cells routinely [110]. However, recent work has shown that stem cells in the hair follicle do not contribute to epidermal homeostasis, but points instead to the presence of a distinct population of interfollicular epidermal stem cells [111]. This idea is supported by several observations [112], including the visualisation of discrete slow-cycling cells that are interspersed throughout the basal layer in mouse interfollicular epidermis [113] and in monkey palm epidermis [114], as well as the visualisation of clonal regenerating units of interfollicular epidermis (Figure 6B) [115]. In human skin, however, our inability to mark slow-cycling cells using *in situ* approaches, coupled with a lack of specific stem cell markers, has meant that the identification and location of interfollicular epidermal stem cells remains uncertain. Although it is generally accepted that eccrine sweat duct-derived cells can repair damaged epidermis [116], little attention has been paid to the sweat gland as an additional location of epidermal stem cells. Recently, stem cell populations in sweat glands and ducts have however been identified by lineage-tracing (Figure 6B) [117].

7. Examples of Papillomavirus Niche-Adaptation and Tropisms

Given the genotype-specific tropisms of papillomaviruses and the varied pathologies associated with infection, it is not unreasonable to think that different papillomaviruses have become well adapted to their respective epithelial niches. Although this has already been overviewed to some extent in the preceding paragraphs, the diversity of niche-adaptation can perhaps be appreciated by reference to particular examples that are outlined below.

Schmitt *et al.* [103] have shown that the primary target cells of CRPV co-localize with the hair follicle stem cells. In this study, E6 and E7 transcripts were detected at 11 days post-infection, despite the absence of apparent disease, with expression of these genes representing an early marker of infection. The subsequent finding that viable keratinocyte clones isolated from these infected follicles were CRPV mRNA-positive, suggests infection of a replication competent cell, which given the location, is suspected to be a bulge stem cell. A similar cellular target and epithelial tropism has also been suggested for the HPV types associated with epidermodysplasia verruciformis (EV). These HPV types usually come from the Genus Betapapillomavirus, and can be detected in plucked hair from different body sites. EV is a rare autosomal recessive genodermatosis associated with a high risk of skin carcinoma [11,118,119]. In many cases, the disease is linked to mutations in two genes (EVER1 and EVER2) that are expressed in the endoplasmic reticulum, and which form a complex with zinc-transporter-1 (ZnT-1), thereby controlling zinc balance [120]. Lesions generally start on the dorsa of the hands and the forehead, and spread progressively to the limb, neck and trunk. However, the mucous epithelium is not involved. EV skin lesions are hardly seen on the palmoplantar skin, even though they are frequently productive in nearby epithelium. At a clinical level, it is very clear that the Betapapillomavirus types associated with EV precisely "select" their preferential anatomical sites. The frequent detection of EV-PV in plucked hairs points to the hair follicle as an important site of infection for EV-PV [121,122].

HPV1 and 63 are members of the Genus Mupapillomavirus, and appear to target eccrine ducts in the palmoplantar skin. HPV1 causes typical myrmecia warts on the palms and soles [27,123], and it is quite rare for this HPV type to infect and cause lesions at extra-palmoplantar sites. The most characteristic histological feature is eosinophilic granular intracytoplasmic inclusion bodies (Gr-ICBs), which are larger and more numerous in the upper layer of the epidermis, and which are pathognomonic of infection by this virus. HPV1 infection provided a good example of how the site of infection influences clinical features and can affect viral gene expression [43]. Lesion development occurs on the skin surface ridges on the palms or soles. Histopathological studies and *in situ* hybridization analysis of early lesions have revealed that both HPV1-associated disease-pathology and *in situ* DNA-positivity are closely associated with eccrine ducts (Figure 7A, and our unpublished data).

HPV63, which is the second Mupapillomavirus type, induces tiny punctate warts (0.5 to 2 mm) characterised by the presence of heavily-stained tonofibril-like structures referred to as filamentous inclusion bodies [27,29]. To date, HPV63-induced warts have only been reported on the sole, with HPV63 DNA being detected both in keratinocytes around acrosyringiums (Figure 7B) and also in the uppermost portion of the eccrine dermal duct, which was hypertrophied with the HPV infection [124].

The final example focuses on HPV6 and 11, which share 85% sequence identity and belong to same HPV species (Alphapapillomavirus 2). Interestingly, in contrast to EV-HPVs from the Genus Betapapillomavirus, and which are both detectable in plucked pubic and eyebrow hairs, HPV6 and 11 could be detected only in hair plucked from pubic and perianal regions of patients with genital warts, but not in eyebrow hairs of the same patients. It has been suggested that HPV6 and 11 have a more limited epithelial distribution [125]. In contrast to this, Kocjan *et al.* [126], showed the presence of four genital HPVs (HPV6, 11 31 and 58) in plucked eyebrow hairs from immunocompetent individuals, suggesting that these viruses may sometimes be more widely distributed, but that they develop lesions preferentially at anogenital or oral sites. Since HPV6 and 11 are well-known causative agents for condyloma acuminatum, and can be detected in pubic and perianal hairs, a follicular distribution for this virus may exist. In fact, hair follicle-associated papules are sometimes observed in conjunction with the larger more prominent lesions at anogenital sites, with hair follicles being present widely at many epithelial sites. Interestingly, histological studies show that HPV6/11 DNA can also be detected in association with hair follicles as well as the vicinity of sweat apparatus or inter-appendageal epidermis, along with associated histological changes (Figure 7C, and our unpublished data). HPV6 and 11 are also causative agents of laryngeal papillomatosis, although in this instance the target cells remain unclear. The vocal cord, where HPV6 and 11 can also produce papillomatosis consists of squamous epithelium, pseudostratified epithelium and a transitional zone, which may facilitate access to stem-like cells at this site.

8. High-Risk Mucosal HPV Types

The high-risk mucosal HPVs, such as HPV16, 18, 31 and 33 appear to have evolved a number of additional characteristics that are not shared by low-risk papillomaviruses. These relate primarily to the function of the E5, E6 and E7 gene products, and the regulatory mechanisms that govern their expression (reviewed in [2]). Many functional studies have however focused on the contribution of these viral proteins to cancer development rather than life-cycle completion, and a detailed comparison of the role of these proteins in the life cycles of high and low-risk papillomaviruses remains to be done. A prominent characteristic of high-risk E6 proteins is the presence of a C-terminal PDZ-binding motif [127–131], along with an ability to bind and degrade the cellular p53 protein [132–134]. In addition, the high-risk HPV types have been shown to express their E6 and E7 proteins from a single promoter, with the levels of these two proteins being regulated by differential splicing [1,135]. This contrasts with the presence of separate promoters that regulate E6 and E7 expression in low-risk HPV types, such as those from the Alphpapillomavirus genus. Although there are additional key differences between these broad groups of papillomaviruses, these properties in themselves are known to have a major impact on high-risk virus pathogenesis. Although the precise advantage that the high-risk HPV types gain from these additional functions remains unclear, it is very likely to be related to particular sites of infection and their different strategies for productive infection at these sites. Clearly, the high-risk viruses have an ability to drive cell cycle entry as well as cell proliferation, which may be an important part of their normal life cycle at mucosal sites [83]. High-risk HPV-associated cancers occur predominantly within the cervical and anal transformation zone, as well as in the crypts of the oropharynx (Figure 6C) [136], which are thought to be sites where viral gene expression is poorly controlled. Given the preferential tissue distributions of different papillomaviruses, it is not surprising that particular HPV types may sometimes infect sites where full productive infection is either not supported or supported only poorly. The different cancer risk at the penis, vagina and vulva, when compared to the cervix or anus, illustrates this well [137]. In addition, the different risk of cancer progression associated with different HPV types (e.g., HPV16, 18 and 45) must reflect the different basic tropisms or tissue-specificities of the species Alphapapillomavirus 7 and Alphapapillomaviru 9 [138], and the different ways in which their genes are expressed and achieve their effects at these different epithelial sites. In addition to infection at these genital sites, some high-risk types can be found at other cutaneous sites where they cause "Bowenoid papulosis" [139,140].

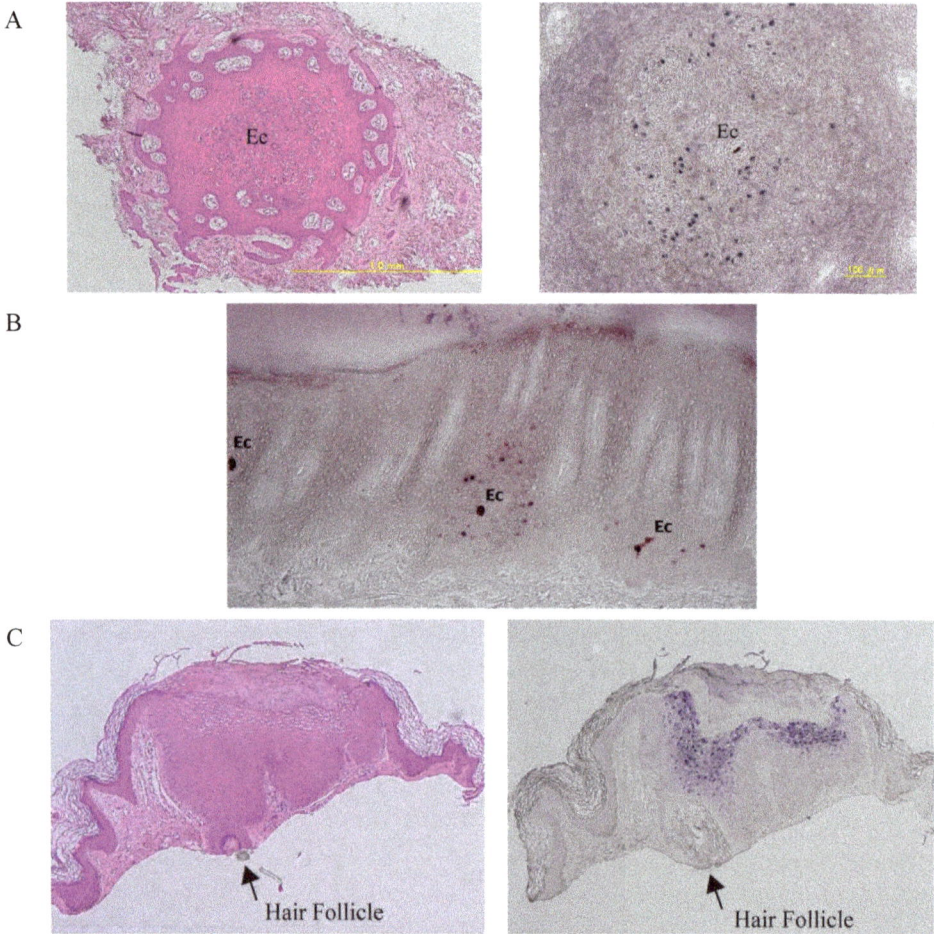

Figure 7. Examples for HPV targeting cutaneous appendages. (**A**) Haemotoxylin & eosin stain (**left**) of a horizontal section of a tiny wart. An eccrine (Ec)-centered distribution of histological changes is observed. HPV1 DNAs are identified within the pathology changes following DNA *in situ* hybridization (**right**); (**B**) HPV63 DNA is identified in resident keratinocytes in the vicinity of eccrine ducts (Ec) in a ridge of the plantar skin following DNA *in situ* hybridization; (**C**) HPV6/11 histopathological changes are identified in the resident keratinocytes in and around the hair follicle (arrow) by haemotoxylin & eosin staining (**left**) and DNA *in situ* hybridization (**right**).

9. Conclusions

Papillomaviruses have evolved over many millions of years to propagate themselves in a range of different species including humans. The evolution of skin glands, hair follicles and other ecological niches has provided papillomaviruses with new opportunities for infection. These have been followed by niche-adaptation, and it is thought that through this route, the great diversity of papillomaviruses has arisen. In general, β- and Gammapapillomavirus types have evolved a strategy of causing chronic asymptomatic infections accompanied by long-term virion-production and shedding from infected epidermis. This strategy minimises the risk of immune clearance, and depends on a balance between immune suppression and virus synthesis without any marked effect on host-fitness. By contrast, other papillomaviruses, such as those from the Genus Alphapapillomavirus, have evolved an array of sophisticated immune evasion strategies that allow them to cause prominent and persistent papillomas, even in immune competent individuals. Immune clearance of such lesions can however lead to asymptomatic or latent infections, with the possibility of an increase in viral copy number upon immunosuppression. Thus, both the cellular environment and the site of infection are important determinants of viral gene expression and virus activity. HPV pathogenicity, however, also reflects differences in viral gene function, which is evidenced most significantly when high and low-risk HPV types are compared. The genotype-specific association with disease, and more specifically with neoplasia, provides the basis for the recent introduction of DNA testing in cervical screening, and underlies the introduction of HPV 16/18 vaccines to protect against cervical cancer. High-risk mucosal HPVs do however have different cancer associations at different sites, prompting a detailed analysis of site-specific patterns of gene expression and gene function at different epithelial sites, such as the endocervix, the cervical transformation zone and the ectocervix, as well as the tonsillar crypts and other oropharyngeal sites where HPV-associated neoplasia can develop. An understanding of the site-specific aspects of HPV infection should facilitate the development of better strategies for disease treatment (e.g., target antivirals, immunotherapeutics) and disease management (e.g., risk assessment), and is very likely to provide new insight into epithelial biology and immunology.

Acknowledgments

Thanks to Ignacio Bravo (Catalan Institute of Oncology, Barcelona, Spain) for the evolutionary tree, which is modified from [26]. Our understanding of HPV immunology has been significantly enhanced through discussion with Margaret Stanley (University of Cambridge, UK) and Merilyn Hibma (University of Otago, New Zealand). The support and encouragement of Geoffrey Smith, Head of Department (Department of Pathology, University of Cambridge, UK) are greatly acknowledged. Thanks to Isao Murakami and Christian Kranjec (University of Cambridge, UK) for helpful discussion. The Human Papillomavirus Research Group at the University of Cambridge is funded by the UK Medical Research Council.

Author Contributions

Nagayasu Egawa and John Doorbar wrote the manuscript. Kiyofumi Egawa wrote the section of "Examples of Papillomavirus Niche-Adaptation and Tropisms" and provided the figures. Heather Griffin wrote the section relating to β-papillomavirus infection and deregulation and provided the figure.

Conflicts of Interest

The authors declare no conflict of interest.

References

1. PaVE: Papillomavirus Episteme. Available online: http://pave.niaid.nih.gov/ (accessed on 11 July 2015).
2. Doorbar, J.; Egawa, N.; Griffin, H.; Kranjec, C.; Murakami, I. Human papillomavirus molecular biology and disease association. *Rev. Med. Virol.* **2015**, *25* (Suppl. 1), 2–23. [CrossRef] [PubMed]
3. Bottalico, D.; Chen, Z.; Dunne, A.; Ostoloza, J.; McKinney, S.; Sun, C.; Schlecht, N.F.; Herrero, R.; Fatahzadeh, M.; Schiffman, M.; *et al.* The oral cavity contains abundant known and novel human papillomaviruses from the betapapillomavirus and gammapapillomavirus genera. *J. Infect. Dis.* **2011**, *204*, 787–792. [CrossRef] [PubMed]
4. De Koning, M.N.; Quint, K.D.; Bruggink, S.C.; Gussekloo, J.; Bouwes Bavinck, J.N.; Quint, W.G.; Feltkamp, M.C.; Eekhof, J.A. High prevalence of cutaneous warts in elementary school children and the ubiquitous presence of wart-associated human papillomavirus on clinically normal skin. *Br. J. Dermatol.* **2015**, *172*, 196–201. [CrossRef] [PubMed]
5. Poljak, M.; Kocjan, B.J.; Potocnik, M.; Seme, K. Anogenital hairs are an important reservoir of α-papillomaviruses in patients with genital warts. *J. Infect. Dis.* **2009**, *199*, 1270–1274. [CrossRef] [PubMed]
6. Potocnik, M.; Kocjan, B.J.; Seme, K.; Luzar, B.; Babic, D.Z.; Poljak, M. β-papillomaviruses in anogenital hairs plucked from healthy individuals. *J. Med. Virol.* **2006**, *78*, 1673–1678. [CrossRef] [PubMed]
7. De Koning, M.N.; Struijk, L.; Bavinck, J.N.; Kleter, B.; ter Schegget, J.; Quint, W.G.; Feltkamp, M.C. Betapapillomaviruses frequently persist in the skin of healthy individuals. *J. Gen. Virol.* **2007**, *88*, 1489–1495. [CrossRef] [PubMed]
8. Antonsson, A.; Erfurt, C.; Hazard, K.; Holmgren, V.; Simon, M.; Kataoka, A.; Hossain, S.; Hakangard, C.; Hansson, B.G. Prevalence and type spectrum of human papillomaviruses in healthy skin samples collected in three continents. *J. Gen. Virol.* **2003**, *84*, 1881–1886. [CrossRef] [PubMed]
9. Antonsson, A.; Karanfilovska, S.; Lindqvist, P.G.; Hansson, B.G. General acquisition of human papillomavirus infections of skin occurs in early infancy. *J. Clin. Microbiol.* **2003**, *41*, 2509–2514. [CrossRef] [PubMed]

10. Jablonska, S.; Orth, G. Epidermodysplasia verruciformis. *Clin. Dermatol.* **1985**, *3*, 83–96. [CrossRef]

11. Quint, K.D.; Genders, R.E.; de Koning, M.N.; Borgogna, C.; Gariglio, M.; Bouwes Bavinck, J.N.; Doorbar, J.; Feltkamp, M.C. Human β-papillomavirus infection and keratinocyte carcinomas. *J. Pathol.* **2015**, *235*, 342–354. [CrossRef] [PubMed]

12. Goon, P.; Sonnex, C.; Jani, P.; Stanley, M.; Sudhoff, H. Recurrent respiratory papillomatosis: An overview of current thinking and treatment. *Eur. Arch. Oto-Rhino-Laryngol.* **2008**, *265*, 147–151. [CrossRef] [PubMed]

13. Cubie, H.A. Diseases associated with human papillomavirus infection. *Virology* **2013**, *445*, 21–34. [CrossRef] [PubMed]

14. Harwood, C.A.; Spink, P.J.; Surentheran, T.; Leigh, I.M.; de Villiers, E.M.; McGregor, J.M.; Proby, C.M.; Breuer, J. Degenerate and nested PCR: A highly sensitive and specific method for detection of human papillomavirus infection in cutaneous warts. *J. Clin. Microbiol.* **1999**, *37*, 3545–3555. [PubMed]

15. Rogers, H.D.; Macgregor, J.L.; Nord, K.M.; Tyring, S.; Rady, P.; Engler, D.E.; Grossman, M.E. Acquired epidermodysplasia verruciformis. *J. Am. Acad. Dermatol.* **2009**, *60*, 315–320. [CrossRef] [PubMed]

16. Bouwes Bavinck, J.N.; Feltkamp, M.; Struijk, L.; ter Schegget, J. Human papillomavirus infection and skin cancer risk in organ transplant recipients. *J. Investig. Dermatol. Symp. Proc. Soc.* **2001**, *6*, 207–211. [CrossRef] [PubMed]

17. Van Doorslaer, K. Evolution of the papillomaviridae. *Virology* **2013**, *445*, 11–20. [CrossRef] [PubMed]

18. Bravo, I.G.; Felez-Sanchez, M. Papillomaviruses: Viral evolution, cancer and evolutionary medicine. *Evol. Med. Public Health* **2015**, *2015*, 32–51. [CrossRef] [PubMed]

19. Van Doorslaer, K.; Burk, R.D. Evolution of human papillomavirus carcinogenicity. *Adv. Virus Res.* **2010**, *77*, 41–62. [PubMed]

20. Burk, R.D.; Chen, Z.; van Doorslaer, K. Human papillomaviruses: Genetic basis of carcinogenicity. *Public Health Genomics* **2009**, *12*, 281–290. [CrossRef] [PubMed]

21. Bernard, H.U.; Burk, R.D.; Chen, Z.; van Doorslaer, K.; zur Hausen, H.; de Villiers, E.M. Classification of papillomaviruses (PVS) based on 189 PV types and proposal of taxonomic amendments. *Virology* **2010**, *401*, 70–79. [CrossRef] [PubMed]

22. De Villiers, E.M.; Fauquet, C.; Broker, T.R.; Bernard, H.U.; zur Hausen, H. Classification of papillomaviruses. *Virology* **2004**, *324*, 17–27. [CrossRef] [PubMed]

23. Drolet, M.; Benard, E.; Boily, M.C.; Ali, H.; Baandrup, L.; Bauer, H.; Beddows, S.; Brisson, J.; Brotherton, J.M.; Cummings, T.; *et al.* Population-level impact and herd effects following human papillomavirus vaccination programmes: A systematic review and meta-analysis. *Lancet. Infect. Dis.* **2015**, *15*, 565–580. [CrossRef]

24. Buck, C.B.; Day, P.M.; Trus, B.L. The papillomavirus major capsid protein L1. *Virology* **2013**, *445*, 169–174. [CrossRef] [PubMed]

25. Dasgupta, J.; Bienkowska-Haba, M.; Ortega, M.E.; Patel, H.D.; Bodevin, S.; Spillmann, D.; Bishop, B.; Sapp, M.; Chen, X.S. Structural basis of oligosaccharide receptor recognition by human papillomavirus. *J. Biol. Chem.* **2011**, *286*, 2617–2624. [CrossRef] [PubMed]

26. Doorbar, J.; Quint, W.; Banks, L.; Bravo, I.G.; Stoler, M.; Broker, T.R.; Stanley, M.A. The biology and life-cycle of human papillomaviruses. *Vaccine* **2012**, *30*, F55–F70. [CrossRef] [PubMed]

27. Egawa, K. New types of human papillomaviruses and intracytoplasmic inclusion bodies: A classification of inclusion warts according to clinical features, histology and associated HPV types. *Br. J. Dermatol.* **1994**, *130*, 158–166. [CrossRef] [PubMed]

28. Egawa, K.; Honda, Y.; Inaba, Y.; Ono, T. Pigmented viral warts: A clinical and histopathological study including human papillomavirus typing. *Br. J. Dermatol.* **1998**, *138*, 381–389. [CrossRef] [PubMed]

29. Egawa, K.; Delius, H.; Matsukura, T.; Kawashima, M.; de Villiers, E.M. Two novel types of human papillomavirus, HPV 63 and HPV 65: Comparisons of their clinical and histological features and DNA sequences to other HPV types. *Virology* **1993**, *194*, 789–799. [CrossRef] [PubMed]

30. Egawa, K.; Kimmel, R.; de Villiers, E.M. A novel type of human papillomavirus (HPV 95): Comparison with infections of closely related human papillomavirus types. *Br. J. Dermatol.* **2005**, *153*, 688–689. [CrossRef] [PubMed]

31. Ball, S.L.; Winder, D.M.; Vaughan, K.; Hanna, N.; Levy, J.; Sterling, J.C.; Stanley, M.A.; Goon, P.K. Analyses of human papillomavirus genotypes and viral loads in anogenital warts. *J. Med. Virol.* **2011**, *83*, 1345–1350. [CrossRef] [PubMed]

32. Dickens, P.; Srivastava, G.; Loke, S.L.; Larkin, S. Human papillomavirus 6, 11, and 16 in laryngeal papillomas. *J. Pathol.* **1991**, *165*, 243–246. [CrossRef] [PubMed]

33. Syrjanen, S. Human papillomavirus infections and oral tumors. *Med. Microbiol. Immunol. Berl* **2003**, *192*, 123–128. [CrossRef] [PubMed]

34. De Villiers, E.M. Human pathogenic papillomavirus types: An update. *Curr. Top. Microbiol. Immunol.* **1994**, *186*, 1–12. [PubMed]

35. Orth, G.; Jablonska, S.; Favre, M.; Croissant, O.; Obalek, S.; Jarzabek-Chorzelska, M.; Jibard, N. Identification of papillomaviruses in butchers' warts. *J. Investig. Dermatol.* **1981**, *76*, 97–102. [CrossRef] [PubMed]

36. International Agency for the Research on Cancer. Avilable online: http://monographs.iarc.fr/ENG/ Monographs/vol90/ (acessed on 11 July 2015).

37. Munoz, N.; Bosch, F.X.; de Sanjose, S.; Herrero, R.; Castellsague, X.; Shah, K.V.; Snijders, P.J.; Meijer, C.J.; International Agency for Research on Cancer Multicenter Cervical Cancer Study Group. Epidemiologic classification of human papillomavirus types associated with cervical cancer. *N. Engl. J. Med.* **2003**, *348*, 518–527. [CrossRef] [PubMed]

38. Herrero, R.; Castellsague, X.; Pawlita, M.; Lissowska, J.; Kee, F.; Balaram, P.; Rajkumar, T.; Sridhar, H.; Rose, B.; Pintos, J.; *et al.* Human papillomavirus and oral cancer: The international agency for research on cancer multicenter study. *J. Natl. Cancer Inst.* **2003**, *95*, 1772–1783. [CrossRef] [PubMed]

39. Day, P.M.; Schelhaas, M. Concepts of papillomavirus entry into host cells. *Curr. Opin. Virol.* **2014**, *4*, 24–31. [CrossRef] [PubMed]

40. Steinberg, B.M.; Auborn, K.J.; Brandsma, J.L.; Taichman, L.B. Tissue site-specific enhancer function of the upstream regulatory region of human papillomavirus type 11 in cultured keratinocytes. *J. Virol.* **1989**, *63*, 957–960. [PubMed]

41. Ottinger, M.; Smith, J.A.; Schweiger, M.R.; Robbins, D.; Powell, M.L.; You, J.; Howley, P.M. Cell-type specific transcriptional activities among different papillomavirus long control regions and their regulation by E2. *Virology* **2009**, *395*, 161–171. [CrossRef] [PubMed]

42. Mistry, N.; Wibom, C.; Evander, M. Cutaneous and mucosal human papillomaviruses differ in net surface charge, potential impact on tropism. *Virol. J.* **2008**, *5* [CrossRef] [PubMed]

43. Egawa, K.; Inaba, Y.; Yoshimura, K.; Ono, T. Varied clinical morphology of HPV-1-induced warts, depending on anatomical factors. *Br. J. Dermatol.* **1993**, *128*, 271–276. [CrossRef] [PubMed]

44. Trenfield, K.; Spradbrow, P.B.; Vanselow, B. Sequences of papillomavirus DNA in equine sarcoids. *Equine Vet. J.* **1985**, *17*, 449–452. [CrossRef] [PubMed]

45. Otten, N.; von Tscharner, C.; Lazary, S.; Antczak, D.F.; Gerber, H. DNA of bovine papillomavirus type 1 and 2 in equine sarcoids: Pcr detection and direct sequencing. *Arch. Virol.* **1993**, *132*, 121–131. [CrossRef] [PubMed]

46. Angelos, J.A.; Marti, E.; Lazary, S.; Carmichael, L.E. Characterization of BPV-like DNA in equine sarcoids. *Arch. Virol.* **1991**, *119*, 95–109. [CrossRef] [PubMed]

47. Chambers, G.; Ellsmore, V.A.; O'Brien, P.M.; Reid, S.W.; Love, S.; Campo, M.S.; Nasir, L. Association of bovine papillomavirus with the equine sarcoid. *J. Gen. Virol.* **2003**, *84*, 1055–1062. [CrossRef] [PubMed]

48. Rous, P.; Beard, J.W. The progression to carcinoma of virus-induced rabbit papillomas (shope). *J. Exp. Med.* **1935**, *62*, 523–548. [CrossRef] [PubMed]

49. Syverton, J.T. The pathogenesis of the rabbit papilloma-to-carcinoma sequence. *Ann. N. Y. Acad. Sci.* **1952**, *54*, 1126–1140. [CrossRef] [PubMed]

50. Jeckel, S.; Huber, E.; Stubenrauch, F.; Iftner, T. A transactivator function of cottontail rabbit papillomavirus E2 is essential for tumor induction in rabbits. *J. Virol.* **2002**, *76*, 11209–11215. [CrossRef] [PubMed]

51. Meyers, C.; Harry, J.; Lin, Y.L.; Wettstein, F.O. Identification of three transforming proteins encoded by cottontail rabbit papillomavirus. *J. Virol.* **1992**, *66*, 1655–1664. [PubMed]

52. Gjoerup, O.; Chang, Y. Update on human polyomaviruses and cancer. *Adv. Cancer Res.* **2010**, *106*, 1–51. [PubMed]

53. Schlesinger, R.W. Adenoviruses: The nature of the virion and of controlling factors in productive or abortive infection and tumorigenesis. *Adv. Virus Res.* **1969**, *14*, 1–61. [PubMed]

54. Pipas, J.M. SV40: Cell transformation and tumorigenesis. *Virology* **2009**, *384*, 294–303. [CrossRef] [PubMed]

55. Endter, C.; Dobner, T. Cell transformation by human adenoviruses. *Curr. Top. Microbiol. Immunol.* **2004**, *273*, 163–214. [PubMed]

56. Vande Pol, S.B.; Klingelhutz, A.J. Papillomavirus E6 oncoproteins. *Virology* **2013**, *445*, 115–137. [CrossRef] [PubMed]

57. Roman, A.; Munger, K. The papillomavirus E7 proteins. *Virology* **2013**, *445*, 138–168. [CrossRef] [PubMed]

58. Middleton, K.; Peh, W.; Southern, S.; Griffin, H.; Sotlar, K.; Nakahara, T.; El-Sherif, A.; Morris, L.; Seth, R.; Hibma, M.; *et al.* Organization of human papillomavirus productive cycle during neoplastic progression provides a basis for selection of diagnostic markers. *J. Virol.* **2003**, *77*, 10186–10201. [CrossRef] [PubMed]

59. Griffin, H.; Soneji, Y.; van Baars, R.; Arora, R.; Jenkins, D.; van de Sandt, M.; Wu, Z.; Quint, W.; Jach, R.; Okon, K.; *et al.* Stratification of HPV-induced cervical pathology using the virally encoded molecular marker E4 in conbination with p16 or mcm. *Mod. Pathol.* **2015**. in press [CrossRef] [PubMed]

60. Khare, S.; Pater, M.M.; Tang, S.C.; Pater, A. Effect of glucocorticoid hormones on viral gene expression, growth, and dysplastic differentiation in HPV16-immortalized ectocervical cells. *Exp. Cell Res.* **1997**, *232*, 353–360. [CrossRef] [PubMed]

61. Mittal, R.; Tsutsumi, K.; Pater, A.; Pater, M.M. Human papillomavirus type 16 expression in cervical keratinocytes: Role of progesterone and glucocorticoid hormones. *Obstet. Gynecol.* **1993**, *81*, 5–12. [PubMed]

62. Kyo, S.; Inoue, M.; Hayasaka, N.; Inoue, T.; Yutsudo, M.; Tanizawa, O.; Hakura, A. Regulation of early gene expression of human papillomavirus type 16 by inflammatory cytokines. *Virology* **1994**, *200*, 130–139. [CrossRef] [PubMed]

63. Antonsson, A.; Forslund, O.; Ekberg, H.; Sterner, G.; Hansson, B.G. The ubiquity and impressive genomic diversity of human skin papillomaviruses suggest a commensalic nature of these viruses. *J. Virol.* **2000**, *74*, 11636–11641. [CrossRef] [PubMed]

64. Antonsson, A.; Hansson, B.G. Healthy skin of many animal species harbors papillomaviruses which are closely related to their human counterparts. *J. Virol.* **2002**, *76*, 12537–12542. [CrossRef] [PubMed]

65. Maglennon, G.A.; McIntosh, P.; Doorbar, J. Persistence of viral DNA in the epithelial basal layer suggests a model for papillomavirus latency following immune regression. *Virology* **2011**, *414*, 153–163. [CrossRef] [PubMed]

66. Maglennon, G.A.; McIntosh, P.B.; Doorbar, J. Immunosuppression facilitates the reactivation of latent papillomavirus infections. *J. Virol.* **2014**, *88*, 710–716. [CrossRef] [PubMed]

67. Doorbar, J. Latent papillomavirus infections and their regulation. *Curr. Opin. Virol.* **2013**, *3*, 416–421. [CrossRef] [PubMed]

68. Gravitt, P.E. Evidence and impact of human papillomavirus latency. *Open Virol. J.* **2012**, *6*, 198–203. [CrossRef] [PubMed]

69. De Villiers, E.M. Cross-roads in the classification of papillomaviruses. *Virology* **2013**, *445*, 2–10. [CrossRef] [PubMed]

70. Delius, H.; van Ranst, M.A.; Jenson, A.B.; zur Hausen, H.; Sundberg, J.P. Canine oral papillomavirus genomic sequence: A unique 1.5-kb intervening sequence between the E2 and L2 open reading frames. *Virology* **1994**, *204*, 447–452. [CrossRef] [PubMed]

71. Terai, M.; Burk, R.D. Felis domesticus papillomavirus, isolated from a skin lesion, is related to canine oral papillomavirus and contains a 1.3 kb non-coding region between the E2 and L2 open reading frames. *J. Gen. Virol.* **2002**, *83*, 2303–2307. [PubMed]

72. Bergvall, M.; Melendy, T.; Archambault, J. The E1 proteins. *Virology* **2013**, *445*, 35–56. [CrossRef] [PubMed]

73. McBride, A.A. The papillomavirus E2 proteins. *Virology* **2013**, *445*, 57–79. [CrossRef] [PubMed]

74. Wang, J.W.; Roden, R.B. L2, the minor capsid protein of papillomavirus. *Virology* **2013**, *445*, 175–186. [CrossRef] [PubMed]

75. DiMaio, D.; Petti, L.M. The E5 proteins. *Virology* **2013**, *445*, 99–114. [CrossRef] [PubMed]

76. Doorbar, J. The E4 protein; structure, function and patterns of expression. *Virology* **2013**, *445*, 80–98. [CrossRef] [PubMed]

77. Chen, X.S.; Garcea, R.L.; Goldberg, I.; Casini, G.; Harrison, S.C. Structure of small virus-like particles assembled from the L1 protein of human papillomavirus 16. *Mol. Cell* **2000**, *5*, 557–567. [CrossRef]

78. Holmgren, S.C.; Patterson, N.A.; Ozbun, M.A.; Lambert, P.F. The minor capsid protein L2 contributes to two steps in the human papillomavirus type 31 life cycle. *J. Virol.* **2005**, *79*, 3938–3948. [CrossRef] [PubMed]

79. Doorbar, J. The papillomavirus life cycle. *J. Clin. Virol.* **2005**, *32*, S7–S15. [CrossRef] [PubMed]

80. Kines, R.C.; Thompson, C.D.; Lowy, D.R.; Schiller, J.T.; Day, P.M. The initial steps leading to papillomavirus infection occur on the basement membrane prior to cell surface binding. *Proc. Natl. Acad. Sci. USA* **2009**, *106*, 20458–20463. [CrossRef] [PubMed]

81. Pyeon, D.; Pearce, S.M.; Lank, S.M.; Ahlquist, P.; Lambert, P.F. Establishment of human papillomavirus infection requires cell cycle progression. *PLoS Pathog.* **2009**, *5*, e1000318. [CrossRef] [PubMed]

82. Parish, J.L.; Bean, A.M.; Park, R.B.; Androphy, E.J. Chlr1 is required for loading papillomavirus E2 onto mitotic chromosomes and viral genome maintenance. *Mol. Cell* **2006**, *24*, 867–876. [CrossRef] [PubMed]

83. Doorbar, J. Molecular biology of human papillomavirus infection and cervical cancer. *Clin. Sci. Lond.* **2006**, *110*, 525–541. [CrossRef] [PubMed]

84. Melendy, T.; Sedman, J.; Stenlund, A. Cellular factors required for papillomavirus DNA replication. *J. Virol.* **1995**, *69*, 7857–7867. [PubMed]

85. White, E.A.; Howley, P.M. Proteomic approaches to the study of papillomavirus-host interactions. *Virology* **2013**, *435*, 57–69. [CrossRef] [PubMed]

86. Yu, J.H.; Lin, B.Y.; Deng, W.; Broker, T.R.; Chow, L.T. Mitogen-activated protein kinases activate the nuclear localization sequence of human papillomavirus type 11 E1 DNA helicase to promote efficient nuclear import. *J. Virol.* **2007**, *81*, 5066–5078. [CrossRef] [PubMed]

87. Fradet-Turcotte, A.; Moody, C.; Laimins, L.A.; Archambault, J. Nuclear export of human papillomavirus type 31 E1 is regulated by CDK2 phosphorylation and required for viral genome maintenance. *J. Virol.* **2010**, *84*, 11747–11760. [CrossRef] [PubMed]

88. McIntosh, P.B.; Laskey, P.; Sullivan, K.; Davy, C.; Wang, Q.; Jackson, D.J.; Griffin, H.M.; Doorbar, J. E1–E4-mediated keratin phosphorylation and ubiquitylation: A mechanism for keratin depletion in HPV16-infected epithelium. *J. Cell Sci.* **2010**, *123*, 2810–2822. [CrossRef] [PubMed]

89. McIntosh, P.B.; Martin, S.R.; Jackson, D.J.; Khan, J.; Isaacson, E.R.; Calder, L.; Raj, K.; Griffin, H.M.; Wang, Q.; Laskey, P.; *et al.* Structural analysis reveals an amyloid form of the human papillomavirus type 16 E1–E4 protein and provides a molecular basis for its accumulation. *J. Virol.* **2008**, *82*, 8196–8203. [CrossRef] [PubMed]

90. Nicholls, P.; Klaunberg, B.; Moore, R.A.; Santos, E.B.; Parry, N.R.; Gough, G.W.; Stanley, M.A. Naturally occurring, nonregressing canine oral papillomavirus infection: Host immunity, virus characterization, and experimental infection. *Virology* **1999**, *265*, 365–374. [CrossRef] [PubMed]

91. Wilgenburg, B.J.; Budgeon, L.R.; Lang, C.M.; Griffith, J.W.; Christensen, N.D. Characterization of immune responses during regression of rabbit oral papillomavirus infections. *Comp. Med.* **2005**, *55*, 431–439. [PubMed]

92. Nicholls, P.K.; Moore, P.F.; Anderson, D.M.; Moore, R.A.; Parry, N.R.; Gough, G.W.; Stanley, M.A. Regression of canine oral papillomas is associated with infiltration of CD4+ and CD8+ lymphocytes. *Virology* **2001**, *283*, 31–39. [CrossRef] [PubMed]

93. Coleman, N.; Birley, H.D.; Renton, A.M.; Hanna, N.F.; Ryait, B.K.; Byrne, M.; Taylor-Robinson, D.; Stanley, M.A. Immunological events in regressing genital warts. *Am. J. Clin. Pathol.* **1994**, *102*, 768–774. [PubMed]

94. Monnier-Benoit, S.; Mauny, F.; Riethmuller, D.; Guerrini, J.S.; Capilna, M.; Felix, S.; Seilles, E.; Mougin, C.; Pretet, J.L. Immunohistochemical analysis of CD4+ and CD8+ T-cell subsets in high risk human papillomavirus-associated pre-malignant and malignant lesions of the uterine cervix. *Gynecol. Oncol.* **2006**, *102*, 22–31. [CrossRef] [PubMed]

95. Stanley, M.A. Epithelial cell responses to infection with human papillomavirus. *Clin. Microbiol. Rev.* **2012**, *25*, 215–222. [CrossRef] [PubMed]

96. Kanodia, S.; Fahey, L.M.; Kast, W.M. Mechanisms used by human papillomaviruses to escape the host immune response. *Curr. Cancer Drug Targets* **2007**, *7*, 79–89. [CrossRef] [PubMed]

97. Egawa, K. Do human papillomaviruses target epidermal stem cells? *Dermatology* **2003**, *207*, 251–254. [CrossRef] [PubMed]

98. Herfs, M.; Vargas, S.O.; Yamamoto, Y.; Howitt, B.E.; Nucci, M.R.; Hornick, J.L.; McKeon, F.D.; Xian, W.; Crum, C.P. A novel blueprint for "top down" differentiation defines the cervical squamocolumnar junction during development, reproductive life, and neoplasia. *J. Pathol.* **2013**, *229*, 460–468. [CrossRef] [PubMed]

99. Herfs, M.; Yamamoto, Y.; Laury, A.; Wang, X.; Nucci, M.R.; McLaughlin-Drubin, M.E.; Munger, K.; Feldman, S.; McKeon, F.D.; Xian, W.; *et al.* A discrete population of squamocolumnar junction cells implicated in the pathogenesis of cervical cancer. *Proc. Natl. Acad. Sci. USA* **2012**, *109*, 10516–10521. [CrossRef] [PubMed]

100. Watt, F.M. Epidermal stem cells: Markers, patterning and the control of stem cell fate. *Philos. Trans. R. Soc. Lond. Ser. B Biol. Sci.* **1998**, *353*, 831–837. [CrossRef] [PubMed]

101. Potten, C.S. Cell replacement in epidermis (keratopoiesis) via discrete units of proliferation. *Int. Rev. Cytol.* **1981**, *69*, 271–318. [PubMed]

102. Feuerman, E. Verruca vulgaris—Benign tumor of viral etiology and clonal development. *Harefuah* **1972**, *83*, 211–212. [PubMed]

103. Schmitt, A.; Rochat, A.; Zeltner, R.; Borenstein, L.; Barrandon, Y.; Wettstein, F.O.; Iftner, T. The primary target cells of the high-risk cottontail rabbit papillomavirus colocalize with hair follicle stem cells. *J. Virol.* **1996**, *70*, 1912–1922. [PubMed]

104. Martens, J.E.; Smedts, F.M.; Ploeger, D.; Helmerhorst, T.J.; Ramaekers, F.C.; Arends, J.W.; Hopman, A.H. Distribution pattern and marker profile show two subpopulations of reserve cells in the endocervical canal. *Int. J. Gynecol. Pathol.* **2009**, *28*, 381–388. [CrossRef] [PubMed]

105. Mirkovic, J.; Howitt, B.E.; Roncarati, P.; Demoulin, S.; Suarez-Carmona, M.; Hubert, P.; McKeon, F.D.; Xian, W.; Li, A.; Delvenne, P.; *et al.* Carcinogenic HPV infection in the cervical squamo-columnar junction. *J. Pathol.* **2015**, *236*, 265–271. [CrossRef] [PubMed]

106. Isaacson Wechsler, E.; Wang, Q.; Roberts, I.; Pagliarulo, E.; Jackson, D.; Untersperger, C.; Coleman, N.; Griffin, H.; Doorbar, J. Reconstruction of human papillomavirus type 16-mediated early-stage neoplasia implicates E6/E7 deregulation and the loss of contact inhibition in neoplastic progression. *J. Virol.* **2012**, *86*, 6358–6364. [CrossRef] [PubMed]

107. Cotsarelis, G.; Sun, T.T.; Lavker, R.M. Label-retaining cells reside in the bulge area of pilosebaceous unit: Implications for follicular stem cells, hair cycle, and skin carcinogenesis. *Cell* **1990**, *61*, 1329–1337. [CrossRef]

108. Cotsarelis, G. Epithelial stem cells: A folliculocentric view. *J. Investig. Dermatol.* **2006**, *126*, 1459–1468. [CrossRef] [PubMed]

109. Ghazizadeh, S.; Taichman, L.B. Organization of stem cells and their progeny in human epidermis. *J. Investig. Dermatol.* **2005**, *124*, 367–372. [CrossRef] [PubMed]

110. Taylor, G.; Lehrer, M.S.; Jensen, P.J.; Sun, T.T.; Lavker, R.M. Involvement of follicular stem cells in forming not only the follicle but also the epidermis. *Cell* **2000**, *102*, 451–461. [CrossRef]

111. Ito, M.; Liu, Y.; Yang, Z.; Nguyen, J.; Liang, F.; Morris, R.J.; Cotsarelis, G. Stem cells in the hair follicle bulge contribute to wound repair but not to homeostasis of the epidermis. *Nat. Med.* **2005**, *11*, 1351–1354. [CrossRef] [PubMed]

112. Kaur, P. Interfollicular epidermal stem cells: Identification, challenges, potential. *J. Investig. Dermatol.* **2006**, *126*, 1450–1458. [CrossRef] [PubMed]

113. Braun, K.M.; Niemann, C.; Jensen, U.B.; Sundberg, J.P.; Silva-Vargas, V.; Watt, F.M. Manipulation of stem cell proliferation and lineage commitment: Visualisation of label-retaining cells in wholemounts of mouse epidermis. *Development* **2003**, *130*, 5241–5255. [CrossRef] [PubMed]

114. Lavker, R.M.; Sun, T.T. Epidermal stem cells. *J. Investig. Dermatol.* **1983**, *81*, 121s–127s. [CrossRef] [PubMed]

115. Ghazizadeh, S.; Taichman, L.B. Multiple classes of stem cells in cutaneous epithelium: A lineage analysis of adult mouse skin. *EMBO J.* **2001**, *20*, 1215–1222. [CrossRef] [PubMed]

116. Miller, S.J.; Burke, E.M.; Rader, M.D.; Coulombe, P.A.; Lavker, R.M. Re-epithelialization of porcine skin by the sweat apparatus. *J. Investig. Dermatol.* **1998**, *110*, 13–19. [CrossRef] [PubMed]

117. Lu, C.P.; Polak, L.; Rocha, A.S.; Pasolli, H.A.; Chen, S.C.; Sharma, N.; Blanpain, C.; Fuchs, E. Identification of stem cell populations in sweat glands and ducts reveals roles in homeostasis and wound repair. *Cell* **2012**, *150*, 136–150. [CrossRef] [PubMed]

118. Jablonska, S.; Dabrowski, J.; Jakubowicz, K. Epidermodysplasia verruciformis as a model in studies on the role of papovaviruses in oncogenesis. *Cancer Res.* **1972**, *32*, 583–589. [PubMed]

119. Orth, G. Genetics of epidermodysplasia verruciformis: Insights into host defense against papillomaviruses. *Semin. Immunol.* **2006**, *18*, 362–374. [CrossRef] [PubMed]

120. Lazarczyk, M.; Dalard, C.; Hayder, M.; Dupre, L.; Pignolet, B.; Majewski, S.; Vuillier, F.; Favre, M.; Liblau, R.S. Ever proteins, key elements of the natural anti-human papillomavirus barrier, are regulated upon t-cell activation. *PLoS ONE* **2012**, *7*, e39995. [CrossRef] [PubMed]

121. Boxman, I.L.; Berkhout, R.J.; Mulder, L.H.; Wolkers, M.C.; Bouwes Bavinck, J.N.; Vermeer, B.J.; ter Schegget, J. Detection of human papillomavirus DNA in plucked hairs from renal transplant recipients and healthy volunteers. *J. Investig. Dermatol.* **1997**, *108*, 712–715. [CrossRef] [PubMed]

122. Boxman, I.L.; Mulder, L.H.; Russell, A.; Bouwes Bavinck, J.N.; Green, A.; Ter Schegget, J. Human papillomavirus type 5 is commonly present in immunosuppressed and immunocompetent individuals. *Br. J. Dermatol.* **1999**, *141*, 246–249. [CrossRef] [PubMed]

123. Jablonska, S.; Orth, G.; Obalek, S.; Croissant, O. Cutaneous warts. Clinical, histologic, and virologic correlations. *Clin. Dermatol.* **1985**, *3*, 71–82. [CrossRef]

124. Egawa, K. Eccrine-centred distribution of human papillomavirus 63 infection in the epidermis of the plantar skin. *Br. J. Dermatol.* **2005**, *152*, 993–996. [CrossRef] [PubMed]

125. Boxman, I.L.; Hogewoning, A.; Mulder, L.H.; Bouwes Bavinck, J.N.; ter Schegget, J. Detection of human papillomavirus types 6 and 11 in pubic and perianal hair from patients with genital warts. *J. Clin. Microbiol.* **1999**, *37*, 2270–2273. [PubMed]

126. Kocjan, B.J.; Poljak, M.; Seme, K.; Potocnik, M.; Fujs, K.; Babic, D.Z. Distribution of human papillomavirus genotypes in plucked eyebrow hairs from slovenian males with genital warts. *Infect. Genet. Evol. J. Mol. Epidemiol. Evol. Genet. Infect. Dis.* **2005**, *5*, 255–259. [CrossRef] [PubMed]

127. Javier, R.T. Cell polarity proteins: Common targets for tumorigenic human viruses. *Oncogene* **2008**, *27*, 7031–7046. [CrossRef] [PubMed]

128. Kranjec, C.; Banks, L. A systematic analysis of human papillomavirus (HPV) E6 pdz substrates identifies magi-1 as a major target of HPV type 16 (HPV-16) and HPV-18 whose loss accompanies disruption of tight junctions. *J. Virol.* **2011**, *85*, 1757–1764. [CrossRef] [PubMed]

129. Massimi, P.; Narayan, N.; Thomas, M.; Gammoh, N.; Strand, S.; Strand, D.; Banks, L. Regulation of the hdlg/hscrib/hugl-1 tumour suppressor complex. *Exp. Cell Res.* **2008**, *314*, 3306–3317. [CrossRef] [PubMed]

130. Pim, D.; Bergant, M.; Boon, S.S.; Ganti, K.; Kranjec, C.; Massimi, P.; Subbaiah, V.K.; Thomas, M.; Tomaic, V.; Banks, L. Human papillomaviruses and the specificity of pdz domain targeting. *FEBS J.* **2012**, *279*, 3530–3537. [CrossRef] [PubMed]

131. Thomas, M.; Dasgupta, J.; Zhang, Y.; Chen, X.; Banks, L. Analysis of specificity determinants in the interactions of different HPV E6 proteins with their pdz domain-containing substrates. *Virology* **2008**, *376*, 371–378. [CrossRef] [PubMed]

132. Fu, L.; van Doorslaer, K.; Chen, Z.; Ristriani, T.; Masson, M.; Trave, G.; Burk, R.D. Degradation of p53 by human alphapapillomavirus E6 proteins shows a stronger correlation with phylogeny than oncogenicity. *PLoS ONE* **2010**, *5*, e12816. [CrossRef] [PubMed]

133. Zanier, K.; ould M'hamed ould Sidi, A.; Boulade-Ladame, C.; Rybin, V.; Chappelle, A.; Atkinson, A.; Kieffer, B.; Trave, G. Solution structure analysis of the HPV16 E6 oncoprotein reveals a self-association mechanism required for E6-mediated degradation of p53. *Structure* **2012**, *20*, 604–617. [CrossRef] [PubMed]

134. Pim, D.; Banks, L. Interaction of viral oncoproteins with cellular target molecules: Infection with high-risk *vs.* low-risk human papillomaviruses. *Apmis* **2010**, *118*, 471–493. [CrossRef] [PubMed]

135. Zheng, Z.M.; Baker, C.C. Papillomavirus genome structure, expression, and post-transcriptional regulation. *Front. Biosci.* **2006**, *11*, 2286–2302. [CrossRef] [PubMed]

136. Westra, W.H. The morphologic profile of HPV-related head and neck squamous carcinoma: Implications for diagnosis, prognosis, and clinical management. *Head Neck Pathol.* **2012**, *6*, S48–S54. [CrossRef] [PubMed]

137. Parkin, D.M. The global health burden of infection-associated cancers in the year 2002. *Int. J. Cancer J. Int. Cancer* **2006**, *118*, 3030–3044. [CrossRef] [PubMed]

138. Munoz, N.; Bosch, F.X.; Castellsague, X.; Diaz, M.; de Sanjose, S.; Hammouda, D.; Shah, K.V.; Meijer, C.J. Against which human papillomavirus types shall we vaccinate and screen? The international perspective. *Int. J. Cancer J. Int. Cancer* **2004**, *111*, 278–285. [CrossRef] [PubMed]

139. Gross, G.; Hagedorn, M.; Ikenberg, H.; Rufli, T.; Dahlet, C.; Grosshans, E.; Gissmann, L. Bowenoid papulosis. Presence of human papillomavirus (HPV) structural antigens and of HPV 16-related DNA sequences. *Arch. Dermatol.* **1985**, *121*, 858–863. [CrossRef] [PubMed]
140. Tschandl, P.; Rosendahl, C.; Kittler, H. Cutaneous human papillomavirus infection: Manifestations and diagnosis. *Curr. Probl. Dermatol.* **2014**, *45*, 92–97. [PubMed]

Human Papillomavirus and Tonsillar and Base of Tongue Cancer

Torbjörn Ramqvist, Nathalie Grün and Tina Dalianis

Abstract: In 2007, human papillomavirus (HPV) type 16 was recognized as a risk factor by the International Agency for Research on Cancer, for oropharyngeal squamous cell carcinoma (OSCC), where tonsillar and base of tongue cancer (TSCC and BOTSCC) dominate. Furthermore, patients with HPV-positive TSCC and BOTSCC, had a much better clinical outcome than those with corresponding HPV-negative cancer and other head and neck cancer. More specifically, survival was around 80% for HPV-positive TSCC and BOTSCC *vs.* 40% five-year disease free survival, for the corresponding HPV-negative tumors with conventional radiotherapy and surgery, while this could not be observed for HPV-positive OSCC at other sites. In addition, the past 20–40 years in many Western Countries, the incidence of HPV-positive TSCC and BOTSCC has risen, and >70% are men. This has resulted in a relative increase of patients with HPV-positive TSCC and BOTSCC that may not need the intensified chemo-radiotherapy (with many more severe debilitating side effects) often given today to patients with head and neck cancer. However, before tapering therapy, one needs to enable selection of patients for such treatment, by identifying clinical and molecular markers that together with HPV-positive status will better predict patient prognosis and response to therapy. To conclude, there is a new increasing group of patients with HPV-positive TSCC and BOTSCC with good clinical outcome, where options for better-tailored therapy are needed. For prevention, it would be of benefit to vaccinate both girls and boys against HPV16 infection. For potential future screening the ways to do so need optimizing.

Reprinted from *Viruses*. Cite as: Ramqvist, T.; Grün, N.; Dalianis, T. Human Papillomavirus and Tonsillar and Base of Tongue Cancer. *Viruses* **2015**, *7*, 1332-1343.

1. Introduction

Twenty years ago in 1995, the International Agency for Research against Cancer acknowledged an association between human papillomavirus (HPV) type 16 and cancer of the cervix, uteri, and other anogenital cancer [1]. It took another 12 years for HPV16 to be recognized as a risk factor for oropharyngeal squamous cell carcinoma (OSCC) in 2007, dominated tonsillar and base of tongue cancer (TSCC and BOTSCC) and where HPV is mainly found [2–7]. The most notable difference between HPV-positive TSCC and BOTSCC compared to their HPV-negative counterparts and other head and neck squamous cell carcinoma (HNSCC) is the better clinical outcome (80% *vs.* 40% five-year disease specific survival) [2–4,6–9]. This difference in clinical outcome has not been observed between HPV-positive and HPV-negative OSCC at other sites other than TSCC and BOTSCC [6]. In the past decades, the incidence of especially TSCC and BOTSCC (and thereby also OSCC) has increased in many Western countries, mainly due to a rise of HPV-positive TSCC and BOTSCC cases and >70% are men [8,10–21]. HNSCC has poor prognosis in general and is now given more aggressive treatment with more intensified chemotherapy and radiotherapy, leading to more adverse side effects [21]. Such therapy may not be beneficial for most patients with HPV-positive TSCC and BOTSCC, where 80% of the patients survived, before treatment was intensified, and when given only conventional radiotherapy alone, with the addition of surgery if needed [2–4,6–9,19]. Here, differences between HPV-positive and HPV-negative TSCC and BOTSCC and issues of finding biomarkers in HPV-positive TSCC and BOTSCC useful for predicting which patients may have a good response to therapy and be eligible for de-escalated therapy trials are described. Data on oral HPV infection, effects of HPV vaccination, and potential screening for TSCC and BOTSCC are also discussed.

2. Human Papillomavirus (HPV) and Disease and Cancer

There are >170 HPV types, with the majority found in the skin (cutaneous HPV types), but many also found in mucous tissues (mucosal HPV types) and where clearly the vast majority cause only asymptomatic infections [22,23]. A phylogenetic tree based on the homologous nucleotide sequence of the major capsid protein L1 groups the different HPV types into five genera—alpha, beta, gamma, mu and nu [23]. The mucosal types are included in genera alpha and the others mainly consist of cutaneous types [23].

Mucosal HPV types can be divided into high-risk (HR) types that have a clear and well-recognized potential to cause cancer, or low-risk types (LR) that are very rarely observed in cancer [22,23]. The best-known association between HR-HPVs and cancer is that of HPV and cancer of the uterine cervix [22]. However, HPV is also associated with vulvar, vaginal, penile and anal cancer, and since 2007, HPV16 has also been acknowledged to be a risk factor for OSCC, where tonsillar and base of tongue cancer dominate [5,22–25]. Furthermore, in addition to HPV16, HPV33, HPV35 and others (also found in cervical cancer) have been observed to contribute to OSCC [4,7,15,24,26]. LR-HPV types are not associated with cancer development in general, but are often found in benign genital lesions, such as condylomas and recurrent respiratory papillomas [22,23].

Cutaneous HPV types are best known to cause skin warts, but multiple skin cancers may emerge from verruca-like papillomatous lesions in Epidermodyplasia vercucciformis (EV) patients that are especially sensitive to infections with e.g., HPV5 and 8 [1,5,23]. Whether HPV is responsible for other squamous cell carcinoma of the skin is however still a question of debate.

HPVs have double stranded circular DNA genomes of around 7.9 kb. The genome, including a non-coding control region (NCCR), an early and a late coding region, is enclosed together with histones within a 52–55 nm virion [23,24]. The regulatory proteins E1-E2, E4-E7 important for gene regulation, replication and pathogenesis are coded by the early region, while the two structural proteins L1 and L2 responsible for the viral capsid are encoded by the late region [23,24]. E6 and E7 regarded as oncogenes in HR-HPV types and have high affinity to p53 and pRb, respectively, and are without doubt of relevance for immortalization and transformation [23,24]. The binding of E6 to p53 causes its degradation, preventing e.g., control of DNA damage and cell repair or apoptosis, while the binding of E7 to Rb and abrogates deregulation of cell cycle control [1,18,19,23,24]. The latter, also results in an increase in the expression of the cyclin dependent kinase inhibitor p16^{INK4a} [23,27]. Overexpression of p16^{INK4a} was in the past used as a surrogate marker for presence of HPV in OSCC [24,27]. Today, neither overexpression of p16^{INK4a} or the presence of HPV DNA alone, are regarded as sufficient to point that a tumor is caused by HPV [24,28]. However, presence of HPV DNA combined with p16^{INK4a} overexpression is almost as sensitive as using the golden standard, *i.e.* analyzing for presence of HPV E6 and E7 mRNA [24,28]. The L1 major capsid protein contributes to around 80% of the viral capsid and can spontaneously self-assemble into virus-like particles (VLPs) under specific conditions [22,23,29]. Todays' HPV vaccines Cevarix (GMK) and Gardasil (Merk) consist of VLPs from different HPV types and both contain HPV16 and 18 VLPs, while the latter also contains VLPs also of HPV6 and HPV11 [30,31]. These vaccines have been shown to be very efficient against cervical HPV infection, and also very likely efficient against oral HPV infection [30–35].

3. TSCC, BOTSCC and HPV and Its Influence on Clinical Outcome and Number of Cases

OSCC includes not only TSCC and BOTSCC, which accounting for 80% of the cases, but also cancer of the walls of the pharynx and the soft palate [36]. However, it becomes evident that the presence of HPV is highest in TSCC and BOTSCC (79% and 73%, respectively), which both contain lymphoid tissue and together contribute to Waldeyers ring, while in other OSCC the presence of HPV is lower (17%) [6,8,17]. This is not always appreciated, and it is likely that this distinction must also be made more clearly before selecting patients for clinical studies. Furthermore, HPV16 is present in >90% of the HPV-positive TSCC and BOTSCC cases, and even more common than in cervical cancer, where HPV16 accounts for 50%–55% of the cases; moreover, most of the patients are men [1–7,23,24,26]. Taken together, this is also of importance when reflecting over HPV vaccination against these tumors.

The most prominent difference between HPV-positive and HPV-negative TSCC and BOTSCC is that patients with HPV-positive tumors have a much better clinical outcome than those with HPV-negative tumors and other head and neck cancer, including OSCC other than TSCC and BOTSCC (80% *vs.* 40% five-year survival) [3,4,7–9]. In the literature, similar findings with regard to clinical outcome are found for HPV-positive and HPV-negative OSCC, which is not surprising, since the numbers of OSCC cases outside the tonsil and base of tongue at other subsites are limited and do not change the general trend [4,6,36].

Patients with HPV-positive TSCC and BOTSCC as compared to those with HPV-negative tumors are also often somewhat younger, more often non-smokers, have smaller tumors, but have a higher tumor stage, since many patients also have nodal disease [2–4,6–9,37]. Nonetheless, the latter still does not necessarily affect the better survival among patients in the HPV-positive group [2–4,6–9,37].

Most HPV-positive TSCC and BOTSCC independent of episomal/and or integrated HPV genomes, exhibit E6 and E7 mRNA expression; with p53 expression more often, being normal and with 16^{Ink4a} overexpressed in most cases, in contrast to that observed in HPV-negative TSCC and OSCC [23,37–40]. HPV-positive TSCC and BOTSCC is also generally less differentiated; more frequently aneuploid compared to HPV-negative OSCC; and chromosome 3q often amplified similar to cervical cancer, but amplification of chromosome 3q did not further influence clinical outcome [41,42]. HPV has repeatedly been a favorable prognostic marker, independent of tumor stage, age, gender, differentiation, or DNA ploidy, and in never-smokers clinical outcome has generally been found to be even better [2–4,6–9,37,41,42].

Today the golden standard of HPV-positive status in TSCC/BOTSCC/OSCC is the presence of E6 and E7 mRNA expression by RT-PCR, suggested to be associated with functional HPV expression [28]. Notably, the combined presence of HPV DNA tested by PCR and p16 overexpression is now more and more widely recognized as very close to the golden standard in TSCC and BOTSCC and especially useful for analysis in formalin fixed paraffin embedded (FFPE) tumor samples [28]. Nonetheless, in the past many ways of defining HPV-positive status have been used, including different types of PCR assays and primers, which may have given some variations in the results obtained. [4,7,27,28,37,43–48]. Moreover, HPV prevalence varies not only due to the methodology used, but also depending on geographical location and time period of investigation. Numerous studies have indicated an increase in HPV prevalence in TSCC, BOTSCC and OSCC the past decades and this has certainly contributed to the increased incidence of these tumors that has been reported from many Western countries [8,10–21].

For example, between 1970 and 2002, a three-fold increase in both the incidence and prevalence of HPV (both in women and men) was observed in TSCC in Stockholm, Sweden and this lead to the hypothesis that HPV was responsible for the increase in incidence of TSCC [15]. This was followed by additional reports again from Stockholm, showing that HPV-positive TSCC had increased with a seven-fold doubling per decade 1970–2007, while HPV-negative cancer had decreased and analogous changes were also reported for BOTSCC [8,14,17]. The most recent report from Sweden, shows a continued increase in incidence of TSCC and BOTSCC, however, while the number of HPV-positive TSCC cases remains high in the Stockholm area, it has not increased the past five years [49]. Whether this trend is temporary or not needs to be investigated further.

Parallel to the Stockholm studies, an emerging epidemic of HPV associated OSCC was suggested in the US, and in 2011 in the U.S. a rise in the incidence of HPV-positive OSCC and a decline in HPV-negative OPCC was also reported for the past decades [11,12,16,21]. Furthermore, during the same period accumulating reports from many Western countries, such as Scotland, the UK, and the Netherlands, conveyed an increasing incidence of OSCC, while in Eastern Denmark an increase in HPV-positive TSCC was described [10,13,20,50].

As already mentioned, the increase in incidence of TSCC, BOTSCC and OSCC has been suggested due to a rise in HPV-positive cases and changes in life style, and there is a significant correlation between HPV-positive OSCC, early sex debut, and numbers of oral or vaginal partners [51]. However, it has also been shown that oral-to-oral contact and HPV-transmission at birth may also result in oral HPV infection [52,53]. Clearly, in many Western countries, the numbers of HPV-positive TSCC and BOTSCC, where most patients are men, are still increasing, although new trends may be appearing [8,10–21,49,50].

4. TSCC, BOTSCC, HPV and Other Biomarkers and Treatment

HPV-positive TSCC and BOTSCC contribute to an increasing proportion of HNSCC in recent years [19]. Furthermore, this group of patients with a favorable clinical outcome already with conventional radiotherapy, may in most cases not need the intensified chemo-radiotherapy, with more side effects and increasing expenses for society that is given to head- and neck-cancer today [19,24]. Thus, it is imperative to distinguish patients that need intensive therapy from those who do not. For this purpose, it is important to combine positive HPV-status with additional biomarkers in order to better predict response to therapy and only select patients with a very probable good response rate to randomized trials with lessened therapy.

It has been shown that for TSCC and BOTSCC, the presence of HPV DNA/RNA and p16 overexpression are very good prognostic markers especially when combined with accurate data on that the patient being a never smoker [9,28,37]. In fact even the quantity of smoking, *i.e.*, package years, was also important [37]. However, none of these factors distinguish 100% of the patients and additional markers are therefore needed.

Additional biomarkers have been investigated. It has been reported that absent/low expression of MHC class I, CD44, CD98, LMP7, or LMP10 intensity staining, or absence of HLA-A*02, or high LRIG1 expression improved prediction of clinical outcome for patients with HPV-positive TSCC and BOTSCC [54–62]. Absence of HLA class I immunohistochemistry (IHC) staining, e.g., indicated a 95%–100% probability of a three-year disease free survival, but identified only around 20% of the patients with a good clinical outcome [56,57]. In addition, the high CD8+ tumor infiltrating lymphocyte (TIL) counts were also very favorable for patients with HPV-positive TSCC and BOTSCC as compared to the corresponding tumors with low CD8+ TIL counts, but was less sensitive [63,64].

That having high CD8+ TIL counts was favorable was expected, since an efficient immune response may result in a favorable clinical outcome of HPV-positive TSCC and BOTSCC [63,64]. More enigmatic was that HPV-positive TSCC and BOTSCC with absent HLA class I expression had good prognosis since HLA class I down-regulation abrogates the immune response, especially in absence of NK-cells [56,57]. Here, it is possible that HPV E5 and E7 expression contributes to HLA class I downregulation and that treatment increases HLA class I expression this way enhancing the immune response against these tumors [56,57].

Other biomarkers that have been investigated are miRNAs and data are accumulating, but stringent concordance between different studies has not been shown so far [65–67].

Clearly, additional molecular knowledge and ways to combine different markers, both clinical and molecular for prediction of clinical outcome in patients with HPV-positive TSCC and BOTSCC would be of great value.

5. Prevention of HPV-Positive TSCC and BOTSCC

HPV16 is the most common HPV type in the oral cavity in non-vaccinated individuals [68–71]. Oral HPV prevalence, including all HPV types, varies and has been reported to be 3%–9% in studies including individuals at all ages and when limited to unvaccinated youth [68–71]. Furthermore, it has been shown that women with a cervical HPV infection more often have an oral HPV infection [68].

In HPV vaccinated groups, oral HPV prevalence, especially HPV16, seems to be lower than in non-vaccinated groups suggesting a vaccination effect [32–35]. Thus vaccinating both girls and boys against HPV16 may be a good option not only to prevent cervical cancer but also to prevent the majority of HPV16 positive TSCC and BOTSCC in the future, especially since >70% of the latter comprise men.

For non-vaccinated individuals, other approaches may be of interest and screening may not be optimal for several reasons especially since the incidence of HPV-positive TSCC and BOTSCC is still relatively low [12,14,49]. Furthermore, testing HPV-prevalence in the oral cavity could result in an underestimation, e.g., due to saliva production, and some individuals may be reported as falsely negative because the obtained HPV signals are generally lower compared to those obtained for the cervical site [68].

Nevertheless, in the proportion of patients with HPV-positive TSCC and BOTSCC, it has been shown that the viral load of HPV16 in mouthwashes is considerably higher than that obtained in healthy youth and often comparable to that obtained in the cervix [72]. Whether a high viral HPV load is indicative of HPV-positive TSCC and BOTSCC needs to be investigated further. However, there is also the possibility to combine this approach with serology, where certain antibody profiles, especially the appearance of HPV16 E6 have been shown to predict risk for development of HPV-positive OSCC, and such antibodies are very seldom found in non-cancer patients [73,74]. Alternatively, serology could be used to identify patients at risk for OSCC and these patients could be followed using mouthwashes [72,73]. Using cytology has also been attempted, but not found very useful so far [67,75].

6. Conclusions

HPV-positive TSCC and BOTSCC have better clinical outcome than corresponding HPV-negative cancers and are increasing in incidence. Preventing spread of HPV16 infection by vaccination, and deescalating intensive therapy by using additional predictive markers to identify and select HPV-positive TSCC and BOTSCC patients eligible for randomized trials with lessened therapy are issues of importance.

Acknowledgments

This work was supported in part by the Swedish Cancer Foundation, the Stockholm Cancer Society, the Cancer and Allergy Foundation, the Stockholm City Council and the Karolinska Institutet, Sweden.

Author Contributions

TD wrote the review. TR read and modified the review and completed the reference list. NG read and modified the review.

Conflicts of Interest

The authors declare no conflict of interest.

References

1. IARC. *Human Papillomaviruses*; WHO: Lyon, France, 1995; Volume 64.
2. Dahlgren, L.; Dahlstrand, H.M.; Lindquist, D.; Hogmo, A.; Bjornestal, L.; Lindholm, J.; Lundberg, B.; Dalianis, T.; Munck-Wikland, E. Human papillomavirus is more common in base of tongue than in mobile tongue cancer and is a favorable prognostic factor in base of tongue cancer patients. *Int. J. Cancer* **2004**, *112*, 1015–1019.
3. Dahlstrand, H.M.; Dalianis, T. Presence and influence of human papillomaviruses (HPV) in tonsillar cancer. *Adv. Cancer Res.* **2005**, *93*, 59–89.
4. Gillison, M.L.; Koch, W.M.; Capone, R.B.; Spafford, M.; Westra, W.H.; Wu, L.; Zahurak, M.L.; Daniel, R.W.; Viglione, M.; Symer, D.E.; *et al*. Evidence for a causal association between human papillomavirus and a subset of head and neck cancers. *J. Natl. Cancer Inst.* **2000**, *92*, 709–720.
5. IARC. *Human Papillomaviruses*; WHO: Lyon, France, 2007; Volume 90.
6. Marklund, L.; Näsman, A.; Ramqvist, T.; Dalianis, T.; Munck-Wikland, E.; Hammarstedt, L. Prevalence of human papillomavirus and survival in oropharyngeal cancer other than tonsil or base of tongue cancer. *Cancer Med.* **2012**, *1*, 82–88.
7. Mellin, H.; Friesland, S.; Lewensohn, R.; Dalianis, T.; Munck-Wikland, E. Human papillomavirus (HPV) DNA in tonsillar cancer: Clinical correlates, risk of relapse, and survival. *Int. J. Cancer* **2000**, *89*, 300–304.
8. Attner, P.; Du, J.; Nasman, A.; Hammarstedt, L.; Ramqvist, T.; Lindholm, J.; Marklund, L.; Dalianis, T.; Munck-Wikland, E. The role of human papillomavirus in the increased incidence of base of tongue cancer. *Int. J. Cancer* **2010**, *126*, 2879–2884.
9. Lindquist, D.; Romanitan, M.; Hammarstedt, L.; Nasman, A.; Dahlstrand, H.; Lindholm, J.; Onelov, L.; Ramqvist, T.; Ye, W.; Munck-Wikland, E.; *et al*. Human papillomavirus is a favourable prognostic factor in tonsillar cancer and its oncogenic role is supported by the expression of e6 and e7. *Mol. Oncol.* **2007**, *1*, 350–355.
10. Braakhuis, B.J.; Visser, O.; Leemans, C.R. Oral and oropharyngeal cancer in the netherlands between 1989 and 2006: Increasing incidence, but not in young adults. *Oral. Oncol.* 2009, *45*, e85–e89.
11. Chaturvedi, A.K.; Engels, E.A.; Anderson, W.F.; Gillison, M.L. Incidence trends for human papillomavirus-related and -unrelated oral squamous cell carcinomas in the united states. *J. Clin. Oncol.* **2008**, *26*, 612–619.
12. Chaturvedi, A.K.; Engels, E.A.; Pfeiffer, R.M.; Hernandez, B.Y.; Xiao, W.; Kim, E.; Jiang, B.; Goodman, M.T.; Sibug-Saber, M.; Cozen, W.; *et al*. Human papillomavirus and rising oropharyngeal cancer incidence in the united states. *J. Clin. Oncol.* **2011**, *29*, 4294–4301.
13. Conway, D.I.; Stockton, D.L.; Warnakulasuriya, K.A.; Ogden, G.; Macpherson, L.M. Incidence of oral and oropharyngeal cancer in united kingdom (1990–1999)—Recent trends and regional variation. *Oral. Oncol.* **2006**, *42*, 586–592.

14. Hammarstedt, L.; Dahlstrand, H.; Lindquist, D.; Onelov, L.; Ryott, M.; Luo, J.; Dalianis, T.; Ye, W.; Munck-Wikland, E. The incidence of tonsillar cancer in sweden is increasing. *Acta Otolaryngol.* **2007**, *127*, 988–992.

15. Hammarstedt, L.; Lindquist, D.; Dahlstrand, H.; Romanitan, M.; Dahlgren, L.O.; Joneberg, J.; Creson, N.; Lindholm, J.; Ye, W.; Dalianis, T.; *et al.* Human papillomavirus as a risk factor for the increase in incidence of tonsillar cancer. *Int. J. Cancer* **2006**, *119*, 2620–2623.

16. Marur, S.; D'Souza, G.; Westra, W.H.; Forastiere, A.A. HPV-associated head and neck cancer: A virus-related cancer epidemic. *Lancet Oncol.* **2010**, *11*, 781–789.

17. Nasman, A.; Attner, P.; Hammarstedt, L.; Du, J.; Eriksson, M.; Giraud, G.; Ahrlund-Richter, S.; Marklund, L.; Romanitan, M.; Lindquist, D.; *et al.* Incidence of human papillomavirus (HPV) positive tonsillar carcinoma in stockholm, sweden: An epidemic of viral-induced carcinoma? *Int. J. Cancer* **2009**, *124*, 150–156.

18. Ramqvist, T.; Dalianis, T. Oropharyngeal cancer epidemic and human papillomavirus. *Emerg. Infect. Dis.* **2010**, *16*, 1671–1677.

19. Ramqvist, T.; Dalianis, T. An epidemic of oropharyngeal squamous cell carcinoma (oscc) due to human papillomavirus (HPV) infection and aspects of treatment and prevention. *Anticancer Res.* **2011**, *31*, 1515–1519.

20. Robinson, K.L.; Macfarlane, G.J. Oropharyngeal cancer incidence and mortality in scotland: Are rates still increasing? *Oral. Oncol.* **2003**, *39*, 31–36.

21. Sturgis, E.M.; Cinciripini, P.M. Trends in head and neck cancer incidence in relation to smoking prevalence: An emerging epidemic of human papillomavirus-associated cancers? *Cancer* **2007**, *110*, 1429–1435.

22. Tommasino, M. The human papillomavirus family and its role in carcinogenesis. *Semin. Cancer Biol.* **2014**, *26*, 13–21.

23. Zur Hausen, H. Papillomavirus infections: A major cause of human cancer. In *Infections Causing Human Cancer*; Wiley-VCH Verlag: Weinheim, Germany, 2006; pp. 145–243.

24. Dalianis, T. Human papillomavirus and oropharyngeal cancer, the epidemics, and significance of additional clinical biomarkers for prediction of response to therapy (review). *Int. J. Oncol.* **2014**, *44*, 1799–1805.

25. IARC. Human papillomaviruses. In *Iarc Monographs on the Evaluation of Carcinogenic Risk to Humans. A Review of Human Carcinogens. B. Biological Agents*; WHO: Lyon, France, 2011; Volume 100B, pp. 261–301.

26. Du, J.; Nasman, A.; Carlson, J.W.; Ramqvist, T.; Dalianis, T. Prevalence of human papillomavirus (HPV) types in cervical cancer 2003–2008 in stockholm, sweden, before public HPV vaccination. *Acta Oncol.* **2011**, *50*, 1215–1219.

27. Oguejiofor, K.K.; Hall, J.S.; Mani, N.; Douglas, C.; Slevin, N.J.; Homer, J.; Hall, G.; West, C.M. The prognostic significance of the biomarker p16 in oropharyngeal squamous cell carcinoma. *Clin. Oncol.* **2013**, *25*, 630–638.

28. Smeets, S.J.; Hesselink, A.T.; Speel, E.J.; Haesevoets, A.; Snijders, P.J.; Pawlita, M.; Meijer, C.J.; Braakhuis, B.J.; Leemans, C.R.; Brakenhoff, R.H. A novel algorithm for reliable detection of human papillomavirus in paraffin embedded head and neck cancer specimen. *Int. J. Cancer* **2007**, *121*, 2465–2472.

29. Ramqvist, T.; Andreasson, K.; Dalianis, T. Vaccination, immune and gene therapy based on virus-like particles against viral infections and cancer. *Expert Opin. Biol. Ther.* **2007**, *7*, 997–1007.

30. Group, F.I.S. Quadrivalent vaccine against human papillomavirus to prevent high-grade cervical lesions. *N. Engl. J. Med.* **2007**, *356*, 1915–1927.

31. Paavonen, J.; Jenkins, D.; Bosch, F.X.; Naud, P.; Salmeron, J.; Wheeler, C.M.; Chow, S.N.; Apter, D.L.; Kitchener, H.C.; Castellsague, X.; *et al.* Efficacy of a prophylactic adjuvanted bivalent ll virus-like-particle vaccine against infection with human papillomavirus types 16 and 18 in young women: An interim analysis of a phase iii double-blind, randomised controlled trial. *Lancet* **2007**, *369*, 2161–2170.

32. Grun, N.; Ahrlund-Richter, A.; Franzen, J.; Mirzaie, L.; Marions, L.; Ramqvist, T.; Dalianis, T. Oral human papillomavirus (HPV) prevalence in youth and cervical HPV prevalence in women attending a youth clinic in sweden, a follow up-study 2013–2014 after gradual introduction of public HPV vaccination. *Infect. Dis.* **2015**, *47*, 57–61.

33. Herrero, R.; Quint, W.; Hildesheim, A.; Gonzalez, P.; Struijk, L.; Katki, H.A.; Porras, C.; Schiffman, M.; Rodriguez, A.C.; Solomon, D.; *et al.* Reduced prevalence of oral human papillomavirus (HPV) 4 years after bivalent HPV vaccination in a randomized clinical trial in costa rica. *PLOS ONE* **2013**, *8*, e68329.

34. Antonsson, A.; Cornford, M.; Perry, S.; Davis, M.; Dunne, M.P.; Whiteman, D.C. Prevalence and risk factors for oral HPV infection in young australians. *PLOS ONE* **2014**, *9*, e91761.

35. Nordfors, C.; Grun, N.; Haeggblom, L.; Tertipis, N.; Sivars, L.; Mattebo, M.; Larsson, M.; Haggstrom-Nordin, E.; Tyden, T.; Ramqvist, T.; *et al.* Oral human papillomavirus prevalence in high school students of one municipality in sweden. *Scand. J. Infect. Dis.* **2013**, *45*, 878–881.

36. Licitra, L.; Bernier, J.; Grandi, C.; Merlano, M.; Bruzzi, P.; Lefebvre, J.L. Cancer of the oropharynx. *Crit. Rev. Oncol. Hematol.* **2002**, *41*, 107–122.

37. Ang, K.K.; Harris, J.; Wheeler, R.; Weber, R.; Rosenthal, D.I.; Nguyen-Tan, P.F.; Westra, W.H.; Chung, C.H.; Jordan, R.C.; Lu, C.; *et al.* Human papillomavirus and survival of patients with oropharyngeal cancer. *N. Engl. J. Med.* **2010**, *363*, 24–35.

38. Koskinen, W.J.; Chen, R.W.; Leivo, I.; Makitie, A.; Back, L.; Kontio, R.; Suuronen, R.; Lindqvist, C.; Auvinen, E.; Molijn, A.; *et al.* Prevalence and physical status of human papillomavirus in squamous cell carcinomas of the head and neck. *Int. J. Cancer* **2003**, *107*, 401–406.

39. Mellin Dahlstrand, H.; Lindquist, D.; Bjornestal, L.; Ohlsson, A.; Dalianis, T.; Munck-Wikland, E.; Elmberger, G. P16(ink4a) correlates to human papillomavirus presence, response to radiotherapy and clinical outcome in tonsillar carcinoma. *Anticancer Res.* **2005**, *25*, 4375–4383.

40. Mellin, H.; Dahlgren, L.; Munck-Wikland, E.; Lindholm, J.; Rabbani, H.; Kalantari, M.; Dalianis, T. Human papillomavirus type 16 is episomal and a high viral load may be correlated to better prognosis in tonsillar cancer. *Int. J. Cancer* **2002**, *102*, 152–158.

41. Dahlgren, L.; Mellin, H.; Wangsa, D.; Heselmeyer-Haddad, K.; Bjornestal, L.; Lindholm, J.; Munck-Wikland, E.; Auer, G.; Ried, T.; Dalianis, T. Comparative genomic hybridization analysis of tonsillar cancer reveals a different pattern of genomic imbalances in human papillomavirus-positive and -negative tumors. *Int. J. Cancer* **2003**, *107*, 244–249.

42. Mellin, H.; Friesland, S.; Auer, G.; Dalianis, T.; Munck-Wikland, E. Human papillomavirus and DNA ploidy in tonsillar cancer—Correlation to prognosis. *Anticancer Res.* **2003**, *23*, 2821–2828.

43. Clavel, C.; Masure, M.; Bory, J.P.; Putaud, I.; Mangeonjean, C.; Lorenzato, M.; Gabriel, R.; Quereux, C.; Birembaut, P. Hybrid capture II-based human papillomavirus detection, a sensitive test to detect in routine high-grade cervical lesions: A preliminary study on 1518 women. *Br. J. Cancer* **1999**, *80*, 1306–1311.

44. de Roda Husman, A.M.; Walboomers, J.M.; van den Brule, A.J.; Meijer, C.J.; Snijders, P.J. The use of general primers gp5 and gp6 elongated at their 3' ends with adjacent highly conserved sequences improves human papillomavirus detection by PCR. *J. Gen. Virol.* **1995**, *76*, 1057–1062.

45. Gravitt, P.E.; Peyton, C.L.; Apple, R.J.; Wheeler, C.M. Genotyping of 27 human papillomavirus types by using L1 consensus PCR products by a single-hybridization, reverse line blot detection method. *J. Clin. Microbiol.* **1998**, *36*, 3020–3027.

46. Schmitt, M.; Bravo, I.G.; Snijders, P.J.; Gissmann, L.; Pawlita, M.; Waterboer, T. Bead-based multiplex genotyping of human papillomaviruses. *J. Clin. Microbiol.* **2006**, *44*, 504–512.

47. Tieben, L.M.; ter Schegget, J.; Minnaar, R.P.; Bouwes Bavinck, J.N.; Berkhout, R.J.; Vermeer, B.J.; Jebbink, M.F.; Smits, H.L. Detection of cutaneous and genital HPV types in clinical samples by PCR using consensus primers. *J. Virol. Methods* **1993**, *42*, 265–279.

48. van den Brule, A.J.; Pol, R.; Fransen-Daalmeijer, N.; Schouls, L.M.; Meijer, C.J.; Snijders, P.J. Gp5+/6+ pcr followed by reverse line blot analysis enables rapid and high-throughput identification of human papillomavirus genotypes. *J. Clin. Microbiol.* **2002**, *40*, 779–787.

49. Nasman, A.; Nordfors, C.; Holzhauser, S.; Vlastos, A.; Tertipis, N.; Hammar, U.; Hammarstedt-Nordenvall, L.; Marklund, L.; Munck-Wikland, E.; Ramqvist, T.; *et al.* Incidence of human papillomavirus positive tonsillar and base of tongue carcinoma: A stabilisation of an epidemic of viral induced carcinoma? *Eur. J. Cancer* **2015**, *51*, 55–61.

50. Garnaes, E.; Kiss, K.; Andersen, L.; Therkildsen, M.H.; Franzmann, M.B.; Filtenborg-Barnkob, B.; Hoegdall, E.; Krenk, L.; Josiassen, M.; Lajer, C.B.; *et al.* A high and increasing HPV prevalence in tonsillar cancers in eastern denmark, 2000–2010: The largest registry-based study to date. *Int. J. Cancer* **2014**, *136*, 2196–2203.

51. Anaya-Saavedra, G.; Ramirez-Amador, V.; Irigoyen-Camacho, M.E.; Garcia-Cuellar, C.M.; Guido-Jimenez, M.; Mendez-Martinez, R.; Garcia-Carranca, A. High association of human papillomavirus infection with oral cancer: A case-control study. *Arch. Med. Res.* **2008**, *39*, 189–197.

52. D'Souza, G.; Agrawal, Y.; Halpern, J.; Bodison, S.; Gillison, M.L. Oral sexual behaviors associated with prevalent oral human papillomavirus infection. *J. Infect. Dis.* **2009**, *199*, 1263–1269.

53. Syrjanen, S.; Puranen, M. Human papillomavirus infections in children: The potential role of maternal transmission. *Crit. Rev. Oral Biol. Med.* **2000**, *11*, 259–274.

54. Lindquist, D.; Ahrlund-Richter, A.; Tarjan, M.; Tot, T.; Dalianis, T. Intense CD44 expression is a negative prognostic factor in tonsillar and base of tongue cancer. *Anticancer Res.* **2012**, *32*, 153–161.

55. Lindquist, D.; Nasman, A.; Tarjan, M.; Henriksson, R.; Tot, T.; Dalianis, T.; Hedman, H. Expression of LRIG1 is associated with good prognosis and human papillomavirus status in oropharyngeal cancer. *Br. J. Cancer* **2014**, *110*, 1793–1800.

56. Nasman, A.; Andersson, E.; Marklund, L.; Tertipis, N.; Hammarstedt-Nordenvall, L.; Attner, P.; Nyberg, T.; Masucci, G.V.; Munck-Wikland, E.; Ramqvist, T.; *et al.* HLA class I and II expression in oropharyngeal squamous cell carcinoma in relation to tumor HPV status and clinical outcome. *PLOS ONE* **2013**, *8*, e77025.

57. Nasman, A.; Andersson, E.; Nordfors, C.; Grun, N.; Johansson, H.; Munck-Wikland, E.; Massucci, G.; Dalianis, T.; Ramqvist, T. MHC class I expression in HPV positive and negative tonsillar squamous cell carcinoma in correlation to clinical outcome. *Int. J. Cancer* **2013**, *132*, 72–81.

58. Nasman, A.; Nordfors, C.; Grun, N.; Munck-Wikland, E.; Ramqvist, T.; Marklund, L.; Lindquist, D.; Dalianis, T. Absent/weak CD44 intensity and positive human papillomavirus (HPV) status in oropharyngeal squamous cell carcinoma indicates a very high survival. *Cancer Med.* **2013**, *2*, 507–518.

59. Rietbergen, M.M.; Martens-de Kemp, S.R.; Bloemena, E.; Witte, B.I.; Brink, A.; Baatenburg de Jong, R.J.; Leemans, C.R.; Braakhuis, B.J.; Brakenhoff, R.H. Cancer stem cell enrichment marker CD98: A prognostic factor for survival in patients with human papillomavirus-positive oropharyngeal cancer. *Eur. J. Cancer* **2014**, *50*, 765–773.

60. Tertipis, N.; Haeggblom, L.; Grün, N.; Nordfors, C.; Näsman, A.; Dalianis, T.; Ramqvist, T. Reduced expression of the antigen processing machinery components TAP2, LMP2 and LMP7 in tonsillar and base of tongue cancer and implications for clinical outcome. *Transl. Oncol.* **2015**, *8*, 10–17.

61. Tertipis, N.; Haeggblom, L.; Nordfors, C.; Grun, N.; Nasman, A.; Vlastos, A.; Dalianis, T.; Ramqvist, T. Correlation of lmp10 expression and clinical outcome in human papillomavirus (HPV) positive and HPV-negative tonsillar and base of tongue cancer. *PLOS ONE* **2014**, *9*, e95624.

62. Tertipis, N.; Villabona, L.; Nordfors, C.; Nasman, A.; Ramqvist, T.; Vlastos, A.; Masucci, G.; Dalianis, T. Hla-a*02 in relation to outcome in human papillomavirus positive tonsillar and base of tongue cancer. *Anticancer Res.* **2014**, *34*, 2369–2375.

63. Nasman, A.; Romanitan, M.; Nordfors, C.; Grun, N.; Johansson, H.; Hammarstedt, L.; Marklund, L.; Munck-Wikland, E.; Dalianis, T.; Ramqvist, T. Tumor infiltrating CD8+ and Foxp3+ lymphocytes correlate to clinical outcome and human papillomavirus (HPV) status in tonsillar cancer. *PLOS ONE* **2012**, *7*, e38711.

64. Nordfors, C.; Grün, N.; Tertipis, N.; Ährlund-Richter, A.; Haeggblom, L.; Sivars, L.; Du, J.; Nyberg, T.; Marklund, L.; Munck-Wikland, E.; *et al*. CD8+ and CD4+ tumour infiltrating lymphocytes in relation to human papillomavirus status and clinical outcome in tonsillar and base of tongue squamous cell carcinoma. *Eur. J. Cancer* **2013**, *49*, 2522–2530.

65. Hui, A.B.; Lin, A.; Xu, W.; Waldron, L.; Perez-Ordonez, B.; Weinreb, I.; Shi, W.; Bruce, J.; Huang, S.H.; O'Sullivan, B.; *et al*. Potentially prognostic mirnas in HPV-associated oropharyngeal carcinoma. *Clin. Cancer Res.* **2013**, *19*, 2154–2162.

66. Lajer, C.B.; Nielsen, F.C.; Friis-Hansen, L.; Norrild, B.; Borup, R.; Garnaes, E.; Rossing, M.; Specht, L.; Therkildsen, M.H.; Nauntofte, B.; *et al*. Different mirna signatures of oral and pharyngeal squamous cell carcinomas: A prospective translational study. *Br. J. Cancer* **2011**, *104*, 830–840.

67. Miller, J.F. Influence of thymectomy on tumor induction by polyoma virus in C57BL mice. *Proc. Soc. Exp. Biol. Med.* **1964**, *116*, 323–327.

68. Du, J.; Nordfors, C.; Ahrlund-Richter, A.; Sobkowiak, M.; Romanitan, M.; Nasman, A.; Andersson, S.; Ramqvist, T.; Dalianis, T. Prevalence of oral human papillomavirus infection among youth, Sweden. *Emerg. Infect. Dis.* **2012**, *18*, 1468–1471.

69. Kreimer, A.R.; Bhatia, R.K.; Messeguer, A.L.; Gonzalez, P.; Herrero, R.; Giuliano, A.R. Oral human papillomavirus in healthy individuals: A systematic review of the literature. *Sex. Transm. Dis.* **2010**, *37*, 386–391.

70. Rautava, J.; Syrjanen, S. Human papillomavirus infections in the oral mucosa. *J. Am. Dent. Assoc.* **2011**, *142*, 905–914.

71. Steinau, M.; Hariri, S.; Gillison, M.L.; Broutian, T.R.; Dunne, E.F.; Tong, Z.Y.; Markowitz, L.E.; Unger, E.R. Prevalence of cervical and oral human papillomavirus infections among US women. *J. Infect. Dis.* **2014**, *209*, 1739–1743.

72. Nordfors, C.; Vlastos, A.; Du, J.; Ahrlund-Richter, A.; Tertipis, N.; Grun, N.; Romanitan, M.; Haeggblom, L.; Roosaar, A.; Dahllof, G.; *et al*. Human papillomavirus prevalence is high in oral samples of patients with tonsillar and base of tongue cancer. *Oral Oncol.* **2014**, *50*, 491–497.

73. D'Souza, G.; Kreimer, A.R.; Viscidi, R.; Pawlita, M.; Fakhry, C.; Koch, W.M.; Westra, W.H.; Gillison, M.L. Case-control study of human papillomavirus and oropharyngeal cancer. *N. Engl. J. Med.* **2007**, *356*, 1944–1956.

74. Lang Kuhs, K.A.; Anantharaman, D.; Waterboer, T.; Johansson, M.; Brennan, P.; Michel, A.; Willhauck-Fleckenstein, M.; Purdue, M.P.; Holcatova, I.; Ahrens, W. Human papillomavirus 16 E6 antibodies in individuals without diagnosed cancer: A Pooled Analysis. *Cancer Epidemiol. Biomark. Prev.* **2015**, doi:10.1158/1055-9965.EPI-14-1217.

75. Fakhry, C.; Rosenthal, B.T.; Clark, D.P.; Gillison, M.L. Associations between oral HPV16 infection and cytopathology: Evaluation of an oropharyngeal "pap-test equivalent" in high-risk populations. *Cancer Prev. Res.* **2011**, *4*, 1378–1384.

Interaction of Human Tumor Viruses with Host Cell Surface Receptors and Cell Entry

Georgia Schäfer, Melissa J. Blumenthal and Arieh A. Katz

Abstract: Currently, seven viruses, namely Epstein-Barr virus (EBV), Kaposi's sarcoma-associated herpes virus (KSHV), high-risk human papillomaviruses (HPVs), Merkel cell polyomavirus (MCPyV), hepatitis B virus (HBV), hepatitis C virus (HCV) and human T cell lymphotropic virus type 1 (HTLV-1), have been described to be consistently associated with different types of human cancer. These oncogenic viruses belong to distinct viral families, display diverse cell tropism and cause different malignancies. A key to their pathogenicity is attachment to the host cell and entry in order to replicate and complete their life cycle. Interaction with the host cell during viral entry is characterized by a sequence of events, involving viral envelope and/or capsid molecules as well as cellular entry factors that are critical in target cell recognition, thereby determining cell tropism. Most oncogenic viruses initially attach to cell surface heparan sulfate proteoglycans, followed by conformational change and transfer of the viral particle to secondary high-affinity cell- and virus-specific receptors. This review summarizes the current knowledge of the host cell surface factors and molecular mechanisms underlying oncogenic virus binding and uptake by their cognate host cell(s) with the aim to provide a concise overview of potential target molecules for prevention and/or treatment of oncogenic virus infection.

Reprinted from *Viruses*. Cite as: Schäfer, G.; Blumenthal, M.J.; Katz, A.A. Interaction of Human Tumor Viruses with Host Cell Surface Receptors and Cell Entry. *Viruses* **2015**, *7*, 2592-2617.

1. Introduction

Cell entry is a fundamental process of the infectivity of any virus, and cell surface receptors are the critical molecules in target cell recognition determining cell tropism and species specificity [1,2]. To date, there are seven known human oncogenic viruses, namely Epstein-Barr virus (EBV), Kaposi's sarcoma-associated herpes virus (KSHV), high-risk human papillomaviruses (HPVs), Merkel cell polyomavirus (MCPyV), hepatitis B virus (HBV), hepatitis C virus (HCV) and human T cell lymphotropic virus type 1 (HTLV-1). These viruses belong to distinct viral families, display different modes of entry and host cell tropism and lead to distinct types of cancer. They all infect their specific target cells through a two-step process: the first step involves attachment to the cell surface followed by the second step of specific receptor-mediated uptake. With the exception of EBV [3] which uses the cellular receptor CD21 (or CR2) for attachment [4], all oncogenic viruses initially attach to their target cells via ubiquitously expressed cell membrane heparan sulfate proteoglycans (HSPG), also termed the "universal receptor", which contain charged carbohydrate moieties that can interact with several viral surface glycoproteins or protein ligands expressed in the envelope and/or the capsid, respectively. This leads to accumulation of the viral particle at the cell surface and is generally followed by conformational changes ultimately facilitating viral uptake by secondary cell type- and/or virus-specific receptors. Some oncogenic viruses display simple binding and uptake involving only a small number of participating viral and/or host surface molecules, while others involve multiple viral proteins engaging with several specific receptors, co-receptors and co-factors that may vary according to cell type, implicating rather complex, highly regulated and concerted interactions [1,2].

This review summarizes the virus and host cell surface molecular interactions essential for productive viral infection and initiation of the viral life cycle of the individual oncogenic viruses that culminate in the tumorigenic process. While human immunodeficiency virus (HIV) is not considered a tumor virus *per se*, immunodeficiency-related malignancies play an important role in the total number of oncogenic virus-associated cancer types and will be mentioned where appropriate.

2. Epstein-Barr Virus

Epstein-Barr virus (EBV) belongs to the large and diverse family of *Herpesviridae* (subfamily Gammaherpesvirinae). As the first virus that has been identified as an etiologic agent for human cancer, EBV is now known to be causally associated with several B cell malignancies including Burkitt's lymphoma, Hodgkin's lymphoma, immunosuppression-related lymphoma, T and NK cell lymphomas as well as malignancies involving epithelial cells of the upper digestive tract, particularly nasopharyngeal and stomach carcinomas [5]. Found ubiquitously in approximately 95% of the adult population worldwide, most individuals acquire EBV infections during early childhood when EBV establishes a latent infection that persists generally asymptomatically throughout life. However, under certain circumstances when host-virus balance is not achieved such as due to immunosuppression as a result of HIV infection or in response to unrelated infections, EBV can cause malignant diseases [6]. As a predominantly orally transmitted virus, EBV has (unlike other members of the herpesvirus family) a rather narrow spectrum of potential target cells and primarily infects naïve tonsillar B cells [7] and (more rarely) oral epithelial cells [8].

Like all members of the herpesvirus family, the comparatively large EBV double-stranded linear DNA genome is packed inside a capsid which is surrounded by a tegument. This is further enclosed by a lipid envelope consisting of several conserved, as well as EBV-unique, glycoproteins. These glycoproteins play important roles during initial attachment and subsequent viral entry through interaction with specific host cell surface receptors mediating macropinocytosis [9] and lipid raft-dependent endocytosis [9,10]. It has long been known that the initial phase of EBV tethering to the host cell surface of B cells and epithelial cells occurs via the viral envelope glycoprotein gp350 (or its alternative isoform gp220) which interacts with the cellular receptor CD21 (or CR2) [4]. A recent report suggested also the involvement of CD35 as an alternative EBV attachment receptor in certain CD21-negative cells [11]. Unique among herpesviruses, gp350/220 dominates the outer viral membrane and is one of the most abundant EBV glycoproteins [12]. Although the absence of gp350/220 reduces EBV entry into epithelial and B cells, it is not absolutely required for infection [13]. In addition, the EBV transmembrane envelope glycoprotein BMRF2 has been shown to interact with $\beta 1$ and $\alpha 5$ integrins on oral epithelial cells but not on B cells [14].

Initial tethering of EBV to either the B cell or the epithelial cell membrane eventually triggers fusion with the EBV envelope which is considered the second phase of the infection process. This requires three conserved viral glycoproteins comprising the core fusion machinery, namely the gH-gL heterodimer and gB, the latter being crucial to the fusion process as it can insert into target membranes and refold through large conformational changes to bring viral and host membranes into close proximity, resulting in the formation of a fusion pore [15]. However, activation of this core fusion machinery differs significantly for each cell type [15]. While infection of B cells occurs mainly via endocytosis followed by fusion of the virus envelope with the endocytic vesicle membrane, epithelial cells are generally entered through direct fusion with the host cell plasma membrane at the cell surface [15]. In B lymphocytes, it was found that EBV uses the host cell surface human leukocyte antigen class II (HLA class II) through binding to the viral glycoprotein gp42 which associates non-covalently with the complex of the core fusion machinery gH-gL and gB. This interaction eventually triggers fusion of the virus with the endosomal membrane, allowing entry of the tegumented capsid into the cytoplasm [16]. While the interaction between gp42, gH-gL and gB is required for EBV entry into B cells, infection of epithelial cells requires only gH-gL and gB [17]. In fact, gp42 is not only dispensable but was even found to be inhibitory for epithelial cell infection [17]. EBV fusion with the plasma membrane and entry into epithelial cells has recently been shown to involve neuropilin 1 (NRP1) which directly interacts with gB [9], as well as binding of gH-gL to $\alpha V\beta 6$ and $\alpha V\beta 8$ integrins [18] via recruitment and activation of receptor tyrosine kinases and signaling via the epidermal growth factor receptor (EGFR)/Akt and EGFR/extracellular-signal-regulated kinases (ERK) pathways [9]. The presence or absence of gp42 in the EBV envelope renders the virus mutually exclusive for B cell and epithelial cell entry, respectively, therefore pointing towards the role of gp42 in determining (and redirecting) EBV's cell tropism [17].

3. Kaposi's Sarcoma-Associated Herpes Virus

Kaposi's sarcoma-associated herpes virus (KSHV) or human herpesvirus-8 (HHV-8) is a gamma-2 or rhadinovirus subfamily herpesvirus (*Herpesviridae* family) that causes two types of malignancies: the vascular tumor known as Kaposi's Sarcoma (KS), including classical KS, African-endemic KS, iatrogenic KS and epidemic or AIDS-associated KS, as well as B cell malignancies including multicentric Castleman disease and primary effusion lymphoma [19–21]. With the onset of the AIDS epidemic, immunodeficiency-related neoplasias have been steadily on the rise, and KS is now considered the most common AIDS-related malignancy worldwide [22–24]. Unlike EBV, KSHV is not a ubiquitous virus but is most prevalent in African populations [25].

KSHV DNA sequences can be detected in various body fluids such as blood, plasma, semen, and saliva. However, transmission of the infection seems to mainly occur through contaminated saliva [26–31]. KSHV infects a variety of target cells, such as activated B cells, endothelial cells, monocytes (including macrophages and dendritic cells), fibroblasts and epithelial cells [32].

As a member of the herpesvirus family, KSHV structure shows close similarities to EBV as described above, with the KSHV double-stranded linear DNA genome packed inside a capsid which is surrounded by a tegument and further enclosed by a lipid envelope, the latter consisting of five conserved glycoproteins, namely gB, gH, gL, gM and gN, as well as several KSHV-unique lytic cycle associated glycoproteins such as ORF4 and gpK8.1A [32,33]. Although KSHV displays broad cell tropism with the available host cell surface molecules determining the particular cell-type specific mode of entry via engagement with multiple KSHV glycoproteins [34], generally, glycoproteins gB and gpK8.1A function as key molecules in the initial attachment of the virus to the cell surface [35], while the non-covalently linked gH-gL complex is indispensable for subsequent KSHV entry [36].

In the initial phase of the infectious cell entry, KSHV attaches to host cell surface HSPGs mostly via gB and gpK8.1A [35,37], possibly followed by conformational changes in viral glycoproteins allowing access to specific entry receptors which function in the second phase of the infection [38]. According to cell tropism and entry pathway, these entry receptors vary and/or are utilized by KSHV in different combinations. Several studies on human fibroblasts and human endothelial cells demonstrated the necessity of viral gB engagement with the integrins $\alpha3\beta1$, $\alpha V\beta3$, $\alpha V\beta5$ for cell entry [39,40], and recently also the interaction of gB with integrin $\alpha9\beta1$ has been discussed [41]. Multiple integrin receptors can interact with the glutamate/cystine exchange transporter xCT, which has been identified as a fusion-entry receptor mediating KSHV-induced endocytic events [39,42]. However, it is yet unknown which KSHV glycoprotein interacts with xCT. During infection of dendritic cells, macrophages and activated B cells, KSHV has been reported to utilize the C-type lectin DC-SIGN [43,44] via the highly mannosylated glycoprotein gB [45]. Crucial for the entry process of KSHV into endothelial cells and downstream signaling is the recently identified ephrin receptor tyrosine kinase A2 (EPHA2) at the host cell membrane [36,46] that binds the gH-gL dimer. Binding to EPHA2 follows KSHVs initial interaction with integrins, leading to EPHA2 phosphorylation, coordinated integrin-associated downstream signaling and endocytosis, which enables KSHV entry and subsequent infection [47].

Depending on the targeted cell type, KSHV has been reported to induce different routes of endocytic uptake, particularly macropinocytosis and clathrin-mediated endocytosis [32]. In endothelial cells, actin-dependent, dynamin-independent macropinocytosis has been described as a major route for productive KSHV infection [48]. This is in contrast to the route of entry in human fibroblasts, which occurs via dynamin-dependent, clathrin-mediated endocytosis [48,49]. Downstream of the described surface-mediated events, KSHV infection has been found to be dependent on activation of Focal adhesion kinase (FAK) [50], phosphorylated Src [51] and phosphoinositide 3-kinase (PI3K) signaling [52].

4. Human Papillomavirus

Papillomaviruses belong to the diverse family of *Papillomaviridae* comprising predominantly squamous-epitheliotropic, non-enveloped, small DNA viruses, which are ubiquitous throughout the world. There are more than 100 different types of human papillomaviruses (HPVs), which can be broadly divided into the types infecting mucosal epithelium and those infecting cutaneous epithelium [53]. HPV infections particularly affect squamous cells of the anogenital region, the skin and the upper aerodigestive tract. Depending on their ability to cause malignant cell transformation, HPVs have been grouped into low-risk and high-risk genotypes with HPV 16 and 18 being the most common oncogenic types world-wide. Indeed, HPVs are the most prevalent causes of sexually transmitted infections in the world and are known to be associated with the development of multiple carcinomas with virtually all cases of cervical cancer being attributable to infection by oncogenic HPVs [54].

HPV is generally not passed through body fluids but is rather transmitted by direct skin-to-skin contact. Viral infection predominantly occurs in the mitotically active basal layer of keratinocytes attached to the basement membrane of the host's cutaneous and mucosal epithelia. While entry and infection in the upper epithelial layers has not been conclusively ruled out, cell cycle progression through early stages of mitosis was found to be critical for successful HPV infection providing a reason why HPV seems to infect only undifferentiated proliferating cells [55]. As the life cycle of HPV is dependent on the differentiation of basal cells into keratinocytes, the virus is highly tissue-restricted and thought to reach its cellular targets through disruptions of the integrity of the stratified or columnar epithelium that exposes the basal layer of cells for infection [56].

The icosahedral capsid coat of HPV containing the circular double-stranded DNA viral genome is composed of two structural proteins, the major capsid protein L1, and the minor capsid protein L2 which is mostly hidden from the capsid surface [57]. Although both L1 and L2, to different extents, are involved in the attachment of the virus to basal keratinocytes, the exact mechanism and receptors used by HPV to specifically bind to and infect epithelial cells is still highly debated. Differing experimental approaches and types of HPV particles used challenge the consistencies of the observations between studies [58]. The current two-step entry model entails that HPV, like many pathogens, initially binds to HSPGs at the epithelial cell surface or on the basement membrane which serve as primary attachment receptors [56,59–62]. Indeed, cells lacking HSPGs display greatly reduced HPV uptake [63–65]. In addition, laminin-5 secreted onto the extracellular matrix by keratinocytes has been proposed as a high-affinity initial attachment molecule [66]. The initial L1-mediated attachment is thought to induce conformational changes in the virus capsid through sequential engagement of multiple heparan sulfate (HS) binding sites eventually resulting in loss of affinity to the primary attachment receptor and transfer of the virus to an as yet unidentified secondary receptor [67]. The conformational changes in the virus capsid are thought to be brought about by cyclophilin B [68], thereby increasing the exposure of the otherwise hidden amino terminal portion of the L2 protein which contains a highly conserved consensus site for the proprotein convertase furin. Cleavage of L2 by furin is thought to be necessary for infection although it is debated whether it is a prerequisite at the early stage of binding and entry, or for endosome escape at later stages during the infection process [69,70]. Several secondary, predominantly L1-specific, receptors have been proposed, such as α6 integrin [71,72], EGFR and keratinocyte growth factor receptor (KGFR) [73] and tetraspanins [74]. Interestingly, the tetraspanin CD151 was found to be highly expressed in the basal layers of human cervical mucosa which—in association with integrins— may further explain the tissue-tropism of HPV16 infection [75]. Moreover, the annexin A2 heterotetramer (AIIt), a multifunctional protein involved in diverse cellular processes [76], has been implicated to act as an L2-specific receptor [77], regulating entry and intracellular trafficking of the virus [78].

Once bound to its specific uptake receptors, HPV enters the cell through a comparatively very slow asynchronous actin-dependent, and clathrin-, caveolin-, cholesterol-, and dynamin-independent endocytic process, with an average half-time of 12 h [79]. Internalization was found to be dependent on several signaling molecules such as EGFR, protein kinase C (PKC), p21-activated kinase 1 (PAK-1), the PI3K/Akt/mTOR pathway, and other yet unidentified tyrosine and serine/threonine kinases [79,80].

While there is no clear evidence for additive and/or alternative pathways, it is tempting to speculate that multiple internalization routes exist given the multitude of potential molecules involved in HPV binding and uptake. This is particularly interesting in the context of other viral entry pathways using similar entry molecules to HPV, for example cytomegalovirus (CMV) which infects cells via HSPGs (initial attachment), EGFR, integrins and/or annexin A2 (specific uptake) [81]; or Hepatitis C Virus (HCV) which employs HSPGs, EGFR, tetraspanins and similar intracellular signaling molecules ([82–86] and below). Moreover, the ability of different HPV types to species-specifically interact with individual binding and/or uptake molecules adds to the complexity of HPV entry and may provide an explanation of anatomical-site preferences of a given HPV type [87].

5. Merkel Cell Polyomavirus

Merkel cell polyomavirus (MCPyV) is the most recently identified member of the *Polyomaviridae* family which consists of non-enveloped, icosahedral, double-stranded DNA viruses capable of infecting a variety of species [88] including humans and having tumorigenic potential [88,89]. Merkel cell carcinoma (MCC) is generally found in sun-exposed areas of the skin, most commonly the head and neck [90]. MCPyV was first isolated from MCC tumors in 2008 [91] using Digital Transcriptome Subtraction [92]. Since then, a compelling line of evidence has been established for MCPyV as the causative agent of MCC, and 64%–88% of primary MCC tumors as well as metastases harbor MCPyV DNA and express MCPyV gene products [91,93–100]. MCC is a relatively rare, albeit severe, human cancer, predominantly presenting in elderly, Caucasian people [101] with suppressed immune systems and also HIV-infected individuals [102]. The incidence of MCC has drastically increased in the last two decades, presumably due to the increasing incidence of HIV/AIDS [90,101].

MCPyV is common in the population [99,103–105] and has a worldwide distribution [97,99,103–107]. MCPyV has been detected in children and seroprevalence found to gradually increase with age [105,106]. A correlation between siblings as well as mothers and small children was noted, suggesting transmission via close contact of skin and saliva during early childhood [106]. The virus has been detected in a diverse range of human tissues, although at significantly lower levels than in MCC tumors [108]. Keratinocytes are suspected to be the primary cells infected by MCPyV, supported by the chronic shedding of MCPyV from the skin [103]. However, the relationship between the cells that MCPyV infects and those that it transforms to give rise to MCC is currently unclear. MCC was thought to originate from Merkel cells due to similar expression patterns; however, this has come under debate due to differences in histomorphologic growth patterns and the discovery of markers previously thought to be of neuroendocrine cells, in a subset of lymphomas. A recent study proposes pro/pre- or pre-B cells as the cells of MCC due to the expression of terminal deoxynucleotidyl transferase (TdT) and paired box gene 5 (PAX5) in 72.8% and 100% of MCCs tested, respectively [109]. Jankowski *et al.* add that dysregulated expression of transcription factors in an epithelial cell could equally explain this expression pattern [110].

The MCPyV capsid consists of the structural proteins VP1 and VP2 in a ratio of 5:2 for native virions, while the VP3 minor capsid protein found in other polyomaviruses is undetectable in MCPyV [91,111]. Essential for MCPyV entry are pentameric knobs made up of the major capsid protein VP1 which bind to cellular receptors [112]. VP2 is essential for infectious entry in some cell types *in vitro* but others were transduced with MCPyV pseudovirions deficient in VP2. Knockdown of VP2 did not affect virion assembly, packaging of DNA or attachment to a target cell, indicating that the role of VP2 is in post-attachment entry. Myristoylation of VP2 has been shown to be important for infection of some cell types [111]. Current evidence points to a 2-step attachment-and-entry process for MCPyV whereby VP1 initially attaches to sulfated glycosaminoglycans, in particular heparan sulfate [113]. This is followed by transferal to a sialylated glycan which serves as a secondary post-attachment co-receptor and may mediate uptake by the cell. In concordance with the attachment mechanisms of other polyomaviruses [114–117], the ganglioside Gt1b was the first proposed receptor for MCPyV which was shown to interact with the sialic acid moieties on the Gt1b carbohydrate chain [118]. However, this was later disproved as Lec2 cells that lack sialylated glycans were found to bind MCPyV pseudovirions but were not productively infected [113]. The recently solved crystal structure of MCPyV major capsid protein VP1 showed a sialic acid binding site that, through mutation studies, was identified as having a role in post-attachment infection, not initial attachment. This sialic acid binding site differs from those of the previously characterized polyomaviruses BKPyV and murine polyomavirus for which sialylated glycans are primary attachment receptors and which is distinct from the yet unidentified binding site for sulfated glycosaminoglycans [112].

MCPyV seems to enter its target cells very slowly and asynchronously [119]. Eventually, viral replication takes place in the host cell nucleus, establishing productive infection. However, the events following penetration of the cell membrane and delivery of the encapsidated DNA to the nucleus are not yet fully established due to the lack of a reliable system to study MCPyV infection in cultured cells, especially with regard to infectivity and cell type tropism [119].

6. Hepatitis B Virus

Hepatitis B Virus (HBV) is the prototype of the *Hepadnaviridae* family of small, enveloped viruses that are tropic for the liver [120]. The virus is ubiquitous with a global distribution. HBV infection usually persists throughout the lifetime of the host as a result of noncytocidal infection of hepatocytes and their shedding of infectious virus into the blood stream and body fluids which represents the main route of transmission. Host immune response to the infection results in liver damage and scarring, creating a favorable environment for oncogenesis with a lifetime risk of developing liver cancer of 10%–25% in patients with chronic HBV infections [121]. HBV accounts for approximately 50% of hepatocellular carcinoma world-wide, which occurs mainly in less developed regions [122].

HBV is the smallest known enveloped virus. It is comprised of an icosahedral nucleocapsid containing the partially double-stranded, relaxed circular DNA (rcDNA) surrounded by a compact envelope [123,124]. The envelope consists of a host-derived cholesterol-rich lipid bilayer in which three surface glycoproteins, encoded by the same open reading frame, are embedded: L (large), M (middle) and S (small) [123,125,126]. The N-terminal preS1 domain of the L-protein is myristoylated at glycine 2 and this post-translational modification has been shown to be essential for viral infectivity [127–129]. Mutants lacking myristoylation are capable of normal viral assembly but are non-infectious to human hepatocytes [127,129].

Investigations into the entry mechanisms of HBV were hampered by the lack of a workable HBV infection system. Established human hepatoma cell lines were found to be unsusceptible to HBV and primary human hepatocyte cultures (PHHs) obtained from surgically resected liver pieces were susceptible but limited in availability, difficult to work with and variable in their preparation [130]. Primary hepatocyte cultures from the northern treeshrew *Tupaia belangeri* (PTH), which were shown to be susceptible to HBV, and a newly derived HepaRG hepatoma-derived cell line, provided the required infection systems to progress understanding of the mechanisms of HBV entry [131–133].

In 2005, Heparin Chromatography was reported as a method for purifying HBV virions from blood plasma, noting that heparin was a close homologue of liver heparan sulfate [134]. HSPGs were subsequently shown to act as an initial attachment receptor for HBV. The addition of heparin or dextran sulfate inhibits HBV infection, presumably by outcompeting HSPG for HBV binding, whereas a low-sulfated form of chondroitin sulfate does not [135]. Interestingly, liver-HS is also the binding receptor for other liver-targeting pathogens such as HCV (see below), dengue virus and malaria circumsporozoite [136–138].

Following attachment to HSPGs in the initial phase, HBV uptake through a specific entry receptor in the second phase of infection has long been elusive. While other viruses use a multitude of receptors for entry, only one candidate, namely sodium taurocholate co-transporting polypeptide (NTCP) [139–141], has been identified as a functional uptake receptor for HBV [139–141], initially using a synthetic preS1 construct together with PTH hepatocytes as target cells [139]. Importantly, susceptibility to HBV can be conferred on non-susceptible liver cell lines by transfecting them with NTCP [139–141]. NTCP is found exclusively in hepatocytes which explains the highly specific HBV tropism for the liver.

HBV uptake into hepatocytes is thought to be regulated by clathrin-mediated endocytosis of the NTCP-HBV complex. PreS1 of the HBV L-protein interacts with clathrin heavy chain and clathrin adaptor protein AP-2. Employing sh-RNA knockdown of these molecules attenuated infection of a human primary hepatocyte cell line. In addition, treatment with an inhibitor of clathrin mediated endocytosis, chlorpromazine, inhibited HBV infection [142] presumably due to inhibition of NTCP endocytosis [143].

7. Hepatitis C Virus

Hepatitis C Virus (HCV) is a member of the *Flaviviridae* family—enveloped RNA viruses which are characterized by single-stranded RNA of positive polarity embedded in a capsid which is wrapped in a lipid and protein envelope. Like HBV, HCV is a hepatotropic virus and a major cause of chronic hepatitis, progressive liver disease and hepatocellular carcinoma worldwide. While about 3% of the global population is affected by HCV, approximately 1%–4% develop liver cancer, predominantly in Europe and Asia [122,144,145]. HCV is primarily transmitted through blood and infected blood products with very low risk of sexual or vertical transmission [146]. The virus can initiate infection of the host cell as cell-free particle and/or via cell-cell contact [147].

The viral envelope consists of host phosphatidylcholine, cholesterol-rich and triglyceride-rich lipoproteins incorporating host Very-Low Density Lipoprotein (VLDL) components such as apolipoproteins (Apos) AI, B, and E, as well as the viral glycoproteins E1 and E2 that form a heterodimer. These moeities are essential for infectivity of mature circulating HCV particles [148,149]. Unlike HBV entry as described above, infection of hepatocytes by HCV is more complex and involves multiple host cell molecules, suggesting a highly orchestrated process. However, initial attachment and concentrating of the virus to the hepatocyte surface is similar for both HBV and HCV and occurs via HSPGs which is mediated by the HCV glycoprotein E2 and host ApoE present in the viral envelope [136,150]. Subsequent internalization has been found to be mediated by a complex interplay between the HCV glycoproteins E1/E2 and several cellular factors which are important in defining HCV tissue specificity [151] including tetraspanin CD81, Scavenger Receptor B1 (SR-B1), and the tight junction proteins claudin-1 and occludin (reviewed in [152,153]). However, the sequential order of HCV-receptor interactions during the uptake phase is not well understood. It is hypothesized that the lipid-rich nature of the HCV particle may favor an initial interaction with SR-B1 possibly leading to conformational changes in the HCV glycoproteins [153], thereby contributing to HCV entry in an High Density Lipoprotein (HDL)-dependent manner [154,155]. SR-B1 is a lipoprotein receptor which is expressed primarily in liver and nonplacental steroidogenic tissues, and mediates both the selective uptake of cholesteryl esters and the efflux of cholesterol [156]. Moreover, several reports have indicated a role of SR-B1 in microbial recognition and uptake, such as that of *M. tuberculosis* [157] and *P. falciparum* [158]. While SR-B1 might function independently during intermediate steps of HCV infection by preparing the virus for hepatocyte entry, host surface molecules tetraspanin CD81 and claudin-1 form complexes that facilitate HCV uptake later in the entry process [159]. It is thought that HCV directly interacts with CD81 (and only indirectly with claudin-1 [160]) leading to further conformational rearrangement of the HCV E1/E2 glycoproteins [161]. Required for productive HCV infection is also the recently identified tight junction protein occludin [162–164], suggesting that HCV entry predominantly occurs at hepatocyte tight junctions [153]. Interestingly, Niemann-Pick C1-like cholesterol receptor (NPC1L1) [165] and the iron uptake receptor transferrin receptor 1 [166] have been assigned to play accessory but essential roles in HCV uptake. Lastly, the receptor tyrosine kinases EGFR and EPHA2 have recently been found to be involved in mediating HCV entry by regulating CD81-claudin-1 co-receptor associations [85,86], illustrating once more the complex involvement of multiple host surface molecules in HCV uptake.

Following the interactions with the host receptors, HCV enters the cell via clathrin- and dynamin-dependent endocytic pathways [82,84] with downstream activation of Rho GTPase family members Rac, Rho and cell division cycle 42 (CDC42) and mitogen-activated protein kinase (MAPK) signaling cascades [83].

8. Human T Cell Lymphotropic Virus Type 1

Human T cell lymphotropic Virus Type-1 (HTLV-1) is a member of the *Retroviridae* family of single stranded RNA viruses that replicate in the host via reverse transcription [167–169]. As the etiological agent of Adult T cell lymphoma, HTLV-1 was independently described in the early 1980s [167–170] and became the first oncogenic retrovirus to be discovered. Chronic myelopathy and specific types of uveitis (HTLV-uveitis) have also since been associated with HTLV-1 infection [167–172].

HTLV-1 is endemic to south-western Japan [173] with a prevalence, in some areas, of >10% of the population [174] and is also prevalent in sub-Saharan Africa, the Caribbean, Iran and South America (reviewed in [175]). However, most HTLV-1 prevalence studies are based on screening of blood donors, pregnant women or sub-populations within endemic areas which introduces screening bias and limits an accurate estimation of number of infections which is thought to be greater than 20 million people [176].

HTLV-1 is transmitted by the introduction of cell-associated virus from mother to child through breast feeding [177], through sexual intercourse [178,179] or via blood or blood products [180]. The primary target cells of HTLV-1 are CD4$^+$ T cells, however other cell types, such as dendritic cells [181], CD8$^+$ T cells [182] and endothelial cells [183] can be infected. Cell-free virus is thought to be poorly infectious and is found at minimal levels in the peripheral blood even in individuals with a high percentage of T cells containing integrated virus [184]. Dendritic cells have, however, been shown to be susceptible to infection by cell-free virus and capable of rapidly transmitting it to T cells [181,185]. Cell-to-cell transmission accounts for the early spread of HTLV-1 in an individual. At a later stage, HTLV-1 infection persists through mitotic division of infected cells [186].

The HTLV-1 virion, like other retroviruses, is surrounded by a proteo-lipid envelope containing proteins with transmembrane (TM) and surface (SU) subunits. Most important for HTLV-1 infection is the product of the viral *env* gene which is cleaved by cellular proteases into the non-covalently linked TM subunit gp21 and the extracellular SU subunit gp46 [187], the latter consisting of an N-terminal Receptor-Binding Domain (RBD), a mid Proline rich region (PRR) and a C-terminal domain [187].

The current understanding of HTLV-1 entry into host cells is a multi-receptor model involving cellular HSPGs, NRP1 and glucose transporter-1 (GLUT-1) receptors through viral attachment and virus/cell fusion [188]. Initial attachment is mediated by interaction between the C-terminal domain of viral gp46 and HS moieties of cell surface HSPGs [189,190] of activated CD4$^+$ T cells [191] or dendritic cells [181]. The length of HS chains has been found to affect the susceptibility of cells to HTLV-1 entry. Shorter HS chains facilitate closer attachment of the HTLV-1 virion to the target cell [192]. This eventually leads to viral binding to NRP1, a cell surface co-receptor for multiple growth factors such as vascular endothelial growth factor (VEGF), which can itself bind heparan sulfate [193–195]. The viral gp46 RBD contains the KPxR motif that has been identified as the NRP1 binding site [185]. The NRP1 b-domain which binds VEGF is required for HTLV-1 entry, and gp46 SU binding to NRP1 is inhibited by the HSPG binding domain of VEGF, indicating that HTLV-1 utilizes HSPG and NRP1 in complex by molecular mimicry of VEGF, particularly in cell types that normally lack VEGF production [185].

Subsequent to stable binding of HTLV-1 to HSPGs and NRP1, GLUT-1 binding sites on gp46 are exposed, presumably as a result of conformational changes in gp46 [188]. HTLV-1 binds GLUT-1 at its large extracellular loop (ECL1) [196,197]. The amino acid residues in the RBD of gp46 that are essential for GLUT-1 binding have been shown to be distinct from the binding sites of NRP1 and HSPGs, indicative of the formation of a multi-receptor complex [193,197]. Antibodies to GLUT-1 attenuate HTLV-1 fusion and infection but not binding to CD4$^+$ T cells, indicating that binding to GLUT-1 is essential for fusion [196].

Cell-to-cell transmission requires cell contact between an infected cell and an uninfected cell and the formation of a Virological Synapse (VS) [198]. Igakura *et al.* observed that infected T cells formed conjugates with neighboring uninfected cells which induced polarization of the microtubule organization center (MTOC) to the point of contact [198]. Polarization of the MTOC is promoted by the interaction of intercellular adhesion molecule-1 or -3 (ICAM-1, ICAM-3) or vascular cell adhesion molecule-1 (VCAM-1) on infected cells with β-integrins, such as lymphocyte function-associated antigen-1 (LFA-1) on uninfected cells [199–201]. The HTLV-1 Env and Gag proteins and the HTLV-1 genome accumulate at the point of contact and then egress into the VS and interact with receptors on the conjugated, uninfected cell. NRP1 and GLUT-1 have been shown to co-localize at the point of contact and presumably promote the formation of the VS [193,202] which is also facilitated by the HTLV-1 encoded transcriptional transactivator protein Tax (Transcriptional Activator of pX region) [203]. Little is known about the mechanism at play to facilitate uptake of HTLV-1 subsequent to transmission across the VS, although it is speculated that this occurs via endocytosis of the virus as is the case for HIV [186].

9. Conclusion

With the discovery of oncogenic viruses as the etiologic agents of various human cancers, targeting viral entry is particularly relevant in prevention of viral infection, antiviral therapy, vaccine development, and ultimately cancer prevention. The knowledge of the viral particle components and the host cell entry factors and their specific interactions can enable the design of efficient antiviral strategies. These strategies can include vaccines against distinct viral proteins required for viral entry or small molecules that target the host entry factors or the interactions between both. While almost all oncogenic viruses use HSPGs during initial attachment to their target cells, the specific uptake receptors differ considerably between the individual viruses (Table 1). This review summarized the current knowledge of the host factors required for oncogenic virus uptake in order to provide a concise overview of potential target molecules for prevention and/or treatment of oncogenic virus infection.

Table 1. Summary of cellular and molecular factors involved in human oncogenic virus uptake.

Virus	Infected cell type	Associated type of cancer	Attachment receptor(s)	Proposed uptake receptor(s) and/or co-receptors	Uptake mechanism
EBV	B cells, oropharyngeal epithelial cells	Burkitt's, Hodgkin's, immunosuppression-related, T and NK cell lymphomas, nasopharyngeal and stomach carcinomas	CD21 (CR2), CD35, β1 and α5 integrins	HLA class II molecules, αVβ6, αVβ8 integrins, NRP1	macropinocytosis and lipid raft-dependent endocytosis (epithelial cells)
KSHV	endothelial cells, B cells, monocytes (macrophages and dendritic cells), fibroblasts, epithelial cells	KS, B cell malignancies	HSPGs	α3β1, αVβ3, αVβ5, α9β1 integrins, xCT, DC-SIGN EPHA2	actin-dependent, dynamin-independent macro-pinocytosis (endothelial cells), dynamin-dependent, clathrin-mediated endocytosis (fibroblasts)
HPV	epithelial cells (particularly basal keratinocytes)	cervical, anal, head and neck cancer	HSPGs, laminin-5	α6 integrin, EGFR, KGFR, tetraspanins, AIIt	actin-dependent, clathrin- and lipid raft-independent endocytosis
MCPyV	keratinocytes	Merkel cell carcinoma	sulfated glycosamino-glycans (most likely HSPGs)	sialylated glycans	unknown
HBV	hepatocytes	hepatocellular carcinoma	HSPGs	NTCP	clathrin-mediated endocytosis
HCV	hepatocytes	hepatocellular carcinoma	HSPGs	tetraspanin CD81, SR-B1, claudin-1, occludin, NPC1L1, transferrin receptor 1, EGFR, EPHA2	clathrin- and dynamin-dependent endocytosis
HTLV-1	CD4+ T cells (dendritic cells, CD8+ T cells, endothelial cells)	Adult T-cell leukemia	HSPGs/NRP1	GLUT-1	viral synapse formation possibly followed by endocytosis

Listed are all known cellular factors involved in binding and uptake of the seven human tumor viruses described to date. Note that multiple molecules can be engaged by a particular virus in different combinations and not necessarily all at the same time. Please refer to the text for references.

Acknowledgments

G.S. was funded by grants from the Poliomyelitis Research Foundation (PRF), the Cancer Association of South Africa (CANSA), the South African Medical Research Council (SA-MRC) and the National Research Foundation (NRF) of South Africa; M.J.B. received an MSc bursary from the NRF; A.A.K. was funded by the SA-MRC, CANSA and PRF.

Author Contributions

G.S., M.J.B. and A.A.K. wrote the paper.

Conflicts of Interest

The authors declare no conflict of interest.

References

1. Marsh, M.; Helenius, A. Virus entry: Open sesame. *Cell* **2006**, *124*, 729–740.
2. Sieczkarski, S.B.; Whittaker, G.R. Viral entry. *Curr. Top. Microbiol. Immunol.* **2005**, *285*, 1–23.
3. Shukla, D.; Spear, P.G. Herpesviruses and heparan sulfate: An intimate relationship in aid of viral entry. *J. Clin. Investig.* **2001**, *108*, 503–510.
4. Tanner, J.; Weis, J.; Fearon, D.; Whang, Y.; Kieff, E. Epstein-Barr virus gp350/220 binding to the B lymphocyte C3D receptor mediates adsorption, capping, and endocytosis. *Cell* **1987**, *50*, 203–213.
5. Thompson, M.P.; Kurzrock, R. Epstein-Barr virus and cancer. *Clin. Cancer Res. Off. J. Am. Assoc. Cancer Res.* **2004**, *10*, 803–821.
6. Odumade, O.A.; Hogquist, K.A.; Balfour, H.H., Jr. Progress and problems in understanding and managing primary Epstein-Barr virus infections. *Clin. Microbiol. Rev.* **2011**, *24*, 193–209.
7. Klein, G.; Klein, E.; Kashuba, E. Interaction of Epstein-Barr virus (EBV) with human B-lymphocytes. *Biochem. Biophys. Res. Commun.* **2010**, *396*, 67–73.
8. Hutt-Fletcher, L.M. Epstein-Barr virus entry. *J. Virol.* **2007**, *81*, 7825–7832.
9. Wang, H.B.; Zhang, H.; Zhang, J.P.; Li, Y.; Zhao, B.; Feng, G.K.; Du, Y.; Xiong, D.; Zhong, Q.; Liu, W.L.; *et al.* Neuropilin 1 is an entry factor that promotes EBV infection of nasopharyngeal epithelial cells. *Nat. Commun.* **2015**, *6*, 6240.
10. Katzman, R.B.; Longnecker, R. Cholesterol-dependent infection of Burkitt's lymphoma cell lines by Epstein-Barr virus. *J. Gen. Virol.* **2003**, *84*, 2987–2992.
11. Ogembo, J.G.; Kannan, L.; Ghiran, I.; Nicholson-Weller, A.; Finberg, R.W.; Tsokos, G.C.; Fingeroth, J.D. Human complement receptor type 1/CD35 is an Epstein-Barr virus receptor. *Cell Rep.* **2013**, *3*, 371–385.
12. Edson, C.M.; Thorley-Lawson, D.A. Epstein-Barr virus membrane antigens: Characterization, distribution, and strain differences. *J. Virol.* **1981**, *39*, 172–184.

13. Janz, A.; Oezel, M.; Kurzeder, C.; Mautner, J.; Pich, D.; Kost, M.; Hammerschmidt, W.; Delecluse, H.J. Infectious Epstein-Barr virus lacking major glycoprotein BLLF1 (gp350/220) demonstrates the existence of additional viral ligands. *J. Virol.* **2000**, *74*, 10142–10152.

14. Xiao, J.; Palefsky, J.M.; Herrera, R.; Berline, J.; Tugizov, S.M. The Epstein-Barr virus BMRF-2 protein facilitates virus attachment to oral epithelial cells. *Virology* **2008**, *370*, 430–442.

15. Connolly, S.A.; Jackson, J.O.; Jardetzky, T.S.; Longnecker, R. Fusing structure and function: A structural view of the herpesvirus entry machinery. *Nat. Rev. Microbiol.* **2011**, *9*, 369–381.

16. Li, Q.; Spriggs, M.K.; Kovats, S.; Turk, S.M.; Comeau, M.R.; Nepom, B.; Hutt-Fletcher, L.M. Epstein-Barr virus uses HLA class II as a cofactor for infection of B lymphocytes. *J. Virol.* **1997**, *71*, 4657–4662.

17. Wang, X.; Kenyon, W.J.; Li, Q.; Mullberg, J.; Hutt-Fletcher, L.M. Epstein-Barr virus uses different complexes of glycoproteins gH and gL to infect B lymphocytes and epithelial cells. *J. Virol.* **1998**, *72*, 5552–5558.

18. Chesnokova, L.S.; Nishimura, S.L.; Hutt-Fletcher, L.M. Fusion of epithelial cells by Epstein-Barr virus proteins is triggered by binding of viral glycoproteins gHgL to integrins alphavbeta6 or alphavbeta8. *Proc. Natl. Acad. Sci. USA* **2009**, *106*, 20464–20469.

19. Cesarman, E.; Chang, Y.; Moore, P.S.; Said, J.W.; Knowles, D.M. Kaposi's sarcoma-associated herpesvirus-like DNA sequences in AIDS-related body-cavity-based lymphomas. *N. Engl. J. Med.* **1995**, *332*, 1186–1191.

20. Chang, Y.; Cesarman, E.; Pessin, M.S.; Lee, F.; Culpepper, J.; Knowles, D.M.; Moore, P.S. Identification of herpesvirus-like DNA sequences in AIDS-associated Kaposi's sarcoma. *Science* **1994**, *266*, 1865–1869.

21. Soulier, J.; Grollet, L.; Oksenhendler, E.; Cacoub, P.; Cazals-Hatem, D.; Babinet, P.; d'Agay, M.F.; Clauvel, J.P.; Raphael, M.; Degos, L.; *et al.* Kaposi's sarcoma-associated herpesvirus-like DNA sequences in multicentric Castleman's disease. *Blood* **1995**, *86*, 1276–1280.

22. Morris, K. Cancer? In Africa? *Lancet Oncol.* **2003**, *4*, 5.

23. Parkin, D.M.; Bray, F.; Ferlay, J.; Pisani, P. Global cancer statistics, 2002. *CA: Cancer J. Clin.* **2005**, *55*, 74–108.

24. Chadburn, A.; Abdul-Nabi, A.M.; Teruya, B.S.; Lo, A.A. Lymphoid proliferations associated with human immunodeficiency virus infection. *Arch. Pathol. Lab. Med.* **2013**, *137*, 360–370.

25. Sitas, F.; Newton, R. Kaposi's sarcoma in South Africa. *J. Natl. Cancer Inst. Monogr.* **2001**, *28*, 1–4.

26. Dedicoat, M.; Newton, R.; Alkharsah, K.R.; Sheldon, J.; Szabados, I.; Ndlovu, B.; Page, T.; Casabonne, D.; Gilks, C.F.; Cassol, S.A.; *et al.* Mother-to-child transmission of human herpesvirus-8 in South Africa. *J. Infect. Dis.* **2004**, *190*, 1068–1075.

27. Henke-Gendo, C.; Schulz, T.F. Transmission and disease association of Kaposi's sarcoma-associated herpesvirus: Recent developments. *Curr. Opin. Infect. Dis.* **2004**, *17*, 53–57.

28. Matteoli, B.; Broccolo, F.; Scaccino, A.; Cottoni, F.; Angeloni, A.; Faggioni, A.; Ceccherini-Nelli, L. *In vivo* and *in vitro* evidence for an association between the route-specific transmission of HHV-8 and the virus genotype. *J. Med. Virol.* **2012**, *84*, 786–791.

29. Mbulaiteye, S.; Marshall, V.; Bagni, R.K.; Wang, C.D.; Mbisa, G.; Bakaki, P.M.; Owor, A.M.; Ndugwa, C.M.; Engels, E.A.; Katongole-Mbidde, E.; *et al.* Molecular evidence for mother-to-child transmission of Kaposi sarcoma-associated herpesvirus in Uganda and K1 gene evolution within the host. *J. Infect. Dis.* **2006**, *193*, 1250–1257.

30. Vieira, J.; Huang, M.L.; Koelle, D.M.; Corey, L. Transmissible Kaposi's sarcoma-associated herpesvirus (human herpesvirus 8) in saliva of men with a history of Kaposi's sarcoma. *J. Virol.* **1997**, *71*, 7083–7087.

31. Koelle, D.M.; Huang, M.L.; Chandran, B.; Vieira, J.; Piepkorn, M.; Corey, L. Frequent detection of Kaposi's sarcoma-associated herpesvirus (human herpesvirus 8) DNA in saliva of human immunodeficiency virus-infected men: Clinical and immunologic correlates. *J. Infect. Dis.* **1997**, *176*, 94–102.

32. Veettil, M.V.; Bandyopadhyay, C.; Dutta, D.; Chandran, B. Interaction of KSHV with host cell surface receptors and cell entry. *Viruses* **2014**, *6*, 4024–4046.

33. Neipel, F.; Albrecht, J.C.; Fleckenstein, B. Human herpesvirus 8—The first human rhadinovirus. *J. Natl. Cancer Inst. Monogr.* **1998**, *23*, 73–77.

34. Chandran, B. Early events in Kaposi's sarcoma-associated herpesvirus infection of target cells. *J. Virol.* **2010**, *84*, 2188–2199.

35. Wang, F.Z.; Akula, S.M.; Sharma-Walia, N.; Zeng, L.; Chandran, B. Human herpesvirus 8 envelope glycoprotein B mediates cell adhesion via its RGD sequence. *J. Virol.* **2003**, *77*, 3131–3147.

36. Hahn, A.S.; Kaufmann, J.K.; Wies, E.; Naschberger, E.; Panteleev-Ivlev, J.; Schmidt, K.; Holzer, A.; Schmidt, M.; Chen, J.; Konig, S.; *et al.* The ephrin receptor tyrosine kinase A2 is a cellular receptor for Kaposi's sarcoma-associated herpesvirus. *Nat. Med.* **2012**, *18*, 961–966.

37. Birkmann, A.; Mahr, K.; Ensser, A.; Yaguboglu, S.; Titgemeyer, F.; Fleckenstein, B.; Neipel, F. Cell surface heparan sulfate is a receptor for human herpesvirus 8 and interacts with envelope glycoprotein K8.1. *J. Virol.* **2001**, *75*, 11583–11593.

38. Akula, S.M.; Pramod, N.P.; Wang, F.Z.; Chandran, B. Human herpesvirus 8 envelope-associated glycoprotein B interacts with heparan sulfate-like moieties. *Virology* **2001**, *284*, 235–249.

39. Veettil, M.V.; Sadagopan, S.; Sharma-Walia, N.; Wang, F.Z.; Raghu, H.; Varga, L.; Chandran, B. Kaposi's sarcoma-associated herpesvirus forms a multimolecular complex of integrins (alphavbeta5, alphavbeta3, and alpha3beta1) and CD98-xCT during infection of human dermal microvascular endothelial cells, and CD98-xCT is essential for the postentry stage of infection. *J. Virol.* **2008**, *82*, 12126–12144.

40. Akula, S.M.; Pramod, N.P.; Wang, F.Z.; Chandran, B. Integrin alpha3beta1 (CD 49c/29) is a cellular receptor for Kaposi's sarcoma-associated herpesvirus (KSHV/HHV-8) entry into the target cells. *Cell* **2002**, *108*, 407–419.

41. Walker, L.R.; Hussein, H.A.; Akula, S.M. Disintegrin-like domain of glycoprotein B regulates Kaposi's sarcoma-associated herpesvirus infection of cells. *J. Gen. Virol.* **2014**, *95*, 1770–1782.

42. Kaleeba, J.A.; Berger, E.A. Kaposi's sarcoma-associated herpesvirus fusion-entry receptor: Cystine transporter xCT. *Science* **2006**, *311*, 1921–1924.

43. Rappocciolo, G.; Hensler, H.R.; Jais, M.; Reinhart, T.A.; Pegu, A.; Jenkins, F.J.; Rinaldo, C.R. Human herpesvirus 8 infects and replicates in primary cultures of activated B lymphocytes through DC-SIGN. *J. Virol.* **2008**, *82*, 4793–4806.

44. Rappocciolo, G.; Jenkins, F.J.; Hensler, H.R.; Piazza, P.; Jais, M.; Borowski, L.; Watkins, S.C.; Rinaldo, C.R., Jr. DC-SIGN is a receptor for human herpesvirus 8 on dendritic cells and macrophages. *J. Immunol.* **2006**, *176*, 1741–1749.

45. Hensler, H.R.; Tomaszewski, M.J.; Rappocciolo, G.; Rinaldo, C.R.; Jenkins, F.J. Human herpesvirus 8 glycoprotein B binds the entry receptor DC-SIGN. *Virus Res.* **2014**, *190*, 97–103.

46. Chakraborty, S.; Veettil, M.V.; Bottero, V.; Chandran, B. Kaposi's sarcoma-associated herpesvirus interacts with EphrinA2 receptor to amplify signaling essential for productive infection. *Proc. Natl. Acad. Sci. USA* **2012**, *109*, E1163–E1172.

47. Dutta, D.; Chakraborty, S.; Bandyopadhyay, C.; Valiya Veettil, M.; Ansari, M.A.; Singh, V.V.; Chandran, B. EphrinA2 regulates clathrin mediated KSHV endocytosis in fibroblast cells by coordinating integrin-associated signaling and c-Cbl directed polyubiquitination. *PLoS Pathog.* **2013**, *9*, e1003510.

48. Raghu, H.; Sharma-Walia, N.; Veettil, M.V.; Sadagopan, S.; Chandran, B. Kaposi's sarcoma-associated herpesvirus utilizes an actin polymerization-dependent macropinocytic pathway to enter human dermal microvascular endothelial and human umbilical vein endothelial cells. *J. Virol.* **2009**, *83*, 4895–4911.

49. Akula, S.M.; Naranatt, P.P.; Walia, N.S.; Wang, F.Z.; Fegley, B.; Chandran, B. Kaposi's sarcoma-associated herpesvirus (human herpesvirus 8) infection of human fibroblast cells occurs through endocytosis. *J. Virol.* **2003**, *77*, 7978–7990.

50. Krishnan, H.H.; Sharma-Walia, N.; Streblow, D.N.; Naranatt, P.P.; Chandran, B. Focal adhesion kinase is critical for entry of Kaposi's sarcoma-associated herpesvirus into target cells. *J. Virol.* **2006**, *80*, 1167–1180.

51. Veettil, M.V.; Sharma-Walia, N.; Sadagopan, S.; Raghu, H.; Sivakumar, R.; Naranatt, P.P.; Chandran, B. RhoA-GTPase facilitates entry of Kaposi's sarcoma-associated herpesvirus into adherent target cells in a Src-dependent manner. *J. Virol.* **2006**, *80*, 11432–11446.

52. Naranatt, P.P.; Akula, S.M.; Zien, C.A.; Krishnan, H.H.; Chandran, B. Kaposi's sarcoma-associated herpesvirus induces the phosphatidylinositol 3-kinase-PKC-zeta-MEK-ERK signaling pathway in target cells early during infection: Implications for infectivity. *J. Virol.* **2003**, *77*, 1524–1539.

53. de Villiers, E.M.; Fauquet, C.; Broker, T.R.; Bernard, H.U.; zur Hausen, H. Classification of papillomaviruses. *Virology* **2004**, *324*, 17–27.

54. Trottier, H.; Franco, E.L. The epidemiology of genital human papillomavirus infection. *Vaccine* **2006**, *24 Suppl 1*, S1–S15.

55. Pyeon, D.; Pearce, S.M.; Lank, S.M.; Ahlquist, P.; Lambert, P.F. Establishment of human papillomavirus infection requires cell cycle progression. *PLoS Pathog.* **2009**, *5*, e1000318.

56. Schiller, J.T.; Day, P.M.; Kines, R.C. Current understanding of the mechanism of HPV infection. *Gynecol. Oncol.* **2010**, *118*, S12–S17.

57. Buck, C.B.; Cheng, N.; Thompson, C.D.; Lowy, D.R.; Steven, A.C.; Schiller, J.T.; Trus, B.L. Arrangement of L2 within the papillomavirus capsid. *J. Virol.* **2008**, *82*, 5190–5197.

58. Raff, A.B.; Woodham, A.W.; Raff, L.M.; Skeate, J.G.; Yan, L.; Da Silva, D.M.; Schelhaas, M.; Kast, W.M. The evolving field of human papillomavirus receptor research: A review of binding and entry. *J. Virol.* **2013**, *87*, 6062–6072.

59. Kines, R.C.; Thompson, C.D.; Lowy, D.R.; Schiller, J.T.; Day, P.M. The initial steps leading to papillomavirus infection occur on the basement membrane prior to cell surface binding. *Proc. Natl. Acad. Sci. USA* **2009**, *106*, 20458–20463.

60. Abban, C.Y.; Meneses, P.I. Usage of heparan sulfate, integrins, and FAK in HPV16 infection. *Virology* **2010**, *403*, 1–16.

61. Johnson, K.M.; Kines, R.C.; Roberts, J.N.; Lowy, D.R.; Schiller, J.T.; Day, P.M. Role of heparan sulfate in attachment to and infection of the murine female genital tract by human papillomavirus. *J. Virol.* **2009**, *83*, 2067–2074.

62. Culp, T.D.; Budgeon, L.R.; Christensen, N.D. Human papillomaviruses bind a basal extracellular matrix component secreted by keratinocytes which is distinct from a membrane-associated receptor. *Virology* **2006**, *347*, 147–159.

63. Schäfer, G.; Kabanda, S.; van Rooyen, B.; Marusic, M.B.; Banks, L.; Parker, M.I. The role of inflammation in HPV infection of the oesophagus. *BMC Cancer* **2013**, *13*, e185.

64. Selinka, H.C.; Florin, L.; Patel, H.D.; Freitag, K.; Schmidtke, M.; Makarov, V.A.; Sapp, M. Inhibition of transfer to secondary receptors by heparan sulfate-binding drug or antibody induces noninfectious uptake of human papillomavirus. *J. Virol.* **2007**, *81*, 10970–10980.

65. Shafti-Keramat, S.; Handisurya, A.; Kriehuber, E.; Meneguzzi, G.; Slupetzky, K.; Kirnbauer, R. Different heparan sulfate proteoglycans serve as cellular receptors for human papillomaviruses. *J. Virol.* **2003**, *77*, 13125–13135.

66. Culp, T.D.; Budgeon, L.R.; Marinkovich, M.P.; Meneguzzi, G.; Christensen, N.D. Keratinocyte-secreted laminin 5 can function as a transient receptor for human papillomaviruses by binding virions and transferring them to adjacent cells. *J. Virol.* **2006**, *80*, 8940–8950.

67. Richards, K.F.; Bienkowska-Haba, M.; Dasgupta, J.; Chen, X.S.; Sapp, M. Multiple heparan sulfate binding site engagements are required for the infectious entry of human papillomavirus type 16. *J. Virol.* **2013**, *87*, 11426–11437.

68. Bienkowska-Haba, M.; Patel, H.D.; Sapp, M. Target cell cyclophilins facilitate human papillomavirus type 16 infection. *PLoS Pathog.* **2009**, *5*, e1000524.

69. Day, P.M.; Schiller, J.T. The role of furin in papillomavirus infection. *Futur. Microbiol.* **2009**, *4*, 1255–1262.

70. Richards, R.M.; Lowy, D.R.; Schiller, J.T.; Day, P.M. Cleavage of the papillomavirus minor capsid protein, L2, at a furin consensus site is necessary for infection. *Proc. Natl. Acad. Sci. USA* **2006**, *103*, 1522–1527.

71. Evander, M.; Frazer, I.H.; Payne, E.; Qi, Y.M.; Hengst, K.; McMillan, N.A. Identification of the alpha6 integrin as a candidate receptor for papillomaviruses. *J. Virol.* **1997**, *71*, 2449–2456.

72. Yoon, C.S.; Kim, K.D.; Park, S.N.; Cheong, S.W. Alpha(6) integrin is the main receptor of human papillomavirus type 16 VLP. *Biochem. Biophys. Res. Commun.* **2001**, *283*, 668–673.

73. Surviladze, Z.; Dziduszko, A.; Ozbun, M.A. Essential roles for soluble virion-associated heparan sulfonated proteoglycans and growth factors in human papillomavirus infections. *PLoS Pathog.* **2012**, *8*, e1002519.

74. Spoden, G.; Freitag, K.; Husmann, M.; Boller, K.; Sapp, M.; Lambert, C.; Florin, L. Clathrin- and caveolin-independent entry of human papillomavirus type 16--involvement of tetraspanin-enriched microdomains (TEMs). *PLoS ONE* **2008**, *3*, e3313.

75. Scheffer, K.D.; Gawlitza, A.; Spoden, G.A.; Zhang, X.A.; Lambert, C.; Berditchevski, F.; Florin, L. Tetraspanin CD151 mediates papillomavirus type 16 endocytosis. *J. Virol.* **2013**, *87*, 3435–3446.

76. Hitchcock, J.K.; Katz, A.A.; Schäfer, G. Dynamic reciprocity: The role of annexin A2 in tissue integrity. *J. Cell Commun. Signal.* **2014**, *8*, 125–133.

77. Woodham, A.W.; Da Silva, D.M.; Skeate, J.G.; Raff, A.B.; Ambroso, M.R.; Brand, H.E.; Isas, J.M.; Langen, R.; Kast, W.M. The S100A10 subunit of the annexin A2 heterotetramer facilitates L2-mediated human papillomavirus infection. *PLoS ONE* **2012**, *7*, e43519.

78. Dziduszko, A.; Ozbun, M.A. Annexin A2 and S100A10 regulate human papillomavirus type 16 entry and intracellular trafficking in human keratinocytes. *J. Virol.* **2013**, *87*, 7502–7515.

79. Schelhaas, M.; Shah, B.; Holzer, M.; Blattmann, P.; Kuhling, L.; Day, P.M.; Schiller, J.T.; Helenius, A. Entry of human papillomavirus type 16 by actin-dependent, clathrin- and lipid raft-independent endocytosis. *PLoS Pathog.* **2012**, *8*, e1002657.

80. Surviladze, Z.; Sterk, R.T.; DeHaro, S.A.; Ozbun, M.A. Cellular entry of human papillomavirus type 16 involves activation of the phosphatidylinositol 3-kinase/Akt/mTOR pathway and inhibition of autophagy. *J. Virol.* **2013**, *87*, 2508–2517.

81. Compton, T. Receptors and immune sensors: The complex entry path of human cytomegalovirus. *Trends Cell Biol.* **2004**, *14*, 5–8.

82. Blanchard, E.; Belouzard, S.; Goueslain, L.; Wakita, T.; Dubuisson, J.; Wychowski, C.; Rouille, Y. Hepatitis C Virus entry depends on clathrin-mediated endocytosis. *J. Virol.* **2006**, *80*, 6964–6972.

83. Brazzoli, M.; Bianchi, A.; Filippini, S.; Weiner, A.; Zhu, Q.; Pizza, M.; Crotta, S. CD81 is a central regulator of cellular events required for hepatitis C virus infection of human hepatocytes. *J. Virol.* **2008**, *82*, 8316–8329.

84. Farquhar, M.J.; Hu, K.; Harris, H.J.; Davis, C.; Brimacombe, C.L.; Fletcher, S.J.; Baumert, T.F.; Rappoport, J.Z.; Balfe, P.; McKeating, J.A. Hepatitis C Virus induces CD81 and claudin-1 endocytosis. *J. Virol.* **2012**, *86*, 4305–4316.

85. Lupberger, J.; Zeisel, M.B.; Xiao, F.; Thumann, C.; Fofana, I.; Zona, L.; Davis, C.; Mee, C.J.; Turek, M.; Gorke, S.; *et al.* EGFR and EphA2 are host factors for Hepatitis C Virus entry and possible targets for antiviral therapy. *Nat. Med.* **2011**, *17*, 589–595.

86. Zona, L.; Lupberger, J.; Sidahmed-Adrar, N.; Thumann, C.; Harris, H.J.; Barnes, A.; Florentin, J.; Tawar, R.G.; Xiao, F.; Turek, M.; *et al.* HRas signal transduction promotes hepatitis C virus cell entry by triggering assembly of the host tetraspanin receptor complex. *Cell Host Microbe* **2013**, *13*, 302–313.

87. Richards, K.F.; Mukherjee, S.; Bienkowska-Haba, M.; Pang, J.; Sapp, M. Human papillomavirus species-specific interaction with the basement membrane-resident non-heparan sulfate receptor. *Viruses* **2014**, *6*, 4856–4879.

88. Johne, R.; Buck, C.B.; Allander, T.; Atwood, W.J.; Garcea, R.L.; Imperiale, M.J.; Major, E.O.; Ramqvist, T.; Norkin, L.C. Taxonomical developments in the family polyomaviridae. *Arch. Virol.* **2011**, *156*, 1627–1634.

89. Spurgeon, M.E.; Lambert, P.F. Merkel cell polyomavirus: A newly discovered human virus with oncogenic potential. *Virology* **2013**, *435*, 118–130.

90. Kaae, J.; Hansen, A.V.; Biggar, R.J.; Boyd, H.A.; Moore, P.S.; Wohlfahrt, J.; Melbye, M. Merkel cell carcinoma: Incidence, mortality, and risk of other cancers. *J. Natl. Cancer Inst.* **2010**, *102*, 793–801.

91. Feng, H.; Shuda, M.; Chang, Y.; Moore, P.S. Clonal integration of a polyomavirus in human Merkel cell carcinoma. *Science* **2008**, *319*, 1096–1100.

92. Feng, H.; Taylor, J.L.; Benos, P.V.; Newton, R.; Waddell, K.; Lucas, S.B.; Chang, Y.; Moore, P.S. Human transcriptome subtraction by using short sequence tags to search for tumor viruses in conjunctival carcinoma. *J. Virol.* **2007**, *81*, 11332–11340.

93. Houben, R.; Shuda, M.; Weinkam, R.; Schrama, D.; Feng, H.; Chang, Y.; Moore, P.S.; Becker, J.C. Merkel cell polyomavirus-infected Merkel cell carcinoma cells require expression of viral T antigens. *J. Virol.* **2010**, *84*, 7064–7072.

94. Shuda, M.; Arora, R.; Kwun, H.J.; Feng, H.; Sarid, R.; Fernandez-Figueras, M.T.; Tolstov, Y.; Gjoerup, O.; Mansukhani, M.M.; Swerdlow, S.H.; *et al.* Human Merkel cell polyomavirus infection I. MCV T antigen expression in Merkel cell carcinoma, lymphoid tissues and lymphoid tumors. *Int. J. Cancer. J. Int. Cancer* **2009**, *125*, 1243–1249.

95. Shuda, M.; Kwun, H.J.; Feng, H.; Chang, Y.; Moore, P.S. Human Merkel cell polyomavirus small T antigen is an oncoprotein targeting the 4E-BP1 translation regulator. *J. Clin. Investig.* **2011**, *121*, 3623–3634.

96. Andres, C.; Belloni, B.; Puchta, U.; Sander, C.A.; Flaig, M.J. Prevalence of MCPyV in Merkel cell carcinoma and non-MCC tumors. *J. Cutan. Pathol.* **2010**, *37*, 28–34.

97. Becker, J.C.; Houben, R.; Ugurel, S.; Trefzer, U.; Pfohler, C.; Schrama, D. MC polyomavirus is frequently present in Merkel cell carcinoma of European patients. *J. Investig. Dermatol.* **2009**, *129*, 248–250.

98. Kassem, A.; Schopflin, A.; Diaz, C.; Weyers, W.; Stickeler, E.; Werner, M.; Zur Hausen, A. Frequent detection of Merkel cell polyomavirus in human Merkel cell carcinomas and identification of a unique deletion in the VP1 gene. *Cancer Res.* **2008**, *68*, 5009–5013.

99. Wieland, U.; Mauch, C.; Kreuter, A.; Krieg, T.; Pfister, H. Merkel cell polyomavirus DNA in persons without Merkel cell carcinoma. *Emerg. Infect. Dis.* **2009**, *15*, 1496–1498.

100. Laude, H.C.; Jonchere, B.; Maubec, E.; Carlotti, A.; Marinho, E.; Couturaud, B.; Peter, M.; Sastre-Garau, X.; Avril, M.F.; Dupin, N.; *et al.* Distinct Merkel cell polyomavirus molecular features in tumour and non tumour specimens from patients with Merkel cell carcinoma. *PLoS Pathog.* **2010**, *6*, e1001076.

101. Agelli, M.; Clegg, L.X. Epidemiology of primary Merkel cell carcinoma in the United States. *J. Am. Acad. Dermatol.* **2003**, *49*, 832–841.

102. Engels, E.A.; Frisch, M.; Goedert, J.J.; Biggar, R.J.; Miller, R.W. Merkel cell carcinoma and HIV infection. *Lancet* **2002**, *359*, 497–498.

103. Schowalter, R.M.; Pastrana, D.V.; Pumphrey, K.A.; Moyer, A.L.; Buck, C.B. Merkel cell polyomavirus and two previously unknown polyomaviruses are chronically shed from human skin. *Cell Host Microbe* **2010**, *7*, 509–515.

104. Tolstov, Y.L.; Pastrana, D.V.; Feng, H.; Becker, J.C.; Jenkins, F.J.; Moschos, S.; Chang, Y.; Buck, C.B.; Moore, P.S. Human Merkel cell polyomavirus infection II. MCV is a common human infection that can be detected by conformational capsid epitope immunoassays. *Int. J. Cancer. J. Int. Cancer* **2009**, *125*, 1250–1256.

105. van der Meijden, E.; Bialasiewicz, S.; Rockett, R.J.; Tozer, S.J.; Sloots, T.P.; Feltkamp, M.C. Different serologic behavior of MCPyV, TSPyV, HPyV6, HPyV7 and HPyV9 polyomaviruses found on the skin. *PLoS ONE* **2013**, *8*, e81078.

106. Martel-Jantin, C.; Pedergnana, V.; Nicol, J.T.; Leblond, V.; Tregouet, D.A.; Tortevoye, P.; Plancoulaine, S.; Coursaget, P.; Touze, A.; Abel, L.; *et al.* Merkel cell polyomavirus infection occurs during early childhood and is transmitted between siblings. *J. Clin. Virol. Off. Publ. Pan Am. Soc. Clin. Virol.* **2013**, *58*, 288–291.

107. Matsushita, M.; Kuwamoto, S.; Iwasaki, T.; Higaki-Mori, H.; Yashima, S.; Kato, M.; Murakami, I.; Horie, Y.; Kitamura, Y.; Hayashi, K. Detection of Merkel cell polyomavirus in the human tissues from 41 Japanese autopsy cases using polymerase chain reaction. *Intervirology* **2013**, *56*, 1–5.

108. Loyo, M.; Guerrero-Preston, R.; Brait, M.; Hoque, M.O.; Chuang, A.; Kim, M.S.; Sharma, R.; Liegeois, N.J.; Koch, W.M.; Califano, J.A.; *et al.* Quantitative detection of Merkel cell virus in human tissues and possible mode of transmission. *Int. J. Cancer. J. Int. Cancer* **2010**, *126*, 2991–2996.

109. Zur Hausen, A.; Rennspiess, D.; Winnepenninckx, V.; Speel, E.J.; Kurz, A.K. Early B-cell differentiation in Merkel cell carcinomas: Clues to cellular ancestry. *Cancer Res.* **2013**, *73*, 4982–4987.

110. Jankowski, M.; Kopinski, P.; Schwartz, R.; Czajkowski, R. Merkel cell carcinoma: Is this a true carcinoma? *Exp. Dermatol.* **2014**, *23*, 792–794.

111. Schowalter, R.M.; Buck, C.B. The Merkel cell polyomavirus minor capsid protein. *PLoS Pathogens* **2013**, *9*, e1003558.

112. Neu, U.; Hengel, H.; Blaum, B.S.; Schowalter, R.M.; Macejak, D.; Gilbert, M.; Wakarchuk, W.W.; Imamura, A.; Ando, H.; Kiso, M.; *et al.* Structures of Merkel cell polyomavirus VP1 complexes define a sialic acid binding site required for infection. *PLoS Pathogens* **2012**, *8*, e1002738.

113. Schowalter, R.M.; Pastrana, D.V.; Buck, C.B. Glycosaminoglycans and sialylated glycans sequentially facilitate Merkel cell polyomavirus infectious entry. *PLoS Pathogens* **2011**, *7*, e1002161.

114. Campanero-Rhodes, M.A.; Smith, A.; Chai, W.; Sonnino, S.; Mauri, L.; Childs, R.A.; Zhang, Y.; Ewers, H.; Helenius, A.; Imberty, A.; *et al.* N-glycolyl GM1 ganglioside as a receptor for simian virus 40. *J. Virol.* **2007**, *81*, 12846–12858.

115. Low, J.A.; Magnuson, B.; Tsai, B.; Imperiale, M.J. Identification of gangliosides GD1b and GT1b as receptors for BK virus. *J. Virol.* **2006**, *80*, 1361–1366.

116. Smith, A.E.; Lilie, H.; Helenius, A. Ganglioside-dependent cell attachment and endocytosis of murine polyomavirus-like particles. *FEBS Lett.* **2003**, *555*, 199–203.

117. Tsai, B.; Gilbert, J.M.; Stehle, T.; Lencer, W.; Benjamin, T.L.; Rapoport, T.A. Gangliosides are receptors for murine polyoma virus and SV40. *EMBO J.* **2003**, *22*, 4346–4355.

118. Erickson, K.D.; Garcea, R.L.; Tsai, B. Ganglioside GT1b is a putative host cell receptor for the Merkel cell polyomavirus. *J. Virol.* **2009**, *83*, 10275–10279.

119. Schowalter, R.M.; Reinhold, W.C.; Buck, C.B. Entry tropism of BK and Merkel cell polyomaviruses in cell culture. *PLoS ONE* **2012**, *7*, e42181.

120. Gust, I.D.; Burrell, C.J.; Coulepis, A.G.; Robinson, W.S.; Zuckerman, A.J. Taxonomic classification of human hepatitis B virus. *Intervirology* **1986**, *25*, 14–29.

121. Seeger, C.; Mason, W.S. Hepatitis B virus biology. *Microbiol. Mol. Biol. Rev. MMBR* **2000**, *64*, 51–68.

122. El-Serag, H.B. Epidemiology of viral hepatitis and hepatocellular carcinoma. *Gastroenterology* **2012**, *142*, 1264–1273.e1.

123. Seitz, S.; Urban, S.; Antoni, C.; Bottcher, B. Cryo-electron microscopy of hepatitis B virions reveals variability in envelope capsid interactions. *EMBO J.* **2007**, *26*, 4160–4167.

124. Wynne, S.A.; Crowther, R.A.; Leslie, A.G. The crystal structure of the human hepatitis B virus capsid. *Mol. Cell* **1999**, *3*, 771–780.

125. Gavilanes, F.; Gonzalez-Ros, J.M.; Peterson, D.L. Structure of hepatitis B surface antigen. Characterization of the lipid components and their association with the viral proteins. *J. Biol. Chem.* **1982**, *257*, 7770–7777.

126. Heermann, K.H.; Goldmann, U.; Schwartz, W.; Seyffarth, T.; Baumgarten, H.; Gerlich, W.H. Large surface proteins of hepatitis B virus containing the pre-s sequence. *J. Virol.* **1984**, *52*, 396–402.

127. Bruss, V.; Hagelstein, J.; Gerhardt, E.; Galle, P.R. Myristylation of the large surface protein is required for hepatitis B virus *in vitro* infectivity. *Virology* **1996**, *218*, 396–399.

128. Gripon, P.; Cannie, I.; Urban, S. Efficient inhibition of hepatitis B virus infection by acylated peptides derived from the large viral surface protein. *J. Virol.* **2005**, *79*, 1613–1622.

129. Gripon, P.; Le Seyec, J.; Rumin, S.; Guguen-Guillouzo, C. Myristylation of the hepatitis B virus large surface protein is essential for viral infectivity. *Virology* **1995**, *213*, 292–299.

130. Gripon, P.; Diot, C.; Theze, N.; Fourel, I.; Loreal, O.; Brechot, C.; Guguen-Guillouzo, C. Hepatitis B virus infection of adult human hepatocytes cultured in the presence of dimethyl sulfoxide. *J. Virol.* **1988**, *62*, 4136–4143.

131. Gripon, P.; Rumin, S.; Urban, S.; Le Seyec, J.; Glaise, D.; Cannie, I.; Guyomard, C.; Lucas, J.; Trepo, C.; Guguen-Guillouzo, C. Infection of a human hepatoma cell line by hepatitis B virus. *Proc. Natl. Acad. Sci. USA* **2002**, *99*, 15655–15660.

132. Kock, J.; Nassal, M.; MacNelly, S.; Baumert, T.F.; Blum, H.E.; von Weizsacker, F. Efficient infection of primary tupaia hepatocytes with purified human and woolly monkey hepatitis B virus. *J. Virol.* **2001**, *75*, 5084–5089.

133. Walter, E.; Keist, R.; Niederost, B.; Pult, I.; Blum, H.E. Hepatitis B virus infection of tupaia hepatocytes *in vitro* and *in vivo*. *Hepatology* **1996**, *24*, 1–5.

134. Zahn, A.; Allain, J.P. Hepatitis C virus and hepatitis B virus bind to heparin: Purification of largely IgG-free virions from infected plasma by heparin chromatography. *J. Gen. Virol.* **2005**, *86*, 677–685.

135. Leistner, C.M.; Gruen-Bernhard, S.; Glebe, D. Role of glycosaminoglycans for binding and infection of hepatitis B virus. *Cell. Microbiol.* **2008**, *10*, 122–133.

136. Barth, H.; Schafer, C.; Adah, M.I.; Zhang, F.; Linhardt, R.J.; Toyoda, H.; Kinoshita-Toyoda, A.; Toida, T.; Van Kuppevelt, T.H.; Depla, E.; *et al.* Cellular binding of hepatitis C virus envelope glycoprotein E2 requires cell surface heparan sulfate. *J. Biol. Chem.* **2003**, *278*, 41003–41012.

137. Chen, Y.; Maguire, T.; Hileman, R.E.; Fromm, J.R.; Esko, J.D.; Linhardt, R.J.; Marks, R.M. Dengue virus infectivity depends on envelope protein binding to target cell heparan sulfate. *Nat. Med.* **1997**, *3*, 866–871.

138. Rathore, D.; McCutchan, T.F.; Garboczi, D.N.; Toida, T.; Hernaiz, M.J.; LeBrun, L.A.; Lang, S.C.; Linhardt, R.J. Direct measurement of the interactions of glycosaminoglycans and a heparin decasaccharide with the malaria circumsporozoite protein. *Biochemistry* **2001**, *40*, 11518–11524.

139. Yan, H.; Zhong, G.; Xu, G.; He, W.; Jing, Z.; Gao, Z.; Huang, Y.; Qi, Y.; Peng, B.; Wang, H.; *et al.* Sodium taurocholate cotransporting polypeptide is a functional receptor for human hepatitis B and D virus. *eLife* **2012**, *1*, e00049.

140. Ni, Y.; Lempp, F.A.; Mehrle, S.; Nkongolo, S.; Kaufman, C.; Falth, M.; Stindt, J.; Koniger, C.; Nassal, M.; Kubitz, R.; *et al.* Hepatitis B and D viruses exploit sodium taurocholate co-transporting polypeptide for species-specific entry into hepatocytes. *Gastroenterology* **2014**, *146*, 1070–1083.

141. Zhong, G.; Yan, H.; Wang, H.; He, W.; Jing, Z.; Qi, Y.; Fu, L.; Gao, Z.; Huang, Y.; Xu, G.; *et al.* Sodium taurocholate cotransporting polypeptide mediates woolly monkey hepatitis B virus infection of tupaia hepatocytes. *J. Virol.* **2013**, *87*, 7176–7184.

142. Huang, H.C.; Chen, C.C.; Chang, W.C.; Tao, M.H.; Huang, C. Entry of hepatitis B virus into immortalized human primary hepatocytes by clathrin-dependent endocytosis. *J. Virol.* **2012**, *86*, 9443–9453.

143. Stross, C.; Kluge, S.; Weissenberger, K.; Winands, E.; Haussinger, D.; Kubitz, R. A dileucine motif is involved in plasma membrane expression and endocytosis of rat sodium taurocholate cotransporting polypeptide (NTCP). *Am. J. Physiol. Gastrointest. Liver Physiol.* **2013**, *305*, G722–G730.

144. Morgan, R.L.; Baack, B.; Smith, B.D.; Yartel, A.; Pitasi, M.; Falck-Ytter, Y. Eradication of hepatitis C virus infection and the development of hepatocellular carcinoma: A meta-analysis of observational studies. *Ann. Intern. Med.* **2013**, *158*, 329–337.

145. Tang, H.; Grise, H. Cellular and molecular biology of HCV infection and hepatitis. *Clin. Sci.* **2009**, *117*, 49–65.

146. Shepard, C.W.; Finelli, L.; Alter, M.J. Global epidemiology of hepatitis C virus infection. *Lancet Infect. Dis.* **2005**, *5*, 558–567.

147. Brimacombe, C.L.; Grove, J.; Meredith, L.W.; Hu, K.; Syder, A.J.; Flores, M.V.; Timpe, J.M.; Krieger, S.E.; Baumert, T.F.; Tellinghuisen, T.L.; *et al.* Neutralizing antibody-resistant hepatitis C virus cell-to-cell transmission. *J. Virol.* **2011**, *85*, 596–605.

148. Catanese, M.T.; Uryu, K.; Kopp, M.; Edwards, T.J.; Andrus, L.; Rice, W.J.; Silvestry, M.; Kuhn, R.J.; Rice, C.M. Ultrastructural analysis of hepatitis C virus particles. *Proc. Natl. Acad. Sci. USA* **2013**, *110*, 9505–9510.

149. Merz, A.; Long, G.; Hiet, M.S.; Brugger, B.; Chlanda, P.; Andre, P.; Wieland, F.; Krijnse-Locker, J.; Bartenschlager, R. Biochemical and morphological properties of hepatitis C virus particles and determination of their lipidome. *J. Biol. Chem.* **2011**, *286*, 3018–3032.

150. Jiang, J.; Cun, W.; Wu, X.; Shi, Q.; Tang, H.; Luo, G. Hepatitis C virus attachment mediated by apolipoprotein E binding to cell surface heparan sulfate. *J. Virol.* **2012**, *86*, 7256–7267.

151. Da Costa, D.; Turek, M.; Felmlee, D.J.; Girardi, E.; Pfeffer, S.; Long, G.; Bartenschlager, R.; Zeisel, M.B.; Baumert, T.F. Reconstitution of the entire hepatitis C virus life cycle in nonhepatic cells. *J. Virol.* **2012**, *86*, 11919–11925.

152. Baumert, T.F.; Meredith, L.; Ni, Y.; Felmlee, D.J.; McKeating, J.A.; Urban, S. Entry of hepatitis B and C viruses - recent progress and future impact. *Curr. Opin. Virol.* **2014**, *4*, 58–65.

153. Meredith, L.W.; Wilson, G.K.; Fletcher, N.F.; McKeating, J.A. Hepatitis C virus entry: Beyond receptors. *Rev. Med. Virol.* **2012**, *22*, 182–193.

154. Catanese, M.T.; Graziani, R.; von Hahn, T.; Moreau, M.; Huby, T.; Paonessa, G.; Santini, C.; Luzzago, A.; Rice, C.M.; Cortese, R.; *et al.* High-avidity monoclonal antibodies against the human scavenger class B type I receptor efficiently block hepatitis C virus infection in the presence of high-density lipoprotein. *J. Virol.* **2007**, *81*, 8063–8071.

155. Voisset, C.; Callens, N.; Blanchard, E.; Op De Beeck, A.; Dubuisson, J.; Vu-Dac, N. High density lipoproteins facilitate hepatitis C virus entry through the scavenger receptor class B type I. *J. Biol. Chem.* **2005**, *280*, 7793–7799.

156. Acton, S.; Rigotti, A.; Landschulz, K.T.; Xu, S.; Hobbs, H.H.; Krieger, M. Identification of scavenger receptor SR-BI as a high density lipoprotein receptor. *Science* **1996**, *271*, 518–520.

157. Schäfer, G.; Guler, R.; Murray, G.; Brombacher, F.; Brown, G.D. The role of scavenger receptor B1 in infection with mycobacterium tuberculosis in a murine model. *PLoS ONE* **2009**, *4*, e8448.

158. Rodrigues, C.D.; Hannus, M.; Prudencio, M.; Martin, C.; Goncalves, L.A.; Portugal, S.; Epiphanio, S.; Akinc, A.; Hadwiger, P.; Jahn-Hofmann, K.; *et al.* Host scavenger receptor SR-BI plays a dual role in the establishment of malaria parasite liver infection. *Cell Host Microbe* **2008**, *4*, 271–282.

159. Evans, M.J.; von Hahn, T.; Tscherne, D.M.; Syder, A.J.; Panis, M.; Wolk, B.; Hatziioannou, T.; McKeating, J.A.; Bieniasz, P.D.; Rice, C.M. Claudin-1 is a hepatitis C virus co-receptor required for a late step in entry. *Nature* **2007**, *446*, 801–805.

160. Davis, C.; Harris, H.J.; Hu, K.; Drummer, H.E.; McKeating, J.A.; Mullins, J.G.; Balfe, P. In silico directed mutagenesis identifies the CD81/claudin-1 hepatitis C virus receptor interface. *Cell. Microbiol.* **2012**, *14*, 1892–1903.

161. Sharma, N.R.; Mateu, G.; Dreux, M.; Grakoui, A.; Cosset, F.L.; Melikyan, G.B. Hepatitis C virus is primed by CD81 protein for low pH-dependent fusion. *J. Biol. Chem.* **2011**, *286*, 30361–30376.

162. Benedicto, I.; Molina-Jimenez, F.; Bartosch, B.; Cosset, F.L.; Lavillette, D.; Prieto, J.; Moreno-Otero, R.; Valenzuela-Fernandez, A.; Aldabe, R.; Lopez-Cabrera, M.; *et al.* The tight junction-associated protein occludin is required for a postbinding step in hepatitis C virus entry and infection. *J. Virol.* **2009**, *83*, 8012–8020.

163. Liu, S.; Yang, W.; Shen, L.; Turner, J.R.; Coyne, C.B.; Wang, T. Tight junction proteins claudin-1 and occludin control hepatitis C virus entry and are downregulated during infection to prevent superinfection. *J. Virol.* **2009**, *83*, 2011–2014.

164. Ploss, A.; Evans, M.J.; Gaysinskaya, V.A.; Panis, M.; You, H.; de Jong, Y.P.; Rice, C.M. Human occludin is a hepatitis C virus entry factor required for infection of mouse cells. *Nature* **2009**, *457*, 882–886.

165. Sainz, B., Jr.; Barretto, N.; Martin, D.N.; Hiraga, N.; Imamura, M.; Hussain, S.; Marsh, K.A.; Yu, X.; Chayama, K.; Alrefai, W.A.; *et al.* Identification of the Niemann-Pick C1-like 1 cholesterol absorption receptor as a new hepatitis C virus entry factor. *Nat. Med.* **2012**, *18*, 281–285.

166. Martin, D.N.; Uprichard, S.L. Identification of transferrin receptor 1 as a hepatitis C virus entry factor. *Proc. Natl. Acad. Sci. USA* **2013**, *110*, 10777–10782.

167. Hinuma, Y.; Nagata, K.; Hanaoka, M.; Nakai, M.; Matsumoto, T.; Kinoshita, K.I.; Shirakawa, S.; Miyoshi, I. Adult T-cell leukemia: Antigen in an ATL cell line and detection of antibodies to the antigen in human sera. *Proc. Natl. Acad. Sci. USA* **1981**, *78*, 6476–6480.

168. Poiesz, B.J.; Ruscetti, F.W.; Gazdar, A.F.; Bunn, P.A.; Minna, J.D.; Gallo, R.C. Detection and isolation of type C retrovirus particles from fresh and cultured lymphocytes of a patient with cutaneous T-cell lymphoma. *Proc. Natl. Acad. Sci. USA* **1980**, *77*, 7415–7419.

169. Yoshida, M.; Miyoshi, I.; Hinuma, Y. Isolation and characterization of retrovirus from cell lines of human adult T-cell leukemia and its implication in the disease. *Proc. Natl. Acad. Sci. USA* **1982**, *79*, 2031–2035.

170. Watanabe, T.; Seiki, M.; Yoshida, M. HTLV type I (U. S. Isolate) and ATLV (Japanese isolate) are the same species of human retrovirus. *Virology* **1984**, *133*, 238–241.

171. Kwok, S.; Kellogg, D.; Ehrlich, G.; Poiesz, B.; Bhagavati, S.; Sninsky, J.J. Characterization of a sequence of human T cell leukemia virus type I from a patient with chronic progressive myelopathy. *J. Infect. Dis.* **1988**, *158*, 1193–1197.

172. Mochizuki, M.; Tajima, K.; Watanabe, T.; Yamaguchi, K. Human T lymphotropic virus type 1 uveitis. *Br. J. Ophthalmol.* **1994**, *78*, 149–154.

173. Satake, M.; Yamaguchi, K.; Tadokoro, K. Current prevalence of HTLV-1 in Japan as determined by screening of blood donors. *J. Med. Virol.* **2012**, *84*, 327–335.

174. Kishihara, Y.; Furusyo, N.; Kashiwagi, K.; Mitsutake, A.; Kashiwagi, S.; Hayashi, J. Human T lymphotropic virus type 1 infection influences hepatitis C virus clearance. *J. Infect. Dis.* **2001**, *184*, 1114–1119.

175. Proietti, F.A.; Carneiro-Proietti, A.B.; Catalan-Soares, B.C.; Murphy, E.L. Global epidemiology of HTLV-I infection and associated diseases. *Oncogene* **2005**, *24*, 6058–6068.

176. Hlela, C.; Shepperd, S.; Khumalo, N.P.; Taylor, G.P. The prevalence of human T-cell lymphotropic virus type 1 in the general population is unknown. *AIDS Rev.* **2009**, *11*, 205–214.

177. Li, H.C.; Biggar, R.J.; Miley, W.J.; Maloney, E.M.; Cranston, B.; Hanchard, B.; Hisada, M. Provirus load in breast milk and risk of mother-to-child transmission of human T lymphotropic virus type I. *J. Infect. Dis.* **2004**, *190*, 1275–1278.

178. Mueller, N.; Okayama, A.; Stuver, S.; Tachibana, N. Findings from the Miyazaki cohort study. *J. Acquir. Immune Defic. Syndr. Hum. Retrovirology : Off. Publ. Int. Retrovirology Assoc.* **1996**, *13 Suppl 1*, S2–S7.

179. Stuver, S.O.; Tachibana, N.; Okayama, A.; Shioiri, S.; Tsunetoshi, Y.; Tsuda, K.; Mueller, N.E. Heterosexual transmission of human T cell leukemia/lymphoma virus type I among married couples in southwestern Japan: An initial report from the Miyazaki cohort study. *J. Infect. Dis.* **1993**, *167*, 57–65.

180. Manns, A.; Murphy, E.L.; Wilks, R.; Haynes, G.; Figueroa, J.P.; Hanchard, B.; Barnett, M.; Drummond, J.; Waters, D.; Cerney, M.; *et al.* Detection of early human T-cell lymphotropic virus type I antibody patterns during seroconversion among transfusion recipients. *Blood* **1991**, *77*, 896–905.

181. Jones, K.S.; Petrow-Sadowski, C.; Huang, Y.K.; Bertolette, D.C.; Ruscetti, F.W. Cell-free HTLV-1 infects dendritic cells leading to transmission and transformation of CD4(+) T cells. *Nat. Med.* **2008**, *14*, 429–436.

182. Hanon, E.; Stinchcombe, J.C.; Saito, M.; Asquith, B.E.; Taylor, G.P.; Tanaka, Y.; Weber, J.N.; Griffiths, G.M.; Bangham, C.R. Fratricide among CD8(+) T lymphocytes naturally infected with human T cell lymphotropic virus type I. *Immunity* **2000**, *13*, 657–664.

183. Hoxie, J.A.; Matthews, D.M.; Cines, D.B. Infection of human endothelial cells by human T-cell leukemia virus type I. *Proc. Natl. Acad. Sci. USA* **1984**, *81*, 7591–7595.

184. Derse, D.; Hill, S.A.; Lloyd, P.A.; Chung, H.; Morse, B.A. Examining human T-lymphotropic virus type 1 infection and replication by cell-free infection with recombinant virus vectors. *J. Virol.* **2001**, *75*, 8461–8468.

185. Lambert, S.; Bouttier, M.; Vassy, R.; Seigneuret, M.; Petrow-Sadowski, C.; Janvier, S.; Heveker, N.; Ruscetti, F.W.; Perret, G.; Jones, K.S.; *et al.* HTLV-1 uses HSPG and neuropilin-1 for entry by molecular mimicry of VEGF165. *Blood* **2009**, *113*, 5176–5185.

186. Pique, C.; Jones, K.S. Pathways of cell-cell transmission of HTLV-1. *Front. Microbiol.* **2012**, *3*, e378.

187. Pique, C.; Pham, D.; Tursz, T.; Dokhelar, M.C. Human T-cell leukemia virus type I envelope protein maturation process: Requirements for syncytium formation. *J. Virol.* **1992**, *66*, 906–913.

188. Jones, K.S.; Lambert, S.; Bouttier, M.; Benit, L.; Ruscetti, F.W.; Hermine, O.; Pique, C. Molecular aspects of HTLV-1 entry: Functional domains of the HTLV-1 surface subunit (SU) and their relationships to the entry receptors. *Viruses* **2011**, *3*, 794–810.

189. Pinon, J.D.; Klasse, P.J.; Jassal, S.R.; Welson, S.; Weber, J.; Brighty, D.W.; Sattentau, Q.J. Human T-cell leukemia virus type 1 envelope glycoprotein gp46 interacts with cell surface heparan sulfate proteoglycans. *J. Virol.* **2003**, *77*, 9922–9930.

190. Jones, K.S.; Fugo, K.; Petrow-Sadowski, C.; Huang, Y.; Bertolette, D.C.; Lisinski, I.; Cushman, S.W.; Jacobson, S.; Ruscetti, F.W. Human T-cell leukemia virus type 1 (HTLV-1) and HTLV-2 use different receptor complexes to enter T cells. *J. Virol.* **2006**, *80*, 8291–8302.

191. Jones, K.S.; Petrow-Sadowski, C.; Bertolette, D.C.; Huang, Y.; Ruscetti, F.W. Heparan sulfate proteoglycans mediate attachment and entry of human T-cell leukemia virus type 1 virions into CD4+ T cells. *J. Virol.* **2005**, *79*, 12692–12702.

192. Tanaka, A.; Jinno-Oue, A.; Shimizu, N.; Hoque, A.; Mori, T.; Islam, S.; Nakatani, Y.; Shinagawa, M.; Hoshino, H. Entry of human T-cell leukemia virus type 1 is augmented by heparin sulfate proteoglycans bearing short heparin-like structures. *J. Virol.* **2012**, *86*, 2959–2969.

193. Ghez, D.; Lepelletier, Y.; Lambert, S.; Fourneau, J.M.; Blot, V.; Janvier, S.; Arnulf, B.; van Endert, P.M.; Heveker, N.; Pique, C.; *et al.* Neuropilin-1 is involved in human T-cell lymphotropic virus type 1 entry. *J. Virol.* **2006**, *80*, 6844–6854.

194. Soker, S.; Takashima, S.; Miao, H.Q.; Neufeld, G.; Klagsbrun, M. Neuropilin-1 is expressed by endothelial and tumor cells as an isoform-specific receptor for vascular endothelial growth factor. *Cell* **1998**, *92*, 735–745.

195. Vander Kooi, C.W.; Jusino, M.A.; Perman, B.; Neau, D.B.; Bellamy, H.D.; Leahy, D.J. Structural basis for ligand and heparin binding to neuropilin B domains. *Proc. Natl. Acad. Sci. USA* **2007**, *104*, 6152–6157.

196. Jin, Q.; Agrawal, L.; VanHorn-Ali, Z.; Alkhatib, G. Infection of CD4+ T lymphocytes by the human T cell leukemia virus type 1 is mediated by the glucose transporter GLUT-1: Evidence using antibodies specific to the receptor's large extracellular domain. *Virology* **2006**, *349*, 184–196.

197. Manel, N.; Kim, F.J.; Kinet, S.; Taylor, N.; Sitbon, M.; Battini, J.L. The ubiquitous glucose transporter GLUT-1 is a receptor for HTLV. *Cell* **2003**, *115*, 449–459.

198. Igakura, T.; Stinchcombe, J.C.; Goon, P.K.; Taylor, G.P.; Weber, J.N.; Griffiths, G.M.; Tanaka, Y.; Osame, M.; Bangham, C.R. Spread of HTLV-I between lymphocytes by virus-induced polarization of the cytoskeleton. *Science* **2003**, *299*, 1713–1716.

199. Barnard, A.L.; Igakura, T.; Tanaka, Y.; Taylor, G.P.; Bangham, C.R. Engagement of specific T-cell surface molecules regulates cytoskeletal polarization in HTLV-1-infected lymphocytes. *Blood* **2005**, *106*, 988–995.

200. Hildreth, J.E.; Subramanium, A.; Hampton, R.A. Human T-cell lymphotropic virus type 1 (HTLV-1)-induced syncytium formation mediated by vascular cell adhesion molecule-1: Evidence for involvement of cell adhesion molecules in HTLV-1 biology. *J. Virol.* **1997**, *71*, 1173–1180.

201. Daenke, S.; McCracken, S.A.; Booth, S. Human T-cell leukaemia/lymphoma virus type 1 syncytium formation is regulated in a cell-specific manner by ICAM-1, ICAM-3 and VCAM-1 and can be inhibited by antibodies to integrin beta2 or beta7. *J. Gen. Virol.* **1999**, *80*, 1429–1436.

202. Takenouchi, N.; Jones, K.S.; Lisinski, I.; Fugo, K.; Yao, K.; Cushman, S.W.; Ruscetti, F.W.; Jacobson, S. GLUT1 is not the primary binding receptor but is associated with cell-to-cell transmission of human T-cell leukemia virus type 1. *J. Virol.* **2007**, *81*, 1506–1510.

203. Nejmeddine, M.; Barnard, A.L.; Tanaka, Y.; Taylor, G.P.; Bangham, C.R. Human T-lymphotropic virus, type 1, tax protein triggers microtubule reorientation in the virological synapse. *J. Biol. Chem.* **2005**, *280*, 29653–29660.

Papillomavirus Infectious Pathways: A Comparison of Systems

Jennifer Biryukov and Craig Meyers

Abstract: The HPV viral lifecycle is tightly linked to the host cell differentiation, causing difficulty in growing virions in culture. A system that bypasses the need for differentiating epithelium has allowed for generation of recombinant particles, such as virus-like particles (VLPs), pseudovirions (PsV), and quasivirions (QV). Much of the research looking at the HPV life cycle, infectivity, and structure has been generated utilizing recombinant particles. While recombinant particles have proven to be invaluable, allowing for a rapid progression of the HPV field, there are some significant differences between recombinant particles and native virions and very few comparative studies using native virions to confirm results are done. This review serves to address the conflicting data in the HPV field regarding native virions and recombinant particles.

Reprinted from *Viruses*. Cite as: Biryukov, J.; Meyers, C. Papillomavirus Infectious Pathways: A Comparison of Systems. *Viruses* **2015**, *7*, 4303–4325.

1. Introduction

Human papillomaviruses are the etiologic agent of cervical cancer and other anogenital and oral cancers [1–4]. To date, over 150 types have been identified. Cervical cancer is the third most common cancer in women worldwide, causing a quarter of a million deaths per year [5]. All HPV types replicate within epithelium, however, they are subdivided based on their ability to infect either mucosal or cutaneous keratinocytes. HPVs that infect mucosal keratinocytes are further sub-divided into low-risk and high-risk types. Low-risk HPVs cause benign lesions such as condylomas or warts while high-risk types cause malignant neoplasms such as cervical cancer [6,7]. The most significant risk of developing cervical cancer is infection with a high-risk type. High-risk types such as HPV16, HPV18, HPV31, HPV45, and HPV58 have a strong link with malignant progression [6–8].

HPV virions consist of an icosahedral capsid containing histone-associated dsDNA. The viral genome is circular and approximately 8000 bp in length. There is an average of 8 open reading frames (ORFs), which are divided into early and late gene expression classes. HPV replication is tightly linked to the differentiation program of host epithelial cells. Infection occurs in basal epithelial cells via microabrasions. Once infected, early non-structural proteins involved in activities such as genome maintenance and transcription activation are expressed. As the daughter cells divide and become increasingly differentiated in the suprabasal layers, viral genomes are amplified, the late structural proteins are expressed, and new virions assembled [9–13].

Because the HPV life cycle is tightly linked to the host cell differentiation, the virus has been hard to grow in culture. Therefore, systems to create viral particles bypassing epithelial differentiation were developed. Recent research studying the HPV life cycle including infectivity, transmission, and viral structure predominantly relies on the use of recombinant papillomavirus particles. These recombinant particles such as virus-like particles (VLPs), pseudovirions (PsVs), and quasivirions

(QVs) all bypass the need for stratifying and differentiating human epithelium. While data collected using recombinant particles has allowed for a rapid progression of the field of HPV research, there are few comparative studies utilizing native virions produced in organotypic raft culture, xenografts, or native tissue to confirm results in the context of stratifying and differentiating human epithelial tissue.

The main focus of this review will be to compare native virions produced in stratifying epithelium to recombinant particles (Table 1). All aspects of the viral life cycle including particle synthesis, maturation, attachment/entry, and infectivity will be evaluated. While much of the data generated utilizing recombinant particles has served to advance the field, there are many instances of conflicting results when compared to data generated utilizing native virions. A fraction of the conflicting data is likely due to utilization of different cell lines, particle maturation, and the multiplicity of infection used. Additionally, much of the research has been done using recombinant HPV16 with the assumption that what holds true for one HPV type is true for all HPV types. However, some recent data suggests that this may not be the case [14,15]. We hope to highlight not only the necessity for confirmatory studies but also the importance of studying more than one HPV type.

Table 1. A comparison of the advantages and disadvantages of the different types of papillomavirus particles.

HPV Particle Type	Method of Production	Particle and System Advantages	Particle and System Disadvantages
Native Virions	Organotypic raft culture system	Contains the full HPV capsid as well as full HPV genome. Assembles and matures over a period of 10–20 days in tissue	Expensive. Slow production time
Virus Like Particles (VLPs)	Transfection based system	Quick and inexpensive particle production	Codon modification of L1 and L2 is necessary for production. Over-expression system. Particles produced in non-relevant cell lines. Particle assembly within 48 hours. Maturation occurs overnight. Deletion of in frame methionines upstream of the consensus methionine in L1. VLPs contain no genomes, which may alter particle structure
Pseudovirions (PsVs)	Transfection based system	Quick and inexpensive particle production. Ability to track cellular infectivity	
Quasivirions (QVs)	Transfection based system	Quick and inexpensive particle production. Contains L1, L2, and a full HPV genome. Closest to NVs—retains majority of cell surface exposed conformational epitopes	

2. Papillomavirus Particle Production

Organotypic raft culture is the only system outside of the xenograft system that allows for the production of infectious HPV in its natural environment, a differentiating epithelium [9,16–18]. This *in vitro* system preserves the molecular events that occur during differentiation with the convenience of not having to use animals such as in the xenograft system. In the organotypic system, human keratinocytes that stably maintain HPV viral genomes are grown above a dermal equivalent made of fibroblast cells embedded in a collagen matrix. Keratinocytes grow at the air liquid interface and are fed via diffusion from cell culture media below the dermal equivalent [9]. After a period of 10–20 days, tissue is harvested and homogenized to generate a virus stock (Figure 1). This system allows for the study of genetic mutants in the natural life cycle as genomes can be mutated, electroporated into keratinocytes, and grown in raft culture for the production of mutant virions. One drawback to using this system is the increased cost for virion production. Additionally, instead of producing particles in a matter of days, the raft culture system takes 3–4 weeks. Overall, this system allows researchers to examine, temporally and spatially, all parts of the viral life cycle including viral genome amplification, late gene expression, virion assembly, maturation, and infectivity.

The increased cost and technical challenges associated with acquiring virions from either an *in vivo* or *in vitro* system created a need for the development of the recombinant system (Figure 2). The basis of recombinant particle production comes from the fact that expression of either L1 alone or L1 and L2 together results in the self-assembly of proteins into viral capsids called VLPs [19–25]. For production of PsVs, in addition to the capsid proteins, a reporter plasmid encoding GFP or other reporter is co-transfected to function as a mock genome. These particles allow for researchers to easily track cellular infectivity [26,27]. Finally, when the full HPV genome is transfected into cells along with expression vectors for the capsid proteins, the particles are called QVs [26,27]. VLPs, PsVs, and QVs are assembled over a period of 48 h in monolayer culture and do not go through the natural maturation process. QVs represent the closest resembling recombinant particle to native virions, resembling NVs when compared by cryoelectron microscopy. The QV particles also retain the majority of surface exposed conformational dependent epitopes [28–32]. It is unclear if there are any structural differences between recombinant particles and NVs that might affect the biology of the virus. Due to the ease and low cost of production, recombinant particles remain the main tool used to study HPV structure, assembly, entry, and infectivity.

While the use of recombinant particles allows for rapid collection of data, there are some caveats to this system of particle production that should be considered. The first being that expression plasmids utilized to synthesize L1 and L2 have been codon optimized to remove rare codons and negate any possible limitations in protein expression [33,34]. While this does not change the protein sequence, this can affect the speed at which proteins are translated, leading to an increased chance of errors in the final amino acid sequence, or could affect protein folding [35]. Additionally, expression of proteins in non-keratinocytes, and especially in prokaryotic cells, could lead to a change in the post translational modifications in the protein. When comparing HPV6b L1 produced using a recombinant baculovirus in Sf9 insect cells or using a recombinant vaccinia virus in kidney cells, differing post-translation modifications were observed. Specifically, L1 produced in the Sf9 cells presented with several post-translationally modified variants, both threonine and serine residues were phosphorylated compared to just serine residues in the L1 produced in the kidney cells, and Sf9 produced L1 incorporated with both mannose and galactose whereas kidney cell produced L1 incorporated with only galactose [36]. While there are likely differences, studies to look at the differences in post-translational modifications between raft-culture derived virions and recombinant particles have not been done and the possibility of these differences should be kept in mind when drawing conclusions from experiments.

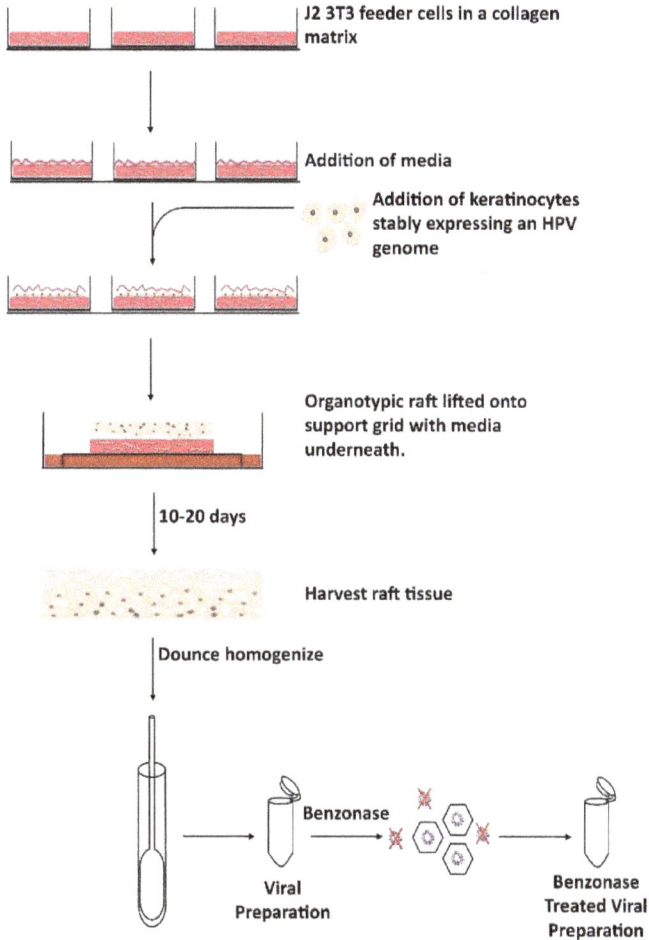

Figure 1. Schematic of native HPV virion production. HPV-positive cells are seeded on top of collagen plugs comprised of collagen and J2 3T3 feeder cells. Once the cells are have grown to confluence, the plugs are lifted onto a support grid. The rafts are then fed via diffusion from media underneath the grid. After 10–20 days, the tissue is harvested, homogenized, and benzonase treated to get a final viral preparation.

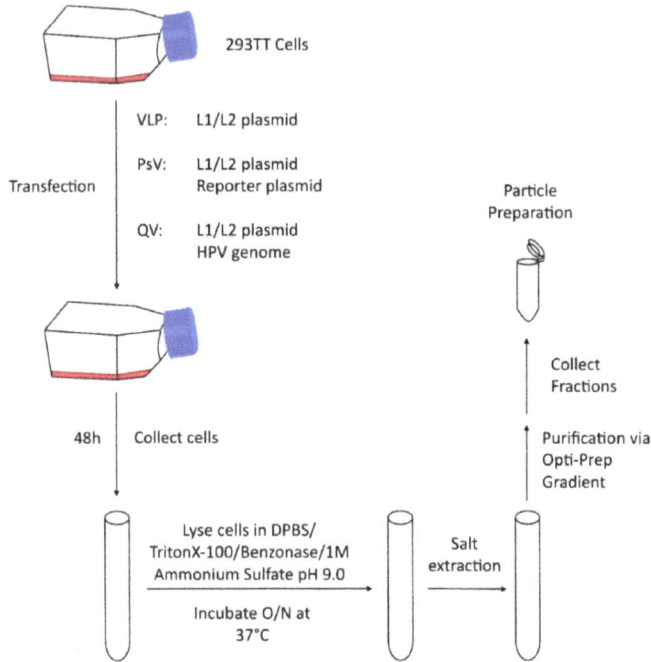

Figure 2. Schematic of recombinant HPV particle production. 293TT cells are transfected with a L1/L2 plasmid (VLPs), a L1/L2 plasmid and a reporter plasmid (PsVs), or a L1/L2 plasmid and a full HPV genome (QVs). Forty-eight hours post transfection, cells are harvested, lysed, and incubated at 37 °C overnight in the presence of cell lysate. Particles are then purified on a gradient.

In the recombinant particle system, the L1 in all recombinant particles begins at a consensus methionine that was found when aligning the L1 N-terminal open reading frame (ORF) of many papillomaviruses. However, some HPV types, such as HPV16 and HPV18 actually have upstream, in frame, methionines that allow for the production of more than one L1 species. Specifically, HPV16 has one upstream methionine that allows for production of an L1 26 amino acids longer then L1 started at the consensus methionine and HPV18 has two upstream methionines enabling production of L1 proteins that are 61 and 26 amino acids longer then the L1 produced from the consensus methionine. These methionines were removed from the expression plasmids utilized in the recombinant system because only production of L1 from the consensus methionine allowed for efficient expression of L1 and production of particles [37]. A HPV18 NV 35 amino acid deletion mutant, deleting the region from the upstream methionine to the consensus methionine, was produced and tested for neutralization with two conformational dependent antibodies created using HPV18 VLPs—H18.K2 and H18.J4. The deletion mutant loses the ability to be neutralized with the H18.K2 antibody but is still neutralized by the conformation dependent H18.J4 antibody. The H18.K2 antibody was produced against VLPs, and therefore, the binding site cannot be contained in the 35 upstream amino acids. This suggests that the 35 upstream amino acids induce structural changes within the HPV18 capsid. When HPV16 virions produced in organotypic raft culture are run on an SDS-PAGE gel, two bands for L1 can be detected on the western blot [38–41]. In contrast, western blots of HPV16 QV appear to only have one form of L1 incorporated into virions, as only one band is visible [38]. This suggests both a potential conformational and structural difference between capsids produced in differentiated epithelium compared to those produced in monolayer cells. While the ability to quickly and easily produce infectious recombinant particles allows for quick experimentation, it is becoming clear that any important results generated should be verified utilizing native particles derived from differentiating epithelium.

3. Maturation and Assembly of Viral Particles

During the process of developing into fully infectious virions, many viruses undergo structural changes, called maturation [42,43]. In the raft culture system, virions mature over a period of 10–20 days within the tissue. In contrast, recombinant particles are matured over a period of 24 h after they are released from cells. After 10 days in foreskin epithelial raft culture, encapsidated and infectious virions are present, however, given another 10 days (20-day tissue) the virions are more mature as displayed by an increase in stability, defined by an enhanced resistance to stress imposed by fractionation in an ultracentrifuge [38]. This is similar to what has been seen for PsVs, with mature capsids having an increased stability [44]. Matured PsVs have also been shown to have an increased resistance to trypsin digestion and chemical reduction, as well as a more ordered structure when viewed by transmission electron microscopy (TEM) [44]. In addition to a more stable and mature phenotype, 20-day native virions were twice as infectious as 10-day virions and had an increased sensitivity to antibody neutralization by both L1 and L2 antibodies [38]. Both immature and mature PsV capsids were equally neutralized by many L1 and L2 antibodies [44]. This suggests that, for NV, the particles may be changing their exposure of key epitopes during the assembly and maturation process.

Analysis of both 10-day and 20-day tissue sections show a difference in virion localization with 10-day virions predominantly localized in the nuclei of suprabasal cells and 20-day virions being found mainly in the cornified layers of the epithelium. Human epithelium has a natural redox gradient with the lower part of the tissue being a reducing environment and the upper layers being an oxidizing environment [38]. This natural redox gradient coincides with the location of immature and mature virions, leading to the conclusion that virion maturation occurs when virions move from a reducing environment to an oxidizing environment. When an oxidizing agent, oxidized glutathione (GSSG) was added to the tissue during virion production, 10-day virions exhibited increased maturation [38,45]. This suggests that movement to an oxidizing environment induces changes in the viral capsid, converting it to a more stable and mature phenotype. PsVs are matured in an oxidizing environment in the presence of cellular lysate after being released from cells. Incubation of particles with cellular lysate, but not clarified lysate produced more mature particles, indicating that currently unknown cellular factors, the same of which are presumably present in tissue, are required for maturation as well [46].

While not absolutely essential for particle formation, a series of disulfide bonds in L1 and L2 molecules form both intra- and interpentameric interactions, stabilizing particles and making them more resistant to environmental influences from nucleases and proteases [44,47]. Within the epithelium, disulfide bond formation is thought to happen slowly during virion maturation in an oxidizing environment [38,44,47,48]. There are 12 conserved cysteine residues across L1 from multiple HPV types [48]. Evaluation of bovine papillomavirus (BPV) via cryoelectron microscopy showed that two separate disulfide bonds are present. However, only one of these is conserved among the human papillomaviruses [49]. Native and recombinant particles have both been reported to contain disulfide bonds. In HPV16, the disulfide bonding occurs at C175 and C428, forming an interpentameric bond between two neighboring capsomeres [50]. Based on mass spectrometry data for HPV18 VLPs, the homologous cysteines that participate in disulfide bonding are C175 and C429. Additionally, both genetic and biochemical analysis of HPV16 and HPV33 PsVs have shown the importance of these conserved cysteines in the structural integrity of the particles as mutation of the cysteines in HPV16 PsV prevents maturation of the particles [44,48]. Mutation and analysis of all 12 conserved cysteines in HPV16 VLPs identified three cysteines potentially involved in disulfide bonding—C175, C185, and C428. Mutation of C428 was especially detrimental to the capsid formation as only capsomeres were visible by EM. The C175S mutation produced tube like structures and the C185S mutation made smaller capsids [48]. These cysteine mutations were also evaluated in the context of NVs grown in differentiating epithelium [45]. Only the C175S mutation severely reduced production of infectious virions in both 10-day and 20-day tissue while C428S, C185S, and the C175S, C185S double mutation only affected 20-day virions. In contrast to the C428S PsV mutant, particle formation was observed for the C428S NV mutant. This suggests that in the context of differentiating epithelium, these cysteines are important for the formation of mature NV. Other conserved cysteines in HPV16 that do not play a role in interpentameric disulfide bonding were also evaluated. In HPV16 VLPs, cysteine to serine mutations of C161, C229, and C379, produced particles with a high susceptibility to tryptic proteolysis [48]. The same cysteine to serine substitutions in HPV16 NV hindered the accumulation of endonuclease resistant genomes in a stratifying epithelium with only the C229S mutant forming non-infectious virions. This suggests that these cysteines may be involved in forming transient disulfide bonds early in the assembly process to guide the capsid to its mature form [51]. Determining the precise roles of these cysteines in the HPV capsid may lead to be much better understanding of HPV capsid stability.

In addition to disulfide bonds in L1, L2 plays an important role in capsid formation and stabilization. L2 forms heterotypic interactions with L1 on the inside of the capsid as well as homotypic interactions with other L2 molecules. When L1 disulfide bonds were unable to form, L1 only capsomeres did not assemble into capsids. However when L2 was present, VLPs were formed [52]. Within the L2 terminus exists two highly conserved cysteines which, when mutated in PsVs, leads to production of non-infectious particles [53,54]. The importance of the L2 cysteines in particle formation was evaluated in the context of NVs produced in differentiating epithelium [55]. In contrast to previous studies using HPV16 PsV and QV, when one or both of the cysteines were mutated, infectious NVs were produced in both 10-day and 20-day tissue [53,54]. Evaluation of the mutant virions produced compared to wild-type showed that the mutants were more infectious but less stable. This could be due to enhanced presentation of a favored binding site on the capsid or a more effective release of viral genomes after host entry due to reduced capsid stability [53]. These L2 cysteines are in close proximity of the proposed external loop, which has been implicated as a feasible candidate for a cross neutralizing epitope across many HPV types for vaccine development. This loop is likely exposed in the final stages of virion maturation as only 20-day virions more efficiently neutralized with the L2 specific antibody RG-1 [38]. Other antibodies targeting this loop also only efficiently neutralized 20-day virions [38,39]. This data provides evidence for the importance of L2 in capsid stability and highlights a significant structural difference between virions assembled in monolayer *vs.* differentiating epithelium.

One of the remaining unknowns of HPV assembly and structure is the concentration of L2 in the viral capsid. L1 itself is able to self-assemble into capsids, however, including L2 into recombinant particles has been shown to increase yield, capsid stability, DNA encapsidation, and infectivity of particles [31,39,52,56–58]. Cryoelectron microscopy of native BPV1 suggests that there are 12 copies of L2 per capsid—one at the center of each of the pentameric capsomeres [59]. However SDS-PAGE analysis from native HPV1 particles, co-immunoprecipitation of HPV11 L1 pentamers with L2, as well as cryoelectron microscopy of HPV16 PsV suggests that up to 72 copies of L2 are present with one being at the center of each capsomere [59,60]. In the natural system, there is regulation as to the quantity of L1 and L2 produced. However, in an over expression system such as the ones used to produce recombinant particles, there would be no regulation of protein production and thus the ratio of L1 and L2 available for particle formation is likely skewed. This could allow for additional L2 to be incorporated into recombinant particles. Additional studies to identify the number of copies of L2 in native particles need to be done.

Organotypic raft culture allows for the generation of not only single base mutant virions but also for the generation of chimeric HPVs composed of different components of different HPV types in the natural environment of differentiating tissue. This whole gene replacement method has aimed at highlighting areas of conservation among different HPV types via complementation experiments. A chimeric virus, named HPV18/16, was created with the L1 and L2 ORF of HPV16 swapped into HPV18 in place of HPV18 L1 and L2 [61]. Substitution of both late gene ORFs did not affect viral genome maintenance in cells and raft cultures were found to have late gene functions such as capsid gene expression and virion morphogenesis. Additionally, chimeric virions purified from raft culture were able to infect keratinocytes [61]. As expected, the chimeric HPV18/16 was unable to be neutralized by HPV18 polyclonal antiserum but was neutralized by a HPV16 polyclonal antiserum. Additional chimeras created by swapping the L1s and L2s of more genetically diverse HPV types such as HPV45, 39, 31, 33, 11, 6b, 1a, CRPV, and BPV1 were also engineered. All the chimeras produced infectious virions, however, titers were much lower then those of wild-type HPV18 [62]. These data demonstrate that there are conserved biological functions between different HPV types. Similar studies illustrate that both capsid proteins play a role in the structure of each other and that this can ultimately affect the overall conformation and behavior of the virion [62,63]. Obviously, both L1 and L2 play an integral role in the biology of HPV.

In addition to L1 and L2, a variety of components comprise the viral capsid including viral DNA and histones. There are other possible viral and cellular components that either directly or indirectly affect the final structure and/or yield of capsids. These components could include HPV early proteins such as E2, E4, E5, and E7 or cellular proteins such as chaperones or karyopherins. There are undoubtedly unknown viral and/or cellular factors necessary for the assembly of native virions. These proteins could act by direction of proper subcellular localization, initial interaction of capsid proteins, mediation of the correct temporal formation of disulfide bonds, and regulation of capsid protein expression.

4. Attachment and Internalization

Our current understanding of the PV infection strategy proposes that, via microabrasions, virus particles gain access to the basal cells as well as the basement membrane. Utilizing recombinant particles, most papillomaviruses have been observed to infect cells by first binding to a glycosaminoglycan (GAG), heparan sulfate (HS), via L1 on either the basement membrane or the cell surface [64–67]. This binding event induces a conformational change allowing for the L2 N-terminus to be cleaved by a proprotein convertase (PC) such as furin or PC5/6 [68,69]. Additionally, these conformational changes may allow for interactions with secondary HS sites on the capsid, transfer the virion to an uptake receptor or receptor complex, and allow for exposure of hidden epitopes that may be important for the interaction with other cellular proteins in the entry process [69–71]. The involvement of GAGs as an attachment receptor was first demonstrated utilizing HPV11 L1 only VLPs and HaCaT cells. Not only was HPV shown to bind, but experiments suggested that L1 interactions with GAGs were GAG type specific [72]. Shortly thereafter, dependence on HS engagement was shown for both HPV16 and HPV33 PsVs [66]. However, in contrast to what was seen for HPV11 VLPs, both HPV16 and HPV33 PsV attachment was to a different subset of GAG molecules. This suggests either a difference between VLPs and PsVs or a difference between HPV types. While primary attachment to heparan sulfate proteoglycans (HSPGs) has been suggested to be a universal step for all PVs, noticeably, tissue derived HPV31 NV and HPV16 NV attachment to and infection of human keratinocytes was shown to be able to occur in the absence of HSPGs [15,73]. HPV31 NV infection of HaCaT cells, n-TERT-1 cells, and primary keratinocytes was unable to be blocked by exogenous heparin competition and could not be blocked by the enzymatic removal of cell surface HS. However, infection of Cos-7 cells and transformed C-33A cells was efficiently blocked by both methods [73]. These data indicate that the virus may use more than one receptor and/or entry pathway to enter a host cell depending on the cell type [73]. It is interesting to note that contradictory findings were observed for HPV31, where HPV31 NV was not dependent on HSPG for attachment, and HPV31 PsV was shown to be dependent on HS for attachment and infection of cells [67]. To complicate the story more, a recent report suggests that both HPV16 and HPV31 PsV attachment and infection can be independent of HSPGs if the non-HSPG molecule laminin-332 is present on the cell surface [14]. Possible explanations for these contradictions include the HPV type being studied, a possible structural variation between NV and PsV due to differences in maturation and assembly, and the cell types being utilized.

Due to minor structural differences between different virus types, it is possible that initial attachment to and interaction with the ECM and cell surface might be HPV type specific. Though the

L1 protein displays overall sequence homology and structural conformation, both structural diversity and conformational differences in the loop structures of the pentamer surface have been demonstrated for HPV types 11, 16, 18 and 35 [74]. It is feasible that minor structural differences might have an impact on receptor engagement, including binding to attachment receptors as well as putative entry receptors. In a study looking at VLP binding for four different HPV types to the ECM and cell surface, some diversity was demonstrated [75]. Two more recent studies, one utilizing native virions [15] and one utilizing PsVs [14] also found HPV type dependent differences in ECM and cell surface binding. It is likely that slightly varying structures of different HPV types facilitate a preferred attachment to a specific GAG/GAG modification, laminin-332, or other unidentified receptors to initially bind to cells and trigger subsequent entry events. For the closely related polyomavirus, different types have been shown to utilize different receptors for infection [76–78]. Thus, a general hypothesis for attachment that encompasses all HPV types is not probable.

The entry process for HPV is slow, with cell surface events such as interaction with several receptors and conformational changes thought to be responsible [79,80]. HPV16 PsV and HPV18 PsV infection have been reported to have an average entry half time of 12 h in HaCaT cells [66,81]. Similarly, entry time for HPV31 NV as well as HPV31 QV was reported to be slow with a half time of about 14 h. In contrast, HPV16 QV was comparably rapid with an average entry half time of only 4 h in HaCaT cells [82]. This data suggests differences in HPV infections to be both HPV type and model system dependent. Several reports suggest that post engagement with cell surface HSPGs, HPV16 PsV particles require interaction with one or more non-HSPG receptors for internalization [69,81,83]. Post-attachment, there is initial co-localization of capsids with HSPGs on the cell surface. However, this co-localization is lost as the particles are internalized into intracellular vesicles, suggesting that particles have been transferred to a non-HSPG receptor [81]. Also giving strong evidence for the existence of other receptors is that furin pre-cleaved virions can bind to and infect HSPG negative cells [83]. Using biochemical inhibitors of endocytosis, it was suggested that HPV16 PsV and HPV58 PsV were internalized via clathrin-mediated endocytosis in COS-7 cells, while HPV31 PsV internalization was dependent on caveolae [84]. This is in contrast to a more recent study whereby HPV31 PsV was also reported to use clathrin mediated endocytosis in both COS-7 and 293TT cells [85]. When utilizing QV, caveolae mediated entry was demonstrated for HPV31 QV and HPV16 QV entry was again suggested to be mediated by clathrin coated pits [82]. HPV16 PsV infection has also been shown to be dependent on entry via caveolin-1 followed by particles trafficking to the ER upon entering cells [86,87]. In another system, this time using dominant negative inhibitors and siRNA knockdown in HeLa, 293TT, and HaCaT cells, the entry pathway of HPV16, HPV18, and HPV31 PsVs was described as a "clatharin, caveolin, lipid raft, flotillin, cholesterol, and dynamin independent mechanism distinct from macropinocytosis" [88,89]. The pathway was defined by a requirement for actin polymerization and tetraspanin microdomains with all particles trafficking to the late endosomal compartment with similar kinetics [89]. Taking into account the multiple HPV types studied, the virus system and cell type utilized, as well as the experimental approach taken, multiple pathways have been identified as playing a role in HPV entry. Considering the number of diverse HPV types, their virus tropisms, and the various HPV related diseases, it is possible that there is both overlapping and non-overlapping receptor usage that could feasibly shunt different

HPV types into different entry pathways depending on cell type. It is also important to consider that when attachment assays are done utilizing recombinant particles, often a high multiplicity of infection (MOI) is used—typically in the thousands—which could be shunting particles into cells via pathways the virus would not typically use. In contrast, work with native virions typically uses MOIs of less than 100. Side-by-side comparisons should be done prior to making broad conclusions regarding the mechanism used for infectious entry. A more detailed analysis of the various HPV types using PsV, QV, and NV in relevant keratinocyte cell lines as well as in primary keratinocytes would both show similarities and highlight differences between the types that would help settle the growing amount of conflicting data in the literature.

To date, a multitude of receptors have been identified as potential primary and secondary HPV internalization receptors. One of the first identified was α6-integrin, an epithelial adhesion protein [90]. An α6-integrin specific monoclonal antibody was able to block HPV6b L1 only VLP binding to the cell surface by 60%. Additionally, HPV16 L1 only VLP binding correlated with α6-integrin in a study of 10 different cell lines [91]. However, cells deficient in α6-integrin can bind HPV16 L1 only VLPs and HPV11 NV infects α6-integrin negative cells at levels similar to an α6-integrin expressing cells [92]. In contrast to what is seen for VLPs, β4-integrin along with α6-integrin processing was found to be essential for infection of HaCaT cells with HPV16 PsV [93]. While it appears not to be absolutely essential for all HPV types, α6-integrin may contribute to infection efficiency, especially with its involvement in the Ras/MAPK and PI3K intracellular signaling pathways that are activated upon binding, which may function to create a more proliferative cellular environment to support infection [94]. A role for growth factor receptors (GFRs), both epidermal growth factor receptor (EGFR) and keratinocyte growth factor receptor (KGFR), has also been suggested [95]. Though not yet demonstrated, it is proposed that a HPV-HSPG complex binds directly to GFRs. While blocking EGFR inhibits infection of HaCaT cells with both HPV16 PsV, HPV31 PsV and HPV31 QV, the actual role for EGFR has yet to be determined [88,95]. It is possible that EGFR is directly involved in entry or it is possible that EGFR is activated and involved only in a signaling cascade. Another candidate, Annexin-A2 is a calcium and phospholipid binding protein that is expressed on the cell surface as a part of a heterotetramer, A2t, consisting of two Annexin-A2 monomers associated with a S100A10 dimer [96]. Both HPV16 VLPs and HPV16 PsVs directly interact with A2t, as shown by co-immunoprecipitation. Knockdown of Annexin-A2 by shRNA yields a significant decrease in internalization by HPV16 VLPs and infection by 16 PsV. Further, mutation of a site in L1 shown to be important for A2t binding reduces both binding and infection of HPV16 PsV [97]. HPV16 PsV is still able to infect Annexin-A2 deficient HepG2 cells, suggesting the use of an alternative infectious pathway [98]. While it will be important to discover the internalization receptor(s) for HPV, there is again a necessity for the consideration of HPV type, viral system, and cell line being utilized.

5. Subsequent Steps in HPV Infection

After endocytosis, acidification of the endosomal lumen allows for disassembly of the viral capsid and cellular sorting allows for transport to the nucleus where viral replication occurs [28,65,88,99,100]. Much of the L1 is dissociated from the L2-DNA complex with the help of cyclophillins [101]. Utilizing both HaCaT and HeLa cells, it has been shown that the L1-L2-DNA complex then travels to the Golgi via the retromer, a complex involved in transport of cellular cargo from the endosome to the Golgi [102–107]. This retrograde trafficking is required for HPV infection and HPV16 L2 has been shown to bind directly to the retromer to mediate escape from the endosome [103,108]. Utilizing BPV PsV, Syntaxin 18, which mediates trafficking of vesicles between the ER and the Golgi apparatus, has also been implicated in playing a role in trafficking the L2-DNA complex to the nucleus [109,110]. The L2-DNA complex then travels through the ER, requiring γ-secretase [87]. Inhibition of γ-secretase blocks infection of HPV after endosomal exit but prior to arrival in the Golgi [87]. Two separate HPV entry siRNA screens have identified ER components that are enriched during entry [102,103]. Additionally, both knockdown of specific ER proteins and use of chemical inhibitors of ER function inhibit HPV infection [102,111]. However, there are conflicting studies regarding ER co-localization with one study reporting co-localization of HPV with ER markers during infection [86] and the other failing to observe ER co-localization [104]. Finally, the L2-DNA complex trafficks to the nucleus, gaining entry during nuclear envelope breakdown during mitosis [27,102,112]. The L2 minor protein is essential for infection, as it facilitates transport of the HPV genome to the nucleus. Specifically, a transmembrane domain along with three GxxxG motifs within L2 have been shown to be essential for infectivity [113]. Much of the research looking at viral infection has been done utilizing PsV, which harbors a non-viral pseudogenome. While this allows for a much faster method of screening, results should be confirmed utilizing native virions.

In a natural infection, HPV infects keratinocytes. However, experiments *in vitro* rarely utilize primary keratinocytes. This is in part due to the extremely low infection efficiency of recombinant particles in primary cells [83]. Conversely, native virions infect primary cells quite well. HPV31 and HPV45 have been shown to infect primary cells just as efficiently as HaCaT cells and HPV16 infects primary cells with a greater efficiency compared to HaCaT cells [15]. HPV18 infects primary cells, however, with lower efficiency compared to HaCaT cell infection. This could be due to a difference in the expression of attachment or internalization receptors present on primary cells compared to other cells. Much of the data for HPV attachment and infection, whether using recombinant particles or native virions, is done in a variety of cell lines including, but not limited to, HaCaT cells, HeLa cells, 293TT cells, and CHO cells. The lack of utilizing a consistent cell line between experiments is likely responsible for some of the conflicting data when looking at both attachment and infectivity.

Of equal importance when looking at infection is the ability to look at virus neutralization. This is especially important due to the use of VLPs in the HPV vaccine that is now being administered to prevent infection. Almost all of the antibodies generated for HPV have been made using recombinant particles. Many of these antibodies generated, such as H16.V5 and H16.E70, have been shown to neutralize both recombinant particles and native virions [16,32,114,115]. However, some antibodies shown to be highly effective in neutralizing recombinant particles, such as H16.U4, have no neutralizing capabilities against native virions [32]. Also of significance are studies highlighting the mechanisms with which different antibodies neutralize particles. For example, H16.V5 and H16.E70 allow attachment of particles to the cell surface but blocks association with the ECM as well as internalization. H16.U4, however, prevented binding to the cell surface but not to the ECM [115]. These neutralization studies again highlight the probable structural differences between recombinant particles and native virions and the importance of confirming data using native virions.

6. Translational Research

There are two approved vaccines against the high-risk HPV types HPV16 and HPV18—Gardasil and Cervarix. Gardasil also provides protection against low-risk HPV types HPV6 and HPV11. These vaccines, however, are not efficiently cross protective against other high-risk HPV types. Therefore, another HPV vaccine, Gardasil 9 was recently approved that, in addition to the four types of HPV previously mentioned also provides protection against high-risk types HPV31, HPV33, HPV45, HP52, and HPV58. While these vaccines have been shown to be highly effective at preventing infection and the development of lesions, their high cost limits their use in developing countries. Additionally, the vaccine rate is low in some developed countries. Thus, the development of less expensive microbicides that would offer protection against a multitude of HPV types would offer women an additional protection against contracting HPV infection.

Due to the importance of GAGs, specifically HS, in HPV attachment to and infection of host cells, agents targeting GAGs are of great interest. Both high molecular weight sulfated or sulfonated polysaccharides and polymers such as cellulose sulfate, dextran sulfate, and polystyrene sulfonate showed microbicidal activity against BPV1 in mouse cells and HPV11 and HPV40 in human A431 cells without showing cellular toxicity [116]. More recently, Carrageenan, a highly sulfated polysaccharide derived from red algae, was identified in an *in vitro* screen of compounds that may effectively block infection by high-risk HPV types [117]. In this study, PsVs were used to determine whether there was a block in infection. However, when tested with NV, carrageenan failed to inhibit infection by HPV16 at concentrations up to 100 µg/mL. In contrast, significant levels of inhibition were observed at concentrations as low as 1 µg/mL for native HPV18. The IC_{50} for carrageenan inhibition of various HPV PsV types was in the ng/mL range [117]. In contrast, utilizing similar titers of NV, infection with HPV18 took at least 10 µg/mL of carrageenan for 50% inhibition and HPV16 was not inhibited. HPV31 NV was also found to be sensitive to inhibition by carrageenan and HPV45 showed no dose-dependent decrease in infection in the presence of carrageenan [15]. These results suggest that not only is there a difference in resistance and susceptibility of NV and PsV to the effects of carrageenan, but that different HPV types are very selective in their requirements for the type of GAG required for infection. Virion attachment to the cell surface in the presence of carrageenan was then analyzed. Attachment of HPV16 NV to the cell surface was unaffected in the presence of carrageenan. Conversely, HPV18 NV attachment was completely ablated in the presence of carrageenan. Additionally, HPV31 NV attachment was slightly reduced, whereas HPV45 attachment was not significantly reduced [15]. Taken together, these data suggest that when testing microbicides, confirmation of data utilizing native virions is essential. In addition, confirmation of results utilizing multiple HPV types should be done, as these data again confirm that not every HPV type behaves in the same manner.

Preventing the incidence of HPV infection through the use of vaccinations and microbicides is invaluable. Of equal importance is the use of chemical disinfectants to keep surfaces and medical equipment free of virions that could be spread between patients. Determining the efficacy of chemical disinfectants against HPV is important because it includes real world scenarios of disinfection protocols on equipment in hospital settings that are possibly contaminated with HPV. Currently, what little information that is available on HPV disinfection has been based mainly on studies from surrogate viruses, such as hepatitis B virus (HBV), which were deemed, without experimental evidence, to have similar resistance to HPV. Additionally, the United States Centers for Disease Control and Prevention (CDC) currently acknowledges the lack of research regarding effective HPV disinfectants [118]. Disinfectant information from studies that have included HPV has all been acquired utilizing recombinant HPV particles. Until recently, no studies had been done using native virions and there existed no functional assays to determine HPVs susceptibility to clinical disinfectants. A recent study tested the susceptibility of both HPV16 recombinant particles (QV) and native HPV16 grown in organotypic raft culture to a variety of common clinical disinfectants. Viral particles were incubated with common clinical disinfectants such as isopropanol, ethanol, triple phenolic, paracetic acid silver-based disinfectant (PAA), gluteraldehyde, hypochlorite, and ortho-phthalaldehyde. The disinfectant was then washed away and virus was incubated with cells to determine its ability to infect cells compared to untreated virus. HPV16 QV and HPV16 NV had some similarities in their resistance/susceptibility profiles to the disinfectants tested. Both types of particles were resistant to gluteraldehyde and ortho-phthalaldehyde and both were susceptible to hypochlorite and high concentration PAA. Of importance, however, HPV16 QV was also susceptible to isopropanol, triple phenolic, and a lower concentration of PAA [119]. This study highlights the necessity of utilizing native HPV particles when investigating clinical aspects of viral infection and transmission.

7. Concluding Remarks

Taken together, all of the data has provided great insight into the structure of HPV as well as the method of binding to, entering, and infecting cells. The use of recombinant particle technology is invaluable, as it allows for rapid data generation in studies looking at multiple parts of the HPV life cycle. However, as this review points out, there are some significant differences between native virions and recombinant particles that necessitate conformational studies utilizing native virions.

Acknowledgments

This work was supported by a PHS grant for the National Institute of Allergy and Infectious Disease (RO1AI57988). Jennifer Biryukov is supported by a grant from the National Cancer Institute (U01CA179724-02).

Author Contributions

Jennifer Biryukov and Craig Meyers discussed the manuscript content and illustrations. Jennifer Biryukov wrote the review.

Conflicts of Interest

The authors declare no conflict of interest.

References

1. Zur Hausen, H. Papillomaviruses in the causation of human cancers—A brief historical account. *Virology* **2009**, *384*, 260–265.
2. Syrjanen, S.; Lodi, G.; von Bültzingslöwen, I.; Aliko, A.; Arduino, P.; Campisi, G.; Challacombe, S.; Ficarra, G.; Flaitz, C.; Zhou, H.M.; *et al.* Human papillomaviruses in oral carcinoma and oral potentially malignant disorders: A systematic review. *Oral Dis.* **2011**, *17*, S58–S72. [CrossRef] [PubMed]
3. D'Souza, G.; Kreimer, A.R.; Viscidi, R.; Pawlita, M.; Fakhry, C.; Koch, W.M.; Westra, W.H.; Gillison, M.L. Case-control study of human papillomavirus and oropharyngeal cancer. *N. Engl. J. Med.* **2007**, *356*, 1944–1956. [CrossRef] [PubMed]
4. Schiffman, M.; Kjaer, S.K. Chapter 2: Natural history of anogenital human papillomavirus infection and neoplasia. *J. Natl. Cancer Inst. Monogr.* **2003**, *31*, 14–19. [CrossRef] [PubMed]
5. Arbyn, M.; Castellsagué, X.; de Sanjosé, S.; Bruni, L.; Saraiya, M.; Bray, F.; Ferlay, J. Worldwide burden of cervical cancer in 2008. *Ann. Oncol.* **2011**, *22*, 2675–2686. [CrossRef] [PubMed]
6. De Villiers, E.M.; Fauquet, C.; Broker, T.R.; Bernard, H.U.; zur Hausen, H. Classification of papillomaviruses. *Virology* **2004**, *324*, 17–27. [CrossRef] [PubMed]
7. Longworth, M.S.; Laimins, L.A. Pathogenesis of human papillomaviruses in differentiating epithelia. *Microbiol. Mol. Biol. Rev.* **2004**, *68*, 362–372. [CrossRef] [PubMed]
8. Madsen, B.S.; Jensen, H.L.; van den Brule, A.J.; Wohlfahrt, J.; Frisch, M. Risk factors for invasive squamous cell carcinoma of the vulva and vagina—Population-based case-control study in Denmark. *Int. J. Cancer* **2008**, *122*, 2827–2834. [CrossRef] [PubMed]
9. Meyers, C.; Frattini, M.G.; Hudson, J.B.; Laimins, L.A. Biosynthesis of human papillomavirus from a continuous cell line upon epithelial differentiation. *Science* **1992**, *257*, 971–973. [CrossRef] [PubMed]
10. Hummel, M.; Hudson, J.B.; Laimins, L.A. Differentiation-induced and constitutive transcription of human papillomavirus type 31b in cell lines containing viral episomes. *J. Virol.* **1992**, *66*, 6070–6080. [PubMed]
11. Bedell, M.A.; Hudson, J.B.; Golub, T.R.; Turyk, M.E.; Hosken, M.; Wilbanks, G.D.; Laimins, L.A. Amplification of human papillomavirus genomes *in vitro* is dependent on epithelial differentiation. *J. Virol.* **1991**, *65*, 2254–2260. [PubMed]

12. Grassmann, K.; Rapp, B.; Maschek, H.; Petry, K.U.; Iftner, T. Identification of a differentiation-inducible promoter in the E7 open reading frame of human papillomavirus type 16 (HPV-16) in raft cultures of a new cell line containing high copy numbers of episomal HPV-16 DNA. *J. Virol.* **1996**, *70*, 2339–2349. [PubMed]

13. Ozbun, M.A.; Meyers, C. Characterization of late gene transcripts expressed during vegetative replication of human papillomavirus type 31b. *J. Virol.* **1997**, *71*, 5161–5172. [PubMed]

14. Richards, K.F.; Mukherjee, S.; Bienkowska-Haba, M.; Pang, J.; Sapp, M. Human papillomavirus species-specific interaction with the basement membrane-resident non-heparan sulfate receptor. *Viruses* **2014**, *6*, 4856–4879. [CrossRef] [PubMed]

15. Cruz, L.; Meyers, C. Differential dependence on host cell glycosaminoglycans for infection of epithelial cells by high-risk HPV types. *PLoS ONE* **2013**, *8*, e68379. [CrossRef] [PubMed]

16. McLaughlin-Drubin, M.E.; Christensen, N.D.; Meyers, C. Propagation, infection, and neutralization of authentic HPV16 virus. *Virology* **2004**, *322*, 213–219. [CrossRef] [PubMed]

17. McLaughlin-Drubin, M.E.; Wilson, S.; Mullikin, B.; Suzich, J.; Meyers, C. Human papillomavirus type 45 propagation, infection, and neutralization. *Virology* **2003**, *312*, 1–7. [CrossRef]

18. Meyers, C.; Mayer, T.J.; Ozbun, M.A. Synthesis of infectious human papillomavirus type 18 in differentiating epithelium transfected with viral DNA. *J. Virol.* **1997**, *71*, 7381–7386. [PubMed]

19. Kirnbauer, R.; Booy, F.; Cheng, N.; Lowy, D.R.; Schiller, J.T. Papillomavirus L1 major capsid protein self-assembles into virus-like particles that are highly immunogenic. *Proc. Natl. Acad. Sci. USA* **1992**, *89*, 12180–12184. [CrossRef] [PubMed]

20. Kirnbauer, R.; Taub, J.; Greenstone, H.; Roden, R.; Dürst, M.; Gissmann, L.; Lowy, D.R.; Schiller, J.T. Efficient self-assembly of human papillomavirus type 16 L1 and L1-L2 into virus-like particles. *J. Virol.* **1993**, *67*, 6929–6936. [PubMed]

21. Rose, R.C.; Bonnez, W.; Reichman, R.C.; Garcea, R.L. Expression of human papillomavirus type 11 L1 protein in insect cells: *In vivo* and *in vitro* assembly of viruslike particles. *J. Virol.* **1993**, *67*, 1936–1944. [PubMed]

22. Hagensee, M.E.; Yaegashi, N.; Galloway, D.A. Self-assembly of human papillomavirus type 1 capsids by expression of the L1 protein alone or by coexpression of the L1 and L2 capsid proteins. *J. Virol.* **1993**, *67*, 315–322. [PubMed]

23. Hagensee, M.E.; Olson, N.H.; Baker, T.S.; Galloway, D.A. Three-dimensional structure of vaccinia virus-produced human papillomavirus type 1 capsids. *J. Virol.* **1994**, *68*, 4503–4505. [PubMed]

24. Chen, X.S.; Garcea, R.L.; Goldberg, I.; Casini, G.; Harrison, S.C. Structure of small virus-like particles assembled from the L1 protein of human papillomavirus 16. *Mol. Cell* **2000**, *5*, 557–567. [CrossRef]

25. Zhang, W.; Carmichael, J.; Ferguson, J.; Inglis, S.; Ashrafian, H.; Stanley, M. Expression of human papillomavirus type 16 L1 protein in *Escherichia coli*: Denaturation, renaturation, and self-assembly of virus-like particles *in vitro*. *Virology* **1998**, *243*, 423–431. [CrossRef] [PubMed]

26. Culp, T.D.; Christensen, N.D. Kinetics of *in vitro* adsorption and entry of papillomavirus virions. *Virology* **2004**, *319*, 152–161. [CrossRef] [PubMed]

27. Pyeon, D.; Lambert, P.F.; Ahlquist, P. Production of infectious human papillomavirus independently of viral replication and epithelial cell differentiation. *Proc. Natl. Acad. Sci. USA* **2005**, *102*, 9311–9316. [CrossRef] [PubMed]

28. Day, P.M.; Lowy, D.R.; Schiller, J.T. Papillomaviruses infect cells via a clathrin-dependent pathway. *Virology* **2003**, *307*, 1–11. [CrossRef]

29. Embers, M.E.; Budgeon, L.R.; Culp, T.D.; Reed, C.A.; Pickel, M.D.; Christensen, N.D. Differential antibody responses to a distinct region of human papillomavirus minor capsid proteins. *Vaccine* **2004**, *22*, 670–680. [CrossRef] [PubMed]

30. Gambhira, R.; Karanam, B.; Jagu, S.; Roberts, J.N.; Buck, C.B.; Bossis, I.; Alphs, H.; Culp, T.; Christensen, N.D.; Roden, R.B. A protective and broadly cross-neutralizing epitope of human papillomavirus L2. *J. Virol.* **2007**, *81*, 13927–13931. [CrossRef] [PubMed]

31. Buck, C.B.; Cheng, N.; Thompson, C.D.; Lowy, D.R.; Steven, A.C.; Schiller, J.T.; Trus, B.L. Arrangement of L2 within the papillomavirus capsid. *J. Virol.* **2008**, *82*, 5190–5197. [CrossRef] [PubMed]

32. White, W.I.; Wilson, S.D.; Parmer-Hill, F.J.; Woods, R.M.; Ghim, S.; Hewitt, L.A.; Goldman, D.M.; Burke, S.J.; Jerson, A.B.; Koening, S.; *et al.* Characterization of a major neutralizing epitope on human papillomavirus type 16 L1. *J. Virol.* **1999**, *73*, 4882–4889. [PubMed]

33. Zhou, J.; Liu, W.J.; Peng, S.W.; Sun, X.Y.; Frazer, I. Papillomavirus capsid protein expression level depends on the match between codon usage and tRNA availability. *J. Virol.* **1999**, *73*, 4972–4982. [PubMed]

34. Leder, C.; Kleinschmidt, J.A.; Wiethe, C.; Müller, M. Enhancement of capsid gene expression: Preparing the human papillomavirus type 16 major structural gene L1 for DNA vaccination purposes. *J. Virol.* **2001**, *75*, 9201–9209. [CrossRef] [PubMed]

35. Spencer, P.S.; Barral, J.M. Genetic code redundancy and its influence on the encoded polypeptides. *Comput. Struct. Biotechnol. J.* **2012**, *1*, e201204006. [CrossRef] [PubMed]

36. Fang, N.X.; Frazer, I.H.; Fernando, G.J. Differences in the post-translational modifications of human papillomavirus type 6b major capsid protein expressed from a baculovirus system compared with a vaccinia virus system. *Biotechnol. Appl. Biochem.* **2000**, *32*, 27–33. [CrossRef] [PubMed]

37. Zhou, J.; Sun, X.Y.; Stenzel, D.J.; Frazer, I.H. Expression of vaccinia recombinant HPV 16 L1 and L2 ORF proteins in epithelial cells is sufficient for assembly of HPV virion-like particles. *Virology* **1991**, *185*, 251–257. [CrossRef]

38. Conway, M.J.; Alam, S.; Ryndock, E.J.; Cruz, L.; Christensen, N.D.; Roden, R.B.; Meyers, C. Tissue-spanning redox gradient-dependent assembly of native human papillomavirus type 16 virions. *J. Virol.* **2009**, *83*, 10515–10526. [CrossRef] [PubMed]

39. Holmgren, S.C.; Patterson, N.A.; Ozbun, M.A.; Lambert, P.F. The minor capsid protein L2 contributes to two steps in the human papillomavirus type 31 life cycle. *J. Virol.* **2005**, *79*, 3938–3948. [CrossRef] [PubMed]

40. Fey, S.J.; Nielsen, T.; Vetner, M.; Storgaard, L.; Smed, V.; Larsen, P.M. Demonstration of *in vitro* synthesis of human papilloma viral proteins from hand and foot warts. *J. Investig. Dermatol.* **1989**, *92*, 817–824. [CrossRef] [PubMed]

41. Larsen, P.M.; Storgaard, L.; Fey, S.J. Proteins present in bovine papillomavirus particles. *J. Virol.* **1987**, *61*, 3596–3601. [PubMed]

42. Joshi, A.; Nagashima, K.; Freed, E.O. Mutation of dileucine-like motifs in the human immunodeficiency virus type 1 capsid disrupts virus assembly, gag-gag interactions, gag-membrane binding, and virion maturation. *J. Virol.* **2006**, *80*, 7939–7951. [CrossRef] [PubMed]

43. Perez-Berna, A.J.; Ortega-Esteban, A.; Menéndez-Conejero, R.; Winkler, D.C.; Menéndez, M.; Steven, A.C.; Flint, S.J.; de Pablo, P.J.; San Martín, C. The role of capsid maturation on adenovirus priming for sequential uncoating. *J. Biol. Chem.* **2012**, *287*, 31582–31595. [CrossRef] [PubMed]

44. Buck, C.B.; Thompson, C.D.; Pang, Y.Y.; Lowy, D.R.; Schiller, J.T. Maturation of papillomavirus capsids. *J. Virol.* **2005**, *79*, 2839–2846. [CrossRef] [PubMed]

45. Conway, M.J.; Cruz, L.; Alam, S.; Christensen, N.D.; Meyers, C. Differentiation-dependent interpentameric disulfide bond stabilizes native human papillomavirus type 16. *PLoS ONE* **2011**, *6*, e22427. [CrossRef] [PubMed]

46. Cardone, G.; Moyer, A.L.; Cheng, N.; Thompson, C.D.; Dvoretzky, I.; Lowy, D.R.; Schiller, J.T.; Steven, A.C.; Buck, C.B.; Trus, B.L. Maturation of the human papillomavirus 16 capsid. *MBio* **2014**, *5*, e01104-14. [CrossRef] [PubMed]

47. Fligge, C.; Schäfer, F.; Selinka, H.C.; Sapp, C.; Sapp, M. DNA-induced structural changes in the papillomavirus capsid. *J. Virol.* **2001**, *75*, 7727–7731. [CrossRef] [PubMed]

48. Ishii, Y.; Tanaka, K.; Kanda, T. Mutational analysis of human papillomavirus type 16 major capsid protein L1: The cysteines affecting the intermolecular bonding and structure of L1-capsids. *Virology* **2003**, *308*, 128–136. [CrossRef]

49. Wolf, M.; Garcea, R.L.; Grigorieff, N.; Harrison, S.C. Subunit interactions in bovine papillomavirus. *Proc. Natl. Acad. Sci. USA* **2010**, *107*, 6298–6303. [CrossRef] [PubMed]

50. Modis, Y.; Trus, B.L.; Harrison, S.C. Atomic model of the papillomavirus capsid. *EMBO J.* **2002**, *21*, 4754–4762. [CrossRef] [PubMed]

51. Ryndock, E.J.; Conway, M.J.; Alam, S.; Gul, S.; Murad, S.; Christensen, N.D.; Meyers, C. Roles for human papillomavirus type 16 l1 cysteine residues 161, 229, and 379 in genome encapsidation and capsid stability. *PLoS ONE* **2014**, *9*, e99488. [CrossRef] [PubMed]

52. Ishii, Y.; Ozaki, S.; Tanaka, K.; Kanda, T. Human papillomavirus 16 minor capsid protein L2 helps capsomeres assemble independently of intercapsomeric disulfide bonding. *Virus Genes* **2005**, *31*, 321–328. [CrossRef] [PubMed]

53. Campos, S.K.; Ozbun, M.A. Two highly conserved cysteine residues in HPV16 L2 form an intramolecular disulfide bond and are critical for infectivity in human keratinocytes. *PLoS ONE* **2009**, *4*, e4463. [CrossRef] [PubMed]

54. Gambhira, R.; Jagu, S.; Karanam, B.; Day, P.M.; Roden, R. Role of L2 cysteines in papillomavirus infection and neutralization. *Virol. J.* **2009**, *6*, e176. [CrossRef] [PubMed]

55. Conway, M.J.; Alam, S.; Christensen, N.D.; Meyers, C. Overlapping and independent structural roles for human papillomavirus type 16 L2 conserved cysteines. *Virology* **2009**, *393*, 295–303. [CrossRef] [PubMed]

56. Kondo, K.; Ishii, Y.; Ochi, H.; Matsumoto, T.; Yoshikawa, H.; Kanda, T. Neutralization of HPV16, 18, 31, and 58 pseudovirions with antisera induced by immunizing rabbits with synthetic peptides representing segments of the HPV16 minor capsid protein L2 surface region. *Virology* **2007**, *358*, 266–272. [CrossRef] [PubMed]

57. Roden, R.B.; Greenstone, H.L.; Kirnbauer, R.; Booy, F.P.; Jessie, J.; Lowy, D.R.; Schiller, J.T. *In vitro* generation and type-specific neutralization of a human papillomavirus type 16 virion pseudotype. *J. Virol.* **1996**, *70*, 5875–5883. [PubMed]

58. Zhao, K.N.; Sun, X.Y.; Frazer, I.H.; Zhou, J. DNA packaging by L1 and L2 capsid proteins of bovine papillomavirus type 1. *Virology* **1998**, *243*, 482–491. [CrossRef] [PubMed]

59. Trus, B.L.; Roden, R.B.; Greenstone, H.L.; Vrhel, M.; Schiller, J.T.; Booy, F.P. Novel structural features of bovine papillomavirus capsid revealed by a three-dimensional reconstruction to 9 Å resolution. *Nat. Struct. Biol.* **1997**, *4*, 413–420. [CrossRef] [PubMed]

60. Doorbar, J.; Gallimore, P.H. Identification of proteins encoded by the L1 and L2 open reading frames of human papillomavirus 1a. *J. Virol.* **1987**, *61*, 2793–2799. [PubMed]

61. Meyers, C.; Bromberg-White, J.L.; Zhang, J.; Kaupas, M.E.; Bryan, J.T.; Lowe, R.S.; Jansen, K.U. Infectious virions produced from a human papillomavirus type 18/16 genomic DNA chimera. *J. Virol.* **2002**, *76*, 4723–4733. [CrossRef] [PubMed]

62. Bowser, B.S.; Chen, H.S.; Conway, M.J.; Christensen, N.D.; Meyers, C. Human papillomavirus type 18 chimeras containing the L2/L1 capsid genes from evolutionarily diverse papillomavirus types generate infectious virus. *Virus Res.* **2011**, *160*, 246–255. [CrossRef] [PubMed]

63. Chen, H.S.; Conway, M.J.; Christensen, N.D.; Alam, S.; Meyers, C. Papillomavirus capsid proteins mutually impact structure. *Virology* **2011**, *412*, 378–383. [CrossRef] [PubMed]

64. Kines, R.C.; Thompson, C.D.; Lowy, D.R.; Schiller, J.T.; Day, P.M. The initial steps leading to papillomavirus infection occur on the basement membrane prior to cell surface binding. *Proc. Natl. Acad. Sci. USA* **2009**, *106*, 20458–20463. [CrossRef] [PubMed]

65. Sapp, M.; Bienkowska-Haba, M. Viral entry mechanisms: Human papillomavirus and a long journey from extracellular matrix to the nucleus. *FEBS J.* **2009**, *276*, 7206–7216. [CrossRef] [PubMed]

66. Giroglou, T.; Florin, L.; Schäfer, F.; Streeck, R.E.; Sapp, M. Human papillomavirus infection requires cell surface heparan sulfate. *J. Virol.* **2001**, *75*, 1565–1570. [CrossRef] [PubMed]

67. Johnson, K.M.; Kines, R.C.; Roberts, J.N.; Lowy, D.R.; Schiller, J.T.; Day, P.M. Role of heparan sulfate in attachment to and infection of the murine female genital tract by human papillomavirus. *J. Virol.* **2009**, *83*, 2067–2074. [CrossRef] [PubMed]

68. Richards, R.M.; Lowy, D.R.; Schiller, J.T.; Day, P.M. Cleavage of the papillomavirus minor capsid protein, L2, at a furin consensus site is necessary for infection. *Proc. Natl. Acad. Sci. USA* **2006**, *103*, 1522–1527. [CrossRef] [PubMed]

69. Day, P.M.; Gambhira, R.; Roden, R.B.; Lowy, D.R.; Schiller, J.T. Mechanisms of human papillomavirus type 16 neutralization by 12 cross-neutralizing and 11 type-specific antibodies. *J. Virol.* **2008**, *82*, 4638–4646. [CrossRef] [PubMed]

70. Knappe, M.; Bodevin, S.; Selinka, H.C.; Spillmann, D.; Streeck, R.E.; Chen, X.S.; Lindahl, U.; Sapp, M. Surface-exposed amino acid residues of HPV16 L1 protein mediating interaction with cell surface heparan sulfate. *J. Biol. Chem.* **2007**, *282*, 27913–27922. [CrossRef] [PubMed]

71. Kumar, A.; Jacob, T.; Abban, C.Y.; Meneses, P.I. Intermediate heparan sulfate binding during HPV-16 infection in HaCaTs. *Am. J. Ther.* **2014**, *21*, 331–342. [CrossRef] [PubMed]

72. Joyce, J.G.; Tung, J.S.; Przysiecki, C.T.; Cook, J.C.; Lehman, E.D.; Sands, J.A.; Jansen, K.U.; Keller, P.M. The L1 major capsid protein of human papillomavirus type 11 recombinant virus-like particles interacts with heparin and cell-surface glycosaminoglycans on human keratinocytes. *J. Biol. Chem.* **1999**, *274*, 5810–5822. [CrossRef] [PubMed]

73. Patterson, N.A.; Smith, J.L.; Ozbun, M.A. Human papillomavirus type 31b infection of human keratinocytes does not require heparan sulfate. *J. Virol.* **2005**, *79*, 6838–6847. [CrossRef] [PubMed]

74. Bishop, B.; Dasgupta, J.; Klein, M.; Garcea, R.L.; Christensen, N.D.; Zhao, R.; Chen, X.S. Crystal structures of four types of human papillomavirus L1 capsid proteins: Understanding the specificity of neutralizing monoclonal antibodies. *J. Biol. Chem.* **2007**, *282*, 31803–31811. [CrossRef] [PubMed]

75. Broutian, T.R.; Brendle, S.A.; Christensen, N.D. Differential binding patterns to host cells associated with particles of several human alphapapillomavirus types. *J. Gen. Virol.* **2010**, *91*, 531–540. [CrossRef] [PubMed]

76. Sapp, M.; Day, P.M. Structure, attachment and entry of polyoma- and papillomaviruses. *Virology* **2009**, *384*, 400–409. [CrossRef] [PubMed]

77. Bauer, P.H.; Cui, C.; Liu, W.R.; Stehle, T.; Harrison, S.C.; DeCaprio, J.A.; Benjamin, T.L. Discrimination between sialic acid-containing receptors and pseudoreceptors regulates polyomavirus spread in the mouse. *J. Virol.* **1999**, *73*, 5826–5832. [CrossRef] [PubMed]

78. Neu, U.; Maginnis, M.S.; Palma, A.S.; Ströh, L.J.; Nelson, C.D.; Feizi, T.; Atwood, W.J.; Stehle, T. Structure-function analysis of the human JC polyomavirus establishes the LSTc pentasaccharide as a functional receptor motif. *Cell Host Microbe* **2010**, *8*, 309–319. [CrossRef] [PubMed]

79. Horvath, C.A.; Boulet, G.A.; Renoux, V.M.; Delvenne, P.O.; Bogers, J.P. Mechanisms of cell entry by human papillomaviruses: An overview. *Virol. J.* **2010**, *7*, e11. [CrossRef] [PubMed]

80. Spoden, G.; Freitag, K.; Husmann, M.; Boller, K.; Sapp, M.; Lambert, C.; Florin, L. Clathrin- and caveolin-independent entry of human papillomavirus type 16—Involvement of tetraspanin-enriched microdomains (TEMs). *PLoS ONE* **2008**, *3*, e3313. [CrossRef] [PubMed]

81. Selinka, H.C.; Florin, L.; Patel, H.D.; Freitag, K.; Schmidtke, M.; Makarov, V.A.; Sapp, M. Inhibition of transfer to secondary receptors by heparan sulfate-binding drug or antibody induces noninfectious uptake of human papillomavirus. *J. Virol.* **2007**, *81*, 10970–10980. [CrossRef] [PubMed]

82. Smith, J.L.; Campos, S.K.; Ozbun, M.A. Human papillomavirus type 31 uses a caveolin 1- and dynamin 2-mediated entry pathway for infection of human keratinocytes. *J. Virol.* **2007**, *81*, 9922–9931. [CrossRef] [PubMed]

83. Day, P.M.; Lowy, D.R.; Schiller, J.T. Heparan sulfate-independent cell binding and infection with furin-precleaved papillomavirus capsids. *J. Virol.* **2008**, *82*, 12565–12568. [CrossRef] [PubMed]

84. Bousarghin, L.; Touzé, A.; Sizaret, P.Y.; Coursaget, P. Human papillomavirus types 16, 31, and 58 use different endocytosis pathways to enter cells. *J. Virol.* **2003**, *77*, 3846–3850. [CrossRef] [PubMed]

85. Hindmarsh, P.L.; Laimins, L.A. Mechanisms regulating expression of the HPV 31 L1 and L2 capsid proteins and pseudovirion entry. *Virol. J.* **2007**, *4*, e19. [CrossRef] [PubMed]

86. Laniosz, V.; Dabydeen, S.A.; Havens, M.A.; Meneses, P.I. Human papillomavirus type 16 infection of human keratinocytes requires clathrin and caveolin-1 and is brefeldin a sensitive. *J. Virol.* **2009**, *83*, 8221–8232. [CrossRef] [PubMed]

87. Zhang, W.; Kazakov, T.; Popa, A.; DiMaio, D. Vesicular trafficking of incoming human papillomavirus 16 to the Golgi apparatus and endoplasmic reticulum requires gamma-secretase activity. *MBio* **2014**, *5*, e01777-14. [CrossRef] [PubMed]

88. Schelhaas, M.; Shah, B.; Holzer, M.; Blattmann, P.; Kühling, L.; Day, P.M.; Schiller, J.T.; Helenius, A. Entry of human papillomavirus type 16 by actin-dependent, clathrin- and lipid raft-independent endocytosis. *PLoS Pathog.* **2012**, *8*, e1002657. [CrossRef] [PubMed]

89. Spoden, G.; Kühling, L.; Cordes, N.; Frenzel, B.; Sapp, M.; Boller, K.; Florin, L.; Schelhaas, M. Human papillomavirus types 16, 18, and 31 share similar endocytic requirements for entry. *J. Virol.* **2013**, *87*, 7765–7773. [CrossRef] [PubMed]

90. Evander, M.; Frazer, I.H.; Payne, E.; Qi, Y.M.; Hengst, K.; McMillan, N.A. Identification of the alpha6 integrin as a candidate receptor for papillomaviruses. *J. Virol.* **1997**, *71*, 2449–2456. [PubMed]

91. Yoon, C.S.; Kim, K.D.; Park, S.N.; Cheong, S.W. Alpha(6) integrin is the main receptor of human papillomavirus type 16 VLP. *Biochem. Biophys. Res. Commun.* **2001**, *283*, 668–673. [CrossRef] [PubMed]

92. Sibbet, G.; Romero-Graillet, C.; Meneguzzi, G.; Campo, M.S. Alpha6 integrin is not the obligatory cell receptor for bovine papillomavirus type 4. *J. Gen. Virol.* **2000**, *81*, 327–334. [PubMed]

93. Aksoy, P.; Abban, C.Y.; Kiyashka, E.; Qiang, W.; Meneses, P.I. HPV16 infection of HaCaTs is dependent on beta4 integrin, and alpha6 integrin processing. *Virology* **2014**, *449*, 45–52. [CrossRef] [PubMed]

94. Fothergill, T.; McMillan, N.A. Papillomavirus virus-like particles activate the PI3-kinase pathway via alpha-6 beta-4 integrin upon binding. *Virology* **2006**, *352*, 319–328. [CrossRef] [PubMed]

95. Surviladze, Z.; Dziduszko, A.; Ozbun, M.A. Essential roles for soluble virion-associated heparan sulfonated proteoglycans and growth factors in human papillomavirus infections. *PLoS Pathog.* **2012**, *8*, e1002519. [CrossRef] [PubMed]

96. Waisman, D.M. Annexin II tetramer: Structure and function. *Mol. Cell. Biochem.* **1995**, *149–150*, 301–322. [CrossRef] [PubMed]

97. Woodham, A.W.; da Silva, D.M.; Skeate, J.G.; Raff, A.B.; Ambroso, M.R.; Brand, H.E.; Isas, J.M.; Langen, R.; Kast, W.M. The S100A10 subunit of the annexin A2 heterotetramer facilitates L2-mediated human papillomavirus infection. *PLoS ONE* **2012**, *7*, e43519. [CrossRef] [PubMed]

98. Raff, A.B.; Woodham, A.W.; Raff, L.M.; Skeate, J.G.; Yan, L.; da Silva, D.M.; Schelhaas, M.; Kast, W.M. The evolving field of human papillomavirus receptor research: A review of binding and entry. *J. Virol.* **2013**, *87*, 6062–6072. [CrossRef] [PubMed]

99. Smith, J.L.; Lidke, D.S.; Ozbun, M.A. Virus activated filopodia promote human papillomavirus type 31 uptake from the extracellular matrix. *Virology* **2008**, *381*, 16–21. [CrossRef] [PubMed]

100. Selinka, H.C.; Giroglou, T.; Sapp, M. Analysis of the infectious entry pathway of human papillomavirus type 33 pseudovirions. *Virology* **2002**, *299*, 279–287. [CrossRef] [PubMed]

101. Bienkowska-Haba, M.; Williams, C.; Kim, S.M.; Garcea, R.L.; Sapp, M. Cyclophilins facilitate dissociation of the human papillomavirus type 16 capsid protein L1 from the L2/DNA complex following virus entry. *J. Virol.* **2012**, *86*, 9875–9887. [CrossRef] [PubMed]

102. Aydin, I.; Weber, S.; Snijder, B.; Samperio Ventayol, P.; Kühbacher, A.; Becker, M.; Day, P.M.; Schiller, J.T.; Kann, M.; Pelkmans, L.; *et al.* Large scale RNAi reveals the requirement of nuclear envelope breakdown for nuclear import of human papillomaviruses. *PLoS Pathog.* **2014**, *10*, e1004162. [CrossRef] [PubMed]

103. Lipovsky, A.; Popa, A.; Pimienta, G.; Wyler, M.; Bhan, A.; Kuruvilla, L.; Guie, M.A.; Poffenberger, A.C.; Nelson, C.D.; Atwood, W.J.; *et al.* Genome-wide siRNA screen identifies the retromer as a cellular entry factor for human papillomavirus. *Proc. Natl. Acad. Sci. USA* **2013**, *110*, 7452–7457. [CrossRef] [PubMed]

104. Day, P.M.; Thompson, C.D.; Schowalter, R.M.; Lowy, D.R.; Schiller, J.T. Identification of a role for the trans-Golgi network in human papillomavirus 16 pseudovirus infection. *J. Virol.* **2013**, *87*, 3862–3870. [CrossRef] [PubMed]

105. DiGiuseppe, S.; Bienkowska-Haba, M.; Hilbig, L.; Sapp, M. The nuclear retention signal of HPV16 L2 protein is essential for incoming viral genome to transverse the trans-Golgi network. *Virology* **2014**, *458–459*, 93–105. [CrossRef] [PubMed]

106. Attar, N.; Cullen, P.J. The retromer complex. *Adv. Enzyme Regul.* **2010**, *50*, 216–236. [CrossRef] [PubMed]

107. Burd, C.; Cullen, P.J. Retromer: A master conductor of endosome sorting. *Cold Spring Harb. Perspect. Biol.* **2014**, *6*, a016774. [CrossRef] [PubMed]

108. Popa, A.; Zhang, W.; Harrison, M.S.; Goodner, K.; Kazakov, T.; Goodwin, E.C.; Lipovsky, A.; Burd, C.G.; DiMaio, D. Direct binding of retromer to human papillomavirus type 16 minor capsid protein L2 mediates endosome exit during viral infection. *PLoS Pathog.* **2015**, *11*, e1004699. [CrossRef] [PubMed]

109. Bossis, I.; Roden, R.B.; Gambhira, R.; Yang, R.; Tagaya, M.; Howley, P.M.; Meneses, P.I. Interaction of tSNARE syntaxin 18 with the papillomavirus minor capsid protein mediates infection. *J. Virol.* **2005**, *79*, 6723–6731. [CrossRef] [PubMed]

110. Laniosz, V.; Nguyen, K.C.; Meneses, P.I. Bovine papillomavirus type 1 infection is mediated by SNARE syntaxin 18. *J. Virol.* **2007**, *81*, 7435–7448. [CrossRef] [PubMed]

111. Campos, S.K.; Chapman, J.A.; Deymier, M.J.; Bronnimann, M.P.; Ozbun, M.A. Opposing effects of bacitracin on human papillomavirus type 16 infection: Enhancement of binding and entry and inhibition of endosomal penetration. *J. Virol.* **2012**, *86*, 4169–4181. [CrossRef] [PubMed]

112. Pyeon, D.; Pearce, S.M.; Lank, S.M.; Ahlquist, P.; Lambert, P.F. Establishment of human papillomavirus infection requires cell cycle progression. *PLoS Pathog.* **2009**, *5*, e1000318. [CrossRef] [PubMed]

113. Bronnimann, M.P.; Chapman, J.A.; Park, C.K.; Campos, S.K. A transmembrane domain and GxxxG motifs within L2 are essential for papillomavirus infection. *J. Virol.* **2013**, *87*, 464–473. [CrossRef] [PubMed]

114. Christensen, N.D.; Cladel, N.M.; Reed, C.A.; Budgeon, L.R.; Embers, M.E.; Skulsky, D.M.; McClements, W.L.; Ludmerer, S.W.; Jansen, K.U. Hybrid papillomavirus L1 molecules assemble into virus-like particles that reconstitute conformational epitopes and induce neutralizing antibodies to distinct HPV types. *Virology* **2001**, *291*, 324–334. [CrossRef] [PubMed]

115. Day, P.M.; Thompson, C.D.; Buck, C.B.; Pang, Y.Y.; Lowy, D.R.; Schiller, J.T. Neutralization of human papillomavirus with monoclonal antibodies reveals different mechanisms of inhibition. *J. Virol.* **2007**, *81*, 8784–8792. [CrossRef] [PubMed]

116. Christensen, N.D.; Reed, C.A.; Culp, T.D.; Hermonat, P.L.; Howett, M.K.; Anderson, R.A.; Zaneveld, L.J. Papillomavirus microbicidal activities of high-molecular-weight cellulose sulfate, dextran sulfate, and polystyrene sulfonate. *Antimicrob. Agents Chemother.* **2001**, *45*, 3427–3432. [CrossRef] [PubMed]

117. Buck, C.B.; Thompson, C.D.; Roberts, J.N.; Müller, M.; Lowy, D.R.; Schiller, J.T. Carrageenan is a potent inhibitor of papillomavirus infection. *PLoS Pathog.* **2006**, *2*, e69. [CrossRef] [PubMed]

118. CDC Guideline for Disinfection and Sterilization in Healthcare Facilities, 2008. Available online: http://www.cdc.gov/hicpac/Disinfection_Sterilization/3_1deLaparoArthro.html (accessed on 13 May 2015).

119. Meyers, J.; Ryndock, E.; Conway, M.J.; Meyers, C.; Robison, R. Susceptibility of high-risk human papillomavirus type 16 to clinical disinfectants. *J. Antimicrob. Chemother.* **2014**, *69*, 1546–1550. [CrossRef] [PubMed]

High-Risk Human Papillomavirus Targets Crossroads in Immune Signaling

Bart Tummers and Sjoerd H. Van Der Burg

Abstract: Persistent infections with a high-risk type human papillomavirus (hrHPV) can progress to cancer. High-risk HPVs infect keratinocytes (KCs) and successfully suppress host immunity for up to two years despite the fact that KCs are well equipped to detect and initiate immune responses to invading pathogens. Viral persistence is achieved by active interference with KCs innate and adaptive immune mechanisms. To this end hrHPV utilizes proteins encoded by its viral genome, as well as exploits cellular proteins to interfere with signaling of innate and adaptive immune pathways. This results in impairment of interferon and pro-inflammatory cytokine production and subsequent immune cell attraction, as well as resistance to incoming signals from the immune system. Furthermore, hrHPV avoids the killing of infected cells by interfering with antigen presentation to antigen-specific cytotoxic T lymphocytes. Thus, hrHPV has evolved multiple mechanisms to avoid detection and clearance by both the innate and adaptive immune system, the molecular mechanisms of which will be dealt with in detail in this review.

Reprinted from *Viruses*. Cite as: Tummers, B.; van Der Burg, S.H. High-Risk Human Papillomavirus Targets Crossroads in Immune Signaling. *Viruses* **2015**, *7*, 2485-2506.

1. Introduction

Human papillomaviruses (HPVs) are small, non-enveloped icosahedral viruses belonging to the *Papillomaviridae* family. HPV is widespread within all human populations and transmitted via the skin, including the genitalia. With a double-stranded episomal DNA genome of only 7–8 kb, containing six non-structural early genes (E6, E7, E1, E2, E4 and E5), and two late genes (L2 and L1) that encode the capsid proteins [1], HPVs induce diseases ranging from warts to cancers [2]. Over 150 HPV types have currently been identified. They are divided into genera α, β, γ, μ and ν, based on the nucleotide sequence of the L1 gene [3]. HPV types of the α genus (~40) infect cutaneous and mucosal epithelia. Based on their oncogenic potential, mucosal HPVs are classified as low-risk, associated with benign warts or epithelial lesions, or high-risk, that can cause oropharyngeal and anogenital malignancies, including cancers of the cervix, vulva, vagina, penis and anus. HPV types of the other genera infect cutaneous epithelium and are associated with cutaneous papillomas and warts. βHPV types can cause non-melanoma skin cancer in immunocompromised individuals [4]. Most HPV infections resolve spontaneously within one (70%) to two (90%) years [5], and in only <1% of cases malignancies develop. Still, HPV causes ~528,000 new cancer cases and ~266,000 deaths each year. High-risk HPV (hrHPV) types are responsible for ~5% of all human cancers and are detected in 99.7% of cervical cancer cases, the fourth most common cancer in women, accounting for 7.5% of all cancer-associated deaths in women worldwide per year [6,7]. HPV types 16, 18, 31, 33, 35, 39, 45, 51, 52, 56, 59, 69, 73 and

82 have been detected in cervical carcinomas, but HPV16 is the most prevalent hrHPV type in cervical cancer and dominant in all other HPV-induced cancers [8,9].

HPVs exclusively infect keratinocytes (KCs) of the basal layer of the epidermis and mucosal epithelia, which they reach via micro-wounds and abrasions. Binding of the L1 protein of HPV to heparan sulfate proteoglycans at the surface of KCs induces endocytosis of the virion. Subsequently the capsid disassembles following acidification of the endosome and then the viral episome, still associated with L2, travels via the Golgi apparatus and Endoplasmic Reticulum to the nucleus [10] where low levels of viral early proteins are produced that reside mainly in the nucleus [11]. E1 and E2 initiate episome replication and, together with the host DNA replication machinery, maintain a low episome copy-number of 50–100 per cell [12]. Furthermore, E6 and E7 are produced to prevent cell growth arrest and apoptosis and delay differentiation, by inactivating p53 and the retinoblastoma protein (pRB). This induces a proliferative, non-differentiating state of the infected KC, resulting in lateral cell division. As the infected KC differentiates and migrates through the suprabasal layers of the epithelium, the expression of all viral genes is induced to enhance viral episome replication, which reaches high copy-numbers of hundreds to thousands per cell. In the higher layers of the epithelium the production of the late proteins L1 and L2, together forming the viral capsid, is induced and virion assembly takes place. With the rupture and shedding of the matured KC the viral particles are released [13]. Sometimes, for yet unknown reasons, hrHPV genomes can spontaneously integrate into the host genome, leading to release of the tight regulation of E6 and E7 expression. The newly transformed cells stably express E6, which binds to p53 and recruits the E3 ligase E6AP to target p53 for proteasomal degradation, as well as E7, which recruits the E3 ligase cullin 2 to target pRb for proteasomal degradation. The loss of these tumor suppressors results in uncontrollable cell growth, host genome mutations and inhibition of apoptosis, ultimately leading to cancer formation [1,13,14].

High-risk HPV infections can persist despite viral activity in keratinocytes. This indicates that HPV has developed mechanisms to effectively evade or suppress the host's innate and/or adaptive immune response. Indeed, several studies on the spontaneous immune response to HPV have shown that HPV-specific cellular immunity develops quite late during persistent HPV infections and often are of dubious quality in people with progressive infections [15].

Viral persistence may be linked to the life cycle of HPV since HPV does not cause viremia, cell death, or cell lysis, and the life-cycle takes place within the boundary of the *lamina basalis*, away from dermal immune cells. Thus, spontaneous contact between the immune system and the virus are minimal and inflammatory responses are not readily elicited. Langerhans cells residing within the epidermis can sense viral presence, but HPV counteracts their recruitment by interfering with the production of immune attractants. In addition, after the infection is established in basal keratinocytes, major viral gene expression is differentiation dependent and as such viral peptide presentation to immune cells is limited. Besides these passive mechanisms to evade the immune system, hrHPV also actively interferes with innate and adaptive immune mechanisms. These are discussed below.

160

2. Viral Recognition by Keratinocytes

Keratinocytes are well equipped to sense pathogens. Basal KCs express pattern-recognition receptors (PRRs), such as Toll-like receptors (TLRs), NOD-like receptors (NLRs), and RNA helicases, to recognize pathogen-associated molecular patterns (PAMPs) on viruses and microbes. PRR ligation leads to activation of inflammatory and proliferative signaling cascades and subsequent production of pro-inflammatory cytokines that can induce innate and adaptive immune responses. *In vitro* studies showed that KCs express TLR 1, 2, 3, 5 and 6 on the cell-surface and the nucleic acid-sensing TLR3 in endosomes. TLR7 and TLR8 are not expressed, but TLR7 expression can be induced upon TLR3 ligation [16]. The expression of TLR4 and TLR9 in basal KCs is still under debate, but TLR9 expression can be induced after terminal differentiation [17]. Cytosolically, KCs express the RNA helicases retinoic acid-inducible gene I (RIG-I; DDX58) and melanoma differentiation-associated protein 5 (MDA5; IFIH1) [18], and the dsDNA sensors gamma-interferon-inducible 16 (IFI16) and absent in melanoma 2 (AIM2) [19]. Expression of TLR3, 7, 9, PKR, RIG-I and MDA5 was confirmed *in situ* [18,20].

Although the vesicle-mediated entry mechanism used by HPV may hide the virus from recognition by cytoplasmic DNA sensors, KCs can produce type I interferon (IFN) and pro-inflammatory cytokines upon viral entry and, therefore, do recognize HPV [17]. Indeed, the episome contains CpG motifs that can be recognized by TLR9 [21] and the viral capsid itself is a potential PAMP. Whether HPV interferes with the expression of TLRs, RIG-I or MDA5 in HPV episome-containing KCs is still under debate [17,18,22]. While *TLR9* expression and function was shown to be abolished in KCs that overexpressed HPV16 E6 and E7 [21], by an E7-induced recruitment of a NFκB1, ERα and HDAC1 inhibitory complex to the TLR9 promotor [23], others concluded that E6 nor E7 influenced TLR9 expression or function [24]. The DNA sensor AIM2 is strongly expressed in HPV16-infected skin lesions, whereas IFI16 expression is not elevated [19]. Hence, it is not yet clear if HPV affects the expression of virus sensory molecules on KCs.

3. HPV Influences Innate Immune Signaling

Keratinocytes produce type I IFNs and pro-inflammatory cytokines upon PRR ligation through signaling via interferon regulatory factor (IRF) and nuclear factor of kappa-light-chain-enhancer of activated B cells (NFκB) activating pathways. Type I IFNs (mainly IFNα (13 subtypes) and IFNβ, but also IFNε, IFNτ, IFNκ, IFNω, IFNδ and IFNζ) stimulate cells to express genes inducing an anti-viral state. They can also stimulate dendritic cells and as such act as a bridge between innate and adaptive immunity [25–27]. Pro-inflammatory cytokines are chemoattractants for immune cells and regulate cell migration, activation, polarization and proliferation. Several genome-wide transcription studies reported that hrHPV types 16, 18 and 31 influence—mainly reduce—basal, TLR3-induced cytokine expression, and type I IFN-induced interferon-stimulated gene (ISG) expression [18,28–30], indicating that hrHPV affects PRR- and type I IFN-induced signaling pathways.

3.1. The Effect of HPV (Proteins) on the IRF Signaling Pathway

All TLRs, except TLR3, convey their signals via the adapter molecule MyD88. This induces the IRAK complex (consisting of IRAK1, 2 and 4) to recruit TRAF3, which stimulates IKKα to phosphorylate IRF7. TLR3 and 4 signal via TRIF, cytosolic RNA sensors via MAVS, and cytosolic DNA sensors signal via the adaptor molecule STING to activate TRAF3, which then induces the TBK1-IKKε complex to phosphorylate IRF3. Phosphorylated IRF3 and IRF7 homo-dimerize and translocate to the nucleus where production of type I IFNs is initiated. Furthermore, PRR ligation can result in IRF1 activation (Figure 1).

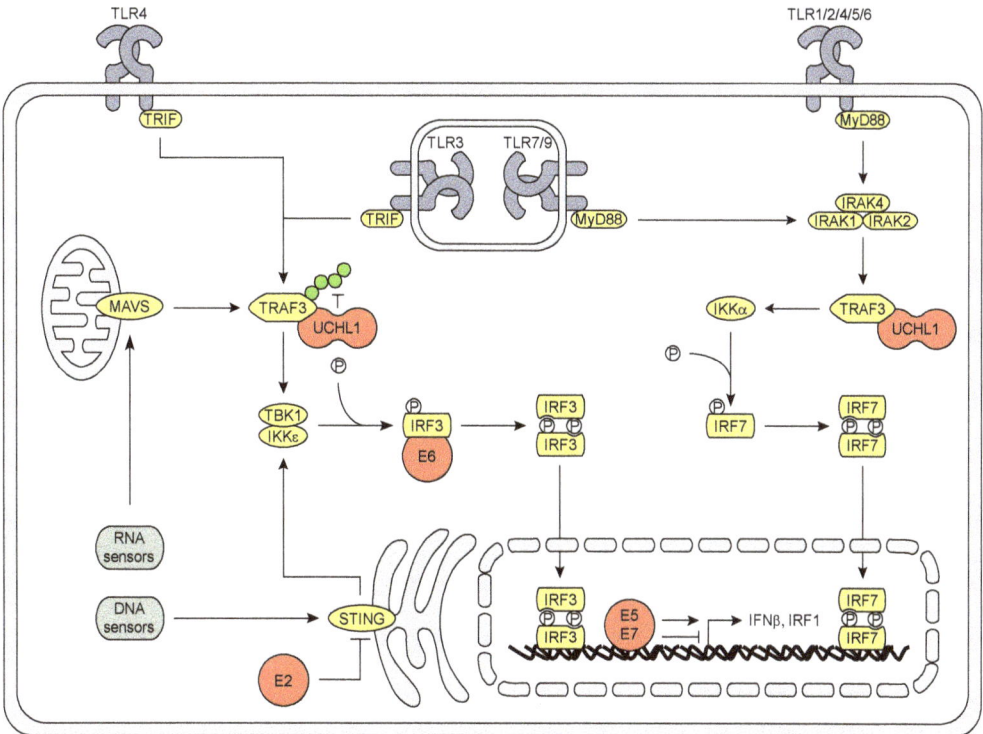

Figure 1. Schematic representation of the effects of hrHPV on IRF signaling. All TLRs, except TLR3, activate IRF7 via signaling through MyD88, the IRAK complex, TRAF3 and IKKα. TLR3 and 4 signal via TRIF, cytosolic RNA sensors through MAVS and cytosolic DNA sensors via STING activate IRF3 through TRAF3, TBK1 and IKKε. Activated IRFs dimerize, translocate to the nucleus and initiate gene transcription. HPV utilizes its own encoded E proteins (red) as well as exploits the cellular protein UCHL1 (red) to interfere with these signaling pathways. Green circles on TRAF3 indicate K63-linked poly-ubiquitin chains.

HrHPV influences type I IFN production by interfering at several points in the signaling cascade. HrHPV E2 proteins reduce the expression of *STING* and *IFNκ* [31], the latter of which its expression is also reduced by E6 [22,32]. HPV16, but not HPV18, E6 protein binds to IRF3 and, thereby, may prevent its transcriptional activity [33]. E7 blocks *IFNβ* transcription by binding to IRF1 and recruiting histone deacetylases (HDACs) to the *IFNβ* promotor site [34,35]. In contrast, E5 enhances *IFNβ* and *IRF1* expression [36]. HrHPV also exploits cellular proteins to interfere with IRF signaling; it upregulates the endogenous deubiquitinase ubiquitin carboxy-terminal hydrolase L1 (UCHL1) to interact with and deubiquitinate K63-linked poly-ubiquitin chains from TRAF3, resulting in reduced TBK1—TRAF3 interaction, IRF3 phosphorylation and *IFNβ* expression [17].

3.2. The Effects of HPV (Proteins) on IFNAR Signaling

The PRR-induced type I IFNs IFNα and IFNβ are secreted and can induce IFN-stimulated gene (ISG) expression in the infected cell itself but also in their uninfected neighbors. IFNα and IFNβ bind to the heterodimeric transmembrane IFNα/β receptor (IFNAR), composed of the IFNAR1 and IFNAR2 subunits. The IFNAR activates the receptor-associated protein tyrosine kinases Janus kinase 1 (JAK1) and tyrosine kinase 2 (TYK2), which recruit and phosphorylate STAT1 and STAT2, causing them to hetero-dimerize, bind IRF9, thereby forming the IFN-stimulated gene factor 3 (ISGF3) complex, and translocate to the nucleus. ISGF3 binds to IFN-stimulated response elements (ISREs) on the DNA and activates ISG transcription. IFNAR ligation can also lead to STAT1 homo-dimerization. STAT1 homo-dimers translocate to the nucleus and bind to γ-activated sequences (GAS) on the DNA, thereby activating ISG transcription more associated with IFNγ signaling (Figure 2) [26,37].

HrHPV also interferes with IFNAR signaling. HPV18 E6 can bind to TYK2 in order to hamper phosphorylation of STAT1 and STAT2 [38]. E6, and to a lesser extend E7, of the hrHPV types 16 and 31 were shown to impair STAT1 transcription and translation, and binding of STAT1 to the ISRE [28,30,39]. However, although hrHPV represses STAT1 protein levels, the IFNβ-induced STAT1 signal cascade is not affected by hrHPV, as phosphorylation of STAT1 still occurs [39]. Expression of STAT2 and IRF9 are not affected, but E7 can interact with cytosolic IRF9, preventing IRF9 to translocate to the nucleus with as a consequence impairment of ISGF3 complex formation [40,41].

3.3. The Effect of HPV (Proteins) on the NFκB Signaling Pathway

PRRs also induce cytokine production through signaling via TRIF, MAVS, STING and the IRAK complex, which leads to the K63-linked poly-ubiquitination of TRAF6. The TAB1-TAB2-TAK1 complex and the IKK complex (consisting of NEMO, IKKα and IKKβ) bind to the poly-ubiquitin chain on TRAF6, resulting in the phosphorylation of IKKβ by TAK1. Activated IKKβ then phosphorylates IκBα, leading to the SCF-βTrCP-mediated K48-linked poly-ubiquitination of IκBα and its subsequent degradation. This releases the NFκB1 complex (consisting of RelA and p50) and allows it to translocate to the nucleus where it is further modified to induce DNA binding and transcriptional activation (Figure 3) [27,42].

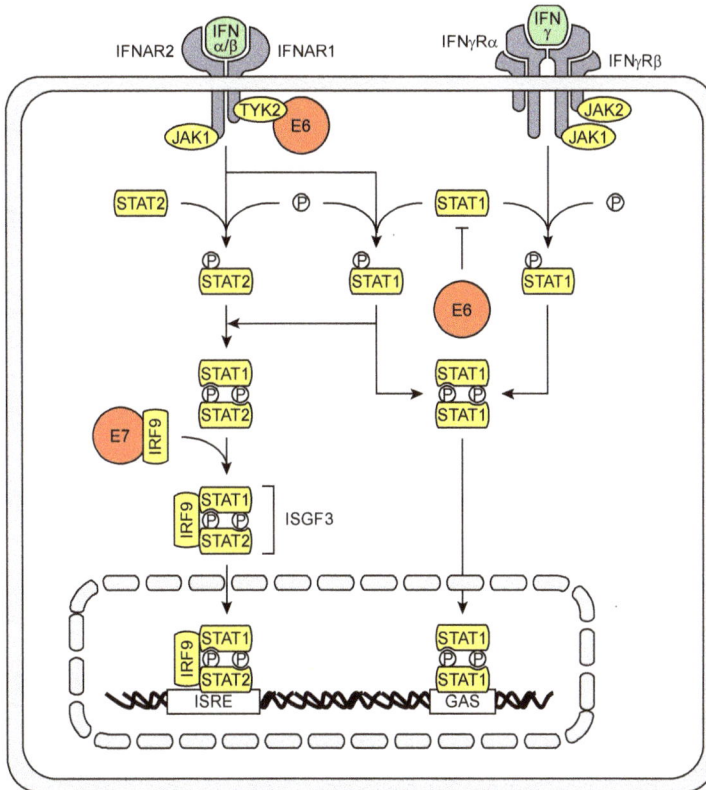

Figure 2. Schematic representation of the effects of hrHPV on IFNAR and IFNγR signaling. Type I IFN binding to the IFNAR leads to signaling via JAK1 and TYK2 to activate STAT1 and STAT2. STAT1 and STAT2 heterodimerize and recruit IRF9, forming the ISGF3 complex, which translocates to the nucleus, binds to ISREs and initiates ISG transcription. Activated STAT1 can also homodimerize, translocate to the nucleus, bind to GAS and initiate ISG transcription. Type II IFN binding to the IFNγR results in the activation of JAK1 and JAK2 and recruitment and phosphorylation of STAT1, which homo-dimerizes, translocates to the nucleus, binds to GAS on the DNA and initiates ISG transcription. HPV proteins (red) interfere with both IFNAR and IFNγR signaling by decreasing STAT1 levels, and hampering TYK2 and IRF9.

By the use of several different model systems hrHPV or its individual proteins have been shown to affect the PRR-induced signaling cascade that leads to NFκB nuclear translocation and to impair the function of NFκB within the nucleus. HrHPV upregulates the NFκB family members RelA, c-Rel, and the precursor proteins p105 and p100, which are processed into p50 and p52, respectively, and sequesters these proteins in the cytoplasm [30,43–46]. The last NFκB family member, RelB, is not reported to be regulated by HPV. HrHPV exploits the endogenous protein UCHL1 to bind TRAF6 and influence the Ub status of TRAF6 and NEMO, resulting in NEMO degradation [17]. Furthermore, UCHL1 can prevent IκBα ubiquitination [47].

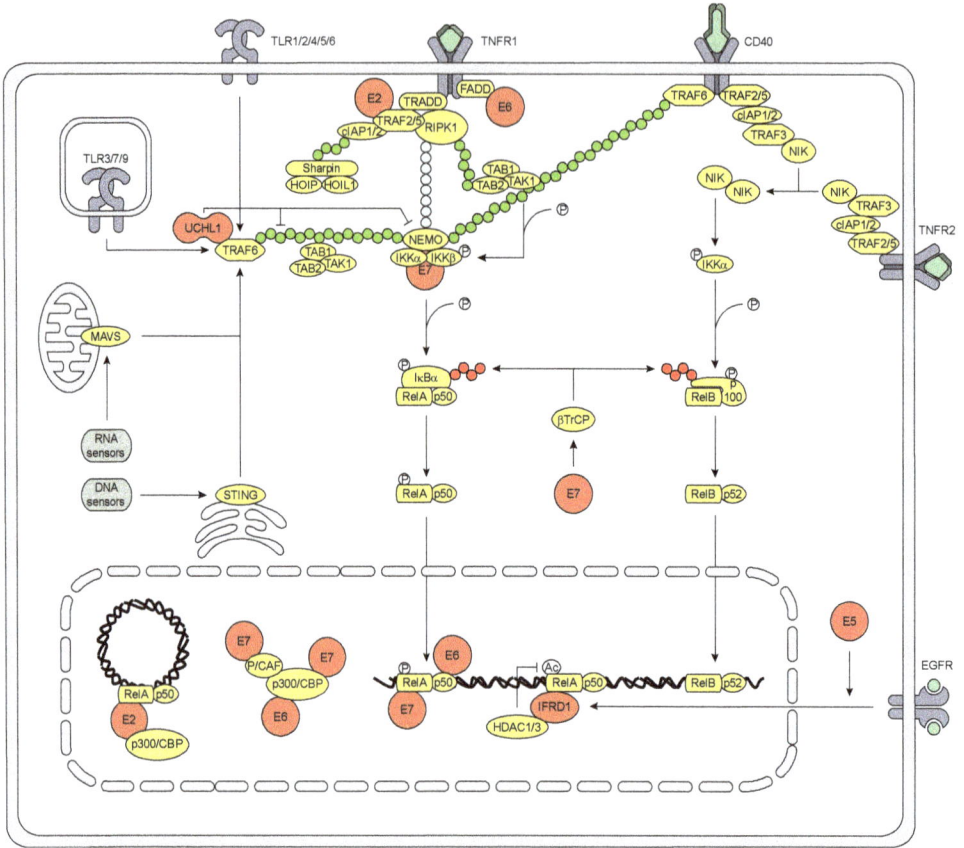

Figure 3. Schematic representation of the effects of hrHPV on NFκB signaling. The canonical NFκB1 pathway is activated by PRRs and CD40 through TRAF6 and TNFR1 through RIP1. Poly-ubiquitination of TRAF6 and RIP1 recruits the TAB1-TAB2-TAK1 and IKK complexes resulting in the phosphorylation of IKKβ by TAK1. IKKβ phosphorylates IκBα, which is then ubiquitinated by SCF-βTrCP and subsequently degraded, and thereby releases the NFκB1 complex to translocate to the nucleus. CD40 and TNFR2 initiate non-canonical NFκB2 signaling by recruitment of TRAF2/5, cIAP1/2 and TRAF3 to the respective receptor, leading to TRAF3 degradation. This causes NIK to accumulate and activate IKKα to phosphorylate p100. This induces SCF-βTrCP to ubiquitinate p100, leading to the proteosomal processing of p100 into p52, and the subsequent nuclear translocation of NFκB2. In the nucleus NFκB binds to the DNA and is aided by coactivators to initiate gene transcription. HPV utilizes its own encoded E proteins (red) as well as exploits the cellular proteins (red) UCHL1 and IFRD1 to interfere with NFκB1 signaling at multiple positions in the pathway. Green circles indicate K63-linked poly-ubiquitin chains, red circles indicate K48-linked poly-ubiquitin chains, and blue circles indicate linear poly-ubiquitin chains.

Within the nucleus, E6 reduces NFκB RelA-dependent transcriptional activity [48], by binding to the C/H1, C/H3 and C terminal domains of CBP/p300 [49,50], thereby competing with RelA and SRC1, which bind the C/H1 and C terminal domain of CBP/p300, respectively [51]. P/CAF can still bind to the C/H3 domain of CBP/p300 in presence of E6, but P/CAF cannot acetylate NFκB since E7 binds to, and thereby blocks, the HAT domain of P/CAF [51]. E7 blocks NFκB DNA binding activity [35] and competes with E2 for binding the C/H1 domain of p300/CBP, thereby hampering E2 transactivation [52]. In contrast, E2 binds to p300/CBP [53,54] and increases NFκB signaling by enhancing RelA expression and transcriptional activation upon TNFα treatment [45].

HrHPV upregulates EGFR gene and surface expression via the E5, E6 and E7 proteins [55,56], and enhances EGFR signaling via E5 and E6 [57–59]. EGFR activation on epithelial cells has been shown to result in a decreased production of pro-inflammatory cytokines [60–62]. HrHPV-induced EGFR signaling, via mTOR, RAF and/or MEK1, increases the expression of the cellular protein interferon-related developmental regulator 1 (IFRD1) which mediates RelA K310 deacetylation by HDAC1/3 and, thereby, attenuates the transcriptional activity of NFκB1 [56].

3.4. The Effect on the Inflammasome Pathway

It is not clear whether the inflammasome pathway is important in the protection against HPV. However, recently it was reported that the production of IL1β, a cytokine that is secreted upon cleavage of pro-IL1β by inflammasome-activated caspase1, is impaired. HPV E6 binds to E6-AP and p53 and this complex induces the inflammasome-independent proteasome-mediated degradation of pro-IL1β and, as such, hampers IL1β formation [63], indicating that hrHPV may suppress immunity by interference with post-translational processes.

Altogether it is clear that hrHPV invested heavily in preventing infected cells to adapt an anti-viral state as well as to suppress the production of cytokines that can induce the attraction of adaptive immune cells which may control HPV infection.

4. The Action of KCs to Secondary Immune Signals is Suppressed by HPV

Cells of the adaptive immune system, in particular T cells, are activated by APCs in the lymph nodes and migrate to infected sites. They produce cytokines and express ligands that can activate signaling cascades in the KC involved in survival and pro-inflammatory cytokine production, leading to killing of KCs and in parallel the reinforcement of adaptive immunity. Despite the infiltration of adaptive immune effector cells the persistence of hrHPV-infected sites suggests that hrHPV has evolved mechanisms to resist this attack. CD4$^+$ T helper 1 (Th1) cells are especially important in controlling hrHPV infections. However, even vaccines that boost viral Th1 immunity during chronic infection are only partially successful [64]. Th1 cells produce IFNγ and TNFα, and express CD40L, which induce cytokine production and proliferative changes in KCs.

TNFα is the ligand for both the TNFα receptor 1 (TNFR1) and TNFR2. TNFR1 activates canonical NFκB1 by recruiting and activating TRADD, leading to the formation of a complex consisting of RIP1, TRAF2 or 5, and cIAP1 or 2. cIAP1/2 is ubiquitinated with a K63-linked poly-ubiquitin chain to which the LUBAC complex (consisting of Sharpin, HOIP and HOIL1) binds. RIP1 is ubiquitinated with both K63-linked and linear poly-ubiquitin chains. The TAB1-TAB2-TAK1 complex binds to the K63-linked poly-ubiquitin chain and phosphorylates the IKK complex that binds to the linear poly-ubiquitin chain of RIP1, leading to NFκB1 release through IKKβ-induced SCF-βTrCP-mediated degradation of IκBα. TNFR2 activates the non-canonical NFκB2 pathway by recruiting TRAF2/5, cIAP1/2 and TRAF3, resulting in TRAF3 degradation. This abrogates TRAF3-induced NIK degradation, causing NIK to accumulate and activate IKKα. IKKα phosphorylates the p100 NFκB precursor protein of the NFκB2 complex, which further consists of RelB. This induces SCF-βTrCP to ubiquitinate p100 with a K48-linked poly-ubiquitin chain, leading to the proteosomal processing of p100 into p52, and the subsequent nuclear translocation of the p52-RelB dimer (Figure 3).

HPV interferes with these cascades in a similar way as it attenuates PRR-induced NFκB signaling by using its own E proteins and endogenous proteins. Additionally, E6 binds to the C terminus of TNFR1 [65], and the N terminus of the death effector domains (DEDs) of FADD, which accelerates the degradation of FADD [66], thereby hampering the induction of apoptosis. E6 does not bind to the TRADD adaptor molecule [66]. Furthermore, E7 binds to the IKK complex and attenuates TNFα-induced kinase activity of IKKα and IKKβ, which hampers IκBα phosphorylation and degradation, and subsequent NFκB nuclear translocation [48]. In contrast to E6, E7, UCHL1 and IFRD1, E2 stimulates TNFα-induced, but not IL1-induced, NFκB signaling [45,67], by directly interacting with TRAF5 and TRAF6, but not TRAF2, thereby stimulating K63-linked ubiquitination of TRAF5 [67].

IFNγ and TNFα are known to synergistically affect gene expression, and also in KCs pro-inflammatory cytokine expression is synergistically higher than expression induced by IFNγ or TNFα alone. Still, hrHPV attenuates IFNγ and TNFα-induced pro-inflammatory cytokine expression and the attraction of PBMCs to KCs that have been stimulated with the combination of IFNγ and TNFα [56]. Ligation of the IFNγR with type II IFN results in the activation of JAK1 and JAK2 and recruitment and phosphorylation of STAT1, which homo-dimerizes, translocates to the nucleus, binds to GAS on the DNA and initiates ISG transcription (Figure 2). The effects of HrHPV on the IFNγ-signaling pathway might be explained by the repressed STAT1 expression and protein levels in HPV infected cells, albeit that STAT1 phosphorylation still is intact [39,68]. However, exposure of hrHPV-infected KCs to IFNγ fails to induce cellular programs associated with a block of proliferation [68]. Furthermore, IFNγ and TNFα stimulation induces processing of the non-canonical NFκB precursor p100 into p52 in hrHPV-infected cells but not uninfected KCs (Tummers, Unpublished data), indicating that hrHPV skews the response of KCs upon stimulation with TNFα and IFNγ towards the non-canonical NFκB pathway. Potentially, this is caused by E7 as this oncoprotein was shown to increase SCF-βTrCP protein levels [69] and in this way might accelerate IκBα degradation and p100 processing [70]. Although unexplored at this point, it is highly likely that this forms another pathway allowing hrHPV-infected cells to resist control of infection by the immune system. Last but not least, epithelial cells express CD40 on their cell surface [71]. Ligation of CD40 induces both canonical and non-canonical NFκB activation, similar to TNFR1 and 2, respectively. Activation of this pathway in epithelial cells results in a very coordinated response by KCs, dominated by the expression of genes involved in leukocyte migration, cell-to-cell signaling and interaction, as well as cell death and survival. The presence of HPV does not affect the gene expression profile of CD40 stimulated KCs, but it does attenuate the extent of the response and reduces the attraction of PBMCs [72]. Based on our previous studies it is likely that the CD40—NFκB1 axes of CD40 signaling is affected via the interaction of UCHL1 and TRAF6, the effects of E7 on the IKK complex, and that of IFRD1 on NFκB1 transcriptional activation. Speculatively, at the non-canonical side signaling could be hampered by abrogation of UCHL1-mediated TRAF2 and/or 5- or E7-mediated IKKα functioning. However, UCHL1-mediated TRAF3 hampering could also lead to constitutive NIK accumulation and subsequent pathway activation. It remains to be determined if hrHPV prefers to skew KCs towards non-canonical NFκB activation after CD40 ligation.

In conclusion, hrHPV does not only try to prevent the attraction of immune cells via the impairment of cytokine secretion but it also hampers the regulation of intracellular growth and apoptosis programs of infected cells that normally are activated as a response to effector molecules of the adaptive immune system.

5. HrHPV Influences MHC Surface Expression and Peptide Presentation

The attack of virus-infected cells by T cells is a highly effective and specific mechanism to prevent the production and spread of virus particles. T cells recognize cells when viral protein-derived peptides are presented in the context of MHC molecules. Literature shows that primary KCs constitute excellent targets for antigen-specific cytotoxic T lymphocytes (CTLs) if their cognate peptide is presented on the KCs cell surface [73]. The overexpression of E5 [74] or E7 [75], however, makes cells more resistant to CTL-mediated lysis. E5 and E7 both reduce MHC-I surface expression, but act on different levels (Figure 4). E7 reduces MHC-I gene expression by physically associating with a putative RXRbeta binding motif (GGTCA) of the proximal promoter of MHC-I genes and recruiting HDAC1, 2 and 8 to this promoter site, leading to repressed chromatin activation. Indeed, E7 knock-down in Caski cells released HDAC1 and 2 from the MHC class-I promoter, and increased histone acetylation and MHC-I expression [75–79]. Furthermore, E7 represses the LMP2 and TAP1 promotors [76,77], two important proteins involved in peptide production and transportation, respectively. E7 also reduces IRF1 expression by suppression of IFNγ-induced STAT1-Tyr701 phosphorylation, repressing IFNγ-mediated upregulation of MHC-I expression via the JAK1/JAK2/STAT1/IRF-1 signal transduction pathway [80,81]. E5 does not influence MHC-I synthesis, but reduces MHC-I surface expression [80] by retaining MHC-I in the Golgi complex via interaction of di-leucine motifs (LL1 and LL3) localized in the N-terminal helical transmembrane (TM1) region of the protein [82]. This E5—MHC-I interaction is not haplotype specific, suggesting that E5 can hamper all MHC-I-dependent antigen presentation [83]. Moreover, binding of the TM1 domain of E5 to the ER chaperone Calnexin retains MHC-I in the ER [84], and down-regulates surface expression of CD1d, a sentinel protein in bridging innate and adaptive immunity [85]. Furthermore, via its C-terminus E5 can bind the B-cell-associated protein 31 (Bap31) [86], a protein involved in the exit of peptide-loaded MHC-I from the ER [87]. Interestingly, E5 selectively downregulates the surface expression of HLA-A and -B, but not that of HLA-C and HLA-E [80]. Under normal conditions expression of HLA class II is not affected but upon IFNγ stimulation E5 does abrogate MHC-II surface expression and blocks peptide-loading of MHC-II and invariant chain degradation [88], by inhibiting endosome acidification [89] or perturbing trafficking from early to late endocytic structures [90].

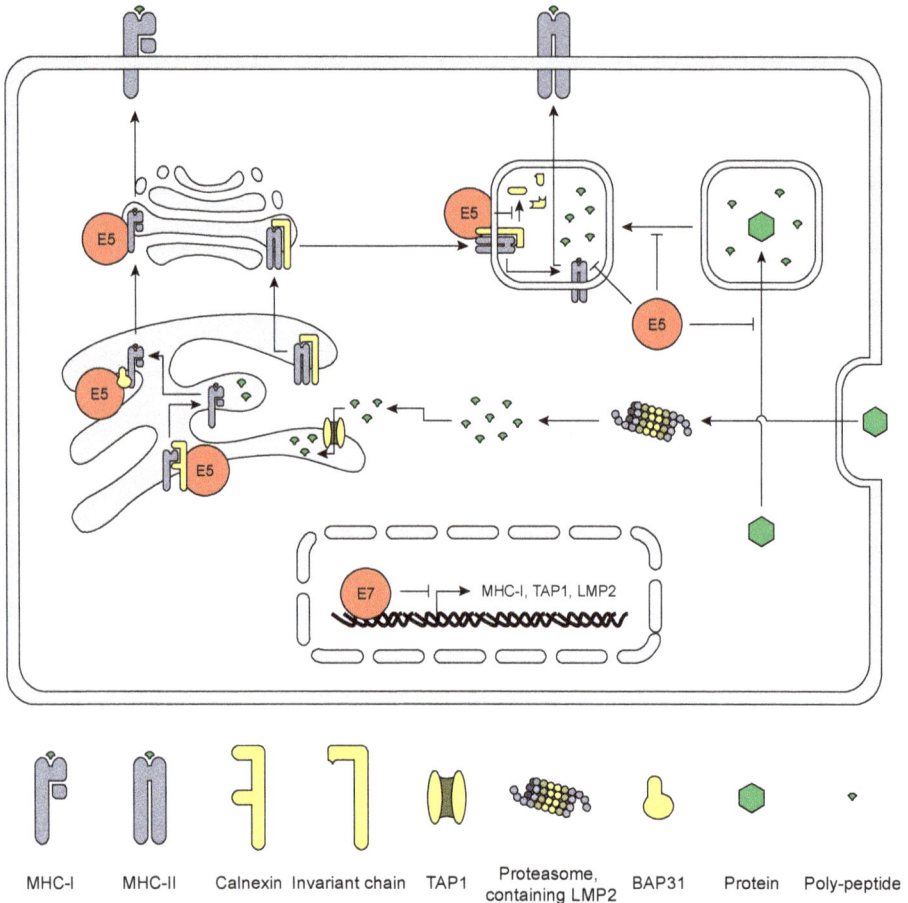

Figure 4. Schematic representation of the effects of hrHPV on antigen presentation. The proteasome processes proteins into peptides, which are transported into the ER via TAP1. Aided by several chaperone proteins, MHC-I is folded and loaded with peptide after which it exits the ER to travel via the Golgi apparatus to the plasma membrane were the peptides are presented to T cells. HPV proteins (red) attenuate gene expression of critical components of this pathway, as well as actively retains MHC-I in the ER and Golgi apparatus. MHC-II forms in the ER and complexes with the invariant chain. The complex travels via the ER and Golgi apparatus to lysosomes where the invariant chain is degraded and MHC-II is loaded with processed peptides from endocytosed proteins. Loaded MHC-II then travels to the plasma membrane to present the peptides. Upon IFNγ stimulation, HPV E5 (red) blocks invariant chain degradation and peptide loading, as well as inhibits endosome acidification and maturation.

Thus, as a last resort, to evade host immunity HPV perturbates the expression of HLA class I and II molecules making the infected cells less visible to the adaptive immune system, slowing down the resolution of infected lesions.

6. Final Comments

Keratinocytes are well equipped to recognize and react to invading pathogens, and hrHPV is no exception to this. However, hrHPV initiates several immune evasion mechanisms soon after infecting the KC. The virus interferes with the innate immune response by affecting several signaling pathways that otherwise would prompt anti-viral mechanisms in the host cell. Furthermore, hrHPV interferes with the production of cytokines that are involved in the attraction of immune cells to the infected epithelium. In addition, the virus hides itself from the immune system by suppressing the antigen presentation machinery normally allowing infected cells to be recognized by adaptive immune cells and, if this is not successful, hrHPV still employs means to hamper the response of KC's to signals from the effector molecules used by adaptive immune cells to exert their antiviral function.

The IFN pathway seems to be centrally attacked through downregulation of STAT1 levels which is observed in hrHPV episome-bearing KCs when compared to uninfected KCs. Downregulation of STAT1 results in attenuated ISG expression, albeit that signaling downstream of the IFNAR and IFNγR still functions [39,68]. Thus, the attenuated type I IFN-induced ISG expression in HPV+ KCs must be due to the basal lowered STAT1 levels. In contrast, in experiments where E6 is overexpressed, E6 was shown to bind TYK2 and to interfere with STAT1 and STAT2 phosphorylation. In addition, overexpressed E7 binds and sequesters IRF9 in the cytosol, so that the ISGF3 complex cannot form in the nucleus. Blocking the type I IFN response is also beneficial for the virus as it allows viral replication. Long-term high-dose IFNβ treatment of HPV-episome bearing KCs results in growth arrest and apoptosis [91,92], thus preventing viral replication. Treatment of KCs with IFN upregulates *IFIT1* (*ISG56*), which can block E1-mediated episome replication by directly interacting with E1, inhibiting E1 DNA helicase activity and causing E1 to translocate from the nucleus to the cytosol [93]. By interfering with IFN signaling through downregulation of STAT1 HPV is able to maintain and amplify its episomes [39]. Interestingly, in virally infected cells p53 was shown to boost type 1 IFN production and signaling resulting in enhanced apoptosis of the infected cells with as consequence limited spread of the infection [94,95]. HPV interferes with the function of p53 and as such with the ability of KCs to boost their antiviral activity.

The canonical NFκB pathway is attacked by hrHPV at multiple positions in the signaling cascade downstream of immune receptors. This indicates that suppression of the NFκB pathway forms a very important target for the virus and implies that this pathway normally would allow the host to resist viral infection. There are several proteins involved in this process. Interestingly, E2 may promote canonical NFκB signaling. It may form an E2-NFκB-p300/CBP transcriptional repressor complex on the LCR of the episome and as such regulates episome transcription which is required for the virus to sustain a low profile. However, as luciferase assays show that the E2 protein renders NFκB more active, the virus thus may prompt E2-mediated NFκB-induced pro-inflammatory cytokine production and immune cell attraction, indicating that the virus needs additional mechanisms in order to regulate the episome while keeping pro-inflammatory cytokine expression in check during infection. Our knock-down experiments in HPV episome-bearing KCs revealed that hrHPV exploits the cellular proteins UCHL1 and IFRD1 to interfere with NFκB signaling. UCHL1 acts on TRAF proteins in the cytosol upstream of NFκB signaling, whereas IFRD1 attenuates the transcriptional activity of NFκB. The combined expression of E2, UCHL1 and IFRD1 during an infection thus might allow hrHPV to regulate its episome while suppressing KCs pro-inflammatory cytokine production (Figure 3). Furthermore, hrHPV seems to skew the response of KCs to IFNγ and TNFα towards the non-canonical NFκB pathway. How and why the virus does this is currently unknown, but it may be that hrHPV utilizes the non-canonical NFκB route to resist the anti-proliferative effects of IFNγ and TNFα [68].

In contrast to hrHPV-infected cells, higher intraepithelial neoplastic lesions and HPV-positive cancers often show overactive canonical NFκB gene expression [96]. Indeed, overexpression experiments showed that E6 and/or E7 can also have pro-NFκB signaling effects and can increase NFκB target gene expression [30]. Mechanistically, E6 targets the NFκB repressor NFX1-91 for degradation [97] and under hypoxic conditions hampers CYLD, a negative regulator of NFκB signaling [98]. E6 also upregulates gene expression of the NFκB signaling components p50, NIK and TRAIP [30]. E7 upregulates SCF-βTrCP protein levels [69], which might lead to accelerated IκBα degradation and p100 processing. The transformed cell may benefit from E6/E7-enhanced NFκB signaling by maintaining a proliferative, anti-apoptotic state, although also pro-inflammatory cytokine expression is increased. Notably, cell type and growth rate are important determinants whether HPV E6 or E6/E7 stimulate or inhibit NFκB activation [99].

Most HPV infections resolve spontaneously, although HPV invests heavily in suppressing host immunity. This indicates that external factors, such as genetic and environmental factors may contribute to the establishment of a persistent infection and progression to cancer. Genetic predisposition to cervical tumors was found [100] and several combinations of single nucleotide polymorphisms (SNPs) were associated with an increased risk to cancer. SNPs in genes of the antigen processing machinery, such as HLA-A, LMP7, TAP2 and ERAP1 [101], and in the FANCA and IRF3 genes [102] were linked to persistent HPV infection and formation of cancer. SNPs in the TLR and NFκB pathways were also studied [103]. Of the thirty-two candidate genes involved in these pathways, including TLR3, NFκB1, NFκB2, RelA, RelB, TRAF3 and TRAF6, only a SNP in the 5' UTR of the lymphotoxin alpha (LTA; TNF superfamily member 1) was significantly associated with increased risks of cervical and vulvar cancers [103]. Based on the interactions between the different proteins in the downstream signaling pathways and their outcomes with respect to activation, splicing, degradation and translocation it might well be that combinations of SNPs, of multiple genes associated with the IRF and NFκB pathways, rather than single SNPs, may confer protection or susceptibility towards persistence of HPV infection and the ultimate progression to cancer.

In conclusion, there is accumulating evidence that hrHPV targets multiple immune-associated pathways. Notably, since viral gene expression considerably differs between hrHPV-infected KCs and hrHPV-transformed cells, data obtained from viral protein overexpression experiments should be carefully interpreted with respect to what their effects are in infection or cancer.

Acknowledgments

This work was supported by the Netherlands Organization for Health Research (NWO/ZonMw) TOP grant 91209012.

Author Contributions

BT and SHvdB wrote the manuscript.

Conflicts of Interest

The authors declare no conflict of interest. The founding sponsors had no role in the writing of the manuscript, and in the decision to publish.

References

1. Zur Hausen, H. Papillomaviruses and cancer: From basic studies to clinical application. *Nat. Rev. Cancer* **2002**, *2*, 342–350.
2. Cubie, H.A. Diseases associated with human papillomavirus infection. *Virology* **2013**, *445*, 21–34.
3. De Villiers, E.M.; Fauquet, C.; Broker, T.R.; Bernard, H.U.; zur Hausen, H. Classification of papillomaviruses. *Virology* **2004**, *324*, 17–27.
4. Akgul, B.; Cooke, J.C.; Storey, A. HPV-associated skin disease. *J. Pathol.* **2006**, *208*, 165–175.
5. Veldhuijzen, N.J.; Snijders, P.J.; Reiss, P.; Meijer, C.J.; van de Wijgert, J.H. Factors affecting transmission of mucosal human papillomavirus. *Lancet Infect. Dis.* **2010**, *10*, 862–874.
6. Arbyn, M.; Anttila, A.; Jordan, J.; Ronco, G.; Schenck, U.; Segnan, N.; Wiener, H.; Herbert, A.; von Karsa, L. European Guidelines for Quality Assurance in Cervical Cancer Screening. Second edition—Summary document. *Ann. Oncol.* **2010**, *21*, 448–458.
7. Ferlay, J.; Soerjomataram, I.; Dikshit, R.; Eser, S.; Mathers, C.; Rebelo, M.; Parkin, D.M.; Forman, D.; Bray, F. Cancer incidence and mortality worldwide: Sources, methods and major patterns in GLOBOCAN 2012. *Int. J. Cancer* **2015**, *136*, E359–E386.
8. Munoz, N.; Bosch, F.X.; de Sanjose, S.; Herrero, R.; Castellsague, X.; Shah, K.V.; Snijders, P.J.; Meijer, C.J. Epidemiologic classification of human papillomavirus types associated with cervical cancer. *N. Engl. J. Med.* **2003**, *348*, 518–527.
9. Smith, J.S.; Lindsay, L.; Hoots, B.; Keys, J.; Franceschi, S.; Winer, R.; Clifford, G.M. Human papillomavirus type distribution in invasive cervical cancer and high-grade cervical lesions: A meta-analysis update. *Int. J. Cancer* **2007**, *121*, 621–632.
10. Sapp, M.; Bienkowska-Haba, M. Viral entry mechanisms: Human papillomavirus and a long journey from extracellular matrix to the nucleus. *FEBS J.* **2009**, *276*, 7206–7216.
11. Doorbar, J. The papillomavirus life cycle. *J. Clin. Virol.* **2005**, *32 (Suppl. 1)*, S7–S15.

12. Bedell, M.A.; Hudson, J.B.; Golub, T.R.; Turyk, M.E.; Hosken, M.; Wilbanks, G.D.; Laimins, L.A. Amplification of human papillomavirus genomes *in vitro* is dependent on epithelial differentiation. *J. Virol.* **1991**, *65*, 2254–2260.

13. Doorbar, J. Molecular biology of human papillomavirus infection and cervical cancer. *Clin. Sci.* **2006**, *110*, 525–541.

14. Lou, Z.; Wang, S. E3 ubiquitin ligases and human papillomavirus-induced carcinogenesis. *J. Int. Med. Res.* **2014**, *42*, 247–260.

15. Van der Burg, S.H.; Melief, C.J. Therapeutic vaccination against human papilloma virus induced malignancies. *Curr. Opin. Immunol.* **2011**, *23*, 252–257.

16. Nestle, F.O.; Di Meglio, P.; Qin, J.Z.; Nickoloff, B.J. Skin immune sentinels in health and disease. *Nat. Rev. Immunol.* **2009**, *9*, 679–691.

17. Karim, R.; Tummers, B.; Meyers, C.; Biryukov, J.L.; Alam, S.; Backendorf, C.; Jha, V.; Offringa, R.; van Ommen, G.J.; Melief, C.J.; *et al.* Human papillomavirus (HPV) upregulates the cellular deubiquitinase UCHL1 to suppress the keratinocyte's innate immune response. *PLOS Pathog.* **2013**, *9*, e1003384.

18. Karim, R.; Meyers, C.; Backendorf, C.; Ludigs, K.; Offringa, R.; van Ommen, G.J.; Melief, C.J.; van der Burg, S.H.; Boer, J.M. Human papillomavirus deregulates the response of a cellular network comprising of chemotactic and proinflammatory genes. *PLOS ONE* **2011**, *6*, e17848.

19. Reinholz, M.; Kawakami, Y.; Salzer, S.; Kreuter, A.; Dombrowski, Y.; Koglin, S.; Kresse, S.; Ruzicka, T.; Schauber, J. HPV16 activates the AIM2 inflammasome in keratinocytes. *Arch. Dermatol. Res.* **2013**, *305*, 723–732.

20. Kalali, B.N.; Kollisch, G.; Mages, J.; Muller, T.; Bauer, S.; Wagner, H.; Ring, J.; Lang, R.; Mempel, M.; Ollert, M. Double-stranded RNA induces an antiviral defense status in epidermal keratinocytes through TLR3-, PKR-, and MDA5/RIG-I-mediated differential signaling. *J. Immunol.* **2008**, *181*, 2694–2704.

21. Hasan, U.A.; Bates, E.; Takeshita, F.; Biliato, A.; Accardi, R.; Bouvard, V.; Mansour, M.; Vincent, I.; Gissmann, L.; Iftner, T.; *et al.* TLR9 expression and function is abolished by the cervical cancer-associated human papillomavirus type 16. *J. Immunol.* **2007**, *178*, 3186–3197.

22. Reiser, J.; Hurst, J.; Voges, M.; Krauss, P.; Munch, P.; Iftner, T.; Stubenrauch, F. High-risk human papillomaviruses repress constitutive kappa interferon transcription via E6 to prevent pathogen recognition receptor and antiviral-gene expression. *J. Virol.* **2011**, *85*, 11372–11380.

23. Hasan, U.A.; Zannetti, C.; Parroche, P.; Goutagny, N.; Malfroy, M.; Roblot, G.; Carreira, C.; Hussain, I.; Muller, M.; Taylor-Papadimitriou, J.; *et al.* The human papillomavirus type 16 E7 oncoprotein induces a transcriptional repressor complex on the Toll-like receptor 9 promoter. *J. Exp. Med.* **2013**, *210*, 1369–1387.

24. Andersen, J.M.; Al-Khairy, D.; Ingalls, R.R. Innate immunity at the mucosal surface: Role of toll-like receptor 3 and toll-like receptor 9 in cervical epithelial cell responses to microbial pathogens. *Biol. Reprod.* **2006**, *74*, 824–831.

25. Le Bon, A.; Tough, D.F. Links between innate and adaptive immunity via type I interferon. *Curr. Opin. Immunol.* **2002**, *14*, 432–436.

26. McNab, F.; Mayer-Barber, K.; Sher, A.; Wack, A.; O'Garra, A. Type I interferons in infectious disease. *Nat. Rev. Immunol.* **2015**, *15*, 87–103.

27. O'Neill, L.A.; Golenbock, D.; Bowie, A.G. The history of Toll-like receptors—Redefining innate immunity. *Nat. Rev. Immunol.* **2013**, *13*, 453–460.

28. Chang, Y.E.; Laimins, L.A. Microarray analysis identifies interferon-inducible genes and Stat-1 as major transcriptional targets of human papillomavirus type 31. *J. Virol.* **2000**, *74*, 4174–4182.

29. Karstensen, B.; Poppelreuther, S.; Bonin, M.; Walter, M.; Iftner, T.; Stubenrauch, F. Gene expression profiles reveal an upregulation of E2F and downregulation of interferon targets by HPV18 but no changes between keratinocytes with integrated or episomal viral genomes. *Virology* **2006**, *353*, 200–209.

30. Nees, M.; Geoghegan, J.M.; Hyman, T.; Frank, S.; Miller, L.; Woodworth, C.D. Papillomavirus type 16 oncogenes downregulate expression of interferon-responsive genes and upregulate proliferation-associated and NF-kappaB-responsive genes in cervical keratinocytes. *J. Virol.* **2001**, *75*, 4283–4296.

31. Sunthamala, N.; Thierry, F.; Teissier, S.; Pientong, C.; Kongyingyoes, B.; Tangsiriwatthana, T.; Sangkomkamhang, U.; Ekalaksananan, T. E2 proteins of high risk human papillomaviruses down-modulate STING and IFN-kappa transcription in keratinocytes. *PLOS One* **2014**, *9*, e91473.

32. Rincon-Orozco, B.; Halec, G.; Rosenberger, S.; Muschik, D.; Nindl, I.; Bachmann, A.; Ritter, T.M.; Dondog, B.; Ly, R.; Bosch, F.X.; *et al.* Epigenetic silencing of interferon-kappa in human papillomavirus type 16-positive cells. *Cancer Res.* **2009**, *69*, 8718–8725.

33. Ronco, L.V.; Karpova, A.Y.; Vidal, M.; Howley, P.M. Human papillomavirus 16 E6 oncoprotein binds to interferon regulatory factor-3 and inhibits its transcriptional activity. *Genes Dev.* **1998**, *12*, 2061–2072.

34. Park, J.S.; Kim, E.J.; Kwon, H.J.; Hwang, E.S.; Namkoong, S.E.; Um, S.J. Inactivation of interferon regulatory factor-1 tumor suppressor protein by HPV E7 oncoprotein. Implication for the E7-mediated immune evasion mechanism in cervical carcinogenesis. *J. Biol. Chem.* **2000**, *275*, 6764–6769.

35. Perea, S.E.; Massimi, P.; Banks, L. Human papillomavirus type 16 E7 impairs the activation of the interferon regulatory factor-1. *Int. J. Mol. Med.* **2000**, *5*, 661–666.

36. Muto, V.; Stellacci, E.; Lamberti, A.G.; Perrotti, E.; Carrabba, A.; Matera, G.; Sgarbanti, M.; Battistini, A.; Liberto, M.C.; Foca, A. Human papillomavirus type 16 E5 protein induces expression of beta interferon through interferon regulatory factor 1 in human keratinocytes. *J. Virol.* **2011**, *85*, 5070–5080.

37. Ivashkiv, L.B.; Donlin, L.T. Regulation of type I interferon responses. *Nat. Rev. Immunol.* **2014**, *14*, 36–49.

38. Li, S.; Labrecque, S.; Gauzzi, M.C.; Cuddihy, A.R.; Wong, A.H.; Pellegrini, S.; Matlashewski, G.J.; Koromilas, A.E. The human papilloma virus (HPV)-18 E6 oncoprotein physically associates with Tyk2 and impairs Jak-STAT activation by interferon-alpha. *Oncogene* **1999**, *18*, 5727–5737.

39. Hong, S.; Mehta, K.P.; Laimins, L.A. Suppression of STAT-1 expression by human papillomaviruses is necessary for differentiation-dependent genome amplification and plasmid maintenance. *J. Virol.* **2011**, *85*, 9486–9494.

40. Barnard, P.; McMillan, N.A. The human papillomavirus E7 oncoprotein abrogates signaling mediated by interferon-alpha. *Virology* **1999**, *259*, 305–313.

41. Barnard, P.; Payne, E.; McMillan, N.A. The human papillomavirus E7 protein is able to inhibit the antiviral and anti-growth functions of interferon-alpha. *Virology* **2000**, *277*, 411–419.

42. Oeckinghaus, A.; Hayden, M.S.; Ghosh, S. Crosstalk in NF-kappaB signaling pathways. *Nat. Immunol.* **2011**, *12*, 695–708.

43. Havard, L.; Delvenne, P.; Frare, P.; Boniver, J.; Giannini, S.L. Differential production of cytokines and activation of NF-kappaB in HPV-transformed keratinocytes. *Virology* **2002**, *298*, 271–285.

44. Havard, L.; Rahmouni, S.; Boniver, J.; Delvenne, P. High levels of p105 (NFKB1) and p100 (NFKB2) proteins in HPV16-transformed keratinocytes: Role of E6 and E7 oncoproteins. *Virology* **2005**, *331*, 357–366.

45. Prabhavathy, D.; Prabhakar, B.N.; Karunagaran, D. HPV16 E2-mediated potentiation of NF-kappaB activation induced by TNF-alpha involves parallel activation of STAT3 with a reduction in E2-induced apoptosis. *Mol. Cell. Biochem.* **2014**, *394*, 77–90.

46. Vancurova, I.; Wu, R.; Miskolci, V.; Sun, S. Increased p50/p50 NF-kappaB activation in human papillomavirus type 6- or type 11-induced laryngeal papilloma tissue. *J. Virol.* **2002**, *76*, 1533–1536.

47. Takami, Y.; Nakagami, H.; Morishita, R.; Katsuya, T.; Cui, T.X.; Ichikawa, T.; Saito, Y.; Hayashi, H.; Kikuchi, Y.; Nishikawa, T.; *et al.* Ubiquitin carboxyl-terminal hydrolase L1, a novel deubiquitinating enzyme in the vasculature, attenuates NF-kappaB activation. *Arterioscler. Thromb. Vasc. Biol.* **2007**, *27*, 2184–2190.

48. Spitkovsky, D.; Hehner, S.P.; Hofmann, T.G.; Moller, A.; Schmitz, M.L. The human papillomavirus oncoprotein E7 attenuates NF-kappa B activation by targeting the Ikappa B kinase complex. *J. Biol. Chem.* **2002**, *277*, 25576–25582.

49. Patel, D.; Huang, S.M.; Baglia, L.A.; McCance, D.J. The E6 protein of human papillomavirus type 16 binds to and inhibits co-activation by CBP and p300. *EMBO J.* **1999**, *18*, 5061–5072.

50. Zimmermann, H.; Degenkolbe, R.; Bernard, H.U.; O'Connor, M.J. The human papillomavirus type 16 E6 oncoprotein can down-regulate p53 activity by targeting the transcriptional coactivator CBP/p300. *J. Virol.* **1999**, *73*, 6209–6219.

51. Huang, S.M.; McCance, D.J. Down regulation of the interleukin-8 promoter by human papillomavirus type 16 E6 and E7 through effects on CREB binding protein/p300 and P/CAF. *J. Virol.* **2002**, *76*, 8710–8721.

52. Bernat, A.; Avvakumov, N.; Mymryk, J.S.; Banks, L. Interaction between the HPV E7 oncoprotein and the transcriptional coactivator p300. *Oncogene* **2003**, *22*, 7871–7881.

53. Lee, D.; Lee, B.; Kim, J.; Kim, D.W.; Choe, J. cAMP response element-binding protein-binding protein binds to human papillomavirus E2 protein and activates E2-dependent transcription. *J. Biol. Chem.* **2000**, *275*, 7045–7051.

54. Marcello, A.; Massimi, P.; Banks, L.; Giacca, M. Adeno-associated virus type 2 rep protein inhibits human papillomavirus type 16 E2 recruitment of the transcriptional coactivator p300. *J. Virol.* **2000**, *74*, 9090–9098.

55. Akerman, G.S.; Tolleson, W.H.; Brown, K.L.; Zyzak, L.L.; Mourateva, E.; Engin, T.S.; Basaraba, A.; Coker, A.L.; Creek, K.E.; Pirisi, L. Human papillomavirus type 16 E6 and E7 cooperate to increase epidermal growth factor receptor (EGFR) mRNA levels, overcoming mechanisms by which excessive EGFR signaling shortens the life span of normal human keratinocytes. *Cancer Res.* **2001**, *61*, 3837–3843.

56. Tummers, B.; Goedemans, R.; Pelascini, L.P.L.; Jordanova, E.S.; van Esch, E.M.G.; Meyers, C.; Melief, C.J.M.; Boer, J.M.; van der Burg, S.H. The interferon-related developmental regulator 1 is used by human papillomavirus to suppress NFκB activation. *Nat. Commun.* **2015**, *6*, 6537.

57. Kim, S.H.; Juhnn, Y.S.; Kang, S.; Park, S.W.; Sung, M.W.; Bang, Y.J.; Song, Y.S. Human papillomavirus 16 E5 up-regulates the expression of vascular endothelial growth factor through the activation of epidermal growth factor receptor, MEK/ ERK1,2 and PI3K/Akt. *Cell. Mol. Life Sci.* **2006**, *63*, 930–938.

58. Spangle, J.M.; Munger, K. The HPV16 E6 oncoprotein causes prolonged receptor protein tyrosine kinase signaling and enhances internalization of phosphorylated receptor species. *PLOS Pathog.* **2013**, *9*, e1003237.

59. Zhang, B.; Srirangam, A.; Potter, D.A.; Roman, A. HPV16 E5 protein disrupts the c-Cbl-EGFR interaction and EGFR ubiquitination in human foreskin keratinocytes. *Oncogene* **2005**, *24*, 2585–2588.

60. Kalinowski, A.; Ueki, I.; Min-Oo, G.; Ballon-Landa, E.; Knoff, D.; Galen, B.; Lanier, L.L.; Nadel, J.A.; Koff, J.L. EGFR activation suppresses respiratory virus-induced IRF1-dependent CXCL10 production. *Am. J. Physiol. Lung Cell. Mol. Physiol.* **2014**, *307*, L186–L196.

61. Mascia, F.; Mariani, V.; Girolomoni, G.; Pastore, S. Blockade of the EGF receptor induces a deranged chemokine expression in keratinocytes leading to enhanced skin inflammation. *Am. J. Pathol.* **2003**, *163*, 303–312.

62. Paul, T.; Schumann, C.; Rudiger, S.; Boeck, S.; Heinemann, V.; Kachele, V.; Steffens, M.; Scholl, C.; Hichert, V.; Seufferlein, T.; *et al.* Cytokine regulation by epidermal growth factor receptor inhibitors and epidermal growth factor receptor inhibitor associated skin toxicity in cancer patients. *Eur. J. Cancer* **2014**, *50*, 1855–1863.

63. Niebler, M.; Qian, X.; Hofler, D.; Kogosov, V.; Kaewprag, J.; Kaufmann, A.M.; Ly, R.; Bohmer, G.; Zawatzky, R.; Rosl, F.; *et al.* Post-translational control of IL-1beta via the human papillomavirus type 16 E6 oncoprotein: A novel mechanism of innate immune escape mediated by the E3-ubiquitin ligase E6-AP and p53. *PLOS Pathog.* **2013**, *9*, e1003536.

64. Van der Burg, S.H.; Arens, R.; Melief, C.J. Immunotherapy for persistent viral infections and associated disease. *Trends Immunol.* **2011**, *32*, 97–103.

65. Filippova, M.; Song, H.; Connolly, J.L.; Dermody, T.S.; Duerksen-Hughes, P.J. The human papillomavirus 16 E6 protein binds to tumor necrosis factor (TNF) R1 and protects cells from TNF-induced apoptosis. *J. Biol. Chem.* **2002**, *277*, 21730–21739.

66. Filippova, M.; Parkhurst, L.; Duerksen-Hughes, P.J. The human papillomavirus 16 E6 protein binds to Fas-associated death domain and protects cells from Fas-triggered apoptosis. *J. Biol. Chem.* **2004**, *279*, 25729–25744.

67. Boulabiar, M.; Boubaker, S.; Favre, M.; Demeret, C. Keratinocyte sensitization to tumour necrosis factor-induced nuclear factor kappa B activation by the E2 regulatory protein of human papillomaviruses. *J. Gen. Virol.* **2011**, *92*, 2422–2427.

68. Tummers, B.; van Esch, E.M.; Ma, W.; Goedemans, R.; Pelascini, L.P.L.; Vilarrasa-Blasi, R.; Torreggiani, S.; Meyers, C.; Melief, C.J.M.; Boer, J.M.; *et al.* Human papillomavirus (HPV) down regulates the expression of IFITM1 to resist the anti-proliferative effects of IFNγ and TNFα. **2015**, submitted.

69. Spardy, N.; Covella, K.; Cha, E.; Hoskins, E.E.; Wells, S.I.; Duensing, A.; Duensing, S. Human papillomavirus 16 E7 oncoprotein attenuates DNA damage checkpoint control by increasing the proteolytic turnover of claspin. *Cancer Res.* **2009**, *69*, 7022–7029.

70. Fukushima, H.; Matsumoto, A.; Inuzuka, H.; Zhai, B.; Lau, A.W.; Wan, L.; Gao, D.; Shaik, S.; Yuan, M.; Gygi, S.P.; *et al.* SCF(Fbw7) modulates the NFkB signaling pathway by targeting NFkB2 for ubiquitination and destruction. *Cell Rep.* **2012**, *1*, 434–443.

71. Galy, A.H.; Spits, H. CD40 is functionally expressed on human thymic epithelial cells. *J. Immunol.* **1992**, *149*, 775–782.

72. Tummers, B.; Goedemans, R.; Jha, V.; Meyers, C.; Melief, C.J.M.; van der Burg, S.H.; Boer, J.M. CD40-mediated amplification of local immunity by epithelial cells is impaired by HPV. *J. Investig. Dermatol.* **2014**, *134*, 2918–2927.

73. Zhou, F.; Frazer, I.H.; Leggatt, G.R. Keratinocytes efficiently process endogenous antigens for cytotoxic T-cell mediated lysis. *Exp. Dermatol.* **2009**, *18*, 1053–1059.

74. Campo, M.S.; Graham, S.V.; Cortese, M.S.; Ashrafi, G.H.; Araibi, E.H.; Dornan, E.S.; Miners, K.; Nunes, C.; Man, S. HPV-16 E5 down-regulates expression of surface HLA class I and reduces recognition by CD8 T cells. *Virology* **2010**, *407*, 137–142.

75. Zhou, F.; Leggatt, G.R.; Frazer, I.H. Human papillomavirus 16 E7 protein inhibits interferon-gamma-mediated enhancement of keratinocyte antigen processing and T-cell lysis. *FEBS J.* **2011**, *278*, 955–963.

76. Georgopoulos, N.T.; Proffitt, J.L.; Blair, G.E. Transcriptional regulation of the major histocompatibility complex (MHC) class I heavy chain, TAP1 and LMP2 genes by the human papillomavirus (HPV) type 6b, 16 and 18 E7 oncoproteins. *Oncogene* **2000**, *19*, 4930–4935.

77. Heller, C.; Weisser, T.; Mueller-Schickert, A.; Rufer, E.; Hoh, A.; Leonhardt, R.M.; Knittler, M.R. Identification of key amino acid residues that determine the ability of high risk HPV16-E7 to dysregulate major histocompatibility complex class I expression. *J. Biol. Chem.* **2011**, *286*, 10983–10997.

78. Li, H.; Ou, X.; Xiong, J.; Wang, T. HPV16E7 mediates HADC chromatin repression and downregulation of MHC class I genes in HPV16 tumorigenic cells through interaction with an MHC class I promoter. *Biochem. Biophys. Res. Commun.* **2006**, *349*, 1315–1321.

79. Li, H.; Zhan, T.; Li, C.; Liu, M.; Wang, Q.K. Repression of MHC class I transcription by HPV16E7 through interaction with a putative RXRbeta motif and NF-kappaB cytoplasmic sequestration. *Biochem. Biophys. Res. Commun.* **2009**, *388*, 383–388.

80. Ashrafi, G.H.; Haghshenas, M.R.; Marchetti, B.; O'Brien, P.M.; Campo, M.S. E5 protein of human papillomavirus type 16 selectively downregulates surface HLA class I. *Int. J. Cancer* **2005**, *113*, 276–283.

81. Zhou, F.; Chen, J.; Zhao, K.N. Human papillomavirus 16-encoded E7 protein inhibits IFN-gamma-mediated MHC class I antigen presentation and CTL-induced lysis by blocking IRF-1 expression in mouse keratinocytes. *J. Gen. Virol.* **2013**, *94*, 2504–2514.

82. Cortese, M.S.; Ashrafi, G.H.; Campo, M.S. All 4 di-leucine motifs in the first hydrophobic domain of the E5 oncoprotein of human papillomavirus type 16 are essential for surface MHC class I downregulation activity and E5 endomembrane localization. *Int. J. Cancer* **2010**, *126*, 1675–1682.

83. Ashrafi, G.H.; Haghshenas, M.; Marchetti, B.; Campo, M.S. E5 protein of human papillomavirus 16 downregulates HLA class I and interacts with the heavy chain via its first hydrophobic domain. *Int. J. Cancer* **2006**, *119*, 2105–2112.

84. Gruener, M.; Bravo, I.G.; Momburg, F.; Alonso, A.; Tomakidi, P. The E5 protein of the human papillomavirus type 16 down-regulates HLA-I surface expression in calnexin-expressing but not in calnexin-deficient cells. *Virol. J.* **2007**, *4*, e116.

85. Miura, S.; Kawana, K.; Schust, D.J.; Fujii, T.; Yokoyama, T.; Iwasawa, Y.; Nagamatsu, T.; Adachi, K.; Tomio, A.; Tomio, K.; *et al.* CD1d, a sentinel molecule bridging innate and adaptive immunity, is downregulated by the human papillomavirus (HPV) E5 protein: A possible mechanism for immune evasion by HPV. *J. Virol.* **2010**, *84*, 11614–11623.

86. Regan, J.A.; Laimins, L.A. Bap31 is a novel target of the human papillomavirus E5 protein. *J. Virol.* **2008**, *82*, 10042–10051.

87. Abe, F.; van Prooyen, N.; Ladasky, J.J.; Edidin, M. Interaction of Bap31 and MHC class I molecules and their traffic out of the endoplasmic reticulum. *J. Immunol.* **2009**, *182*, 4776–4783.

88. Zhang, B.; Li, P.; Wang, E.; Brahmi, Z.; Dunn, K.W.; Blum, J.S.; Roman, A. The E5 protein of human papillomavirus type 16 perturbs MHC class II antigen maturation in human foreskin keratinocytes treated with interferon-gamma. *Virology* **2003**, *310*, 100–108.

89. Straight, S.W.; Herman, B.; McCance, D.J. The E5 oncoprotein of human papillomavirus type 16 inhibits the acidification of endosomes in human keratinocytes. *J. Virol.* **1995**, *69*, 3185–3192.

90. Thomsen, P.; van Deurs, B.; Norrild, B.; Kayser, L. The HPV16 E5 oncogene inhibits endocytic trafficking. *Oncogene* **2000**, *19*, 6023–6032.

91. Chang, Y.E.; Pena, L.; Sen, G.C.; Park, J.K.; Laimins, L.A. Long-term effect of interferon on keratinocytes that maintain human papillomavirus type 31. *J. Virol.* **2002**, *76*, 8864–8874.

92. Herdman, M.T.; Pett, M.R.; Roberts, I.; Alazawi, W.O.; Teschendorff, A.E.; Zhang, X.Y.; Stanley, M.A.; Coleman, N. Interferon-beta treatment of cervical keratinocytes naturally infected with human papillomavirus 16 episomes promotes rapid reduction in episome numbers and emergence of latent integrants. *Carcinogenesis* **2006**, *27*, 2341–2353.

93. Terenzi, F.; Saikia, P.; Sen, G.C. Interferon-inducible protein, P56, inhibits HPV DNA replication by binding to the viral protein E1. *EMBO J.* **2008**, *27*, 3311–3321.

94. Munoz-Fontela, C.; Macip, S.; Martinez-Sobrido, L.; Brown, L.; Ashour, J.; Garcia-Sastre, A.; Lee, S.W.; Aaronson, S.A. Transcriptional role of p53 in interferon-mediated antiviral immunity. *J. Exp. Med.* **2008**, *205*, 1929–1938.

95. Munoz-Fontela, C.; Pazos, M.; Delgado, I.; Murk, W.; Mungamuri, S.K.; Lee, S.W.; Garcia-Sastre, A.; Moran, T.M.; Aaronson, S.A. p53 serves as a host antiviral factor that enhances innate and adaptive immune responses to influenza A virus. *J. Immunol.* **2011**, *187*, 6428–6436.

96. Branca, M.; Giorgi, C.; Ciotti, M.; Santini, D.; di Bonito, L.; Costa, S.; Benedetto, A.; Bonifacio, D.; Di Bonito, P.; Paba, P.; *et al.* Upregulation of nuclear factor-kappaB (NF-kappaB) is related to the grade of cervical intraepithelial neoplasia, but is not an independent predictor of high-risk human papillomavirus or disease outcome in cervical cancer. *Diagn. Cytopathol.* **2006**, *34*, 555–563.

97. Xu, M.; Katzenellenbogen, R.A.; Grandori, C.; Galloway, D.A. NFX1 plays a role in human papillomavirus type 16 E6 activation of NFkappaB activity. *J. Virol.* **2010**, *84*, 11461–11469.

98. An, J.; Mo, D.; Liu, H.; Veena, M.S.; Srivatsan, E.S.; Massoumi, R.; Rettig, M.B. Inactivation of the CYLD deubiquitinase by HPV E6 mediates hypoxia-induced NF-kappaB activation. *Cancer Cell* **2008**, *14*, 394–407.

99. Vandermark, E.R.; Deluca, K.A.; Gardner, C.R.; Marker, D.F.; Schreiner, C.N.; Strickland, D.A.; Wilton, K.M.; Mondal, S.; Woodworth, C.D. Human papillomavirus type 16 E6 and E 7 proteins alter NF-kB in cultured cervical epithelial cells and inhibition of NF-kB promotes cell growth and immortalization. *Virology* **2012**, *425*, 53–60.

100. Magnusson, P.K.; Lichtenstein, P.; Gyllensten, U.B. Heritability of cervical tumours. *Int. J. Cancer* **2000**, *88*, 698–701.

101. Mehta, A.M.; Jordanova, E.S.; van Wezel, T.; Uh, H.W.; Corver, W.E.; Kwappenberg, K.M.; Verduijn, W.; Kenter, G.G.; van der Burg, S.H.; Fleuren, G.J. Genetic variation of antigen processing machinery components and association with cervical carcinoma. *Genes Chromosom. Cancer* **2007**, *46*, 577–586.

102. Wang, S.S.; Bratti, M.C.; Rodriguez, A.C.; Herrero, R.; Burk, R.D.; Porras, C.; Gonzalez, P.; Sherman, M.E.; Wacholder, S.; Lan, Z.E.; *et al.* Common variants in immune and DNA repair genes and risk for human papillomavirus persistence and progression to cervical cancer. *J. Infect. Dis.* **2009**, *199*, 20–30.

103. Bodelon, C.; Madeleine, M.M.; Johnson, L.G.; Du, Q.; Galloway, D.A.; Malkki, M.; Petersdorf, E.W.; Schwartz, S.M. Genetic variation in the TLR and NF-kappaB pathways and cervical and vulvar cancer risk: A population-based case-control study. *Int. J. Cancer* **2014**, *134*, 437–444.

Modulation of DNA Damage and Repair Pathways by Human Tumour Viruses

Robert Hollingworth and Roger J. Grand

Abstract: With between 10% and 15% of human cancers attributable to viral infection, there is great interest, from both a scientific and clinical viewpoint, as to how these pathogens modulate host cell functions. Seven human tumour viruses have been identified as being involved in the development of specific malignancies. It has long been known that the introduction of chromosomal aberrations is a common feature of viral infections. Intensive research over the past two decades has subsequently revealed that viruses specifically interact with cellular mechanisms responsible for the recognition and repair of DNA lesions, collectively known as the DNA damage response (DDR). These interactions can involve activation and deactivation of individual DDR pathways as well as the recruitment of specific proteins to sites of viral replication. Since the DDR has evolved to protect the genome from the accumulation of deleterious mutations, deregulation is inevitably associated with an increased risk of tumour formation. This review summarises the current literature regarding the complex relationship between known human tumour viruses and the DDR and aims to shed light on how these interactions can contribute to genomic instability and ultimately the development of human cancers.

Reprinted from *Viruses*. Cite as: Hollingworth, R.; Grand, R.J. Modulation of DNA Damage and Repair Pathways by Human Tumour Viruses. *Viruses* **2015**, *7*, 2542-2591.

1. Introduction

Over the past two decades interest in the relationship between viruses and the DNA damage response (DDR) has increased exponentially. For all the commonly studied viruses, at least some investigations have been undertaken into the effect of viral infection, or expression of individual viral genes, on activation of the host DDR. Similarly, the ways in which viruses impinge on pathways responsible for DNA repair have been the subject of multiple investigations. While most attention has focused on DNA viruses [1–3], there is also a growing body of literature concerning both RNA viruses and retroviruses, whose lifecycle includes a DNA intermediate [4–6].

Historically, interest in the relationship between viruses and DNA damage originates from observations made half a century ago, that infection with a number of virus species could lead to extensive chromosomal damage [7]. It was subsequently found that, in the case of adenovirus, herpes simplex virus 1 (HSV-1), and human cytomegalovirus (HCMV), chromosomal breaks induced by viruses frequently occur at specific sites [8]. More recently, as knowledge of the DDR has developed, increasing layers of complexity concerning viral/DDR interactions have been revealed. It is now clear that some aspects of the DDR can represent a potent antiviral defence that may be disabled following entry to the host cell. In other cases, viruses take advantage of DDR activation to modulate the cell cycle while specific proteins are hijacked to aid viral replication.

Over the past few years a number of excellent reviews of the virus/DDR relationship have been published [1,9–11]. Some of these have focused on particular aspects of the DDR whereas others have concentrated on particular viral species. Here we have tried to draw together the relevant published literature specific to viruses associated with human tumours *in vivo*, including relevant RNA viruses as well as those with a DNA genome. Thus we have summarised the literature concerning human papilloma virus (HPV), Merkel cell polyomavirus (MCPyV), human T-cell leukemia virus type 1 (HTLV-1), Epstein-Barr virus (EBV), Kaposi's sarcoma-associated herpesvirus (KSHV), hepatitis B virus (HBV) and hepatitis C virus (HCV) (Table 1). By adopting this approach it is hoped that an appreciation can be gained of how the interaction between a virus and the DDR may contribute to its oncogenic potential and, ultimately, to cancer development.

2. Human Tumour Viruses

Human oncogenic viruses have been defined as necessary but not sufficient to initiate cancer. It is now apparent that only in relatively rare cases do oncogenic viruses in isolation give rise to tumours in otherwise healthy individuals. For example, while the majority of the population carries latent persistent EBV infection, EBV-derived tumours are comparatively rare. In the case of KSHV, although most infected individuals remain asymptomatic, when the immune system is incapacitated, this herpesvirus can induce Kaposi's sarcoma in a significant proportion of patients. In the case of other viruses, which are present in a latent state, it seems that multiple genetic events, often attributable to environmental factors, are required to initiate tumour development. As outlined here, it is probable that genomic instability induced during the viral lifecycle plays a significant role in tumour initiation. Regardless of the mechanisms of tumourigenesis, it is estimated that between 10% and 15%, of human malignancies are attributable to viral infection [12] and viruses are now second only to smoking as a leading cause of theoretically preventable cancer. This review will focus on genetic instability introduced by infection with oncogenic viruses, and how these pathogens subsequently modulate pathways responsible for recognition and repair of DNA damage.

3. The DNA Damage Response (DDR)

The DDR is the response of the cell to lesions in its genome that occur during normal cellular processes or are introduced by exogenous agents. The inability to repair this damage can result in a variety of acquired clinical conditions such as cancer and neurodegeneration [13,14]. In addition, several inherited disorders are associated with defects in genes crucial for proper functioning of the DDR. Here we include a brief outline of a very complex and fast developing area of biological research; mention is made of only a limited number of proteins involved in the DDR which are relevant to the subsequent discussion.

3.1. Sensors and Transducers

The cellular response to DNA damage is primarily regulated by the activities of three phosphatidylinositol 3-kinase-like kinases (PIKKs): Ataxia telangiectasia mutated (ATM), ATM and Rad3-related (ATR) and DNA-dependent protein kinase (DNA-PK) [15–17]. ATM and DNA-PK are principally activated in response to double-strand breaks (DSBs) while ATR can be activated following stalled replication forks and by DNA lesions that result in persistent single-stranded DNA (ssDNA). Subsequent phosphorylation of transducer and effector molecules by these kinases results in cell cycle arrest, DNA repair and, potentially, apoptosis or senescence.

DSBs are recognised by the MRN complex (comprising Mre11, Rad50 and NBS1), which mediates the recruitment and activation of ATM [18–20] (Figure 1). The activation of ATM also requires acetylation by TIP60, and autophosphorylation that results in disassociation of the inactive dimer to an active monomer [21,22]. ATM can phosphorylate a large number of downstream targets that include CHK2 and H2AX. The phosphorylation of histone H2AX (giving rise to γH2AX) at sites of DSBs allows binding of MDC1, which acts as a scaffold for the formation of large protein complexes spreading appreciable distances from the site of the actual break [23]. MDC1 promotes recruitment of ubiquitin ligases, such as RNF8 and RNF168, which in turn facilitate binding of 53BP1 and BRCA1 [24–26]. DSBs also lead to activation of DNA-PK, which is involved in the non-homologous end joining (NHEJ) repair pathway summarised below.

The stalling of replication forks and resection of DSBs can generate stretches of ssDNA that provide a signal for activation of the ATR kinase (Figure 2) [16]. The ssDNA is bound by RPA which recruits ATR via its binding partner ATRIP. The generation of ds/ssDNA junctions can promote loading of the Rad9-Rad1-Hus1 (9-1-1) complex by the Rad17-RFC complex [27]. The recruitment of TOPBP1 then facilitates ATR activation which subsequently phosphorylates CHK1 with the help of the adaptor protein Claspin [28]. The Timeless-Tipin complex also contributes to CHK1 activation by interacting with both RPA and Claspin [29]. Activation of the ATR-CHK1 pathway is able to slow replication by inhibition of replication origin firing and initiate cell cycle checkpoints by phosphorylation of the Cdc25 phosphatases [16,30].

3.2. Activation of Cell Cycle Checkpoints

A typical consequence of DDR activation is the triggering of cell cycle arrest that prevents replication of potentially mutagenic lesions and provides opportunity for DNA repair [31]. Three major checkpoints are involved in monitoring the integrity of the genome under normal circumstances. The G1/S checkpoint is activated to prevent the initiation of cellular replication in the presence of DNA damage. The intra-S-phase checkpoint can slow or arrest DNA synthesis following replication stress or DNA damage that occurs during DNA replication. Finally, the G2/M checkpoint is activated to ensure that damage incurred during S and G2 is not segregated during mitosis (Figure 3).

Activation of ATM and ATR leads to phosphorylation of CHK1 and CHK2, which in turn regulate downstream targets, such as p53, Cdc25 and Wee1. The phosphorylation of the Cdc25 phosphatases leads to their degradation, which prevents the activation of cyclin-dependent kinases (CDKs) causing cell cycle delay. The key transcriptional target following p53 stabilization is p21, which can impede G1/S transition through inhibition of the cyclin E/Cdk2 complex. The retinoblastoma protein (Rb) also has a key role in cell cycle progression by binding the E2F transcription factors and inhibiting progression into S-phase. In addition, activation of the tyrosine kinase Wee1 following DNA damage leads to inhibition of the cyclin B/Cdk1 complex preventing G2/M transition.

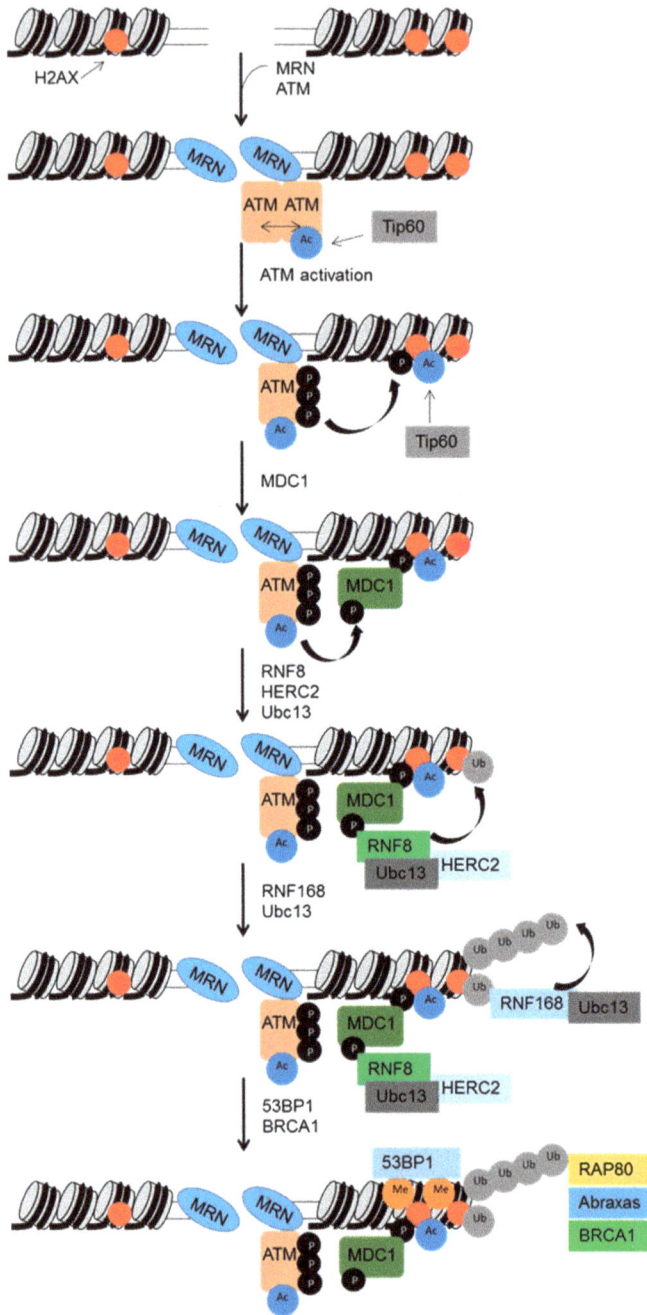

Figure 1. ATM activation in response to DSBs. The MRN complex rapidly migrates to sites of DSBs and, along with the acetyltransferase Tip60, contributes to activation of ATM kinase activity. Phosphorylation of histone H2AX by ATM results in binding of MDC1 which subsequently mediates recruitment of factors, such as 53BP1 and BRCA1, which participate in DSB repair and regulation of cell cycle checkpoints.

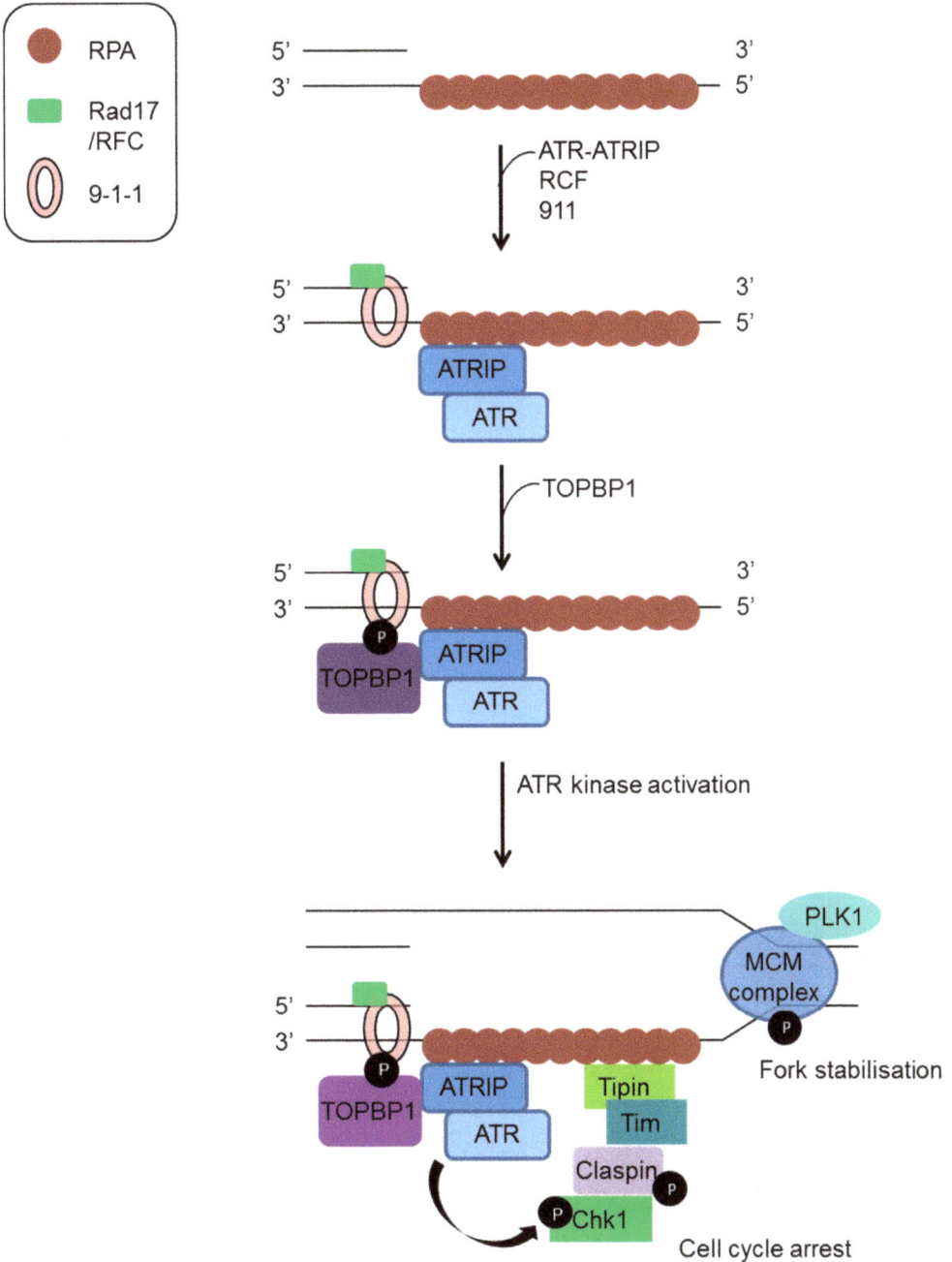

Figure 2. ATR activation in response to stalled replication forks. ATR and ATRIP bind to stretches of ssDNA coated with RPA while the Rad17-RFC complex independently loads the 9-1-1 checkpoint clamp onto ssDNA/dsDNA junctions. Subsequent recruitment of TOPBP1 mediates activation of ATR while Claspin and the Timeless (Tim)/Tipin complex facilitate CHK1 activation.

Many viruses have evolved mechanisms of disrupting the G1/S checkpoint to allow S-phase progression, which is often required for replication of their genetic material. The best characterised of these is disassociation of the Rb/E2F complex by the small DNA tumour viruses (adenovirus, HPV, simian virus 40 [SV40]) allowing E2F-mediated transcription of viral genes [32]. Several viruses, such as human immunodeficiency virus (HIV), can also activate the G2/M checkpoint to prevent mitotic entry or to maintain a pseudo S-phase state [33]. This can be achieved via several mechanisms including prevention of nuclear entry of the cyclin B/Cdk1 complex, activation of the ATR-CHK1 pathway or manipulation of Wee1 and Cdc25C activities [33].

3.3. Apoptosis and Senescence

While cell cycle arrest and DNA repair are typical consequences of DDR activation, failure to properly correct DNA lesions can lead to cellular senescence or apoptosis. The activation of apoptotic pathways following DNA damage typically occurs via p53 and defects in this response are associated with tumourigenesis. ATM, ATR, CHK1, and CHK2 can all directly phosphorylate p53 preventing its degradation and increasing its transcriptional activity [34]. While p53 can play a protective role by arresting the cell cycle and upregulating DNA repair pathways, it can also induce transcription of pro-apoptotic genes, such as PUMA, NOXA and BAX [35]. Since apoptosis of the host cell can reduce the opportunity for viral propagation, many viral species have evolved mechanisms to inhibit programmed cell death and promote cellular survival [36]. For example, oncogenic herpesviruses encode homologs of BCL-2, an anti-apoptotic protein that prevents the release of cytochrome c from mitochondria [37,38]. HPV E6 meanwhile promotes the ubiquitination and degradation of p53, as well as other pro-apoptotic proteins, such as BAK, FADD and c-Myc [39,40]. Lymphocytes expressing the HTLV-1 Tax protein have elevated levels of c-FLIP, a protein that can inhibit apoptosis mediated by the CD95 death receptor [41].

Persistent activation of the DDR can also lead to cellular senescence whereby the cell remains viable but replicative capacity is lost. A DDR that results in cellular senescence can be triggered by the progressive shortening of telomeres, repetitive DNA sequences that protect chromosome ends, following successive cell divisions [42]. As this provides an important safeguard against malignant proliferation, the interference of tumour viruses with cellular mechanisms involved in the maintenance of telomeres could play a role virally-induced oncogenesis. To promote elongation of telomeres, and thus replicative immortality, viruses can both activate telomerase enzymes and stimulate a more error-prone alternative lengthening of telomere (ALT) pathway [43]. Cellular proliferation induced by viruses can lead to progressive shortening of telomeres and this has been observed in HPV-associated cervical carcinoma and in hepatocarcinomas containing HBV and HCV [44,45]. To counteract this effect, HPV and HBV oncoproteins have been shown to increase expression and activity of hTERT, the catalytic subunit of telomerase [46,47]. There is also evidence of activation of the ALT pathway in human embryonic fibroblasts immortalised by HPV E6 and E7 and also in endothelial cells immortalised by KSHV vGPCR [48,49]. It is also conceivable that DNA damage induced by viruses could cause irreparable damage to telomeric DNA that, if the senescence response is compromised due to deregulation of cell cycle checkpoints, may lead to telomere dysfunction and genomic instability [43].

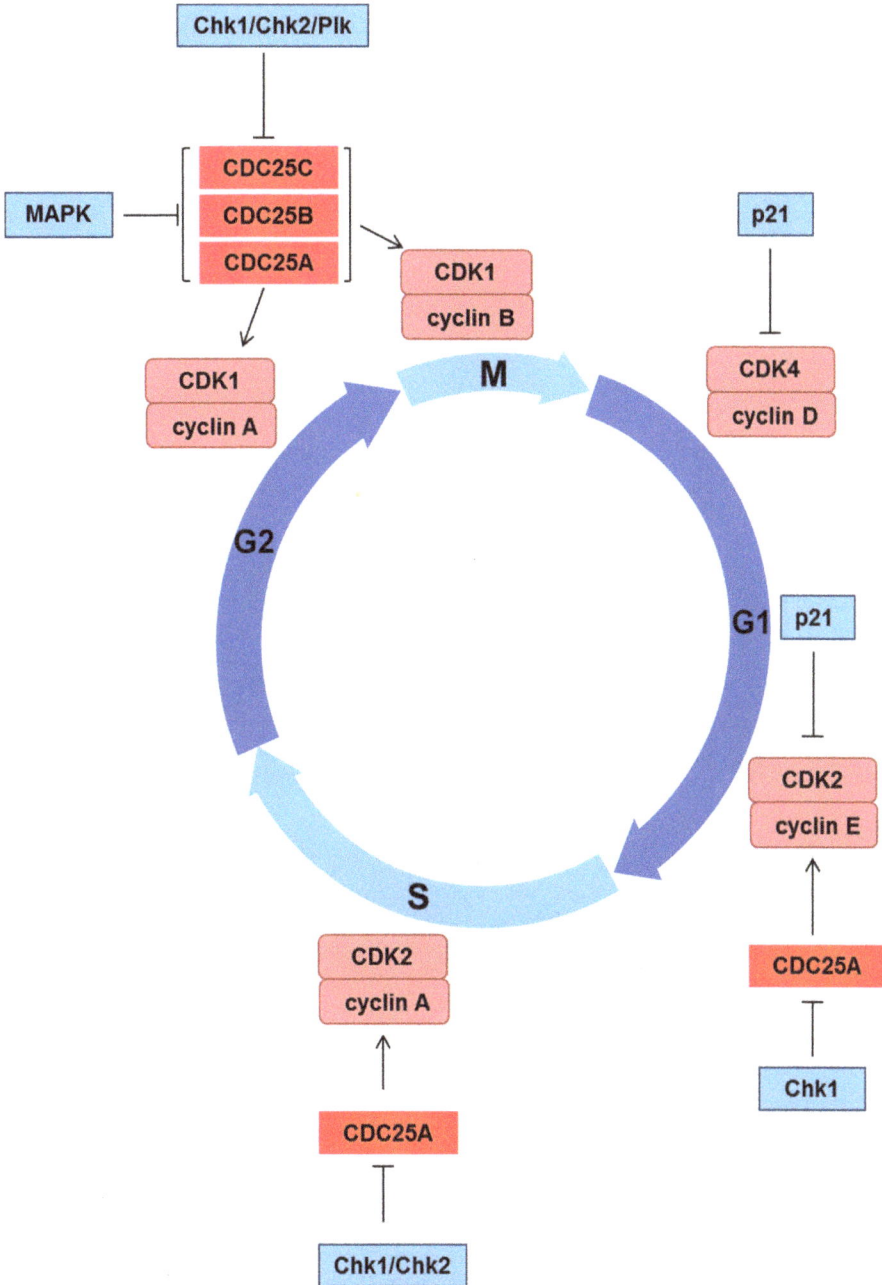

Figure 3. Regulation of the cell cycle by cyclins and CDKs. In response to DNA damage phosphorylation of CHK1 and CHK2 can lead to degradation of Cdc25 phosphatases. Cdc25 degradation inhibits activation of CDKs and delays progression of the cell cycle. Increased expression of p21 by p53 following DNA damage can also mediate cell cycle arrest through inhibition of CDKs.

Table 1. Known human tumour viruses. * Includes well-characterised proteins rather than an exhaustive list.

Virus	Genome	Viral Oncoproteins *	Associated Cancer
Human papilloma virus (HPV)	dsDNA	E6 and E7	Cervical cancer, penile cancer, anogenital carcinoma, head and neck cancer
Merkel cell polyoma virus (MCPyV)	dsDNA	Large T antigen	Merkel cell carcinoma (MCC)
Human T cell leukaemia virus-1 (HTLV-1)	ssRNA	Tax	Adult T cell leukaemia (ATL)
Epstein Barr virus (EBV)	dsDNA	EBNA2, EBNA3C, LMP-1, LMP-2	Nasopharyngeal carcinoma (NPC), Burkitt's lymphoma, Hodgkin's lymphoma, post-transplant lymphoproliferative disease (PTLD), T cell lymphoma, gastric cancer
Kaposi's sarcoma-associated herpesvirus (KSHV)	dsDNA	LANA, v-cyclin, vGPCR, vIL-6, vBcl-2, vFLIP, Kaposin B	Kaposi's sarcoma (KS), primary effusion lymphoma (PEL), multicentric Castleman's disease (MCD)
Hepatitis B virus (HBV)	ssDNA + ssRNA	HBx	Hepatocellular carcinoma (HCC)
Hepatitis C virus (HCV)	ssRNA	Core, NS3, NS5A	Hepatocellular carcinoma (HCC), B-cell lymphoma

4. DNA Damage Repair Pathways

As DNA lesions can take many forms, multiple DNA repair pathways have evolved to correct specific type of DNA damage. A brief overview of the principle mechanisms of DNA repair pathways are included below. In addition, the interactions between human tumour viruses and key components of these repair pathways are summarised in Table 2.

4.1. Non-Homologues End Joining (NHEJ)

Double-strand breaks in DNA can be repaired by non-homologous end joining (NHEJ) that operates throughout the cell cycle (Figure 4). DNA ends are first recognised and bound by the Ku70/80 heterodimer which recruits the catalytic subunit of DNA-PK (DNA-PKcs) [68,69]. Subsequent phosphorylation by DNA-PK results in its disassociation from the site of damage. End processing involves Artemis, an endonuclease, together with polynucleotide kinase/phosphatase (PNKP), DNA polymerase μ (Pol μ) and DNA polymerase λ (Pol λ). These are required to synthesise complementary nucleotides to fill in single-strand overhangs [70]. DNA-PK also facilitates recruitment of the NHEJ ligation complex, comprising DNA Ligase IV, XRCC4 and XLF, which is required for the final joining of the DNA strands [71,72].

Table 2. Viral interactions with components of core DNA repair pathways. This list is not exhaustive but includes observations mentioned in this article. VRCs—Viral replication centres.

Repair Pathway	DDR Target Protein	Virus/Viral Protein	Reference
Direct repair	MGMT	HPV E6	[50]
Base excision repair (BER)	XRCC1	HPV E6	[51]
	B polymerase	HTLV-1 Tax	[52]
Nucleotide excision repair (GG-NER)	XPC	MCPyV LT	[53]
	PCNA	HTLV-1 Tax	[54]
	DDB1	EBV BPLF1	[55]
	XPB/XPD	HBx	[56–59]
Mismatch repair (MMR)	MSH2	KSHV VRCs	[60]
	MSH6	HTLV-1 Tax	[61]
	MLH1	EBV VRCs	[62]
	hPSM2		
Single-strand break repair	PARP-1	KSHV VRCs	[60]
	XRCC1	HPV E6	[51]
Homologous recombination (HR)	Rad51	HPV16 E7	[63]
	BRCA1	EBV VRCs	[64]
	Rad52	HPV31 VRCs	[65]
		HTLV-1 VRCs	[66]
Non-homologous end joining (NHEJ)	DNA-PKcs	KSHV VRCs	[60]
	Ku70/Ku80	HTLV-1 VRCs	[66]
		KSHVorf59	[67]
		KSHV VRCs	[60]

4.2. Homologous Recombination (HR)

Repair of double-strand breaks by homologous recombination (HR), which occurs in the S and G2 phases of the cell cycle, is considered more accurate than NHEJ because an undamaged sister chromatid is used as a template (Figure 5). During HR, DNA ends are processed to form 3′ ssDNA tails that are bound by RPA. This requires the endonuclease activity of the MRN component Mre11 together with CtIP [73,74]. Additional resection involves the combined activities of exonuclease-1 (EXO1), DNA replication helicase 2 (DNA2) and Bloom Syndrome protein (BLM) [24]. The recombinase Rad51, together with BRCA2 and Rad52, displaces RPA and regions of homology on the sister chromatid are located through the action of Rad52 [70,75,76]. A Rad51 nucleoprotein complex with Rad54 locates homologous regions in the undamaged dsDNA template. Strand invasion which leads to the formation of a displacement loop is followed by DNA synthesis by DNA polymerases such as Pol δ using the homologous DNA as a template. HR can be completed either by the synthesis-dependent strand annealing pathway (SDSA) or the DSBR pathway through a process of "second end capture" facilitated by Rad52 [77]. During SDSA the newly synthesised strand is displaced by RTEL helicase, annealing with ssDNA on the other side of the break. Second end capture gives rise to double Holliday junctions which are resolved to separate sister chromatids. This resolution requires the endonucleases Mus1/Eme1 and SLX1/SLX4 and the resolvase GEN1 [78].

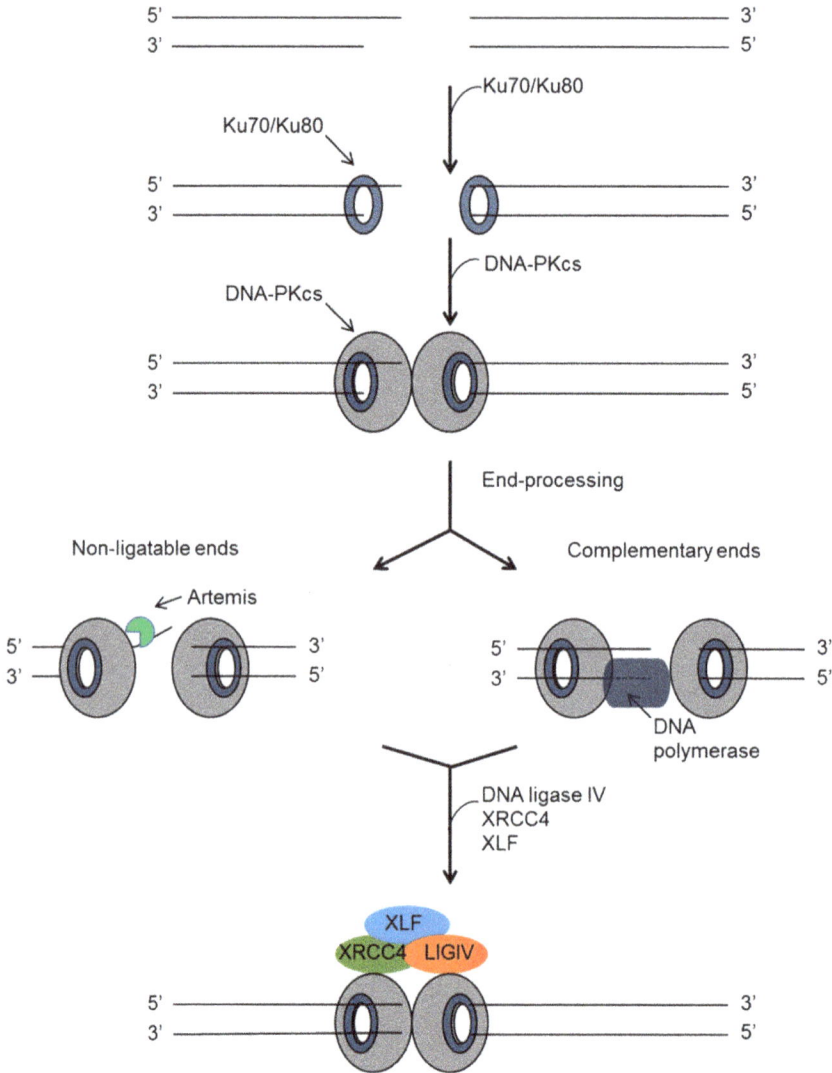

Figure 4. Non-homologous end joining repair of DSBs. Ku70/80 binds to DNA ends and provides a scaffold for the recruitment of additional proteins such as DNA-PKcs. If required, DNA ends may be cleaved by the Artemis nuclease and gaps filled by DNA polymerases. Finally, the ligation complex consisting of XLF, XRCC4 and DNA Ligase IV ligates the DNA ends to complete the repair.

4.3. Single-Strand Break Repair

Single-strand breaks (SSBs) are much more common than DSBs although they are often converted to DSBs which are more deleterious to the cell [79]. SSBs can arise from a variety of sources such as the action of reactive oxygen species (ROS) and the erroneous incorporation of ribonucleotides into DNA. They can also arise indirectly during the process of excision repair and as a result of the activity of DNA topoisomerase 1. SSBs are generally recognised by poly (ADP-ribose) polymerase 1 (PARP1), which is activated by DNA strand breaks and then ribosylates both itself and a number of target proteins [80]. Poly (ADP-ribose) chains are rapidly degraded by poly (ADP-ribose) glycohydrolase (PARG) after SSB repair is completed. PARP1 is responsible for recruitment of XRCC1, to the site of damage. This acts as a scaffold for the accumulation of further factors such as polynucleotide kinase/phosphatase (PNKP), aprataxin, and Polβ. Depending on how the SSB has arisen, and therefore the structure of the 3' and 5' DNA ends, different enzymes are involved in processing and repairing the break. For example, PNKP and AP endonuclease 1 (APE1) modify 3' ends whereas 5' termini are substrates for DNA polymerase β [81]. After processing of the damaged termini, gaps in the DNA are filled by Polβ, Polδ and Polε. However, other proteins, such as XRCC1, PARP1, FEN-1 and proliferating cell nuclear antigen (PCNA) may also contribute to this process [79]. As a final step, the DNA is ligated by Ligase 3α (short patch repair) or Ligase 1 (long patch repair) [82–84].

4.4. Base Excision Repair (BER)

Lesions and mutations, which affect single bases, are removed by the base excision repair pathway (BER). Adducts in DNA tend to occur following exposure to alkylating agents and ROS. Certain modifications, such as O^6-methyl guanine, 1-methyl adenine and 3-methyl cytosine can be repaired directly by alkyltransferases and DNA dioxygenases [85,86]. During BER, damaged DNA bases resulting from single base loss or base oxidation are recognised and removed by one of a number of DNA glycosylases which cleave the bond linking the base to the sugar-phosphate backbone [87]. Removal of the base and end processing by the AP endonuclease 1 (APE1) results in formation of an apyrimidinic/apurinic (AP) site and a SSB. The single nucleotide gap is filled by Polβ and the DNA religated by Ligase 3 and XRCC1 [88]. Alternative BER pathway components, such as PNKP, may come into play depending on the specific DNA glycosylases involved [89].

4.5. Nucleotide Excision Repair (NER)

Nucleotide excision repair (NER) is used to repair DNA which contains large helix-distorting adducts such as cyclo-butane pyrimidine dimers (CPD) and 6, 4 pyrimidine-pyrimidone photoproducts formed as a result of UV irradiation [90]. Two forms of NER are recognised: Global genome NER (GG-NER), which is used to generally repair lesions, and transcription-coupled NER (TC-NER), which is used for repair of transcriptionally active DNA [91]. During GG-NER, lesions are recognised by the Xeroderma-pigmentosum C (XPC)-RAD23B-centrin 2 complex. Recruitment of TFIIH, containing the helicases XPB and XPD and the endonucleases XPG and ERCC1-XPF results in opening of the DNA double helix and cleavage and removal of the aberrant base. The gap is filled by Polδ and Polε in the presence of PCNA and the DNA is religated by Ligase 1 or Ligase 3 [92]. During TC-NER, the lesion is detected by RNA polymerase II (Pol II) when it becomes stalled during elongation [93]. This results in the recruitment of a large protein complex which includes Cockayne Syndrome proteins, ERCC6 and ERCC8, certain core NER factors as well as TC-NER specific proteins UVSSA, USP7 and HMG14 [91]. Repair then proceeds by a similar mechanism to that of GG-NER.

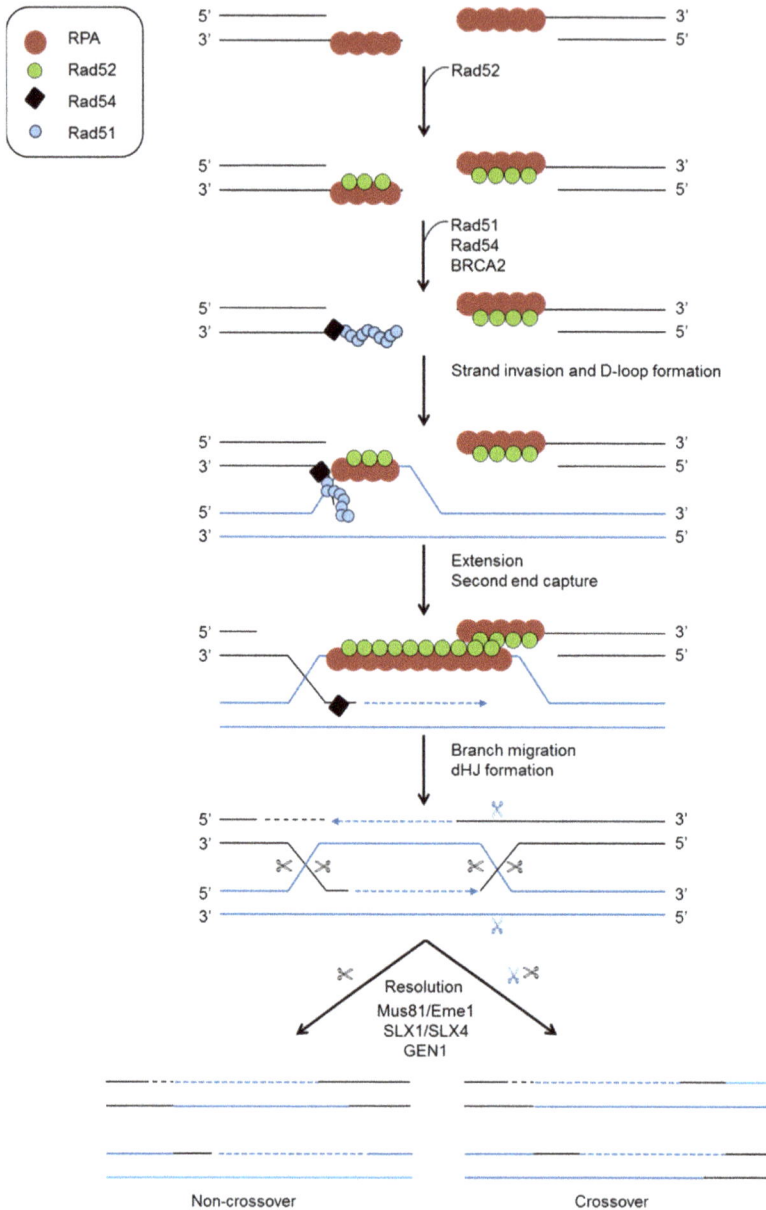

Figure 5. Homologous recombination repair of DBSs. DNA end resection results in ssDNA that is first coated by RPA. Rad51, in conjunction with Rad52 and BRCA2, then displaces RPA. Rad51 and Rad54 catalyse strand invasion and homology search with the undamaged template. Following DNA synthesis via polymerases, the resulting Holliday junctions are resolved.

4.6. Mismatch Repair (MMR)

The mismatch repair (MMR) pathways are activated to deal with DNA base mispairings which arise during DNA replication, generally due to errors of DNA polymerases [94]. Two MMR complexes detect mismatched DNA: MutSα, comprising MSH2 and MSH6, recognises mismatched insertions/deletions 1–2 bases long and MutSβ, comprising MSH2 and MSH3, recognises longer insertion/deletion mispairs [95]. The MutLα-ATPase complex (MLH1 and PMS2) is recruited to the MutSα complex on the mismatched DNA with PCNA and RPA [96]. PMS2 is an endonuclease which, together with EXO1, is required for excision of the mismatched region [97]. The gap is filled by DNA Pol δ using the sister strand as a template and the DNA religated by Ligase 1, together with PCNA [96]. Whilst this simplified account explains MMR *in vitro*, it appears that *in vivo,* other chromatin remodelling/modification factors and epigenetic modifications of histones are involved [98].

4.7. Fanconi Anaemia (FA) Pathway

The Fanconi anaemia (FA) pathway comprises between 15 and 20 proteins and is activated in response to interstrand cross-links (ICLs) [99,100]. ICLs arise following exposure to chemicals such as cisplatin and mitomycin C and result in the covalent cross linking of two DNA strands, inhibiting transcription and replication. The ICL is recognised by a FA anchor complex containing a number of proteins such as FANCM [99]. Subsequent recruitment of the FA core complex, comprising eight proteins, leads to monoubiquitination of FANCD2 and FANCI. This monoubiquitination results in the recruitment and activation of nucleases, such as FANCP and FANCQ, which cleave the DNA and "unhook" the cross-link. HR proteins are engaged in the later stages of ICL repair to resolve DSBs which are generated [99].

4.8. DNA Repair Pathways and the Cell Cycle

Although the DNA repair pathways outlined above have evolved to deal with specific types of DNA damage, their activity can vary significantly during different phases of the cell cycle [101]. As detailed above, DSB repair by HR is restricted to S and G2 phases due to the requirement of a template sister chromatid that allows faithful repair of the damaged region. The other primary DSB repair pathway, NHEJ, can occur throughout the cell cycle but is more prevalent in G1 due to the unavailability of HR. The CtIP protein has been identified as an important factor in stimulating HR since its phosphorylation by CDKs in S and G2 phases promotes its role in the initial resection step of HR [102]. Proteasomal degradation of CtIP in G1 may also contribute to inhibiting HR during this phase [103].

While the principle excision repair pathways can operate throughout the cell cycle, activity can also vary in different phases. For example, MMR is more prevalent during S-phase to correct replication errors while NER plays a key role in G1 to remove bulky lesions that could block DNA polymerases [101,104]. It has also been shown that the activities of key BER enzymes are higher in G1 following IR-induced DNA damage compared with the G2 phase [105]. S-phase is also associated with DNA damage tolerance pathways (DDT) that allow replication to proceed in the presence of unrepaired DNA damage. Lesions can be bypassed in an error-prone manner using specialised translesion synthesis (TLS) DNA polymerases, or more accurately by template switching (TS) which employs the sister chromatid as a template [106]. Since viruses can specifically interact with proteins involved in both DNA repair and cell cycle regulation, it is worthwhile to consider possible cell cycle effects when evaluating the efficiency of DNA repair pathways during viral infection.

The following text summarises the published literature concerning how the pathways detailed above are activated or subverted by viruses known to cause tumours in humans (Table 2).

5. Human Papillomaviruses (HPV)

Human papillomaviruses (HPV) are small double-stranded DNA viruses of approximately 8 kb that target the mucosal and cutaneous epithelium. HPV infection is associated with malignancies of the anogenital tract and the oropharynx and is a particular risk factor for the development of cervical cancer [107]. Over 100 HPV strains have been identified although only a limited number have been classified as high-risk based on their potential to cause disease. Among these high risk types, type 16, 18, 31, and 33 are responsible for approximately 90% of all cervical cancers [108].

HPV initially establishes infection in undifferentiated and actively proliferating cells in the basal layer of epithelium where progression of the viral lifecycle is tightly linked to cellular differentiation. In undifferentiated cells, the viral genomes are maintained as extra-chromosomal nuclear episomes, which are replicated in synchrony with cellular DNA. Upon cellular differentiation, the HPV genome is amplified to produce infectious particles and the cell cycle is deregulated to aid viral replication. In HPV-containing cancer cells, the viral episome is often lost and in high grade lesions HPV DNA is typically found integrated into the host genome [109].

The HPV genome can be divided into an early region that comprises six open reading frames (ORFs) designated E1 to E7, a noncoding region and a late region comprising two ORFs, L1 and L2, that transcribe major and minor capsid proteins respectively [110]. The E1 and E2 viral proteins are necessary for initiation of HPV replication while E6 and E7, highly expressed during genome amplification, are the primary oncoproteins required for malignant transformation. E6 and E7 are known to promote degradation of the tumour suppressors, p53 and retinoblastoma protein (pRB) respectively, while disruption of the E2F/RB complex by E7 drives progression of the cell cycle [111,112].

5.1. Activation of the DDR by HPV

Expression of the viral helicase E1 in several high-risk HPV types can induce a DDR that principally comprises activation of the ATM-CHK2 pathway [113–116]. Activation of ATM has been attributed to the induction of DSBs in cellular DNA caused by E1 which specifically requires the ATPase and dsDNA melting activity of the protein [116]. ATM activation by E1 has also been shown to induce cell cycle arrest in S and G2 phases leading to suppression of cell growth [114,116]. Nuclear export of E1 prevents DDR activation and cell cycle arrest in undifferentiated keratinocytes and the formation of a complex with E2 limits DDR activation while E1 is in the nucleus [114]. In several studies a subset of E1-expressing cells were found to contain phosphorylated CHK1 suggesting that HPV may induce cell cycle-dependent activation of the ATR pathway [113,115,117].

While a number of studies have focused on the ability of E1 to induce DNA damage directly, it has also been demonstrated that the coexistence of HPV episomes and integrated HPV DNA in some cells can lead to unscheduled DNA replication and host genomic alterations [117,118]. The re-replication of integrated HPV18 DNA can generate heterogeneous replication intermediates that recruit DNA repair proteins and activate the ATM-CHK2 pathway [117]. In this study, DDR activation was ascribed to the recognition of viral DNA structures rather than the direct induction of DNA damage by viral proteins, although E1 expression was still required to initiate HPV replication.

Expression of E7 has also been shown to cause DNA damage, which can subsequently activate the Fanconi anemia (FA) repair pathway in squamous cell carcinomas (SCCs) [119–121]. FANCD2 foci, a marker of FA activation, have been observed in SSC tissues while FANCD2, FANCD1 and BRCA2 are recruited to chromatin in HPV16 E7-expressing cells [120]. Consequently, expression of E7 in cells deficient in the FA pathway leads to unrepaired DNA damage and increases the risk of genetic instability.

5.2. Involvement of DDR Factors in HPV Replication

Several studies have examined a role for DDR activation and individual DDR proteins in replication of HPV DNA [65,113,115]. It has been suggested that HPV-induced ATM activation and subsequent cell cycle arrest can provide a more suitable environment for viral replication in differentiating cells [113]. ATM inhibition during HPV31 infection was shown to adversely affect late viral genome amplification but not episomal maintenance [113]. In the same study it was demonstrated that CHK2 activity is required for caspase activation, which subsequently plays a role in viral replication through cleavage of E1.

Numerous DNA damage and repair proteins have been observed associated with HPV replication centres suggesting a role for these factors in the viral lifecycle. Levels of the DSB markers γH2AX and 53BP1 are elevated in HPV31-positive cells compared with uninfected cells and both can localize to HPV DNA foci along with phosphorylated ATM and CHK2 [65]. In addition, HR factors such as Rad51, BRCA1 and phosphorylated RPA have been seen associated with HPV31 replication centres suggesting that HR may be involved in HPV replication [65]. ATR pathway proteins ATRIP and TOPBP1 have also been observed at HPV18 replication sites [116,117]. TOPBP1 interacts directly with HPV E2 and suppression of this interaction results in an inability of the virus to establish episomes and reduces overall viral DNA replication [122].

5.3. Deregulation of DDR Signalling by E6 and E7

As well as activating the cellular DDR and recruiting DDR factors to replication centres, HPV also interferes with several DNA repair pathways, principally through the actions of E6 and E7. Expression of HPV16 E7 can increase persistence of γH2AX and Rad51 foci following IR-induced DNA damage in head and neck cancer cells [63]. While E7 expression can increase overall levels of the HR protein Rad51, the normal kinetics of DNA repair including the resolution of γH2AX foci are impeded.

E6 from at least three HPV types interacts with XRCC1, a scaffold protein involved in BER, impairing the ability of E6-expressing cells to repair SSBs [51]. E6 also interacts with O^6methylguanine-DNA methyl transferase (MGMT), a protein that protects against harmful mutations by repairing DNA adducts, and promotes its ubiquitin-dependent degradation [50]. Expression of HPV16 E6 was found to interfere with the recovery of fibroblasts from UV radiation by deregulating CHK1 activity [123] while HPV5 and 8 E6 proteins can promote degradation of the histone acetyl transferase p300. Degradation of p300 by E6 was shown to reduce ATR protein levels and subsequently increase thymine dimer persistence and DSB formation following UV exposure [124].

5.4. Summary

HPV activates the ATM pathway primarily through the expression of E1, although other viral proteins as well as viral DNA structures can also simulate DDR signalling. Several components of the ATM and ATR signalling pathways, as well as DNA repair factors, are localised to HPV DNA foci and may contribute to viral replication. Despite the presence of DNA damage, E6 and E7 promote cellular proliferation and survival by deregulating the cell cycle and impairing the apoptotic response.

By driving cellular proliferation in the presence of unrepaired DNA damage and abrogating repair mechanisms, the introduction of genomic instability is an inevitable consequence of the HPV lifecycle that increases the chances of malignant transformation of the host cell. In addition, it has been demonstrated that the introduction of DSBs in cellular DNA leads to an increase in HPV16 integration events [125]. Since viral integration is frequently observed in HPV-related malignancies, it is conceivable that DNA damage induced by HPV proteins can facilitate viral integration which subsequently drives HPV-mediated carcinogenesis [126].

6. Merkel Cell Polyomavirus (MCPyV)

Merkel cell polyomavirus (MCPyV), the most recent human tumour virus to be identified, is responsible for approximately 80% of cases of the aggressive skin cancer known as Merkel cell carcinoma (MCC) [127]. As with other polyomaviruses, MCPyV has a small double-stranded DNA genome divided into early and late coding regions separated by a non-coding control region (NCCR). The early region produces a transcript whose alternative splicing results in the translation of three proteins; a large T antigen (LT), a small T antigen (sT) and the 57 kT antigen. Three capsid proteins, VP1, VP2 and VP3, are encoded by the late region. In MCC tumours positive for MCPyV, the viral genome is integrated prior to clonal expansion of the cancer cells, strongly suggesting a viral driver of oncogenesis [128].

Similar to other polyomaviruses, MCPyV sT is involved in viral replication and targets the protein phosphatase 2A (PP2A). MCPyV sT may also play a role in regulation of the innate immune response and has been implicated in the development and progression of MCC [129,130]. Most attention, however, has focused on the multifunctional LT protein which contributes to initiating replication of the viral genome and also modulates a range of host cellular processes, some of which are analogous to the actions of SV40 LT [131]. In MCPyV-positive MCC, mutations in the C-terminus of LT result in production of a truncated LT protein that lacks the helicase activity required for viral replication but retains oncogenic properties, such as an ability to inactivate Rb [131,132]. Since exposure to UV is a known risk factor for the development of MCC and pyrimidine dimer substitutions are common among LT mutations, it is speculated that UV-induced DNA damage plays a role in carcinogenesis [131].

6.1. Activation of the DDR by MCPyV

It has been demonstrated that activation of both the ATM and ATR pathways occurs following MCPyV infection [133]. Activation of both pathways was observed following infection with native MCPyV virions and after viral genomes were directly introduced into cells via transfection. Expression of LT alone led to activation of ATR but not ATM while comet assays revealed that LT expression can cause DNA damage in the host cell. Use of various LT mutants indicated that the C-terminal region of the protein was responsible for both ATR activation and DNA damage. Activation of ATR by LT C-terminal region led to p53 phosphorylation, cell cycle arrest and suppression of cellular proliferation. The authors concluded that the C-terminal region of LT can act as a brake on cellular proliferation via DDR activation which offers a possible explanation as to why mutations in this region can contribute to development of MCC.

6.2. Involvement of DDR Factors in MCPyV Replication

It was recently shown that in MCPyV-infected cells, several DDR proteins associated with both the ATM and ATR pathways localise to foci containing MCPyV LT [134]. Using immuno-FISH and BrdU staining to visualise viral DNA, it was found that DDR factors including ATR, phosphorylated CHK2, and γH2AX co-localised with LT at sites of active MCPyV replication. This localisation was abolished following introduction of a replication-defective virus or a virus containing a mutated origin of replication. Experiments using DDR kinase inhibitors and small interfering RNA (siRNA) suggested that the ATR kinase is important for efficient viral replication.

6.3. Interference with DNA Repair by LT

It has been demonstrated that MCC cells containing MCPyV have impaired ability to respond to UV-induced DNA damage compared to uninfected MCC cells [53]. These effects were attributed to a defect in the GG-NER pathway, in particular reduced expression of the XPC protein in the MCPyV-positive cells. Ectopic expression of both wild type and mutated LT inhibited G1 arrest following UV exposure but only expression of the mutated protein led to defects in DNA repair. The results suggest that mutated LT could increase genomic instability leading to MCC and provide a possible explanation for the higher incidence of MCC on sun-exposed skin.

6.4. Summary

Similar to SV40 LT [135], expression of MCPyV LT can activate a cellular DDR. But while SV40 LT interferes with downstream activation of p53, the DDR elicited by MCPyV LT leads to cell cycle arrest in a p53-dependent manner. In this way, checkpoint activation may provide selective pressure for generation of the truncated LT protein that fails to activate a growth-suppressive DDR. Truncated LT can also interfere with DNA repair of UV damage that could further contribute to genomic instability and tumourigenesis. During MCPyV infection, DDR factors are localised to sites of viral replication centres while ATR activation plays a positive role in viral replication. It has also been reported that ATM can directly phosphorylate LT although the functional significance of this has yet to be fully determined [136].

Investigations into the interaction between MCPyV and the DDR are clearly at an early stage although findings from closely related polyomaviruses, as well as other tumour viruses, have provided a basis for current research projects. Future studies may uncover how exactly the virus benefits from activation of the DDR and assess the relevant contribution of DDR dysregulation and other factors to the development of MCPyV-associated MCC.

7. Epstein-Barr Virus (EBV)

Epstein-Barr virus (EBV) is a ubiquitous gammaherpesvirus carried by over 90% of the human population [137]. EBV has a dsDNA genome of approximately 170 kb, primarily targets resting B lymphocytes and epithelial cells and is implicated in the development of human malignancies that include Burkitt's lymphoma, Hodgkin's lymphoma, nasopharyngeal carcinoma and gastric carcinoma [138].

The EBV lifecycle consists of two distinct states: Latent infection and lytic replication. Following initial infection, the virus can establish life-long latency in the host where it is typically maintained as an extra-chromosomal episome that is replicated synchronously with cellular DNA by host polymerases. Latency is characterised by the expression of a tightly restricted subset of genes required for viral genome persistence. EBV latency transcripts include three membrane proteins (LMPs), six nuclear proteins (EBNAs), two non-coding RNAs (EBERs) and several microRNAs (miRNAs) [139]. The expression pattern of latent viral genes is variable among EBV-associated tumours and has led to the classification of four latency programmes known as Latency 0, I, II and III [139].

In a subset of cells, EBV can enter the lytic replication programme in which the viral genome can be amplified 100- to 1000-fold. The method of genome replication is distinct from latency in that there is greater requirement for replication proteins encoded by the virus. The immediate-early genes BZLF1 and BRLF1 are expressed first and encode transactivator proteins. These proteins then mediate expression of an ordered cascade of early and late viral genes that culminates in release of progeny virus from the host cell. Proteins expressed in both the latent and lytic stages have unique interactions with the host DDR.

7.1. Genetic Instability Induced by EBV Proteins

Several reports have demonstrated that EBV infection can lead to genetic instability in host cells through the introduction of DNA damage and dysregulation of DNA repair mechanisms. EBV-infected Burkitt's lymphoma cell lines have elevated levels of γH2AX in the absence of exogenous damaging agents as well as an increased number of chromosomal abnormalities [140].

EBV nuclear antigen 1 (EBNA1), expressed during latency and present in all EBV-associated malignancies, has been shown to induce chromosomal aberrations and DSBs through the production of ROS via NADPH oxidase activation [141,142]. Latent membrane protein 1 (LMP1), known to be essential for EBV-induced transformation of lymphocytes, can induce micronucleus formation and repress DNA repair in epithelial cells [143,144]. It was shown that LMP1 can stimulate the PI3K/Akt pathway resulting in phosphorylation of FOXO3a [144]. FOXO3a phosphorylation causes its retention in the nucleus impairing its ability to activate target genes such as DDB1, a component of the NER pathway.

Several EBV lytic proteins have also been shown to contribute to genetic instability in host cells. In nasopharyngeal carcinoma cells, EBV terminase BALF3 can induce DNA strand breaks and micronuclei formation and recurrent expression leads to tumourigenic features such as enhanced cell migration and invasiveness [145]. The EBV-encoded kinase BGLF4 is also associated with genetic instability [146]. BGLF4 expression leads to DDR activation that interferes with host DNA replication and delays the progression through S-phase [146]. Expression of the EBV DNase BGLF5 in epithelial cells can induce DSBs leading to an increase in microsatellite instability and genetic mutations [147]. BGLF5 expression also results in the down-regulation of several DNA repair genes involved in pathways such as MMR, NER, BER, and HR. Finally, BPLF1, a late tegument protein, deubiquitinates PCNA resulting in reduced recruitment of DNA polymerase η to sites of stalled replication where it would normally play a role in the translesion synthesis (TLS) DNA damage tolerance pathway [55].

7.2. Involvement of DDR Factors in EBV Replication

The role of the DDR proteins in EBV lytic replication has been studied by several research groups although the findings have not always been consistent. During lytic replication in B cells, EBV can elicit an ATM-mediated DDR with phosphorylation of H2AX, CHK2 and p53 observed following induction of the lytic cycle [148]. Phosphorylated ATM, NBS1 and Mre11 were found localised to sites of viral replication while downstream p53 signalling was inhibited. Inhibition of ATM activation by caffeine (a general PIKK inhibitor) during this investigation did not affect production of infectious virus. Two subsequent studies by the same group found that MMR proteins, such as MSH2, MSH6, MLH1, and hPSM2, as well as HR factors, including RPA, Rad51 and Rad52, are also localised to EBV replication compartments [62,64]. The HR factors, as well as PCNA and Mre11, were shown to be loaded onto newly synthesised viral DNA while knockdown of RPA and Rad51 significantly reduced viral replication. These findings point to a key role for individual DNA repair factors and the HR pathway in synthesis of EBV DNA. An interaction has also been reported between the early lytic protein BZLF1 and 53BP1, a protein that usually forms foci at the sites of DSBs [149]. Knockdown of 53BP1 inhibited production of progeny virus indicating that the protein is involved in EBV replication, possibly by protecting exposed DNA ends during genome amplification.

In a mechanism believed to be conserved among herpesviruses, the EBV lytic protein BGLF4 activates the acetyl transferase TIP60, a known regulator of the DDR through acetylation of ATM [150]. This interaction subsequently triggers an ATM-mediated DDR that plays a positive role in lytic replication of EBV. In contrast to previous reports, viral replication was suppressed in a dose-dependent manner following application of a specific ATM inhibitor. A further study provided evidence that ATM activation is required for efficient expression of EBV early lytic genes but not for viral DNA replication [151]. This study presented a model in which cellular stress activates the ATM pathway leading to activation of the EBV early lytic BZLF1 promoter by ATM substrates that can modify viral chromatin. More recently it has been shown that ATM activation during EBV lytic replication leads to phosphorylation of the transcription factor Sp1 [152]. Sp1 is subsequently localised to sites of replicating viral DNA along with other DDR factors associated with the ATM pathway. Phosphorylation of Sp1 was shown to mediate the recruitment of EBV replication proteins to viral DNA, highlighted by the fact that knockdown of Sp1 dramatically suppressed viral replication.

7.3. Dysregulation of Cell Cycle Checkpoints during EBV Infection

It has also been demonstrated that EBV-induced cellular proliferation can activate cell cycle checkpoints that lead to DDR signalling [153,154]. Activation of the ATM-CHK2 pathway has been observed following hyper-proliferation of primary B cells containing EBV [153]. In this context the DDR can act as an anti-tumour barrier as inhibition of ATM and CHK2 significantly increased the efficiency of B cell transformation by EBV. The study also demonstrated that expression of the EBV nuclear antigen 3C (EBNA3C) latent protein could attenuate DDR signalling that would otherwise arrest the growth of EBV-infected cells.

EBNA3C can also interact directly with CHK2 and disrupt the G2/M cell cycle checkpoint induced by treatment with nocodazole which blocks cells in mitosis [155]. The binding of EBNA3C to CHK2 leads to phosphorylation of Cdc25c at Ser^{216}. This causes Cdc25c to be localised to the cytoplasm allowing cyclin B/Cdc2 activation and progression through the G2/M checkpoint. EBNA3C has subsequently been shown to form a complex with H2AX facilitating its degradation via the ubiquitin proteasome pathway [156]. In lymphoblastoid cell lines (LCLs), H2AX knockdown resulted in down-regulation of the tumour suppressor p53 and increased expression of the oncoprotein Bub1. In addition to the actions of EBNA3C, expression of LMP1 in nasopharyngeal epithelial cells impairs the G2 checkpoint through defective CHK1 activation resulting in increased chromosomal instability [157].

Activation of the ATR pathway has also been demonstrated following oncogene-driven cellular replication induced by EBV infection [154]. EBV can interrupt downstream ATR signalling by increasing expression of STAT3, which leads to the loss of Claspin due to caspase 7 activity. Since Claspin is involved in mediating CHK1 phosphorylation by ATR, its loss leads to relaxation of the intra S-phase checkpoint, which would normally act as a barrier against EBV-driven cellular proliferation.

7.4. Summary

Activation of the DDR has been observed in EBV-infected B cells and epithelial cells, during both the latent and lytic stages of the viral lifecycle. The cause of this DDR activation has been ascribed to several sources including the introduction of damage and inhibition of DNA repair, recognition of viral DNA as host cell damaged DNA, and aberrant cellular proliferation induced by viral proteins.

In addition, the virus can interfere with cell cycle checkpoints to allow continued cellular proliferation in the presence of DNA damage. EBV also appears to recruit several DDR factors, including HR proteins, to replication sites where they may participate in viral DNA synthesis.

While several studies have noted the activation of ATM following lytic reactivation of EBV, there have been diverging opinions regarding the role of this kinase in viral replication. Contradictory observations could be the result of using different cell types, varying methods of inducing viral replication or the use of non-specific ATM inhibitors. While question marks remain over the role of the DDR during the lytic cycle, it has been shown that DDR factors contribute to viral-genome replication during latency. For example, Mre11 and NBS1 are recruited to the origin of plasmid replication (OriP) in S-phase and play a role in maintenance of viral episomes through the formation of recombination junctions [158]. In summary, EBV appears to take advantage of DDR activation and utilise certain DDR proteins while disabling other aspects of the DDR that could lead to growth suppression and apoptosis. In view of the prevalence of EBV infection, the multiple interactions of the virus with the DDR and its ability to damage host cell DNA, it is perhaps surprising that EBV-associated tumours are not more common.

8. Kaposi's Sarcoma Related Virus (KSHV)

Kaposi's sarcoma-associated herpesvirus (KSHV), also known as human herpesvirus 8 (HHV8), is the second gammaherpesvirus after EBV to be identified as a causative agent of human cancers. KSHV can infect and transform endothelial cells leading to the development of the angioproliferative malignancy known as Kaposi's sarcoma [159]. In addition, KSHV is associated with B cell malignancies, primary effusion lymphoma (PEL), and multicentric Castleman's disease (MCD) [160,161]. Despite its association with cancer, KSHV infection is largely asymptomatic, with disease most likely to occur only following suppression of the host immune system.

As with EBV, the KSHV lifecycle is biphasic and consists of latent and lytic stages. KSHV latency-associated nuclear antigen (LANA) is a multifunctional viral protein expressed in all latently infected cells and present in all KSHV-positive KS tumours [162]. LANA plays a vital role in tethering the KSHV episome to host DNA ensuring its distribution to daughter cells following cellular division [163]. LANA can also interact with multiple host cell proteins and has been implicated in tumourigenesis through interference with cellular pathways associated with cell cycle control, apoptosis, gene expression and immune regulation [164]. In contrast to the latent programme, KSHV lytic replication involves expression of a large set of viral genes necessary for production of new infectious progeny. Both latent and lytic genetic programmes have been implicated in the development of KSHV-related malignancies [165].

8.1. DDR Activation during Latent KSHV Infection

KSHV infection induces phosphorylation of H2AX and elevated levels of γH2AX have been observed in latently infected B cells [166,167]. LANA can interact with H2AX at its C-terminus leading to formation of γH2AX, which is important for episomal persistence [166]. KSHV infection of primary endothelial cells was shown to result in phosphorylation of H2AX and ATM as early as

30 min after viral entry with γH2AX co-localising with viral genomes [167]. Transient phosphorylation of DDR effector kinases CHK1 and CHK2 was observed but downstream phosphorylation of Cdc25c was unaffected. Total H2AX levels also increased during KSHV infection, which was attributed to a decrease in K^{48}-linked polyubiquitination and an increase in K^{63}-linked polyubiquitination. Inhibition of ATM or depletion of H2AX has a negative effect on the ability of the virus to establish latency, suggesting that KSHV may mediate activation of the ATM arm of the DDR while inhibiting downstream signalling.

Ectopic expression of v-Cyclin, a homolog of mammalian D-type cyclins expressed during latency, leads to activation of the ATM pathway and S-phase arrest in endothelial cells [168]. This DDR acts as a block to cellular proliferation and could create selective pressure for the acquisition of mutations that abrogate this barrier while providing a growth advantage to cells defective in DDR components. Expression of v-Cyclin also leads to the introduction of chromosomal instability through disruption of the centrosome cycle resulting in multinucleation and aneuploidy [168–170]. Cells latently infected with KSHV can also bypass a G2/M checkpoint block induced by nocodazole [171]. It was shown that LANA can interact directly with CHK2 leading to activation of the cyclin B/Cdc2 complex that mediates progression through the G2/M checkpoint.

8.2. Involvement of DDR Factors in KSHV Replication

DNA affinity purification combined with mass spectrometry has been used to identify host cell factors that bind to KSHV origins of lytic replication (*ori-Lyt*) [60]. The proteins identified included the MSH2/MSH6 heterodimer involved in MMR and DNA-PKcs, PARP-1, and Ku80/70, all involved in NHEJ. Several of these proteins were also found to localise to viral replication compartments suggesting a role in viral replication. Treatment of KSHV-infected B cells with a PARP-1 inhibitor has been shown to enhance lytic replication but further analysis has also suggested that PARP-1 may play a positive role in replication of viral DNA, although it can negatively regulate late assembly of infectious virions [60,172].

8.3. DDR Activation during Lytic Replication of KSHV

Elevated levels of phosphorylated H2AX have been observed following initiation of KSHV lytic replication [67,173]. This observation has been attributed to the introduction of DSBs in cellular DNA by the viral mRNA export factor orf57 [173]. It was shown that sequestration of the human Transcription and Export complex (hTREX) by orf57 can cause newly transcribed mRNA to form R-loops through annealing to the DNA template strand which in turn leads to formation of DSBs. Another lytic protein, the processivity factor orf59, was shown to interact with Ku70 and Ku80 and impair NHEJ repair of DSBs [67]. Although KSHV lytic replication can activate the DDR, the lytic protein viral interferon regulatory factor 1 (vIRF1) has been shown to interact with ATM and impair DDR activation following etoposide treatment [174]. In addition, vIRF1 can interact with p53 and facilitate its proteasomal degradation.

8.4. Summary

As with EBV, KSHV infection activates the ATM pathway and latently-infected cells display elevated markers of DNA damage and chromosomal aberrations. DDR activation appears to play a positive role in establishing and maintaining latency but does not result in cell cycle arrest or apoptosis that would negatively impact the viral lifecycle. Latent KSHV proteins can also activate the DDR by inducing cellular proliferation while LANA plays a role in abrogating cell cycle checkpoints. KSHV also induces DNA damage during lytic replication and several DDR factors associate with replicating viral DNA.

In both EBV and KSHV, it is problematic to link DNA damage during lytic replication with cellular transformation, since this phase of the viral lifecycle inevitably results in destruction of the host cell. However, several lytic proteins are also expressed briefly following initial viral entry and could conceivably have a detrimental impact on chromosomal stability at an earlier stage of the viral lifecycle [175]. In addition, expression of early lytic proteins during abortive lytic replication of EBV has been linked to the development of lymphomas in mice [176]. Being the most recent human herpesvirus to be identified, there are fewer reports concerning the interaction between KSHV and the DDR compared with other members of the same family. Future studies may focus on identifying a role for DDR activation during lytic replication and the relative contribution of DDR modulation to development of KSHV-related malignancies.

9. Human T-cell Leukemia Virus Type 1 (HTLV-1)

Human T-cell leukemia virus (HTLV-1), the only known human retrovirus that can directly transform human cells, is responsible for the rare and aggressive cancer known as adult T-cell-leukemia/lymphoma (ATL) [177]. HTLV-1 is an enveloped virus with a genome composed of two identical plus-sense single-stranded RNA molecules that are synthesised into a DNA provirus that integrates into the host genome. Like all retroviruses, the HTLV-1 genome contains gag, pol, and ENV structural genes encompassed by two long terminal repeat (LTR) sequences. Bordering the 3' LTR is the unique pX region that encodes regulatory proteins Tax, Rex, p12, p13, p30, p21, and HTLV-1 basic Zip factor (HBZ) [178].

Like HIV, HTLV-1 predominately targets CD4+ T cells but, while HIV infection can lead to CD4 depletion, HTLV-1-related diseases are characterised by the unregulated proliferation of these lymphocytes. The 40 kDa Tax protein is a transcriptional activator that is essential for viral replication and implicated in HTLV-1-induced cellular transformation [179–181]. Tax can stimulate cellular proliferation and survival through modulation of cell cycle checkpoints, stimulation of the NF-kB pathway and activation of the hTERT promoter in quiescent T cells [182]. In addition, Tax can interfere with tumour suppressor function through the inactivation of p53 [183] and proteasomal degradation of Rb [184]. Tax can also form foci, known as Tax Speckled Structures (TSS) [185] or Tax nuclear bodies [186], in the nuclei of infected cells.

9.1. Genetic Instability in Tax-Expressing Cells

Multiple groups have reported markers of genomic instability in host cells following expression of Tax, characterised by the prevalence of DNA damage markers, formation of micronuclei, and occurrence of aneuploidy [187–189]. Tax was shown to sensitise cells with deleted or mutant p53 to a range of DNA damaging agents including mitomycin C, etoposide and UV light [190]. Expression of Tax can also increase levels of intracellular ROS, DNA damage and the senescence marker SEN1 in both fibroblasts and T cells [191]. All of these markers were reduced following siRNA depletion of Tax or treatment with the ROS scavenger *N*-acetyl-L-cysteine (NAC), indicating that ROS production was the primary source of Tax-induced DNA damage in these cells. Tax has also been shown to impair DNA replication fork progression causing formation of DSBs during S-phase [192]. In the same study, NF-kB activation by Tax also led to DSBs via the production of nitric oxide (NO).

9.2. Interference with DNA Repair by HTLV-1 Proteins

Tax expression has been shown to increase the frequency of mutations in the cellular genome [193]. The random nature of these mutations indicates that this viral protein could interfere with the repair of DNA damage that accumulates during normal cellular processes. Tax has since been shown to interfere directly and indirectly with several different repair pathways that include BER, NER, MMR, NHEJ and HR.

The presence of Tax reduces the expression of human polymerase β, an enzyme involved in BER [52]. Using a range of DNA-damaging agents and a plasmid reactivation assay it was shown that base-excision repair of oxidative damage was the primary repair pathway suppressed in Tax-expressing cells [188]. Tax has also been demonstrated to repress NER through transactivation of PCNA, a cofactor for DNA polymerase δ, that plays a central role in DNA replication and repair [54,194]. In cells containing *wt* p53, Tax has a dose-dependent dual effect on NER [195]. Low levels of Tax stimulated NER through increased transcriptional activity of p53 while high levels of Tax increased p53 levels but functionally inactivated the protein leading to inhibition of p53-dependent NER. Altered expression of several MMR genes has also been recorded in primary leukemic cells derived from patients with ATL [61]. Expression of MSH2 and PMS1 was decreased in all cases examined and the reduction in the latter was ascribed to increased methylation of the PMS1 promoter.

Following IR, Tax-expressing cells display defects in the formation of DNA repair foci including disruption of the association between MDC1 and γH2AX and reduced association between ATM and chromatin [196]. As a result of impaired ATM activity, Tax-expressing cells progressed more rapidly into S-phase despite the presence of unrepaired DNA damage. A subsequent study demonstrated that MDC1 is bound by Tax and recruited to nuclear foci that also contain DNA-PK and BRCA1 [66]. Formation of these Tax-containing foci led to a reduction in IR-induced foci containing NBS1 suggesting that viral sequestration of DDR factors impedes their involvement in the normal DDR. Tax has also been found to enhance the expression of the cellular phosphatase WIP1 leading to a reduction in levels of γH2AX following UV radiation [197]. Interference with γH2AX accumulation allows Tax expressing cells to bypass the G1/S checkpoint and enter S-phase with unrepaired DNA damage.

Expression of Tax has been shown to directly induce DSBs during S-phase while NF-kB activation by Tax leads to suppression of HR [189]. Tax also represses transcription of the NHEJ protein Ku80 [198,199]. Reduced Ku80 expression increased the number of unprotected DNA breaks as well as the occurrence of micronuclei and nucleoplasmic bridges in Tax-expressing cells. Although the majority of studies investigating genetic instability during HTLV-1 infection have focused on the activities of Tax, there is evidence that HTLV-1 RNA binding protein p30 can also interfere with repair of DSBs [200]. Following DNA damage, p30 alters its distribution from the nucleolus to the nucleoplasm and binds to NBS1 and Rad50, which interferes with the formation of the MRN complex. The disruption of the MRN complex by p30 results in impaired HR during S-phase and facilitates a switch to the more error-prone NHEJ.

9.3. Interaction between Tax and Checkpoint Kinases

Expression of Tax has been shown to induce ATM-CHK2 signalling resulting in the accumulation of cells in the G2 cell cycle phase [201]. Tax can interact directly with CHK2 and, following DDR activation, they, along with 53BP1, have been observed co-localised in nuclear foci [201]. Another study confirmed the interaction between Tax and CHK2 and demonstrated that Tax can also bind to CHK1 [202]. Through this interaction, Tax was shown to inhibit the kinase activity of CHK1 preventing the degradation of Cdc25A and impairing G2 arrest following IR. A subsequent publication attempted to address these contradictory findings by demonstrating that Tax can interact with the kinase domain of CHK2 stimulating oligomerization, autophosphorylation, and stabilisation of the protein [203]. Following IR, Tax sequesters the phosphorylated form of CHK2 within chromatin and impairs its role in the DDR following IR treatment. This study concluded that following Tax expression, CHK2 maintains the ability to orchestrate cell cycle arrest in G2 but is impaired in its role in the DDR following IR. DNA-PK has also been observed associated with nuclear foci containing Tax and CHK2 [204]. Tax can mediate an interaction between DNA-PK and CHK2 leading to increased DNA-PK activity, which impairs the cellular response to IR.

9.4. Summary

Multiple groups have shown that expression of HTLV-1 Tax can cause DNA damage and that this can occur via several mechanisms including the generation of free radicals and interference with DNA replication. There is evidence that at least five key DNA repair pathways are compromised during HTLV infection and that the proper formation of repair complexes is disrupted. In addition, checkpoint kinases are targeted and the virus modulates cell cycle progression to maximise viral replication efficiency.

Following infection, HTLV-1 can enter a long period of asymptomatic latency with only 3%–4% of infected individuals eventually developing ATL [205]. This suggests that cancer development and progression following HTLV-1 infection requires the accumulation of multiple genetic changes [206]. By introducing DNA damage and interfering with repair pathways and cell cycle checkpoints, HTLV-1 can induce cellular proliferation in the presence of genetic aberrations that could ultimately result in cellular transformation.

10. Hepatitis B Virus (HBV)

Hepatitis B virus (HBV) is a small enveloped DNA virus with a 3.2 kb genome belonging to the *Hepadnaviridae* family. HBV targets hepatocytes and, while transient infection can cause acute hepatitis, chronic infection is a major risk factor for the development of hepatocellular carcinoma (HCC). The circular HBV genome is partially double-stranded and contains four ORFs named S, C, P, and X. The S gene, also known as HBsAg, is divided into Pre-S1, Pre-S2, and S domains and encodes three envelope proteins often referred to as large (L), middle (M), and small (S). The three remaining ORFs encode a DNA polymerase (Pol), a capsid protein (core) and the 154 amino acid X protein (HBx) [207].

HBV has a relatively complex lifecycle that involves reverse transcription. Following entry to the cell, the HBV relaxed circular DNA (rcDNA) is transported to the nucleus where it is converted into covalently closed circular DNA (cccDNA). The cccDNA provides a template for production of viral RNA that is converted by reverse transcriptase to HBV rcDNA in the cytoplasm. Although not an essential step in the viral lifecycle, the HBV genome is often found integrated into the host genome and the process has been proposed to play a role in hepatocarcinogenesis [208].

The multifunctional HBx is involved in viral replication and is implicated in the development of HCC due to its ability to interfere with multiple cellular processes [209]. In the cytoplasm, HBx can interact with mitochondria and activate mitogenic signalling cascades [209]. In the nucleus, HBx can influence gene expression through interaction with cellular transcription machinery, although it does not bind DNA directly [210]. HBx can also interact with p53 and impair its transcriptional activation of target genes, thus interfering with its role in apoptotic and DNA repair pathways [211–215].

10.1. Introduction of Oxidative DNA Damage by HBV

Chronic HBV infection can result in increased oxidative stress that has been linked to development of HBV-associated liver disease [216]. Using transgenic mice that expressed the HBV large envelope protein in hepatocytes, it was shown that development of chronic liver disease is associated with an increase in oxidative DNA damage [217]. Mutations in the Pre-S region of the HBV surface antigen (HBsAg) are associated with increased ROS production via endoplasmic reticulum (ER) stress [218,219]. Expression of pre-S mutant antigens in hepatoma cell lines was shown to cause increased oxidative DNA damage and increase expression of the DNA repair gene ogg1 [218]. However, a subsequent study failed to find a correlation between HBV pre-S mutations and increased oxidative DNA damage in patients with HCC [220].

It has also been demonstrated that DNA damage caused by oxidative stress can promote HBV integration and this has been linked to the development of HCC [221]. Inhibition of PARP-1, involved in SSB repair, also increased the occurrence of HPV genome integration indicating that functional DNA repair pathways, as well as limiting the impact of oxidative DNA damage, may protect against tumour development by restricting viral integration frequency.

10.2. Interference with DNA Repair by HBV

As well as promoting oxidative damage, HBV proteins have been shown to interfere with DNA repair pathways including those associated with processing of ROS-induced DNA lesions. HBx is known to bind to DNA binding protein 1 (DDB1) and the interaction has been shown to be important for HBV replication [56,57,222,223]. As well as being a subunit of an E3 ubiquitin ligase complex, DDB1 also plays a role in both GG-NER and TC-NER by forming complexes with DDB2 and Cockayne syndrome group A protein, respectively [224]. Using a panel of HBx mutants it was shown that, while HBx can impede repair of UV-damage, the interaction between HBx-DDB1 was not absolutely required for this effect [225]. Although it has not been directly linked to impaired NER, the interaction between HBx and DDB1 interferes with normal S-phase progression and increases the incidence of lagging chromosomes during mitosis that can subsequently cause multi-nucleation [226].

HBx, through its interaction with the Sp1 transcription factor, can also reduce expression of the XPD and XPB subunits of the TFIIH transcription factor involved in NER [227]. In addition, HBx has been shown to interact directly with XPD and XPB and sensitise cells to UV-induced DNA damage [58,59]. The interaction between HBx and p53 has been linked to defects in NER and expression of HBx can inhibit p53-dependent GG-NER in primary mouse hepatocytes [228]. Following UV exposure, levels of HBx and p53 were shown to increase in a hepatocellular carcinoma cell line and both proteins co-localized in the nucleus [229]. The HBx-expressing cells also displayed increased G2/M arrest and apoptosis, as well as a reduced capacity to repair DNA damage compared with controls. HBx expression also impedes the other NER sub-pathway, TC-NER, in both *wt* and p53-null cell lines [214].

Although the majority of studies regarding HBV and DNA repair have focused on the NER pathway, there is also evidence that HBx can interfere with BER. The structure of HBx has been shown to be similar to the human thymine DNA glycosylase (TDG) involved in initiating BER [230]. While the presence of TDG does not affect viral replication, HBx expression strongly supresses BER of G/T mismatches normally initiated by TDG. A more recent study assessed the individual capacity of HBsAg, core protein and HBx to induce DNA damage and interfere with DNA repair [231]. Accumulation of HBsAg was shown to enhance degradation of PML, which led to delayed repair of DSBs and increased resistance to apoptosis following IR. The results also showed that while HBx can induce DNA damage and apoptosis, it did not impede the repair of IR-induced lesions.

10.3. Activation of the ATR Pathway by HBx

HBx expression has been demonstrated to activate the ATR pathway [232–234]. HBV infection can activate the ATR-CHK1 pathway with minimal effect on ATM-CHK2 [233]. Expression of HBx alone has been shown to induce DNA re-replication and polyploidy through increased expression of Cdc6 and Cdt1 and down-regulation of geminin [234]. DNA re-replication resulted in increased DNA damage and the activation of ATR leading to phosphorylation of Rad17 and H2AX but not CHK1. It was also demonstrated that HBx-expressing cells are able to progress to mitosis despite the presence of DNA damage [235]. Activation of Polo-like kinase 1 (Plk1) in the G2 phase of the cell cycle by HBx reduced levels of Claspin, leading to CHK1 inactivation. This led to attenuation of the G2/M cell cycle checkpoint and subsequently suppressed DNA repair and p53-mediated apoptosis. Treatment of HBV infected cells with theophylline, a compound that inhibits ATR and ATM signalling, significantly reduces HBV replication suggesting a role for DDR signalling in the HBV lifecycle [236].

10.4. Summary

It is clear that HBV infection can increase oxidative stress leading to elevated levels of DNA damage in infected hepatocytes. HBV also appears to interfere with several DNA repair pathways, primarily through the actions of HBx, potentially exacerbating the increase in DNA damage. In addition, HBx can activate the ATR pathway, interfere with checkpoint activation and promote cell cycle progression in the presence of DNA damage that could ultimately result in genetic aberrations.

Despite BER being the primary pathway for repair of oxidative DNA damage, the majority of reports concerning HBV and DNA repair have focused on NER. Although NER can play a part in repair of oxidative damage, it is still unclear whether modulation of this pathway has a substantial role in tumour development following viral infection. As with HPV, DNA damage and abrogated repair can increase the likelihood of HBV integration, which has been linked to deleterious genetic alterations and altered expression of cellular genes. The fact that 80%–90% of HBV-related HCCs contain integrated HBV sequences highlights the link between viral integration and tumourigenesis [237].

11. Hepatitis C Virus (HCV)

Belonging to the *Flaviviridae* family, the Hepatitis C virus (HCV) chronically infects over 170 million people worldwide and is responsible for around a third of cases of HCC [238]. While primarily a hepatotrophic virus, HCV can also replicate in lymphocytes and has been linked to several lymphoproliferative disorders including B-cell non-Hodgkin's lymphoma (NHL) [239].

HCV has a positive-sense single-stranded RNA genome of 9.6 kb that contains 5′ and 3′ untranslated regions (UTRs) that border a single ORF. An internal ribosome entry site (IRES) in the 5′ UTR initiates translation of the HCV genome. The HCV ORF encodes a viral polyprotein of approximately 3010 amino acids that is subsequently cleaved by host proteases to produce a core protein, the E1 and E2 glycoproteins and the p7 ion channel protein. Additional cleavage by viral proteases produces six non-structural proteins NS2, NS3, NS4A, NS4B, NS5A, and NS5B.

Like all positive-strand RNA viruses, HCV reorganises intracellular membranes, such as those associated with mitochondria or endoplasmic reticulum, to form replication structures in the cytoplasm known as "membranous webs" [240,241]. Remodelling of endoplasmic reticulum has been observed following expression of NS4B suggesting that this protein is primarily responsible for the formation of HCV replication complexes [240].

11.1. Introduction of Oxidative DNA Damage by HCV

As with HBV, infection with HCV can lead to elevated levels of oxidative stress that subsequently increase the occurrence of DNA damage. Expression of HCV core and NS3 proteins have been shown to stimulate the inducible NO synthase (iNOS) gene leading to increased production of nitric oxide (NO) and formation of DSBs in cellular DNA [242]. Expression of E1 during the study also caused a moderate increase in DSB formation but by a mechanism independent of NO production. It was subsequently demonstrated that HCV infection leads to a reduction in mitochondrial membrane potential and an increase in ROS levels [243]. Individual expression of core, E1, and NS3 confirmed that each protein is able to induce ROS that subsequently leads to an increase in DSB formation.

HCV infection can also lead to overexpression of 3β-hydroxysterol Δ24-reductase (DHCR24) which has been linked to attenuation of the p53-mediated response to oxidative stress [244]. Activation of Akt by NS5A can increase c-Myc transcription through stabilisation of the transcription factor β-catenin [245]. This increased expression of c-Myc enhances ROS production and results in increased DNA damage and cell cycle arrest.

11.2. Interference with DNA Repair by HCV

Several studies have demonstrated that interference with DNA repair pathways by HCV could contribute to the development of HCC. Reduced expression of several genes involved in the MMR pathway including MSH2, MLH1, GTBP and PMS2, has been observed in HCV-positive tissues from patients with HCC [246]. Another study demonstrated that ROS levels were elevated by 30–60 fold in HCV-infected HCC cells compared to uninfected controls and that expression of the DNA glycosylase NEIL1, involved in the initial steps of BER, was suppressed [247]. Treatment with the antioxidant NAC or antiviral interferon (IFN) led to a partial restoration in NEIL1 expression indicating that virally-induced ROS could potentially impair BER. Down-regulation of Gadd45β, that plays a role in NER, was observed in HCV-infected cell lines and tissues, as well as transgenic mice expressing the entire HCV ORF [248]. Hypermethylation of the Gadd45β promoter in the presence of HCV was found to be responsible for the defect and led to impaired cell cycle arrest and reduced DNA excision repair. It has also been shown that expression of the HCV core protein in hepatocellular carcinoma cells leads to an impaired ability to repair UV-induced DNA damage [249].

As well as attenuating excision repair pathways, HCV can also interfere with DSB repair. Infection of lymphocytes with HCV was found to cause chromosomal aberrations, which are at least partially attributed to elevated NO and ROS levels [250]. In the same study, expression of Core and NS3 HCV proteins sensitised cells to DNA damaging agents by inhibiting NHEJ. In addition, HCV core protein was shown to interact with NBS1 and prevent formation of the MRN complex, which impairs ATM-mediated repair of DSBs.

11.3. Interaction between HCV Proteins and the ATM Pathway

In addition to inducing DNA damage, HCV can also modulate the DDR through interaction with the ATM-CHK2 pathway. NS3/4A can interact directly with ATM while NS5B interacts with both ATM and CHK2 [251]. In the same study, knockdown of ATM and CHK2 reduced replication of

HCV RNA suggesting a functional ATM pathway is required for efficient viral replication. The interaction between NS3/4A and ATM was confirmed in a separate study and shown to sensitise cells to IR via retention of ATM in the cytoplasm that subsequently impairs phosphorylation of H2AX following DNA damage [252]. In addition, expression of the HCV non-structural transmembrane protein NS2 can enhance cellular proliferation and activate the ATM-CHK2 pathway [253]. Following DDR activation, p53 was retained in the cytoplasm in NS2-expressing cells resulting in impaired downstream signalling through p21 activation.

11.4. Summary

As with HBV, infection with HCV can lead to elevated intracellular oxidative stress that increases the occurrence of DNA damage. HCV proteins can also interact with several DDR factors resulting in a reduced capacity to repair DNA lesions. Expression of HCV proteins activates the ATM pathway although the virus appears to deregulate downstream signalling while using key components for viral replication. There is also evidence that HCV exploits the DDR to modulate cell cycle progression [254,255]. NS5B was shown to interact with cyclin-dependent kinase 2-interacting protein (CINP), involved in the ATR-CHK1 pathway, preventing its localisation to the nucleus and ultimately causing a delay in S-phase progression [255].

Despite having an RNA genome and replicating in the cytoplasm, HCV can still impact significantly on host DDR pathways. While the HCV lifecycle does not include a nuclear stage, truncated or mutated versions of several HCV proteins have been observed in the nucleus where they can interact with cellular proteins including those associated with the DDR [250,256,257]. There is some doubt, however, about the nuclear localisation of these proteins during realistic models of infection [258]. To this end, the development of superior murine models of HCV pathogenesis will be invaluable in assessing the relative contribution of DDR deregulation and other factors to the progression of HCC.

12. Conclusions

It is clear from the literature summarised here that the interaction between human tumour viruses and the DDR can contribute to an increase in genomic instability during viral infection. As is apparent in high-risk HPV types, the accumulation of deleterious mutations in the presence of viral oncoproteins that promote cellular survival and proliferation will inevitably increase the likelihood of cellular transformation. While the DDR can represent an antiviral defence and barrier to tumourigenesis, chronic activation also creates selection pressure for the deactivation of key components. In all the viruses covered here, DDR activation is an inevitable consequence of viral infection and can be the direct result of the recognition of foreign DNA or expression of viral proteins, or can occur more indirectly through elevated oxidative stress or aberrant cellular proliferation. Since cell cycle arrest and apoptosis are consequences of DDR activation, viruses typically inhibit downstream signalling that would have negative consequences for viral propagation. It is also notable that, in HBV and HPV, DNA damage inflicted during infection can increase the chances of viral integration. This, in turn, can contribute to malignant transformation through host genomic alterations and elevated expression of viral oncogenes [259].

While there have been many excellent studies in this field, in certain cases, experimental limitations have led to some contradictory findings that have impeded progress. The use of different cell culture models and non-specific kinase inhibitors has resulted in inconsistent conclusions regarding the role of DDR activation during replication of viruses such as EBV. A more detailed mechanistic understanding of the precise roles of DDR proteins during synthesis of viral genomes will help confirm their importance during the viral lifecycle. While the ectopic expression of viral genes is a typical way of assessing their contribution to DDR deregulation, it may be misleading in cell types that are not natural viral targets *in vivo*. In the case of EBV, for example, the BZLF1 lytic protein has contrasting effects on cell cycle progression when expressed in different cell lines [260]. It is also apparent that viral proteins may have dual effects on the DDR depending on levels of expression; an example mentioned above is the effect of HTLV-1 Tax on p53-dependent NER [195]. In these cases, results must be considered in the context of gene expression levels during natural viral infection. Furthermore, possible contributory effects of other viral proteins should be borne in mind.

Reports demonstrating the involvement of DDR factors in viral replication have led to suggestions that these proteins could be targeted as a novel therapeutic strategy [251]. This is particularly appealing in the case of viruses with RNA genomes where higher rates of mutation have limited efforts to target the virus directly [261]. In addition, it has also been proposed that inactivation of key DNA repair pathways by viruses could open the door for the use of synthetic lethality to specifically target tumours containing viral genomes [262]. While these are attractive possibilities, the DDR is a rapidly evolving area of research with new components and novel regulatory mechanisms being identified frequently. Virologists must keep abreast of the most recent findings to gain a more complete understanding of how viruses modulate these pathways. While it remains to be seen whether the study of viruses and the DDR will yield effective therapeutic interventions, at the very least it has shed new light on how these pathogens modulate host cell functions and uncovered new facets of DDR regulation.

Acknowledgments

We are most grateful to Anoushka Thomas for the illustrations used here and to Cancer Research UK for funding.

Conflicts of Interest

The authors declare no conflict of interest.

References

1. Weitzman, M.D.; Lilley, C.E.; Chaurushiya, M.S. Genomes in conflict: Maintaining genome integrity during virus infection. *Annu. Rev. Microbiol.* **2010**, *64*, 61–81.
2. Nikitin, P.A.; Luftig, M.A. At a crossroads: Human DNA tumor viruses and the host DNA damage response. *Future Virol.* **2011**, *6*, 813–830.

3. Turnell, A.S.; Grand, R.J. DNA viruses and the cellular DNA-damage response. *J. Gen. Virol.* **2012**, *93*, 2076–2097.

4. Marriott, S.J.; Semmes, O.J. Impact of HTLV-I tax on cell cycle progression and the cellular DNA damage repair response. *Oncogene* **2005**, *24*, 5986–5995.

5. Smith, J.A.; Daniel, R. Following the path of the virus: The exploitation of host DNA repair mechanisms by retroviruses. *ACS Chem. Biol.* **2006**, *1*, 217–226.

6. Higgs, M.R.; Chouteau, P.; Lerat, H. 'Liver let die': Oxidative DNA damage and hepatotropic viruses. *J. Gen. Virol.* **2014**, *95*, 991–1004.

7. Nichols, W.W.; Levan, A.; Aula, P.; Norrby, E. Chromosome damage associated with the measles virus *in vitro*. *Hereditas* **1965**, *54*, 101–118.

8. Fortunato, E.A.; Spector, D.H. Viral induction of site-specific chromosome damage. *Rev. Med. Virol.* **2003**, *13*, 21–37.

9. Sinclair, A.; Yarranton, S.; Schelcher, C. DNA-damage response pathways triggered by viral replication. *Expert Rev. Mol. Med.* **2006**, *8*, 1–11.

10. Lilley, C.E.; Chaurushiya, M.S.; Weitzman, M.D. Chromatin at the intersection of viral infection and DNA damage. *Biochim. Biophys. Acta* **2010**, *1799*, 319–327.

11. Weitzman, M.D.; Lilley, C.E.; Chaurushiya, M.S. Changing the ubiquitin landscape during viral manipulation of the DNA damage response. *FEBS Lett.* **2011**, *585*, 2897–2906.

12. Parkin, D.M. The global health burden of infection-associated cancers in the year 2002. *Int. J. Cancer* **2006**, *118*, 3030–3044.

13. Jackson, S.P.; Bartek, J. The DNA-damage response in human biology and disease. *Nature* **2009**, *461*, 1071–1078.

14. Reynolds, J.J.; Stewart, G.S. A single strand that links multiple neuropathologies in human disease. *Brain* **2013**, *136*, 14–27.

15. Jeggo, P.A.; Lobrich, M. DNA double-strand breaks: Their cellular and clinical impact? *Oncogene* **2007**, *26*, 7717–7719.

16. Cimprich, K.A.; Cortez, D. Atr: An essential regulator of genome integrity. *Nat. Rev. Mol. Cell Biol.* **2008**, *9*, 616–627.

17. Davis, A.J.; Chen, B.P.; Chen, D.J. DNA-pk: A dynamic enzyme in a versatile dsb repair pathway. *DNA Repair* **2014**, *17*, 21–29.

18. Falck, J.; Coates, J.; Jackson, S.P. Conserved modes of recruitment of atm, atr and DNA-pkcs to sites of DNA damage. *Nature* **2005**, *434*, 605–611.

19. Petrini, J.H.; Stracker, T.H. The cellular response to DNA double-strand breaks: Defining the sensors and mediators. *Trends Cell Biol.* **2003**, *13*, 458–462.

20. Bekker-Jensen, S.; Lukas, C.; Kitagawa, R.; Melander, F.; Kastan, M.B.; Bartek, J.; Lukas, J. Spatial organization of the mammalian genome surveillance machinery in response to DNA strand breaks. *J. Cell Biol.* **2006**, *173*, 195–206.

21. Bakkenist, C.J.; Kastan, M.B. DNA damage activates atm through intermolecular autophosphorylation and dimer dissociation. *Nature* **2003**, *421*, 499–506.

22. Sun, Y.; Jiang, X.; Chen, S.; Fernandes, N.; Price, B.D. A role for the tip60 histone acetyltransferase in the acetylation and activation of atm. *Proc. Natl. Acad. Sci. USA* **2005**, *102*, 13182–13187.

23. Lou, Z.; Minter-Dykhouse, K.; Franco, S.; Gostissa, M.; Rivera, M.A.; Celeste, A.; Manis, J.P.; van Deursen, J.; Nussenzweig, A.; Paull, T.T.; *et al.* Mdc1 maintains genomic stability by participating in the amplification of atm-dependent DNA damage signals. *Mol. Cell* **2006**, *21*, 187–200.

24. Ciccia, A.; Elledge, S.J. The DNA damage response: Making it safe to play with knives. *Mol. Cell* **2010**, *40*, 179–204.

25. Stewart, G.S.; Panier, S.; Townsend, K.; Al-Hakim, A.K.; Kolas, N.K.; Miller, E.S.; Nakada, S.; Ylanko, J.; Olivarius, S.; Mendez, M.; *et al.* The riddle syndrome protein mediates a ubiquitin-dependent signaling cascade at sites of DNA damage. *Cell* **2009**, *136*, 420–434.

26. Bekker-Jensen, S.; Rendtlew Danielsen, J.; Fugger, K.; Gromova, I.; Nerstedt, A.; Lukas, C.; Bartek, J.; Lukas, J.; Mailand, N. Herc2 coordinates ubiquitin-dependent assembly of DNA repair factors on damaged chromosomes. *Nat. Cell Biol.* **2010**, *12*, 80–86.

27. Ellison, V.; Stillman, B. Biochemical characterization of DNA damage checkpoint complexes: Clamp loader and clamp complexes with specificity for 5′ recessed DNA. *PLoS Biol.* **2003**, *1*, E33.

28. Kumagai, A.; Kim, S.M.; Dunphy, W.G. Claspin and the activated form of atr-atrip collaborate in the activation of chk1. *J. Biol. Chem.* **2004**, *279*, 49599–49608.

29. Kemp, M.G.; Akan, Z.; Yilmaz, S.; Grillo, M.; Smith-Roe, S.L.; Kang, T.H.; Cordeiro-Stone, M.; Kaufmann, W.K.; Abraham, R.T.; Sancar, A.; *et al.* Tipin-replication protein a interaction mediates chk1 phosphorylation by atr in response to genotoxic stress. *J. Biol. Chem.* **2010**, *285*, 16562–16571.

30. Nam, E.A.; Cortez, D. Atr signalling: More than meeting at the fork. *Biochem. J.* **2011**, *436*, 527–536.

31. Lukas, J.; Lukas, C.; Bartek, J. Mammalian cell cycle checkpoints: Signalling pathways and their organization in space and time. *DNA Repair* **2004**, *3*, 997–1007.

32. DeCaprio, J.A. How the rb tumor suppressor structure and function was revealed by the study of adenovirus and sv40. *Virology* **2009**, *384*, 274–284.

33. Davy, C.; Doorbar, J. G2/m cell cycle arrest in the life cycle of viruses. *Virology* **2007**, *368*, 219–226.

34. Sancar, A.; Lindsey-Boltz, L.A.; Unsal-Kacmaz, K.; Linn, S. Molecular mechanisms of mammalian DNA repair and the DNA damage checkpoints. *Annu. Rev. Biochem.* **2004**, *73*, 39–85.

35. Fridman, J.S.; Lowe, S.W. Control of apoptosis by p53. *Oncogene* **2003**, *22*, 9030–9040.

36. Fuentes-Gonzalez, A.M.; Contreras-Paredes, A.; Manzo-Merino, J.; Lizano, M. The modulation of apoptosis by oncogenic viruses. *Virol. J.* **2013**, *10*, e182.

37. Cheng, E.H.; Nicholas, J.; Bellows, D.S.; Hayward, G.S.; Guo, H.G.; Reitz, M.S.; Hardwick, J.M. A bcl-2 homolog encoded by kaposi sarcoma-associated virus, human herpesvirus 8, inhibits apoptosis but does not heterodimerize with bax or bak. *Proc. Natl. Acad. Sci. USA* **1997**, *94*, 690–694.

38. Henderson, S.; Rowe, M.; Gregory, C.; Croom-Carter, D.; Wang, F.; Longnecker, R.; Kieff, E.; Rickinson, A. Induction of bcl-2 expression by epstein-barr virus latent membrane protein 1 protects infected b cells from programmed cell death. *Cell* **1991**, *65*, 1107–1115.

39. Gross-Mesilaty, S.; Reinstein, E.; Bercovich, B.; Tobias, K.E.; Schwartz, A.L.; Kahana, C.; Ciechanover, A. Basal and human papillomavirus e6 oncoprotein-induced degradation of myc proteins by the ubiquitin pathway. *Proc. Natl. Acad. Sci. USA* **1998**, *95*, 8058–8063.

40. Howley, P.M. Warts, cancer and ubiquitylation: Lessons from the papillomaviruses. *Trans. Am. Clin. Climatol. Assoc.* **2006**, *117*, 113–126, discussion 126–117.

41. Krueger, A.; Fas, S.C.; Giaisi, M.; Bleumink, M.; Merling, A.; Stumpf, C.; Baumann, S.; Holtkotte, D.; Bosch, V.; Krammer, P.H.; *et al.* Htlv-1 tax protects against cd95-mediated apoptosis by induction of the cellular flice-inhibitory protein (c-flip). *Blood* **2006**, *107*, 3933–3939.

42. Hewitt, G.; Jurk, D.; Marques, F.D.; Correia-Melo, C.; Hardy, T.; Gackowska, A.; Anderson, R.; Taschuk, M.; Mann, J.; Passos, J.F. Telomeres are favoured targets of a persistent DNA damage response in ageing and stress-induced senescence. *Nat. Commun.* **2012**, *3*, 708.

43. Chen, X.; Kamranvar, S.A.; Masucci, M.G. Tumor viruses and replicative immortality—Avoiding the telomere hurdle. *Semin. Cancer Biol.* **2014**, *26*, 43–51.

44. Zhang, A.; Wang, J.; Zheng, B.; Fang, X.; Angstrom, T.; Liu, C.; Li, X.; Erlandsson, F.; Bjorkholm, M.; Nordenskjord, M.; *et al.* Telomere attrition predominantly occurs in precursor lesions during *in vivo* carcinogenic process of the uterine cervix. *Oncogene* **2004**, *23*, 7441–7447.

45. Miura, N.; Horikawa, I.; Nishimoto, A.; Ohmura, H.; Ito, H.; Hirohashi, S.; Shay, J.W.; Oshimura, M. Progressive telomere shortening and telomerase reactivation during hepatocellular carcinogenesis. *Cancer Genet. Cytogenet.* **1997**, *93*, 56–62.

46. Oh, S.T.; Kyo, S.; Laimins, L.A. Telomerase activation by human papillomavirus type 16 e6 protein: Induction of human telomerase reverse transcriptase expression through myc and gc-rich sp1 binding sites. *J. Virol.* **2001**, *75*, 5559–5566.

47. Zhang, X.; Dong, N.; Zhang, H.; You, J.; Wang, H.; Ye, L. Effects of hepatitis b virus x protein on human telomerase reverse transcriptase expression and activity in hepatoma cells. *J. Lab. Clin. Med.* **2005**, *145*, 98–104.

48. Bais, C.; Van Geelen, A.; Eroles, P.; Mutlu, A.; Chiozzini, C.; Dias, S.; Silverstein, R.L.; Rafii, S.; Mesri, E.A. Kaposi's sarcoma associated herpesvirus g protein-coupled receptor immortalizes human endothelial cells by activation of the vegf receptor-2/ kdr. *Cancer Cell* **2003**, *3*, 131–143.

49. Yamamoto, A.; Kumakura, S.; Uchida, M.; Barrett, J.C.; Tsutsui, T. Immortalization of normal human embryonic fibroblasts by introduction of either the human papillomavirus type 16 e6 or e7 gene alone. *Int. J. Cancer* **2003**, *106*, 301–309.

50. Srivenugopal, K.S.; Ali-Osman, F. The DNA repair protein, o(6)-methylguanine-DNA methyltransferase is a proteolytic target for the e6 human papillomavirus oncoprotein. *Oncogene* **2002**, *21*, 5940–5945.

51. Iftner, T.; Elbel, M.; Schopp, B.; Hiller, T.; Loizou, J.I.; Caldecott, K.W.; Stubenrauch, F. Interference of papillomavirus e6 protein with single-strand break repair by interaction with xrcc1. *EMBO J.* **2002**, *21*, 4741–4748.

52. Jeang, K.T.; Widen, S.G.; Semmes, O.J.T.; Wilson, S.H. Htlv-i trans-activator protein, tax, is a trans-repressor of the human beta-polymerase gene. *Science* **1990**, *247*, 1082–1084.

53. Demetriou, S.K.; Ona-Vu, K.; Sullivan, E.M.; Dong, T.K.; Hsu, S.W.; Oh, D.H. Defective DNA repair and cell cycle arrest in cells expressing merkel cell polyomavirus t antigen. *Int. J. Cancer* **2012**, *131*, 1818–1827.

54. Kao, S.Y.; Marriott, S.J. Disruption of nucleotide excision repair by the human t-cell leukemia virus type 1 tax protein. *J. Virol.* **1999**, *73*, 4299–4304.

55. Whitehurst, C.B.; Vaziri, C.; Shackelford, J.; Pagano, J.S. Epstein-barr virus bplf1 deubiquitinates pcna and attenuates polymerase eta recruitment to DNA damage sites. *J. Virol.* **2012**, *86*, 8097–8106.

56. Lee, T.H.; Elledge, S.J.; Butel, J.S. Hepatitis b virus x protein interacts with a probable cellular DNA repair protein. *J. Virol.* **1995**, *69*, 1107–1114.

57. Sitterlin, D.; Lee, T.H.; Prigent, S.; Tiollais, P.; Butel, J.S.; Transy, C. Interaction of the uv-damaged DNA-binding protein with hepatitis b virus x protein is conserved among mammalian hepadnaviruses and restricted to transactivation-proficient x-insertion mutants. *J. Virol.* **1997**, *71*, 6194–6199.

58. Qadri, I.; Conaway, J.W.; Conaway, R.C.; Schaack, J.; Siddiqui, A. Hepatitis b virus transactivator protein, hbx, associates with the components of tfiih and stimulates the DNA helicase activity of tfiih. *Proc. Natl. Acad. Sci. USA* **1996**, *93*, 10578–10583.

59. Qadri, I.; Fatima, K.; Abde, L.H.H. Hepatitis b virus x protein impedes the DNA repair via its association with transcription factor, tfiih. *BMC Microbiol.* **2011**, *11*, e48.

60. Wang, Y.; Li, H.; Tang, Q.; Maul, G.G.; Yuan, Y. Kaposi's sarcoma-associated herpesvirus ori-lyt-dependent DNA replication: Involvement of host cellular factors. *J. Virol.* **2008**, *82*, 2867–2882.

61. Morimoto, H.; Tsukada, J.; Kominato, Y.; Tanaka, Y. Reduced expression of human mismatch repair genes in adult t-cell leukemia. *Am. J. Hematol.* **2005**, *78*, 100–107.

62. Daikoku, T.; Kudoh, A.; Sugaya, Y.; Iwahori, S.; Shirata, N.; Isomura, H.; Tsurumi, T. Postreplicative mismatch repair factors are recruited to epstein-barr virus replication compartments. *J. Biol. Chem.* **2006**, *281*, 11422–11430.

63. Park, J.W.; Nickel, K.P.; Torres, A.D.; Lee, D.; Lambert, P.F.; Kimple, R.J. Human papillomavirus type 16 e7 oncoprotein causes a delay in repair of DNA damage. *Radiother. Oncol.* **2014**, *113*, 337–344.

64. Kudoh, A.; Iwahori, S.; Sato, Y.; Nakayama, S.; Isomura, H.; Murata, T.; Tsurumi, T. Homologous recombinational repair factors are recruited and loaded onto the viral DNA genome in epstein-barr virus replication compartments. *J. Virol.* **2009**, *83*, 6641–6651.

65. Gillespie, K.A.; Mehta, K.P.; Laimins, L.A.; Moody, C.A. Human papillomaviruses recruit cellular DNA repair and homologous recombination factors to viral replication centers. *J. Virol.* **2012**, *86*, 9520–9526.

66. Belgnaoui, S.M.; Fryrear, K.A.; Nyalwidhe, J.O.; Guo, X.; Semmes, O.J. The viral oncoprotein tax sequesters DNA damage response factors by tethering mdc1 to chromatin. *J. Biol. Chem.* **2010**, *285*, 32897–32905.

67. Xiao, Y.; Chen, J.; Liao, Q.; Wu, Y.; Peng, C.; Chen, X. Lytic infection of Kaposi's sarcoma-associated herpesvirus induces DNA double-strand breaks and impairs non-homologous end joining. *J. Gen. Virol.* **2013**, *94*, 1870–1875.

68. Gottlieb, T.M.; Jackson, S.P. The DNA-dependent protein kinase: Requirement for DNA ends and association with ku antigen. *Cell* **1993**, *72*, 131–142.

69. Nick McElhinny, S.A.; Snowden, C.M.; McCarville, J.; Ramsden, D.A. Ku recruits the xrcc4-ligase iv complex to DNA ends. *Mol. Cell. Biol.* **2000**, *20*, 2996–3003.

70. Hartlerode, A.J.; Scully, R. Mechanisms of double-strand break repair in somatic mammalian cells. *Biochem. J.* **2009**, *423*, 157–168.

71. Grawunder, U.; Wilm, M.; Wu, X.; Kulesza, P.; Wilson, T.E.; Mann, M.; Lieber, M.R. Activity of DNA ligase iv stimulated by complex formation with xrcc4 protein in mammalian cells. *Nature* **1997**, *388*, 492–495.

72. Ahnesorg, P.; Smith, P.; Jackson, S.P. Xlf interacts with the xrcc4-DNA ligase iv complex to promote DNA nonhomologous end-joining. *Cell* **2006**, *124*, 301–313.

73. Sartori, A.A.; Lukas, C.; Coates, J.; Mistrik, M.; Fu, S.; Bartek, J.; Baer, R.; Lukas, J.; Jackson, S.P. Human ctip promotes DNA end resection. *Nature* **2007**, *450*, 509–U506.

74. Symington, L.S.; Gautier, J. Double-strand break end resection and repair pathway choice. *Annu. Rev. Genet.* **2011**, *45*, 247–271.

75. Nimonkar, A.V.; Sica, R.A.; Kowalczykowski, S.C. Rad52 promotes second-end DNA capture in double-stranded break repair to form complement-stabilized joint molecules. *Proc. Natl. Acad. Sci. USA* **2009**, *106*, 3077–3082.

76. Nimonkar, A.V.; Genschel, J.; Kinoshita, E.; Polaczek, P.; Campbell, J.L.; Wyman, C.; Modrich, P.; Kowalczykowski, S.C. Blm-dna2-rpa-mrn and exo1-blm-rpa-mrn constitute two DNA end resection machineries for human DNA break repair. *Genes Dev.* **2011**, *25*, 350–362.

77. McIlwraith, M.J.; West, S.C. DNA repair synthesis facilitates rad52-mediated second-end capture during dsb repair. *Mol. Cell* **2008**, *29*, 510–516.

78. Nimonkar, A.V.; Kowalczykowski, S.C. Second-end DNA capture in double-strand break repair: How to catch a DNA by its tail. *Cell Cycle* **2009**, *8*, 1816–1817.

79. Caldecott, K.W. DNA single-strand break repair. *Exp. Cell Res.* **2014**, *329*, 2–8.

80. D'Amours, D.; Desnoyers, S.; D'Silva, I.; Poirier, G.G. Poly(adp-ribosyl)ation reactions in the regulation of nuclear functions. *Biochem. J.* **1999**, *342*, 249–268.

81. Caldecott, K.W. Single-strand break repair and genetic disease. *Nat. Rev. Genet.* **2008**, *9*, 619–631.

82. Caldecott, K.W.; Aoufouchi, S.; Johnson, P.; Shall, S. Xrcc1 polypeptide interacts with DNA polymerase beta and possibly poly(adp-ribose) polymerase, and DNA ligase iii is a novel molecular "nick-sensor" *in vitro*. *Nucleic Acids Res.* **1996**, *24*, 4387–4394.

83. Whitehouse, C.J.; Taylor, R.M.; Thistlethwaite, A.; Zhang, H.; Karimi-Busheri, F.; Lasko, D.D.; Weinfeld, M.; Caldecott, K.W. Xrcc1 stimulates human polynucleotide kinase activity at damaged DNA termini and accelerates DNA single-strand break repair. *Cell* **2001**, *104*, 107–117.

84. Caldecott, K.W. Mammalian single-strand break repair: Mechanisms and links with chromatin. *DNA Repair* **2007**, *6*, 443–453.

85. Sedgwick, B. Repairing DNA-methylation damage. *Nat. Rev. Mol. Cell Biol.* **2004**, *5*, 148–157.

86. Sedgwick, B.; Bates, P.A.; Paik, J.; Jacobs, S.C.; Lindahl, T. Repair of alkylated DNA: Recent advances. *DNA Repair* **2007**, *6*, 429–442.

87. Parsons, J.L.; Dianov, G.L. Co-ordination of base excision repair and genome stability. *DNA Repair* **2013**, *12*, 326–333.

88. Dutta, A.; Yang, C.; Sengupta, S.; Mitra, S.; Hegde, M.L. New paradigms in the repair of oxidative damage in human genome: Mechanisms ensuring repair of mutagenic base lesions during replication and involvement of accessory proteins. *Cell. Mol. Life Sci.* **2015**, *72*, 1679–1698.

89. Wallace, S.S. Base excision repair: A critical player in many games. *DNA Repair* **2014**, *19*, 14–26.

90. Alekseev, S.; Coin, F. Orchestral maneuvers at the damaged sites in nucleotide excision repair. *Cell. Mol. Life Sci.* **2015**, *72*, 2177–2186.

91. Vermeulen, W.; Fousteri, M. Mammalian transcription-coupled excision repair. *Cold Spring Harb. Perspect. Biol.* **2013**, *5*, a012625.

92. De Laat, W.L.; Jaspers, N.G.; Hoeijmakers, J.H. Molecular mechanism of nucleotide excision repair. *Genes Dev.* **1999**, *13*, 768–785.

93. Marteijn, J.A.; Lans, H.; Vermeulen, W.; Hoeijmakers, J.H. Understanding nucleotide excision repair and its roles in cancer and ageing. *Nat. Rev. Mol. Cell Biol.* **2014**, *15*, 465–481.

94. Li, G.M. New insights and challenges in mismatch repair: Getting over the chromatin hurdle. *DNA Repair* **2014**, *19*, 48–54.

95. Kunkel, T.A.; Erie, D.A. DNA mismatch repair. *Annu. Rev. Biochem.* **2005**, *74*, 681–710.

96. Jiricny, J. The multifaceted mismatch-repair system. *Nat. Rev. Mol. Cell Biol.* **2006**, *7*, 335–346.

97. Genschel, J.; Modrich, P. Mechanism of 5′-directed excision in human mismatch repair. *Mol. Cell* **2003**, *12*, 1077–1086.

98. Williams, J.S.; Kunkel, T.A. Ribonucleotides in DNA: Origins, repair and consequences. *DNA Repair* **2014**, *19*, 27–37.

99. Walden, H.; Deans, A.J. The fanconi anemia DNA repair pathway: Structural and functional insights into a complex disorder. *Annu. Rev. Biophys.* **2014**, *43*, 257–278.

100. Kim, H.; D'Andrea, A.D. Regulation of DNA cross-link repair by the fanconi anemia/brca pathway. *Genes Dev.* **2012**, *26*, 1393–1408.

101. Branzei, D.; Foiani, M. Regulation of DNA repair throughout the cell cycle. *Nat. Rev. Mol. Cell Biol.* **2008**, *9*, 297–308.

102. You, Z.; Bailis, J.M. DNA damage and decisions: Ctip coordinates DNA repair and cell cycle checkpoints. *Trends Cell Biol.* **2010**, *20*, 402–409.

103. Germani, A.; Prabel, A.; Mourah, S.; Podgorniak, M.P.; di Carlo, A.; Ehrlich, R.; Gisselbrecht, S.; Varin-Blank, N.; Calvo, F.; Bruzzoni-Giovanelli, H. Siah-1 interacts with ctip and promotes its degradation by the proteasome pathway. *Oncogene* **2003**, *22*, 8845–8851.

104. Schroering, A.G.; Edelbrock, M.A.; Richards, T.J.; Williams, K.J. The cell cycle and DNA mismatch repair. *Exp. Cell Res.* **2007**, *313*, 292–304.

105. Chaudhry, M.A. Base excision repair of ionizing radiation-induced DNA damage in g1 and g2 cell cycle phases. *Cancer Cell Int.* **2007**, *7*, e15.

106. Ghosal, G.; Chen, J. DNA damage tolerance: A double-edged sword guarding the genome. *Transl. Cancer Res.* **2013**, *2*, 107–129.

107. Walboomers, J.M.; Jacobs, M.V.; Manos, M.M.; Bosch, F.X.; Kummer, J.A.; Shah, K.V.; Snijders, P.J.; Peto, J.; Meijer, C.J.; Munoz, N. Human papillomavirus is a necessary cause of invasive cervical cancer worldwide. *J. Pathol.* **1999**, *189*, 12–19.

108. Clifford, G.M.; Rana, R.K.; Franceschi, S.; Smith, J.S.; Gough, G.; Pimenta, J.M. Human papillomavirus genotype distribution in low-grade cervical lesions: Comparison by geographic region and with cervical cancer. *Cancer Epidemiol. Biomarkers Prev.* **2005**, *14*, 1157–1164.

109. Cooper, K.; Herrington, C.S.; Stickland, J.E.; Evans, M.F.; McGee, J.O. Episomal and integrated human papillomavirus in cervical neoplasia shown by non-isotopic *in situ* hybridisation. *J. Clin. Pathol.* **1991**, *44*, 990–996.

110. Zheng, Z.M.; Baker, C.C. Papillomavirus genome structure, expression, and post-transcriptional regulation. *Front. Biosci.* **2006**, *11*, 2286–2302.

111. Crook, T.; Tidy, J.A.; Vousden, K.H. Degradation of p53 can be targeted by hpv e6 sequences distinct from those required for p53 binding and trans-activation. *Cell* **1991**, *67*, 547–556.

112. Jones, D.L.; Munger, K. Interactions of the human papillomavirus e7 protein with cell cycle regulators. *Semin. Cancer Biol.* **1996**, *7*, 327–337.

113. Moody, C.A.; Laimins, L.A. Human papillomaviruses activate the atm DNA damage pathway for viral genome amplification upon differentiation. *PLoS Pathog.* **2009**, *5*, e1000605.

114. Fradet-Turcotte, A.; Bergeron-Labrecque, F.; Moody, C.A.; Lehoux, M.; Laimins, L.A.; Archambault, J. Nuclear accumulation of the papillomavirus e1 helicase blocks s-phase progression and triggers an atm-dependent DNA damage response. *J. Virol.* **2011**, *85*, 8996–9012.

115. Sakakibara, N.; Mitra, R.; McBride, A.A. The papillomavirus e1 helicase activates a cellular DNA damage response in viral replication foci. *J. Virol.* **2011**, *85*, 8981–8995.

116. Reinson, T.; Toots, M.; Kadaja, M.; Pipitch, R.; Allik, M.; Ustav, E.; Ustav, M. Engagement of the atr-dependent DNA damage response at the human papillomavirus 18 replication centers during the initial amplification. *J. Virol.* **2013**, *87*, 951–964.

117. Kadaja, M.; Isok-Paas, H.; Laos, T.; Ustav, E.; Ustav, M. Mechanism of genomic instability in cells infected with the high-risk human papillomaviruses. *PLoS Pathog.* **2009**, *5*, e1000397.

118. Kadaja, M.; Sumerina, A.; Verst, T.; Ojarand, M.; Ustav, E.; Ustav, M. Genomic instability of the host cell induced by the human papillomavirus replication machinery. *EMBO J.* **2007**, *26*, 2180–2191.

119. Duensing, S.; Munger, K. The human papillomavirus type 16 e6 and e7 oncoproteins independently induce numerical and structural chromosome instability. *Cancer Res.* **2002**, *62*, 7075–7082.

120. Spardy, N.; Duensing, A.; Charles, D.; Haines, N.; Nakahara, T.; Lambert, P.F.; Duensing, S. The human papillomavirus type 16 e7 oncoprotein activates the fanconi anemia (fa) pathway and causes accelerated chromosomal instability in fa cells. *J. Virol.* **2007**, *81*, 13265–13270.

121. Park, J.W.; Pitot, H.C.; Strati, K.; Spardy, N.; Duensing, S.; Grompe, M.; Lambert, P.F. Deficiencies in the fanconi anemia DNA damage response pathway increase sensitivity to hpv-associated head and neck cancer. *Cancer Res.* **2010**, *70*, 9959–9968.

122. Donaldson, M.M.; Mackintosh, L.J.; Bodily, J.M.; Dornan, E.S.; Laimins, L.A.; Morgan, I.M. An interaction between human papillomavirus 16 e2 and topbp1 is required for optimum viral DNA replication and episomal genome establishment. *J. Virol.* **2012**, *86*, 12806–12815.

123. Chen, B.; Simpson, D.A.; Zhou, Y.; Mitra, A.; Mitchell, D.L.; Cordeiro-Stone, M.; Kaufmann, W.K. Human papilloma virus type16 e6 deregulates chk1 and sensitizes human fibroblasts to environmental carcinogens independently of its effect on p53. *Cell Cycle* **2009**, *8*, 1775–1787.

124. Wallace, N.A.; Robinson, K.; Howie, H.L.; Galloway, D.A. Hpv 5 and 8 e6 abrogate atr activity resulting in increased persistence of uvb induced DNA damage. *PLoS Pathog.* **2012**, *8*, e1002807.

125. Winder, D.M.; Pett, M.R.; Foster, N.; Shivji, M.K.; Herdman, M.T.; Stanley, M.A.; Venkitaraman, A.R.; Coleman, N. An increase in DNA double-strand breaks, induced by ku70 depletion, is associated with human papillomavirus 16 episome loss and *de novo* viral integration events. *J. Pathol.* **2007**, *213*, 27–34.

126. Williams, V.M.; Filippova, M.; Filippov, V.; Payne, K.J.; Duerksen-Hughes, P. Human papillomavirus type 16 e6* induces oxidative stress and DNA damage. *J. Virol.* **2014**, *88*, 6751–6761.

127. Feng, H.; Shuda, M.; Chang, Y.; Moore, P.S. Clonal integration of a polyomavirus in human merkel cell carcinoma. *Science* **2008**, *319*, 1096–1100.

128. Sastre-Garau, X.; Peter, M.; Avril, M.F.; Laude, H.; Couturier, J.; Rozenberg, F.; Almeida, A.; Boitier, F.; Carlotti, A.; Couturaud, B.; *et al.* Merkel cell carcinoma of the skin: Pathological and molecular evidence for a causative role of mcv in oncogenesis. *J. Pathol.* **2009**, *218*, 48–56.

129. Shuda, M.; Kwun, H.J.; Feng, H.; Chang, Y.; Moore, P.S. Human merkel cell polyomavirus small t antigen is an oncoprotein targeting the 4e-bp1 translation regulator. *J. Clin. Investig.* **2011**, *121*, 3623–3634.

130. Griffiths, D.A.; Abdul-Sada, H.; Knight, L.M.; Jackson, B.R.; Richards, K.; Prescott, E.L.; Peach, A.H.; Blair, G.E.; Macdonald, A.; Whitehouse, A. Merkel cell polyomavirus small t antigen targets the nemo adaptor protein to disrupt inflammatory signaling. *J. Virol.* **2013**, *87*, 13853–13867.

131. Shuda, M.; Feng, H.; Kwun, H.J.; Rosen, S.T.; Gjoerup, O.; Moore, P.S.; Chang, Y. T antigen mutations are a human tumor-specific signature for merkel cell polyomavirus. *Proc. Natl. Acad. Sci. USA* **2008**, *105*, 16272–16277.

132. Kwun, H.J.; Guastafierro, A.; Shuda, M.; Meinke, G.; Bohm, A.; Moore, P.S.; Chang, Y. The minimum replication origin of merkel cell polyomavirus has a unique large t-antigen loading architecture and requires small t-antigen expression for optimal replication. *J. Virol.* **2009**, *83*, 12118–12128.

133. Li, J.; Wang, X.; Diaz, J.; Tsang, S.H.; Buck, C.B.; You, J. Merkel cell polyomavirus large t antigen disrupts host genomic integrity and inhibits cellular proliferation. *J. Virol.* **2013**, *87*, 9173–9188.

134. Tsang, S.H.; Wang, X.; Li, J.; Buck, C.B.; You, J. Host DNA damage response factors localize to merkel cell polyomavirus DNA replication sites to support efficient viral DNA replication. *J. Virol.* **2014**, *88*, 3285–3297.

135. Boichuk, S.; Hu, L.; Hein, J.; Gjoerup, O.V. Multiple DNA damage signaling and repair pathways deregulated by simian virus 40 large T antigen. *J. Virol.* **2010**, *84*, 8007–8020.

136. Li, J.; Diaz, J.; Wang, X.; Tsang, S.H.; You, J. Phosphorylation of merkel cell polyomavirus large tumor antigen at serine 816 by atm kinase induces apoptosis in host cells. *J. Biol. Chem.* **2015**, *290*, 1874–1884.

137. Chang, C.M.; Yu, K.J.; Mbulaiteye, S.M.; Hildesheim, A.; Bhatia, K. The extent of genetic diversity of epstein-barr virus and its geographic and disease patterns: A need for reappraisal. *Virus Res.* **2009**, *143*, 209–221.

138. Maeda, E.; Akahane, M.; Kiryu, S.; Kato, N.; Yoshikawa, T.; Hayashi, N.; Aoki, S.; Minami, M.; Uozaki, H.; Fukayama, M.; *et al.* Spectrum of epstein-barr virus-related diseases: A pictorial review. *Jpn. J. Radiol.* **2009**, *27*, 4–19.

139. Young, L.S.; Rickinson, A.B. Epstein-barr virus: 40 years on. *Nat. Rev. Cancer* **2004**, *4*, 757–768.

140. Kamranvar, S.A.; Gruhne, B.; Szeles, A.; Masucci, M.G. Epstein-barr virus promotes genomic instability in burkitt's lymphoma. *Oncogene* **2007**, *26*, 5115–5123.

141. Gruhne, B.; Kamranvar, S.A.; Masucci, M.G.; Sompallae, R. Ebv and genomic instability—A new look at the role of the virus in the pathogenesis of burkitt's lymphoma. *Semin. Cancer Biol.* **2009**, *19*, 394–400.

142. Kamranvar, S.A.; Masucci, M.G. The epstein-barr virus nuclear antigen-1 promotes telomere dysfunction via induction of oxidative stress. *Leukemia* **2011**, *25*, 1017–1025.

143. Liu, M.T.; Chen, Y.R.; Chen, S.C.; Hu, C.Y.; Lin, C.S.; Chang, Y.T.; Wang, W.B.; Chen, J.Y. Epstein-barr virus latent membrane protein 1 induces micronucleus formation, represses DNA repair and enhances sensitivity to DNA-damaging agents in human epithelial cells. *Oncogene* **2004**, *23*, 2531–2539.

144. Chen, Y.R.; Liu, M.T.; Chang, Y.T.; Wu, C.C.; Hu, C.Y.; Chen, J.Y. Epstein-barr virus latent membrane protein 1 represses DNA repair through the pi3k/akt/foxo3a pathway in human epithelial cells. *J. Virol.* **2008**, *82*, 8124–8137.

145. Chiu, S.H.; Wu, C.C.; Fang, C.Y.; Yu, S.L.; Hsu, H.Y.; Chow, Y.H.; Chen, J.Y. Epstein-barr virus balf3 mediates genomic instability and progressive malignancy in nasopharyngeal carcinoma. *Oncotarget* **2014**, *5*, 8583–8601.

146. Chang, Y.H.; Lee, C.P.; Su, M.T.; Wang, J.T.; Chen, J.Y.; Lin, S.F.; Tsai, C.H.; Hsieh, M.J.; Takada, K.; Chen, M.R. Epstein-barr virus bglf4 kinase retards cellular s-phase progression and induces chromosomal abnormality. *PLoS ONE* **2012**, *7*, e39217.

147. Wu, C.C.; Liu, M.T.; Chang, Y.T.; Fang, C.Y.; Chou, S.P.; Liao, H.W.; Kuo, K.L.; Hsu, S.L.; Chen, Y.R.; Wang, P.W.; *et al.* Epstein-barr virus dnase (bglf5) induces genomic instability in human epithelial cells. *Nucleic Acids Res.* **2010**, *38*, 1932–1949.

148. Kudoh, A.; Fujita, M.; Zhang, L.; Shirata, N.; Daikoku, T.; Sugaya, Y.; Isomura, H.; Nishiyama, Y.; Tsurumi, T. Epstein-barr virus lytic replication elicits atm checkpoint signal transduction while providing an s-phase-like cellular environment. *J. Biol. Chem.* **2005**, *280*, 8156–8163.

149. Bailey, S.G.; Verrall, E.; Schelcher, C.; Rhie, A.; Doherty, A.J.; Sinclair, A.J. Functional interaction between epstein-barr virus replication protein zta and host DNA damage response protein 53bp1. *J. Virol.* **2009**, *83*, 11116–11122.

150. Li, R.; Zhu, J.; Xie, Z.; Liao, G.; Liu, J.; Chen, M.R.; Hu, S.; Woodard, C.; Lin, J.; Taverna, S.D.; *et al.* Conserved herpesvirus kinases target the DNA damage response pathway and tip60 histone acetyltransferase to promote virus replication. *Cell Host Microbe* **2011**, *10*, 390–400.

151. Hagemeier, S.R.; Barlow, E.A.; Meng, Q.; Kenney, S.C. The cellular ataxia telangiectasia-mutated kinase promotes epstein-barr virus lytic reactivation in response to multiple different types of lytic reactivation-inducing stimuli. *J. Virol.* **2012**, *86*, 13360–13370.

152. Hau, P.M.; Deng, W.; Jia, L.; Yang, J.; Tsurumi, T.; Chiang, A.K.; Huen, M.S.; Tsao, S.W. Role of atm in the formation of the replication compartment during lytic replication of epstein-barr virus in nasopharyngeal epithelial cells. *J. Virol.* **2015**, *89*, 652–668.

153. Nikitin, P.A.; Yan, C.M.; Forte, E.; Bocedi, A.; Tourigny, J.P.; White, R.E.; Allday, M.J.; Patel, A.; Dave, S.S.; Kim, W.; *et al.* An atm/chk2-mediated DNA damage-responsive signaling pathway suppresses epstein-barr virus transformation of primary human b cells. *Cell Host Microbe* **2010**, *8*, 510–522.

154. Koganti, S.; Hui-Yuen, J.; McAllister, S.; Gardner, B.; Grasser, F.; Palendira, U.; Tangye, S.G.; Freeman, A.F.; Bhaduri-McIntosh, S. Stat3 interrupts atr-chk1 signaling to allow oncovirus-mediated cell proliferation. *Proc. Natl. Acad. Sci. USA* **2014**, *111*, 4946–4951.

155. Choudhuri, T.; Verma, S.C.; Lan, K.; Murakami, M.; Robertson, E.S. The atm/atr signaling effector chk2 is targeted by epstein-barr virus nuclear antigen 3c to release the g2/m cell cycle block. *J. Virol.* **2007**, *81*, 6718–6730.

156. Jha, H.C.; A, J.M.; Saha, A.; Banerjee, S.; Lu, J.; Robertson, E.S. Epstein-Barr virus essential antigen EBNA3C attenuates H2AX expression. *J. Virol.* **2014**, *88*, 3776–3788.

157. Deng, W.; Pang, P.S.; Tsang, C.M.; Hau, P.M.; Yip, Y.L.; Cheung, A.L.; Tsao, S.W. Epstein-barr virus-encoded latent membrane protein 1 impairs g2 checkpoint in human nasopharyngeal epithelial cells through defective chk1 activation. *PLoS ONE* **2012**, *7*, e39095.

158. Dheekollu, J.; Deng, Z.; Wiedmer, A.; Weitzman, M.D.; Lieberman, P.M. A role for mre11, nbs1, and recombination junctions in replication and stable maintenance of ebv episomes. *PLoS ONE* **2007**, *2*, e1257.

159. Chang, Y.; Cesarman, E.; Pessin, M.S.; Lee, F.; Culpepper, J.; Knowles, D.M.; Moore, P.S. Identification of herpesvirus-like DNA sequences in aids-associated Kaposi's sarcoma. *Science* **1994**, *266*, 1865–1869.

160. Cesarman, E.; Chang, Y.; Moore, P.S.; Said, J.W.; Knowles, D.M. Kaposi's sarcoma-associated herpesvirus-like DNA sequences in aids-related body-cavity-based lymphomas. *N. Engl. J. Med.* **1995**, *332*, 1186–1191.

161. Soulier, J.; Grollet, L.; Oksenhendler, E.; Cacoub, P.; Cazals-Hatem, D.; Babinet, P.; D'agay, M.F.; Clauvel, J.P.; Raphael, M.; Degos, L.; *et al.* Kaposi's sarcoma-associated herpesvirus-like DNA sequences in multicentric castleman's disease. *Blood* **1995**, *86*, 1276–1280.

162. Dittmer, D.P. Transcription profile of Kaposi's sarcoma-associated herpesvirus in primary Kaposi's sarcoma lesions as determined by real-time pcr arrays. *Cancer Res.* **2003**, *63*, 2010–2015.

163. Ballestas, M.E.; Chatis, P.A.; Kaye, K.M. Efficient persistence of extrachromosomal kshv DNA mediated by latency-associated nuclear antigen. *Science* **1999**, *284*, 641–644.

164. Uppal, T.; Banerjee, S.; Sun, Z.; Verma, S.C.; Robertson, E.S. Kshv lana—The master regulator of kshv latency. *Viruses* **2014**, *6*, 4961–4998.

165. Mesri, E.A.; Cesarman, E.; Boshoff, C. Kaposi's sarcoma and its associated herpesvirus. *Nat. Rev. Cancer* **2010**, *10*, 707–719.

166. Jha, H.C.; Upadhyay, S.K.; A. J. Prasad, M.; Lu, J.; Cai, Q.; Saha, A.; Robertson, E.S. H2ax phosphorylation is important for lana-mediated Kaposi's sarcoma-associated herpesvirus episome persistence. *J. Virol.* **2013**, *87*, 5255–5269.

167. Singh, V.V.; Dutta, D.; Ansari, M.A.; Dutta, S.; Chandran, B. Kaposi's sarcoma-associated herpesvirus induces the atm and h2ax DNA damage response early during *de novo* infection of primary endothelial cells, which play roles in latency establishment. *J. Virol.* **2014**, *88*, 2821–2834.

168. Koopal, S.; Furuhjelm, J.H.; Jarviluoma, A.; Jaamaa, S.; Pyakurel, P.; Pussinen, C.; Wirzenius, M.; Biberfeld, P.; Alitalo, K.; Laiho, M.; *et al.* Viral oncogene-induced DNA damage response is activated in kaposi sarcoma tumorigenesis. *PLoS Pathog.* **2007**, *3*, 1348–1360.

169. Verschuren, E.W.; Klefstrom, J.; Evan, G.I.; Jones, N. The oncogenic potential of Kaposi's sarcoma-associated herpesvirus cyclin is exposed by p53 loss *in vitro* and *in vivo*. *Cancer Cell* **2002**, *2*, 229–241.

170. Pan, H.; Zhou, F.; Gao, S.J. Kaposi's sarcoma-associated herpesvirus induction of chromosome instability in primary human endothelial cells. *Cancer Res.* **2004**, *64*, 4064–4068.

171. Kumar, A.; Sahu, S.K.; Mohanty, S.; Chakrabarti, S.; Maji, S.; Reddy, R.R.; Jha, A.K.; Goswami, C.; Kundu, C.N.; Rajasubramaniam, S.; *et al.* Kaposi sarcoma herpes virus latency associated nuclear antigen protein release the g2/m cell cycle blocks by modulating atm/atr mediated checkpoint pathway. *PLoS ONE* **2014**, *9*, e100228.

172. Gwack, Y.; Nakamura, H.; Lee, S.H.; Souvlis, J.; Yustein, J.T.; Gygi, S.; Kung, H.J.; Jung, J.U. Poly(adp-ribose) polymerase 1 and ste20-like kinase hkfc act as transcriptional repressors for gamma-2 herpesvirus lytic replication. *Mol. Cell. Biol.* **2003**, *23*, 8282–8294.

173. Jackson, B.R.; Noerenberg, M.; Whitehouse, A. A novel mechanism inducing genome instability in Kaposi's sarcoma-associated herpesvirus infected cells. *PLoS Pathog.* **2014**, *10*, e1004098.

174. Shin, Y.C.; Nakamura, H.; Liang, X.; Feng, P.; Chang, H.; Kowalik, T.F.; Jung, J.U. Inhibition of the atm/p53 signal transduction pathway by Kaposi's sarcoma-associated herpesvirus interferon regulatory factor 1. *J. Virol.* **2006**, *80*, 2257–2266.

175. Krishnan, H.H.; Naranatt, P.P.; Smith, M.S.; Zeng, L.; Bloomer, C.; Chandran, B. Concurrent expression of latent and a limited number of lytic genes with immune modulation and antiapoptotic function by Kaposi's sarcoma-associated herpesvirus early during infection of primary endothelial and fibroblast cells and subsequent decline of lytic gene expression. *J. Virol.* **2004**, *78*, 3601–3620.

176. Ma, S.D.; Hegde, S.; Young, K.H.; Sullivan, R.; Rajesh, D.; Zhou, Y.; Jankowska-Gan, E.; Burlingham, W.J.; Sun, X.; Gulley, M.L.; *et al.* A new model of epstein-barr virus infection reveals an important role for early lytic viral protein expression in the development of lymphomas. *J. Virol.* **2011**, *85*, 165–177.

177. Franchini, G.; Nicot, C.; Johnson, J.M. Seizing of t cells by human t-cell leukemia/lymphoma virus type 1. *Adv. Cancer Res.* **2003**, *89*, 69–132.

178. Matsuoka, M.; Jeang, K.T. Human t-cell leukaemia virus type 1 (htlv-1) infectivity and cellular transformation. *Nat. Rev. Cancer* **2007**, *7*, 270–280.

179. Korber, B.; Okayama, A.; Donnelly, R.; Tachibana, N.; Essex, M. Polymerase chain reaction analysis of defective human t-cell leukemia virus type i proviral genomes in leukemic cells of patients with adult t-cell leukemia. *J. Virol.* **1991**, *65*, 5471–5476.

180. Grassmann, R.; Dengler, C.; Muller-Fleckenstein, I.; Fleckenstein, B.; McGuire, K.; Dokhelar, M.C.; Sodroski, J.G.; Haseltine, W.A. Transformation to continuous growth of primary human t lymphocytes by human t-cell leukemia virus type i x-region genes transduced by a herpesvirus saimiri vector. *Proc. Natl. Acad. Sci. USA* **1989**, *86*, 3351–3355.

181. Grassmann, R.; Berchtold, S.; Radant, I.; Alt, M.; Fleckenstein, B.; Sodroski, J.G.; Haseltine, W.A.; Ramstedt, U. Role of human t-cell leukemia virus type 1 x region proteins in immortalization of primary human lymphocytes in culture. *J. Virol.* **1992**, *66*, 4570–4575.

182. Hara, T.; Matsumura-Arioka, Y.; Ohtani, K.; Nakamura, M. Role of human t-cell leukemia virus type i tax in expression of the human telomerase reverse transcriptase (htert) gene in human t-cells. *Cancer Sci.* **2008**, *99*, 1155–1163.

183. Pise-Mason, C.A.; Mahieux, R.; Jiang, H.; Ashcroft, M.; Radonovich, M.; Duvall, J.; Guillerm, C.; Brady, J.N. Inactivation of p53 by human t-cell lymphotropic virus type 1 tax requires activation of the nf-kappab pathway and is dependent on p53 phosphorylation. *Mol. Cell. Biol.* **2000**, *20*, 3377–3386.

184. Kehn, K.; Fuente Cde, L.; Strouss, K.; Berro, R.; Jiang, H.; Brady, J.; Mahieux, R.; Pumfery, A.; Bottazzi, M.E.; Kashanchi, F. The htlv-i tax oncoprotein targets the retinoblastoma protein for proteasomal degradation. *Oncogene* **2005**, *24*, 525–540.

185. Semmes, O.J.; Jeang, K.T. Localization of human t-cell leukemia virus type 1 tax to subnuclear compartments that overlap with interchromatin speckles. *J. Virol.* **1996**, *70*, 6347–6357.

186. Bex, F.; McDowall, A.; Burny, A.; Gaynor, R. The human t-cell leukemia virus type 1 transactivator protein tax colocalizes in unique nuclear structures with nf-kappab proteins. *J. Virol.* **1997**, *71*, 3484–3497.

187. Majone, F.; Semmes, O.J.; Jeang, K.T. Induction of micronuclei by htlv-i tax: A cellular assay for function. *Virology* **1993**, *193*, 456–459.

188. Philpott, S.M.; Buehring, G.C. Defective DNA repair in cells with human t-cell leukemia/bovine leukemia viruses: Role of tax gene. *J. Natl. Cancer Inst.* **1999**, *91*, 933–942.

189. Baydoun, H.H.; Bai, X.T.; Shelton, S.; Nicot, C. Htlv-i tax increases genetic instability by inducing DNA double strand breaks during DNA replication and switching repair to nhej. *PLoS ONE* **2012**, *7*, e42226.

190. Mihaylova, V.T.; Green, A.M.; Khurgel, M.; Semmes, O.J.; Kupfer, G.M. Human t-cell leukemia virus i tax protein sensitizes p53-mutant cells to DNA damage. *Cancer Res.* **2008**, *68*, 4843–4852.

191. Kinjo, T.; Ham-Terhune, J.; Peloponese, J.M., Jr.; Jeang, K.T. Induction of reactive oxygen species by human t-cell leukemia virus type 1 tax correlates with DNA damage and expression of cellular senescence marker. *J. Virol.* **2010**, *84*, 5431–5437.

192. Chaib-Mezrag, H.; Lemacon, D.; Fontaine, H.; Bellon, M.; Bai, X.T.; Drac, M.; Coquelle, A.; Nicot, C. Tax impairs DNA replication forks and increases DNA breaks in specific oncogenic genome regions. *Mol. Cancer* **2014**, *13*, e205.

193. Miyake, H.; Suzuki, T.; Hirai, H.; Yoshida, M. Trans-activator tax of human t-cell leukemia virus type 1 enhances mutation frequency of the cellular genome. *Virology* **1999**, *253*, 155–161.

194. Cannon, J.S.; Hamzeh, F.; Moore, S.; Nicholas, J.; Ambinder, R.F. Human herpesvirus 8-encoded thymidine kinase and phosphotransferase homologues confer sensitivity to ganciclovir. *J. Virol.* **1999**, *73*, 4786–4793.

195. Schavinsky-Khrapunsky, Y.; Priel, E.; Aboud, M. Dose-dependent dual effect of htlv-1 tax oncoprotein on p53-dependent nucleotide excision repair in human t-cells. *Int. J. Cancer* **2008**, *122*, 305–316.

196. Chandhasin, C.; Ducu, R.I.; Berkovich, E.; Kastan, M.B.; Marriott, S.J. Human t-cell leukemia virus type 1 tax attenuates the atm-mediated cellular DNA damage response. *J. Virol.* **2008**, *82*, 6952–6961.

197. Dayaram, T.; Lemoine, F.J.; Donehower, L.A.; Marriott, S.J. Activation of wip1 phosphatase by htlv-1 tax mitigates the cellular response to DNA damage. *PLoS ONE* **2013**, *8*, e55989.

198. Majone, F.; Luisetto, R.; Zamboni, D.; Iwanaga, Y.; Jeang, K.T. Ku protein as a potential human T-cell leukemia virus type 1 (HTLV-1) tax target in clastogenic chromosomal instability of mammalian cells. *Retrovirology* **2005**, *2*, e45.

199. Ducu, R.I.; Dayaram, T.; Marriott, S.J. The htlv-1 tax oncoprotein represses ku80 gene expression. *Virology* **2011**, *416*, 1–8.

200. Baydoun, H.H.; Pancewicz, J.; Nicot, C. Human t-lymphotropic type 1 virus p30 inhibits homologous recombination and favors unfaithful DNA repair. *Blood* **2011**, *117*, 5897–5906.

201. Haoudi, A.; Daniels, R.C.; Wong, E.; Kupfer, G.; Semmes, O.J. Human t-cell leukemia virus-i tax oncoprotein functionally targets a subnuclear complex involved in cellular DNA damage-response. *J. Biol. Chem.* **2003**, *278*, 37736–37744.
202. Park, H.U.; Jeong, J.H.; Chung, J.H.; Brady, J.N. Human t-cell leukemia virus type 1 tax interacts with chk1 and attenuates DNA-damage induced g2 arrest mediated by chk1. *Oncogene* **2004**, *23*, 4966–4974.
203. Gupta, S.K.; Guo, X.; Durkin, S.S.; Fryrear, K.F.; Ward, M.D.; Semmes, O.J. Human t-cell leukemia virus type 1 tax oncoprotein prevents DNA damage-induced chromatin egress of hyperphosphorylated chk2. *J. Biol. Chem.* **2007**, *282*, 29431–29440.
204. Durkin, S.S.; Guo, X.; Fryrear, K.A.; Mihaylova, V.T.; Gupta, S.K.; Belgnaoui, S.M.; Haoudi, A.; Kupfer, G.M.; Semmes, O.J. Htlv-1 tax oncoprotein subverts the cellular DNA damage response via binding to DNA-dependent protein kinase. *J. Biol. Chem.* **2008**, *283*, 36311–36320.
205. Chlichlia, K.; Khazaie, K. Htlv-1 tax: Linking transformation, DNA damage and apoptotic t-cell death. *Chem. Biol. Interact.* **2010**, *188*, 359–365.
206. Bellon, M.; Baydoun, H.H.; Yao, Y.; Nicot, C. Htlv-i tax-dependent and -independent events associated with immortalization of human primary t lymphocytes. *Blood* **2010**, *115*, 2441–2448.
207. Miller, R.H.; Kaneko, S.; Chung, C.T.; Girones, R.; Purcell, R.H. Compact organization of the hepatitis b virus genome. *Hepatology* **1989**, *9*, 322–327.
208. Edman, J.C.; Gray, P.; Valenzuela, P.; Rall, L.B.; Rutter, W.J. Integration of hepatitis b virus sequences and their expression in a human hepatoma cell. *Nature* **1980**, *286*, 535–538.
209. Wei, Y.; Neuveut, C.; Tiollais, P.; Buendia, M.A. Molecular biology of the hepatitis b virus and role of the x gene. *Pathol. Biol.* **2010**, *58*, 267–272.
210. Bouchard, M.J.; Schneider, R.J. The enigmatic x gene of hepatitis b virus. *J. Virol.* **2004**, *78*, 12725–12734.
211. Feitelson, M.A.; Zhu, M.; Duan, L.X.; London, W.T. Hepatitis b x antigen and p53 are associated *in vitro* and in liver tissues from patients with primary hepatocellular carcinoma. *Oncogene* **1993**, *8*, 1109–1117.
212. Wang, X.W.; Forrester, K.; Yeh, H.; Feitelson, M.A.; Gu, J.R.; Harris, C.C. Hepatitis b virus x protein inhibits p53 sequence-specific DNA binding, transcriptional activity, and association with transcription factor ercc3. *Proc. Natl. Acad. Sci. USA* **1994**, *91*, 2230–2234.
213. Lee, S.G.; Rho, H.M. Transcriptional repression of the human p53 gene by hepatitis b viral x protein. *Oncogene* **2000**, *19*, 468–471.
214. Mathonnet, G.; Lachance, S.; Alaoui-Jamali, M.; Drobetsky, E.A. Expression of hepatitis b virus x oncoprotein inhibits transcription-coupled nucleotide excision repair in human cells. *Mutat. Res.* **2004**, *554*, 305–318.
215. Knoll, S.; Furst, K.; Thomas, S.; Villanueva Baselga, S.; Stoll, A.; Schaefer, S.; Putzer, B.M. Dissection of cell context-dependent interactions between hbx and p53 family members in regulation of apoptosis: A role for hbv-induced hcc. *Cell Cycle* **2011**, *10*, 3554–3565.

216. Bolukbas, C.; Bolukbas, F.F.; Horoz, M.; Aslan, M.; Celik, H.; Erel, O. Increased oxidative stress associated with the severity of the liver disease in various forms of hepatitis b virus infection. *BMC Infect. Dis.* **2005**, *5*, e95.

217. Hagen, T.M.; Huang, S.; Curnutte, J.; Fowler, P.; Martinez, V.; Wehr, C.M.; Ames, B.N.; Chisari, F.V. Extensive oxidative DNA damage in hepatocytes of transgenic mice with chronic active hepatitis destined to develop hepatocellular carcinoma. *Proc. Natl. Acad. Sci. USA* **1994**, *91*, 12808–12812.

218. Hsieh, Y.H.; Su, I.J.; Wang, H.C.; Chang, W.W.; Lei, H.Y.; Lai, M.D.; Chang, W.T.; Huang, W. Pre-s mutant surface antigens in chronic hepatitis b virus infection induce oxidative stress and DNA damage. *Carcinogenesis* **2004**, *25*, 2023–2032.

219. Wang, H.C.; Huang, W.; Lai, M.D.; Su, I.J. Hepatitis b virus pre-s mutants, endoplasmic reticulum stress and hepatocarcinogenesis. *Cancer Sci.* **2006**, *97*, 683–688.

220. Gwak, G.Y.; Lee, D.H.; Moon, T.G.; Choi, M.S.; Lee, J.H.; Koh, K.C.; Paik, S.W.; Park, C.K.; Joh, J.W.; Yoo, B.C. The correlation of hepatitis b virus pre-s mutation with cellular oxidative DNA damage in hepatocellular carcinoma. *Hepatogastroenterology* **2008**, *55*, 2028–2032.

221. Dandri, M.; Burda, M.R.; Burkle, A.; Zuckerman, D.M.; Will, H.; Rogler, C.E.; Greten, H.; Petersen, J. Increase in *de novo* hbv DNA integrations in response to oxidative DNA damage or inhibition of poly(adp-ribosyl)ation. *Hepatology* **2002**, *35*, 217–223.

222. Leupin, O.; Bontron, S.; Schaeffer, C.; Strubin, M. Hepatitis b virus x protein stimulates viral genome replication via a ddb1-dependent pathway distinct from that leading to cell death. *J. Virol.* **2005**, *79*, 4238–4245.

223. Hodgson, A.J.; Hyser, J.M.; Keasler, V.V.; Cang, Y.; Slagle, B.L. Hepatitis b virus regulatory hbx protein binding to ddb1 is required but is not sufficient for maximal hbv replication. *Virology* **2012**, *426*, 73–82.

224. Iovine, B.; Iannella, M.L.; Bevilacqua, M.A. Damage-specific DNA binding protein 1 (ddb1): A protein with a wide range of functions. *Int. J. Biochem. Cell Biol.* **2011**, *43*, 1664–1667.

225. Becker, S.A.; Lee, T.H.; Butel, J.S.; Slagle, B.L. Hepatitis b virus x protein interferes with cellular DNA repair. *J. Virol.* **1998**, *72*, 266–272.

226. Martin-Lluesma, S.; Schaeffer, C.; Robert, E.I.; van Breugel, P.C.; Leupin, O.; Hantz, O.; Strubin, M. Hepatitis b virus x protein affects s phase progression leading to chromosome segregation defects by binding to damaged DNA binding protein 1. *Hepatology* **2008**, *48*, 1467–1476.

227. Jaitovich-Groisman, I.; Benlimame, N.; Slagle, B.L.; Perez, M.H.; Alpert, L.; Song, D.J.; Fotouhi-Ardakani, N.; Galipeau, J.; Alaoui-Jamali, M.A. Transcriptional regulation of the tfiih transcription repair components xpb and xpd by the hepatitis b virus x protein in liver cells and transgenic liver tissue. *J. Biol. Chem.* **2001**, *276*, 14124–14132.

228. Prost, S.; Ford, J.M.; Taylor, C.; Doig, J.; Harrison, D.J. Hepatitis b x protein inhibits p53-dependent DNA repair in primary mouse hepatocytes. *J. Biol. Chem.* **1998**, *273*, 33327–33332.

229. Lee, A.T.; Ren, J.; Wong, E.T.; Ban, K.H.; Lee, L.A.; Lee, C.G. The hepatitis b virus x protein sensitizes hepg2 cells to uv light-induced DNA damage. *J. Biol. Chem.* **2005**, *280*, 33525–33535.

230. Van de Klundert, M.A.; van Hemert, F.J.; Zaaijer, H.L.; Kootstra, N.A. The hepatitis b virus x protein inhibits thymine DNA glycosylase initiated base excision repair. *PLoS ONE* **2012**, *7*, e48940.

231. Chung, Y.L. Defective DNA damage response and repair in liver cells expressing hepatitis b virus surface antigen. *FASEB J.* **2013**, *27*, 2316–2327.

232. Wang, W.H.; Hullinger, R.L.; Andrisani, O.M. Hepatitis b virus x protein via the p38mapk pathway induces e2f1 release and atr kinase activation mediating p53 apoptosis. *J. Biol. Chem.* **2008**, *283*, 25455–25467.

233. Zhao, F.; Hou, N.B.; Yang, X.L.; He, X.; Liu, Y.; Zhang, Y.H.; Wei, C.W.; Song, T.; Li, L.; Ma, Q.J.; *et al.* Ataxia telangiectasia-mutated-rad3-related DNA damage checkpoint signaling pathway triggered by hepatitis b virus infection. *World J. Gastroenterol.* **2008**, *14*, 6163–6170.

234. Rakotomalala, L.; Studach, L.; Wang, W.H.; Gregori, G.; Hullinger, R.L.; Andrisani, O. Hepatitis b virus x protein increases the cdt1-to-geminin ratio inducing DNA re-replication and polyploidy. *J. Biol. Chem.* **2008**, *283*, 28729–28740.

235. Studach, L.; Wang, W.H.; Weber, G.; Tang, J.; Hullinger, R.L.; Malbrue, R.; Liu, X.; Andrisani, O. Polo-like kinase 1 activated by the hepatitis b virus x protein attenuates both the DNA damage checkpoint and DNA repair resulting in partial polyploidy. *J. Biol. Chem.* **2010**, *285*, 30282–30293.

236. Zheng, Z.; Li, J.; Sun, J.; Song, T.; Wei, C.; Zhang, Y.; Rao, G.; Chen, G.; Li, D.; Yang, G.; *et al.* Inhibition of hbv replication by theophylline. *Antiviral Res.* **2011**, *89*, 149–155.

237. Sung, W.K.; Zheng, H.; Li, S.; Chen, R.; Liu, X.; Li, Y.; Lee, N.P.; Lee, W.H.; Ariyaratne, P.N.; Tennakoon, C.; *et al.* Genome-wide survey of recurrent hbv integration in hepatocellular carcinoma. *Nat. Genet.* **2012**, *44*, 765–769.

238. Shepard, C.W.; Finelli, L.; Alter, M.J. Global epidemiology of hepatitis c virus infection. *Lancet Infect. Dis.* **2005**, *5*, 558–567.

239. Zignego, A.L.; Macchia, D.; Monti, M.; Thiers, V.; Mazzetti, M.; Foschi, M.; Maggi, E.; Romagnani, S.; Gentilini, P.; Brechot, C. Infection of peripheral mononuclear blood cells by hepatitis c virus. *J. Hepatol.* **1992**, *15*, 382–386.

240. Egger, D.; Wolk, B.; Gosert, R.; Bianchi, L.; Blum, H.E.; Moradpour, D.; Bienz, K. Expression of hepatitis c virus proteins induces distinct membrane alterations including a candidate viral replication complex. *J. Virol.* **2002**, *76*, 5974–5984.

241. Gosert, R.; Egger, D.; Lohmann, V.; Bartenschlager, R.; Blum, H.E.; Bienz, K.; Moradpour, D. Identification of the hepatitis c virus rna replication complex in huh-7 cells harboring subgenomic replicons. *J. Virol.* **2003**, *77*, 5487–5492.

242. Machida, K.; Cheng, K.T.; Sung, V.M.; Lee, K.J.; Levine, A.M.; Lai, M.M. Hepatitis c virus infection activates the immunologic (type ii) isoform of nitric oxide synthase and thereby enhances DNA damage and mutations of cellular genes. *J. Virol.* **2004**, *78*, 8835–8843.

243. Machida, K.; Cheng, K.T.; Lai, C.K.; Jeng, K.S.; Sung, V.M.; Lai, M.M. Hepatitis c virus triggers mitochondrial permeability transition with production of reactive oxygen species, leading to DNA damage and stat3 activation. *J. Virol.* **2006**, *80*, 7199–7207.

244. Nishimura, T.; Kohara, M.; Izumi, K.; Kasama, Y.; Hirata, Y.; Huang, Y.; Shuda, M.; Mukaidani, C.; Takano, T.; Tokunaga, Y.; *et al.* Hepatitis c virus impairs p53 via persistent overexpression of 3beta-hydroxysterol delta24-reductase. *J. Biol. Chem.* **2009**, *284*, 36442–36452.

245. Higgs, M.R.; Lerat, H.; Pawlotsky, J.M. Hepatitis c virus-induced activation of beta-catenin promotes c-myc expression and a cascade of pro-carcinogenetic events. *Oncogene* **2013**, *32*, 4683–4693.

246. Zekri, A.R.; Sabry, G.M.; Bahnassy, A.A.; Shalaby, K.A.; Abdel-Wahabh, S.A.; Zakaria, S. Mismatch repair genes (hmlh1, hpms1, hpms2, gtbp/hmsh6, hmsh2) in the pathogenesis of hepatocellular carcinoma. *World J. Gastroenterol.* **2005**, *11*, 3020–3026.

247. Pal, S.; Polyak, S.J.; Bano, N.; Qiu, W.C.; Carithers, R.L.; Shuhart, M.; Gretch, D.R.; Das, A. Hepatitis c virus induces oxidative stress, DNA damage and modulates the DNA repair enzyme neil1. *J. Gastroenterol. Hepatol.* **2010**, *25*, 627–634.

248. Higgs, M.R.; Lerat, H.; Pawlotsky, J.M. Downregulation of gadd45beta expression by hepatitis c virus leads to defective cell cycle arrest. *Cancer Res.* **2010**, *70*, 4901–4911.

249. Van Pelt, J.F.; Severi, T.; Crabbe, T.; Eetveldt, A.V.; Verslype, C.; Roskams, T.; Fevery, J. Expression of hepatitis c virus core protein impairs DNA repair in human hepatoma cells. *Cancer Lett.* **2004**, *209*, 197–205.

250. Machida, K.; McNamara, G.; Cheng, K.T.; Huang, J.; Wang, C.H.; Comai, L.; Ou, J.H.; Lai, M.M. Hepatitis c virus inhibits DNA damage repair through reactive oxygen and nitrogen species and by interfering with the atm-nbs1/mre11/rad50 DNA repair pathway in monocytes and hepatocytes. *J. Immunol.* **2010**, *185*, 6985–6998.

251. Ariumi, Y.; Kuroki, M.; Dansako, H.; Abe, K.; Ikeda, M.; Wakita, T.; Kato, N. The DNA damage sensors ataxia-telangiectasia mutated kinase and checkpoint kinase 2 are required for hepatitis c virus rna replication. *J. Virol.* **2008**, *82*, 9639–9646.

252. Lai, C.K.; Jeng, K.S.; Machida, K.; Cheng, Y.S.; Lai, M.M. Hepatitis c virus ns3/4a protein interacts with atm, impairs DNA repair and enhances sensitivity to ionizing radiation. *Virology* **2008**, *370*, 295–309.

253. Bittar, C.; Shrivastava, S.; Bhanja Chowdhury, J.; Rahal, P.; Ray, R.B. Hepatitis c virus ns2 protein inhibits DNA damage pathway by sequestering p53 to the cytoplasm. *PLoS ONE* **2013**, *8*, e62581.

254. Naka, K.; Dansako, H.; Kobayashi, N.; Ikeda, M.; Kato, N. Hepatitis c virus ns5b delays cell cycle progression by inducing interferon-beta via toll-like receptor 3 signaling pathway without replicating viral genomes. *Virology* **2006**, *346*, 348–362.

255. Wang, Y.; Xu, Y.; Tong, W.; Pan, T.; Li, J.; Sun, S.; Shao, J.; Ding, H.; Toyoda, T.; Yuan, Z. Hepatitis c virus ns5b protein delays s phase progression in human hepatocyte-derived cells by relocalizing cyclin-dependent kinase 2-interacting protein (cinp). *J. Biol. Chem.* **2011**, *286*, 26603–26615.

256. Qadri, I.; Iwahashi, M.; Simon, F. Hepatitis c virus ns5a protein binds tbp and p53, inhibiting their DNA binding and p53 interactions with tbp and ercc3. *Biochim. Biophys. Acta* **2002**, *1592*, 193–204.

257. Maqbool, M.A.; Imache, M.R.; Higgs, M.R.; Carmouse, S.; Pawlotsky, J.M.; Lerat, H. Regulation of hepatitis c virus replication by nuclear translocation of nonstructural 5a protein and transcriptional activation of host genes. *J. Virol.* **2013**, *87*, 5523–5539.

258. Levin, A.; Neufeldt, C.J.; Pang, D.; Wilson, K.; Loewen-Dobler, D.; Joyce, M.A.; Wozniak, R.W.; Tyrrell, D.L. Functional characterization of nuclear localization and export signals in hepatitis c virus proteins and their role in the membranous web. *PLoS ONE* **2014**, *9*, e114629.

259. Chen, Y.; Williams, V.; Filippova, M.; Filippov, V.; Duerksen-Hughes, P. Viral carcinogenesis: Factors inducing DNA damage and virus integration. *Cancers* **2014**, *6*, 2155–2186.

260. Mauser, A.; Holley-Guthrie, E.; Zanation, A.; Yarborough, W.; Kaufmann, W.; Klingelhutz, A.; Seaman, W.T.; Kenney, S. The epstein-barr virus immediate-early protein bzlf1 induces expression of e2f-1 and other proteins involved in cell cycle progression in primary keratinocytes and gastric carcinoma cells. *J. Virol.* **2002**, *76*, 12543–12552.

261. Drake, J.W.; Holland, J.J. Mutation rates among rna viruses. *Proc. Natl. Acad. Sci. USA* **1999**, *96*, 13910–13913.

262. Carpentier, A.; Barez, P.Y.; Boxus, M.; Willems, L. Checkpoints modulation by the human T-lymphotropic virus type 1 tax protein. *Retrovirology* **2014**, *11*, eP90.

The Role of the DNA Damage Response throughout the Papillomavirus Life Cycle

Caleb C. McKinney, Katherine L. Hussmann and Alison A. McBride

Abstract: The DNA damage response (DDR) maintains genomic integrity through an elaborate network of signaling pathways that sense DNA damage and recruit effector factors to repair damaged DNA. DDR signaling pathways are usurped and manipulated by the replication programs of many viruses. Here, we review the papillomavirus (PV) life cycle, highlighting current knowledge of how PVs recruit and engage the DDR to facilitate productive infection.

Reprinted from *Viruses*. Cite as: McKinney, C.C.; Hussmann, K.L.; McBride, A.A. The Role of the DNA Damage Response throughout the Papillomavirus Life Cycle. *Viruses* **2015**, *7*, 2450-2469.

1. Introduction

Papillomaviruses (PVs) are an ancient group of small, double-stranded (ds) DNA viruses that infect the highly adapted niche of the stratified epithelium of the skin or mucosa in specific host species. Although HPV infection can be asymptomatic, HPVs are also the etiological agent of a wide range of benign papillomas or warts [1]. A subset of HPVs cause infections that can transition to cancer after long-term, persistent infection [2,3]. As the causative agent of over 99% of cervical cancers and an increasing number of anogenital and oropharyngeal cancers, HPVs remain a serious health threat [4].

There are several hundred different PVs that infect a multitude of individual host species, yet the basic genomic scheme and life cycle strategy of each virus are remarkably consistent. Each virus contains a small dsDNA circular genome of approximately 8 kb that encodes just six to eight genes [5] (Figure 1).

Because of this limited coding capacity, PVs retain a strict dependence on host factors for viral replication [6]. Elucidation of the different replication mechanisms of HPV could identify novel targets that may prove effective in the development of therapies for preexisting HPV infections, where the multivalent vaccines are not readily useful. Recent studies have shown that the DNA damage response and repair (DDR) machinery is necessary for efficient HPV replication. Here, we review the mechanisms of genome replication in the HPV life cycle and summarize how the virus recruits and engages the DDR to facilitate viral DNA synthesis.

2. Replication Phases in the Papillomavirus Life Cycle

2.1. Overview of Phases of Replication in Papillomavirus Infection

Upon initial infection of cells in the basal epithelium, the HPV genome enters the nucleus and undergoes a limited number of rounds of DNA replication to establish a low copy number of genomes per infected cell. Subsequently, when the infected cells in the basal layer replicate and divide, the HPV genome replicates in synchrony as extrachromosomal elements tethered to the host genome.

When these infected cells detach from the basement membrane and enter the suprabasal epithelial layer as part of the differentiation process, the HPV E6 and E7 proteins promote unscheduled cell cycle progression by manipulating cell cycle regulation and cellular differentiation programs (see Figure 2). As cells continue to advance through the epithelial layer, there is a switch in the mode of genome replication to productive viral DNA amplification concomitant with increased levels of the E1 and E2 replication proteins. In the terminally differentiated layers of the epithelium, L1 and L2, the viral capsid proteins, are synthesized, and viral particles are assembled.

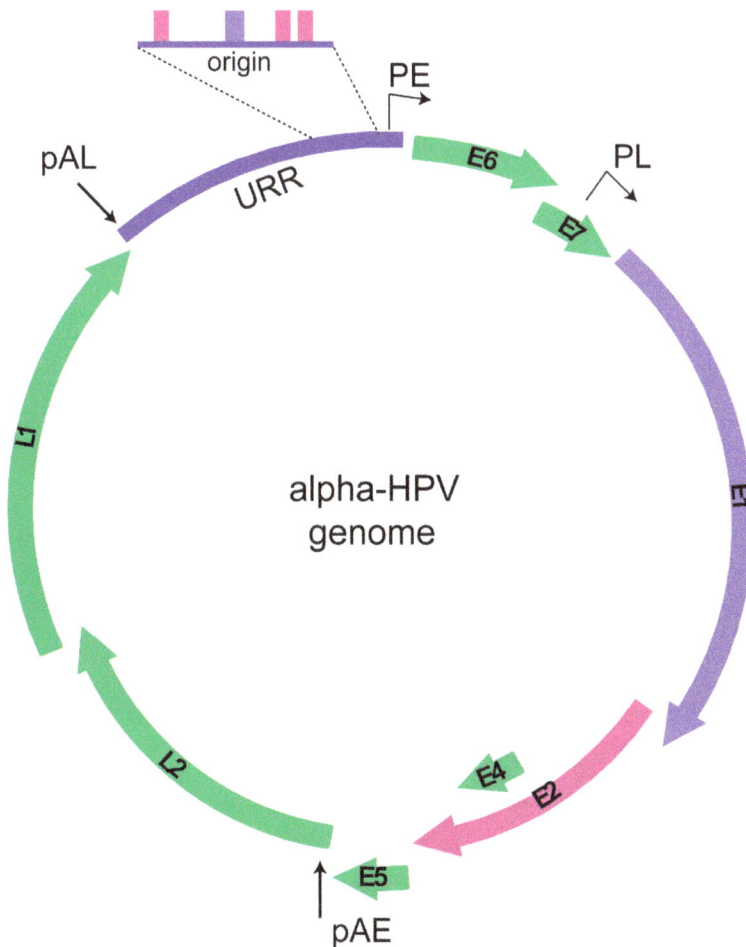

Figure 1. Viral genome. The circular dsDNA genome of an alpha-HPV genome is shown. Viral open reading frames are depicted as curved arrows. The URR (upstream regulatory region) is expanded to show the replication origin containing binding sites for the E2 protein (magenta boxes) and the E1 binding site (purple rectangle) are shown. The early promoter (PE), late promoter (PL) and early and late polyadenylation sites (pAE and pAL) are indicated.

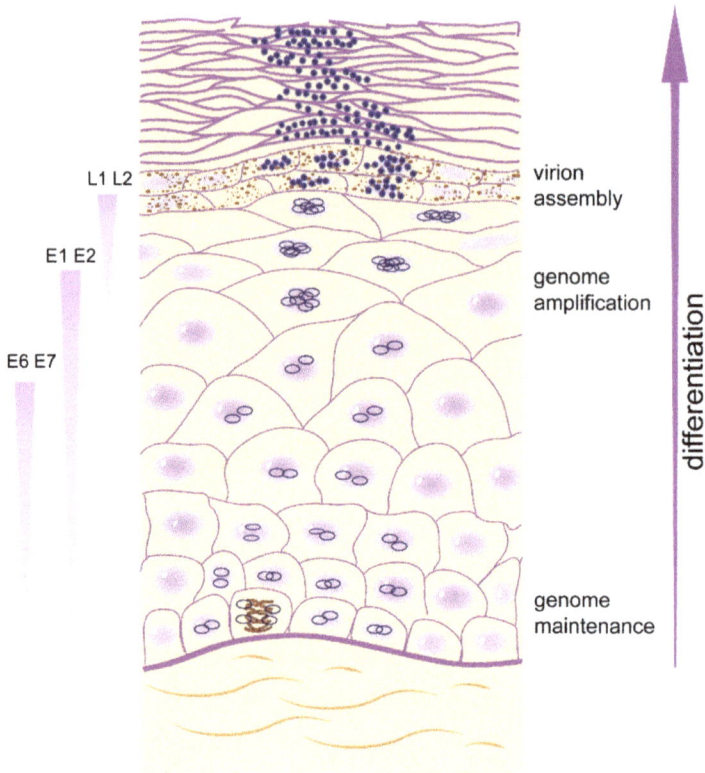

Figure 2. Model of a stratified epithelium. The layers of epithelium shown from bottom to top are stratum basale (with a mitotic cell), stratum spinosum, stratum granulosum (with brown keratohyalin granules), and stratum corneum. The dividing cells in the lowest layer of the epithelium maintain the viral genome as a plasmid (dark blue circles). As these cells progress upwards during the process of differentiation, viral genomes are amplified and packaged in viral particles (dark blue particles). The levels of viral proteins also increase with differentiation as shown on the left.

2.2. HPV Replication Proteins

The relatively small intrinsic coding capacity of HPV genomes warrants exquisite dependence on host factors, as well as viral factors, for replication. HPV encodes two proteins directly involved in replication: E1, a replicative helicase, and E2, a multifunctional protein that recruits E1 to the replication origin of the viral genome [7–9]. The replication origin, located in the Upstream Regulatory Region (URR) of the viral genome, contains binding sites for both E1 and E2 proteins [9]. In cooperation with E2, E1 binds specifically to the replication origin of HPV, but E2 is then displaced and E1 converts to a double hexamer that unwinds the DNA in an ATP-dependent manner [10]. The host DNA replication machinery is then recruited to synthesize the viral DNA. Both E1 and E2 are required for initial amplification and establishment of viral DNA and late vegetative replication [11]. However, in certain circumstances, E1 is dispensable for maintenance replication [11,12].

Additional HPV proteins are required to maintain an environment conducive to viral replication. The oncogenic HPV E6 and E7 proteins promote cellular proliferation, delay cellular differentiation, and promote immune evasion (reviewed in [2]). Both E6 and E7 are required for productive replication [13–15] and it was long thought that E6 and E7 were required to sustain differentiated cells in a pseudo-S phase to provide access to host DNA replication machinery. However, in a stratified epithelium, the differentiated cells that amplify viral genomes are in a G2 like phase that is also dependent on E7 expression [16,17]. As described below, E6 and E7 play essential roles in modulating the cellular DDR for viral replication.

2.3. Initial Amplification and Establishment

HPVs infect the dividing keratinocytes of the basal layer by entering the stratified epithelium through a microabrasion (reviewed in [18]). Virions initially bind through interactions with heparin sulfate proteoglycans [19], and the viral capsid is processed and trafficked through the endosomal pathway. The minor capsid protein, L2, is bound to the viral genome within the virion and, once the virion is processed through the endosome, delivers the genome into the nucleus where it localizes to ND10 bodies [20]. Nuclear access most likely occurs in mitotic cells, where nuclear envelope breakdown allows nuclear access, rather than entry through a nuclear pore [21,22].

Once in the nucleus, the viral genome replicates to a low level number of copies per cell. The genome copy number in the basal cells is low, as the viral genome cannot be detected by *in situ* techniques during this stage [23,24]. The infected cell maintains this low copy number as the viral DNA replicates during subsequent cell divisions in the maintenance replication phase (Figure 3).

A time course of HPV31 infection in HaCaT cells demonstrated that spliced messages which could encode E1 and E2 were present just four hours post-infection [25], consistent with the requirement of E1 and E2 for initial replication. Other viral transcripts appear at eight hours post-infection [25]. The E8^E2 transcript was also detected at four hours post-infection and this is very consistent with its role as a repressor of viral transcription and replication [26]. Expression of E8^E2 at this stage of infection could temper runaway replication and promote the switch to maintenance replication.

2.4. Maintenance Replication and Persistence

During the maintenance phase of replication, the viral genome is replicated in S-phase along with the host genome. Initial evidence using bovine papillomavirus type 1 (BPV1) suggested that viral genome replication is licensed during the maintenance phase [27,28], but later studies demonstrated that in S-phase cells, replication is by a random choice mechanism, whereby some genomes undergo multiple rounds of replication and others remain unreplicated [29,30]. A more recent study of cell lines maintaining HPV genomes revealed that, depending on the HPV type and cell line, both replication mechanisms could be detected in dividing cells [31]. High levels of E1 expression promoted random choice replication, which likely represents the unscheduled DNA synthesis characteristic of vegetative amplification [31].

The level of replication must be tightly controlled during the maintenance phase of the viral life cycle and this may occur via several mechanisms. The E1 protein is retained in the cytoplasm of cells that are not undergoing S-phase [32,33] and in certain circumstances the E1 protein might even be dispensable for maintenance replication [11,12]. The E8^E2 transcriptional repressor tightly regulates both transcription and replication and is required to regulate maintenance replication [26,34]. HPVs also regulate microRNA 145, which in turn binds to the E1 and E2 genes and down-regulates their expression [35]. In BPV1, E2 is a limiting factor in maintenance replication and phosphorylation of the E2 protein regulates its stability and modulates genome copy number in dividing cells [36].

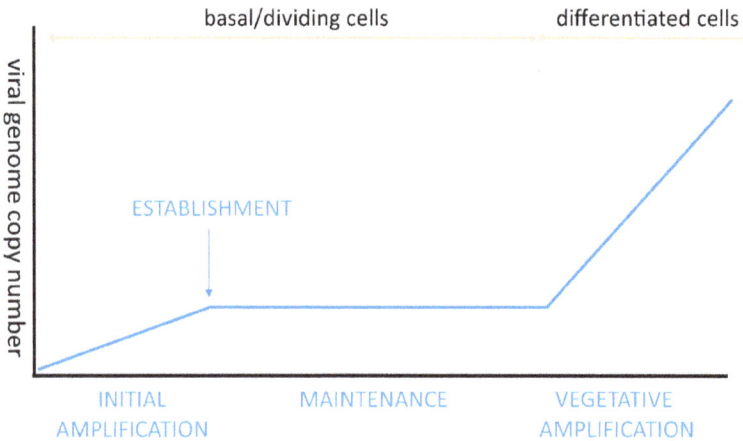

Figure 3. Viral Genome Copy Number during the Different Phases of Replication.

For HPV to persist in dividing cells, not only must the viral genomes be replicated in synchrony with host DNA, but nascent genomes must be efficiently partitioned to daughter cells. The partitioning model, best defined for BPV1, is that the E2 protein binds to multiple E2 binding motifs within the viral URR and tethers the viral genome to host chromatin, thus ensuring it is perpetuated within the basal layer of cells [37–39]. While this model likely applies to all PVs, it is probable that there are differences in the details [40]. For example, most HPVs do not have the large number of E2 binding sites found in BPV1 (and other delta PVs) and all PV E2 proteins do not bind tightly to host mitotic chromatin [41]. Nevertheless, the tethering strategy is likely to be universal for all PVs, as a similar mechanism is used by other persistent viruses that maintain their genomes as extrachromosomal plasmids.

Attachment of PV genomes to host chromatin is important for more than just partitioning of genomes to daughter cells because tethering to active, inactive or even genetically unstable regions of host chromatin could have important outcomes for the infection [42–44]. Extensive studies of the E2 protein have shown its interaction with several well characterized proteins involved in replication, transcription, and host cell cycle regulation [45] and reviewed in [46], and several proteins have been proposed to be the chromatin target responsible for partitioning the E2-PVgenome complex (reviewed in [46]). One of the best-studied (and most debated) targets is BRD4, a chromatin adaptor protein essential for transcriptional initiation and elongation, as well as mitotic bookmarking of host chromatin (reviewed in [47,48]). E2 binds to BRD4 and stabilizes its association with host chromatin [49–51], and BRD4 is a key regulator of PV transcription [52–56]. However, the role of BRD4 in the partitioning of the alpha-PV genomes seems to be more complex since these E2 proteins bind with lower affinity to BRD4 (and to host chromatin) [45,52]. Alpha-E2s and BRD4 are not observed on host mitotic chromosomes under conditions in which other E2 proteins are readily detected; however, the HPV16 E2-BRD4 complex has recently been discovered on chromosomes by bimolecular fluorescence complementation [57] or when E2 is expressed as a GFP-fusion protein [58]. However, HPV31 genomes that encode an E2 protein defective for BRD4 binding are maintained in stable cell lines [59,60], suggesting either that BRD4 is not essential for genome maintenance of all HPV strains, or that the interaction of BRD4 and E2 is more complex in the background of a viral infection. Accordingly, there is evidence that the latter situation is true, since BRD4 is associated with replication foci formed both by E1 and E2 overexpression and in HPV31 genome-containing differentiated cells [61–63]. In the latter situation BRD4 surrounds the replication foci in a satellite pattern [61], but its exact mechanistic role is not clear. BRD4 could facilitate viral transcription in replication factories, could directly function in replication, or might interface with components of the DDR. There are hints that BRD4 might function in the DDR; it interacts with the DDR associated RFC1 subunit, ATAD5 [53], a minor isoform of BRD4 protects adjacent chromatin from DNA damage signaling [64], and we find that BRD4 is associated with aphidicolin inducible fragile sites in C-33A cells [43,64]. Thus, BRD4, and possibly other host proteins required for maintenance replication, may play complex roles during the transition from maintenance replication to vegetative amplification.

2.5. Differentiation-Dependent Viral DNA Amplification

The third phase of HPV replication is tightly linked to the differentiation program of the stratified epithelium. HPVs must transition out of maintenance phase to induce high level viral DNA synthesis, late gene expression, and production and perpetuation of infectious particles. As proliferating cells that harbor HPV genomes transition through the differentiation process, the late promoter is activated, high levels of the E1 and E2 proteins are expressed and the mode of viral genome replication switches to support productive viral genome amplification (Figures 2 and 3 and [65]). Using laser microdissection to analyze viral genome copy number of OcPV1 (ROPV; rabbit oral papillomavirus) in the stratified layers of oral lesions, Maglennon *et al.* demonstrate a five log amplification of viral DNA in the superficial layers compared to the basal layers [66].

The exact mechanism of genome amplification has not been elucidated, but there has long been evidence that it might differ from the bidirectional theta mode characteristic of maintenance replication [67,68]. A breakthrough in the study of HPV replication was the observation that differentiation-dependent genome amplification required the cellular ATM DNA damage response and repair (DDR) pathway [69]. This finding opened up the possibility that genomes might be amplified in differentiated cells by a recombination-dependent replication mode supported by the DDR response.

The remainder of this review will focus on recent developments that uncover how HPV contests and hijacks the carefully crafted DDR network to recruit, engage, and manipulate the host DDR machinery for viral DNA replication at various stages of the life cycle.

3. Role of the DNA Damage Response (DDR) in HPV Replication

3.1. DNA Damage Response Overview

Cellular DNA constantly sustains endogenous and exogenous insults that result in mutations, crosslinks and single and double stranded breaks [70]. However, cells have evolved an elaborate pathway to correct these lesions known as the DNA damage and repair response (DDR) (Figure 4 and reviewed in [71]). The DDR consists of sensors, such as the MRE11-Rad50-Nbs1 complex (MRN) or replication protein A (RPA), that detect a wide range of DNA lesions and recruit the Ataxia telangiectasia mutated (ATM) and Ataxia telangiectasia and Rad3 related (ATR) transducer kinases to the site of damage. With the help of mediator proteins, these transducer kinases induce a signaling cascade through the effector kinases Chk1 and Chk2, activating cell cycle checkpoints to prevent cell cycle progression and ultimately recruiting the DNA repair machinery to the damaged DNA. Once the damage is repaired, the checkpoints are mitigated and cells return to their normal state. However, extensive DNA damage may warrant the initiation of cellular death pathways (Figure 4).

Some of the most severe types of damage are dsDNA breaks that result from ionizing radiation, replication stress, reactive metabolic intermediates, or exogenous chemicals and result in activation of the ATM pathway. dsDNA breaks are bound by the MRN complex that both bridges the gap and activates the ATM signaling cascade. This results in activation of a Chk2 dependent cell cycle checkpoint. dsDNA breaks can be repaired either through the low fidelity, non-homologous end joining (NHEJ) pathway or the high fidelity, homologous recombination (HR) pathway. In HR, a section of DNA adjacent to the 5' end of the break is resectioned by an exonuclease complex recruited by the MRN complex. This ssDNA becomes coated with Rad51 and invades the undamaged sister chromatid, which is used to provide a template for DNA synthesis to faithfully restore the sequence on the damaged strand. Since a homologous DNA sequence is required for proper repair, HR is restricted to S and G2 phases of the cell cycle, when a homologous sister chromatid is present to provide a template.

ATR is activated in response to the presence of persistent ssDNA that results from replication stress. RPA (replication protein A) binds this ssDNA and recruits ATRIP (ATR interacting partner) and ATR. The ring-shaped 9-1-1 complex (consisting of Rad9, Hus1, Rad1) is loaded onto collapsed replication forks and recruits TopBP1 (Topoisomerase II-binding protein 1), a multifaceted factor essential for maintaining genomic stability and facilitating DNA replication by recruiting replication factors to replication forks. 9-1-1 further enlists claspin, which recruits the kinase, Chk1. ATR-dependent phosphorylation of Chk1 activates a cell cycle checkpoint and facilitates stabilization of replication forks. The DDR pathway not only functions to accurately repair DNA, but must also regulate the cell cycle, pausing it to allow repair to be completed. The DDR activates specific downstream cell cycle checkpoints during each phase of the cell cycle to maintain DNA integrity. For example, a DDR mediated increase in stability of p53 can lead to increased expression of CDK inhibitor p21, which arrests cells in G1. Likewise, Chk1 and Chk2 phosphorylation inhibits Cdc25 family members, which are important for progression at several stages of the cell cycle (reviewed in [72]).

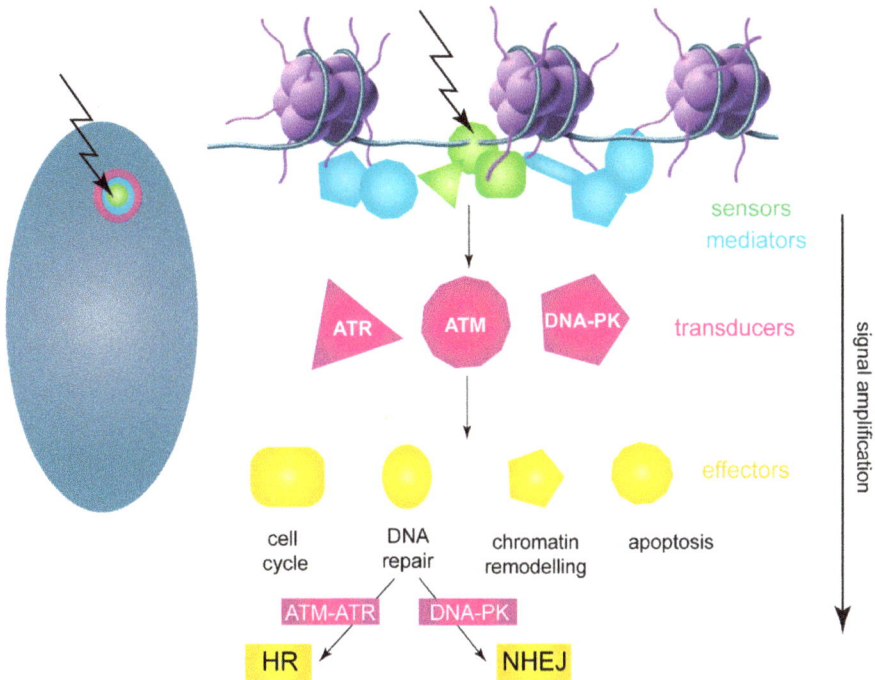

Figure 4. Diagram of the DNA Damage and Repair Response Pathway. DNA breaks or collapsed replication forks are detected by sensors that mark the site of damage. Three kinase signaling cascades (ATM, ATR, and DNA-PK) regulate and transduce the response to damage. With the help of mediator proteins, transducer and effector proteins transmit and greatly amplify the signal throughout the cell, resulting in a huge influx of factors to repair damage and remodel chromatin. Two major repair pathways, HR and NHEJ, are regulated by the ATM/ATR and DNA-PK pathways, respectively. The nucleus on the left contains an HPV replication focus that mimics a DDR focus. Adapted from [71].

The unusual structure of viral DNA (e.g., extrachromosomal linear or small circular molecules), unscheduled viral DNA synthesis, viral protein expression or aberrant cellular proliferation can induce a DDR in infected cells. While this can be detrimental to infection, many viruses disable certain components of the DDR and take advantage of others to synthesize and process viral DNA in productive infection [73,74]. Viral DNA synthesis can also result in additional unique structures that require processing. For example, studies in SV40 show that ATM and ATR are required for quality control of viral replication products and for maintenance of the replication foci; ATM signaling reduces unidirectional replication and ATR signaling prevents DNA breakage at the converging forks that result from bidirectional replication [75]. ATM also ensures that HR factors, and not NHEJ factors, are recruited to SV40 replication centers [76]. The use of repair pathways to synthesize viral

DNA also enables replication in cells that are outwith S-phase. Not only does this reduce competition from host cell synthesis, but in the case of HPVs, the use of repair pathways to synthesize DNA allows replication in differentiated cells. A landmark study in the HPV field demonstrated that the ATM arm of the DDR response was required for amplification of HPV31 genomes in differentiated cells [69]. Subsequent studies have shown that several viral proteins, including E1 and E7, are involved in the recruitment of DSB repair factors to HPV replication factories [61,77–80]. HPV replication foci also recruit factors from both the ATM and ATR pathways, such as TopBP1, Rad51, pNBS1, MRN, RPA, BRCA1, and 53BP1 [61,63,69,77–81].

In the context of the life cycle, E6 and E7 play important roles in both promoting DDR pathways as well as inhibiting the downstream consequences. In general, E7 promotes the DDR to provide an environment conducive for viral replication while E6 mitigates the downstream effects of the DDR cascade to allow cells to tolerate chronic DDR activation, growth arrest and/or apoptosis. Thus, HPVs have usurped DNA damage sensing and repair strategies to promote viral DNA replication. The sections below will review current knowledge of the role of the DDR in the different phases of HPV DNA replication.

3.2. HPV Replication Foci Formation

High level amplification of viral DNA occurs as infected cells differentiate and progress toward the epithelial surface. Most viruses replicate their genomes in defined cellular regions that are designated replication foci, compartments, or factories. These regions serve to nucleate and concentrate the components required for viral DNA synthesis and other processes to specific regions of the cell, presumably to facilitate viral replication. Nuclear replication foci can be detected in differentiated cells that contain replicating HPV [61,69,82].

Studies from our lab as well as others have shown that HPV viral DNA replication occurs in nuclear foci and is dependent on the E1 and E2 proteins. Expression of the E1 and E2 replication proteins (along with an origin containing replicon) is sufficient for the formation of replication foci that recruit components of the DDR [77,78,80,83]. The E1 protein induces the DDR and this is concentrated into nuclear foci by the E2 protein [77,78]. Viral DNA is amplified to high levels in these foci and this does not require keratinocyte differentiation, most likely because E1 and E2 are provided from heterologous promoters. Remarkably, viral replication factories mimic cellular DNA damage foci, whereby a single DNA lesion results in a huge influx of factors to repair damage (Figure 4). Thus, by inducing and usurping the DDR pathways, HPVs can take advantage of major cellular pathways and resources

3.3. The Role of the DNA Damage Response during Initial Amplification of HPV

Difficulties in obtaining large numbers of virions have greatly hampered the study of the early events of HPV infection. Viral DNA (cleaved from plasmids propagated in bacteria and recircularized) can be transfected into keratinocytes or other cells as a surrogate for early infection. Such studies may not completely mimic delivery of a viral genome in a virion particle, but they can provide some insight into the early events required for infection. On the other hand, replication obtained with viral genomes deficient in E8^E2 repressor expression, or when viral genomes are co-transfected with E1 and E2 expression vectors, most likely represent the runaway replication observed in vegetative amplification.

It has been shown that PVs enter the nucleus and initiate their replication and transcription program adjacent to ND10 bodies, like many other viruses [20]. ND10 bodies are important for anti-viral defense and one of the components, Sp100, represses transcription and replication of incoming virions [84]. ND10 bodies are associated with regions of DNA damage in uninfected cells and may provide a link between early events in HPV entry and the initial rounds of unscheduled viral DNA synthesis required to initiate infection [85]. As HPV enters the cell it is likely that this unscheduled viral DNA synthesis will activate or use components of the DDR for limited amplification of the viral genome. A recent study from our laboratory demonstrated that the HPV E2 proteins bind to host chromatin in complex with BRD4 at regions of the genome undergoing replication stress [43], and we proposed that these interactions are important for both genome tethering during maintenance replication as well as nucleation of replication foci in vegetative amplification. We predict that the interaction of viral genomes with host chromatin becomes established at very early stages of infection and involves the DNA damage response.

Although it has been shown in some studies that transient replication of HPV16 is not inhibited by DNA damage signaling or by p53 induction [86,87], it is likely that these responses have to be tempered for the infected cell to transition into the maintenance phase. A quantitative colony-forming assay is a useful measure of genome establishment [84,88], but it indicates that this is a rare event. This is very similar to findings that establishment of Epstein-Barr virus derived replicons is infrequent [89]. Thus, the virus must evade intrinsic immune responses, modulate DDR responses, and tether to beneficial regions of host chromatin to ensure a long-term, persistent infection.

3.4. The Role of the DNA Damage Response during Maintenance Replication.

During the maintenance phase of replication, HPV genomes are replicated in S-phase in synchrony with host DNA replication. Cells containing HPV genomes have increased markers of ATM and ATR signaling, but pharmacological inhibition of the ATM pathway does not affect genome maintenance [69]. However, it has been reported that siRNA-mediated reduction of ATM, ATR or several other proteins in the DDR pathway, as well as Chk1 inhibition results in 40%–50% reduction in HPV16 copy number in W12 cells [90,91].

Many DNA tumor viruses inactivate pRB and p21 and derepress expression of the E2F transcription factor family and for HMCV this is important for ATM-driven viral replication [92]. Furthermore, deregulation of E2F1 by E7 induces the expression of Chk2 [93], implying that this function of E7 could promote HPV replication. E7 deregulation of E2F, and subsequent nucleotide deficiency, further promotes replication stress and genomic instability [94]. In addition, E7 can promote mitotic entry in the presence of a DDR response by accelerating the degradation of the Chk1 binding protein, claspin [95]. Therefore, HPVs integrate DNA damage signaling with anti-apoptotic and proliferative functions to create a unique cellular environment that supports viral replication.

Activation of ATM and ATR signaling results in activation of p53 with a concomitant growth arrest that would be detrimental for viral replication in dividing cells. However, several studies have shown that the ability of E6 to degrade and/or inactivate p53 is crucial for long-term genome maintenance [14,96,97]. Furthermore, inactivation of p53 or expression of a dominant negative p53 protein can complement genomes defective for E6 expression [87,98].

3.5. The Role of the DNA Damage Response during Vegetative Amplification of Viral Genomes

Most of what we know about the role of the DDR in HPV biology has been elucidated from studies of late viral DNA replication. The E1 and E2 replication proteins are induced to high levels by a differentiation-dependent induction of the late promoter [99] and this coincides with activation of a DDR [69]. Many studies have also shown that E1 can specifically upregulate and contribute to the activation of the DDR in non-differentiated cells, and when coexpressed with E2, many DDR factors such as pATM, pATR, γH2AX, pChk2, pChk1, BRCA1, RAD51, TopBP1 and pNBS1 colocalize within nuclear foci that replicate viral DNA to high levels [63,77,78,80].

In the more natural condition of a stratified epithelium, E7 plays an important role in inducing the ATM arm of the DDR in differentiated cells [16] and ATM signaling is required for genome amplification [69]. E7 specifically interacts with NBS1 (which is also required for viral replication), and this association can be separated from activation of the ATM pathway [100]. Further, studies using inhibitors of signal transducer and transactivator 5 (STAT5), a member of the JAK-STAT pathway and an important regulator of the immune response, displayed that STAT5 positively regulates vegetative amplification of HPV31 via potentiation of the ATM pathway through activation of Chk2 [101]. STAT5 activity was dependent on E7 expression, further demonstrating the role of E7 during vegetative amplification. Although the extent of the landscape of E7-driven regulation of cellular protein expression and signaling pathways is not fully understood, E7-mediated upregulation and accumulation of MRN and HR factors as well as stimulation of ATM through the activity of STAT5 is required to properly engage the DDR machinery to facilitate differentiation-dependent genome amplification, bridging the nexus of innate immunity and DNA repair.

The dysregulation of p53 via E6 during vegetative amplification is also necessary to provide an environment conducive for viral DNA amplification. Viruses that are unable to degrade or inactivate p53 are unable to amplify viral DNA in a stratified epithelium [102]. Separate from its role in checkpoint control, p53 can directly inhibit amplification replication in several PVs [103–105], perhaps by directly binding to the E2 protein [106]. Therefore, like other viruses, HPVs take advantage of some aspects of the DDR, but inactivate others.

The presence of several HR factors within HPV replication foci indicates that the transition from maintenance to amplification might also involve a switch to a distinct replication mechanism, such as recombination-dependent replication [67,68,82,107,108]. This has several advantages such as enabling replication in the G2 phase of the cell cycle, as well as generating large amounts of viral DNA without the need for replication initiator proteins to reinitiate DNA synthesis [109].

3.6. The Role of the DNA Damage Response in HPV Integration and Carcinogenesis

As described above, E7 and E6 have important roles in promoting the DDR to support viral DNA replication and render cells resistant to the consequences of checkpoint activation. However, this inadvertently leaves cells highly susceptible to mutation and genetic instability. This has been well studied for the oncogenic alpha-HPVs, but there is also evidence that the cutaneous beta-HPVs might manipulate the host DDR pathways to promote viral replication. Indeed, expression of beta-HPV E6 interferes with DDR signaling at many levels, rendering cellular genomes vulnerable to UV damage and carcinogenic genetic instability [110–116].

In many HPV associated cancers, the viral genome has become integrated into the host genome [117] often in regions of genetic instability called common fragile sites [118,119]. Could this be related to the reliance of the virus on the DDR for replication? We have shown recently that HPV E2 proteins associate with genetically unstable regions of the host chromosomes and also that late replication foci form adjacent to these sites [43]. It could be beneficial to the virus to associate with regions of host chromatin that are highly susceptible to replication stress and prone to late replication. However, the close association of viral and host replication (especially regions undergoing recombination-directed repair) could inadvertently lead to integration of the viral genome. Notably, common fragile sites and extrachromosomal HPV genomes have similar susceptibilities to ATR inhibitors, and to the replication inhibitor aphidicolin [90]. Recent studies have shown that HPV integration sites are highly unstable and involve many rearrangements and duplications of virus and host sequences [120]. It is generally thought that the E1 and E2 replication proteins are not expressed from integrated genomes but expression of these proteins from co-replicating, extrachromosomal HPV genomes induces recruitment of DDR proteins to the integration loci, resulting in onion skin replication and promoting genetic instability [121,122]. In the absence of the viral replication proteins, genetic instability can be further increased by homologous recombination-related looping mechanisms [120]. Any modification that results in increased expression of the E6 and E7 oncoproteins will further stimulate genetic instability and promote carcinogenesis.

4. Conclusions

DDR signaling upon infection is a common strategy utilized by many DNA viruses to recruit factors necessary for replication [74], and, as described here, activation of the DDR plays an important role in the HPV life cycle. As a small virus with limited coding capacity and few replication proteins, HPV needs to utilize the host DNA synthesis machinery to amplify itself. However, it must also reprogram and coordinate multiple cellular signaling pathways to establish an environment conducive to viral replication. Manipulation of the DDR provides a means to efficiently modulate cell cycle checkpoints as well as recruit and usurp replication factors, especially in differentiated cells. The rapid expansion of the genome copy number during the amplification phase of the viral life cycle requires a replication mechanism with high fidelity and there is evidence that homologous recombination repair pathways may provide a means to both initiate replication and resolve DNA intermediates that occur during this phase of rapid DNA synthesis. Ongoing research in many laboratories should provide further mechanistic insight into the pathways and events that result in successful HPV propagation.

Acknowledgments

The authors' research is funded by the Intramural Research Program of the National Institute of Allergy and Infectious Disease of the National Institutes of Health.

Author Contributions

C.M., K.H., and A.M. wrote the manuscript.

Conflicts of Interest

The authors declare no conflict of interest.

References

1. Cubie, H.A. Diseases associated with human papillomavirus infection. *Virology* **2013**, *445*, 21–34.
2. McLaughlin-Drubin, M.E.; Meyers, J.; Munger, K. Cancer associated human papillomaviruses. *Curr. Opin. Virol.* **2012**, *2*, 459–466.
3. Zur Hausen, H. Papillomavirus infections—A major cause of human cancers. *Biochim. Biophys. Acta* **1996**, *1288*, F55–F78.
4. HPV-Associated Cancers Statistics. Available online: http://www.cdc.gov/cancer/hpv/statistics/ (accessed on 13 March 2015).
5. NIAID The Papillomavirus Episteme. Available online: http://pave.niaid.nih.gov/ (accessed on 13 March 2015).
6. McBride, A.A. Replication and partitioning of papillomavirus genomes. *Adv. Virus Res.* **2008**, *72*, 155–205.
7. Mohr, I.J.; Clark, R.; Sun, S.; Androphy, E.J.; MacPherson, P.; Botchan, M.R. Targeting the E1 replication protein to the papillomavirus origin of replication by complex formation with the E2 transactivator. *Science* **1990**, *250*, 1694–1699.
8. Ustav, M.; Stenlund, A. Transient replication of BPV-1 requires two viral polypeptides encoded by the E1 and E2 open reading frames. *EMBO J.* **1991**, *10*, 449–457.
9. Ustav, M.; Ustav, E.; Szymanski, P.; Stenlund, A. Identification of the origin of replication of bovine papillomavirus and characterization of the viral origin recognition factor E1. *EMBO J.* **1991**, *10*, 4321–4329.
10. Sanders, C.M.; Stenlund, A. Recruitment and loading of the E1 initiator protein: An ATP-dependent process catalysed by a transcription factor. *EMBO J.* **1998**, *17*, 7044–7055.
11. Egawa, N.; Nakahara, T.; Ohno, S.; Narisawa-Saito, M.; Yugawa, T.; Fujita, M.; Yamato, K.; Natori, Y.; Kiyono, T. The E1 protein of human papillomavirus type 16 is dispensable for maintenance replication of the viral genome. *J. Virol.* **2012**, *86*, 3276–3283.
12. Kim, K.; Lambert, P.F. E1 protein of bovine papillomavirus 1 is not required for the maintenance of viral plasmid DNA replication. *Virology* **2002**, *293*, 10–14.

13. McLaughlin-Drubin, M.E.; Bromberg-White, J.L.; Meyers, C. The role of the human papillomavirus type 18 E7 oncoprotein during the complete viral life cycle. *Virology* **2005**, *338*, 61–68.

14. Flores, E.R.; Allen-Hoffmann, B.L.; Lee, D.; Lambert, P.F. The human papillomavirus type 16 E7 oncogene is required for the productive stage of the viral life cycle. *J. Virol.* **2000**, *74*, 6622–6631.

15. Chow, L.T.; Duffy, A.A.; Wang, H.K.; Broker, T.R. A highly efficient system to produce infectious human papillomavirus: Elucidation of natural virus-host interactions. *Cell Cycle* **2009**, *8*, 1319–1323.

16. Banerjee, N.S.; Wang, H.K.; Broker, T.R.; Chow, L.T. Human papillomavirus (HPV) E7 induces prolonged G2 following S phase reentry in differentiated human keratinocytes. *J. Biol. Chem.* **2011**, *286*, 15473–15482.

17. Nakahara, T.; Peh, W.L.; Doorbar, J.; Lee, D.; Lambert, P.F. Human papillomavirus type 16 E1^E4 contributes to multiple facets of the papillomavirus life cycle. *J. Virol.* **2005**, *79*, 13150–13165.

18. Day, P.M.; Schelhaas, M. Concepts of papillomavirus entry into host cells. *Curr. Opin. Virol.* **2014**, *4*, 24–31.

19. Culp, T.D.; Budgeon, L.R.; Christensen, N.D. Human papillomaviruses bind a basal extracellular matrix component secreted by keratinocytes which is distinct from a membrane-associated receptor. *Virology* **2006**, *347*, 147–159.

20. Day, P.M.; Baker, C.C.; Lowy, D.R.; Schiller, J.T. Establishment of papillomavirus infection is enhanced by promyelocytic leukemia protein (PML) expression. *Proc. Natl. Acad. Sci. USA* **2004**, *101*, 14252–14257.

21. Pyeon, D.; Pearce, S.M.; Lank, S.M.; Ahlquist, P.; Lambert, P.F. Establishment of human papillomavirus infection requires cell cycle progression. *PLoS Pathog.* **2009**, *5*, e1000318.

22. Aydin, I.; Weber, S.; Snijder, B.; Samperio Ventayol, P.; Kuhbacher, A.; Becker, M.; Day, P.M.; Schiller, J.T.; Kann, M.; Pelkmans, L.; *et al.* Large scale RNAi reveals the requirement of nuclear envelope breakdown for nuclear import of human papillomaviruses. *PLoS Pathog.* **2014**, *10*, e1004162.

23. Evans, M.F.; Aliesky, H.A.; Cooper, K. Optimization of biotinyl-tyramide-based *in situ* hybridization for sensitive background-free applications on formalin-fixed, paraffin-embedded tissue specimens. *BMC Clin. Pathol.* **2003**, *3*, e2.

24. Peh, W.L.; Middleton, K.; Christensen, N.; Nicholls, P.; Egawa, K.; Sotlar, K.; Brandsma, J.; Percival, A.; Lewis, J.; Liu, W.J.; *et al.* Life cycle heterogeneity in animal models of human papillomavirus-associated disease. *J. Virol.* **2002**, *76*, 10401–10416.

25. Ozbun, M.A. Human papillomavirus type 31b infection of human keratinocytes and the onset of early transcription. *J. Virol.* **2002**, *76*, 11291–11300.

26. Stubenrauch, F.; Hummel, M.; Iftner, T.; Laimins, L.A. The E8E2C protein, a negative regulator of viral transcription and replication, is required for extrachromosomal maintenance of human papillomavirus type 31 in keratinocytes. *J. Virol.* **2000**, *74*, 1178–1186.

27. Botchan, M.; Berg, L.; Reynolds, J.; Lusky, M. The bovine papillomavirus replicon. In *Papillomaviruses*; John Wiley & Sons: New York, NY, USA, 1986; pp. 53–67.

28. Roberts, J.M.; Weintraub, H. *Cis*-acting negative control of DNA replication in eukaryotic cells. *Cell* **1988**, *52*, 397–404.

29. Gilbert, D.M.; Cohen, S.N. Bovine papilloma virus plasmids replicate randomly in mouse fibroblasts throughout S phase of the cell cycle. *Cell* **1987**, *50*, 59–68.

30. Ravnan, J.-B.; Gilbert, D.M.; Ten Hagen, K.G.; Cohen, S.N. Random-choice replication of extrachromosomal bovine papillomavirus (BPV) molecules in heterogeneous, clonally derived BPV-infected cell lines. *J. Virol.* **1992**, *66*, 6946–6952.

31. Hoffmann, R.; Hirt, B.; Bechtold, V.; Beard, P.; Raj, K. Different modes of human papillomavirus DNA replication during maintenance. *J. Virol.* **2006**, *80*, 4431–4439.

32. Fradet-Turcotte, A.; Moody, C.; Laimins, L.A.; Archambault, J. Nuclear export of human papillomavirus type 31 E1 is regulated by Cdk2 phosphorylation and required for viral genome maintenance. *J. Virol.* **2010**, *84*, 11747–11760.

33. Yu, J.H.; Lin, B.Y.; Deng, W.; Broker, T.R.; Chow, L.T. Mitogen-activated protein kinases activate the nuclear localization sequence of human papillomavirus type 11 E1 DNA helicase to promote efficient nuclear import. *J. Virol.* **2007**, *81*, 5066–5078.

34. Straub, E.; Dreer, M.; Fertey, J.; Iftner, T.; Stubenrauch, F. The viral E8^E2C repressor limits productive replication of human papillomavirus 16. *J. Virol.* **2014**, *88*, 937–947.

35. Gunasekharan, V.; Laimins, L.A. Human papillomaviruses modulate microRNA 145 expression to directly control genome amplification. *J. Virol.* **2013**, *87*, 6037–6043.

36. Penrose, K.J.; McBride, A.A. Proteasome-mediated degradation of the papillomavirus E2-TA protein is regulated by phosphorylation and can modulate viral genome copy number. *J. Virol.* **2000**, *74*, 6031–6038.

37. Skiadopoulos, M.H.; McBride, A.A. Bovine papillomavirus type 1 genomes and the E2 transactivator protein are closely associated with mitotic chromatin. *J. Virol.* **1998**, *72*, 2079–2088.

38. Ilves, I.; Kivi, S.; Ustav, M. Long-term episomal maintenance of bovine papillomavirus type 1 plasmids is determined by attachment to host chromosomes, which is mediated by the viral E2 protein and its binding sites. *J. Virol.* **1999**, *73*, 4404–4412.

39. Bastien, N.; McBride, A.A. Interaction of the papillomavirus E2 protein with mitotic chromosomes. *Virology* **2000**, *270*, 124–134.

40. McBride, A.A.; Oliveira, J.G.; McPhillips, M.G. Partitioning viral genomes in mitosis: Same idea, different targets. *Cell Cycle* **2006**, *5*, 1499–1502.

41. Oliveira, J.G.; Colf, L.A.; McBride, A.A. Variations in the association of papillomavirus E2 proteins with mitotic chromosomes. *Proc. Natl. Acad. Sci. USA* **2006**, *103*, 1047–1052.

42. Jang, M.K.; Kwon, D.; McBride, A.A. Papillomavirus E2 proteins and the host BRD4 protein associate with transcriptionally active cellular chromatin. *J. Virol.* **2009**, *83*, 2592–2600.

43. Jang, M.K.; Shen, K.; McBride, A.A. Papillomavirus genomes associate with BRD4 to replicate at fragile sites in the host genome. *PLoS Pathog.* **2014**, *10*, e1004117.

44. Helfer, C.M.; Yan, J.; You, J. The cellular bromodomain protein Brd4 has multiple functions in E2-mediated papillomavirus transcription activation. *Viruses* **2014**, *6*, 3228–3249.

45. Jang, M.K.; Anderson, D.E.; van Doorslaer, K.; McBride, A.A. A Proteomic approach to discover and compare interacting partners of Papillomavirus E2 proteins from diverse phylogenetic groups. *Proteomics* **2015**, doi:10.1002/pmic.201400613.

46. McBride, A.A. The Papillomavirus E2 proteins. *Virology* **2013**, *445*, 57–79.

47. Devaiah, B.N.; Singer, D.S. Two faces of brd4: Mitotic bookmark and transcriptional lynchpin. *Transcription* **2013**, *4*, 13–17.

48. McBride, A.A.; Jang, M.K. Current understanding of the role of the Brd4 protein in the papillomavirus lifecycle. *Viruses* **2013**, *5*, 1374–1394.

49. You, J.; Croyle, J.L.; Nishimura, A.; Ozato, K.; Howley, P.M. Interaction of the bovine papillomavirus E2 protein with Brd4 tethers the viral DNA to host mitotic chromosomes. *Cell* **2004**, *117*, 349–360.

50. Baxter, M.K.; McPhillips, M.G.; Ozato, K.; McBride, A.A. The mitotic chromosome binding activity of the papillomavirus E2 protein correlates with interaction with the cellular chromosomal protein, Brd4. *J. Virol.* **2005**, *79*, 4806–4818.

51. McPhillips, M.G.; Ozato, K.; McBride, A.A. Interaction of bovine papillomavirus E2 protein with Brd4 stabilizes its association with chromatin. *J. Virol.* **2005**, *79*, 8920–8932.

52. McPhillips, M.G.; Oliveira, J.G.; Spindler, J.E.; Mitra, R.; McBride, A.A. Brd4 is required for e2-mediated transcriptional activation but not genome partitioning of all papillomaviruses. *J. Virol.* **2006**, *80*, 9530–9543.

53. Rahman, S.; Sowa, M.E.; Ottinger, M.; Smith, J.A.; Shi, Y.; Harper, J.W.; Howley, P.M. The Brd4 extraterminal domain confers transcription activation independent of pTEFb by recruiting multiple proteins, including NSD3. *Mol. Cell. Biol.* **2011**, *31*, 2641–2652.

54. Smith, J.A.; White, E.A.; Sowa, M.E.; Powell, M.L.; Ottinger, M.; Harper, J.W.; Howley, P.M. Genome-wide siRNA screen identifies SMCX, EP400, and Brd4 as E2-dependent regulators of human papillomavirus oncogene expression. *Proc. Natl. Acad. Sci. USA* **2010**, *107*, 3752–3757.

55. Schweiger, M.R.; Ottinger, M.; You, J.; Howley, P.M. Brd4 independent transcriptional repression function of the papillomavirus E2 proteins. *J. Virol.* **2007**, *81*, 9612–9622.

56. Wu, S.Y.; Lee, A.Y.; Hou, S.Y.; Kemper, J.K.; Erdjument-Bromage, H.; Tempst, P.; Chiang, C.M. Brd4 links chromatin targeting to HPV transcriptional silencing. *Genes Dev.* **2006**, *20*, 2383–2396.

57. Helfer, C.M.; Wang, R.; You, J. Analysis of the papillomavirus E2 and bromodomain protein Brd4 interaction using bimolecular fluorescence complementation. *PLoS ONE* **2013**, *8*, e77994.

58. Chang, S.-W.; Liu, W.-C.; Liao, K.-Y.; Tsao, Y.-P.; Hsu, P.-H.; Chen, S.-L. Phosphorylation of HPV-16 E2 at Serine 243 Enables Binding to Brd4 and Mitotic Chromosomes. *PLoS ONE* **2014**, *9*, e110882.

59. Stubenrauch, F.; Colbert, A.M.; Laimins, L.A. Transactivation by the E2 protein of oncogenic human papillomavirus type 31 is not essential for early and late viral functions. *J. Virol.* **1998**, *72*, 8115–8123.

60. Senechal, H.; Poirier, G.G.; Coulombe, B.; Laimins, L.A.; Archambault, J. Amino acid substitutions that specifically impair the transcriptional activity of papillomavirus E2 affect binding to the long isoform of Brd4. *Virology* **2007**, *358*, 10–17.

61. Sakakibara, N.; Chen, D.; Jang, M.K.; Kang, D.W.; Luecke, H.F.; Wu, S.Y.; Chiang, C.M.; McBride, A.A. Brd4 Is Displaced from HPV Replication Factories as They Expand and Amplify Viral DNA. *PLoS Pathog.* **2013**, *9*, e1003777.

62. Wang, X.; Helfer, C.M.; Pancholi, N.; Bradner, J.E.; You, J. Recruitment of brd4 to the human papillomavirus type 16 DNA replication complex is essential for replication of viral DNA. *J. Virol.* **2013**, *87*, 3871–3884.

63. Gauson, E.J.; Donaldson, M.M.; Dornan, E.S.; Wang, X.; Bristol, M.; Bodily, J.M.; Morgan, I.M. Evidence supporting a role for TopBP1 and Brd4 in the initiation but not continuation of human papillomavirus 16 E1/E2 mediated DNA replication. *J. Virol.* **2015**, *89*, 4980–4991.

64. Floyd, S.R.; Pacold, M.E.; Huang, Q.; Clarke, S.M.; Lam, F.C.; Cannell, I.G.; Bryson, B.D.; Rameseder, J.; Lee, M.J.; Blake, E.J.; *et al.* The bromodomain protein Brd4 insulates chromatin from DNA damage signalling. *Nature* **2013**, *498*, 246–250.

65. Bedell, M.A.; Hudson, J.B.; Golub, T.R.; Turyk, M.E.; Hosken, M.; Wilbanks, G.D.; Laimins, L.A. Amplification of human papillomavirus genomes *in vitro* is dependent on epithelial differentiation. *J. Virol.* **1991**, *65*, 2254–2260.

66. Maglennon, G.A.; McIntosh, P.; Doorbar, J. Persistence of viral DNA in the epithelial basal layer suggests a model for papillomavirus latency following immune regression. *Virology* **2011**, *414*, 153–163.

67. Flores, E.R.; Lambert, P.F. Evidence for a switch in the mode of human papillomavirus type 16 DNA replication during the viral life cycle. *J. Virol.* **1997**, *71*, 7167–7179.

68. Burnett, S.; Zabielski, J.; Moreno-Lopez, J.; Pettersson, U. Evidence for multiple vegetative DNA replication origins and alternative replication mechanisms of bovine papillomavirus type 1. *J. Mol. Biol.* **1989**, *206*, 239–244.

69. Moody, C.A.; Laimins, L.A. Human Papillomaviruses Activate the ATM DNA Damage Pathway for Viral Genome Amplification upon Differentiation. *PLoS Pathog.* **2009**, *5*, e1000605.

70. Nakamura, J.; Mutlu, E.; Sharma, V.; Collins, L.; Bodnar, W.; Yu, R.; Lai, Y.; Moeller, B.; Lu, K.; Swenberg, J. The endogenous exposome. *DNA Repair* **2014**, *19*, 3–13.

71. Jackson, S.P.; Bartek, J. The DNA-damage response in human biology and disease. *Nature* **2009**, *461*, 1071–1078.

72. Stracker, T.H.; Usui, T.; Petrini, J.H. Taking the time to make important decisions: The checkpoint effector kinases Chk1 and Chk2 and the DNA damage response. *DNA Repair* **2009**, *8*, 1047–1054.

73. Chaurushiya, M.S.; Weitzman, M.D. Viral manipulation of DNA repair and cell cycle checkpoints. *DNA Repair* **2009**, *8*, 1166–1176.

74. Luftig, M, A. Viruses and the DNA Damage Response: Activation and Antagonism. *Annu. Rev. Virol.* **2014**, *1*, 605–625.

75. Sowd, G.A.; Li, N.Y.; Fanning, E. ATM and ATR activities maintain replication fork integrity during SV40 chromatin replication. *PLoS Pathog.* **2013**, *9*, e1003283.

76. Sowd, G.A.; Mody, D.; Eggold, J.; Cortez, D.; Friedman, K.L.; Fanning, E. SV40 utilizes ATM kinase activity to prevent non-homologous end joining of broken viral DNA replication products. *PLoS Pathog.* **2014**, *10*, e1004536.

77. Sakakibara, N.; Mitra, R.; McBride, A.A. The papillomavirus E1 helicase activates a cellular DNA damage response in viral replication foci. *J. Virol.* **2011**, *85*, 8981–8995.

78. Fradet-Turcotte, A.; Bergeron-Labrecque, F.; Moody, C.A.; Lehoux, M.; Laimins, L.A.; Archambault, J. Nuclear accumulation of the papillomavirus E1 helicase blocks S-phase progression and triggers an ATM-dependent DNA damage response. *J. Virol.* **2011**, *85*, 8996–9012.

79. Gillespie, K.A.; Mehta, K.P.; Laimins, L.A.; Moody, C.A. Human papillomaviruses recruit cellular DNA repair and homologous recombination factors to viral replication centers. *J. Virol.* **2012**, *86*, 9520–9526.

80. Reinson, T.; Toots, M.; Kadaja, M.; Pipitch, R.; Allik, M.; Ustav, E.; Ustav, M. Engagement of the ATR-Dependent DNA Damage Response at the Human Papillomavirus 18 Replication Centers during the Initial Amplification. *J. Virol.* **2013**, *87*, 951–964.

81. Kanginakudru, S.; DeSmet, M.; Thomas, Y.; Morgan, I.M.; Androphy, E.J. Levels of the E2 interacting protein TopBP1 modulate papillomavirus maintenance stage replication. *Virology* **2015**, *478*, 135–142.

82. Sakakibara, N.; Chen, D.; McBride, A.A. Papillomaviruses use recombination-dependent replication to vegetatively amplify their genomes in differentiated cells. *PLoS Pathog.* **2013**, *9*, e1003321.

83. Swindle, C.S.; Zou, N.; van Tine, B.A.; Shaw, G.M.; Engler, J.A.; Chow, L.T. Human papillomavirus DNA replication compartments in a transient DNA replication system. *J. Virol.* **1999**, *73*, 1001–1009.

84. Stepp, W.H.; Meyers, J.M.; McBride, A.A. Sp100 provides intrinsic immunity against human papillomavirus infection. *MBio* **2013**, *4*, e00845–13.

85. Everett, R.D. Interactions between DNA viruses, ND10 and the DNA damage response. *Cell. Microbiol.* **2006**, *8*, 365–374.

86. King, L.E.; Fisk, J.C.; Dornan, E.S.; Donaldson, M.M.; Melendy, T.; Morgan, I.M. Human papillomavirus E1 and E2 mediated DNA replication is not arrested by DNA damage signalling. *Virology* **2010**, *406*, 95–102.

87. Lorenz, L.D.; Rivera Cardona, J.; Lambert, P.F. Inactivation of p53 rescues the maintenance of high risk HPV DNA genomes deficient in expression of E6. *PLoS Pathog.* **2013**, *9*, e1003717.

246

88. Lace, M.J.; Turek, L.P.; Anson, J.R.; Haugen, T.H. Analyzing the Human Papillomavirus (HPV) Life Cycle in Primary Keratinocytes with a Quantitative Colony-Forming Assay. *Curr. Protoc. Microbiol.* **2014**, *33*, 14b.2.1–14b.2.13.

89. Leight, E.R.; Sugden, B. Establishment of an oriP replicon is dependent upon an infrequent, epigenetic event. *Mol. Cell. Biol.* **2001**, *21*, 4149–4161.

90. Edwards, T.G.; Helmus, M.J.; Koeller, K.; Bashkin, J.K.; Fisher, C. Human papillomavirus episome stability is reduced by aphidicolin and controlled by DNA damage response pathways. *J. Virol.* **2013**, *87*, 3979–3989.

91. Edwards, T.G.; Vidmar, T.J.; Koeller, K.; Bashkin, J.K.; Fisher, C. DNA damage repair genes controlling human papillomavirus (HPV) episome levels under conditions of stability and extreme instability. *PLoS ONE* **2013**, *8*, e75406.

92. E, X.; Pickering, M.T.; Debatis, M.; Castillo, J.; Lagadinos, A.; Wang, S.; Lu, S.; Kowalik, T.F. An E2F1-mediated DNA damage response contributes to the replication of human cytomegalovirus. *PLoS Pathog.* **2011**, *7*, e1001342.

93. Rogoff, H.A.; Pickering, M.T.; Frame, F.M.; Debatis, M.E.; Sanchez, Y.; Jones, S.; Kowalik, T.F. Apoptosis associated with deregulated E2F activity is dependent on E2F1 and Atm/Nbs1/Chk2. *Mol. Cell. Biol.* **2004**, *24*, 2968–2977.

94. Bester, A.C.; Roniger, M.; Oren, Y.S.; Im, M.M.; Sarni, D.; Chaoat, M.; Bensimon, A.; Zamir, G.; Shewach, D.S.; Kerem, B. Nucleotide deficiency promotes genomic instability in early stages of cancer development. *Cell* **2011**, *145*, 435–446.

95. Spardy, N.; Covella, K.; Cha, E.; Hoskins, E.E.; Wells, S.I.; Duensing, A.; Duensing, S. Human papillomavirus 16 E7 oncoprotein attenuates DNA damage checkpoint control by increasing the proteolytic turnover of claspin. *Cancer Res.* **2009**, *69*, 7022–7029.

96. Thomas, J.T.; Hubert, W.G.; Ruesch, M.N.; Laimins, L.A. Human papillomavirus type 31 oncoproteins E6 and E7 are required for the maintenance of episomes during the viral life cycle in normal human keratinocytes. *Proc. Natl. Acad. Sci. USA* **1999**, *96*, 8449–8454.

97. Park, R.B.; Androphy, E.J. Genetic analysis of high-risk e6 in episomal maintenance of human papillomavirus genomes in primary human keratinocytes. *J. Virol.* **2002**, *76*, 11359–11364.

98. Brimer, N.; Vande Pol, S.B. Papillomavirus E6 PDZ interactions can be replaced by repression of p53 to promote episomal human papillomavirus genome maintenance. *J. Virol.* **2014**, *88*, 3027–3030.

99. Klumpp, D.J.; Laimins, L.A. Differentiation-induced changes in promoter usage for transcripts encoding the human papillomavirus type 31 replication protein E1. *Virology* **1999**, *257*, 239–246.

100. Anacker, D.C.; Gautam, D.; Gillespie, K.A.; Chappell, W.H.; Moody, C.A. Productive replication of human papillomavirus 31 requires DNA repair factor Nbs1. *J. Virol.* **2014**, *88*, 8528–8544.

101. Hong, S.; Laimins, L.A. The JAK-STAT Transcriptional Regulator, STAT-5, Activates the ATM DNA Damage Pathway to Induce HPV 31 Genome Amplification upon Epithelial Differentiation. *PLoS Pathog.* **2013**, *9*, e1003295.

102. Kho, E.Y.; Wang, H.K.; Banerjee, N.S.; Broker, T.R.; Chow, L.T. HPV-18 E6 mutants reveal p53 modulation of viral DNA amplification in organotypic cultures. *Proc. Natl. Acad. Sci. USA* **2013**, *110*, 7542–7549.

103. Lepik, D.; Ilves, I.; Kristjuhan, A.; Maimets, T.; Ustav, M. p53 protein is a suppressor of papillomavirus DNA amplificational replication. *J. Virol.* **1998**, 72, 6822–6831.

104. Ilves, I.; Kadaja, M.; Ustav, M. Two separate replication modes of the bovine papillomavirus BPV1 origin of replication that have different sensitivity to p53. *Virus Res.* **2003**, *96*, 75–84.

105. Akgul, B.; Karle, P.; Adam, M.; Fuchs, P.G.; Pfister, H.J. Dual role of tumor suppressor p53 in regulation of DNA replication and oncogene E6-promoter activity of epidermodysplasia verruciformis-associated human papillomavirus type 8. *Virology* **2003**, *308*, 279–290.

106. Brown, C.; Kowalczyk, A.M.; Taylor, E.R.; Morgan, I.M.; Gaston, K. P53 represses human papillomavirus type 16 DNA replication via the viral E2 protein. *Virol. J.* **2008**, *5*, e5.

107. Dasgupta, S.; Zabielski, J.; Simonsson, M.; Burnett, S. Rolling-circle replication of a high-copy BPV-1 plasmid. *J. Mol. Biol.* **1992**, *228*, 1–6.

108. Auborn, K.J.; Little, R.D.; Platt, T.H.; Vaccariello, M.A.; Schildkraut, C.L. Replicative intermediates of human papillomavirus type 11 in laryngeal papillomas: Site of replication initiation and direction of replication. *Proc. Natl. Acad. Sci. USA* **1994**, *91*, 7340–7344.

109. Lydeard, J.R.; Lipkin-Moore, Z.; Sheu, Y.J.; Stillman, B.; Burgers, P.M.; Haber, J.E. Break-induced replication requires all essential DNA replication factors except those specific for pre-RC assembly. *Genes Dev.* **2010**, *24*, 1133–1144.

110. Wallace, N.A.; Robinson, K.; Howie, H.L.; Galloway, D.A. HPV 5 and 8 E6 abrogate ATR activity resulting in increased persistence of UVB induced DNA damage. *PLoS Pathog.* **2012**, *8*, e1002807.

111. Wallace, N.A.; Galloway, D.A. Manipulation of cellular DNA damage repair machinery facilitates propagation of human papillomaviruses. *Semin. Cancer Biol.* **2014**, *26*, 30–42.

112. Wallace, N.A.; Robinson, K.; Galloway, D.A. Beta human papillomavirus E6 expression inhibits stabilization of p53 and increases tolerance of genomic instability. *J. Virol.* **2014**, *88*, 6112–6127.

113. Wallace, N.A.; Robinson, K.; Howie, H.L.; Galloway, D.A. beta-HPV 5 and 8 E6 Disrupt Homology Dependent Double Strand Break Repair by Attenuating BRCA1 and BRCA2 Expression and Foci Formation. *PLoS Pathog.* **2015**, *11*, e1004687.

114. Jackson, S.; Storey, A. E6 proteins from diverse cutaneous HPV types inhibit apoptosis in response to UV damage. *Oncogene* **2000**, *19*, 592–598.

115. Giampieri, S.; Storey, A. Repair of UV-induced thymine dimers is compromised in cells expressing the E6 protein from human papillomaviruses types 5 and 18. *Br. J. Cancer* **2004**, *90*, 2203–2209.

116. Holloway, A.; Simmonds, M.; Azad, A.; Fox, J.L.; Storey, A. Resistance to UV-induced apoptosis by beta-HPV5 E6 involves targeting of activated BAK for proteolysis by recruitment of the HERC1 ubiquitin ligase. *Int. J. Cancer J. Int. Cancer* **2015**, *136*, 2831–2843.

117. Durst, M.; Gissmann, L.; Ikenberg, H.; zur Hausen, H. A papillomavirus DNA from a cervical carcinoma and its prevalence in cancer biopsy samples from different geographic regions. *Proc. Natl. Acad. Sci. USA* **1983**, *80*, 3812–3815.

118. Thorland, E.C.; Myers, S.L.; Gostout, B.S.; Smith, D.I. Common fragile sites are preferential targets for HPV16 integrations in cervical tumors. *Oncogene* **2003**, *22*, 1225–1237.

119. Smith, P.P.; Friedman, C.L.; Bryant, E.M.; McDougall, J.K. Viral integration and fragile sites in human papillomavirus-immortalized human keratinocyte cell lines. *Genes Chromosomes Cancer* **1992**, *5*, 150–157.

120. Akagi, K.; Li, J.; Broutian, T.R.; Padilla-Nash, H.; Xiao, W.; Jiang, B.; Rocco, J.W.; Teknos, T.N.; Kumar, B.; Wangsa, D.; *et al.* Genome-wide analysis of HPV integration in human cancers reveals recurrent, focal genomic instability. *Genome Res.* **2013**, doi:10.1101/gr.164806.113.

121. Kadaja, M.; Sumerina, A.; Verst, T.; Ojarand, M.; Ustav, E.; Ustav, M. Genomic instability of the host cell induced by the human papillomavirus replication machinery. *EMBO J.* **2007**, *26*, 2180–2191.

122. Kadaja, M.; Isok-Paas, H.; Laos, T.; Ustav, E.; Ustav, M. Mechanism of genomic instability in cells infected with the high-risk human papillomaviruses. *PLoS Pathog.* **2009**, *5*, e1000397.

Activation of DNA Damage Response Pathways during Lytic Replication of KSHV

Robert Hollingworth, George L. Skalka, Grant S. Stewart, Andrew D. Hislop, David J. Blackbourn and Roger J. Grand

Abstract: Kaposi's sarcoma-associated herpesvirus (KSHV) is the causative agent of several human malignancies. Human tumour viruses such as KSHV are known to interact with the DNA damage response (DDR), the molecular pathways that recognise and repair lesions in cellular DNA. Here it is demonstrated that lytic reactivation of KSHV leads to activation of the ATM and DNA-PK DDR kinases resulting in phosphorylation of multiple downstream substrates. Inhibition of ATM results in the reduction of overall levels of viral replication while inhibition of DNA-PK increases activation of ATM and leads to earlier viral release. There is no activation of the ATR-CHK1 pathway following lytic replication and CHK1 phosphorylation is inhibited at later times during the lytic cycle. Despite evidence of double-strand breaks and phosphorylation of H2AX, 53BP1 foci are not consistently observed in cells containing lytic virus although RPA32 and MRE11 localise to sites of viral DNA synthesis. Activation of the DDR following KSHV lytic reactivation does not result in a G1 cell cycle block and cells are able to proceed to S-phase during the lytic cycle. KSHV appears then to selectively activate DDR pathways, modulate cell cycle progression and recruit DDR proteins to sites of viral replication during the lytic cycle.

Reprinted from *Viruses*. Cite as: Hollingworth, R.; Skalka, G.L.; Stewart, G.S.; Hislop, A.D.; Blackbourn, D.J.; Grand, R.J. Activation of DNA Damage Response Pathways during Lytic Replication of KSHV. *Viruses* **2015**, *7*, 2908–2927.

1. Introduction

Kaposi's sarcoma-associated herpesvirus (KSHV) is a γ-2 herpesvirus identified as the etiological agent of several human malignancies. Transformation of endothelial cells by KSHV is associated with the development of Kaposi's sarcoma (KS) [1], while infection of B cells can lead to the lymphoproliferative diseases primary effusion lymphoma (PEL) and multicentric Castleman's disease (MCD) [2,3].

Like all herpesviruses, KSHV has a double-stranded DNA genome and a biphasic lifecycle that consists of latent and lytic stages. During latent infection, the virus persists as an extrachromosomal episome that is replicated by host replication machinery. Latency is characterised by expression of a limited subset of viral genes required for episomal maintenance and immune evasion that include the latency-associated nuclear antigen (LANA). In order to propagate, the virus must switch to the lytic phase, during which the majority of viral genes are expressed, viral DNA is amplified and infectious virus is released following lysis of the host cell. The viral replication and transcription activator (RTA), encoded by open reading frame 50 (ORF50), is a lytic switch protein as expression is necessary and sufficient for initiation of the full lytic replication programme [4]. Lytic viral genes are expressed in an ordered cascade and the lytic cycle can be divided into several stages that include immediate-early gene expression, delayed-early gene expression, viral DNA amplification, late gene expression and finally, assembly and release of infectious virions [5]. Significantly, lytic replication has been shown to play a major role in the progression of KSHV-related malignancies [6,7].

The DNA damage response (DDR) has evolved to respond to the multitude of lesions inflicted on cellular DNA by both endogenous biological processes and exogenous agents. Three phosphoinositide-3-kinase-related kinases (PIKKs), ataxia telangiectasia mutated (ATM), ATM- and Rad3-related (ATR) and DNA-dependent protein kinase (DNA-PK), can orchestrate activation of the DDR via phosphorylation of multiple target proteins involved in cell cycle regulation, apoptosis and DNA repair. ATM and DNA-PK are primarily activated following formation of double-strand breaks (DSBs) in DNA. DSBs are recognised and bound by the MRE11-RAD50-NBS1 (MRN) complex that facilitates ATM recruitment and activation. ATM subsequently activates effector molecules that include H2AX, CHK2, KAP1, SMC1 and NBS1 that mediate cell cycle arrest and DNA repair. DNA-PK is a core component of the non-homologous end-joining (NHEJ) pathway that acts throughout the cell cycle to repair DSBs [8]. DNA-PK is activated after the Ku70/Ku80 heterodimer binds DNA ends and recruits the DNA-PK catalytic subunit (DNA-PKcs) to form the DNA-PK holoenzyme that is critical for completion of NHEJ [9]. In conjunction with ATR, DNA-PK also plays a role in maintaining genome stability following replication stress [10]. ATR is activated by a number of different lesions that result in persistent single-stranded DNA (ssDNA) that is first coated by replication protein A (RPA) before recruitment of ATR via its binding partner, ATR-interacting protein (ATRIP) [11]. The recruitment of DNA topoisomerase 2-binding protein 1 (TOPBP1) contributes to activation of ATR which subsequently orchestrates activation of cell cycle checkpoints via phosphorylation of CHK1 [12,13].

Several species of herpesvirus are known to interact with the DDR during their lifecycles. These interactions can involve the selective activation and deactivation of DDR pathways and the use of specific DDR proteins in the replication of viral genomes. Since the DDR has evolved to protect the genome from the accumulation of harmful mutations, the modulation of these pathways by human tumour viruses may contribute to the malignant transformation of host cells. KSHV has previously been shown to activate the DDR during *de novo* infection of primary endothelial cells and this plays a role in establishing latency [14]. More recently, it has been demonstrated that lytic replication of KSHV in B cells results in increased phosphorylation of H2AX, a sensitive marker for the presence of DNA damage [15,16]. It has also been demonstrated that expression of immediate-early lytic protein ORF57 alone can cause DNA damage through sequestration of the hTREX complex leading to R-loop formation and ultimately DSBs [16]. Here a more detailed assessment of DDR pathways activated during lytic replication of KSHV is presented and the effect of inhibition of the major DDR kinases on replication of viral DNA is examined. In addition, changes in the localisation of several DDR proteins in cells containing lytic virus is assessed.

2. Materials and Methods

2.1. Cell Culture

TRE-BCBL-1-RTA cells (generously provided by Jae Jung, USC, Los Angeles, CA, USA) and BCBL-1 cells were cultured in RPMI (Sigma, St. Louis, MO, USA) supplemented with 10% fetal bovine serum (FBS) (Sigma) and 1% penicillin-streptomycin (Gibco, Grand Island, NY, USA). TRE-BCBL-1-RTA cells were also cultured in the presence of 100 µg/mL of Hygromycin B (Roche, Burgess Hill, UK). The endothelial cell line, EA.hy926 (purchased from ATCC, Manassas, VA, USA), was grown in DMEM (Sigma) supplemented with 10% FBS and 1% penicillin-streptomycin. EA.hy926-RTA cells, transduced with an Inducer 20 lentivirus [17] which expresses RTA under the control of the tetracycline promoter, were cultured in the presence of 250 µg/mL of G418 (Sigma). rKSHV-EA.hy926-RTA cells, which contain the RTA expression construct and are also infected with recombinant rKSHV.219 virus [18], were cultured in the presence of 250 µg/mL of G418 and 1 µg/mL Puromycin (Sigma).

2.2. Induction of Lytic Reactivation in KSHV-Infected Cell Lines

To assess DDR activation in response to KSHV lytic reactivation, TRE-BCBL-1-RTA cells and rKSHV-EA.hy926-RTA cells, as well as corresponding controls, were treated with 0.5 µg/mL doxycycline (Sigma) and subsequently harvested at the indicated times for western blot analysis. To generate positive controls for DDR activation, TRE-BCBL-1-RTA cells were either exposed to 6 Gy ionising radiation (IR) and harvested after 1 h or exposed to 20 Jm^{-2} ultraviolet light (UV) and harvested after 6 h. To inhibit viral DNA synthesis, TRE-BCBL-1-RTA cells were first treated with 100 µM ganciclovir (Cayman Chemical, Ann Arbor, MI, USA) for 2 h prior to the addition of 0.5 µg/mL doxycycline.

2.3. Inhibition of DDR Kinases during Lytic Replication

The ATR inhibitor VE-821, ATM inhibitor KU55933 and DNA-PK inhibitor NU7441 were purchased from Tocris Bioscience (Bristol, UK). TRE-BCBL-1-RTA cells were treated with specified concentrations of kinase inhibitors, or equivalent DMSO control, 1 h prior to the addition of 0.5 µg/mL doxycycline. Cells were harvested after 24 and 48 h for western blot or immunofluorescence microscopy analysis while supernatants were collected and stored at 4 °C for assessment of infectious virus production.

2.4. Infection of EA.hy926 Cells with TRE-BCBL-1-RTA-Derived KSHV Virus

Supernatants collected from TRE-BCBL-1-RTA cells were added to EA.hy926 cells cultured in 6-well plates or on coverslips in 24-well plates. Cells were centrifuged (330× g, 20 min, ambient temperature) and following 4 h incubation at 37 °C, medium containing virus was removed and replaced with supplemented DMEM medium. After 48 h, cells were harvested for western blot analysis or fixed for immunofluorescence microscopy analysis of LANA expression.

2.5. Cell Cycle Analysis

Propidium iodide (PI) staining was used to determine cell cycle distribution in TRE-BCBL-1-RTA cells. Cells were treated with 100 µM ganciclovir for 1 h followed by DDR kinase inhibitors or DMSO for 1 h prior to addition of 0.5 µg/mL doxycycline. TRE-BCBL-1-RTA cells treated with ganciclovir and DMSO but not doxycycline were used as un-reactivated controls. At each time point specified, cells were washed in phosphate buffered saline (PBS) and fixed in cold 70% ethanol before storage at −20 °C. Following two PBS washes, cells were treated with PBS containing RNase A (20 µg/mL) (Sigma) and PI (10 µg/mL) (Sigma) before being analysed using an Accuri C6 flow cytometer (BD, Oxford, UK).

2.6. Immunofluorescence Microscopy (IF)

Following treatment with 0.5 µg/mL doxycycline for 24 h, TRE-BCBL-1-RTA cells were washed in PBS and allowed to adhere on poly-L-lysine coated glass slides (Sigma). EA.hy926-RTA cells, cultured on coverslips, were infected with KSHV derived from TRE-BCBL-1-RTA cells as described above. After 48 h, EA.hy926-RTA cells were treated with 0.5 µg/mL doxycycline for 24 h. Both cell lines were fixed with 4% paraformaldehyde and permeabilised with 0.5% Triton X-100. Cells were then blocked in 10% heat-inactivated goat serum (HINGS) before addition of primary antibodies diluted in 10% HINGS for 1 h. Cells were washed 3 times in PBS before addition of appropriate secondary antibodies diluted in 10% HINGS for 1 h. Following three PBS washes, cells were stained with DAPI nucleic acid stain (Life Technologies, Grand Island, NY, USA) and cover slides were applied using ProLong Gold antifade reagent (Life Technologies). Images were taken on a Leica DM6000B epifluorescence microscope with Leica Application Suite Advanced Fluorescence software (LAS AF) (Leica, Milton Keynes, UK).

For BrdU staining, TRE-BCBL-1-RTA cells were treated with 10 µl BrdU (Sigma) for 1 h prior to adherence on poly-L-lysine coated glass slides. Cells were then fixed in 70% ethanol for 30 min followed by treatment with 0.07 M NaOH for 2 min prior to addition of FITC-conjugated BrdU antibody for 30 min. Visualisation of viral proteins was then carried out as above.

The following primary antibodies were used for immunofluorescence microscopy: anti-γH2AX (S139) (05-636, Merck Millipore, Billerica, MA, USA), anti-RPA32 (NA19L, Calbiochem, Billerica, MA, USA), anti-MRE11 (GTX70212, Genetex, Irvine, USA), anti-53BP1 (ab36823, Abcam, Cambridge, UK), anti-K8α (SAB5300152, Sigma), anti-ORF6 (Provided by Gary Hayward) anti-ORF59 (in house), anti-LANA (NCL-HHV8-LNA, Novacastra, Newcastle, UK), anti-BrdU (347583, BD).

2.7. Western Blot Analysis

Lysates were prepared by re-suspending cell pellets in 8 M urea, 50 mM Tris HCl (pH 7.4) buffer containing 1% β-mercaptoethanol. Samples were then sonicated, centrifuged (16,300× g, 20 min, 4 °C) and stored at −80 °C. Protein determination was carried using Bradford reagent (Bio-Rad,

Hercules, CA, USA) and protein standards of known concentration. Lysates were then subjected to Sodium Dodecyl Sulphate-Polyacrylamide Gel Electrophoresis (SDS-PAGE).

The following primary antibodies were used for Western blotting: anti-DNA-PKcs (04-1024, Merck Millipore), anti-phospho-DNA-PKcs (S2056) (ab124918, Abcam), anti-ATM (2873, Cell Signaling, Danvers, MA, USA), anti-phospho-ATM (S1981) (AF1655, R&D Systems, Minneapolis, MN, USA), anti-SMC1 (A300-055A, Bethyl, Montgomery, AL, USA) anti-phospho-SMC1 (S966) (A300-050A, Bethyl), anti-KAP1 (A300-274A, Bethyl), anti-phospho-KAP1 (S824) (A300-767A, Bethyl), anti-NBS1 (ab23996, Abcam), anti-phospho-NBS1 (S343) (ab47272, Abcam), anti-CHK1 (sc-8408, Santa Cruz, Dallas, TX, USA), anti-phospho-CHK1 (S345) (2341, Cell Signaling), anti-phospho-CHK1 (S317) (2344, Cell signaling), anti-CHK2 (2662, Cell Signaling), anti-phospho-CHK2 (T68) (2661, Cell Signaling), anti-RPA32 (NA19L, Calbiochem), anti-phospho-RPA32 (S4/S8) (A300-245A, Bethyl), anti-phospho-RPA32 (S33) (ab87278, Abcam), anti-H2AX (7631, Cell Signaling), anti-γH2AX (S139) (05-636, Merck Millipore), anti-β-Actin (A2228, Sigma), anti-LANA (NCL-HHV8-LNA, Novacastra), anti-RTA (in house), anti-ORF57 (AP15004a, Abgent, San Diego, CA, USA), anti-K8.1A (in house).

The following HRP conjugated secondary antibodies were also used: Polyclonal Goat Anti-Mouse (Dako Laboratories, Glostrup, Denmark), polyclonal Swine Anti-Rabbit (Dako Laboratories) and polyclonal Rabbit Anti-Goat (Dako Laboratories). Proteins were visualised using a Fusion SL chemiluminescence imaging system (Vilber Lourmat, Marne-la-Vallée, France).

3. Results

3.1. Lytic Reactivation of KSHV in B Cells Activates DDR Pathways

Analysis of DDR pathway activation during KSHV lytic replication was initially performed on a derivative of the BCBL-1 PEL line that is latently infected with KSHV. These cells, called TRE-BCBL-1-RTA, have been modified to include the ORF50/RTA gene under the control of a tetracycline promoter [19]. Addition of doxycycline to TRE-BCBL-1-RTA cells reactivates KSHV into the lytic cycle resulting in expression of immediate-early lytic proteins such as ORF57 followed by late proteins such as envelope glycoprotein K8.1A and eventually release of infectious virions (Figure 1A). Using this cell line and control BCBL-1 cells lacking the RTA-expression construct, levels of phosphorylated DDR proteins were assessed by western blotting at 5 time points following addition of doxycycline. TRE-BCBL1-RTA cells exposed to either IR or UV were used as positive controls for DDR activation. Addition of doxycycline to control BCBL-1 cells did not result in phosphorylation of any of the DDR proteins examined. While addition of doxycycline to TRE-BCBL-1-RTA cells resulted in minimal phosphorylation of SMC1 and CHK1 compared with IR and UV treated controls, after 12 h there was a notable increase in levels of phosphorylated ATM, DNA-PKcs, CHK2, KAP1, NBS1, RPA32 and H2AX indicating widespread activation of the DDR during the KSHV lytic cycle (Figure 1B).

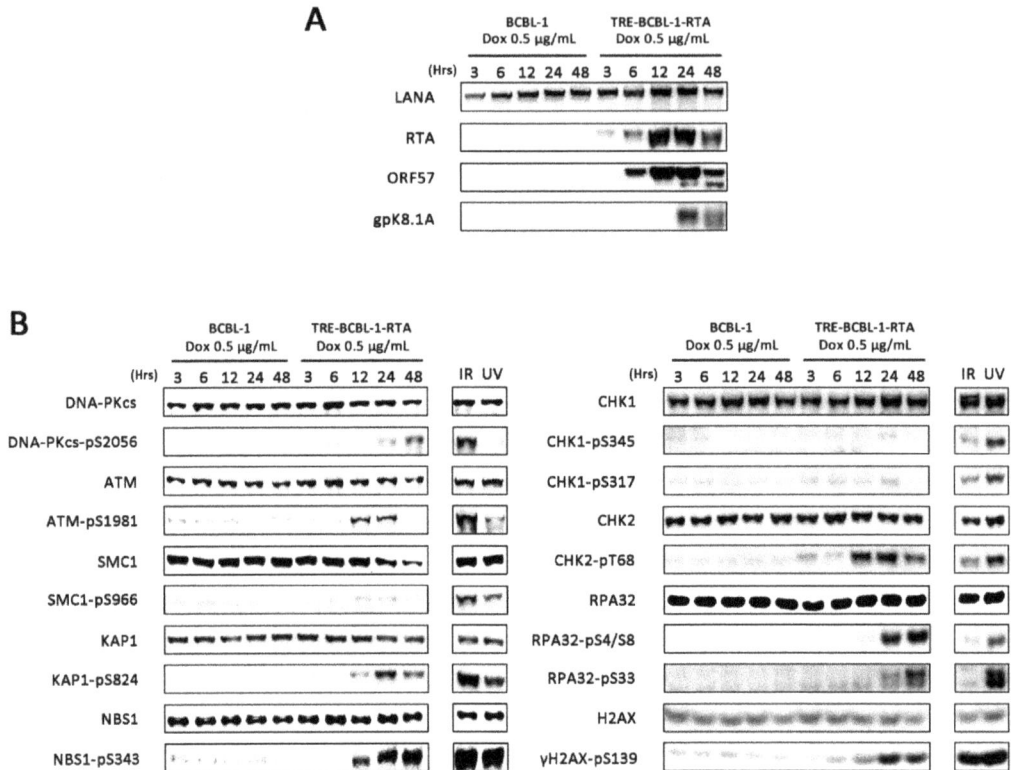

Figure 1. Activation of the DDR during lytic replication of KSHV in TRE-BCBL-1-RTA cells. (**A**) Expression of immediate-early lytic proteins RTA and ORF57 and late lytic glycoprotein K8.1A in TRE-BCBL-1-RTA cells following addition of 0.5 μg/mL doxycycline; (**B**) Phosphorylation of DDR proteins in TRE-BCBL-1-RTA cells following addition of 0.5 μg/mL doxycycline. TRE-BCBL-1-RTA cells treated with ionising radiation (IR) and ultraviolet light (UV) represent positive controls for DDR activation.

3.2. Amplification of Viral DNA and Late Viral Gene Expression Is Not Required for DDR Activation

To determine whether amplification of viral DNA or late viral gene expression is required for DDR activation in TRE-BCBL-1-RTA cells, the antiviral compound ganciclovir was employed. Ganciclovir inhibits viral DNA synthesis without affecting the expression of early lytic proteins or replication of host DNA [20]. TRE-BCBL-1-RTA cells were treated with and without 100 μM ganciclovir prior to the addition of doxycycline. Cells were also treated with ganciclovir alone to ensure that this compound in isolation did not activate the DDR. To ensure that ganciclovir was efficiently inhibiting production of infectious virus, TRE-BCBL-1-RTA supernatants from each condition were collected after 48 h and used to infect EA.hy926 cells. Expression of LANA was assessed by immunofluorescence microscopy as a marker of KSHV infection. Minimal LANA expression was observed in EA.hy926 cells infected with supernatants from TRE-BCBL-1-RTA cells treated with doxycycline and ganciclovir or ganciclovir alone (Figure 2A). Western blots confirmed that the cells treated with ganciclovir and doxycycline expressed the immediate-early lytic proteins RTA and ORF57 but not the late envelope glycoprotein K8.1A (Figure 2B). Levels of phosphorylated H2AX and RPA32 were unchanged in the doxycycline-treated cells despite the presence of ganciclovir suggesting that DDR activation is mediated by early viral gene expression and does not require amplification of viral DNA or expression of late lytic proteins.

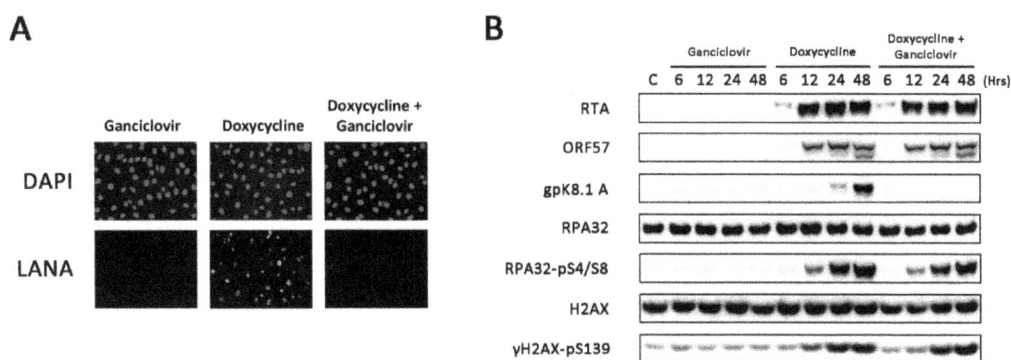

Figure 2. DDR activation following inhibition of viral DNA synthesis and late viral gene expression. (**A**) Immunofluorescence microscopy analysis of LANA expression in EA.hy926 cells following infection with supernatants obtained from TRE-BCBL-1-RTA cells collected 48 h after treatment with 100 µM ganciclovir alone, 0.5 µg/mL doxycycline alone or both compounds in combination; (**B**) Phosphorylation of RPA32 and H2AX in TRE-BCBL-1-RTA cells treated with 100 µM ganciclovir alone, 0.5 µg/mL doxycycline alone or both compounds in combination.

3.3. Inhibition of DDR Kinases Results in Alterations in Production of Infectious Virus

To determine what effect inhibition of DDR kinases has on production of infectious virus, TRE-BCBL-1-RTA cells were treated with two concentrations of ATR, ATM and DNA-PK inhibitors or an equivalent concentration of DMSO 1 h prior to the addition of 0.5 µg/mL doxycycline. After 24 and 48 h, expression of late lytic glycoprotein K8.1A was initially assessed as an indicator of completed viral replication (Figure 3A). At 24 h, expression of K8.1A was relatively low compared to 48 h but was clearly elevated following inhibition of DNA-PK. After 48 h, expression was similar between the DMSO-treated and DNA-PK inhibitor-treated cells but clearly reduced in the cells treated with the higher concentrations of ATR and ATM inhibitors.

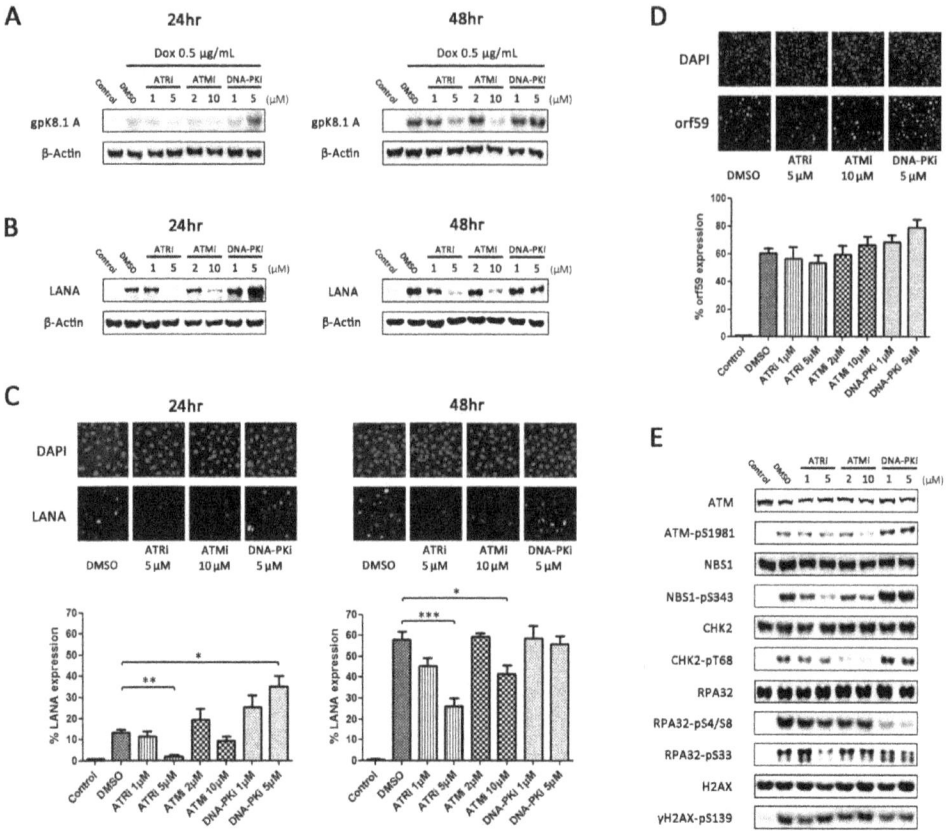

Figure 3. Effect of DDR inhibition on production of infectious virus. (**A**) Expression of late lytic glycoprotein K8.1A in TRE-BCBL-1-RTA cells 24 and 48 h following treatment with DDR kinase inhibitors and doxycycline; (**B**) Expression of LANA in EA.hy926 cells following infection with supernatants collected from TRE-BCBL-1-RTA cells 24 and 48 h after treatment with DDR kinase inhibitors and doxycycline; (**C**) Immunofluorescence microscopy analysis of the percentage of EA.hy926 cells expressing LANA following infection with supernatants collected from TRE-BCBL-1-RTA cells 24 and 48 h after treatment with DDR kinase inhibitors and doxycycline. Each column represents the mean of three independent experiments while the error bars represent the standard error of the mean (SEM). A minimum of 500 cells were analysed for each repetition. * $p < 0.05$; ** $p < 0.01$; *** $p < 0.001$ (statistical analyses were performed using a two-tailed and unpaired Student's t-test); (**D**) Percentage of TRE-BCBL-1-RTA cells expressing early lytic protein ORF59 24 h following treatment with DDR kinase inhibitors and doxycycline. Each column represents the mean of three independent experiments while the error bars represent the standard error of the mean (SEM). A minimum of 500 cells were analysed for each repetition; (**E**) Levels of phosphorylated DDR proteins in TRE-BCBL-1-RTA cells 24 h following treatment with DDR kinase inhibitors and doxycycline.

To ensure that levels of K8.1A expression were representative of the amount of infectious virus produced from TRE-BCBL1-RTA cells, the medium supernatants were collected from the cells after 24 and 48 h and used to infect the EA.hy926 endothelial cell line. After 48 h, western blotting was used to assess expression of LANA as a marker of KSHV infection (Figure 3B). The expression levels of LANA in the EA.hy926 cells correlated well with the levels of K8.1A expression in TRE-BCBL-1-RTA cells at both 24 and 48 h. To quantify further the level of virus contained in the supernatants, the proportion of EA.hy926 cells expressing LANA was calculated using immunofluorescence microscopy (Figure 3C). In the EA.hy926 cells treated with the 24 h supernatants there was a significant increase in the proportion of cells expressing LANA following infection with the 5 μM DNA-PKi supernatants and a significant decrease following infection with the 5 μM ATRi supernatants. In the EA.hy926 cells treated with the 48 h supernatants there was a significant decrease in the proportion of cells expressing LANA following infection with the 5 μM ATRi and 10 μM ATMi supernatants.

Since not all TRE-BCBL-1-RTA cells contain lytically replicating virus following treatment with doxycycline, it is possible that the differences observed in infectious virus production are due to changes in the percentages of cells containing lytic virus rather than changes in the amount of virus produced from individual cells. To quantify the percentage of cells containing lytic virus following treatment with DDR inhibitors and doxycycline, expression of delayed-early lytic protein ORF59 was assessed using immunofluorescence microscopy (Figure 3D). Although the percentage of cells expressing ORF59 appears elevated following the addition of DNA-PK inhibitor there were no significant differences between any of the conditions. This suggests that the changes in production of infectious virus are not because of differing numbers of cells containing lytic virus but because of changes in the efficiency of viral replication.

Several proteins involved in DDR signalling can be phosphorylated by more than one of the three key DDR kinases. To examine which kinase was responsible for the phosphorylation events observed during lytic replication of KSHV, changes in phosphorylation level of several DDR proteins in TRE-BCBL-1-RTA cells was assessed 24 h following addition of DDR inhibitors and doxycycline (Figure 3E). Inhibition of DNA-PK markedly reduced phosphorylation of RPA32 at S4/S8 and moderately reduced phosphorylation of H2AX at S139 and RPA32 at S33. DNA-PK inhibition also led to an increase in ATM, NBS1 and CHK2 phosphorylation. Inhibition of ATM strongly reduced CHK2 phosphorylation as expected but only partially reduced phosphorylation of NBS1. Inhibition of ATR reduced phosphorylation of RPA32 at S33 and reduced NBS1 phosphorylation to a greater extent than the ATM inhibitor. This indicates that DNA-PK and ATM are activating downstream targets during lytic replication but also suggests active ATR despite the lack of CHK1 phosphorylation.

3.4. CHK1 Activation is Inhibited at Later Times during Lytic Replication

While there is evidence of ATR activation during lytic replication, substantial CHK1 phosphorylation is not observed (Figure 1B). To determine whether ATR-CHK1 signalling is functional during lytic replication, TRE-BCBL-1-RTA cells were exposed to 20 Jm^{-2} UV at several time points following initiation of lytic replication (Figure 4). Expression of RTA and ORF57 were used to confirm lytic reactivation of KSHV. Exposure to UV at 3 and 6 h post-doxycycline treatment led to phosphorylation of CHK1 at S345 and S317 which was comparable to the un-reactivated control. However, when TRE-BCBL-1-RTA cells were treated with UV at 12, 24 and 48 h following addition of doxycycline, levels of phosphorylated CHK1 were noticeably lower than in the un-reactivated control. This suggests that at later times in the lytic cycle, when phosphorylation of other DDR substrates occurs, activation of CHK1 is specifically inhibited by the virus.

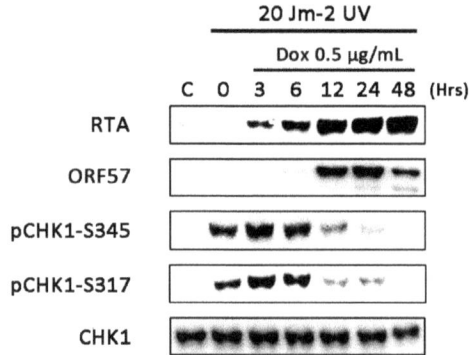

Figure 4. Inhibition of CHK1 signalling at late times during KSHV lytic replication. Levels of phosphorylated CHK1 (S317 and S345) following UV treatment of TRE-BCBL-1-RTA cells treated with doxycycline for varying lengths of time. At each time point following the addition of doxycycline, cells were treated with 20 Jm^{-2} UV and levels of phosphorylated CHK1 were assessed by western blot.

3.5. DDR Activation Does Not Result in a G1 Block Following Lytic Reactivation

Since viruses are known to modulate the cell cycle to facilitate viral replication and DDR activation can vary throughout the cell cycle, an assessment of cell cycle changes during lytic replication and in the presence of DDR kinase inhibitors was undertaken (Figure 5). TRE-BCBL-1-RTA cells were treated with DMSO or the higher concentrations of DDR kinase inhibitors 1 h prior to the addition of doxycycline. Cells were harvested after 6, 12, 18 and 24 h and subjected to PI staining. Because production of viral DNA by herpesviruses has been shown to interfere with PI staining [21], all TRE-BCBL-1-RTA cells were also treated with ganciclovir 1 h prior to the addition of DDR inhibitors or DMSO. At the 6 and 12 h time points the experiment was also conducted without ganciclovir to confirm that this compound alone has no significant effects on the cell cycle distribution following addition of doxycycline (data not shown). Cell counts also confirmed that ganciclovir alone had no significant effect on the growth rate of TRE-BCBL-1-RTA cells over 24 h (data not shown).

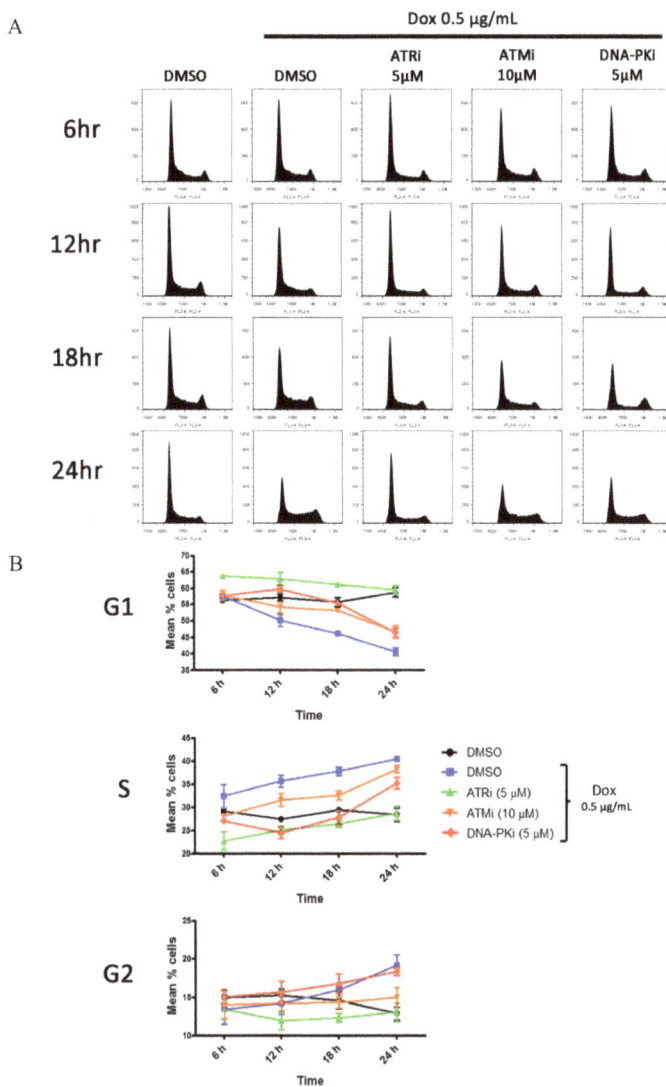

Figure 5. Cell cycle changes during lytic replication of KSHV. (**A**) Representative cell cycle profiles following PI staining of TRE-BCBL-1-RTA cells after addition of doxycycline in the presence and absence of DDR kinase inhibitors; (**B**) Percentage of TRE-BCBL-1-RTA cells in different phases of the cell cycle at 4 different time points following addition of doxycycline in the presence and absence of DDR kinase inhibitors. Each data point represents the mean of three independent experiments while the error bars represent the standard error of the mean (SEM).

The addition of doxycycline to cells treated with DMSO resulted in a decrease in the proportion of cells in G1 and an increase in the proportion in S-phase during the time course compared with the un-reactivated control. By 24 h there was also a notable increase in the proportion of cells in G2 following doxycycline treatment. Addition of the ATR inhibitor prior to doxycycline increased the proportion of cells in G1 and decreased the proportion in S and G2 compared with the cells treated with doxycycline and DMSO. By 24 h, the cell cycle profile in the presence of the ATR inhibitor and doxycycline was almost identical to the un-reactivated control. While treatment with the ATM and DNA-PK inhibitors also initially increased the proportion of cells in G1, by 24 h the difference between these conditions and the cells treated with doxycycline and DMSO was minimal. It appears then that DDR activation following lytic activation does not cause a G1 cell cycle block and instead cells are able to proceed to S and G2 phases. Addition of the ATR inhibitor, however, appears to inhibit the cell cycle changes normally induced during the lytic cycle.

3.6. RPA32 and MRE11 Localise to Sites of Viral Replication

Several DDR proteins are known to form discrete foci at sites of DNA damage [22]. In addition, several herpesviruses are known to recruit DDR proteins to sites of viral replication where they may be involved in viral DNA synthesis [23–25]. An assessment of the localisation of several DDR proteins was undertaken 24 h after induction of KSHV lytic replication (Figure 6). As well as using TRE-BCBL-1-RTA cells (Figure 6A), EA.hy926-RTA cells were examined as their larger size and flat morphology allowed for clearer visualisation of protein localization (Figure 6B). Expression of lytic proteins ORF6 and K8α were used as markers of KSHV lytic replication. The delayed-early lytic protein ORF6 has been shown previously to localise to sites of viral DNA replication in BCBL-1 cells [26]. BrdU staining was used to confirm that ORF6-positive domains were sites of viral DNA synthesis (Figure 6A).

Phosphorylated H2AX is a sensitive marker for the presence of DNA damage and γH2AX foci were consistently observed in TRE-BCBL-1-RTA and EA.hy926 cells positive for ORF6 expression. When ORF6 was localised to discrete replication bodies in EA.hy926-RTA cells, γH2AX foci typically formed outside of these bodies in cellular DNA while in TRE-BCBL-1-RTA cells γH2AX foci appear to form on the margins of these replication compartments. In cells with larger replication bodies, γH2AX staining could be seen on the periphery of replication domains as cellular DNA is marginalised to the edge of the nucleus. RPA32, which binds to single-stranded DNA, also forms foci following DNA damage. In this case, RPA32 was consistently observed localised in the same areas of the nucleus as ORF6 in both TRE-BCBL-1-RTA and EA.hy926-RTA cells, suggesting an association with viral replication domains. In addition, MRE11, part of the MRN complex that localises to DSBs, was also found to localise to viral replication compartments in both cell lines.

The localisation of 53BP1 was also examined as this protein is known to form foci at sites of DSBs [27]. Despite the activation of ATM and DNA-PK in TRE-BCBL-1-RTA cells, consistent formation of 53BP1 foci was not observed in cells positive for K8α after 24 h. In EA.hy926-RTA cells, where it is possible to accurately quantify the number of foci present, less than 30% of cells positive for K8α contained more than ten 53BP1 foci. As K8α did not appear to localise to replication domains in EA.hy926-RTA cells, and since RPA32 was consistently observed in association with

ORF6, RPA32 was used as surrogate marker for KSHV replication domains when staining for 53BP1. When 53BP1 did form foci there was no evidence of localisation to KSHV replication domains.

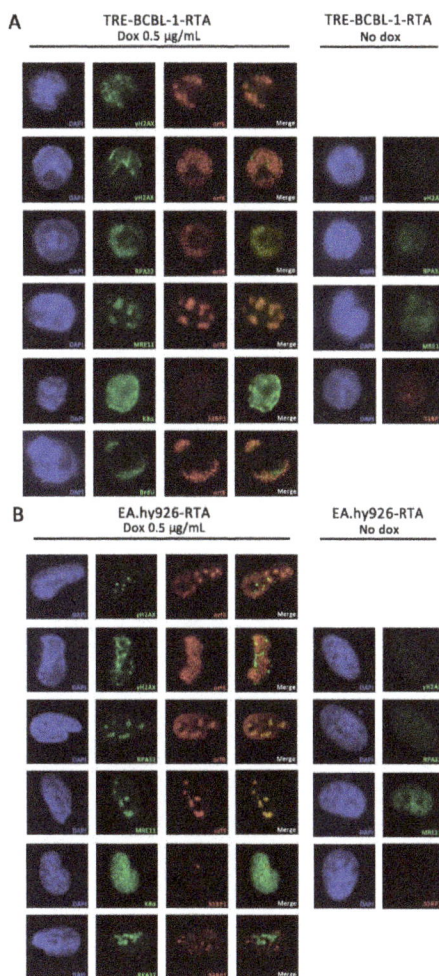

Figure 6. Localisation of DDR proteins during lytic replication of KSHV. (**A**) Localisation of γH2AX, RPA32, MRE11 and 53BP1 in TRE-BCBL-1-RTA cells 24 h following addition of doxycycline and in cells without doxycycline treatment. KSHV lytic proteins K8α and ORF6 were used as markers of lytic reactivation. In the bottom panels, BrdU was used to confirm that ORF6 localises to sites of viral replication; (**B**) Localisation of γH2AX, RPA32, MRE11 and 53BP1 in KSHV-infected EA.hy926-RTA cells 24 h following addition of doxycycline and in cells without doxycycline treatment. KSHV lytic proteins K8α and ORF6 were used as markers of lytic reactivation. In the bottom panels, RPA32 was used as a surrogate marker for sites of viral replication. DAPI staining is only shown in the left hand set of panels in each case for clarity.

3.7. Lytic Reactivation of KSHV in an Endothelial Cell Line Activates the DDR

Since KSHV targets both B cells and endothelial cells *in vivo*, activation of the DDR during KSHV lytic replication was also assessed in the EA.hy926 endothelial cell line (Figure 7). As withthe TRE-BCBL-1-RTA cells, DDR activation was compared between uninfected EA.hy926 cells and KSHV-infected EA.hy926 cells containing an RTA expression construct (termed rKSHV-EA.hy926-RTA) following treatment with doxycycline. In addition, uninfected EA.hy926 cells containing the RTA expression construct (termed EA.hy926-RTA) were also used to assess the effect of RTA expression alone on DDR activation. All three cell lines were treated with doxycycline and levels of phosphorylated H2AX and RPA32 were used to assess DDR activation at the same times examined in TRE-BCBL-1-RTA cells. Expression of LANA and ORF57 were used to confirm KSHV infection and lytic reactivation in the rKSHV-EA.hy926-RTA cell line. Induction of lytic replication in rKSHV-EA.hy926-RTA cells led to DDR activation with increased phosphorylation of RPA32 and H2AX. Expression of RTA alone in EA.hy926-RTA cells also increased phosphorylation of H2AX but to a limited extent compared to cells containing KSHV.

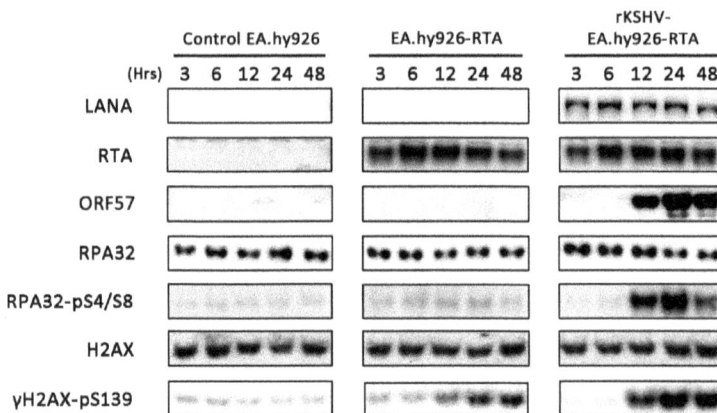

Figure 7. Activation of the DNA damage response during lytic replication of KSHV in EA.hy926 cells and following RTA expression alone. Levels of phosphorylated H2AX and RPA32 following addition of doxycyline to control EA.hy926 cells, EA.hy926 cells containing an ORF50 expression construct (EA.hy926-RTA) and KSHV-infected EA.hy926 cells containing an ORF50 expression construct (rKSHV-EA.hy926-RTA). Expression of LANA and ORF57 confirm presence of KSHV and lytic reactivation of the virus respectively.

4. Discussion

It has been demonstrated here that during lytic replication of KSHV, the ATM and DNA-PK kinases are activated and phosphorylate several downstream targets. The timing of some of these phosphorylation events suggests that immediate early or delayed-early lytic gene expression is responsible for DDR activation. The use of ganciclovir to inhibit viral DNA synthesis and late viral gene expression confirms that these stages of the lytic cycle are not required for the observed activation of the DDR. In accordance with this, it was recently reported that expression of KSHV immediate early lytic protein ORF57 can cause DNA damage and phosphorylation of H2AX when expressed in isolation [16].

It has also been shown here that inhibition of ATR and ATM results in reduced production of infectious virus following induction of lytic replication in B cells. Inhibition of ATM signalling has also been shown to reduce replication of the related gamma herpesvirus Epstein-Barr virus (EBV) in several studies [28–30]. Although the reduction in viral replication following ATM inhibition was moderate, DNA-PK inhibition led to increased phosphorylation of ATM and downstream substrates and increased production of infectious virus after 24 h. This observation also suggests that activation of the ATM-CHK2 arm of the DDR can promote efficient KSHV replication in B cells. The abrogation of ATM signalling was not completely efficient even at the higher concentration of ATM inhibitor. While increased concentrations of inhibitor were avoided due to concerns about toxicity and off-target effects, it would be interesting to see if a more complete knockdown of ATM could result in a more dramatic inhibitory effect on viral replication. This could potentially be achieved using Crispr/Cas in further investigations.

The inhibition of ATR resulted in the largest reduction in virus production although the lack of CHK1 phosphorylation prompts the question of whether this kinase is activated during the lytic cycle. While ATR inhibition reduced phosphorylation of NBS1 and RPA32, both of which are known ATR substrates [31,32], it is possible that cell cycle changes induced by the ATR inhibitor could contribute to a reduction in phosphorylation of these proteins by ATM and DNA-PK as well as the observed decline in virus production, assuming that progression into S-phase is advantageous (Figure 5). Whereas phosphorylation of CHK1 at serines 317 and 345 are typical markers for ATR activation in response to replication stress [33], the ATR-CHK1 pathway can also be activated when repair of DSBs during S and G2 phases generates RPA-coated ssDNA [34,35]. It was interesting therefore to observe that there was minimal CHK1 phosphorylation even at late times during lytic replication. In herpes simplex virus 1 (HSV-1), ATR signalling is disabled during viral replication and proteins involved in the ATR-CHK1 pathway are recruited to viral replication centres [25]. It appears that ATR-CHK1 signalling is disrupted at later times during KSHV lytic replication following treatment with UV. It is not clear at this stage whether inhibition of CHK1 is a deliberate viral strategy or merely a consequence of wider DDR deregulation during lytic replication.

It was notable that a G1 block was not observed during the first 24 h following lytic replication considering that this has been reported in several studies of herpesviruses including KSHV [21,36–39]. KSHV lytic proteins RTA and K8α have both been shown to induce a G1 block when expressed in isolation [37–39]. However, KSHV also encodes a D-cyclin homolog, known as v-cyclin, that has been reported to induce cell cycle progression via inactivation of p27^{KIP1} [40]. v-cyclin is expressed throughout the viral lifecycle but expression is increased following lytic reactivation [41]. It has also been shown to activate the DDR in endothelial cells by promoting deregulated S-phase entry [42]. Considering the apparent opposing effects of these viral proteins on cell cycle progression, and the large number of genes expressed during the lytic cycle, it is clear that single gene expression studies could lead to contradictory finding in this area. While it has been postulated that replication in G1 is ideal during the lytic cycle as the virus avoids competition for DNA replication resources [37], there is also evidence that S-phase can provide KSHV with a favourable environment for viral replication [43]. The data presented here suggest that, despite the presence of DNA damage, cells containing lytic virus can still progress into S-phase and that when a G1 block is induced by inhibition of ATR it has a negative effect on the efficient production of viral DNA.

The activation of ATM and DNA-PK suggests the formation of DSBs during KSHV replication. Increased phosphorylation of H2AX and the presence of γH2AX foci have previously been used to confirm the presence of DSBs in B cells following induction of the lytic cycle [15,16]. The formation of γH2AX foci was also observed during this study but since γH2AX foci can also form at single-stranded DNA breaks and stalled replication forks [44], the presence of 53BP1 foci which form rapidly at sites of DSBs was also examined. Curiously, we did not observe consistent formation of 53BP1 foci in cells positive for lytic virus. Although 53BP1 localisation has been implicated in ATM activation it is not strictly required following high levels of DNA damage [45]. In EBV, 53BP1 has been shown to be important for viral replication [46] while in human cytomegalovirus (HCMV), overall 53BP1 levels decrease during infection [24]. In HSV-1, degradation of the E3 ubiquitin ligases RNF8 and RNF168 inhibits localisation of 53BP1 to incoming viral genomes [47]. It will be interesting to assess further whether expression and localisation of this protein is deregulated during the KSHV lifecycle.

RPA32 and MRE11 can also form foci following DNA damage and both proteins were consistently observed localised to areas of viral replication; this was particularly evident in EA.hy926 cells where ORF6 staining can be seen as discrete bodies that occupy a smaller proportion of the nucleus than in B cells. Both RPA32 and MRE11 have been shown to localise to EBV replication compartments where they are loaded onto newly synthesised viral DNA along with HR factors Rad51 and Rad52 [23]. Further work will be required to determine if any other DDR factors are recruited to KSHV replication compartments and what role they play in viral replication.

Overall, it appears that lytic reactivation of KSHV leads to selective activation of the DDR and this plays a positive role in the viral lifecycle. In addition, the localisation of DDR proteins to areas of viral replication suggests that individual DDR proteins may contribute to viral DNA synthesis, as is the case with other herpesviruses. Although the data presented here do not contradict the demonstration that expression of ORF57 can elicit a DDR, our observations using an endothelial cell line suggest that RTA alone can also modestly increase phosphorylation of H2AX. It is therefore possible that expression of several early lytic proteins could be responsible for DDR activation which then contributes to efficient production of viral progeny.

Acknowledgments

We are most grateful to Cancer Research UK (CRUK) for funding this study. We would also like to thank Jae Jung for kindly providing the TRE-BCBL-1-RTA cells and Gary Hayward for generously supplying the ORF6 antibody.

Author Contributions

Roger J. Grand and Robert Hollingworth conceived and designed the experiments. Robert Hollingworth conducted all experiments in B cells and wrote the manuscript. George L. Skalka conducted all experiments in endothelial cells. David J. Blackbourn, Andrew D. Hislop and Grant S. Stewart provided materials and reagents as well as critical evaluation of experimental procedures. All authors reviewed and approved the final manuscript.

Conflict of Interest

The authors have no conflicting interests.

References

1. Chang, Y.; Cesarman, E.; Pessin, M.S.; Lee, F.; Culpepper, J.; Knowles, D.M.; Moore, P.S. Identification of herpesvirus-like DNA sequences in aids-associated kaposi's sarcoma. *Science* **1994**, *266*, 1865–1869. [CrossRef] [PubMed]
2. Cesarman, E.; Chang, Y.; Moore, P.S.; Said, J.W.; Knowles, D.M. Kaposi's sarcoma-associated herpesvirus-like DNA sequences in aids-related body-cavity-based lymphomas. *N. Engl. J. Med.* **1995**, *332*, 1186–1191. [CrossRef] [PubMed]
3. Soulier, J.; Grollet, L.; Oksenhendler, E.; Cacoub, P.; Cazals-Hatem, D.; Babinet, P.; D'Agay, M.F.; Clauvel, J.P.; Raphael, M.; Degos, L.; *et al.* Kaposi's sarcoma-associated herpesvirus-like DNA sequences in multicentric castleman's disease. *Blood* **1995**, *86*, 1276–1280. [PubMed]
4. Sun, R.; Lin, S.F.; Gradoville, L.; Yuan, Y.; Zhu, F.; Miller, G. A viral gene that activates lytic cycle expression of kaposi's sarcoma-associated herpesvirus. *Proc. Natl. Acad. Sci. USA* **1998**, *95*, 10866–10871. [CrossRef] [PubMed]
5. Guito, J.; Lukac, D.M. Kshv reactivation and novel implications of protein isomerization on lytic switch control. *Viruses* **2015**, *7*, 72–109. [CrossRef] [PubMed]
6. Martin, D.F.; Kuppermann, B.D.; Wolitz, R.A.; Palestine, A.G.; Li, H.; Robinson, C.A. Oral ganciclovir for patients with cytomegalovirus retinitis treated with a ganciclovir implant. Roche ganciclovir study group. *N. Engl. J. Med.* **1999**, *340*, 1063–1070. [CrossRef] [PubMed]
7. Grundhoff, A.; Ganem, D. Inefficient establishment of kshv latency suggests an additional role for continued lytic replication in kaposi sarcoma pathogenesis. *J. Clin. Investig.* **2004**, *113*, 124–136. [CrossRef] [PubMed]

8. Rothkamm, K.; Kruger, I.; Thompson, L.H.; Lobrich, M. Pathways of DNA double-strand break repair during the mammalian cell cycle. *Mol. Cell. Biol.* **2003**, *23*, 5706–5715. [CrossRef] [PubMed]

9. Lieber, M.R. The mechanism of double-strand DNA break repair by the nonhomologous DNA end-joining pathway. *Annu. Rev. Biochem.* **2010**, *79*, 181–211. [CrossRef] [PubMed]

10. Ashley, A.K.; Shrivastav, M.; Nie, J.; Amerin, C.; Troksa, K.; Glanzer, J.G.; Liu, S.; Opiyo, S.O.; Dimitrova, D.D.; Le, P.; *et al.* DNA-pk phosphorylation of rpa32 ser4/ser8 regulates replication stress checkpoint activation, fork restart, homologous recombination and mitotic catastrophe. *DNA Repair Amst.* **2014**, *21*, 131–139. [CrossRef] [PubMed]

11. Zou, L.; Elledge, S.J. Sensing DNA damage through atrip recognition of rpa-ssdna complexes. *Science* **2003**, *300*, 1542–1548. [CrossRef] [PubMed]

12. Chini, C.C.; Chen, J. Human claspin is required for replication checkpoint control. *J. Biol. Chem.* **2003**, *278*, 30057–30062. [CrossRef] [PubMed]

13. Kumagai, A.; Lee, J.; Yoo, H.Y.; Dunphy, W.G. Topbp1 activates the atr-atrip complex. *Cell* **2006**, *124*, 943–955. [CrossRef] [PubMed]

14. Singh, V.V.; Dutta, D.; Ansari, M.A.; Dutta, S.; Chandran, B. Kaposi's sarcoma-associated herpesvirus induces the ATM and H2AX DNA damage response early during de novo infection of primary endothelial cells, which play roles in latency establishment. *J. Virol.* **2014**, *88*, 2821–2834. [CrossRef] [PubMed]

15. Xiao, Y.; Chen, J.; Liao, Q.; Wu, Y.; Peng, C.; Chen, X. Lytic infection of kaposi's sarcoma-associated herpesvirus induces DNA double-strand breaks and impairs non-homologous end joining. *J. Gen. Virol.* **2013**, *94*, 1870–1875. [CrossRef] [PubMed]

16. Jackson, B.R.; Noerenberg, M.; Whitehouse, A. A novel mechanism inducing genome instability in kaposi's sarcoma-associated herpesvirus infected cells. *PLoS Pathog.* **2014**, *10*, e1004098. [CrossRef] [PubMed]

17. Meerbrey, K.L.; Hu, G.; Kessler, J.D.; Roarty, K.; Li, M.Z.; Herschkowitz, J.I.; Burrows, A.E.; Fang, J.E.; Ciccia, A.; Sun, T.; *et al.* The pinducer lentiviral toolkit for inducible rna interference *in vitro* and *in vivo*. *Proc. Natl. Acad. Sci. USA* **2011**, *108*, 3665–3670. [CrossRef] [PubMed]

18. Vieira, J.; O'Hearn, P.M. Use of the red fluorescent protein as a marker of kaposi's sarcoma-associated herpesvirus lytic gene expression. *Virology* **2004**, *325*, 225–240. [CrossRef] [PubMed]

19. Nakamura, H.; Lu, M.; Gwack, Y.; Souvlis, J.; Zeichner, S.L.; Jung, J.U. Global changes in kaposi's sarcoma-associated virus gene expression patterns following expression of a tetracycline-inducible rta transactivator. *J. Virol.* **2003**, *77*, 4205–4220. [CrossRef] [PubMed]

20. Matthews, T.; Boehme, R. Antiviral activity and mechanism of action of ganciclovir. *Rev. Infect. Dis.* **1988**, *10*, S490–S494. [CrossRef] [PubMed]

21. Bresnahan, W.A.; Boldogh, I.; Thompson, E.A.; Albrecht, T. Human cytomegalovirus inhibits cellular DNA synthesis and arrests productively infected cells in late g1. *Virology* **1996**, *224*, 150–160. [CrossRef] [PubMed]

22. Polo, S.E.; Jackson, S.P. Dynamics of DNA damage response proteins at DNA breaks: A focus on protein modifications. *Genes Dev.* **2011**, *25*, 409–433. [CrossRef] [PubMed]

23. Kudoh, A.; Iwahori, S.; Sato, Y.; Nakayama, S.; Isomura, H.; Murata, T.; Tsurumi, T. Homologous recombinational repair factors are recruited and loaded onto the viral DNA genome in epstein-barr virus replication compartments. *J. Virol.* **2009**, *83*, 6641–6651. [CrossRef] [PubMed]

24. Luo, M.H.; Rosenke, K.; Czornak, K.; Fortunato, E.A. Human cytomegalovirus disrupts both ataxia telangiectasia mutated protein (ATM)- and ATM-rad3-related kinase-mediated DNA damage responses during lytic infection. *J. Virol.* **2007**, *81*, 1934–1950. [CrossRef] [PubMed]

25. Mohni, K.N.; Dee, A.R.; Smith, S.; Schumacher, A.J.; Weller, S.K. Efficient herpes simplex virus 1 replication requires cellular atr pathway proteins. *J. Virol.* **2013**, *87*, 531–542. [CrossRef] [PubMed]

26. Wang, Y.; Li, H.; Tang, Q.; Maul, G.G.; Yuan, Y. Kaposi's sarcoma-associated herpesvirus ori-lyt-dependent DNA replication: Involvement of host cellular factors. *J. Virol.* **2008**, *82*, 2867–2882. [CrossRef] [PubMed]

27. Anderson, L.; Henderson, C.; Adachi, Y. Phosphorylation and rapid relocalization of 53bp1 to nuclear foci upon DNA damage. *Mol. Cell. Biol.* **2001**, *21*, 1719–1729. [CrossRef] [PubMed]

28. Li, R.; Zhu, J.; Xie, Z.; Liao, G.; Liu, J.; Chen, M.R.; Hu, S.; Woodard, C.; Lin, J.; Taverna, S.D.; *et al.* Conserved herpesvirus kinases target the DNA damage response pathway and tip60 histone acetyltransferase to promote virus replication. *Cell Host Microbe* **2011**, *10*, 390–400. [CrossRef] [PubMed]

29. Hagemeier, S.R.; Barlow, E.A.; Meng, Q.; Kenney, S.C. The cellular ataxia telangiectasia-mutated kinase promotes epstein-barr virus lytic reactivation in response to multiple different types of lytic reactivation-inducing stimuli. *J. Virol.* **2012**, *86*, 13360–13370. [CrossRef] [PubMed]

30. Hau, P.M.; Deng, W.; Jia, L.; Yang, J.; Tsurumi, T.; Chiang, A.K.; Huen, M.S.; Tsao, S.W. Role of atm in the formation of the replication compartment during lytic replication of epstein-barr virus in nasopharyngeal epithelial cells. *J. Virol.* **2015**, *89*, 652–668. [CrossRef] [PubMed]

31. Liu, S.; Bekker-Jensen, S.; Mailand, N.; Lukas, C.; Bartek, J.; Lukas, J. Claspin operates downstream of TOPBP1 to direct atr signaling towards chk1 activation. *Mol. Cell. Biol.* **2006**, *26*, 6056–6064. [CrossRef] [PubMed]

32. Liu, S.; Opiyo, S.O.; Manthey, K.; Glanzer, J.G.; Ashley, A.K.; Amerin, C.; Troksa, K.; Shrivastav, M.; Nickoloff, J.A.; Oakley, G.G. Distinct roles for DNA-pk, atm and atr in rpa phosphorylation and checkpoint activation in response to replication stress. *Nucleic Acids Res.* **2012**, *40*, 10780–10794. [CrossRef] [PubMed]

33. Cimprich, K.A.; Cortez, D. ATR: An essential regulator of genome integrity. *Nat. Rev. Mol. Cell. Biol.* **2008**, *9*, 616–627. [CrossRef] [PubMed]

34. Jazayeri, A.; Falck, J.; Lukas, C.; Bartek, J.; Smith, G.C.; Lukas, J.; Jackson, S.P. ATM- and cell cycle-dependent regulation of atr in response to DNA double-strand breaks. *Nat. Cell Biol.* **2006**, *8*, 37–45. [CrossRef] [PubMed]

35. Shibata, A.; Barton, O.; Noon, A.T.; Dahm, K.; Deckbar, D.; Goodarzi, A.A.; Lobrich, M.; Jeggo, P.A. Role of ATM and the damage response mediator proteins 53bp1 and mdc1 in the

maintenance of G_2/M checkpoint arrest. *Mol. Cell. Biol.* **2010**, *30*, 3371–3383. [CrossRef] [PubMed]

36. Rodriguez, A.; Jung, E.J.; Flemington, E.K. Cell cycle analysis of epstein-barr virus-infected cells following treatment with lytic cycle-inducing agents. *J. Virol.* **2001**, *75*, 4482–4489. [CrossRef] [PubMed]

37. Wu, F.Y.; Tang, Q.Q.; Chen, H.; ApRhys, C.; Farrell, C.; Chen, J.; Fujimuro, M.; Lane, M.D.; Hayward, G.S. Lytic replication-associated protein (RAP) encoded by kaposi sarcoma-associated herpesvirus causes p21CIP-1-mediated G1 cell cycle arrest through ccaat/enhancer-binding protein-alpha. *Proc. Natl. Acad. Sci. USA* **2002**, *99*, 10683–10688. [CrossRef] [PubMed]

38. Izumiya, Y.; Lin, S.F.; Ellison, T.J.; Levy, A.M.; Mayeur, G.L.; Izumiya, C.; Kung, H.J. Cell cycle regulation by kaposi's sarcoma-associated herpesvirus k-bzip: Direct interaction with cyclin-CDK2 and induction of G1 growth arrest. *J. Virol.* **2003**, *77*, 9652–9661. [CrossRef] [PubMed]

39. Kumar, P.; Wood, C. Kaposi's sarcoma-associated herpesvirus transactivator rta induces cell cycle arrest in G0/G1 phase by stabilizing and promoting nuclear localization of p27kip. *J. Virol.* **2013**, *87*, 13226–13238. [CrossRef] [PubMed]

40. Jarviluoma, A.; Koopal, S.; Rasanen, S.; Makela, T.P.; Ojala, P.M. Kshv viral cyclin binds to p27kip1 in primary effusion lymphomas. *Blood* **2004**, *104*, 3349–3354. [CrossRef] [PubMed]

41. Arias, C.; Weisburd, B.; Stern-Ginossar, N.; Mercier, A.; Madrid, A.S.; Bellare, P.; Holdorf, M.; Weissman, J.S.; Ganem, D. Kshv 2.0: A comprehensive annotation of the kaposi's sarcoma-associated herpesvirus genome using next-generation sequencing reveals novel genomic and functional features. *PLoS Pathog.* **2014**, *10*, e1003847. [CrossRef] [PubMed]

42. Koopal, S.; Furuhjelm, J.H.; Jarviluoma, A.; Jaamaa, S.; Pyakurel, P.; Pussinen, C.; Wirzenius, M.; Biberfeld, P.; Alitalo, K.; Laiho, M.; *et al.* Viral oncogene-induced DNA damage response is activated in kaposi sarcoma tumorigenesis. *PLoS Pathog.* **2007**, *3*, 1348–1360. [CrossRef] [PubMed]

43. Bryan, B.A.; Dyson, O.F.; Akula, S.M. Identifying cellular genes crucial for the reactivation of kaposi's sarcoma-associated herpesvirus latency. *J. Gen. Virol.* **2006**, *87*, 519–529. [CrossRef] [PubMed]

44. Ward, I.M.; Chen, J. Histone H2AX is phosphorylated in an atr-dependent manner in response to replicational stress. *J. Biol. Chem.* **2001**, *276*, 47759–47762. [PubMed]

45. Fernandez-Capetillo, O.; Chen, H.T.; Celeste, A.; Ward, I.; Romanienko, P.J.; Morales, J.C.; Naka, K.; Xia, Z.; Camerini-Otero, R.D.; Motoyama, N.; *et al.* DNA damage-induced g2-m checkpoint activation by histone h2ax and 53bp1. *Nat. Cell Biol.* **2002**, *4*, 993–997. [CrossRef] [PubMed]

46. Bailey, S.G.; Verrall, E.; Schelcher, C.; Rhie, A.; Doherty, A.J.; Sinclair, A.J. Functional interaction between epstein-barr virus replication protein zta and host DNA damage response protein 53bp1. *J. Virol.* **2009**, *83*, 11116–11122. [CrossRef] [PubMed]

47. Lilley, C.E.; Chaurushiya, M.S.; Boutell, C.; Everett, R.D.; Weitzman, M.D. The intrinsic antiviral defense to incoming HSV-1 genomes includes specific DNA repair proteins and is counteracted by the viral protein ICP0. *PLoS Pathog.* **2011**, *7*, e1002084. [CrossRef] [PubMed]

Targeting CTCF to Control Virus Gene Expression: A Common Theme amongst Diverse DNA Viruses

Ieisha Pentland and Joanna L. Parish

Abstract: All viruses target host cell factors for successful life cycle completion. Transcriptional control of DNA viruses by host cell factors is important in the temporal and spatial regulation of virus gene expression. Many of these factors are recruited to enhance virus gene expression and thereby increase virus production, but host cell factors can also restrict virus gene expression and productivity of infection. CCCTC binding factor (CTCF) is a host cell DNA binding protein important for the regulation of genomic chromatin boundaries, transcriptional control and enhancer element usage. CTCF also functions in RNA polymerase II regulation and in doing so can influence co-transcriptional splicing events. Several DNA viruses, including Kaposi's sarcoma-associated herpesvirus (KSHV), Epstein-Barr virus (EBV) and human papillomavirus (HPV) utilize CTCF to control virus gene expression and many studies have highlighted a role for CTCF in the persistence of these diverse oncogenic viruses. CTCF can both enhance and repress virus gene expression and in some cases CTCF increases the complexity of alternatively spliced transcripts. This review article will discuss the function of CTCF in the life cycle of DNA viruses in the context of known host cell CTCF functions.

Reprinted from *Viruses*. Cite as: Pentland, I.; Parish, J.L. Targeting CTCF to Control Virus Gene Expression: A Common Theme amongst Diverse DNA Viruses. *Viruses* **2015**, *7*, 3574–3585.

1. Introduction

CCCTC-binding factor (CTCF) is a ubiquitously expressed DNA binding protein that is highly conserved in bilaterian metazoans [1]. The protein contains 11 zinc fingers, 4 of which (zinc fingers 4–7) bind strongly to the core 12 base pair DNA sequence that is common in most CTCF binding sites [2,3]. It is thought that the remaining zinc fingers differentially bind to sequences up- and downstream of this central motif, allowing a large degree of flexibility and extension of the canonical binding site [4,5]. The exact number of CTCF binding sites within the human genome is not clear, but studies have shown up to 26,000 binding sites in human cells [3,6,7]. However, a more recent study has highlighted the potential for many more CTCF binding sites within the human genome (in the region of 300,000) whose occupancy depends on specific cell type and differentiation status [8]. CTCF binding appears to be enriched within intergenic spaces or in intronic regions of genes. Approximately 12% of CTCF binding sites are located within proximal promoters [8].

The association of CTCF with the human genome is important for genomic organisation and the control of gene expression. In particular, CTCF is essential for genomic imprinting, chromatin insulation, and transcriptional activation and repression. Furthermore, association of CTCF with specific regions of DNA has been linked to DNA looping, nucleosome organization, and the control of RNA Polymerase II (RNA Pol II) progression to co-ordinate co-transcriptional gene splicing events. Adding to the complexity of CTCF-mediated chromatin organization and the regulation of gene expression, DNA CpG methylation can influence CTCF DNA binding. Up to 41% of CTCF binding sites have been shown to be sensitive to CpG methylation [9]. Interestingly, inhibition of CTCF binding by DNA methylation has also been shown to alter gene-splicing events [10].

These important functions of CTCF in the control of host cell gene expression have been extensively reviewed elsewhere. We will therefore focus this article on emerging evidence that CTCF is utilised by diverse DNA viruses to control viral gene expression, genome organisation, virus replication and persistence.

2. CTCF-Mediated Virus Transcription Activation and Repression

The role of CTCF in transcriptional silencing and activation was first described in the 1990s [11–13]. These early studies demonstrated that CTCF binds to multiple sites within the *c-myc* gene promoter to repress transcription. A subsequent study showed that binding of CTCF to a core sequence located at the 5′ end of the chicken β-globin locus conferred strong enhancer blocking activity [14]. Furthermore, transcriptional control of the imprinted Igf2/H19 locus is mediated by CTCF binding to a differentially methylated region (DMR) within the imprinted control region (ICR). Methylation of the paternal ICR prevents CTCF binding, thus allowing downstream enhancers to act on the Igf2 promoter to facilitate Igf2 expression. Conversely, bound CTCF is present at the unmethylated maternal ICR, which blocks enhancers acting on the Igf2 promoter [15,16]. In this particular example, CTCF binding within the maternal allele blocks downstream enhancers from activating Igf2 expression by forming loops within the DNA that prevent interaction of the enhancer elements with the Igf2 promoter, thus promoting H19 expression from the maternal allele only [15,17,18].

The involvement of CTCF in the control of viral gene transcription has been demonstrated in several DNA viruses. In studies of Kaposi's sarcoma-associated herpesvirus (KSHV), CTCF was shown to associate with several regions within the viral genome, the strongest of these binding regions was located at an intergenic site between the divergent ORF73 and K14 open reading frames (ORFs), which are active in the latent and lytic phases of the virus life cycle, respectively. The association of CTCF at this strong binding region occurs in a cell cycle dependent manner, specifically during mid-S phase to repress transcription of lytic genes [19]. Mutation of this CTCF binding cluster disrupted the recruitment of the cohesin complex (described in more detail below) and caused an increase in lytic gene expression due to derepression of the promoter which drives K14 expression [20]. This result was later confirmed by siRNA-mediated depletion of CTCF which showed a specific increase in the early lytic gene expression including K14 and ORF74, but a greater increase in ORF57 and ORF6 was noted [21]. This increase in lytic gene expression caused by depletion of CTCF resulted in a 20–25 fold increase in virion production, leading the authors to propose CTCF as a host cell restriction factor for KSHV lytic replication. Interestingly, it has also been shown that cohesin and CTCF binding at the promoter region of ORF50/RTA in KSHV represses ORF50 expression which is required for latent reactivation, providing further evidence that CTCF and cohesin behave as repressors of lytic transcription [22]. The idea that CTCF may function as a host cell restriction factor for viral infections may also be true for human papillomavirus (HPV) as mutation of a single conserved CTCF binding site in HPV type 18 results in an increase in viral oncoprotein E6 and E7 transcription, causing hyperproliferation of epithelial tissues [23]. In addition, CTCF has been proposed as a restriction factor for human cytomegalovirus (hCMV) infection as it plays a major role in limiting major immediate early (MIE) gene expression [24].

Interestingly, dynamic binding of CTCF at some sites within the KSHV genome has been demonstrated, and rather than a global eviction of CTCF upon lytic cycle activation, CTCF was gradually reduced at the majority of binding sites but maintained at others [21]. In this study, only a subset of lytic KSHV genes were transcriptionally enhanced upon CTCF knockdown, illustrating the use of site-specific CTCF binding. The authors conclude that this contributes to a mechanism whereby CTCF initially acts as a stimulator of lytic gene expression and then subsequently acts as an inhibitor of the lytic gene expression. Although the precise mechanism of this is unclear, it is interesting that CTCF binding is so intimately linked to the switch in latent to lytic gene expression in the KSHV life cycle.

In the Epstein-Barr virus (EBV) genome, CTCF binds to a site between the viral origin of replication (OriP) and the C promoter (Cp). Deletion of this CTCF binding site results in an increase in EBNA2 transcription levels, which is interesting considering EBV latency types are distinguished by their expression of EBNA2 levels. Furthermore, there was more total CTCF protein and mRNA detected in the type-I EBV cells compared to type-III. Overall, the presence of CTCF binding in EBV was shown to negatively affect transcription at Cp [25]. Additionally, when the functional CTCF binding site upstream of another EBV promoter termed Qp was abrogated there was a reduction in initiation of Qp transcription, and instead an alternative promoter (Fp) upstream was preferentially utilized causing overall disruption of latency transcript expression [26]. Mutation of a CTCF site positioned within the EBV LMP1 and LMP2A region also resulted in a decrease in LMP1 and LMP2A transcript expression [27].

There is also evidence to support a role for CTCF in regulating adenovirus replication and late gene expression as siRNA-mediated depletion of CTCF represses late gene expression but had no effect on early gene expression [28]. In addition, a recent study investigating the interaction of CTCF with human cytomegalovirus (hCMV) showed that CTCF binding in the first intron of the major intermediate early (MIE) gene caused repression of MIE gene expression, likely through a transcriptional mechanism believed to involve the manipulation of RNA pol II function [24].

3. Chromatin Barrier Formation

CTCF is the major transcriptional insulator in mammals and, by binding to specific sites within the cellular genome, creates a physical barrier between active and repressive chromatin, forming chromatin boundaries. Genome-wide chromatin immunoprecipitation (ChIP)-Seq analysis has demonstrated a significant enrichment of CTCF binding at regions located between active and repressive chromatin domains [7] and depletion of CTCF appears to cause heterochromatin spread [29,30]. The exact mechanism of how CTCF creates these boundaries remains elusive but it has been suggested that CTCF recruits a variety of binding partners and is differentially post-translationally modified to exert these effects (reviewed by [31]). During Herpes simplex virus type 1 (HSV-1) latent infection, the viral genome is organized in distinct transcriptionally repressed and active chromatin domains characterized by histone H3 hypo- and hyper-acetylation, respectively [32,33]. This separation of the active latency associated transcript (LAT) region and inactive ICP0 lytic promoter is controlled by a cluster of CTCF binding sites at the 3′ end of the LAT region [34,35]. During latent infection, CTCF binding the LAT region insulates the LAT enhancer from exerting effects on the adjacent ICP0 lytic promoter [34]. Binding of CTCF to the LAT region therefore creates a boundary which separates inactive chromatin in the ICP0 lytic gene region from an active LAT promoter region [35]. In the study by Tempera *et al.* [26] described above, mutation of the high affinity CTCF binding site immediately upstream of Qp in EBV resulted in reduced Qp activity but increased Cp and Fp activity. However, long-term culture of cells harboring this mutant bacmid resulted in no detectable Qp activity. This loss of Qp activity was associated with increased histone H3 lysine 9 trimethylation (H3K9me3) and CpG DNA methylation, indicating spread of inactive chromatin marks from the repressed region situated upstream of the mutated CTCF binding sites. Similarly, when the CTCF binding site in the EBV LMP1 and LMP2A region was abrogated there was a noticeable increase in CpG methylation at the corresponding promoter control sites [27]. Conversely, despite CTCF binding between the alternative replication initiation site (rep*) and Cp in the EBV genome of type III Burkitt's lymphoma (BL) cell lines, a high level of CpG methylation has been detected in this region, suggesting CTCF may not be able to prevent methylation across all target sequences [36].

These data elegantly highlight the ability of CTCF to function as a chromatin barrier within the context of a viral genome. Whether CTCF functions as a barrier to prevent the spread of inactive chromatin in small DNA viruses remains to be determined but it is interesting to note that loss of CpG methylation within the HPV16 genome has been demonstrated upon differentiation of host cells [37]. While the HPV life cycle does not have a lytic phase as such, capsid protein production is only induced in differentiated epithelial cells, suggesting that a molecular switch in early to late (capsid) gene expression controls completion of the virus life cycle. Whether this is linked to CTCF binding and epigenetic boundary formation that is altered upon host cell differentiation has yet to be determined.

4. Chromatin Loop Formation

One of the ways in which CTCF contributes to genome wide organization of chromatin is in the formation of loops that mediate long-range interactions between distant loci. Exactly how CTCF co-ordinates loop formation is not entirely understood. Evidence suggests that CTCF is able to form homodimers and complexes with proteins known to associate with insulator sites, such as the nucleolar protein nucleophosmin, providing evidence that loop formation is mediated by CTCF-associated protein complexes [38]. In addition, a number of studies have demonstrated co-localization of CTCF and the cohesion complex at thousands of genomic sites [39–41]. The cohesin complex is composed of four core subunits (Smc1, Smc3, Scc1/Rad21, Scc3), which form a ring-like structure that encircles DNA strands to facilitate sister chromatid cohesion during mitotic segregation (reviewed by [42]). Aside from its role in sister chromatid cohesion, there is evidence that co-localization of CTCF and cohesin at specific sites can facilitate the formation and stabilization of chromatin loops and thereby influence gene expression [43,44]. The role of CTCF in mediating long-range chromatin interactions at the Igf2/H19 locus (described above) and the β-globin locus has been well described. In the β-globin locus, multiple CTCF binding sites are known to interact [45] and a large (>50 kilobasepair) chromatin loop is formed between two CTCF binding sites [46], thus highlighting the possibility of CTCF mediated looping in the control of virus transcription.

In support of this hypothesis, CTCF has been shown to co-localize with cohesin in the KSHV episome on a region of the major latency control transcript during latent infection. Subsequent deletion of this CTCF binding site resulted in a loss of cohesin binding and a reduction in stable colony formation [20]. Co-incidentally, deletion of the CTCF binding site and loss of cohesin binding within the major latency control region of KSHV caused derepression of lytic gene expression, notably the K14/ORF74 transcript. This is similar to the role of Scc1 in transcriptional repression at the *c-myc* locus [20]. Interestingly, knockdown of Scc1 alone was shown to enhance KSHV virion production even more than CTCF knockdown alone, suggesting that cohesin has an even greater influence than CTCF on the control of lytic gene expression and that while these proteins co-locate in some areas of the viral genome, distinct localization of CTCF and cohesin may also occur [21]. Similarly, CTCF and cohesin are highly associated within the control region of the EBV latency membrane proteins LMP-1 and LMP-2A, with this co-occupancy strongly linked with DNA loop formation with the enhancer with the origin of replication (OriP) [27]. These studies highlight the role of CTCF and cohesin co-localization to regulate the complex process of gene expression and chromatin organization, through mediating DNA looping and structural changes of the chromatin.

To address the role of CTCF in mediating chromatin loop formation in EBV, chromatin conformation capture (3C) was used to reveal loop formation between CTCF binding sites at the type I latency promoter Cp and type III latency promoter Qp with the OriP enhancer. As predicted, loop formation was demonstrated between OriP and Qp during type I latency and between OriP and Cp during type III latency [47]. Subsequent deletion of the Qp CTCF binding site led to a loss in loop formation with OriP and a switch to Cp transcription instead. Furthermore, abrogation of CTCF binding at Cp led to a loss in both Qp-OriP and Cp-OriP loop associations, thus demonstrating a critical role for CTCF in this looping function and subsequent regulation of gene expression [47]. Similarly, CTCF-mediated loop formation between OriP and the LMP1/2 gene region of EBV has been described [48]. Abrogation of CTCF binding within OriP disrupts looping with LMP1 and LMP2A control regions leading to an increase in H3K9me3 and CpG methylation in the LMP1 and LMP2A promoter regions and upregulation of LMP2B transcription [27]. It is plausible that the ability of CTCF to confer loop formation between different sequences and regulatory elements can in part explain its varying effects on the control of gene transcription in the EBV life cycle. Furthermore, whilst a DNA looping mechanism may be advantageous for the genomic regulation of larger DNA viruses in order to assemble distal elements, the potential role of looping in much smaller DNA viruses such as adenovirus and HPV remains to be determined. It is likely however that DNA looping may not be required for smaller genome viruses and instead CTCF confers genomic regulation through alternative mechanisms. One potential counter-argument to this, however, has been provided by a study by Mehta and colleagues [49] in which the cohesion subunit Smc1 was shown to bind to the late gene region of the HPV31 genome in a CTCF-dependent manner. At least a subset of the Smc1 protein associated with HPV31 genomes is phosphorylated as part of the Ataxia-Telangiectasia Mutated (ATM)-dependent DNA damage response and is recruited to the HPV31 genome to support viral genome amplification. Whether recruitment of Smc1 to HPV genomes results in loop formation to regulate viral gene expression and/or genome amplification is an interesting question.

5. Nucleosome Positioning and RNA Polymerase II Progression

It has been demonstrated that CTCF can bind to the large subunit of RNA Pol II, which is needed for transcriptional initiation and elongation, and recruit RNA Pol II to target genes [50]. A recent study has also shown that, in co-operation with the general transcription factor TFII-I, CTCF influences the regulation of RNA pol II progression, downstream of recruitment to a pre-initiation complex. TFII-I and CTCF appear to facilitate the recruitment of the cyclin dependent kinase 8 (CDK8) complex, which phosphorylates RNA Pol II within its C-terminal domain at serine 5 to initiate mRNA synthesis [51]. In addition, binding of CTCF within exons has been shown to mediate RNA Pol II pausing and promote inclusion of weak upstream exons *via* co-transcriptional RNA splicing [10]. The potential of CTCF to regulate nucleosome positioning and RNA Pol II function as a mechanism for transcriptional control in DNA viruses has been investigated using a KSHV model system. Here, abrogation of CTCF binding in the first intron of the KSHV latency-associated multicistronic transcript resulted in an elevation or ORF73, ORF72, ORF71 and viral miRNA production but a concomitant decrease in K12 and ORF69 production, consistent with a previous study demonstrating loop formation between the CTCF binding site and the 3′ end of the K12 transcript [52,53]. Experiments also showed that CTCF was required for RNA Pol II programming, since loss of CTCF binding resulted in an increase in phosphorylation of RNA Pol II at serine 5 at the 5′ end of the latency transcript and a decrease in RNA Pol II S5 at the K12 ORF while causing the displacement of nucleosomes within the intron upstream of ORF73 [52]. It appears from these studies that rather than acting as a physical roadblock to RNA Pol II, CTCF may serve to regulate RNA Pol II within the intron upstream of ORF73 and mediate promoter selection by loop formation within the latency transcript unit.

Studies in KSHV have also shown that CTCF binding influences alternative splicing of the major latency transcripts [52], although an earlier study showed that CTCF recruitment to the KSHV genome was cell cycle dependent while alternative transcript splicing was not [19]. It has been proposed that CTCF binding to the EBV LMP region may facilitate RNA Pol II pausing and subsequent gene splicing events [27], although this is yet to be formally proven. Abrogation of CTCF binding by mutation of a binding site within the early gene region of the small DNA virus HPV type 18, in which transcripts are extensively alternatively spliced in order to generate multiple mRNAs and increase the repertoire of expressed proteins, causes a significant alteration of major splicing events important for the expression of early proteins [23]. However, the mechanistic underpinnings of CTCF function in the control of HPV transcript splicing remain to be determined.

6. Viral Genome Persistence

Cohesin and CTCF organize genomic DNA to create discrete nuclear regions to support efficient DNA replication [54]. Whether this is also true for viral DNA replication remains to be determined but the replication and stability of some DNA viruses is affected by CTCF recruitment. In some instances the mutation of specific CTCF binding sites has been shown to alter episomal maintenance. The CTCF binding site in Qp in the EBV genome and in the ORF73/LANA promoter region of the primate gamma-2 herpesvirus, herpesvirus saimiri (HVS) is important for maintaining viral episomes, as mutation to prevent CTCF binding led to a greater loss of episomes compared to the wild type genome for both of these viruses [26,55], although the effect of viral genome loss in both of these studies is more likely due to a reduction in the expression of viral transcripts required for latency. Similarly, disruption of CTCF binding sites in the KSHV latency control region also had a negative impact on stable episome maintenance during latency [20]. On the other hand lymphoblastoid cell lines (LCLs) harboring EBV bacmids with mutated CTCF binding sites at the LMP1 and LMP2 overlapping region revealed increased episome copy number compared to the cells containing wild type EBV bacmids [27]. While abrogation of CTCF binding within the early E2 ORF of HPV18 does not affect the maintenance of viral episomes (Pentland, Parish unpublished), mutating CTCF binding sites within the late L2 region of HPV31 caused a loss of viral episomes, indicating that CTCF may play a role in the replication or maintenance of HPV31 genomes [49]. These contrasting findings may demonstrate that episome maintenance is affected when specific CTCF binding sites are mutated in HPV and the location of these sites is likely to be important for determining the fate of viral maintenance. It has yet to be determined whether CTCF mediates integration of viruses into the host cell genome, or indeed whether CTCF participates in the reactivation of persistent, latent viral infections.

7. Conclusions

It is clear that CTCF plays an integral role in virus genome organization and gene regulation, in both large and small DNA viruses. Studies have demonstrated that CTCF organizes viral genomes by creating boundaries between active and inactive chromatin and can regulate viral gene expression by altering RNA pol II recruitment and progression, and nucleosome positioning. The co-localization of CTCF with cohesin in EBV and KSHV has highlighted the complex regulation of chromatin structure and the CTCF mediated DNA looping required to confer different genomic outcomes. The ability of CTCF to alter nucleosome positioning and RNA Pol II function adds further evidence to its role in altering gene splicing events and chromatin arrangement, in order to facilitate gene expression across the genome. It is unlikely that CTCF mediates separate functions alone but instead uses them in concert to co-ordinate complex genomic processes. It is intriguing that CTCF recruitment to some sites within viral genomes appears to be restrictive to virus replication, while recruitment to other sites results in activation of virus gene expression. The completion of a virus life cycle is without doubt a finely tuned balancing act and it is likely that DNA viruses have evolved to utilize CTCF to maintain this balance for efficient life cycle completion; CTCF thereby behaves as a restriction factor and an activating factor depending on virus life cycle stage and the specific context of CTCF binding.

The continued study of viral interactions with CTCF will allow us to further understand the functions and implications of the vast array of CTCF-mediated events.

Acknowledgments

Joanna L. Parish is funded by an independent Royal Society University Research Fellowship (UF110010). Ieisha Pentland is funded by a Cancer Research UK PhD studentship awarded to the University of Birmingham.

Author Contributions

Ieisha Pentland and Joanna L. Parish co-wrote the manuscript.

Conflicts of Interest

The authors declare no conflict of interest.

References

1. Heger, P.; Marin, B.; Bartkuhn, M.; Schierenberg, E.; Wiehe, T. The chromatin insulator CTCF and the emergence of metazoan diversity. *Proc. Natl. Acad. Sci. USA* **2012**, *109*, 17507–17512. [CrossRef] [PubMed]
2. Renda, M.; Baglivo, I.; Burgess-Beusse, B.; Esposito, S.; Fattorusso, R.; Felsenfeld, G.; Pedone, P.V. Critical DNA binding interactions of the insulator protein CTCF: A small number of zinc fingers mediate strong binding, and a single finger-DNA interaction controls binding at imprinted loci. *J. Biol. Chem.* **2007**, *282*, 33336–33345. [CrossRef] [PubMed]
3. Kim, T.H.; Abdullaev, Z.K.; Smith, A.D.; Ching, K.A.; Loukinov, D.I.; Green, R.D.; Zhang, M.Q.; Lobanenkov, V.V.; Ren, B. Analysis of the vertebrate insulator protein CTCF-binding sites in the human genome. *Cell* **2007**, *128*, 1231–1245. [CrossRef] [PubMed]
4. Schmidt, D.; Schwalie, P.C.; Wilson, M.D.; Ballester, B.; Gonçalves, Â.; Kutter, C.; Brown, G.D.; Marshall, A.; Flicek, P.; Odom, D.T. Waves of retrotransposon expansion remodel genome organization and CTCF binding in multiple mammalian lineages. *Cell* **2012**, *148*, 335–348. [CrossRef] [PubMed]
5. Nakahashi, H.; Kwon, K.-R.K.; Resch, W.; Vian, L.; Dose, M.; Stavreva, D.; Hakim, O.; Pruett, N.; Nelson, S.; Yamane, A.; *et al.* A genome-wide map of CTCF multivalency redefines the CTCF code. *Cell Rep.* **2013**, *3*, 1678–1689. [CrossRef] [PubMed]
6. Jothi, R.; Cuddapah, S.; Barski, A.; Cui, K.; Zhao, K. Genome-wide identification of *in vivo* protein-DNA binding sites from ChIP-Seq data. *Nucleic Acid Res.* **2008**, *36*, 5221–5231. [CrossRef] [PubMed]
7. Cuddapah, S.; Jothi, R.; Schones, D.E.; Roh, T.-Y.; Cui, K.; Zhao, K. Global analysis of the insulator binding protein CTCF in chromatin barrier regions reveals demarcation of active and repressive domains. *Genome Res.* **2009**, *19*, 24–32. [CrossRef] [PubMed]

8. Chen, H.; Tian, Y.; Shu, W.; Bo, X.; Wang, S. Comprehensive identification and annotation of cell type-specific and ubiquitous CTCF-binding sites in the human genome. *PLoS ONE* **2012**, *7*, e41374. [CrossRef] [PubMed]

9. Wang, H.; Maurano, M.T.; Qu, H.; Varley, K.E.; Gertz, J.; Pauli, F.; Lee, K.; Canfield, T.; Weaver, M.; Sandstrom, R.; Thurman, R.E.; Kaul, R.; Myers, R.M.; Stamatoyannopoulos, J.A. Widespread plasticity in CTCF occupancy linked to DNA methylation. *Genome Res.* **2012**, *22*, 1680–1688. [CrossRef] [PubMed]

10. Shukla, S.; Kavak, E.; Gregory, M.; Imashimizu, M.; Shutinoski, B.; Kashlev, M.; Oberdoerffer, P.; Sandberg, R.; Oberdoerffer, S. CTCF-promoted RNA polymerase II pausing links DNA methylation to splicing. *Nature* **2011**, *479*, 74–79. [CrossRef] [PubMed]

11. Lobanenkov, V.V.; Nicolas, R.H.; Adler, V.V.; Paterson, H.; Klenova, E.M.; Polotskaja, A.V.; Goodwin, G.H. A novel sequence-specific DNA binding protein which interacts with three regularly spaced direct repeats of the CCCTC-motif in the 5′-flanking sequence of the chicken *c-myc* gene. *Oncogene* **1990**, *5*, 1743–1753. [PubMed]

12. Klenova, E.M.; Nicolas, R.H.; Paterson, H.F.; Carne, A.F.; Heath, C.M.; Goodwin, G.H.; Neiman, P.E.; Lobanenkov, V.V. CTCF, a conserved nuclear factor required for optimal transcriptional activity of the chicken *c-myc* gene, is an 11-Zn-finger protein differentially expressed in multiple forms. *Mol. Cell Biol.* **1993**, *13*, 7612–7624. [PubMed]

13. Filippova, G.N.; Fagerlie, S.; Klenova, E.M.; Myers, C.; Dehner, Y.; Goodwin, G.; Neiman, P.E.; Collins, S.J.; Lobanenkov, V.V. An exceptionally conserved transcriptional repressor, CTCF, employs different combinations of zinc fingers to bind diverged promoter sequences of avian and mammalian *c-myc* oncogenes. *Mol. Cell Biol.* **1996**, *16*, 2802–2813. [PubMed]

14. Bell, A.; West, A.; Felsenfeld, G. The protein CTCF is required for the enhancer blocking activity of vertebrate insulators. *Cell* **1999**, *98*, 387–396. [CrossRef]

15. Bell, A.C.; Felsenfeld, G. Methylation of a CTCF-dependent boundary controls imprinted expression of the *Igf2* gene. *Nature* **2000**, *405*, 482–485. [PubMed]

16. Hark, A.T.; Schoenherr, C.J.; Katz, D.J.; Ingram, R.S.; Levorse, J.M.; Tilghman, S.M. CTCF mediates methylation-sensitive enhancer-blocking activity at the H19/Igf2 locus. *Nature* **2000**, *405*, 486–489. [PubMed]

17. Murrell, A.; Heeson, S.; Reik, W. Interaction between differentially methylated regions partitions the imprinted genes *Igf2* and *H19* into parent-specific chromatin loops. *Nat. Genet* **2004**, *36*, 889–893. [CrossRef] [PubMed]

18. Ling, J.Q.; Li, T.; Hu, J.F.; Vu, T.H.; Chen, H.L.; Qiu, X.W.; Cherry, A.M.; Hoffman, A.R. CTCF mediates interchromosomal colocalization between *Igf2/H19* and *Wsb1/Nf1*. *Science* **2006**, *312*, 269–272. [CrossRef] [PubMed]

19. Kang, H.; Lieberman, P.M. Cell cycle control of Kaposi's sarcoma-associated herpesvirus latency transcription by CTCF-cohesin interactions. *J. Virol.* **2009**, *83*, 6199–6210. [CrossRef] [PubMed]

20. Stedman, W.; Kang, H.; Lin, S.; Kissil, J.L.; Bartolomei, M.S.; Lieberman, P.M. Cohesins localize with CTCF at the KSHV latency control region and at cellular *c-myc* and *H19/Igf2* insulators. *EMBO J.* **2008**, *27*, 654–666. [CrossRef] [PubMed]

21. Li, D.-J.; Verma, D.; Mosbruger, T.; Swaminathan, S. CTCF and Rad21 act as host cell restriction factors for Kaposi's sarcoma-associated herpesvirus (KSHV) lytic replication by modulating viral gene transcription. *PLoS Pathog.* **2014**, *10*, e1003880. [CrossRef] [PubMed]

22. Chen, H.-S.; Wikramasinghe, P.; Showe, L.; Lieberman, P.M. Cohesins repress Kaposi's sarcoma-associated herpesvirus immediate early gene transcription during latency. *J. Virol.* **2012**, *86*, 9454–9464. [CrossRef] [PubMed]

23. Paris, C.; Pentland, I.; Groves, I.; Roberts, D.C.; Powis, S.J.; Coleman, N.; Roberts, S.; Parish, J.L. CCCTC-binding factor recruitment to the early region of the human papillomavirus type 18 genome regulates viral oncogene expression. *J. Virol.* **2015**, *89*, 4770–4785. [CrossRef] [PubMed]

24. Martinez, F.P.; Cruz, R.; Lu, F.; Plasschaert, R.; Deng, Z.; Rivera-Molina, Y.A.; Bartolomei, M.S.; Lieberman, P.M.; Tang, Q. CTCF binding to the first intron of the major immediate-early (*MIE*) gene of human cytomegalovirus (HCMV) negatively regulates *MIE* gene expression and HCMV replication. *J. Virol.* **2014**, *88*, 7389–7401. [CrossRef] [PubMed]

25. Chau, C.M.; Zhang, X.-Y.; McMahon, S.B.; Lieberman, P.M. Regulation of EPSTEIN-BARR virus latency type by the chromatin boundary factor CTCF. *J. Virol.* **2006**, *80*, 5723–5732. [CrossRef] [PubMed]

26. Tempera, I.; Wiedmer, A.; Dheekollu, J.; Lieberman, P.M. CTCF prevents the epigenetic drift of EBV latency promoter Qp. *PLoS Pathog.* **2010**, *6*, e1001048. [CrossRef] [PubMed]

27. Chen, H.S.; Martin, K.A.; Lu, F.; Lupey, L.N.; Mueller, J.M.; Lieberman, P.M.; Tempera, I. Epigenetic deregulation of the LMP1/LMP2 locus of Epstein-Barr virus by mutation of a single CTCF-cohesin binding site. *J. Virol.* **2014**, *88*, 1703–1713. [CrossRef] [PubMed]

28. Komatsu, T.; Sekiya, T.; Nagata, K. DNA replication-dependent binding of CTCF plays a critical role in adenovirus genome functions. *Sci. Rep.* **2013**, *3*, 2187. [CrossRef] [PubMed]

29. Cho, D.H.; Thienes, C.P.; Mahoney, S.E.; Analau, E.; Filippova, G.N.; Tapscott, S.J. Antisense transcription and heterochromatin at the DM1 CTG repeats are constrained by CTCF. *Mol. Cell* **2005**, *20*, 483–489. [CrossRef] [PubMed]

30. Filippova, G.N.; Thienes, C.P.; Penn, B.H.; Cho, D.H.; Hu, Y.J.; Moore, J.M.; Klesert, T.R.; Lobanenkov, V.V.; Tapscott, S.J. CTCF-binding sites flank CTG/CAG repeats and form a methylation-sensitive insulator at the DM1 locus. *Nat. Genet* **2001**, *28*, 335–343. [CrossRef] [PubMed]

31. Phillips-Cremins, J.E.; Corces, V.G. Chromatin insulators: Linking genome organization to cellular function. *Mol. Cell* **2013**, *50*, 461–474. [CrossRef] [PubMed]

32. Kubat, N.J.; Tran, R.K.; McAnany, P.; Bloom, D.C. Specific histone tail modification and not DNA methylation is a determinant of herpes simplex virus type 1 latent gene expression. *J. Virol.* **2004**, *78*, 1139–1149. [CrossRef] [PubMed]

33. Kubat, N.J.; Amelio, A.L.; Giordani, N.V.; Bloom, D.C. The herpes simplex virus type 1 latency-associated transcript (LAT) enhancer/rcr is hyperacetylated during latency independently of LAT transcription. *J. Virol.* **2004**, *78*, 12508–12518. [CrossRef] [PubMed]

34. Amelio, A.L.; McAnany, P.K.; Bloom, D.C. A Chromatin insulator-like element in the Herpes Simplex virus type 1 latency-associated transcript region binds CCCTC-binding factor and displays enhancer-blocking and silencing activities. *J. Virol.* **2006**, 2358–2368. [CrossRef] [PubMed]

35. Chen, Q.; Lin, L.; Smith, S.; Huang, J.; Berger, S.L.; Zhou, J. CTCF-dependent chromatin boundary element between the latency-associated transcript and ICP0 promoters in the Herpes Simplex virus type 1 genome. *J. Virol.* **2007**, *81*, 5192–5201. [CrossRef] [PubMed]

36. Salamon, D.; Banati, F.; Koroknai, A.; Ravasz, M.; Szenthe, K.; Bathori, Z.; Bakos, A.; Niller, H.H.; Wolf, H.; Minarovits, J. Binding of CCCTC-binding factor *in vivo* to the region located between Rep* and the C promoter of Epstein-Barr virus is unaffected by CpG methylation and does not correlate with Cp activity. *J. Gen. Virol.* **2009**, *90*, 1183–1189. [CrossRef] [PubMed]

37. Kim, K.; Garner-Hamrick, P.A.; Fisher, C.; Lee, D.; Lambert, P.F. Methylation patterns of papillomavirus DNA, its influence on E2 function, and implications in viral infection. *J. Virol.* **2003**, *77*, 12450–12459. [CrossRef] [PubMed]

38. Yusufzai, T.M.; Tagami, H.; Nakatani, Y.; Felsenfeld, G. CTCF tethers an insulator to subnuclear sites, suggesting shared insulator mechanisms across species. *Mol. Cell* **2004**, *13*, 291–298. [CrossRef]

39. Wendt, K.S.; Yoshida, K.; Itoh, T.; Bando, M.; Koch, B.; Schirghuber, E.; Tsutsumi, S.; Nagae, G.; Mishiro, T.; Yahata, K.; *et al.* Cohesin mediates transcriptional insulation by CCCTC-binding factor. *Nature* **2008**, *451*, 796–803. [CrossRef] [PubMed]

40. Parelho, V.; Hadjur, S.; Spivakov, M.; Leleu, M.; Sauer, S.; Gregson, H.C.; Jarmuz, A.; Canzonetta, C.; Webster, Z.; Nesterova, T.; *et al.* Cohesins functionally associate with CTCF on mammalian chromosome arms. *Cell* **2008**, *132*, 422–433. [CrossRef] [PubMed]

41. Rubio, E.D.; Reiss, D.J.; Welcsh, P.L.; Disteche, C.M.; Filippova, G.N.; Baliga, N.S.; Aebersold, R.; Ranish, J.A.; Krumm, A. CTCF physically links cohesin to chromatin. *Proc. Natl. Acad. Sci. USA* **2008**, *105*, 8309–8314. [CrossRef] [PubMed]

42. Feeney, K.M.; Wasson, C.W.; Parish, J.L. Cohesin: A regulator of genome integrity and gene expression. *Biochem. J.* **2010**, *428*, 147–161.

43. Nativio, R.; Wendt, K.S.; Ito, Y.; Huddleston, J.E.; Uribe-Lewis, S.; Woodfine, K.; Krueger, C.; Reik, W.; Peters, J.-M.; Murrell, A. Cohesin is required for higher-order chromatin conformation at the imprinted IGF2-H19 locus. *PLoS Genet.* **2009**, *5*, e1000739. [CrossRef] [PubMed]

44. Hou, C.; Dale, R.; Dean, A. Cell type specificity of chromatin organization mediated by CTCF and cohesin. *Proc. Natl. Acad. Sci. USA* **2010**, *107*, 3651–3656. [CrossRef] [PubMed]

45. Splinter, E.; Heath, H.; Kooren, J.; Palstra, R.-J.; Klous, P.; Grosveld, F.; Galjart, N.; de Laat, W. CTCF mediates long-range chromatin looping and local histone modification in the beta-globin locus. *Genes Dev.* **2006**, *20*, 2349–2354. [CrossRef] [PubMed]

46. Hou, C.; Zhao, H.; Tanimoto, K.; Dean, A. CTCF-dependent enhancer-blocking by alternative chromatin loop formation. *Proc. Natl. Acad. Sci. USA* **2008**, *105*, 20398–20403. [CrossRef] [PubMed]

47. Tempera, I.; Klichinsky, M.; Lieberman, P.M. EBV latency types adopt alternative chromatin conformations. *PLoS Pathog.* **2011**, *7*, e1002180. [CrossRef] [PubMed]

48. Arvey, A.; Tempera, I.; Tsai, K.; Chen, H.-S.; Tikhmyanova, N.; Klichinsky, M.; Leslie, C.; Lieberman, P. An atlas of the Epstein-Barr virus transcriptome and epigenome reveals host-virus regulatory interactions. *Cell Host Microbe* **2012**, *12*, 233–245. [CrossRef] [PubMed]

49. Mehta, K.; Gunasekharan, V.; Satsuka, A.; Laimins, L.A. Human papillomaviruses activate and recruit SMC1 cohesin proteins for the differentiation-dependent life cycle through association with CTCF insulators. *PLoS Pathog.* **2015**, *11*, e1004763. [CrossRef] [PubMed]

50. Chernukhin, I.; Shamsuddin, S.; Kang, S.Y.; Bergström, R.; Kwon, Y.-W.; Yu, W.; Whitehead, J.; Mukhopadhyay, R.; Docquier, F.; Farrar, D.; *et al.* CTCF interacts with and recruits the largest subunit of RNA polymerase II to CTCF target sites genome-wide. *Mol. Cell Biol.* **2007**, *27*, 1631–1648. [CrossRef] [PubMed]

51. Pena-Hernandez, R.; Marques, M.; Hilmi, K.; Zhao, T.; Saad, A.; Alaoui-Jamali, M.A.; del Rincon, S.V.; Ashworth, T.; Roy, A.L.; Emerson, B.M.; Witcher, M. Genome-wide targeting of the epigenetic regulatory protein CTCF to gene promoters by the transcription factor TFII-I. *Proc. Natl. Acad. Sci. USA* **2015**, *112*, E677–E686. [CrossRef] [PubMed]

52. Kang, H.; Cho, H.; Sung, G.H.; Lieberman, P.M. CTCF regulates kaposi's sarcoma-associated herpesvirus latency transcription by nucleosome displacement and rna polymerase programming. *J. Virol.* **2013**, *87*, 1789–1799. [CrossRef] [PubMed]

53. Kang, H.; Wiedmer, A.; Yuan, Y.; Robertson, E.; Lieberman, P.M. Coordination of KSHV latent and lytic gene control by CTCF-cohesin mediated chromosome conformation. *PLoS Pathog.* **2011**, *7*, e1002140. [CrossRef] [PubMed]

54. Guillou, E.; Ibarra, A.; Coulon, V.; Casado-Vela, J.; Rico, D.; Casal, I.; Schwob, E.; Losada, A.; Méndez, J. Cohesin organizes chromatin loops at DNA replication factories. *Genes Dev.* **2010**, *24*, 2812–2822. [CrossRef] [PubMed]

55. Zielke, K.; Full, F.; Teufert, N.; Schmidt, M.; Muller-Fleckenstein, I.; Alberter, B.; Ensser, A. The insulator protein CTCF binding sites in the orf73/LANA promoter region of herpesvirus saimiri are involved in conferring episomal stability in latently infected human T cells. *J. Virol.* **2012**, *86*, 1862–1873. [CrossRef] [PubMed]

Regulation of the Wnt/β-Catenin Signaling Pathway by Human Papillomavirus E6 and E7 Oncoproteins

Jesus Omar Muñoz Bello, Leslie Olmedo Nieva, Adriana Contreras Paredes, Alma Mariana Fuentes Gonzalez, Leticia Rocha Zavaleta and Marcela Lizano

Abstract: Cell signaling pathways are the mechanisms by which cells transduce external stimuli, which control the transcription of genes, to regulate diverse biological effects. In cancer, distinct signaling pathways, such as the Wnt/β-catenin pathway, have been implicated in the deregulation of critical molecular processes that affect cell proliferation and differentiation. For example, changes in β-catenin localization have been identified in Human Papillomavirus (HPV)-related cancers as the lesion progresses. Specifically, β-catenin relocates from the membrane/cytoplasm to the nucleus, suggesting that this transcription regulator participates in cervical carcinogenesis. The E6 and E7 oncoproteins are responsible for the transforming activity of HPV, and some studies have implicated these viral oncoproteins in the regulation of the Wnt/β-catenin pathway. Nevertheless, new interactions of HPV oncoproteins with cellular proteins are emerging, and the study of the biological effects of such interactions will help to understand HPV-related carcinogenesis. This review addresses the accumulated evidence of the involvement of the HPV E6 and E7 oncoproteins in the activation of the Wnt/β-catenin pathway.

Reprinted from *Viruses*. Cite as: Bello, J.O.M.; Nieva, L.O.; Paredes, A.C.; Gonzalez, A.M.F.; Zavaleta, L.R.; Lizano, M. Regulation of the Wnt/β-Catenin Signaling Pathway by Human Papillomavirus E6 and E7 Oncoproteins. *Viruses* **2015**, *7*, 4734–4755.

1. Introduction

Signaling pathways are the mechanisms by which cells decide their fate and communicate with other cells and their environment. The binding of ligands to cell receptors can activate protein cascades and consequently affect gene transcription levels. Via these complex processes, cells transform external stimuli into biochemical signals that control biological effects, such as proliferation, differentiation, and death.

Many signaling pathways have been identified as being deregulated in cancer. Consequently, numerous elements targeting these pathways have been proposed as therapeutic targets. Consistent alterations of some important pathways controlling cell proliferation and apoptosis, such as PI3K/Akt, ERK/MAPK, Notch, and Wnt/β-catenin, have been identified in different types of cancer. In particular, the activation of the Wnt signaling pathway has been implicated in osteosarcoma [1], hepatocellular carcinoma [2], colorectal cancer [3], and breast cancer [4]. More recently, this signaling pathway was also implicated in oral cavity, oropharyngeal [5], and cervical cancers [6,7].

Cervical cancer is the fourth most common cancer in women worldwide and is one of the leading causes of cancer death in women in developing countries [8]. Persistent infection with Human Papillomavirus (HPV) is a necessary factor for cervical cancer development [9]. HPV is also associated with other pathologies, such as head and neck [10] and anal cancers [11]. HPV types linked to cancer are those classified as high-risk HPV (HR-HPV), whose viral oncogenes interact and regulate the function of several cellular proteins.

This review addresses the participation of the Wnt/β-catenin signaling pathway in HPV-related cancers and the possible mechanisms by which HPV E6 and E7 oncoproteins induce the activation of this pathway.

2. Wnt/β-Catenin Cell Signaling Pathway

The Wnt signaling pathway is involved in development, proliferation [12], differentiation [13], adhesion [14], and cellular polarity [15]. The term Wnt, which was adopted in 1991, includes a family of genes that encode secretory glycoproteins. Wnt is an acronym of homologous wingless (wg) and Int-1, which had been described in the fly and mouse, respectively [16]. In 1982, Nusse and Varmus found that the mouse mammary tumor virus (MMTV) promotes mammary carcinogenesis in mice by inserting itself in a specific gene of the host genome [17]. They called this gene Int-1, and its nucleotide and amino acid sequences were obtained in 1984 [18]. Later, in 1987, the wingless gene in *Drosophila melanogaster* proved to be a homologue of Int-1 [19].

Currently, 11 receptors that are members of the Frizzled (Fz) family have been identified in humans. These receptors include Fz1 to Fz10 and Smo, as well as the two co-receptors LRP 5 and 6, and all of these receptors are responsible for Wnt signaling activation. Moreover, 19 Wnt ligands have been described for these receptors: Wnt1, 2, 2b, 3, 3a, 4, 5a, 5b, 6, 7a, 7b, 8a, 8b, 9a, 9b, 10a, 10b, 11, and 16 [20].

At least three signal transduction pathways activated by Wnt ligands are known, namely the canonical Wnt/β-catenin pathway and two non-canonical pathways: the planar cell polarity pathway (Wnt/PCP) and the Wnt/Ca^{2+} pathway. Moreover, the activation of the different pathways is ligand-specific, and the primary ligands that activate the canonical pathway are Wnt1, 2 [21], 3, 3a [22], 7a [23], 8 [24], and 10b [25,26]. The activation of the non-canonical pathways is mediated by Wnt4 [27], 5a [28,29], and 11 [30] ligands. However, diverse Wnt ligands have been shown to elicit various effects when binding to the same Fz receptor [31].

The non-canonical Wnt/PCP, also known as the Wnt/JNK pathway, is important in various processes including wound healing [32], the correct development of the neural tube [33], motility, and the modulation of cellular morphology [34]. These events are all generated by the reorganization of the actin cytoskeleton. Some of the main proteins involved in the transduction of the extracellular signal generated by Wnt/PCP are vangl2, celsr1-3 [35], Dvl, JNK, PKC [36], Rac, and RhoA [37].

In the Wnt/Ca^{2+} pathway, secondary messengers, such as IP3 and DAG, liberate calcium ions from the endoplasmic reticulum [29] and subsequently activate CaMKII [38] and PKC [39].

The processes that are triggered by the activation of this non-canonical pathway include the following: the regulation of convergent extension movements [40], the reorganization of the actin cytoskeleton [41], the modulation of cell motility [42], and the contribution to the inflammatory response [43].

The Wnt canonical signaling pathway is the best understood Wnt signaling cascade. In the absence of Wnt ligands (OFF-STATE), β-catenin is mainly located at cellular junctions.

Nevertheless, a small amount remains in the cytoplasm and binds to a complex responsible for the degradation of β-catenin via the proteasome. This degradation complex consists of the scaffold protein Axin which recruits essential elements during this process such as GSK3β [44], CK1 [45], APC [46], YAP/TAZ, and β-TrCP [47]. CK1 phosphorylates β-catenin at the Ser45 residue, whereas GSK3β phosphorylates this protein at the Ser33, Ser37, and Thr41 residues [48,49]. Moreover, APC impedes the β-catenin dephosphorylation mediated by PP2A phosphatase [50]. Subsequently, the YAP/TAZ complex recruits the E3 ubiquitin ligase β-TrCP, which recognizes Ser/Thr phosphorylation, to promote β-catenin ubiquitination and its subsequent proteosomal degradation [47,51] (Figure 1A).

As a consequence of the Wnt ligand binding to the Fz receptor and LPR5/6 co-receptor [52] (ON-STATE), β-catenin delocalizes, accumulating in the cytoplasm [22] and nucleus [53,54]. When the Fz receptor dimerizes with the LRP5/6 co-receptor, the intracellular motifs of the Fz receptor recruits Disheveled (Dvl) protein [55], whereas CK1 phosphorylates LPR5/6 to allow Axin binding [56,57], which results in the disassembly of the β-catenin destruction complex. This process permits the accumulation and translocation of β-catenin to the nucleus. Moreover, the binding of FOXM1, a member of the Forkhead box (Fox) transcription factor family, to β-catenin promotes its nuclear translocation [58]. In the nucleus, β-catenin binds to transcriptional factors members of the TCF/LEF family [53,59], inducing the dissociation of co-repressors, such as Groucho/TLE [60], which allows the interaction with co-activators including CREPT [61], FHL2, and CBP/p300 [62,63] and remodelers of chromatin such as Brg-1 [64] (Figure 1B). These interactions with β-catenin promote the expression of diverse genes that regulate cellular polarity, proliferation, and differentiation, such as c-jun, c-myc, Cyclin D1, Axin-2, Tcf-1 [20], and β-catenin itself [65].

3. Human Papillomavirus

Persistent infection with Human Papillomaviruses (HPVs) has been implicated in the carcinogenesis of the uterine cervix [9]. Oropharyngeal [10] and anal cancers have also been related to HPV [11]. In fact, almost 70% of cervical cancer cases are associated with HPV16 and HPV18 [66].

The carcinogenic potential of HPV is mainly due to the expression of E6 and E7 viral proteins, which are directly involved in cellular transformation [67,68]. The E6 and E7 oncoproteins interfere with cell cycle regulators and induce genomic instability, which results in a malignant phenotype.

More than 170 HPV types have been identified [69]. HPVs can infect the differentiating squamous epithelium and are classified in two main groups: cutaneotropic and mucosotropic. The majority of cutaneous HPVs belong to the beta and gamma genus, whereas the alpha genus contains all known mucosal HPV types, and at least 40 members of this genus infect the anogenital region [70]. The mucosal HPVs are further divided according to the outcome of infection into low-risk HPVs (LR-HPVs), which are associated with benign and self-limiting benign warts, and high-risk HPVs (HR-HPVs), which are linked to pre-malignant lesions (low- and high-grade cervical intraepithelial neoplasia) and cancer. The most frequent HR-HPV types are: 16, 18, 58, 33, 45, 31, 52, 35, 59, 39, 51, and 56 [71].

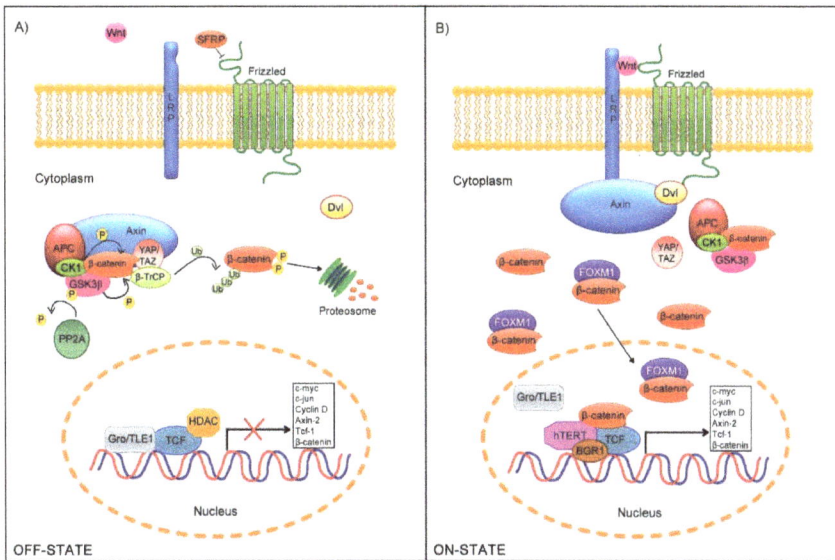

Figure 1. Wnt/β-catenin cell signaling pathway. (**A**) In the absence of stimuli (OFF-STATE), the Fz receptors are regulated by a group of antagonist proteins, such as SFRP, which prevent further receptor-ligand interaction. In the cytoplasm, a degradation complex is formed, to which β-catenin is recruited and phosphorylated at specific residues by the GSK3β and CK1 kinases. These phosphorylated sites are recognized by βTrCP ubiquitin ligase, which mediates β-catenin proteosomal degradation. In the nucleus, the Groucho/TLE repressor binds to TCF/LEF, avoiding its transcriptional activation; (**B**) In the presence of Wnt ligands (ON-STATE), LRP5/6 and Fz dimerize; subsequently, Axin binds to LRP5/6, whereas Disheveled (Dvl) interacts with Fz, allowing Axin-Dvl binding and the disassembly of the β-catenin degradation complex. Finally, β-catenin is released in the cytoplasm and translocated to the nucleus, aided by its binding partner FOXM1, where it binds to TCF/LEF and detaches the Groucho/TLE repressor.

Persistent HR-HPV infection is a crucial event in cellular transformation, but additional events are required to complete the malignant phenotype. Other mechanisms implicated in HPV-related cancers include the activation of multiple cellular pathways such as the Hedgehog [72], Erk/MAPK [73], Notch [74], and Wnt signaling pathways [75], which are involved in embryonic processes, differentiation, survival, proliferation, cell cycle progression, and self-renewal in stem cells.

HPV Genome

The HPV genome consists of a double-stranded circular DNA of approximately 8000 bp that contains genes that are expressed early (E) or late (L) during the viral life cycle and whose transcription and replication are mediated by the long control region (LCR) [76].

L1 and L2 are the HPV structural proteins [77,78]. Specifically, L1 is the major capsid protein and constitutes approximately 80% of the viral capsid [77].

E1 is the viral DNA helicase, and E2 a transcriptional activator and repressor that also complexes with E1 as a critical component of the HPV replisome [79]. E2 protein plays a crucial role in the HPV life cycle due to its ability to regulate viral DNA replication and the transcription of E6 and E7 oncogenes [80].

The E4 coding sequence is contained within the E2 open reading frame (ORF). Although E4 is located in the early region, it is expressed as a late gene and is regulated by a promoter that is responsive to differentiation transcription factors. Moreover, the properties of E4 have not been fully characterized, but several studies implicate E4 in virion release via its association with keratin filaments [81].

E6 and E7 are considered the most important viral oncoproteins: they play a clear role in cellular transformation [82]. Among several cellular interactions, E6 oncoprotein binds to the tumor suppressor protein p53 and to the E3-ubiquitin ligase E6AP, promoting p53 degradation via the proteasome and facilitating DNA damage and mutation [83]. Furthermore, E7 oncoprotein associates with a complex that contains Cullin 2, an E2 ubiquitin ligase, leading to the degradation of the tumor suppressor pRB, to promote cell cycle progression [84].

During the normal viral life cycle, the HPV genome exists in host cells as an episome. However, the viral genome may be incorporated into the host genome in rare cases. Viral genome integration is closely tied to the development of cancer because most HPV-induced cervical cancer cases contain an integrated form of the HPV genome. Viral episome rupture during integration frequently occurs in a zone that includes E1 and E2. The consequent loss of E2 causes the uncontrolled expression of the E6 and E7 oncoproteins, which increases the likelihood of HPV-induced carcinogenesis [85,86].

4. E6 and E7 in Cellular Transformation

E6 and E7 are small proteins that localize to the nucleus and cytoplasm and the interaction of both E6 and E7 immortalizes primary cells in a highly efficient manner [87].

Moreover, the expression of E6 and E7 in organotypic raft cultures results in cellular changes that are similar to those observed in high-grade squamous intraepithelial lesions [88]. Accordingly, transgenic mice expressing HR-HPV E6 and E7 developed basal epithelial squamous carcinomas upon low-dose estrogen treatment [89]. In this model, E7 alone is sufficient to induce high-grade cervical lesions and invasive cervical neoplasia; nevertheless, the inclusion of E6 resulted in larger and more extended tumors. These data demonstrate the cooperative effect of E6 and E7 in promoting the development of cancer [67]. Even when E6 and E7 can immortalize cells in culture, these cells do not form tumors in nude mice models in which the co-expression of supplementary oncogenes, such as v-ras [90] or v-fos [91], is required for tumorigenesis [92].

During HPV infection, E6 and E7 induce the proliferation of undifferentiated and differentiated suprabasal cells and also inhibit apoptosis. These actions promote the accumulation of DNA damage and mutations that can result in cell transformation and the development of cancer [92]. Table 1 summarizes the known E6 and E7 cellular targets and their biological consequences.

Table 1. Cell biological effects induced by HPV E6 and E7 oncoproteins via interactions with cellular elements.

E6-Interactions	Biological Effects
Protein PDZ-domain	Degradation of proteins harboring PDZ domains, with a loss of cell architecture and polarity [93].
E6AP	Degradation of targets such as p53 [83]. Activation of hTERT transcription, inducing immortalization [94].
Bak, FADD Procaspase 8	Induction of respective protein degradation, suppressing apoptosis [95,96].
BRCA1	Activation of estrogen receptor ER signaling pathway [97].
Tyk2	Impairment of Tyk2 activation thereby inhibiting IFN-induced signaling [98].
CBP/p300	Down-regulation of p53 activity by targeting the transcriptional coactivator CBP [99].
NFX1-91	Degradation of NFX1-91 and activation of hTERT [100].
c-Myc	Increased hTERT gene expression [101].
Dvl2	β-catenin stabilization and Wnt signaling activation [102].
E7-Interactions	**Biological effects**
pRb family proteins	Disruption of pRb-E2F complexes thereby initiating the E2F-mediated transcription [103].
AP1	Transactivation of members of AP1 family [104].
Cyclin A/CDK2	Regulation of cell cycle [105].
Cyclin E/CDK2	Regulation of cell cycle (binding through p107) [106].
p21	Inactivation of p21, modulating CDK and PCNA inhibitory functions [107].
MPP2	Enhancement of MPP2-specific transcriptional activity [108].
p600	Contribution to anchorage-independent growth and transformation [109].
Mi2	Complexes with HDAC to promote the E2F2-mediated transcription [110].
IRF1	Abrogation of transactivation function of IRF1 [111].
p48	Down-regulation of IFN α-mediated signal transduction [112].
p27	Abolishment of p27's cell cycle inhibitory function, which endows the cell with invasive properties [113].
PP2A	Inhibition of PP2A catalytic activity [114].

The E6 protein consists of approximately 150 amino acids and contains an LXLL motif in the amino terminal region that is required to interact with the ubiquitin ligase E6AP. Moreover, several proteins also bind to E6 via its LXLL motif, such as E6BP, IRF3, Tuberin, and Paxillin. Another critical E6 motif found in the carboxyl terminus is the S/TXV PBM (PDZ-binding motif), which mediates the interaction with specific domains on cellular proteins known as PDZ domains, specialized in protein-protein interactions [115]. E6 interactions with PDZ-containing proteins commonly induce their proteasome-mediated degradation [116]. PBM is present only in E6 of HR-HPV, suggesting a possible role for this motif in HPV-induced oncogenesis [117] (See Table 1).

The E7 protein consists of 98 amino acids separated in three conserved regions: CR1, CR2, and CR3. CR2 includes a conserved LXCXE motif that mediates high-affinity binding to pRB [118]. The CR3 region contains two CXXC motifs separated by 29 or 30 residues, forming a zinc-binding domain. This region is critical for interaction with cellular proteins, including pRB [119], p21 [107], p27 [120], TBP [121], and E2F [122] (See Table 1).

Growing evidence suggests that HPV oncoproteins can modulate cell signaling pathways to contribute to the carcinogenesis. Upon initial infection, this modulation may be necessary to complete the viral cell cycle and form infective viral particles. Nevertheless, the consistent high-level expression of viral oncoproteins may eventually alter the normal functions of the cell, triggering an uncontrolled transformation process. Via their different cellular interactions, E6 and E7 may be deregulating the different cell signaling pathways implicated in HPV-related cancers. Some of these pathways are involved in cell proliferation and apoptosis, such as PI3K/Akt, Ras/Raf, Notch, and Wnt/β-catenin [123].

5. Wnt/β-Catenin Signaling in HPV-Related Cancers

Several mutations in different components of the Wnt/β-catenin pathway have been described in various types of cancer [124]. In contrast, in HPV-related neoplasias, mutations in Wnt pathway members such as the CTNNB1 and AXIN1 genes are uncommon [125]. However, in cervical cancer biopsies and oropharyngeal squamous carcinoma cells, membrane β-catenin is lost, whereas cytoplasmic and nuclear β-catenin accumulation is observed during cancer progression [6,75]. Some studies have shown that LGR5, a member of the G protein-coupled receptor family, is progressively expressed in cervical neoplasia, promoting the proliferation and tumorigenesis of cervical cancer cells via the activation of the Wnt/β-catenin pathway [126]. Furthermore, post-transcriptional modifications have been identified in components that negatively regulate the pathway; for instance, in cervical cancer samples, GSK3β is inactivated by the phosphorylation of its Ser9 residue, inducing the over-activation of the Wnt signaling pathway [127].

Furthermore, patients with oral and lung cancers that express high levels of the β-catenin-binding partner FOXM1 exhibit worse overall and relapse-free survival than patients with tumors that express low levels of FOXM1; interestingly, this effect is significantly enhanced by the presence of HPV DNA sequences [128]. Therefore, alterations in Wnt cell signaling pathway regulatory elements are associated with cancer progression and poor prognosis in HPV-related cancers.

Epigenetic changes that suppress the activity of the negative regulator of the Wnt pathway have also been identified. Specifically, methylation markers in the APC and SFRP3 promoters have been found in ovarian cancer samples, but only in cases in which HR-HPV genomic sequences were detected [129]. Moreover, in cervical cancer samples, methylation markers have been found in the SRFP2 and DKK3 promoters [130].

Microarray expression studies of cervical cancer-derived tumors and cell lines have identified the over-expression of genes involved in Wnt pathway maintenance and regulation, such as JUN, MYC, FZD2, RAC1, GSK3β, Dvl-1, and CTNNB1 [131,132]. Specifically, Wnt/β-catenin elements are differentially expressed in HPV-positive cervical cancer cell lines (HeLa and SiHa) compared with a non-tumorigenic immortalized cell line (HaCaT) [133]. In this study, 38 genes were identified to be deregulated. Specifically, 15 genes were up-regulated (including CCND3, LRP5, TCF7, and FDZ9), and 23 gene were down-regulated (including CCND2, WNT10A, WNT7A, TCF3, WNT1, FZD4, and BTRC). Because these authors found that WNT7A expression was also significantly reduced in cervical cancer samples, they restored WNT7A expression in HeLa cells, which resulted in a strong decrease in cell viability, proliferation, and migration. In addition, aberrant hypermethylation in the CpG islands within the WNT7A promoter was found in HeLa and SiHa cells but not in HaCaT cells; this event suggests as a possible mechanism by which WNT7A is repressed.

Additionally, a systematic study in cervical cancer samples showed an alteration in the expression of miRNAs involved in Wnt/β-catenin pathway modulation [134]. In this study, miR-21-5p, miR-34c, and miR-96a were up-regulated, whereas miR-99b, miR-497-5p, and miR-617 were down-regulated. Although functional analysis was not performed, these expression patterns were hypothesized to modify the levels of their targets, *i.e.*, WNT5A, FZD1, FAS, MYC, FZD6, CCND1, and PDGFRA, to facilitate cell proliferation and invasion.

Clear evidence indicates that the Wnt pathway is hyperactivated in HPV-related cancers. Currently, HPV oncoproteins are known to bind and alter the function of several cellular targets associated with Wnt pathway regulation, including hTERT, p53, p300/CBP, Dvl, and PP2A (see Table 1), and information about the possible viral regulatory mechanisms in this pathway is emerging.

6. Wnt/β-Catenin Cell Signaling Regulation by E6 and E7 Oncoproteins

The activation of the canonical Wnt pathway represents a second requirement for the malignant transformation of the HPV-infected epithelium [75,135]. Specifically, several findings support the direct or indirect participation of HPV oncoproteins in this pathway.

In HPV-positive oropharyngeal cells, β-catenin expression is strongly localized in the cytoplasm and nucleus, whereas β-catenin is mainly detected in the membranes of HPV-negative cells [5].

In these HPV16-positive oropharyngeal cancer cell lines, E6 and E7 repression was shown to significantly decrease the β-catenin cytoplasmic and nuclear protein levels as well as the β-catenin mRNA levels. Moreover, both E6 and E7 expression were confirmed to up-regulate β-catenin expression and to enhance TCF-mediated transcription. This effect was attributed to a decrease in the ubiquitin ligase type 3 Siah-1 protein (seven in absentia homologue-1 protein), which acts as β-TrCP to induce β-catenin degradation. Because p53 mediates Siah-1 transcriptional activation [136], the down-regulation of p53 induces a decrease in the Siah-1 mRNA and protein levels in HPV-positive cells that are E6-mediated, avoiding β-catenin degradation. However, the activation of Wnt/β-catenin by the E7 oncoprotein is currently poorly understood [5].

An *in vitro* study showed that the HR and LR-HPV E6 proteins can distinctly augment the TCF response, with the highest activity observed for the HR-HPV E6 proteins [137]. In contrast to the previous research, E6 augmented the Wnt/β-catenin/TCF signaling response, although it did not significantly alter β-catenin stability and expression. This process did not depend on p53 degradation, the E6 PDZ-binding motif, APC/Axin/GSK3β complex activity, or β-catenin nuclear localization;

instead, the presence of the E6/E6AP complex enhanced the TCF transcriptional activity mediated by the proteasome, independent of changes in β-catenin levels.

Subsequently, E6AP was confirmed to act as a novel Wnt signaling regulator that cooperates with E6 [138]. Specifically, the levels of E6 decrease in a proteasome-dependent manner in cells in which the Wnt pathway is activated; however, E6 is restored and stabilized in the presence of E6AP, suggesting that E6 requires E6AP to function in Wnt-activated cells. The participation of E6AP in the induction of the TCF response is independent of its catalytic activity. In contrast, β-catenin is stabilized by E6/E6AP, a process that requires the E6AP catalytic domain. Subsequently, β-catenin nuclear accumulation depends on its phosphorylation by GSK3β [138]. The mechanism by which E6/E6AP stabilizes β-catenin is not clear. To date, a direct interaction of E6 or E6AP with β-catenin has not been proven, but the E6/E6AP complex could alternatively participate in the sequestration of a negative regulator of the Wnt pathway.

Another element participating in the Wnt signaling pathway is FOXM1, which is also regulated by the HPV E6 oncogene. FOXM1 can induce β-catenin nuclear translocation by directly binding to β-catenin [58]. In cells harboring the HPV genome, E6 but not E7 oncoprotein was associated with FOXM1 overexpression [128]. This regulation is mediated by the MZF1/NKX2-1 transcriptional factors axis. E6 induces MZF1 expression, and MZF1 consequently activates NKX2-1 transcription. Because the FOXM1 promoter contains three putative sites for NKX2-1, E6 indirectly enhances FOXM1 transcription. In E6-expressing cells, the high levels of FOXM1 increase β-catenin translocation, which promotes TCF transcriptional activation and the expression of Wnt/β-catenin targets such as c-Myc and Cyclin D1 and stemness genes such as Nanog and Oct4. Thus, via the MZF1/NKX2-1 axis, E6 is responsible for metastasis, invasiveness, and stemness induced by the FOXM1-mediated activation of the Wnt/β-catenin pathway.

In vivo studies of transgenic mice support the role of the HPV E6 oncogene in the Wnt signaling pathway. In the K14E6 transgenic mice, the nuclear accumulation of β-catenin depends on the E6 PDZ-binding motif [102]. In this model, Wnt target genes (MYC, BIRC5, and CCND1) were up-regulated in the presence of full-length E6, but these genes were not up-regulated in mice expressing a truncated version of E6 lacking the PDZ-binding motif. Nevertheless, *in vitro* studies showed that both E6 forms enhanced TCF transcriptional activity due to the interaction of E6 with Dvl2, which is responsible for the disassembly of the β-catenin degradation complex [102]. These results suggest that the ability of E6 to activate the TCF response can be both dependent and independent of β-catenin translocation.

Other assays of double-transgenic mice expressing E7 and a constitutively active β-catenin indicated that the co-expression of both proteins promotes invasive cervical cancer, supporting that the activation of the Wnt/β-catenin pathway in premalignant lesions may be partly due to HPV [135].

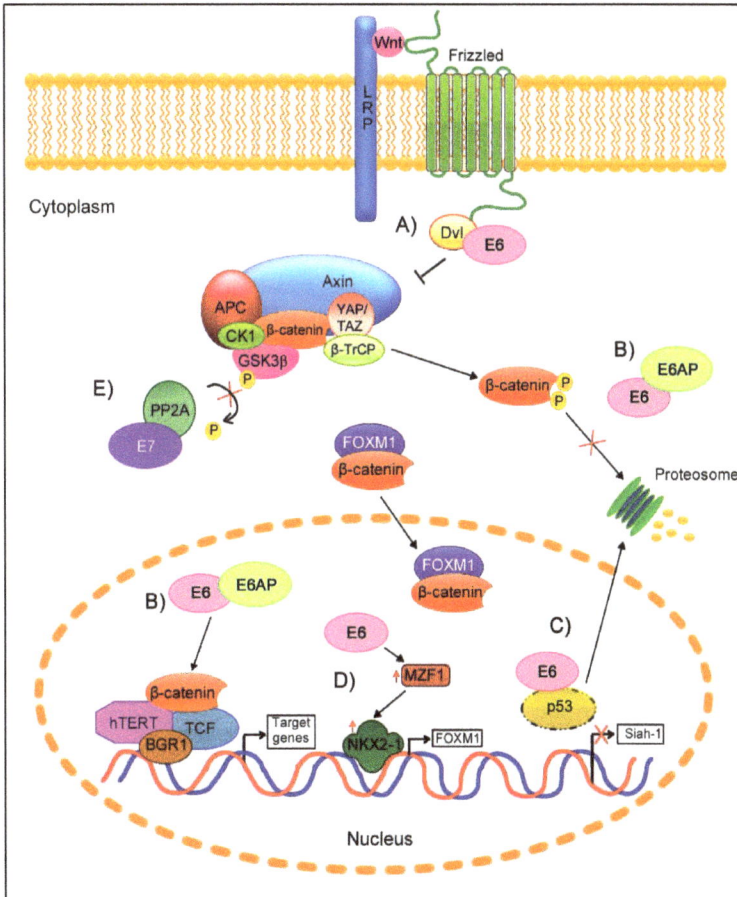

Figure 2. Participation of HPV oncogenes at different levels of Wnt/β-catenin cell signaling regulation. (**A**) The binding of E6-Dvl can disrupt the β-catenin degradation complex, releasing β-catenin which then accumulates in the cytoplasm; (**B**) The E6/E6AP complex stabilizes β-catenin, avoiding its proteasomal degradation and promoting its nuclear translocation, which results in an increase in TCF transcriptional activity; (**C**) E6-induced p53 degradation inhibits Siah-1 expression, which reduces β-catenin degradation; (**D**) E6 induces FOXM1 expression via the MZF1/NKX2-1 axis, which promotes FOXM1/β-catenin nuclear translocation and TCF transcriptional activation; (**E**) E7 binds to PP2A in the structural and catalytic domain, which may avoid the GSK3β activation and consequently β-catenin is stabilized.

Although the role of the E7 oncoprotein in the regulation of Wnt signaling has not been studied as well as that of E6, some findings suggest that this protein is involved in this pathway. Specifically, PP2A participates as a negative regulator of Wnt signaling. This phosphatase induces GSK3β activation via its dephosphorylation at the Ser 9 residue, which results in β-catenin degradation [139]. In a model based on primary human foreskin keratinocytes immortalized with E6 and E7 oncoproteins and transformed with SV40 small T antigen (smt), the smt antigen directly binds to the PP2A catalytic domain, preventing its activation to consequently induce Wnt signaling [75]. Moreover, cell line studies have demonstrated that the functions of E7 and smt are similar: they both strongly bind to the catalytic subunit of PP2A to inhibits its activity [114]. This role of E7 may contribute to β-catenin stabilization in the cytoplasm.

Notably, the above-described findings strongly support a role for HPV oncoproteins in the Wnt canonical pathway. Nevertheless, HPV has not been conclusively linked to the Wnt non-canonical pathway regulation, although some E6 targets, such as WNT7A and Dvl, are known to participate in the activation of Wnt non-canonical pathways.

Evidence supporting the role of HPV in the modulation of the Wnt signaling pathway is shown in Figure 2, which depicts the possible contribution of HPV oncoproteins at different levels in the activation of Wnt signaling.

7. Conclusions

Persistent infection with high-risk HPV types is clearly a main factor in cervical cancer development, and such infections are also implicated in the development of other types of cancer. HPV infection and the activation of diverse cellular processes such as signaling pathways are required to induce a malignant phenotype. Moreover, the Wnt/β-catenin pathway is deregulated in various neoplasias, and has been implicated in HPV-related cancers.

Several studies support the role of HPV oncoproteins in the activation of the canonical Wnt/β-catenin pathway, which may be involved in the onset, progression and maintenance of transformed cells.

Deciphering the precise mechanisms by which HPV oncogenes participate in Wnt/β-catenin modulation will help to elucidate HPV-related carcinogenesis. This information could eventually aid in identifying biomarkers of prognosis and contribute to the design of more effective targeted therapeutics.

Acknowledgments

This manuscript was partially supported by the Consejo Nacional de Ciencia y Tecnología (CONACyT) CB-166808. Jesus Omar Muñoz Bello and Leslie Olmedo Nieva are Ph.D. students from the following programs: "Doctorado en Ciencias Biomédicas" and "Doctorado en Ciencias Bioquímicas", respectively, at the Universidad Nacional Autónoma de México and are recipients of scholarships from CONACyT, México (444223 and 289892, respectively).

Author Contributions

J.O.M.B., L.O.N., A.C.P., and A.M.F.G. performed the bibliographic review and wrote the manuscript; L.R.Z. reviewed the manuscript; and M.L. directed and wrote the manuscript.

Conflicts of Interest

The authors declare no conflict of interest.

References

1. Lin, C.H.; Ji, T.; Chen, C.F.; Hoang, B.H. Wnt signaling in osteosarcoma. *Adv. Exp. Med. Biol.* **2014**, *804*, 33–45. [PubMed]
2. Takigawa, Y.; Brown, A.M. Wnt signaling in liver cancer. *Curr. Drug Targets* **2008**, *9*, 1013–1024. [CrossRef] [PubMed]
3. Krausova, M.; Korinek, V. Wnt signaling in adult intestinal stem cells and cancer. *Cell Signal.* **2014**, *26*, 570–579. [CrossRef] [PubMed]
4. Khramtsov, A.I.; Khramtsova, G.F.; Tretiakova, M.; Huo, D.; Olopade, O.I.; Goss, K.H. Wnt/β-catenin pathway activation is enriched in basal-like breast cancers and predicts poor outcome. *Am. J. Pathol.* **2010**, *176*, 2911–2920. [CrossRef] [PubMed]
5. Rampias, T.; Boutati, E.; Pectasides, E.; Sasaki, C.; Kountourakis, P.; Weinberger, P.; Psyrri, A. Activation of Wnt signaling pathway by human papillomavirus E6 and E7 oncogenes in HPV16-positive oropharyngeal squamous carcinoma cells. *Mol. Cancer Res.* **2010**, *8*, 433–443. [CrossRef] [PubMed]
6. Rodriguez-Sastre, M.A.; Gonzalez-Maya, L.; Delgado, R.; Lizano, M.; Tsubaki, G.; Mohar, A.; Garcia-Carranca, A. Abnormal distribution of E-cadherin and β-catenin in different histologic types of cancer of the uterine cervix. *Gynecol. Oncol.* **2005**, *97*, 330–336. [CrossRef] [PubMed]
7. Clevers, H.; Nusse, R. Wnt/β-catenin signaling and disease. *Cell* **2012**, *149*, 1192–1205. [CrossRef] [PubMed]
8. Globocan 2012: Estimated Cancer Incidence, Mortality and Prevalence Worldwide in 2012. Available online: http://globocan.iarc.fr (accessed on 6 May 2015).
9. Zur Hausen, H. Papillomaviruses and cancer: From basic studies to clinical application. *Nat. Rev. Cancer* **2002**, *2*, 342–350. [CrossRef] [PubMed]
10. D'Souza, G.; Kreimer, A.R.; Viscidi, R.; Pawlita, M.; Fakhry, C.; Koch, W.M.; Westra, W.H.; Gillison, M.L. Case-control study of human papillomavirus and oropharyngeal cancer. *N. Engl. J. Med.* **2007**, *356*, 1944–1956. [CrossRef] [PubMed]
11. Daling, J.R.; Madeleine, M.M.; Johnson, L.G.; Schwartz, S.M.; Shera, K.A.; Wurscher, M.A.; Carter, J.J.; Porter, P.L.; Galloway, D.A.; McDougall, J.K. Human papillomavirus, smoking, and sexual practices in the etiology of anal cancer. *Cancer* **2004**, *101*, 270–280. [CrossRef] [PubMed]
12. Bilir, B.; Kucuk, O.; Moreno, C.S. Wnt signaling blockage inhibits cell proliferation and migration, and induces apoptosis in triple-negative breast cancer cells. *J. Transl. Med.* **2013**, *11*, 280. [CrossRef] [PubMed]
13. Peng, X.; Yang, L.; Chang, H.; Dai, G.; Wang, F.; Duan, X.; Guo, L.; Zhang, Y.; Chen, G. Wnt/β-catenin signaling regulates the proliferation and differentiation of mesenchymal progenitor cells through the p53 pathway. *PLoS ONE* **2014**, *9*, e97283. [CrossRef] [PubMed]

14. Gradl, D.; Kuhl, M.; Wedlich, D. The Wnt/Wg signal transducer β-catenin controls fibronectin expression. *Mol. Cell. Biol.* **1999**, *19*, 5576–5587. [PubMed]

15. Dollar, G.L.; Weber, U.; Mlodzik, M.; Sokol, S.Y. Regulation of Lethal giant larvae by Dishevelled. *Nature* **2005**, *437*, 1376–1380. [CrossRef] [PubMed]

16. Nusse, R.; Brown, A.; Papkoff, J.; Scambler, P.; Shackleford, G.; McMahon, A.; Moon, R.; Varmus, H. A new nomenclature for *int*-1 and related genes: The *Wnt* gene family. *Cell* **1991**, *64*, 231. [CrossRef]

17. Nusse, R.; Varmus, H.E. Many tumors induced by the mouse mammary tumor virus contain a provirus integrated in the same region of the host genome. *Cell* **1982**, *31*, 99–109. [CrossRef]

18. Van Ooyen, A.; Nusse, R. Structure and nucleotide sequence of the putative mammary oncogene *int*-1; proviral insertions leave the protein-encoding domain intact. *Cell* **1984**, *39*, 233–240. [CrossRef]

19. Rijsewijk, F.; Schuermann, M.; Wagenaar, E.; Parren, P.; Weigel, D.; Nusse, R. The drosophila homolog of the mouse mammary oncogene *int*-1 is identical to the segment polarity gene *wingless*. *Cell* **1987**, *50*, 649–657. [CrossRef]

20. The Wnt Homepage. Available online: http://web.stanford.edu/group/nusselab/cgi-bin/wnt/ (accessed on 6 May 2015).

21. Bafico, A.; Gazit, A.; Wu-Morgan, S.S.; Yaniv, A.; Aaronson, S.A. Characterization of Wnt-1 and Wnt-2 induced growth alterations and signaling pathways in NIH3T3 fibroblasts. *Oncogene* **1998**, *16*, 2819–2825. [CrossRef] [PubMed]

22. Shimizu, H.; Julius, M.A.; Giarre, M.; Zheng, Z.; Brown, A.M.; Kitajewski, J. Transformation by Wnt family proteins correlates with regulation of β-catenin. *Cell Growth Differ.* **1997**, *8*, 1349–1358. [PubMed]

23. Ohira, T.; Gemmill, R.M.; Ferguson, K.; Kusy, S.; Roche, J.; Brambilla, E.; Zeng, C.; Baron, A.; Bemis, L.; Erickson, P.; *et al.* WNT7a induces E-cadherin in lung cancer cells. *Proc. Natl. Acad. Sci. USA* **2003**, *100*, 10429–10434. [CrossRef] [PubMed]

24. Fagotto, F.; Guger, K.; Gumbiner, B.M. Induction of the primary dorsalizing center in *Xenopus* by the Wnt/GSK/β-catenin signaling pathway, but not by Vg1, Activin or Noggin. *Development* **1997**, *124*, 453–460. [PubMed]

25. Ye, J.; Yang, T.; Guo, H.; Tang, Y.; Deng, F.; Li, Y.; Xing, Y.; Yang, L.; Yang, K. Wnt10b promotes differentiation of mouse hair follicle melanocytes. *Int. J. Med. Sci.* **2013**, *10*, 691–698. [CrossRef] [PubMed]

26. Li, Y.H.; Zhang, K.; Ye, J.X.; Lian, X.H.; Yang, T. Wnt10b promotes growth of hair follicles via a canonical Wnt signalling pathway. *Clin. Exp. Dermatol.* **2011**, *36*, 534–540. [CrossRef] [PubMed]

27. Louis, I.; Heinonen, K.M.; Chagraoui, J.; Vainio, S.; Sauvageau, G.; Perreault, C. The signaling protein Wnt4 enhances thymopoiesis and expands multipotent hematopoietic progenitors through β-catenin-independent signaling. *Immunity* **2008**, *29*, 57–67. [CrossRef] [PubMed]

28. Liu, A.; Chen, S.; Cai, S.; Dong, L.; Liu, L.; Yang, Y.; Guo, F.; Lu, X.; He, H.; Chen, Q.; *et al.* Wnt5a through noncanonical Wnt/JNK or Wnt/PKC signaling contributes to the differentiation of mesenchymal stem cells into type II alveolar epithelial cells *in vitro.* *PLoS ONE* **2014**, *9*, e90229. [CrossRef] [PubMed]

29. Slusarski, D.C.; Corces, V.G.; Moon, R.T. Interaction of Wnt and a Frizzled homologue triggers G-protein-linked phosphatidylinositol signalling. *Nature* **1997**, *390*, 410–413. [PubMed]

30. Flaherty, M.P.; Dawn, B. Noncanonical Wnt11 signaling and cardiomyogenic differentiation. *Trends Cardiovasc. Med.* **2008**, *18*, 260–268. [CrossRef] [PubMed]

31. Kilander, M.B.; Dahlstrom, J.; Schulte, G. Assessment of Frizzled 6 membrane mobility by FRAP supports G protein coupling and reveals WNT-Frizzled selectivity. *Cell Signal.* **2014**, *26*, 1943–1949. [CrossRef] [PubMed]

32. Caddy, J.; Wilanowski, T.; Darido, C.; Dworkin, S.; Ting, S.B.; Zhao, Q.; Rank, G.; Auden, A.; Srivastava, S.; Papenfuss, T.A.; *et al.* Epidermal wound repair is regulated by the planar cell polarity signaling pathway. *Dev. Cell* **2010**, *19*, 138–147. [CrossRef] [PubMed]

33. Murdoch, J.N.; Damrau, C.; Paudyal, A.; Bogani, D.; Wells, S.; Greene, N.D.; Stanier, P.; Copp, A.J. Genetic interactions between planar cell polarity genes cause diverse neural tube defects in mice. *Dis. Models Mech.* **2014**, *7*, 1153–1163. [CrossRef] [PubMed]

34. Babayeva, S.; Zilber, Y.; Torban, E. Planar cell polarity pathway regulates actin rearrangement, cell shape, motility, and nephrin distribution in podocytes. *Am. J. Physiol. Ren. Physiol.* **2011**, *300*, F549–F560. [CrossRef] [PubMed]

35. Boutin, C.; Labedan, P.; Dimidschstein, J.; Richard, F.; Cremer, H.; Andre, P.; Yang, Y.; Montcouquiol, M.; Goffinet, A.M.; Tissir, F. A dual role for planar cell polarity genes in ciliated cells. *Proc. Natl. Acad. Sci. USA* **2014**, *111*, E3129–E3138. [CrossRef] [PubMed]

36. Kinoshita, N.; Iioka, H.; Miyakoshi, A.; Ueno, N. PKCδ is essential for Dishevelled function in a noncanonical Wnt pathway that regulates *Xenopus* convergent extension movements. *Genes Dev.* **2003**, *17*, 1663–1676. [CrossRef] [PubMed]

37. Fanto, M.; Weber, U.; Strutt, D.I.; Mlodzik, M. Nuclear signaling by Rac and Rho GTPases is required in the establishment of epithelial planar polarity in the *Drosophila* eye. *Curr. Biol.* **2000**, *10*, 979–988. [CrossRef]

38. Kuhl, M.; Sheldahl, L.C.; Malbon, C.C.; Moon, R.T. Ca^{2+}/Calmodulin-dependent protein kinase II is stimulated by Wnt and Frizzled homologs and promotes ventral cell fates in *Xenopus. J. Biol. Chem.* **2000**, *275*, 12701–12711. [CrossRef] [PubMed]

39. Sheldahl, L.C.; Park, M.; Malbon, C.C.; Moon, R.T. Protein Kinase C is differentially stimulated by Wnt and Frizzled homologs in a G-protein-dependent manner. *Curr. Biol.* **1999**, *9*, 695–698. [CrossRef]

40. Kuhl, M.; Geis, K.; Sheldahl, L.C.; Pukrop, T.; Moon, R.T.; Wedlich, D. Antagonistic regulation of convergent extension movements in Xenopus by Wnt/β-catenin and Wnt/Ca^{2+} signaling. *Mech. Dev.* **2001**, *106*, 61–76. [CrossRef]

41. Okamoto, K.; Narayanan, R.; Lee, S.H.; Murata, K.; Hayashi, Y. The role of CaMKII as an F-actin-bundling protein crucial for maintenance of dendritic spine structure. *Proc. Natl. Acad. Sci. USA* **2007**, *104*, 6418–6423. [CrossRef] [PubMed]

42. Wang, Q.; Symes, A.J.; Kane, C.A.; Freeman, A.; Nariculam, J.; Munson, P.; Thrasivoulou, C.; Masters, J.R.; Ahmed, A. A novel role for Wnt/Ca^{2+} signaling in actin cytoskeleton remodeling and cell motility in prostate cancer. *PLoS ONE* **2010**, *5*, e10456. [CrossRef] [PubMed]

43. Pereira, C.; Schaer, D.J.; Bachli, E.B.; Kurrer, M.O.; Schoedon, G. Wnt5A/CaMKII signaling contributes to the inflammatory response of macrophages and is a target for the antiinflammatory action of activated protein C and interleukin-10. *Arterioscler. Thromb. Vasc. Biol.* **2008**, *28*, 504–510. [CrossRef] [PubMed]

44. Ikeda, S.; Kishida, S.; Yamamoto, H.; Murai, H.; Koyama, S.; Kikuchi, A. Axin, a negative regulator of the Wnt signaling pathway, forms a complex with GSK-3beta and beta-catenin and promotes GSK-3beta-dependent phosphorylation of beta-catenin. *EMBO J.* **1998**, *17*, 1371–1384. [CrossRef] [PubMed]

45. Sobrado, P.; Jedlicki, A.; Bustos, V.H.; Allende, C.C.; Allende, J.E. Basic region of residues 228–231 of protein kinase CK1α is involved in its interaction with axin: Binding to axin does not affect the kinase activity. *J. Cell. Biochem.* **2005**, *94*, 217–224. [CrossRef] [PubMed]

46. Papkoff, J.; Rubinfeld, B.; Schryver, B.; Polakis, P. Wnt-1 regulates free pools of catenins and stabilizes APC-catenin complexes. *Mol. Cell. Biol.* **1996**, *16*, 2128–2134. [PubMed]

47. Azzolin, L.; Panciera, T.; Soligo, S.; Enzo, E.; Bicciato, S.; Dupont, S.; Bresolin, S.; Frasson, C.; Basso, G.; Guzzardo, V.; et al. YAP/TAZ incorporation in the β-catenin destruction complex orchestrates the Wnt response. *Cell* **2014**, *158*, 157–170. [CrossRef] [PubMed]

48. Liu, C.; Li, Y.; Semenov, M.; Han, C.; Baeg, G.H.; Tan, Y.; Zhang, Z.; Lin, X.; He, X. Control of β-catenin phosphorylation/degradation by a dual-kinase mechanism. *Cell* **2002**, *108*, 837–847. [CrossRef]

49. Sakanaka, C. Phosphorylation and regulation of β-catenin by casein kinase Iε. *J. Biochem.* **2002**, *132*, 697–703. [CrossRef] [PubMed]

50. Su, Y.; Fu, C.; Ishikawa, S.; Stella, A.; Kojima, M.; Shitoh, K.; Schreiber, E.M.; Day, B.W.; Liu, B. APC is essential for targeting phosphorylated β-catenin to the SCF$^{β-TrCP}$ ubiquitin ligase. *Mol. Cell* **2008**, *32*, 652–661. [CrossRef] [PubMed]

51. Winston, J.T.; Strack, P.; Beer-Romero, P.; Chu, C.Y.; Elledge, S.J.; Harper, J.W. The SCF$^{β-TrCP}$—Ubiquitin ligase complex associates specifically with phosphorylated destruction motifs in IκBα and β-catenin and stimulates IκBα ubiquitination *in vitro*. *Genes Dev.* **1999**, *13*, 270–283. [CrossRef] [PubMed]

52. Cong, F.; Schweizer, L.; Varmus, H. Wnt signals across the plasma membrane to activate the β-catenin pathway by forming oligomers containing its receptors, Frizzled and LRP. *Development* **2004**, *131*, 5103–5115. [CrossRef] [PubMed]

53. Hsu, S.C.; Galceran, J.; Grosschedl, R. Modulation of transcriptional regulation by LEF-1 in response to Wnt-1 signaling and association with β-catenin. *Mol. Cell. Biol.* **1998**, *18*, 4807–4818. [PubMed]

54. Schneider, S.; Steinbeisser, H.; Warga, R.M.; Hausen, P. β-catenin translocation into nuclei demarcates the dorsalizing centers in frog and fish embryos. *Mech. Dev.* **1996**, *57*, 191–198. [CrossRef]

55. Tauriello, D.V.; Jordens, I.; Kirchner, K.; Slootstra, J.W.; Kruitwagen, T.; Bouwman, B.A.; Noutsou, M.; Rudiger, S.G.; Schwamborn, K.; Schambony, A.; *et al.* Wnt/β-catenin signaling requires interaction of the Dishevelled DEP domain and C terminus with a discontinuous motif in Frizzled. *Proc. Natl. Acad. Sci. USA* **2012**, *109*, E812–E820. [CrossRef] [PubMed]

56. Mao, J.; Wang, J.; Liu, B.; Pan, W.; Farr, G.H., 3rd; Flynn, C.; Yuan, H.; Takada, S.; Kimelman, D.; Li, L.; *et al.* Low-density lipoprotein receptor-related protein-5 binds to Axin and regulates the canonical Wnt signaling pathway. *Mol. Cell* **2001**, *7*, 801–809. [CrossRef]

57. Davidson, G.; Wu, W.; Shen, J.; Bilic, J.; Fenger, U.; Stannek, P.; Glinka, A.; Niehrs, C. Casein kinase 1 γ couples Wnt receptor activation to cytoplasmic signal transduction. *Nature* **2005**, *438*, 867–872. [CrossRef] [PubMed]

58. Gong, A.; Huang, S. FoxM1 and Wnt/β-catenin signaling in glioma stem cells. *Cancer Res.* **2012**, *72*, 5658–5662. [CrossRef] [PubMed]

59. Korinek, V.; Barker, N.; Willert, K.; Molenaar, M.; Roose, J.; Wagenaar, G.; Markman, M.; Lamers, W.; Destree, O.; Clevers, H. Two members of the Tcf family implicated in Wnt/β-catenin signaling during embryogenesis in the mouse. *Mol. Cell. Biol.* **1998**, *18*, 1248–1256. [PubMed]

60. Daniels, D.L.; Weis, W.I. β-catenin directly displaces Groucho/TLE repressors from Tcf/Lef in Wnt-mediated transcription activation. *Nat. Struct. Mol. Biol.* **2005**, *12*, 364–371. [CrossRef] [PubMed]

61. Zhang, Y.; Liu, C.; Duan, X.; Ren, F.; Li, S.; Jin, Z.; Wang, Y.; Feng, Y.; Liu, Z.; Chang, Z. CREPT/RPRD1B, a recently identified novel protein highly expressed in tumors, enhances the β-catenin·TCF4 transcriptional activity in response to Wnt signaling. *J. Biol. Chem.* **2014**, *289*, 22589–22599. [CrossRef] [PubMed]

62. Labalette, C.; Renard, C.A.; Neuveut, C.; Buendia, M.A.; Wei, Y. Interaction and functional cooperation between the LIM protein FHL2, CBP/p300, and β-catenin. *Mol. Cell. Biol.* **2004**, *24*, 10689–10702. [CrossRef] [PubMed]

63. Takemaru, K.I.; Moon, R.T. The transcriptional coactivator CBP interacts with β-catenin to activate gene expression. *J. Cell Biol.* **2000**, *149*, 249–254. [CrossRef] [PubMed]

64. Barker, N.; Hurlstone, A.; Musisi, H.; Miles, A.; Bienz, M.; Clevers, H. The chromatin remodelling factor Brg-1 interacts with β-catenin to promote target gene activation. *EMBO J.* **2001**, *20*, 4935–4943. [CrossRef] [PubMed]

65. Bandapalli, O.R.; Dihlmann, S.; Helwa, R.; Macher-Goeppinger, S.; Weitz, J.; Schirmacher, P.; Brand, K. Transcriptional activation of the β-*catenin* gene at the invasion front of colorectal liver metastases. *J. Pathol.* **2009**, *218*, 370–379. [CrossRef] [PubMed]

66. Smith, J.S.; Lindsay, L.; Hoots, B.; Keys, J.; Franceschi, S.; Winer, R.; Clifford, G.M. Human papillomavirus type distribution in invasive cervical cancer and high-grade cervical lesions: A meta-analysis update. *Int. J. Cancer* **2007**, *121*, 621–632. [CrossRef] [PubMed]

67. Riley, R.R.; Duensing, S.; Brake, T.; Munger, K.; Lambert, P.F.; Arbeit, J.M. Dissection of human papillomavirus E6 and E7 function in transgenic mouse models of cervical carcinogenesis. *Cancer Res.* **2003**, *63*, 4862–4871. [PubMed]

68. Song, S.; Pitot, H.C.; Lambert, P.F. The human papillomavirus type 16 E6 gene alone is sufficient to induce carcinomas in transgenic animals. *J. Virol.* **1999**, *73*, 5887–5893. [PubMed]

69. De Villiers, E.M. Cross-roads in the classification of papillomaviruses. *Virology* **2013**, *445*, 2–10. [CrossRef] [PubMed]

70. Bzhalava, D.; Guan, P.; Franceschi, S.; Dillner, J.; Clifford, G. A systematic review of the prevalence of mucosal and cutaneous human papillomavirus types. *Virology* **2013**, *445*, 224–231. [CrossRef] [PubMed]

71. Munoz, N.; Bosch, F.X.; de Sanjose, S.; Herrero, R.; Castellsague, X.; Shah, K.V.; Snijders, P.J.; Meijer, C.J.; International Agency for Research on Cancer Multicenter Cervical Cancer Study Group. Epidemiologic classification of human papillomavirus types associated with cervical cancer. *N. Engl. J. Med.* **2003**, *348*, 518–527. [CrossRef] [PubMed]

72. Samarzija, I.; Beard, P. Hedgehog pathway regulators influence cervical cancer cell proliferation, survival and migration. *Biochem. Biophys. Res. Commun.* **2012**, *425*, 64–69. [CrossRef] [PubMed]

73. Branca, M.; Ciotti, M.; Santini, D.; Bonito, L.D.; Benedetto, A.; Giorgi, C.; Paba, P.; Favalli, C.; Costa, S.; Agarossi, A.; *et al.* Activation of the ERK/MAP kinase pathway in cervical intraepithelial neoplasia is related to grade of the lesion but not to high-risk human papillomavirus, virus clearance, or prognosis in cervical cancer. *Am. J. Clin. Pathol.* **2004**, *122*, 902–911. [CrossRef] [PubMed]

74. Weijzen, S.; Zlobin, A.; Braid, M.; Miele, L.; Kast, W.M. HPV16 E6 and E7 oncoproteins regulate Notch-1 expression and cooperate to induce transformation. *J. Cell. Physiol.* **2003**, *194*, 356–362. [CrossRef] [PubMed]

75. Uren, A.; Fallen, S.; Yuan, H.; Usubutun, A.; Kucukali, T.; Schlegel, R.; Toretsky, J.A. Activation of the canonical Wnt pathway during genital keratinocyte transformation: A model for cervical cancer progression. *Cancer Res.* **2005**, *65*, 6199–6206. [CrossRef] [PubMed]

76. Doorbar, J. Molecular biology of human papillomavirus infection and cervical cancer. *Clin. Sci. Lond.* **2006**, *110*, 525–541. [CrossRef] [PubMed]

77. Buck, C.B.; Day, P.M.; Trus, B.L. The papillomavirus major capsid protein L1. *Virology* **2013**, *445*, 169–174. [CrossRef] [PubMed]

78. Wang, J.W.; Roden, R.B. L2, the minor capsid protein of papillomavirus. *Virology* **2013**, *445*, 175–186. [CrossRef] [PubMed]

79. Bergvall, M.; Melendy, T.; Archambault, J. The E1 proteins. *Virology* **2013**, *445*, 35–56. [CrossRef] [PubMed]

80. McBride, A.A. The papillomavirus E2 proteins. *Virology* **2013**, *445*, 57–79. [CrossRef] [PubMed]

81. Doorbar, J. The E4 protein; structure, function and patterns of expression. *Virology* **2013**, *445*, 80–98. [CrossRef] [PubMed]

82. McLaughlin-Drubin, M.E.; Munger, K. Oncogenic activities of human papillomaviruses. *Virus Res.* **2009**, *143*, 195–208. [CrossRef] [PubMed]

83. Scheffner, M.; Huibregtse, J.M.; Vierstra, R.D.; Howley, P.M. The HPV-16 E6 and E6-AP complex functions as a ubiquitin-protein ligase in the ubiquitination of p53. *Cell* **1993**, *75*, 495–505. [CrossRef]

84. Huh, K.; Zhou, X.; Hayakawa, H.; Cho, J.Y.; Libermann, T.A.; Jin, J.; Harper, J.W.; Munger, K. Human papillomavirus type 16 E7 oncoprotein associates with the cullin 2 ubiquitin ligase complex, which contributes to degradation of the retinoblastoma tumor suppressor. *J. Virol.* **2007**, *81*, 9737–9747. [CrossRef] [PubMed]

85. Pett, M.; Coleman, N. Integration of high-risk human papillomavirus: A key event in cervical carcinogenesis? *J. Pathol.* **2007**, *212*, 356–367. [CrossRef] [PubMed]

86. Jeon, S.; Lambert, P.F. Integration of human papillomavirus type 16 DNA into the human genome leads to increased stability of E6 and E7 mRNAs: Implications for cervical carcinogenesis. *Proc. Natl. Acad. Sci. USA* **1995**, *92*, 1654–1658. [CrossRef] [PubMed]

87. Sashiyama, H.; Shino, Y.; Kawamata, Y.; Tomita, Y.; Ogawa, N.; Shimada, H.; Kobayashi, S.; Asano, T.; Ochiai, T.; Shirasawa, H. Immortalization of human esophageal keratinocytes by E6 and E7 of human papillomavirus type 16. *Int. J. Oncol.* **2001**, *19*, 97–103. [CrossRef] [PubMed]

88. McCance, D.J.; Kopan, R.; Fuchs, E.; Laimins, L.A. Human papillomavirus type 16 alters human epithelial cell differentiation *in vitro*. *Proc. Natl. Acad. Sci. USA* **1988**, *85*, 7169–7173. [CrossRef] [PubMed]

89. Arbeit, J.M.; Howley, P.M.; Hanahan, D. Chronic estrogen-induced cervical and vaginal squamous carcinogenesis in human papillomavirus type 16 transgenic mice. *Proc. Natl. Acad. Sci. USA* **1996**, *93*, 2930–2935. [CrossRef] [PubMed]

90. Durst, M.; Gallahan, D.; Jay, G.; Rhim, J.S. Glucocorticoid-enhanced neoplastic transformation of human keratinocytes by human papillomavirus type 16 and an activated *ras* oncogene. *Virology* **1989**, *173*, 767–771. [CrossRef]

91. Pei, X.F.; Meck, J.M.; Greenhalgh, D.; Schlegel, R. Cotransfection of HPV-18 and *v-fos* DNA induces tumorigenicity of primary human keratinocytes. *Virology* **1993**, *196*, 855–860. [CrossRef] [PubMed]

92. Moody, C.A.; Laimins, L.A. Human papillomavirus oncoproteins: Pathways to transformation. *Nat. Rev. Cancer* **2010**, *10*, 550–560. [CrossRef] [PubMed]

93. Watson, R.A.; Thomas, M.; Banks, L.; Roberts, S. Activity of the human papillomavirus E6 PDZ-binding motif correlates with an enhanced morphological transformation of immortalized human keratinocytes. *J. Cell Sci.* **2003**, *116*, 4925–4934. [CrossRef] [PubMed]

94. Liu, X.; Dakic, A.; Zhang, Y.; Dai, Y.; Chen, R.; Schlegel, R. HPV E6 protein interacts physically and functionally with the cellular telomerase complex. *Proc. Natl. Acad. Sci. USA* **2009**, *106*, 18780–18785. [CrossRef] [PubMed]

95. Thomas, M.; Banks, L. Human papillomavirus (HPV) E6 interactions with Bak are conserved amongst E6 proteins from high and low risk HPV types. *J. Gen. Virol.* **1999**, *80*, 1513–1517. [PubMed]

96. Garnett, T.O.; Filippova, M.; Duerksen-Hughes, P.J. Accelerated degradation of FADD and procaspase 8 in cells expressing human papilloma virus 16 E6 impairs TRAIL-mediated apoptosis. *Cell Death Differ.* **2006**, *13*, 1915–1926. [CrossRef] [PubMed]

97. Zhang, Y.; Fan, S.; Meng, Q.; Ma, Y.; Katiyar, P.; Schlegel, R.; Rosen, E.M. BRCA1 interaction with human papillomavirus oncoproteins. *J. Biol. Chem.* **2005**, *280*, 33165–33177. [CrossRef] [PubMed]

98. Li, S.; Labrecque, S.; Gauzzi, M.C.; Cuddihy, A.R.; Wong, A.H.; Pellegrini, S.; Matlashewski, G.J.; Koromilas, A.E. The human papilloma virus (HPV)-18 E6 oncoprotein physically associates with Tyk2 and impairs Jak-STAT activation by interferon-α. *Oncogene* **1999**, *18*, 5727–5737. [CrossRef] [PubMed]

99. Zimmermann, H.; Degenkolbe, R.; Bernard, H.U.; O'Connor, M.J. The human papillomavirus type 16 E6 oncoprotein can down-regulate p53 activity by targeting the transcriptional coactivator CBP/p300. *J. Virol.* **1999**, *73*, 6209–6219. [PubMed]

100. Gewin, L.; Myers, H.; Kiyono, T.; Galloway, D.A. Identification of a novel telomerase repressor that interacts with the human papillomavirus type-16 E6/E6-AP complex. *Genes Dev.* **2004**, *18*, 2269–2282. [CrossRef] [PubMed]

101. Veldman, T.; Liu, X.; Yuan, H.; Schlegel, R. Human papillomavirus E6 and Myc proteins associate *in vivo* and bind to and cooperatively activate the telomerase reverse transcriptase promoter. *Proc. Natl. Acad. Sci. USA* **2003**, *100*, 8211–8216. [CrossRef] [PubMed]

102. Bonilla-Delgado, J.; Bulut, G.; Liu, X.; Cortes-Malagon, E.M.; Schlegel, R.; Flores-Maldonado, C.; Contreras, R.G.; Chung, S.H.; Lambert, P.F.; Uren, A.; *et al.* The E6 oncoprotein from HPV16 enhances the canonical Wnt/β-catenin pathway in skin epidermis *in vivo*. *Mol. Cancer Res.* **2012**, *10*, 250–258. [CrossRef] [PubMed]

103. Dyson, N.; Howley, P.M.; Munger, K.; Harlow, E. The human papilloma virus-16 E7 oncoprotein is able to bind to the retinoblastoma gene product. *Science* **1989**, *243*, 934–937. [CrossRef] [PubMed]

104. Antinore, M.J.; Birrer, M.J.; Patel, D.; Nader, L.; McCance, D.J. The human papillomavirus type 16 E7 gene product interacts with and *trans*-activates the AP1 family of transcription factors. *EMBO J.* **1996**, *15*, 1950–1960. [PubMed]

105. Nguyen, C.L.; Munger, K. Direct association of the HPV16 E7 oncoprotein with cyclin A/CDK2 and cyclin E/CDK2 complexes. *Virology* **2008**, *380*, 21–25. [CrossRef] [PubMed]

106. McIntyre, M.C.; Ruesch, M.N.; Laimins, L.A. Human papillomavirus E7 oncoproteins bind a single form of cyclin E in a complex with cdk2 and p107. *Virology* **1996**, *215*, 73–82. [CrossRef] [PubMed]

107. Funk, J.O.; Waga, S.; Harry, J.B.; Espling, E.; Stillman, B.; Galloway, D.A. Inhibition of CDK activity and PCNA-dependent DNA replication by p21 is blocked by interaction with the HPV-16 E7 oncoprotein. *Genes Dev.* **1997**, *11*, 2090–2100. [CrossRef] [PubMed]

108. Luscher-Firzlaff, J.M.; Westendorf, J.M.; Zwicker, J.; Burkhardt, H.; Henriksson, M.; Muller, R.; Pirollet, F.; Luscher, B. Interaction of the fork head domain transcription factor MPP2 with the human papilloma virus 16 E7 protein: Enhancement of transformation and transactivation. *Oncogene* **1999**, *18*, 5620–5630. [CrossRef] [PubMed]

109. Huh, K.W.; DeMasi, J.; Ogawa, H.; Nakatani, Y.; Howley, P.M.; Munger, K. Association of the human papillomavirus type 16 E7 oncoprotein with the 600-kDa retinoblastoma protein-associated factor, p600. *Proc. Natl. Acad. Sci. USA* **2005**, *102*, 11492–11497. [CrossRef] [PubMed]

110. Brehm, A.; Nielsen, S.J.; Miska, E.A.; McCance, D.J.; Reid, J.L.; Bannister, A.J.; Kouzarides, T. The E7 oncoprotein associates with Mi2 and histone deacetylase activity to promote cell growth. *EMBO J.* **1999**, *18*, 2449–2458. [CrossRef] [PubMed]

111. Um, S.J.; Rhyu, J.W.; Kim, E.J.; Jeon, K.C.; Hwang, E.S.; Park, J.S. Abrogation of IRF-1 response by high-risk HPV E7 protein *in vivo*. *Cancer Lett.* **2002**, *179*, 205–212. [CrossRef]

112. Barnard, P.; McMillan, N.A. The human papillomavirus E7 oncoprotein abrogates signaling mediated by interferon-α. *Virology* **1999**, *259*, 305–313. [CrossRef] [PubMed]

113. Yan, X.; Shah, W.; Jing, L.; Chen, H.; Wang, Y. High-risk human papillomavirus type 18 E7 caused p27 elevation and cytoplasmic localization. *Cancer Biol. Ther.* **2010**, *9*, 728–735. [CrossRef] [PubMed]

114. Pim, D.; Massimi, P.; Dilworth, S.M.; Banks, L. Activation of the protein kinase B pathway by the HPV-16 E7 oncoprotein occurs through a mechanism involving interaction with PP2A. *Oncogene* **2005**, *24*, 7830–7838. [CrossRef] [PubMed]

115. Howie, H.L.; Katzenellenbogen, R.A.; Galloway, D.A. Papillomavirus E6 proteins. *Virology* **2009**, *384*, 324–334. [CrossRef] [PubMed]

116. Zhang, Y.; Dasgupta, J.; Ma, R.Z.; Banks, L.; Thomas, M.; Chen, X.S. Structures of a human papillomavirus (HPV) E6 polypeptide bound to MAGUK proteins: Mechanisms of targeting tumor suppressors by a high-risk HPV oncoprotein. *J. Virol.* **2007**, *81*, 3618–3626. [CrossRef] [PubMed]

117. Lee, C.; Laimins, L.A. Role of the PDZ domain-binding motif of the oncoprotein E6 in the pathogenesis of human papillomavirus type 31. *J. Virol.* **2004**, *78*, 12366–12377. [CrossRef] [PubMed]

118. Munger, K.; Werness, B.A.; Dyson, N.; Phelps, W.C.; Harlow, E.; Howley, P.M. Complex formation of human papillomavirus E7 proteins with the retinoblastoma tumor suppressor gene product. *EMBO J.* **1989**, *8*, 4099–4105. [PubMed]

119. Liu, X.; Clements, A.; Zhao, K.; Marmorstein, R. Structure of the human papillomavirus E7 oncoprotein and its mechanism for inactivation of the retinoblastoma tumor suppressor. *J. Biol. Chem.* **2006**, *281*, 578–586. [CrossRef] [PubMed]

120. Zerfass-Thome, K.; Zwerschke, W.; Mannhardt, B.; Tindle, R.; Botz, J.W.; Jansen-Durr, P. Inactivation of the cdk inhibitor p27KIP1 by the human papillomavirus type 16 E7 oncoprotein. *Oncogene* **1996**, *13*, 2323–2330. [PubMed]

121. Massimi, P.; Pim, D.; Banks, L. Human papillomavirus type 16 E7 binds to the conserved carboxy-terminal region of the TATA box binding protein and this contributes to E7 transforming activity. *J. Gen. Virol.* **1997**, *78*, 2607–2613. [PubMed]

122. Hwang, S.G.; Lee, D.; Kim, J.; Seo, T.; Choe, J. Human papillomavirus type 16 E7 binds to E2F1 and activates E2F1-driven transcription in a retinoblastoma protein-independent manner. *J. Biol. Chem.* **2002**, *277*, 2923–2930. [CrossRef] [PubMed]

123. Manzo-Merino, J.; Contreras-Paredes, A.; Vazquez-Ulloa, E.; Rocha-Zavaleta, L.; Fuentes-Gonzalez, A.M.; Lizano, M. The role of signaling pathways in cervical cancer and molecular therapeutic targets. *Arch. Med. Res.* **2014**, *45*, 525–539. [CrossRef] [PubMed]

124. Webster, M.T.; Rozycka, M.; Sara, E.; Davis, E.; Smalley, M.; Young, N.; Dale, T.C.; Wooster, R. Sequence variants of the axin gene in breast, colon, and other cancers: An analysis of mutations that interfere with GSK3 binding. *Genes Chromosomes Cancer* **2000**, *28*, 443–453. [CrossRef]

125. Su, T.H.; Chang, J.G.; Yeh, K.T.; Lin, T.H.; Lee, T.P.; Chen, J.C.; Lin, C.C. Mutation analysis of CTNNB1 (β-catenin) and AXIN1, the components of Wnt pathway, in cervical carcinomas. *Oncol. Rep.* **2003**, *10*, 1195–1200. [CrossRef] [PubMed]

126. Chen, Q.; Cao, H.Z.; Zheng, P.S. LGR5 promotes the proliferation and tumor formation of cervical cancer cells through the Wnt/β-catenin signaling pathway. *Oncotarget* **2014**, *5*, 9092–9105. [PubMed]

127. Rath, G.; Jawanjal, P.; Salhan, S.; Nalliah, M.; Dhawan, I. Clinical significance of inactivated glycogen synthase kinase 3β in HPV-associated cervical cancer: Relationship with Wnt/β-catenin pathway activation. *Am. J. Reprod. Immunol.* **2015**, *73*, 460–478. [CrossRef] [PubMed]

128. Chen, P.M.; Cheng, Y.W.; Wang, Y.C.; Wu, T.C.; Chen, C.Y.; Lee, H. Up-regulation of FOXM1 by E6 oncoprotein through the MZF1/NKX2-1 axis is required for human papillomavirus-associated tumorigenesis. *Neoplasia* **2014**, *16*, 961–971. [CrossRef] [PubMed]

129. Al-Shabanah, O.A.; Hafez, M.M.; Hassan, Z.K.; Sayed-Ahmed, M.M.; Abozeed, W.N.; Alsheikh, A.; Al-Rejaie, S.S. Methylation of SFRPS and APC genes in ovarian cancer infected with high risk human papillomavirus. *Asian Pac. J. Cancer Prev.* **2014**, *15*, 2719–2725. [CrossRef] [PubMed]

130. Van der Meide, W.F.; Snellenberg, S.; Meijer, C.J.; Baalbergen, A.; Helmerhorst, T.J.; van der Sluis, W.B.; Snijders, P.J.; Steenbergen, R.D. Promoter methylation analysis of Wnt/β-catenin signaling pathway regulators to detect adenocarcinoma or its precursor lesion of the cervix. *Gynecol. Oncol.* **2011**, *123*, 116–122. [CrossRef] [PubMed]

131. Perez-Plasencia, C.; Vazquez-Ortiz, G.; Lopez-Romero, R.; Pina-Sanchez, P.; Moreno, J.; Salcedo, M. Genome wide expression analysis in HPV16 cervical cancer: Identification of altered metabolic pathways. *Infect. Agent Cancer* **2007**, *2*, 16. [CrossRef] [PubMed]

132. Fragoso-Ontiveros, V.; Maria Alvarez-Garcia, R.; Contreras-Paredes, A.; Vaca-Paniagua, F.; Alonso Herrera, L.; Lopez-Camarillo, C.; Jacobo-Herrera, N.; Lizano-Soberon, M.; Perez-Plasencia, C. Gene expression profiles induced by E6 from non-european HPV18 variants reveals a differential activation on cellular processes driving to carcinogenesis. *Virology* **2012**, *432*, 81–90. [CrossRef] [PubMed]

133. Ramos-Solano, M.; Meza-Canales, I.D.; Torres-Reyes, L.A.; Alvarez-Zavala, M.; Alvarado-Ruiz, L.; Rincon-Orozco, B.; Garcia-Chagollan, M.; Ochoa-Hernandez, A.B.; Ortiz-Lazareno, P.C.; Rosl, F.; *et al.* Expression of *WNT* genes in cervical cancer-derived cells: Implication of WNT7A in cell proliferation and migration. *Exp. Cell Res.* **2015**, *335*, 39–50. [CrossRef] [PubMed]

134. He, Y.; Lin, J.; Ding, Y.; Liu, G.; Luo, Y.; Huang, M.; Xu, C.; Kim, T.K.; Etheridge, A.; Lin, M.; *et al.* A systematic study on dysregulated microRNAs in cervical cancer development. *Int. J. Cancer* **2015**. [CrossRef] [PubMed]

135. Bulut, G.; Fallen, S.; Beauchamp, E.M.; Drebing, L.E.; Sun, J.; Berry, D.L.; Kallakury, B.; Crum, C.P.; Toretsky, J.A.; Schlegel, R.; *et al.* β-catenin accelerates human papilloma virus type-16 mediated cervical carcinogenesis in transgenic mice. *PLoS ONE* **2011**, *6*, e27243. [CrossRef] [PubMed]

136. Matsuzawa, S.I.; Reed, J.C. Siah-1, SIP, and Ebi collaborate in a novel pathway for β-catenin degradation linked to p53 responses. *Mol. Cell* **2001**, *7*, 915–926. [CrossRef]

137. Lichtig, H.; Gilboa, D.A.; Jackman, A.; Gonen, P.; Levav-Cohen, Y.; Haupt, Y.; Sherman, L. HPV16 E6 augments Wnt signaling in an E6AP-dependent manner. *Virology* **2010**, *396*, 47–58. [CrossRef] [PubMed]

138. Sominsky, S.; Kuslansky, Y.; Shapiro, B.; Jackman, A.; Haupt, Y.; Rosin-Arbesfeld, R.; Sherman, L. HPV16 E6 and E6AP differentially cooperate to stimulate or augment Wnt signaling. *Virology* **2014**, *468–470*, 510–523. [CrossRef] [PubMed]

139. Li, X.; Yost, H.J.; Virshup, D.M.; Seeling, J.M. Protein phosphatase 2A and its B56 regulatory subunit inhibit Wnt signaling in Xenopus. *EMBO J.* **2001**, *20*, 4122–4131. [CrossRef] [PubMed]

The Human Papillomavirus E6 PDZ Binding Motif: From Life Cycle to Malignancy

Ketaki Ganti, Justyna Broniarczyk, Wiem Manoubi, Paola Massimi, Suruchi Mittal,
David Pim, Anita Szalmas, Jayashree Thatte, Miranda Thomas, Vjekoslav Tomaić and
Lawrence Banks

Abstract: Cancer-causing HPV E6 oncoproteins are characterized by the presence of a PDZ binding motif (PBM) at their extreme carboxy terminus. It was long thought that this region of E6 had a sole function to confer interaction with a defined set of cellular substrates. However, more recent studies have shown that the E6 PBM has a complex pattern of regulation, whereby phosphorylation within the PBM can regulate interaction with two classes of cellular proteins: those containing PDZ domains and the members of the 14-3-3 family of proteins. In this review, we explore the roles that the PBM and its ligands play in the virus life cycle, and subsequently how these can inadvertently contribute towards the development of malignancy. We also explore how subtle alterations in cellular signal transduction pathways might result in aberrant E6 phosphorylation, which in turn might contribute towards disease progression.

Reprinted from *Viruses*. Cite as: Ganti, K.; Broniarczyk, J.; Manoubi, W.; Massimi, P.; Mittal, S.; Pim, D.; Szalmas, A.; Thatte, J.; Thomas, M.; Tomaić, V.; Banks, L. The Human Papillomavirus E6 PDZ Binding Motif: From Life Cycle to Malignancy. *Viruses* **2015**, *7*, 3530–3551.

1. Human Papillomaviruses and Cancer

Human Papillomaviruses (HPV) are the primary causal agents of cervical cancer and are linked to a growing number of other malignancies [1]. Of these cancers, the most important is cervical cancer of which there are over 500,000 cases worldwide annually [1,2]. The cancer causing HPV types are termed high risk types and the most frequently occurring HPV types in cervical cancer are HPV 16 and HPV 18 [3]. HPV is also the causative agent of cancers at a number of other anogenital sites [4] and in a rapidly increasing proportion of head and neck cancers [5]. The hallmark of HPV induced cancer is the persistent expression of the early proteins E6 and E7, which function as oncogenes throughout development of the malignancy [6,7]. Indeed, ablation of E6 and/or E7 expression in cervical tumor derived cell lines [8,9], primary cervical tumor cells [10] and also in tumors in transgenic mice [11] expressing the E7 oncoprotein, results in a cessation of transformed cell growth and tumor repression, manifested through senescence [12] or apoptosis [13]. Therefore, these viral oncoproteins are excellent targets for therapeutic intervention in HPV induced malignancy.

2. Differences between the High and Low Risk Viruses

There are close to 200 HPV types that have been described thus far, but among these only a few are associated with cancer. These high-risk viruses, which include the predominant cancer causing viruses HPV16 and HPV18, also includes 10 other types (31, 33, 35, 39, 45, 51, 52, 56,

58, and 59), which are also defined as cancer causing by the WHO [3]. The HPV types that do not cause malignancy are termed "Low risk" and include the HPV types 6 and 11, which are generally associated with benign epithelial condylomas. The question of why some viruses are cancer causing, while others are not has garnered a great deal of attention. Detailed analyses of various viral genomes demonstrate a high degree of conservation between the virus types, as well as conservation in many of the cellular targets of the viral oncoproteins. One of the most obvious differences between the E6 oncoproteins of the high and low risk HPV types is the presence of a C-terminal PDZ (PSD-95/DLG/ZO-1) binding motif (PBM) [14] in the E6 oncoproteins of high risk viruses that is absent in the low risk HPV types. This PBM, therefore, represents a molecular signature for the oncogenic potential of the high risk E6 proteins [15,16].

PDZ domains are stretches of 80–90 amino acids in length and were described from the three proteins that were first shown to harbor these domains *i.e.*, the Post Synaptic Density 95 (PSD95), the Discs Large (Dlg) and the Zona Occludens 1 (ZO-1) proteins. These domains are sites of protein-protein interactions with the ligands containing the so-called PDZ binding motifs (PBMs). Many of these proteins contain multiple copies of these PDZ domains and frequently also contain other protein-protein interaction motifs, thereby allowing these proteins to be involved in a plethora of different activities. Indeed, in the case of many PDZ domain containing proteins, they are considered to act as scaffolding proteins, facilitating the assembly of multi-protein complexes [17].

The recognition of the PDZ domain is dependent upon the precise sequence of the PBM on the PDZ ligand. This PBM usually lies at the extreme C-terminus of the protein, but can also be found at internal sites [14]. There are three broad classes of PBM's defined by their C-terminal consensus sequences: type I PBM ($-X-S/T-X-\Phi_{COOH}$), type II PBM ($-X-\Phi-X-\Phi_{COOH}$), and type III PBM ($-X-D/E-X-\Phi_{COOH}$), where X is any residue and Φ is a hydrophobic residue [14,18,19]. However, there is a wide diversity in the types of PBM's, and 16 distinct subtypes have been proposed [14,17]. Furthermore, PDZ domain-ligand specificity is not solely dependent upon the last four amino acids but involves at least the last seven carboxy terminal amino acids, with both canonical and non-canonical residues in the PBM contributing towards substrate specificity [20]. As can be seen from Figure 1, the high risk HPV E6 oncoproteins, while all possessing the canonical $X-S/T-X-\Phi_{COOH}$ consensus site, nonetheless display significant degrees of variation within this region. The first intimation that these changes might be biologically relevant came from early studies on the interaction of HPV 16 and 18 E6 with two potential PDZ domain substrates, the Discs Large (Dlg) and the Scribble (Scrib) proteins. In this case, a single Leucine/Valine substitution at the C terminal residue of the E6 PBM, was instrumental in determining substrate preference, with a C terminal Valine in 18E6 conferring preference for Dlg interactions, with a C terminal Leucine in 16E6 conferring preference for Scrib [21]. Indeed, both crystallographic and NMR studies performed under conditions where the HPV18 E6 PBM is in complex with two different PDZ domain-containing substrates, Dlg and the Membrane-associated guanylate kinase, WW and PDZ domain-containing protein 1 (MAGI-1), defined the importance of non-canonical residues within the PBM for contributing towards substrate specificity [22–24]. As can be seen from Figure 1, apart from the key residues at p0 and p-2 which define the PBM, residues at p-3, p-4, p-5 and p-6, and in the unique case of HPV-18 E6 at p-11, play essential roles in determining substrate specificity in the recognition of both Dlg and/or MAGI-1. Interestingly, most of these residues are non-canonical and lie outside the core PBM consensus sequence. As can be seen, there is also a great degree of variation in the sequence of these residues among the different high risk E6 oncoproteins. This highlights the fact that there is a considerable variability in the way these E6 proteins recognize diverse PDZ domain substrates.

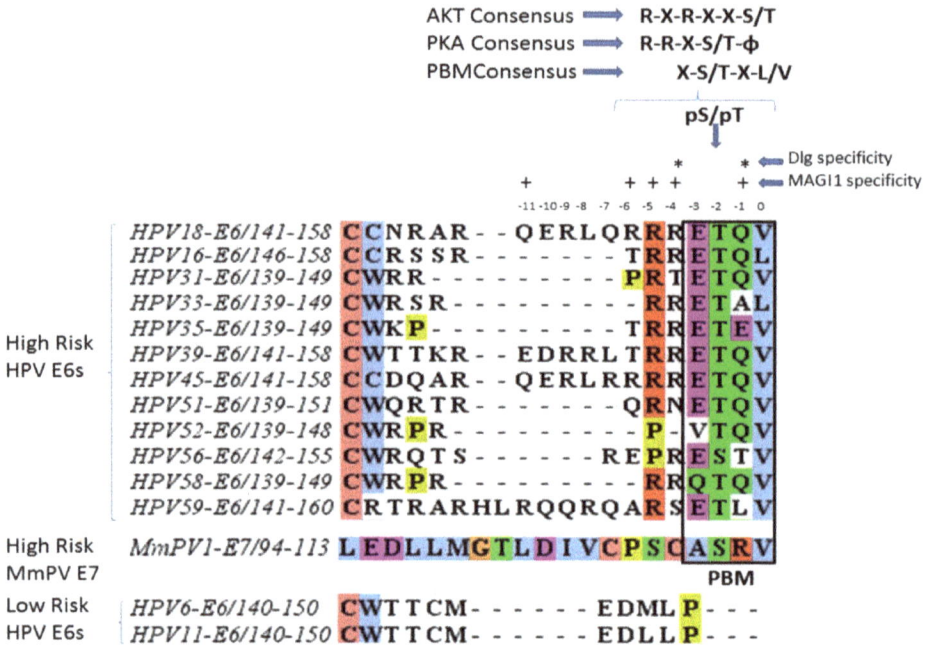

Figure 1. Papillomavirus oncoproteins show diversity in the PBMs and their kinase recognition sequences. The multiple sequence alignment of various HPV E6 proteins from different HPV types and the MmPV E7 using the Clustal X color scheme for the ClustalW sequence alignment program [28] show variation in the sequences of their C-terminus, which includes the PBM in the high risk HPV E6 proteins as well as the MmPV1 E7 protein, which is absent in the low risk HPV E6 types. The 4 boxed amino acids at the extreme C-terminus designate the canonical PBM. Whilst residues at p0 and -2 form the basis of PDZ recognition, the residues marked (*) indicate the amino acids that have been shown to be important for the specificity of E6 interaction with Dlg, while residues marked (+) designate the amino acids crucial for MAGI-1 specificity. The consensus sequences for AKT and PKA recognition of the E6 proteins is also shown.

High resolution NMR studies show that the upstream flanking regions of the PDZ1 domain of MAGI-1 increase the binding affinity to the PBM of E6 and also shows the PBM to have a disordered structure which becomes structured upon binding to the PDZ domain [22]. The number of critical contact sites is reflected directly in the strength of association between E6 and its different PDZ targets, with dissociation constants lying in the micromolar range with regard to MAGI-1, with a

K_D value of about 3.0 μM for 16E6 and 1.5 μM for 18E6 [23], with again the difference in the C-terminal residue playing a significant role. Taken together, these studies indicate conservation of the PBM across high-risk E6 types, although the identity of preferred PDZ domain substrates of these different HPV E6 oncoproteins is likely to vary. As will be seen from the following discussion, this is indeed borne out by the fact that different E6 oncoproteins target a large number of different PDZ domain containing proteins, many of which are involved in the control of the same cell signaling pathways [25] (Table 1). Recent proteomic studies also have identified potential novel interacting partners of HPV E6 that play a role in pathways other than cell polarity [25–27], such as endosomal transport (Table 1).

Table 1. Known PDZ Domain containing targets of E6.

Protein Name	Function	Reference
DLG1	Polarity/tumour suppressor	[15,16,21,29–34]
SCRIB	Polarity/tumour suppressor	[21,35,36]
MAGI1/2/3	Polarity/tumour suppressor	[24,32,37,38]
PSD95	Signaling complex scaffold	[39]
TIP2/GIPC	TGF-β signaling	[40]
NHERF1	PI3K/AKT signaling	[41]
MUPP1	Signaling complex scaffold	[42]
PATJ	Tight junction assembly	[43,44]
PTPH1/PTPN3	Protein tyrosine phosphatase	[45,46]
PTPN13	Non-receptor phosphatase	[47]
PDZRN3/LNX3	RING-containing ubiquitin ligase	[48]
14-3-3	Signaling complex adapter	[49,50]
PAR3	Polarity/tumour suppressor	[51]
SNX27	Endosomal trafficking/signaling	[26,27]
ARHGEF12	RhoGEF	[26]
FRMPD2	Tight junction formation	[26]
LRRC7	Normal synaptic spine function	[26]

A critical aspect of HPV E6 is its multifunctionality. This protein has been reported to interact with at least 30 different cellular substrates [52], and these are involved in all aspects of E6 activity, from roles in the virus life cycle to the development of malignancy. For example, interaction with the cellular ubiquitin ligase, E6AP would appear central to many of E6 activities, by allowing E6 to direct its targets for proteasome mediated degradation [52]. However, this interaction also appears to play a central role in regulating E6 stability and turnover, presumably through holding the E6 structure in a stable conformation [53–56]. Another critical interaction involves the p53 tumor suppressor, and the ability of E6 to target and degrade p53 appears essential for multiple E6 functions [52]. However, it is also clear from the E6 structure that E6 interaction with E6AP and p53, as well as many of its other substrates, is very complex and often involves overlapping interaction sites [54–57]. This makes elucidating the role of these interactions in HPV E6 biological functions very complex, as mutations which abolish one set of interactions are likely to affect other substrates, thereby making interpretations as to which interactions are relevant for each activity very difficult. In contrast, the PBM is a highly defined region of the E6 oncoprotein and structural studies indicate a disordered structure, well away from E6's other protein interaction sites [57]. This makes studies to understand the function of this region of E6 much easier as it allows specific mutations to be introduced in the PBM without affecting any of E6's other activities [58,59]. This is exemplified most clearly in studies on the virus life cycle.

3. The E6 PBM and the Viral Life Cycle

The productive phase of the HPV life cycle is intimately linked with the differentiation program of the host epithelium [60]. The virus infects the basal cells of the epithelium via micro abrasions (see Figure 2), where multiple viral genome copies are maintained as persistent episomes. The division of these infected cells and their movement through the upper strata of the epithelium as they differentiate is coupled with the expression of different viral proteins including E6, and with the amplification of viral DNA [61]. Studies with HPV genomes lacking E6 show that these genomes are unable to establish themselves as episomes within the cell [62,63]. It is speculated that the roles of E6 in the viral life cycle range from the inhibition of apoptosis that is induced by the E7 oncoprotein to the facilitation of viral DNA amplification within the upper epithelial strata [64]. Studies in Human Foreskin Keratinocytes (HFKs) transfected with HPV 31 genomes have shown that the presence of the E6 and E7 oncoproteins is crucial for the stable replication of viral DNA, but not for transient replication [63]. The maintenance of viral copy number and proliferation, as well as episomal maintenance of viral DNA in undifferentiated cells, also requires the presence of the PBM, as demonstrated by the reduced replicative capacity and growth rates of HFKs stably maintaining PBM mutant genomes of HPV 31 [34].

HFKs containing HPV 18 genomes also displayed similar phenotypes, with the cells expressing E6 ΔPBM containing genomes proliferating somewhat more slowly than the cells containing wild type HPV 18 genomes. Furthermore, in organotypic cultures, the E6 ΔPBM genomes exhibited defects in the levels to which the viral genomes could be amplified and a corresponding decrease in the number of suprabasal cells that were capable of replicating DNA. Most interestingly, extended periods of passaging resulted in a loss of E6 ΔPBM genomes, indicating that the E6 PBM plays an essential role in episomal maintenance and in creating an environment favorable for viral DNA replication [65].

Normal Immortal Keratinocytes (NIKS) containing HPV 16 genomes with E6 ΔPBM mutations also exhibited significant defects in the viral life cycle. However, in this particular instance, this was correlated with decreased levels of expression of the HPV 16 E6 protein [66]. This suggested that at least one of the E6's PDZ target proteins might play a role in regulating E6 stability or levels of expression. In support of this, loss of Scrib in HeLa cells was found to greatly reduce the levels of HPV 18E6 protein. However, the molecular basis for this effect remains to be determined [66]. It was also observed that the HPV 16 E6 ΔPBM genomes failed to be maintained as episomes in NIKS cells, compared with wild type genomes.

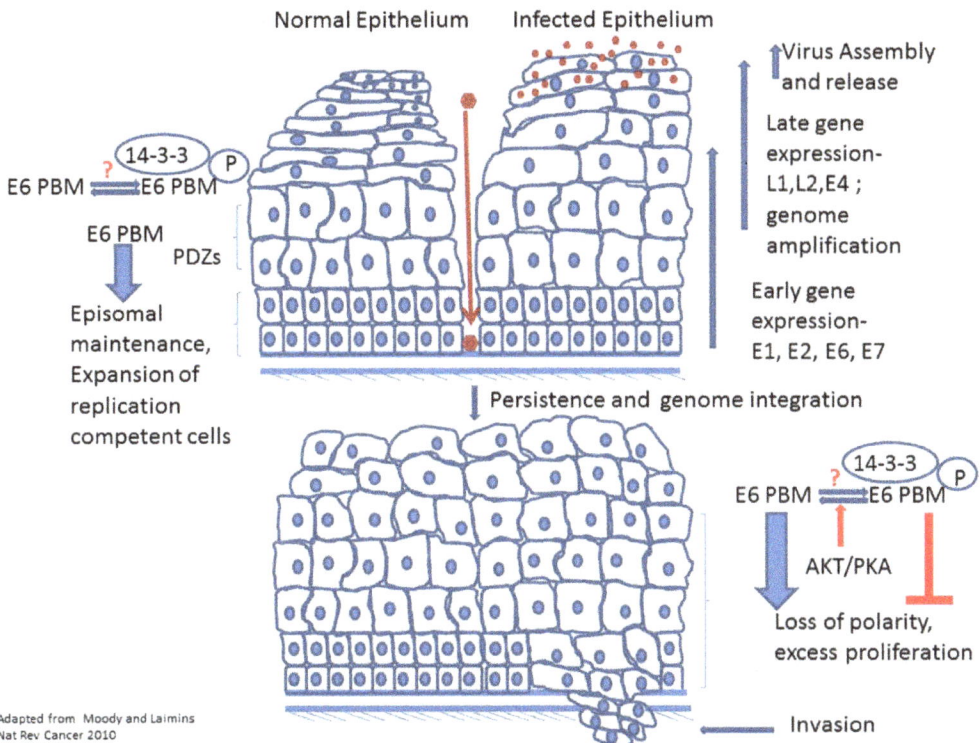

Figure 2. The role of the E6 PBM in the HPV life cycle and malignancy. The figure shows the productive life cycle of the virus after infection of the epithelium with coordinate expression of the different viral gene products during epithelial differentiation, ultimately resulting in the production of new infectious virus particles. The E6 PBM function, most likely through PDZ targeting, is required for expansion of replication competent cells and for maintenance of the viral episomal DNA. During differentiation and viral DNA amplification in the G2M like phase of the cell cycle, E6 will most likely be phosphorylated within the PBM, which could confer interaction with 14-3-3 proteins. Following a persistent infection of up to 20 years, the progression of HPV induced malignancy can occur. The role of the E6 PBM in this stage is unknown, but PDZ targeting might contribute to loss of cell polarity regulators and drive proliferation and invasion. Phosphorylation of the E6 PBM might be a means of negatively regulating this activity of E6.

Recent studies in NIKS containing either wild type HPV16 genomes or genomes with mutant forms of E6 that were either prematurely terminated to ablate all forms of E6 or were lacking just the last two amino acids of the PBM, thus making them unable to interact with PDZ proteins, show that, although the PBM defective mutant is not maintained stably as episomes in these cells through extended passages as described previously [66], this phenotype can be reversed by the repression of p53. The repression of p53, either through shRNA or through the expression of a dominant negative form of the protein, stabilizes the PBM defective 16E6 mutant genome in these NIKS, and the genome is now capable of being maintained as a stable episome [67]. These results are particularly intriguing since the PBM deletion mutants of E6 target p53 as efficiently as wild type E6. This indicates that at least one aspect of E6 PDZ targeting links directly to pathways that are controlled by p53 [67,68]. It is also possible that the PDZ-PBM interaction of wild type E6 contributes towards the abrogation of a p53 function that is independent of the degradation of p53, and this functional ablation is necessary for viral genome maintenance. Clearly, elucidating the molecular basis for the link between PDZ targeting and p53 is a particularly exciting avenue of future research.

Taken together these studies indicate that the E6 PBM plays an essential role in the virus life cycle, being required for expansion of the population of replication-competent cells and thereby increasing the number of cells in which viral genome amplification can occur. A recent intriguing study demonstrated that evolution of the E6 PBM most likely preceded acquisition of an oncogenic phenotype, suggesting the possibility that the PBM may also play an important role in specific tissue tropism or niche selection [69], something that seems to be supported by studies described below in which the biological effects of the E6 PBM varies, depending upon the tissue or origin of cells.

4. The Role of the PBM in Transformation and Cancer

The ability of the high risk HPV E6 oncoproteins to effectively recognize and bind to cellular PDZ domain containing targets has been shown to be important for the transformation potential of E6, both *in vitro* [16,70] and *in vivo* [58], although differences are observed depending on cell type and the anatomical site. Studies in immortalized human keratinocytes that the HPV 18E6 transfected keratinocytes show an exaggerated epithelial to mesenchymal (EMT) phenotype and changes in actin cytoskeletal organization that are significantly reduced in the cells that contain HPV 18E6 PBM mutant cDNA. Furthermore, the wild type E6 expressing cells also show aberrant adherens junction and desmosome formation not observed in the E6 PBM mutant containing cells, implying that the E6 PBM contributes towards the development of transformed characteristics in primary keratinocytes [70]. Likewise, the E6 PBM has been shown to contribute towards anchorage-independent growth in murine and human tonsillar keratinocytes, as well as contributing to their immortalization [47]. The E6 PBM is also necessary for the inhibition of apoptosis in human airway epithelial cells [71]. In the case of rodent cells, the E6 PBM appears to contribute towards the ability of E6 to promote transformation, although this is not observed in mammary epithelial cell or foreskin keratinocyte immortalization assays [16,34,65,72,73]. This indicates that the E6 PBM itself plays a limited role in immortalization *per se*, most likely functioning in the later stages of malignant transformation; however, there are also clear elements of context dependence, where the E6 PBM has a more active role in certain cell types and/or contexts than in others.

Transgenic mice have been used extensively to dissect the role of E6 and E7 in the development of malignancy. This has been based on the expression of the viral oncogenes from a keratin 14 (K14) promoter, which ensures expression of the viral proteins in the correct cell type *i.e.*, basal cells of the cutaneous and squamous epithelia. Mice expressing HPV 16 E6 in the epidermis develop epithelial hyperplasia, which progresses to squamous carcinoma. In contrast, transgenic mice that express an E6 ΔPBM mutant, fail to develop hyperplasia, although they show a radiation response similar to the mice expressing wild type 16E6, as demonstrated by the lack of p53 or p21 induction upon irradiation. This indicates that both the wild type and the E6 ΔPBM mutant are capable of targeting and degrading p53, but that the ability to cause hyperplasia and promote progression towards a transformed phenotype is dependent on the presence of an intact and functional PBM [58]. In the cervix, HPV16 E6 and E7 promote the development of tumors in the presence of chronic estrogen treatment [74,75]. Additionally important to note is that although E6 expressing mice develop tumors in the presence of estrogen, E7 is more efficient in initiating tumor formation in a similar setting, indicating that E6 most likely plays a greater role later in tumor progression [74]. In addition, mice that express wild type 16E6 alone develop cervical tumors at a greater frequency if estrogen treatment is extended to 9 months, whilst the E6 ΔPBM mice have fewer and smaller tumors [59]. There is also a significant reduction in the tumor size and multiplicity of tumors in mice expressing the PBM deletion mutant in co-operation with wild type E7 when compared with mice that express wild type E6 and E7 [59].

Studies with transgenic mouse models of head and neck cancer, as well as models of anal cancer, differ from those of the skin and cervix with respect to the potential role of the E6 PBM in the development of these cancers [76,77]. In the case of head and neck cancer, there is a co-operation between the E6 and E7 oncoproteins in the induction of squamous cell carcinoma, with E7 being the more potent oncogene in the presence of the chemical carcinogen 4-nitroquinoline-1-oxide (4-NQO) [76,78]. The studies with mice that express the E6 ΔPBM, showed that the ability of E6 to co-operate with E7 to cause head and neck squamous cell carcinoma (HNSCC) does not require an intact PBM function [76]. These results indicate that although E6 and E7 synergize to induce HNSCC in transgenic mice, this most likely involves a mechanism that is independent of the E6 PDZ binding potential. Similar phenotypes have been observed in mouse models of anal carcinoma, where HPV16 E7 is more effective in inducing proliferation in the anal epithelium compared with mice expressing 16 E6 alone when treated with the chemical carcinogen DMBA. Additionally important to note is that the co-expression of E6 and E7 in these mice does not seem to increase tumor induction rate or size compared with mice expressing E7 alone [77]. These observations again implicate E7 as the more potent oncogene in the development of anal cancer similar to HNSCC, although further studies with mutant E6 mice and a reduction of the treatment time with DMBA in the case of anal cancer may reveal potential roles of E6 in anal carcinogenesis.

These studies demonstrate a potent contribution of the E6 oncoprotein and the E6 PBM towards the development of certain malignancies in transgenic mice. However, the precise contribution of the PBM varies depending upon the anatomical site, again indicating a degree of context dependence in which the E6 PBM functions. This context dependence is a recurring theme in cell polarity signaling pathways; whose control is largely governed by PDZ domain containing proteins.

5. E6, Cell Adhesion and Polarity

Epithelial tissues have a characteristic polarized cellular architecture and specialized cell-cell junctions, including desmosomes, tight and adherens junctions. There are a number of signaling and polarity complexes that are involved in the recruitment of proteins to the junctions and in the establishment of cell polarity [79]. Cell polarity plays a crucial role in the organization of signaling pathways, which allow the interpretation of signals from the surrounding microenvironment, thus enabling the control of proliferation, metabolism, apoptosis, differentiation and motility [80]. Cell polarity is maintained by an intricate interplay between conserved groups of proteins that form distinct complexes:The Crumbs, Par and Scribble complexes. Apical polarity is controlled by the Crumbs complex, made up of Crumbs3, Pals1 and PatJ [81,82]. The Par complex consists of Par3, Cdc42, Par6 and atypical protein kinase C (aPKC), which is a dynamic complex and which also interacts with the Crumbs complex [80]. The Scribble complex consists of the scaffolding proteins Scribble, Dlg and Hugl, which maintain basolateral polarity. Of these cell polarity proteins, PatJ, Par3, Scrib and Dlg all possess PDZ domains and, hence, are potential targets of E6 [83]. It is important to remember that the control of cell polarity hinges on the correct spatio-temporal regulation of the expression levels of the cell polarity regulators. Alterations in expression levels or mislocalization of any of these components perturbs the function of the complex as a whole and leads to aberrant signaling that may be a driver of neoplastic transformation [84].

One of the primary characteristics of the transition of a benign neoplasm to a malignant phenotype is a major disorganization of cellular architecture, which includes the loss of cell-cell contact and polarity, either through the degradation or mislocalization of the components that regulate these processes. The impact of the loss of polarity can be profound- the perturbation of the trafficking of proteins to the apical or basolateral regions may cause aberrant signaling due to mislocalization of receptors, or a redistribution of cell adhesion molecules that can promote cellular transformation in EMT. Altered polarity can also lead to an inappropriate distribution of degradative enzymes such as matrix metalloproteinases at the cell surface, thus promoting cell invasion and transformation, as well as affecting migration and cytoskeletal organization [85]. Many of the key regulators of cell polarity and adhesion are PDZ domain containing proteins, a number of which have been reported to be targets of high risk HPV E6 oncoproteins, albeit with varying affinities (See Figure 3), and these include the core components of the cell polarity control machinery; the Scribble-Dlg [21,33,36], the Par-aPKC [51,86], and the Crumbs complexes [43,44]. Furthermore, a number of cell junction proteins are targeted by E6 and these include the MAGI group of proteins [37,38], all of which are crucial for maintaining junctional stability and integrity.

The Scribble-Dlg axis appears to be one of the *bona fide* targets of the high risk E6 oncoproteins, with numerous studies showing that both proteins can be bound and degraded by E6 *in vitro* and *in vivo* [33,36,87]. It has also been observed that overexpression of Scrib inhibits the transformation of rodent epithelial cells by high risk E6 and E7 proteins, indicating a potential role for Scribble as a tumor suppressor [21,36,58]. Indeed, deregulation of Scribble function is seen in a wide range of epithelial cancers- including colorectal [88] , breast [89], prostate [90], and endometrial cancers [91]. In the case of cervical cancer, the expression patterns of Scrib and Dlg are severely perturbed during tumor development [92–94]. A general trend during the development of cervical cancer is the unusual cytoplasmic distribution of Dlg in cervical intraepithelial precursor lesions, as opposed to the cell-cell contact localization seen in normal tissue. In contrast, a loss of Dlg is observed only in late stage invasive cervical cancer [92]. This trend is also observed in the case of

Scribble, where a redistribution of Scribble is observed from sites of cell contact in normal squamous cells to the cytoplasm in early dysplasia, followed by a steady reduction in protein levels as the tumor progresses [95]. Whether these perturbations to Dlg and Scrib expression during the progression of cervical cancer are due to the effects of E6 is still an open question and subject of intense research.

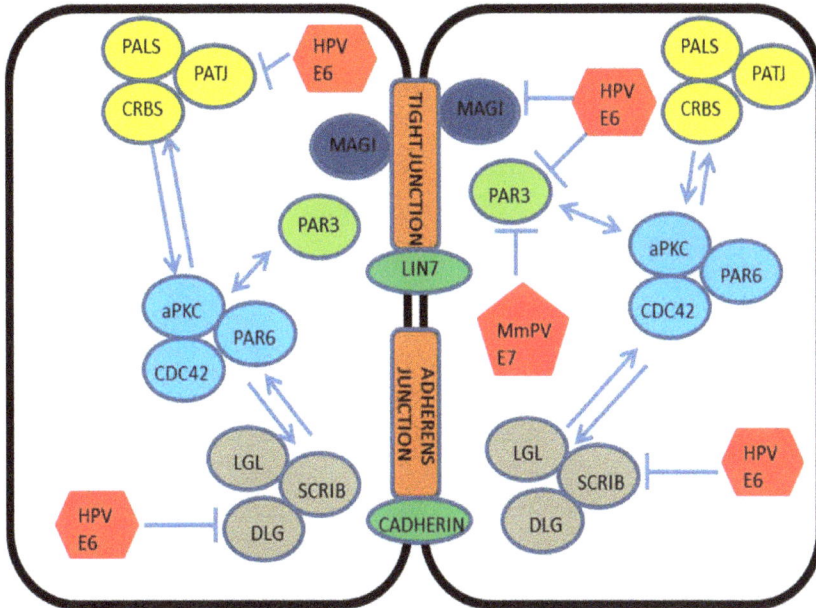

Figure 3. Papillomavirus oncoprotein targeting of cell polarity proteins. The cartoon depicts the various proteins comprising the three major complexes that regulate cell polarity: apical is defined by the Crumbs (CRBS) complex, subapical by the Par complex and basolateral by the Scrib complex. These complexes interact through a series of mutually antagonistic interactions ensuring correct spatial distribution and levels of expression of the individual components. Note the propensity of HPV E6 and MmPV E7 to target diverse components of this cell polarity control network, thereby perturbing their levels of expression or subcellular distribution.

It is important to note that E6 does not induce the degradation of the entire pool of Scribble and Dlg, but instead only targets a specific subset of the proteins. In the case of Dlg and HPV 18E6, only phosphorylated and nuclear fractions of Dlg appear to be targeted for proteasome mediated degradation. Paradoxically, both Dlg and Scribble show pro-oncogenic activity in certain contexts [96,97]. For example, the adenovirus E4 protein triggers Dlg translocation to the plasma membrane of epithelial cells where it goes on to activate PI3K, with the assembly of a ternary complex made up of Dlg1, PI3K, E4 and Akt [98,99]. This membrane associated PI3K is constitutively active and mediates Akt signaling which is associated with tumorigenesis. Scrib has also been found to be overexpressed and mislocalized in breast cancer and has been linked to a gain of pro-oncogenic activity [100]. It is also interesting to hypothesize that such pro-oncogenic functions of PDZ proteins can be manipulated by the HPV E6 protein to create a favorable environment for cellular proliferation.

Indeed, recent studies showed that E6 stimulates RhoG activity by a mechanism that is dependent upon its interaction with Dlg and the RhoG guanine nucleotide exchange factor SGEF. Thus, in HPV positive cells, there are high levels of RhoG activity, resulting in increased invasive potential, which is directly dependent upon continued expression of E6, Dlg and SGEF [101]. These findings demonstrate a hitherto unknown pro-oncogenic function of Dlg in HPV transformed cells. It is thus tempting to speculate that the mislocalization or overexpression of Dlg or Scrib may induce a switch from tumor suppressor to a pro-oncogenic function, possibly due to the alteration in the pool of their interacting partners, thus modulating their function, especially in the case of intermediate grade tumors, thus increasing their potential for progression to invasive cancers [102].

The MAGI family of proteins has been shown to be one of the most strongly bound and most susceptible groups of proteins to be targeted by the high risk E6 proteins [22,37,38,103,104]. MAGI-1 in particular is a major degradation target of both HPV 16E6 and HPV 18E6 [87]. As in the case of Dlg, only certain pools of the protein are targeted, with the membrane and nuclear pools of MAGI-1 being susceptible to proteasome-mediated degradation by both HPV 16E6 and HPV 18E6. The loss of TJ integrity is a direct consequence of the loss of MAGI-1 in cervical cancer cell lines HeLa and CaSKi, which is reinstated upon E6 ablation and the re-emergence of MAGI-1 expression in these cells [87]. It is also interesting to note that the reintroduction of a mutant MAGI-1 that is resistant to E6 induced degradation in HPV positive cells leads to the accumulation of ZO-1 and Par3 at cell contact sites, as well as a significant reduction in cell proliferation and an increase in the number of apoptotic cells [105]. This finding sheds light upon the pathological consequence of the loss of MAGI-1 in HPV positive cells, which includes loss of tight junction stability, an increase in proliferation and a suppression of apoptosis, all of which can be expected to enhance the progression of hyperplastic lesions into metastatic cancer. Interestingly, from a life cycle point of view, this dissociation of the control of cell proliferation by contact inhibition might be means by which E6 targeting of these junctional complexes can be expected to induce proliferation of the suprabasal epithelial cells during the virus life cycle (See Figure 2).

The only other Papillomavirus known to cause cervical cancer in its natural host is the *Macaca mulatta* Papillomavirus type 1 (MmPV1). This virus is very similar to HPV16, is sexually transmitted and causes cervical cancer in rhesus macaques [106–108]. However, unlike the high risk HPVs, the E6 protein of MmPV1 does not contain a C-terminal PBM. Instead, a class I PBM (ASRV) is present on the C-terminus of the E7 protein of this virus. As can be seen from Figure 1, the sequence of the E7 PBM is quite distinct from that found in the high risk HPV E6 proteins [86], however it nevertheless fits perfectly within the consensus sequence. Not surprisingly, such a difference in the PBM is reflected in differences in the PDZ proteins bound by MmPV1 E7, with only very weak levels of interaction seen with Scribble and Dlg. Indeed, the preferred PDZ substrate of the MmPV1 E7 protein is Par3, which, as noted above, is a critical component in the cell polarity control pathway. Par3 defines the sub-apical region of the cell and maintains apico-basal polarity in conjunction with the Crumbs and Scribble complexes and therefore belongs to the same polarity control pathway as Scrib and Dlg [86,109,110]. Whilst there is no information available on the status of Par3 in

Rhesus macaque cervical cancers, it is clear that MmPV1 E7 perturbs the pattern of Par3 expression in cultured cells, and can therefore be expected to perturb the correct functioning of the cell polarity network. This interaction of MmPV1 E7 with Par3 is also biologically significant, as Par3 binding-defective mutants of MmPV1 E7 lose their ability to transform primary rodent cells [86]. It thus appears that the high risk HPVs and MmPV1 target the same pathway of polarity control using a similar mechanism of PDZ recognition, although they target different components of the pathway through the action of different viral oncoproteins, indicating a high degree of evolutionary conservation across these different cancer causing Papillomaviruses.

6. The Multifunctionality of the E6 Oncoprotein

As noted above, the E6 oncoproteins are multifunctional with many different interacting partners required for their multiplicity of function, both during the changing environment of the virus life cycle and during cancer development. Not surprisingly, the E6 PBM is equally multifunctional. Embedded within the PBM of all of the high risk HPV E6 oncoproteins is a potential phospho acceptor site (see Figure 1). It was first shown for HPV 18E6 that Protein Kinase A (PKA) could very efficiently phosphorylate the Threonine residue at position 156 within the PBM, and this in turn resulted in a dramatic inhibition of the ability of E6 to interact with its PDZ substrate Dlg [111]. This is in accordance with observations that phosphorylation of the PBM generally abrogates PDZ-PBM interactions and this is because the phospho moiety cannot be accommodated in the PDZ binding pocket [112]. Not surprisingly, this phospho-dependent inhibition of PDZ recognition is also true for other HPV E6 oncoproteins and other PDZ domain containing substrates [49]. What is more surprising is that the regulation of the different HPV E6 PBMs is not controlled by the same kinase. For example, HPV 18E6 is only phosphorylated by PKA, whilst HPV 16E6 can be phosphorylated by PKA or AKT [49]. A similar situation also holds true for some HPV E6 proteins from other virus types. This raises the surprising possibility that there are significant differences in how the different HPV E6 oncoproteins are regulated. For example, AKT levels are high in proliferating cells [113], whilst PKA levels are high during differentiation [65,70], and this suggests subtle differences in the mechanisms by which the interaction of the high risk HPV E6 oncoproteins with the different PDZ substrates are modulated.

So, does the phosphorylation of the PBM simply act to negatively regulate PDZ recognition or can it confer an additional function upon the E6 protein? Indeed, recent studies have shown that phosphorylation of the E6 PBM confers a strong direct interaction with members of the 14-3-3 family of proteins [49]. 14-3-3 proteins are a group of highly conserved acidic proteins. There are seven known isoforms of 14-3-3 present in mammals that are encoded by seven distinct genes. 14-3-3 proteins bind to a large repertoire of proteins mainly in a phospho-specific manner [114–117]. 14-3-3 proteins function as adapter proteins and interact with a plethora of cellular proteins involved in a wide variety of processes, including signal transduction, apoptosis, metabolic control, cytoskeletal maintenance, tumor suppression, and transcription [118]. The interaction of E6 with 14-3-3 in a phospho-specific manner thus raises intriguing questions as to its role in the modulation of 14-3-3 function or vice versa. The phosphorylation of E6 confers preferred association with 14-3-3 zeta, which in turn seems to be important for maintaining E6 stability in HeLa cells [49].

It therefore seems likely that the E6 PBM function will be differentially regulated through the progression of the viral life cycle, both in the context of recognition of different PDZ containing substrates as well as its interaction with phosphorylation-dependent cellular proteins, such as 14-3-3. Indeed, mutation of the PKA consensus recognition site in HPV 18E6, in the context of the whole genome in organotypic cultures, leads to a more hyperplastic and stratified phenotype, most likely as a result of conferring constitutive interaction with PDZ domain containing substrates [65]. So, what do these observations mean in the context of the fine regulation of E6 PBM function during the course of the viral life cycle and in the development of malignancy? Certainly it is plausible that changes in cellular signal transduction pathways might be reflected in altered patterns of phosphorylation of E6. This in turn could promote or restrict malignant progression, depending upon the specific situation. Since 14-3-3 proteins are heavily involved in the regulation of the cell cycle, it also raises the possibility that the interaction of E6 with 14-3-3 may be modulating its function so as to maintain an environment favorable for viral genome amplification, as for example in the G2/M phase of the cell cycle even in the absence of appropriate signals. It is also tempting to speculate that this intricate phospho-regulation may have arisen as a way of allowing compartmentalization of E6 function during the various stages of the viral life cycle. In this way, phosphorylated E6 will be sequestered by the 14-3-3 proteins and will thus be unable to target the PDZ proteins that are crucial for maintaining structural integrity of the infected cell and appear to be required for promoting cellular proliferation. It is also plausible that the aberrant regulation of E6 phosphorylation may be a prognostic marker for the predisposition of benign lesions to progress into invasive cancer. Whether this is due to a lack or gain of phosphorylation is obviously an aspect requiring further investigation. Future studies involving the dissection of the role of phospho-E6 and its possible effects on 14-3-3 activity, both during the virus life cycle and during the progression to malignancy, should yield fascinating insights into the function of this highly dynamic and multifunctional region of the E6 oncoprotein.

Acknowledgements

L.B. gratefully acknowledges research support from the Associazione Italiana per la Ricerca sul Cancro (AIRC) and the Wellcome Trust. K.G. and J.T. are recipients of the ICGEB Arturo Falaschi Fellowships and are registered students of the Open University, UK.

Conflicts of Interest

The authors declare no conflict of interest.

References

1. Zur Hausen, H. Papillomaviruses and cancer: From basic studies to clinical application. *Nat. Rev. Cancer* **2002**, *2*, 342–350. [CrossRef] [PubMed]
2. Parkin, D.M.; Bray, F. Chapter 2: The burden of HPV-related cancers. *Vaccine* **2006**, *24*. [CrossRef] [PubMed]
3. Bouvard, V.; Baan, R.; Straif, K.; Grosse, Y.; Secretan, B.; El Ghissassi, F.; Benbrahim-Tallaa, L.; Guha, N.; Freeman, C.; Galichet, L.; *et al.* A review of human carcinogens—Part B: Biological agents. *Lancet Oncol.* **2009**, *10*, 321–322. [CrossRef]
4. Wakeham, K.; Kavanagh, K. The Burden of HPV-Associated Anogenital Cancers. *Curr. Oncol. Rep.* **2014**, *16*, 402. [CrossRef] [PubMed]
5. Garbuglia, A.R. Human papillomavirus in head and neck cancer. *Cancers* **2014**, *6*, 1705–1726. [CrossRef] [PubMed]
6. Smotkin, D.; Wettstein, F.O. Transcription of human papillomavirus type 16 early genes in a cervical cancer and a cancer-derived cell line and identification of the E7 protein. *Proc. Natl. Acad. Sci. USA* **1986**, *83*, 4680–4684. [CrossRef] [PubMed]
7. Androphy, E.J.; Hubbert, N.L.; Schiller, J.T.; Lowy, D.R. Identification of the HPV-16 E6 protein from transformed mouse cells and human cervical carcinoma cell lines. *EMBO J.* **1987**, *6*, 989–992. [PubMed]
8. Butz, K.; Ristriani, T.; Hengstermann, A.; Denk, C.; Scheffner, M.; Hoppe-Seyler, F. siRNA targeting of the viral E6 oncogene efficiently kills human papillomavirus-positive cancer cells. *Oncogene* **2003**, *22*, 5938–5945. [CrossRef] [PubMed]
9. Yoshinouchi, M.; Yamada, T.; Kizaki, M.; Fen, J.; Koseki, T.; Ikeda, Y.; Nishihara, T.; Yamato, K. *In vitro* and *in vivo* growth suppression of human papillomavirus 16-positive cervical cancer cells by E6 siRNA. *Mol. Ther.* **2003**, *8*, 762–768. [CrossRef] [PubMed]
10. Magaldi, T.G.; Almstead, L.L.; Bellone, S.; Prevatt, E.G.; Santin, A.D.; DiMaio, D. Primary human cervical carcinoma cells require human papillomavirus E6 and E7 expression for ongoing proliferation. *Virology* **2012**, *422*, 114–124. [CrossRef] [PubMed]
11. Jabbar, S.F.; Abrams, L.; Glick, A.; Lambert, P.F. Persistence of high-grade cervical dysplasia and cervical cancer requires the continuous expression of the human papillomavirus type 16 E7 oncogene. *Cancer Res.* **2009**, *69*, 4407–4414. [CrossRef] [PubMed]

12. Goodwin, E.C.; DiMaio, D. Repression of human papillomavirus oncogenes in HeLa cervical carcinoma cells causes the orderly reactivation of dormant tumor suppressor pathways. *Proc. Natl. Acad. Sci. USA* **2000**, *97*, 12513–12518. [CrossRef] [PubMed]

13. DeFilippis, R.A.; Goodwin, E.C.; Wu, L.; DiMaio, D. Endogenous human papillomavirus E6 and E7 proteins differentially regulate proliferation, senescence, and apoptosis in HeLa cervical carcinoma cells. *J. Virol.* **2003**, *77*, 1551–1563. [CrossRef] [PubMed]

14. Songyang, Z.; Fanning, A.S.; Fu, C.; Xu, J.; Marfatia, S.M.; Chishti, A.H.; Crompton, A.; Chan, A.C.; Anderson, J.M.; Cantley, L.C. Recognition of unique carboxyl-terminal motifs by distinct PDZ domains. *Science* **1997**, *275*, 73–77. [CrossRef] [PubMed]

15. Lee, S.S.; Weiss, R.S.; Javier, R.T. Binding of human virus oncoproteins to hDlg/SAP97, a mammalian homolog of the Drosophila discs large tumor suppressor protein. *Proc. Natl. Acad. Sci. USA* **1997**, *94*, 6670–6675. [CrossRef] [PubMed]

16. Kiyono, T.; Hiraiwa, A.; Fujita, M.; Hayashi, Y.; Akiyama, T.; Ishibashi, M. Binding of high-risk human papillomavirus E6 oncoproteins to the human homologue of the Drosophila discs large tumor suppressor protein. *Proc. Natl. Acad. Sci. USA* **1997**, *94*, 11612–11616. [CrossRef] [PubMed]

17. Ye, F.; Zhang, M. Structures and target recognition modes of PDZ domains: Recurring themes and emerging pictures. *Biochem. J.* **2013**, *455*, 1–14. [CrossRef] [PubMed]

18. Hung, A.Y.; Sheng, M. PDZ domains: Structural modules for protein complex assembly. *J. Biol. Chem.* **2002**, *277*, 5699–5702. [CrossRef] [PubMed]

19. Nourry, C.; Grant, S.G.N.; Borg, J.-P. PDZ domain proteins: Plug and play! *Sci. STKE* **2003**, *2003*, RE7. [CrossRef] [PubMed]

20. Tonikian, R.; Zhang, Y.; Sazinsky, S.L.; Currell, B.; Yeh, J.-H.; Reva, B.; Held, H.A.; Appleton, B.A.; Evangelista, M.; Wu, Y.; *et al.* A specificity map for the PDZ domain family. *PLoS Biol.* **2008**, *6*, e239. [CrossRef] [PubMed]

21. Thomas, M.; Massimi, P.; Navarro, C.; Borg, J.-P.; Banks, L. The hScrib/Dlg apico-basal control complex is differentially targeted by HPV-16 and HPV-18 E6 proteins. *Oncogene* **2005**, *24*, 6222–6230. [CrossRef] [PubMed]

22. Charbonnier, S.; Nominé, Y.; Ramírez, J.; Luck, K.; Chapelle, A.; Stote, R.H.; Travé, G.; Kieffer, B.; Atkinson, R.A. The structural and dynamic response of MAGI-1 PDZ1 with noncanonical domain boundaries to the binding of human papillomavirus E6. *J. Mol. Biol.* **2011**, *406*, 745–763. [CrossRef] [PubMed]

23. Fournane, S.; Charbonnier, S.; Chapelle, A.; Kieffer, B.; Orfanoudakis, G.; Travé, G.; Masson, M.; Nominé, Y. Surface plasmon resonance analysis of the binding of high-risk mucosal HPV E6 oncoproteins to the PDZ1 domain of the tight junction protein MAGI-1. *J. Mol. Recognit.* **2011**, *24*, 511–523. [CrossRef] [PubMed]

24. Zhang, Y.; Dasgupta, J.; Ma, R.Z.; Banks, L.; Thomas, M.; Chen, X.S. Structures of a Human Papillomavirus (HPV) E6 Polypeptide Bound to MAGUK Proteins: Mechanisms of Targeting Tumor Suppressors by a High-Risk HPV Oncoprotein. *J. Virol.* **2007**, *81*, 3618–3626. [CrossRef] [PubMed]

25. Vincentelli, R.; Luck, k.; Poirson, J.; Abdat, J.; Blémont, M.; Turchetto, J.; Iv, F.; Ricquier, K.; Straub, M.-L.; *et al.* Quantifying domain-ligand affinities and specificities by high-throughput holdup assay. *Nat. Methods* **2015**. [CrossRef]

26. Belotti, E.; Polanowska, J.; Daulat, A.M.; Audebert, S.; Thomé, V.; Lissitzky, J.-C.; Lembo, F.; Blibek, K.; Omi, S.; Lenfant, N.; *et al.* The human PDZome: A gateway to PSD95-Disc large-zonula occludens (PDZ)-mediated functions. *Mol. Cell. Proteomics* **2013**, *12*, 2587–2603. [CrossRef] [PubMed]

27. Rozenblatt-Rosen, O.; Deo, R.C.; Padi, M.; Adelmant, G.; Calderwood, M.A.; Rolland, T.; Grace, M.; Dricot, A.; Askenazi, M.; Tavares, M.; *et al.* Interpreting cancer genomes using systematic host network perturbations by tumour virus proteins. *Nature* **2012**, *487*, 491–495. [CrossRef] [PubMed]

28. Van Doorslaer, K.; Tan, Q.; Xirasagar, S.; Bandaru, S.; Gopalan, V.; Mohamoud, Y.; Huyen, Y.; McBride, A.A. The Papillomavirus Episteme: A central resource for papillomavirus sequence data and analysis. *Nucleic Acids Res.* **2013**, *41*, D571–D578. [CrossRef] [PubMed]

29. Du, M.; Fan, X.; Hanada, T.; Gao, H.; Lutchman, M.; Brandsma, J.L.; Chishti, A.H.; Chen, J.J. Association of cottontail rabbit papillomavirus E6 oncoproteins with the hDlg/SAP97 tumor suppressor. *J. Cell. Biochem.* **2005**, *94*, 1038–1045. [CrossRef] [PubMed]

30. Liu, Y.; Henry, G.D.; Hegde, R.S.; Baleja, J.D. Solution structure of the hDlg/SAP97 PDZ2 domain and its mechanism of interaction with HPV-18 papillomavirus E6 protein. *Biochemistry* **2007**, *46*, 10864–10874. [CrossRef] [PubMed]

31. Gardiol, D.; Galizzi, S.; Banks, L. Mutational analysis of the discs large tumour suppressor identifies domains responsible for human papillomavirus type 18 E6-mediated degradation. *J. Gen. Virol.* **2002**, *83*, 283–289. [PubMed]

32. Pim, D.; Thomas, M.; Banks, L. Chimaeric HPV E6 proteins allow dissection of the proteolytic pathways regulating different E6 cellular target proteins. *Oncogene* **2002**, *21*, 8140–8148. [CrossRef] [PubMed]

33. Gardiol, D.; Kühne, C.; Glaunsinger, B.; Lee, S.S.; Javier, R.; Banks, L. Oncogenic human papillomavirus E6 proteins target the discs large tumour suppressor for proteasome-mediated degradation. *Oncogene* **1999**, *18*, 5487–5496. [CrossRef] [PubMed]

34. Lee, C.; Laimins, L.A. Role of the PDZ domain-binding motif of the oncoprotein E6 in the pathogenesis of human papillomavirus type 31. *J. Virol.* **2004**, *78*, 12366–12377. [CrossRef] [PubMed]

35. Takizawa, S.; Nagasaka, K.; Nakagawa, S.; Yano, T.; Nakagawa, K.; Yasugi, T.; Takeuchi, T.; Kanda, T.; Huibregtse, J.M.; Akiyama, T.; *et al.* Human scribble, a novel tumor suppressor identified as a target of high-risk HPV E6 for ubiquitin-mediated degradation, interacts with adenomatous polyposis coli. *Genes Cells* **2006**, *11*, 453–464. [CrossRef] [PubMed]

36. Nakagawa, S.; Huibregtse, J.M. Human scribble (Vartul) is targeted for ubiquitin-mediated degradation by the high-risk papillomavirus E6 proteins and the E6AP ubiquitin-protein ligase. *Mol. Cell. Biol.* **2000**, *20*, 8244–8253. [CrossRef] [PubMed]

37. Glaunsinger, B.A.; Lee, S.S.; Thomas, M.; Banks, L.; Javier, R. Interactions of the PDZ-protein MAGI-1 with adenovirus E4-ORF1 and high-risk papillomavirus E6 oncoproteins. *Oncogene* **2000**, *19*, 5270–5280. [CrossRef] [PubMed]

38. Thomas, M.; Laura, R.; Hepner, K.; Guccione, E.; Sawyers, C.; Lasky, L.; Banks, L. Oncogenic human papillomavirus E6 proteins target the MAGI-2 and MAGI-3 proteins for degradation. *Oncogene* **2002**, *21*, 5088–5096. [CrossRef] [PubMed]

39. Handa, K.; Yugawa, T.; Narisawa-Saito, M.; Ohno, S.-I.; Fujita, M.; Kiyono, T. E6AP-dependent degradation of DLG4/PSD95 by high-risk human papillomavirus type 18 E6 protein. *J. Virol.* **2007**, *81*, 1379–1389. [CrossRef] [PubMed]

40. Favre-Bonvin, A.; Reynaud, C.; Kretz-Remy, C.; Jalinot, P. Human papillomavirus type 18 E6 protein binds the cellular PDZ protein TIP-2/GIPC, which is involved in transforming growth factor β signaling and triggers its degradation by the proteasome. *J. Virol.* **2005**, *79*, 4229–4237. [CrossRef] [PubMed]

41. Accardi, R.; Rubino, R.; Scalise, M.; Gheit, T.; Shahzad, N.; Thomas, M.; Banks, L.; Indiveri, C.; Sylla, B.S.; Cardone, R.A.; *et al.* E6 and E7 from human papillomavirus type 16 cooperate to target the PDZ protein Na/H exchange regulatory factor 1. *J. Virol.* **2011**, *85*, 8208–8216. [CrossRef] [PubMed]

42. Lee, S.S.; Glaunsinger, B.; Mantovani, F.; Banks, L.; Javier, R.T. Multi-PDZ domain protein MUPP1 is a cellular target for both adenovirus E4-ORF1 and high-risk papillomavirus type 18 E6 oncoproteins. *J. Virol.* **2000**, *74*, 9680–9693. [CrossRef] [PubMed]

43. Latorre, I.J.; Roh, M.H.; Frese, K.K.; Weiss, R.S.; Margolis, B.; Javier, R.T. Viral oncoprotein-induced mislocalization of select PDZ proteins disrupts tight junctions and causes polarity defects in epithelial cells. *J. Cell Sci.* **2005**, *118*, 4283–4293. [CrossRef] [PubMed]

44. Storrs, C.H.; Silverstein, S.J. PATJ, a tight junction-associated PDZ protein, is a novel degradation target of high-risk human papillomavirus E6 and the alternatively spliced isoform 18 E6. *J. Virol.* **2007**, *81*, 4080–4090. [CrossRef] [PubMed]

45. Jing, M.; Bohl, J.; Brimer, N.; Kinter, M.; Vande Pol, S.B. Degradation of tyrosine phosphatase PTPN3 (PTPH1) by association with oncogenic human papillomavirus E6 proteins. *J. Virol.* **2007**, *81*, 2231–2239. [CrossRef] [PubMed]

46. Töpffer, S.; Müller-Schiffmann, A.; Matentzoglu, K.; Scheffner, M.; Steger, G. Protein tyrosine phosphatase H1 is a target of the E6 oncoprotein of high-risk genital human papillomaviruses. *J. Gen. Virol.* **2007**, *88*, 2956–2965. [CrossRef] [PubMed]

47. Spanos, W.C.; Geiger, J.; Anderson, M.E.; Harris, G.F.; Bossler, A.D.; Smith, R.B.; Klingelhutz, A.J.; Lee, J.H. Deletion of the PDZ motif of HPV16 E6 preventing immortalization and anchorage-independent growth in human tonsil epithelial cells. *Head Neck* **2008**, *30*, 139–147. [CrossRef] [PubMed]

48. Thomas, M.; Banks, L. PDZRN3/LNX3 is a novel target of human papillomavirus type 16 (HPV-16) and HPV-18 E6. *J. Virol.* **2015**, *89*, 1439–1444. [CrossRef] [PubMed]

49. Boon, S.S.; Banks, L. High-Risk Human Papillomavirus E6 Oncoproteins Interact with 14-3-3 in a PDZ Binding Motif-Dependent Manner. *J. Virol.* **2012**, *87*, 1586–1595. [CrossRef] [PubMed]

50. Boon, S.S.; Tomai, V.; Thomas, M.; Roberts, S.; Banks, L. Cancer-Causing Human Papillomavirus E6 Proteins Display Major Differences in the Phospho-Regulation of their PDZ Interactions. *J. Virol.* **2015**, *89*, 1579–1586. [CrossRef] [PubMed]

51. Facciuto, F.; Bugnon Valdano, M.; Marziali, F.; Massimi, P.; Banks, L.; Cavatorta, A.L.; Gardiol, D. Human papillomavirus (HPV)-18 E6 oncoprotein interferes with the epithelial cell polarity Par3 protein. *Mol. Oncol.* **2014**, *8*, 533–543. [CrossRef] [PubMed]

52. Vande Pol, S.B.; Klingelhutz, A.J. Papillomavirus E6 oncoproteins. *Virology* **2013**, *445*, 115–137. [CrossRef] [PubMed]

53. Tomaić, V.; Pim, D.; Banks, L. The stability of the human papillomavirus E6 oncoprotein is E6AP dependent. *Virology* **2009**, *393*, 7–10. [CrossRef] [PubMed]

54. Zanier, K.; Charbonnier, S.; Baltzinger, M.; Nominé, Y.; Altschuh, D.; Travé, G. Kinetic analysis of the interactions of human papillomavirus E6 oncoproteins with the ubiquitin ligase E6AP using surface plasmon resonance. *J. Mol. Biol.* **2005**, *349*, 401–412. [CrossRef] [PubMed]

55. Zanier, K.; ould M'hamed ould Sidi, A.; Boulade-Ladame, C.; Rybin, V.; Chappelle, A.; Atkinson, A.; Kieffer, B.; Travé, G. Solution structure analysis of the HPV16 E6 oncoprotein reveals a self-association mechanism required for E6-mediated degradation of p53. *Structure* **2012**, *20*, 604–617. [CrossRef] [PubMed]

56. Zanier, K.; Charbonnier, S.; Sidi, A.O.; McEwen, A.G.; Ferrario, M.G.; Poussin-Courmontagne, P.; Cura, V.; Brimer, N.; Babah, K.O.; Ansari, T.; et al. Structural basis for hijacking of cellular LxxLL motifs by papillomavirus E6 oncoproteins. *Science* **2013**, *339*, 694–698. [CrossRef] [PubMed]

57. Nominé, Y.; Masson, M.; Charbonnier, S.; Zanier, K.; Ristriani, T.; Deryckère, F.; Sibler, A.-P.; Desplancq, D.; Atkinson, R.A.; Weiss, E.; et al. Structural and functional analysis of E6 oncoprotein: Insights in the molecular pathways of human papillomavirus-mediated pathogenesis. *Mol. Cell* **2006**, *21*, 665–678. [CrossRef] [PubMed]

58. Nguyen, M.L.; Nguyen, M.M.; Lee, D.; Griep, A.E.; Lambert, P.F. The PDZ ligand domain of the human papillomavirus type 16 E6 protein is required for E6's induction of epithelial hyperplasia *in vivo*. *J. Virol.* **2003**, *77*, 6957–6964. [CrossRef] [PubMed]

59. Shai, A.; Brake, T.; Somoza, C.; Lambert, P.F. The human papillomavirus E6 oncogene dysregulates the cell cycle and contributes to cervical carcinogenesis through two independent activities. *Cancer Res.* **2007**, *67*, 1626–1635. [CrossRef] [PubMed]

60. Howley, P.M.; Lowy, D.R. *Fields Virology*; Wolters Kluwer: Alphen aan den Rijn, The Netherlands; Lippincott Williams & Wilkins: Philadelphia, PA, USA, 2007; pp. 2299–2354.

61. Doorbar, J. Molecular biology of human papillomavirus infection and cervical cancer. *Clin. Sci. (Lond.)* **2006**, *110*, 525–541. [CrossRef] [PubMed]

62. Park, R.B.; Androphy, E.J. Genetic analysis of high-risk e6 in episomal maintenance of human papillomavirus genomes in primary human keratinocytes. *J. Virol.* **2002**, *76*, 11359–11364. [CrossRef] [PubMed]

63. Thomas, J.T.; Hubert, W.G.; Ruesch, M.N.; Laimins, L.A. Human papillomavirus type 31 oncoproteins E6 and E7 are required for the maintenance of episomes during the viral life

cycle in normal human keratinocytes. *Proc. Natl. Acad. Sci. USA* **1999**, *96*, 8449–8454. [CrossRef] [PubMed]

64. Wang, H.-K.; Duffy, A.A.; Broker, T.R.; Chow, L.T. Robust production and passaging of infectious HPV in squamous epithelium of primary human keratinocytes. *Genes Dev.* **2009**, *23*, 181–194. [CrossRef] [PubMed]

65. Delury, C.P.; Marsh, E.K.; James, C.D.; Boon, S.S.; Banks, L.; Knight, G.L.; Roberts, S. The Role of Protein Kinase A Regulation of the E6 PDZ-Binding Domain during the Differentiation-Dependent Life Cycle of Human Papillomavirus Type 18. *J. Virol.* **2013**, *87*, 9463–9472. [CrossRef] [PubMed]

66. Nicolaides, L.; Davy, C.; Raj, K.; Kranjec, C.; Banks, L.; Doorbar, J. Stabilization of HPV16 E6 protein by PDZ proteins, and potential implications for genome maintenance. *Virology* **2011**, *414*, 137–145. [CrossRef] [PubMed]

67. Brimer, N.; Vande Pol, S.B. Papillomavirus E6 PDZ Interactions Can Be Replaced by Repression of p53 To Promote Episomal Human Papillomavirus Genome Maintenance. *J. Virol.* **2013**, *88*, 3027–3030. [CrossRef] [PubMed]

68. Ramirez, J.; Poirson, J.; Foltz, C.; Chebaro, Y.; Schrapp, M.; Meyer, A.; Bonetta, A.; Forster, A.; Jacob, Y.; Masson, M.; *et al.* Targeting the Two Oncogenic Functional Sites of the HPV E6 Oncoprotein with a High-Affinity Bivalent Ligand. *Angew. Chem. Int. Ed. Engl.* **2015**, *54*, 7958–7962. [CrossRef] [PubMed]

69. Van Doorslaer, K.; DeSalle, R.; Einstein, M.H.; Burk, R.D. Degradation of Human PDZ-Proteins by Human Alphapapillomaviruses Represents an Evolutionary Adaptation to a Novel Cellular Niche. *PLoS Pathog.* **2015**, *11*, e1004980. [CrossRef] [PubMed]

70. Watson, R.A.; Thomas, M.; Banks, L.; Roberts, S. Activity of the human papillomavirus E6 PDZ-binding motif correlates with an enhanced morphological transformation of immortalized human keratinocytes. *J. Cell Sci.* **2003**, *116*, 4925–4934. [CrossRef] [PubMed]

71. James, M.A.; Lee, J.H.; Klingelhutz, A.J. Human papillomavirus type 16 E6 activates NF-kappaB, induces cIAP-2 expression, and protects against apoptosis in a PDZ binding motif-dependent manner. *J. Virol.* **2006**, *80*, 5301–5307. [CrossRef] [PubMed]

72. Sherman, L.; Itzhaki, H.; Jackman, A.; Chen, J.J.; Koval, D.; Schlegel, R. Inhibition of Serum- and Calcium-Induced Terminal Differentiation of Human Keratinocytes by HPV 16 E6: Study of the Association with p53 Degradation, Inhibition of p53 Transactivation, and Binding to E6BP. *Virology* **2002**, *292*, 309–320. [CrossRef] [PubMed]

73. Muench, P.; Hiller, T.; Probst, S.; Florea, A.-M.; Stubenrauch, F.; Iftner, T. Binding of PDZ proteins to HPV E6 proteins does neither correlate with epidemiological risk classification nor with the immortalization of foreskin keratinocytes. *Virology* **2009**, *387*, 380–387. [CrossRef] [PubMed]

74. Riley, R.R.; Duensing, S.; Brake, T.; Mu, K.; Lambert, P.F.; Arbeit, J.M. Dissection of Human Papillomavirus E6 and E7 Function in Transgenic Mouse Models of Cervical Carcinogenesis. *Cancer Res.* **2003**, *63*, 4862–4871.

75. Song, S.; Pitot, H.C.; Lambert, P.F. The human papillomavirus type 16 E6 gene alone is sufficient to induce carcinomas in transgenic animals. *J. Virol.* **1999**, *73*, 5887–5893. [PubMed]

76. Jabbar, S.; Strati, K.; Shin, M.K.; Pitot, H.C.; Lambert, P.F. Human papillomavirus type 16 E6 and E7 oncoproteins act synergistically to cause head and neck cancer in mice. *Virology* **2010**, *407*, 60–67. [CrossRef] [PubMed]

77. Thomas, M.K.; Pitot, H.C.; Liem, A.; Lambert, P.F. Dominant role of HPV16 E7 in anal carcinogenesis. *Virology* **2011**, *421*, 114–118. [CrossRef] [PubMed]

78. Strati, K.; Lambert, P.F. Role of Rb-dependent and Rb-independent functions of papillomavirus E7 oncogene in head and neck cancer. *Cancer Res.* **2007**, *67*, 11585–11593. [CrossRef] [PubMed]

79. Suzuki, A.; Ohno, S. The PAR-aPKC system: Lessons in polarity. *J. Cell Sci.* **2006**, *119*, 979–987. [CrossRef] [PubMed]

80. Halaoui, R.; McCaffrey, L. Rewiring cell polarity signaling in cancer. *Oncogene* **2014**, *34*, 1–12. [CrossRef] [PubMed]

81. Pocha, S.M.; Knust, E. Complexities of Crumbs function and regulation in tissue morphogenesis. *Curr. Biol.* **2013**, *23*, R289–R293. [CrossRef] [PubMed]

82. Laprise, P. Emerging role for epithelial polarity proteins of the Crumbs family as potential tumor suppressors. *J. Biomed. Biotechnol.* **2011**, *2011*, 868217. [CrossRef] [PubMed]

83. Pim, D.; Bergant, M.; Boon, S.S.; Ganti, K.; Kranjec, C.; Massimi, P.; Subbaiah, V.K.; Thomas, M.; Tomaić, V.; Banks, L. Human papillomaviruses and the specificity of PDZ domain targeting. *FEBS J.* **2012**, *279*, 3530–3537. [CrossRef] [PubMed]

84. McCaffrey, L.M.; Macara, I.G. Epithelial organization, cell polarity and tumorigenesis. *Trends Cell Biol.* **2011**, *21*, 727–735. [CrossRef] [PubMed]

85. Goldenring, J.R. A central role for vesicle trafficking in epithelial neoplasia: Intracellular highways to carcinogenesis. *Nat. Rev. Cancer* **2013**, *13*, 813–820. [CrossRef] [PubMed]

86. Tomaić, V.; Gardiol, D.; Massimi, P.; Ozbun, M.; Myers, M.; Banks, L. Human and primate tumour viruses use PDZ binding as an evolutionarily conserved mechanism of targeting cell polarity regulators. *Oncogene* **2009**, *28*, 1–8. [CrossRef] [PubMed]

87. Kranjec, C.; Banks, L. A systematic analysis of human papillomavirus (HPV) E6 PDZ substrates identifies MAGI-1 as a major target of HPV type 16 (HPV-16) and HPV-18 whose loss accompanies disruption of tight junctions. *J. Virol.* **2011**, *85*, 1757–1764. [CrossRef] [PubMed]

88. Kamei, Y.; Kito, K.; Takeuchi, T.; Imai, Y.; Murase, R.; Ueda, N.; Kobayashi, N.; Abe, Y. Human scribble accumulates in colorectal neoplasia in association with an altered distribution of β-catenin. *Hum. Pathol.* **2007**, *38*, 1273–1281. [CrossRef] [PubMed]

89. Zhan, L.; Rosenberg, A.; Bergami, K.C.; Yu, M.; Xuan, Z.; Jaffe, A.B.; Allred, C.; Muthuswamy, S.K. Deregulation of scribble promotes mammary tumorigenesis and reveals a role for cell polarity in carcinoma. *Cell* **2008**, *135*, 865–878. [CrossRef] [PubMed]

90. Pearson, H.B.; Perez-Mancera, P.A.; Dow, L.E.; Ryan, A.; Tennstedt, P.; Bogani, D.; Elsum, I.; Greenfield, A.; Tuveson, D.A.; Simon, R.; *et al.* SCRIB expression is deregulated in human prostate cancer, and its deficiency in mice promotes prostate neoplasia. *J. Clin. Investig.* **2011**, *121*, 4257–4267. [CrossRef] [PubMed]

91. Ouyang, Z.; Zhan, W.; Dan, L. hScrib, a human homolog of Drosophila neoplastic tumor suppressor, is involved in the progress of endometrial cancer. *Oncol. Res.* **2010**, *18*, 593–599. [CrossRef] [PubMed]

92. Cavatorta, A.L.; Fumero, G.; Chouhy, D.; Aguirre, R.; Nocito, A.L.; Giri, A.A.; Banks, L.; Gardiol, D. Differential expression of the human homologue of drosophila discs large oncosuppressor in histologic samples from human papillomavirus-associated lesions as a marker for progression to malignancy. *Int. J. Cancer* **2004**, *111*, 373–380. [CrossRef] [PubMed]

93. Watson, R.A.; Rollason, T.P.; Reynolds, G.M.; Murray, P.G.; Banks, L.; Roberts, S. Changes in expression of the human homologue of the Drosophila discs large tumour suppressor protein in high-grade premalignant cervical neoplasias. *Carcinogenesis* **2002**, *23*, 1791–1796. [CrossRef] [PubMed]

94. Lin, H.-T.; Steller, M.A.; Aish, L.; Hanada, T.; Chishti, A.H. Differential expression of human Dlg in cervical intraepithelial neoplasias. *Gynecol. Oncol.* **2004**, *93*, 422–428. [CrossRef] [PubMed]

95. Nakagawa, S.; Yano, T.; Nakagawa, K.; Takizawa, S.; Suzuki, Y.; Yasugi, T.; Huibregtse, J.M.; Taketani, Y. Analysis of the expression and localisation of a LAP protein, human scribble, in the normal and neoplastic epithelium of uterine cervix. *Br. J. Cancer* **2004**, *90*, 194–199. [CrossRef] [PubMed]

96. Roberts, S.; Delury, C.; Marsh, E. The PDZ protein discs-large (DLG): The "Jekyll and Hyde" of the epithelial polarity proteins. *FEBS J.* **2012**, *279*, 3549–3558. [CrossRef] [PubMed]

97. Anastas, J.N.; Biechele, T.L.; Robitaille, M.; Muster, J.; Allison, K.H.; Angers, S.; Moon, R.T. A protein complex of SCRIB, NOS1AP and VANGL1 regulates cell polarity and migration, and is associated with breast cancer progression. *Oncogene* **2012**, *31*, 3696–3708. [CrossRef] [PubMed]

98. Frese, K.K.; Latorre, I.J.; Chung, S.-H.; Caruana, G.; Bernstein, A.; Jones, S.N.; Donehower, L.A.; Justice, M.J.; Garner, C.C.; Javier, R.T. Oncogenic function for the Dlg1 mammalian homolog of the Drosophila discs-large tumor suppressor. *EMBO J.* **2006**, *25*, 1406–1417. [CrossRef] [PubMed]

99. Kong, K.; Kumar, M.; Taruishi, M.; Javier, R.T. The human adenovirus E4-ORF1 protein subverts discs large 1 to mediate membrane recruitment and dysregulation of phosphatidylinositol 3-kinase. *PLoS Pathog.* **2014**, *10*, e1004102. [CrossRef] [PubMed]

100. Feigin, M.E.; Akshinthala, S.D.; Araki, K.; Rosenberg, A.Z.; Muthuswamy, L.B.; Martin, B.; Lehmann, B.D.; Berman, H.K.; Pietenpol, J.A.; Cardiff, R.D.; *et al.* Mislocalization of the cell polarity protein scribble promotes mammary tumorigenesis and is associated with basal breast cancer. *Cancer Res.* **2014**, *74*, 3180–3194. [CrossRef] [PubMed]

101. Krishna Subbaiah, V.; Massimi, P.; Boon, S.S.; Myers, M.P.; Sharek, L.; Garcia-Mata, R.; Banks, L. The invasive capacity of HPV transformed cells requires the hDlg-dependent enhancement of SGEF/RhoG activity. *PLoS Pathog.* **2012**, *8*, e1002543. [CrossRef] [PubMed]

102. Facciuto, F.; Cavatorta, A.L.; Valdano, M.B.; Marziali, F.; Gardiol, D. Differential expression of PDZ domain-containing proteins in human diseases—Challenging topics and novel issues. *FEBS J.* **2012**, *279*, 3538–3548. [CrossRef] [PubMed]

103. Zhang, Y.; Dasgupta, J.; Ma, R.Z.; Banks, L.; Thomas, M.; Chen, X.S. Structures of a human papillomavirus (HPV) E6 polypeptide bound to MAGUK proteins: Mechanisms of targeting tumor suppressors by a high-risk HPV oncoprotein. *J. Virol.* **2007**, *81*, 3618–3626. [CrossRef] [PubMed]

104. Thomas, M.; Glaunsinger, B.; Pim, D.; Javier, R.; Banks, L. HPV E6 and MAGUK protein interactions: Determination of the molecular basis for specific protein recognition and degradation. *Oncogene* **2001**, *20*, 5431–5439. [CrossRef] [PubMed]

105. Kranjec, C.; Massimi, P.; Banks, L. Restoration of MAGI-1 expression in human papillomavirus-positive tumor cells induces cell growth arrest and apoptosis. *J. Virol.* **2014**, *88*, 7155–7169. [CrossRef] [PubMed]

106. Ostrow, R.S.; LaBresh, K.V; Faras, A.J. Characterization of the complete RhPV 1 genomic sequence and an integration locus from a metastatic tumor. *Virology* **1991**, *181*, 424–429. [CrossRef]

107. De Villiers, E.-M.; Fauquet, C.; Broker, T.R.; Bernard, H.-U.; zur Hausen, H. Classification of papillomaviruses. *Virology* **2004**, *324*, 17–27. [CrossRef] [PubMed]

108. Wood, C.E.; Chen, Z.; Cline, J.M.; Miller, B.E.; Burk, R.D. Characterization and experimental transmission of an oncogenic papillomavirus in female macaques. *J. Virol.* **2007**, *81*, 6339–6345. [CrossRef] [PubMed]

109. Bilder, D. Epithelial polarity and proliferation control: Links from the Drosophila neoplastictumor suppressors. *Genes Dev.* **2004**, *18*, 1909–1925. [CrossRef] [PubMed]

110. Humbert, P.O.; Grzeschik, N.A.; Brumby, A.M.; Galea, R.; Elsum, I.; Richardson, H.E. Control of tumourigenesis by the Scribble/Dlg/Lgl polarity module. *Oncogene* **2008**, *27*, 6888–6907. [CrossRef] [PubMed]

111. Kühne, C.; Gardiol, D.; Guarnaccia, C.; Amenitsch, H.; Banks, L. Differential regulation of human papillomavirus E6 by protein kinase A: Conditional degradation of human discs large protein by oncogenic E6. *Oncogene* **2000**, *19*, 5884–5891. [CrossRef] [PubMed]

112. Ivarsson, Y. Plasticity of PDZ domains in ligand recognition and signaling. *FEBS Lett.* **2012**, *586*, 2638–2647. [CrossRef] [PubMed]

113. Contreras-Paredes, A.; De la Cruz-Hernández, E.; Martínez-Ramírez, I.; Dueñas-González, A.; Lizano, M. E6 variants of human papillomavirus 18 differentially modulate the protein kinase B/phosphatidylinositol 3-kinase (akt/PI3K) signaling pathway. *Virology* **2009**, *383*, 78–85. [CrossRef] [PubMed]

114. Rubio, M.P.; Geraghty, K.M.; Wong, B.H.C.; Wood, N.T.; Campbell, D.G.; Morrice, N.; Mackintosh, C. 14-3-3-affinity purification of over 200 human phosphoproteins reveals new links to regulation of cellular metabolism, proliferation and trafficking. *Biochem. J.* **2004**, *379*, 395–408. [CrossRef] [PubMed]

115. Jin, J.; Smith, F.D.; Stark, C.; Wells, C.D.; Fawcett, J.P.; Kulkarni, S.; Metalnikov, P.; O'Donnell, P.; Taylor, P.; Taylor, L.; *et al.* Proteomic, Functional, and Domain-Based Analysis of *In Vivo* 14-3-3 Binding Proteins Involved in Cytoskeletal Regulation and Cellular Organization. *Curr. Biol.* **2004**, *14*, 1436–1450. [CrossRef] [PubMed]

116. Meek, S.E.M.; Lane, W.S.; Piwnica-Worms, H. Comprehensive Proteomic Analysis of Interphase and Mitotic 14-3-3-binding Proteins. *J. Biol. Chem.* **2004**, *279*, 32046–32054. [CrossRef] [PubMed]

117. Benzinger, A. Targeted Proteomic Analysis of 14-3-3, a p53 Effector Commonly Silenced in Cancer. *Mol. Cell. Proteomics* **2005**, *4*, 785–795. [CrossRef] [PubMed]

118. Dougherty, M.K. Unlocking the code of 14-3-3. *J. Cell Sci.* **2004**, *117*, 1875–1884. [CrossRef] [PubMed]

HPV16 E6 Controls the Gap Junction Protein Cx43 in Cervical Tumour Cells

Peng Sun, Li Dong, Alasdair I. MacDonald, Shahrzad Akbari, Michael Edward,

Malcolm B. Hodgins, Scott R. Johnstone and Sheila V. Graham

Abstract: Human papillomavirus type 16 (HPV16) causes a range of cancers including cervical and head and neck cancers. HPV E6 oncoprotein binds the cell polarity regulator hDlg (human homologue of *Drosophila* Discs Large). Previously we showed *in vitro*, and now *in vivo*, that hDlg also binds Connexin 43 (Cx43), a major component of gap junctions that mediate intercellular transfer of small molecules. In HPV16-positive non-tumour cervical epithelial cells (W12G) Cx43 localised to the plasma membrane, while in W12T tumour cells derived from these, it relocated with hDlg into the cytoplasm. We now provide evidence that E6 regulates this cytoplasmic pool of Cx43. E6 siRNA depletion in W12T cells resulted in restoration of Cx43 and hDlg trafficking to the cell membrane. In C33a HPV-negative cervical tumour cells expressing HPV16 or 18 E6, Cx43 was located primarily in the cytoplasm, but mutation of the 18E6 C-terminal hDlg binding motif resulted in redistribution of Cx43 to the membrane. The data indicate for the first time that increased cytoplasmic E6 levels associated with malignant progression alter Cx43 trafficking and recycling to the membrane and the E6/hDlg interaction may be involved. This suggests a novel E6-associated mechanism for changes in Cx43 trafficking in cervical tumour cells.

Reprinted from *Viruses*. Cite as: Sun, P.; Dong, L.; MacDonald, A.I.; Akbari, S.; Edward, M.; Hodgins, M.B.; Johnstone, S.R.; Graham, S.V. HPV16 E6 Controls the Gap Junction Protein Cx43 in Cervical Tumour Cells. *Viruses* **2015**, *7*(10), 5243-5256.

1. Introduction

Gap junctions are specialized cell membrane channels that allow direct intercellular diffusion of critical regulatory ions and small molecules between contiguous cells and are normally present in large aggregates called "plaques" on the cell membrane [1,2]. Gap junction assembly is dependent on the efficient delivery of connexons (half of a gap junction) to the plasma membrane and subsequent docking of these between adjacent cells [3]. Gap junctions are then recycled from the centre of the plaques into the endosomal/lysosomal pathway, but they can also be degraded by the proteasome [4]. Regulation of gap junctional intercellular communication (GJIC) has been demonstrated to produce cellular changes underlying tumour formation. Additionally, connexons have been shown to have gap junction-independent tumour promoting activity [5].

There are 21 human connexin proteins, all of which have four transmembrane α helices anchored in the cell membrane with a short N- and variable length C-terminus in the cytoplasm [3]. For example, Connexin 43 (Cx43), the most widespread connexin and a major component of gap junctions in stratified epithelia, has a 151 amino acid long C-terminus which integrates with intracellular signalling pathways [6]. A body of evidence has accumulated to show that GJIC may be lost during malignant progression, as seen in HPV-positive cervical cancer [7]. Cx43 is often down-regulated in epithelial carcinomas [7] as well as precancerous lesions [8] although in other cases expression may be increased in invasive tumours [9]. Nevertheless, the steps leading to changes in connexin expression and trafficking and how these are related to tumour progression are largely unknown.

Human papillomaviruses (HPVs) are small double-stranded DNA viruses, which infect the stratified epithelia [10]. HPV16 is the most prevalent so-called "high-risk" HPV genotype associated with cervical and other anogenital carcinomas [11], in addition to a subset of head and neck cancers [12]. Progression from the premalignant to malignant phase of high-risk HPV-associated disease is driven by overexpression of the viral oncoproteins E6 and E7 [10]. In the nucleus, E6 binds and targets the tumour suppressor p53 for degradation [13]. However, E6 also contains a highly conserved C-terminal motif [14,15] that can interact with the PDZ (PSD-95/Dlg/ZO-1) domain-containing proteins MAGI-1, 2, 3, MUPP-1, hScrib and hDlg [16,17]. *In vitro* and *in vivo* studies have revealed that the E6 PDZ binding motif is essential for the HPV infectious life cycle and for HPV-associated tumour progression underlining the importance of E6/PDZ protein interactions [15,18].

Proteins of the membrane-associated guanylate kinase homologue (MAGUK) family can form protein scaffolds and comprise macromolecular complexes with protein partners thought to be involved in cell signalling cascades and cell morphology organization [19,20]. hDlg is a MAGUK protein located at intercellular contact sites in epithelial cells [21,22]. Previously we reported an interaction between Cx43 and hDlg in HPV16-positive cervical epithelial cells. The C-terminal domain of Cx43 binds the N- and C-termini of hDlg [23]. hDlg and Cx43 were both located at the plasma membrane in non-tumour cervical epithelial cells (W12G) but were co-localised in the cytoplasm in invasive cervical tumour cells derived from these (W12T; formerly named W12GPXY) [23,24]. Functional studies indicated that hDlg was responsible for maintaining a cytoplasmic pool of Cx43, protected from degradation that may be capable of trafficking to the membrane.

In this study we first demonstrate a physical association between hDlg and Cx43 *in vivo*. The known interaction of E6 with hDlg prompted us to investigate whether HPV16 E6 could interact with and regulate the subcellular location of the hDlg/Cx43 complex. We detected association of E6 with Cx43 at low levels in W12T cells. However, E6 siRNA depletion in W12T cells caused redistribution of Cx43 from the cytoplasm to the membrane. Conversely, overexpression of HPV 16 or 18 E6 in HPV-negative C33a cells resulted in reduced levels of Cx43 and redistribution from the membrane to the cytoplasm. Finally, HPV-negative C33a cells overexpressing a wild type or mutant HPV18 E6 demonstrated the critical role E6 plays in membrane localisation of Cx43. These findings reveal a novel E6-regulated pathway through which Cx43 trafficking, and potentially GJIC, is altered in cervical tumour cells.

2. Materials and Methods

2.1. Cell Culture and Antibodies

W12G cells are non-tumour cervical epithelial cells (clone 20861: [25]). W12T cells are invasive tumour cells derived from W12G cells. They were originally called W12GPXY [24]. W12T, C33a, and HEK-293 cells were cultured in Dulbecco's modified Eagle's medium (DMEM) supplemented with 10% FCS; W12E, W12G cells were cultured in serum-free keratinocyte growth medium (KGM) from Cambrex, UK (cc-3101). All cells were maintained under humidified 5% CO_2 at 37 °C.

Polyclonal antibody raised in rabbits against a synthetic peptide corresponding to residues 363–382 of native Cx43, was kindly provided by Dr E. Rivedal, The Norwegian Radium Hospital, Oslo. Polyclonal antibodies against HPV-16 E6 (sc-1583) and against ZO-1 (61-7300) and a monoclonal antibody against hDlg (sc-9961) were purchased from Santa Cruz Biotechnology, California, CA, USA. Polyclonal antibody against flag (F7425) was purchased from Sigma (Poole, UK). Monoclonal antibody against Cx43 (C-6219) was purchased from Sigma. Antibody dilutions were as described [23].

2.2. Proximity Ligation Assay (PLA)

Archival paraffin-embedded cervical biopsy samples were obtained with ethical permission (Glasgow Royal Infirmary: RN04PC003). Diagnosis was made by two gynaehistopatholgists. HPV presence was confirmed by PCR. Sections on slides were de-paraffined, and antigen retrieval performed using sodium citrate (10 mM, pH6.0). Sections were incubated in blocking solution (PBS, 0.5% BSA, 0.25% TritonX-100) for 1 h and primary antibodies for Cx43 (Rivedal, 1:50) and hDlg (Santa Cruz, 1:50) added in blocking solution overnight at 4 °C. For standard immunofluorescence, samples were washed in PBS three times, blocked for 30 min, then secondary antibodies (Alexa Fluor, Lifetechnologies, Paisley, UK, 1:500) added for 30 min at room temperature. For Proximity Ligation Assay (PLA), following overnight primary antibody incubation, samples were washed in PBS and detection performed using a Duolink kit (Sigma, Poole, UK) as previously described [26]. All samples were mounted using ProLong antifade diamond containing DAPI (Life Techologies, Paisley, UK). Negative controls (no primary antibody) were included in all experiments. Images were taken using a Zeiss LSM510 Meta confocal microscope.

2.3. Co-Immunoprecipitation

Cells at 70% confluence were washed twice with ice-cold PBS and scraped into 5 mL chilled IP buffer containing PBS, 1% Triton X-100, 0.5% CHAPS, 0.1% SDS with one tablet of mini-proteinase inhibitor cocktail and one tablet of PhosSTOP phosphatase inhibitor cocktail per 10 mL (Roche, UK). The cell lysates were incubated on ice for 15 min and the solutions sonicated for 1 min on ice. Cell lysates were then cleared of cellular debris by centrifugation at 12,000 g for 10 min at 4 °C and the protein concentration determined by Bradford's assay (Bio-Rad Laboratories, Hemel Hempstead, UK). Cell lysates were pre-cleared with protein-G sepharose beads (Sigma, Poole, UK) for one hour at 4 °C. Primary antibodies were then added to 100 µg protein of each cell lysate and incubated for

2 h at 4 °C with rotation. Subsequently, 40 μg of reconstituted protein G-Sepharose was added and the volume was adjusted to 750 μL. The samples were mixed by rotation at 4 °C for 4–5 h. Immunocomplexes were then harvested by centrifugation and washed four times with 250 μL ice-cold IP buffer and once with 25 μL ice cold PBS. Proteins were solubilised by sonicating the complexes in PBS mixed with 5 μL 6× protein loading buffer (1× buffer: 125 mM Tris (pH 6.8), 4% SDS, 20% glycerol, 10% mercaptoethanol and 0.006% bromophenol blue) for 5 min on ice and separated by SDS-PAGE.

2.4. Western Blotting

Confluent cells were scraped into protein-loading buffer (125 mM Tris (pH 6.8), 4% SDS, 20% glycerol, 10% mercaptoethanol and 0.006% bromophenol blue, fresh protein inhibitor cocktail (Roche, Welwyn Garden City, UK); 50 μg protein was resolved by polyacrylamide gel electrophoresis and subsequently transferred to a nitrocellulose membrane. The membrane was preincubated for 1 h at room temperature with 5% dried milk in PBS-0.1% Tween, before overnight incubation at 4 °C with diluted primary antibody in PBS-Tween, 1% dried milk. Horseradish peroxidase (HRP)-conjugated secondary antibodies were diluted 1:1000 in PBS-Tween and incubated for 1 h. The blot was developed using Pierce enhanced chemi-luminescence (ECL) kit and exposed to Kodak X-OMAT film.

2.5. Plasmid Construction and Cell Transfection

The HPV18 E6 wild type gene and E6 aa156 Thr → Glu mutation cloned into pcDNA-3 plasmid were provided by Dr Lawrence Banks [27]. The plasmids were sequentially digested with EcoR I and Hind III, then the E6 fragments were purified with phenol:chloroform extraction and ethanol precipitation. The vector p3x-FLAG-CMV™-10 (Sigma, Poole, UK) was cut with EcoR I and Hind III and ligated with the wild type and mutant HPV18 E6 DNA fragments to generate the pN-terminus 3XFLAG-fused wild-type HPV18 E6 (pNFWE6) and pN-terminus 3X FLAG-fused mutant HPV 18 E6 (pNFME6). The recombined plasmids were sequenced to confirm correct insertion.

C33a cells (2×10^5) were transfected with 3 μg of p3x-FLAG-CMV™-10, pNFWE6 or pNFME6 using Lipofectamine (Invitrogen, Paisley, UK) according to the recommended protocol and selected in medium supplemented with 500 μg/mL G418 (Sigma, Poole, UK) to obtain stably transfected cell colonies. The experiments were carried out with 2 independent cell lines selected from each transfection. The plasmid pMAXGFP was used as an expression control plasmid and for calculating transfection efficiency.

2.6. W12T Cell Transfected with siRNA

Purified siRNA against HPV16 E6 (5′GUUACCACAGUUAUGCACATT3′) or a control siRNA (siGLO), were transfected into 1.5×10^5 W12T cells per well in DMEM with 10% FCS using Lipofectamine RNAiMAX (Invitrogen, Paisley, UK) according to the manufacturer's instructions. The final siRNA final concentration was 0.2 μM. Cells were harvested for analysis at 32 h post-transfection.

2.7. Immunofluorescence Microscopy

Cell were grown on sterile 18 × 18 mm coverslips until 90% confluent, washed three times with PBS and fixed and permeabilised with 100% ice-cold methanol for 5 min at −20 °C or with 58 mM sucrose, 4% formaldehyde in PBS for 10 min at room temperature [23]. Coverslips were blocked using 5% (*v/v*) horse serum in PBS for 30 min at room temperature then washed three times with PBS. Cells were incubated with diluted primary antibodies in 1% horse serum in PBS for 1 h at room temperature. Coverslips were washed in PBS six times before incubation for 1 h with secondary antibodies labelled with fluorescein or Texas-Red diluted 1:100 in PBS (Vector laboratories, Peterborough, UK). After washing in PBS six times, the coverslips were mounted with Vectashield mounting medium (with DAPI as a nuclear stain). Negative controls (no primary antibody) were included in all experiments. Images were taken using a Zeiss LSM510 Meta confocal microscope.

3. Results

3.1. hDlg and Cx43 Interact In Vivo

Previously we demonstrated loss of GJIC in a number of HPV16-positive (CaSki, SiHa, W12T (formerly named W12GPXY)) and HPV18-positive (HeLa) cervical cancer cell lines but not in HPV16-positive non-tumour cervical epithelial cells (W12E, W12G) [23,24]. In the W12T cervical tumour cells Cx43 was no longer located in membrane gap junction plaques but was redistributed to the cytoplasm where it colocalised with the PDZ domain protein hDlg. Cx43 was also located in the cytoplasm in two other HPV-16-positive cervical cancer cell lines (CaSki and SiHa cells (Supplementary Figure S2). To determine if the Cx43/hDlg interaction occurs in cervical epithelial tissues *in vivo* we examined location of the proteins in HPV16-positive high grade cervical lesions. Analysis by immunofluorescence showed that hDlg and Cx43 co-localise in epithelial cells in discrete regions of the cells *i.e.*, perinuclear location (Figure 1A,B). To further define this, we probed tissues using a proximity ligation assay (PLA), which identifies protein interactions based on proteins being within 40 nm of each other. Figure 1C,D (red spots) shows detection of the hDlg-Cx43 complex within the epithelial cells of the cervical lesion. Figure 1E shows low level detection of Cx43/hDlg complexes in a limited area of a representative cervical tumour in agreement with the observation that Cx43 levels decline in cervical cancer tissues *in vitro* and *in vivo* [24,28,29]. Two cervical lesions and two cervical cancers were examined and there was evidence that Cx43 and hDlg were in close proximity in all tissues. Figure 1F shows a duolink secondary control. The image is from the outer region of the tissue shown in Figure 1C. We chose this area of the tissue because it represents the only autofluorescence we detected in any of the tissues we examined. Some antibody trapping on the outer surface of the epithelium was detected but there was no staining detected in the cells in the tissue interior. These data confirm our previous *in vitro* findings that Cx43 and hDlg interact and demonstrates the formation of protein complexes in human cervical epithelial cells *in vivo*.

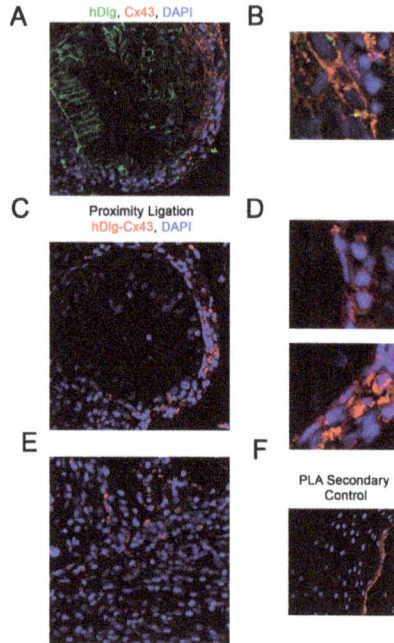

Figure 1. Cx43 and hDlg interact directly in cervical tumour cells *in vivo*. (**A**) Representative immunofluorescence image showing a section of a high grade cervical lesion; (**B**) A region of Cx43 and hDlg colocalisation in (**A**) enlarged 5×. Immunofluorescence shows hDlg (green), Cx43 (red) and DAPI (blue). The yellow arrow indicates one area of Cx43/hDlg colocalisation; (**C**) Immunofluorescence proximity ligation for hDlg and Cx43, where red staining indicates colocalisation of the proteins within 40 nm of each other; (**D**) Regions of Cx43 and hDlg colocalisation detected by PLA in C enlarged 5×; (**E**) Immunofluorescence proximity ligation for hDlg and Cx43 in a region of a cervical tumour; (**F**) Immunofluorescence proximity ligation secondary control on the outer region of the cervical lesion a part of which is shown in (**C**).

3.2. HPV16 E6 Is Present in a Complex Containing hDlg and Cx43 Cervical Tumour Cells

Structural analysis has demonstrated that PDZ domains in hDlg can bind to the X-S/T-X-V/L protein motif of high risk HPV E6 proteins [14]. As we have shown previously that the C-terminal domain of Cx43 interacts with the N- and C- termini of hDlg [23], it is possible that E6, a 17 kDa protein, could form a complex with hDlg while hDlg is also bound to Cx43. Therefore we investigated formation of protein complexes between Cx43, hDlg and E6 by co-immunoprecipitation in W12G (non-tumour cervical epithelial cells) and in W12T (cervical tumour) cells (Figure 2).

Cx43 and hDlg were co-immunoprecipitated from both W12G (top panel, lane 3) and W12T (top panel, lane 5) cell extracts (Figure 2A). Moreover, as expected, E6/hDlg complexes were able to be precipitated from both W12G and W12T cells (Figure 2A middle panel, lanes 3 and 5) but were more abundant in the latter. Co-immunoprecipitation of Cx43 with E6 antibody was detected at low levels in W12T cells (Figure 2A, bottom panel, lane 5) but not in W12G cells. A reverse precipitation of E6 with hDlg or Cx43 was not successful due to poor reactivity of the HPV16 E6 antibody in western blots. These data indicate that in cervical tumour cells E6 can be part of a complex containing hDlg and Cx43, at least at low levels.

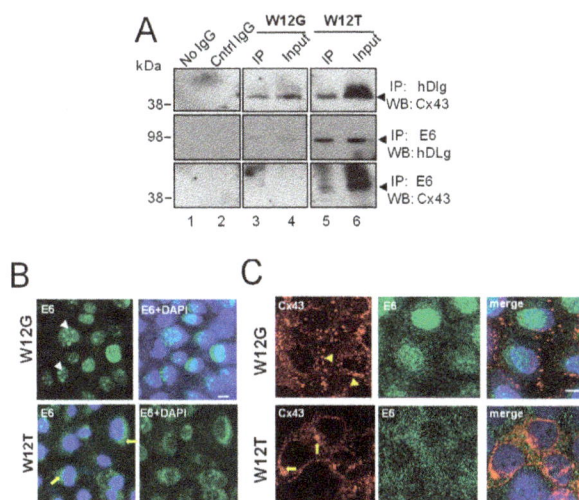

Figure 2. HPV16 E6 associates with Cx43 in W12T cervical cancer cells. (**A**) Co-immunoprecipitation of E6 with hDlg and Cx43 in W12G non-tumour and W12T tumour cells. Antibodies used in immunoprecipitation (IP) and the antibodies used to probe western blots (WB) are indicated on the right hand side. No IgG, beads alone used in the immunoprecipitation. Cntr IgG, a matched antibody isotype negative control immunoprecipitation. Input, 10% of the volume of cellular extract as used in the co-immunoprecipitation experiments; (**B**) Confocal immunofluorescence microscopy imaging showing E6 (green) in the nucleus of W12G cells (arrowheads) but in the cytoplasm of W12T cells (arrows); (**C**) Confocal immunofluorescence microscopy imaging of Cx43 (red) in W12G cells located in large punctuate gap junction plaques (arrowheads) on the membrane. In W12T cells membrane Cx43 (red) is reduced and some perinuclear staining is detected (arrows). E6 is shown in green. Nuclei are stained with DAPI (blue). Bar = 10 µM.

Next we used confocal immunofluorescence microscopy to investigate the subcellular location of endogenous Cx43 and E6 in the W12 non-tumour and tumour cells. In the W12G non-tumour cells E6 was primarily localised within the nucleus as expected (Figure 2B, arrowheads) [30]. However, in the W12T tumour cells E6 was located in both the nucleus and the cytoplasm (Figure 2B, arrows). Immunofluorescence analysis of E6 and Cx43 in the two cell types revealed that Cx43 was located mainly in gap junction plaques (arrowheads) at the plasma membrane in W12G cells. There was no colocalisation of E6 and Cx43 as they were detected in different subcellular compartments (Figure 2C). However, in W12T cells Cx43 was located in a perinuclear location in the cytoplasm as previously observed (arrows) [23,24] and E6 was diffusely located in the cytoplasm (Figure 2C). E6 and Cx43 colocalised around the nucleus of W12T cells.

3.3. HPV16 E6 Controls Cx43 Trafficking in Cervical Tumour Cells

Since our previous study indicated that hDlg was not responsible for Cx43 relocation from the membrane to the cytoplasm [23] next, we investigated the hypothesis that if E6 can potentially form a complex with hDlg and Cx43 it could restrict Cx43 trafficking to the plasma membrane. W12G and W12T are isogenic cell lines where the HPV16 genome is integrated into the host genome. Both lines express only the bicistronic E6E7 viral mRNA [25,31]. Therefore, we depleted the E6E7 mRNA using siRNA and examined changes in the subcellular location of Cx43 in W12T cells. Western blotting showed that specific siRNA treatment caused a marked reduction in total E6 protein levels (Figure 3A). Confocal immunofluorescence microscopy showed that treatment of W12T cells with a control siRNA (siGLO) did not alter the cytoplasmic location of Cx43 (Figure 3B arrowheads). In contrast, treatment of W12T cells with the E6 siRNA resulted in some restoration of Cx43 on the plasma membrane in typical gap junction plaques at points of cell-cell contact (Figure 3C, arrows). hDlg also appeared at the membrane in the E6 siRNA-treated cells (Figure 3C). These data indicate that a protein translated from the E6E7 bicistronic mRNA may be involved in Cx43 (and hDlg) intracellular trafficking. There are a number of possible proteins: E6 full length, E6*I, E6*II, E6*X and E7 [32]. However, only E6 full length can bind hDlg. Next we tested whether this protein was responsible for controlling Cx43.

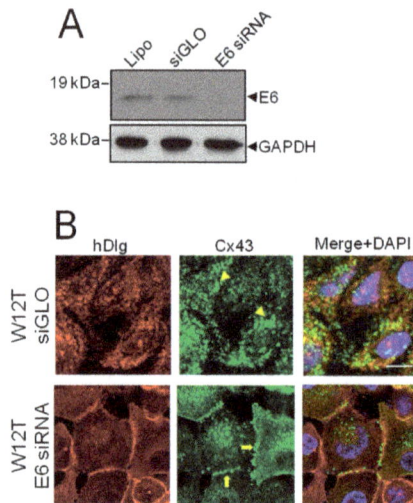

Figure 3. E6 restricts Cx43 trafficking to the plasma membrane in W12T cells. (**A**) Western blot showing depletion of E6 following siRNA treatment of W12T cells. Lipo, mock transfected cells treated with lipofectamine. siGLO, cells transfected with control siRNA. E6 siRNA, cells transfected with HPV16 E6 siRNA; (**B**) Confocal immunofluorescence microscopy imaging showing the location of hDlg (red) and Cx43 (green) in control transfected W12T cells or in W12T cells transfected with siRNA against E6. Arrowheads indicate cytoplasmic Cx43 while arrows indicate Cx43 located in gap junctions plaques on the cell membrane. Nuclei are stained with DAPI. Bar = 10 µM.

3.4. HPV E6 Causes Reduced Levels of Cx43 in C33a Cells

Previously we found that cytoplasmic hDlg protected a pool of Cx43 from degradation: siRNA depletion of hDlg resulted in reduced levels of Cx43 in W12T cells [23]. Although not as efficient as HPV18 E6, HPV16 E6 can target hDlg for degradation via the proteasome in cervical cancer cells [27]. Thus it is possible that the reduction in Cx43 levels following depletion of hDlg could be mediated by HPV E6. To test this we examined Cx43 expression in HPV-negative C33a cervical tumour cells stably transfected with either an empty FLAG vector (C33aV, no E6 expression) or a vector expressing FLAG-tagged HPV16 E6 (C33a16E6). Levels of Cx43 protein were assessed using a protein lysate titration western blot and quantified (Figure 4A,B). As expected FLAG-E6 was detected using an anti-FLAG antibody in C33a16E6 cells but not in C33aV cells (Figure 4A). Expression of E6 in C33a16E6 cells resulted in a significant reduction in Cx43 expression (Figure 4A,B) with no notable changes in the expression of hDlg (Figure 4A). These data suggest that E6 may target Cx43 for degradation independent of its effects on hDlg.

C33a cells normally retain strong GJIC and display Cx43 gap junction plaques on the plasma membrane [24] indicating that HPV expression might regulate GJIC in tumour cells. To confirm our data that E6 regulates Cx43 intracellular trafficking we examined the effect of E6 expression on Cx43 and hDlg in C33a cells using confocal immunofluorescence microscopy. Figure 4C (arrowheads) shows that Cx43 was located primarily on the plasma membranes between adjacent C33aV cells. In contrast, in C33a16E6 cells Cx43 was primarily localised within intracellular regions (Figure 4C arrows), with a marked reduction in membrane gap junction plaques. Figure 4D show that in these cells E6 was located throughout the cytoplasm and the nucleus with some evidence of colocalisation with Cx43 on the plasma membrane (arrowhead) and in the cytoplasm (arrow). In contrast to what we observed in W12T cells, hDlg was found mainly at the cell periphery and there was only limited colocalisation of hDlg with Cx43 in the cytoplasm of C33a16E6 cells (Figure 4E). To test the effect of ectopic expression of HPV16 E6 on Cx43 levels in a non-cervical tumour cell line we transfected the FLAGE6 expression vector into HEK-293 cells. Figure 4F shows that no change in hDlg levels was detected when E6 was expressed and instead of a decrease in Cx43 levels a slight increase was observed. E6 expression did not cause Cx43 relocation from the membrane in these cells (Figure 4G).

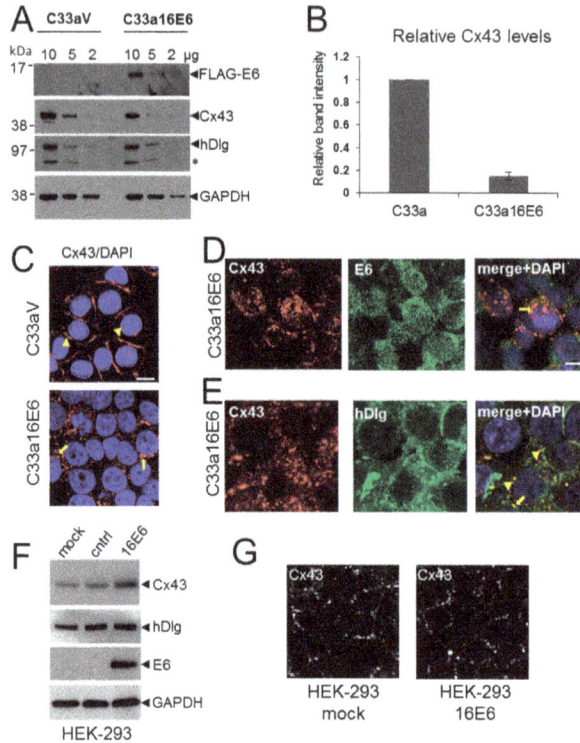

Figure 4. C33a cells expressing HPV16E6 show reduced Cx43 levels and some colocalisation between E6 and Cx43 is seen in the cytoplasm. (**A**) Titration western blot showing the levels of Cx43 and hDlg in C33a cells stably transfected with vector alone (C33aV) or with a vector expressing FLAG-tagged HPV16 E6 (C33a16E6). As indicated, 10, 5 or 2 μg protein extract was applied to the lanes as indicated. hDlg degradation results in a product that appears as an additional protein species of around 70 kDa (asterisk). E6 was detected in C33a16E6 cells using a FLAG antibody. GAPDH is shown as a loading control; (**B**) Quantification of levels of Cx43 relative to GAPDH from 5 separate western blot experiments. The data show the mean and standard error of the mean; (**C**) Confocal immunofluorescence microscopy imaging of Cx43 (red) location in C33aV and C33a16E6 cells; (**D**) Confocal immunofluorescence microscopy imaging of the location of Cx43 (red) and E6 (green) in C33a16E6 cells. Cytoplasmic colocalisation is indicated (arrow). The arrowhead indicates some Cx43 remaining on the plasma membrane where there is also co-staining with E6; (**E**) Location of Cx43 (red) and hDlg (green) in C33a16E6 cells. Co-staining in the cytoplasm is indicated with arrowheads. Nuclei are stained with DAPI. Bar = 10 μM; (**F**) Western blot analysis of Cx43 and hDlg levels in HEK-293 cells expressing HPV16 E6. Mock, mock transfected cells, cntrl, cells transfected with a control plasmid, 16E6, cells transfected with an expression construct for HPV16 E6. (**G**) Confocal microscopy analysis of Cx43 location in 293 cells transfected with the control plasmid (mock) or with an expression construct for HPV16 E6 (16E6).

3.5. E6 PDZ Binding Motif Is Required for Relocation of Cx43 from the Membrane to the Cytoplasm

To investigate the relationship between E6 and hDlg and Cx43 re-location to the cytoplasm in cervical tumour cells, C33a cells were transfected with empty vector (C33aV) or a vector expressing wild type E6 (C33a18E6) or the same vector expressing an HPV18 E6 mutant that does not bind hDlg [27] (Figure 5A). This time HPV18 E6 was used because it binds and degrades hDlg more effectively than HPV16 E6 [33]. Moreover, there is a high degree of conservation between the HPV18 and HPV16 E6 proteins and loss of Cx43 GJIC is observed in HPV18 as well as HPV16-positive cervical tumour cells [24]. HPV18 E6 expression caused a reduction in both Cx43 and hDlg levels but expression of the HPV18 E6 mutant that does not bind hDlg did not affect levels of either protein compared to cells transfected with vector alone (Figure 5B). Similar to what was observed for HPV16 E6, Cx43 relocated from the cell membrane of C33aV cells (Figure 5C) to a perinuclear location in the C33a cells expressing ectopic 18E6 (Figure 5C). hDlg was present at the membrane and in the cytoplasm of both the C33aV and C33a18E6 cells but displayed increased colocalisation with Cx43 in the C33a cells expressing 18E6 (Figure 5C).

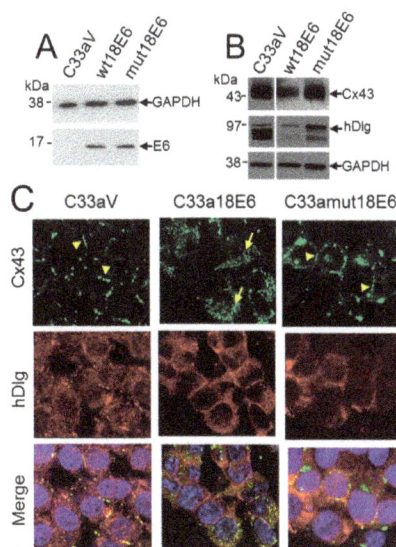

Figure 5. C33a cells expressing a mutant HPV18 E6 that cannot bind hDlg display membrane Cx43. (**A**) Western blot analysis of FLAG E6 expression in C33a cells stably transfected with vector alone (C33aV), vector expressing HPV18E6 (wt18E6) or a mutant HPV18 E6 that does not bind hDlg (mut18E6); (**B**) Western blot analysis of levels of Cx43 and hDlg in C33aV, C33a wt18E6 and C33a mut18E6. GAPDH is used as a loading control for the blots in (**A,B**); (**C**) Confocal immunofluorescence microscopy imaging of Cx43 (green) and hDlg (red) in C33aV cells, C33a18E6 cells and C33amut18E6 cells (cells expressing C-terminal mutated 18E6 that cannot bind hDlg). Membrane gap junction plaques are indicated with arrowheads. The arrow in B indicates cytoplasmic Cx43 staining. Nuclei are stained with DAPI. Bar = 10 μM.

Our data suggest that E6 is involved in Cx43 trafficking and turnover in cervical cancer cells. E6 could control Cx43 via changes in intracellular signalling that impact on the Cx43 cytoplasmic C-terminal domain. Alternatively, E6 could regulate Cx43 via its effects on hDlg. If hDlg-E6 interaction was important in Cx43 recycling from the membrane then a mutant E6 that could no longer bind hDlg might alter Cx43 trafficking. To test this, we examined the effects on Cx43 localisation in C33a cells of the HPV18 E6 mutated in its hDlg binding motif (C33amut18E6) [27] (Figure 5C). In contrast to wild type E6 transfected cells (C33a18E6 cells) where Cx43 and hDlg were located in the cytoplasm and Cx43 was found in a perinuclear location (Figure 5C), C33a cells expressing the mutant E6 displayed some Cx43 on the cell membrane in gap junction plaques. Taken together these data reveal that high risk HPV E6 controls critical steps in Cx43 recycling from the plasma membrane and, at least for HPV18, that E6 interaction with hDlg may underlie the loss of GJIC observed in cervical tumour cells.

4. Discussion

Elucidation of the control of Cx43 trafficking is crucial in understanding gap junction and connexin hemichannel assembly and disassembly, and its impact in a range of skin diseases, cardiovascular disease, diabetes and cancer. Although both reduced levels and overexpression of Cx43 have been reported in various cancers [7], loss of Cx43 from the plasma membrane has been observed in cervical precancers and cancers [8]. These studies suggest that Cx43 trafficking is altered in cervical cancer cells.

Previous studies on Cx43 trafficking have focused on the interaction of Cx43 with the membrane-associated scaffolding protein ZO-1. ZO-1 was shown to control cell migration, adhesion and Cx43 trafficking to and from the membrane in a range of cell types [34–39]. However, we discovered that Cx43 interacts with another scaffolding protein hDlg, in HPV16-positive cervical tumour cells. We have now confirmed this interaction can occur *in vivo* suggesting it has a functional significance. hDlg appears to have a role in Cx43 trafficking by maintaining a cytoplasmic pool of Cx43 protected from lysosomal degradation [23].

GJIC is disrupted and Cx43 is located in the cytoplasm of HPV-positive cervical tumour cells but not in HPV-negative cervical tumour cells [24]. This indicates that one of the viral oncoproteins could control Cx43 trafficking. High risk HPV E6 oncoprotein is the most obvious candidate because it associates with hDlg through its C-terminal PDZ binding motif [14,15], and we have evidence that Cx43 forms a complex with hDlg [23]. Because the amounts and intracellular locations of HPV16 E6 were significantly different between W12G non-tumour and the W12T tumour cells we considered the hypothesis that cytoplasmic E6 in W12T cells could modulate Cx43 and/or hDlg trafficking to the plasma membrane. Our data suggest that E6 does not appear to have a very strong interaction with Cx43 because it was detected only at low levels in co-immunoprecipitation in cervical tumour cells. This result is perhaps not surprising given that HPV16 E6 has been shown to interact with hDlg only at low affinity [40]. Some co-localisation of E6 with Cx43 was detected in the cytoplasm of W12T and in C33a cells ectopically expressing FLAG-tagged E6, but because E6 was distributed widely in the cytoplasm the specificity of this co-localisation requires further investigation. The data suggest that Cx43, hDlg and E6 interact in HPV-positive cervical tumour cells. It seems likely that Cx43 interaction with E6 is mediated by E6 binding hDlg.

siRNA depletion of E6 in W12T cells led to relocation of some Cx43 onto the plasma membrane where it formed gap junction plaques. This suggests that E6 is involved in the trafficking of Cx43. E6 can control intracellular signalling pathways [41–43] that impact upon Cx43 trafficking [6]. Therefore, increased E6 levels could affect Cx43 trafficking indirectly via changes in intracellular signalling sensed by the C-terminal intracellular domain of Cx43. Such changes may include post-translation modifications such as phosphorylation. Indeed dephosphorylation of Cx43 is known to lead to loss of Cx43 from gap junction plaques and intracellular relocation [44–46] and phosphorylation levels of Cx43 are lower in W12T cells than in W12G cells (Supplementary Figure 1). Alternatively, formation of an intracellular Cx43/hDlg/E6 complex in the E6-expressing cervical tumour cells might inhibit Cx43 trafficking to the membrane directly by removing Cx43 from its normal trafficking pathway. In our experiments HPV18 E6/hDlg interaction seemed to be important for Cx43 trafficking because it was found on the plasma membrane in HPV-negative C33a cells expressing the hDlg binding mutant of HPV18 E6 Cx43. It is also possible that hDlg could function in the transport cycle of Cx43 by providing a docking platform for molecules [47] in addition to E6 that are involved in assembling or disassembling connexon hemichannels. Expression or activity of such molecules could be induced by increased expression of E6 and/or by other changes induced by tumour progression.

It has been suggested that changes in Cx43 phosphorylation can be induced by changes in tissue architecture [48]. Compared to W12G cells, W12T cells exhibit alterations in microtubules and microfilaments and drug-induced disruption of the cytoskeleton led to changes in Cx43 cellular location (unpublished data). Interestingly, no relocation of Cx43 was observed in HEK-293 cells expressing HPV16 E6. It is possible that Cx43 post-translation modifications in HEK-293 cells result in a protein that is not able to dock with hDlg (or other trafficking proteins) or that signalling pathways in HEK-293 cells cannot be altered by E6 in a manner similar to those in W12T or C33a cells. It will be interesting to examine Cx43 and hDlg localization in the normal cervix and the HPV-infected cervix. For example, the transformation zone that has altered architecture and appears to be susceptible to HPV infection may exhibit Cx43/hDlg colocalisation. The relative contribution of E6/hDlg interaction, Cx43 phosphorylation and cytoskeletal alterations to Cx43 trafficking remains to be tested in normal and tumour tissues.

Ectopic expression of E6 in C33a cervical tumour cells resulted in reduced levels of Cx43. These data suggest that E6 is involved in targeting Cx43 for degradation. This observation correlates with the fact that in W12T cells that express higher levels of E6, but not in the W12G cells that express lower levels, Cx43 entered, and was degraded by, the lysosomal degradation pathway [23]. A study using clones of HPV-18 positive HeLa cervical tumour cells revealed a spectrum of Cx43 expression with the majority displaying no or low Cx43 levels in agreement with our data [28]. Similarly, in C33a cells expressing HPV18 E6 Cx43 was located in the cytoplasm and few gap junction plaques were observed. In contrast, King et al. showed that overexpression of Cx43 in HeLa cells resulted in appearance of gap junction plaques on the membrane [29]. However, Cx43 was also strongly expressed in the nucleus and cytoplasm, presumably due to the very high levels of Cx43 overexpression [29].

In this case endogenous HPV18 E6 could not mediate full cytoplasmic internalization of Cx43 perhaps due to the large amount of Cx43 in the cell overwhelming trafficking pathways. In HEK-293 cells overexpression of HPV16 E6 did not cause a reduction in Cx43 levels. This suggests that the effect of E6 is cell type-specific and perhaps restricted to cancer cells, for example Cx43 is also not degraded in W12G cells [23]. We have not yet tested whether Cx43 is targeted for lysosomal degradation in C33aE6 cells. Critically, Cx43 can also be degraded by the proteasome. E6 targets p53 for proteasomal degradation via the ubiquitin ligase, E6-associated protein (E6AP). It is possible that via E6, E6AP may also be able to ubiquitinate Cx43 and target it for degradation.

5. Conclusion

In conclusion, our findings highlight a role for high risk HPV E6 in Cx43 trafficking between the plasma membrane and the cytoplasm. Taken together, the present results suggest that when expressed at high levels, for example in cervical cancer cells, high risk HPV E6 may regulate Cx43 trafficking, resulting in decreased delivery of connexon hemichannels to the plasma membrane and inhibition of the formation of gap junctions. Alternatively, E6 could control rapid recycling of gap junction Cx43 into the cytoplasm. Whichever route is correct, reduced gap junctional communication may be another pathway by which E6 overexpression leads to tumour progression.

Acknowledgments

We are grateful to Margaret Stanley for the original W12 cell line, to Paul Lambert for the W12E and W12G clones of this line and to Lawrence Banks for pcDNA-3-E6 plasmids. Peng Sun was supported with a Glasgow University postgraduate scholarship and an Overseas Research Studentship award. This work was supported by grant number 08-0159 from the Association for International Cancer Research (Worldwide Cancer Research).

Author Contributions

P.S., A.I.MacD, L.D. and S.A. carried out the experiments. S.R.J. carried out the PLA assay and contributed discussions. M.B.H., M.E. and S.V.G. conceived the study and designed the experiments. S.V.G., S.R.J. and A.I.MacD. wrote the paper.

Conflicts of Interest

The authors declare no conflict of interest.

References

1. Evans, W.H.; Martin, P.E. Gap junctions: Structure and function. *Mol. Membr. Biol.* **2002**, *19*, 121–136.
2. Saez, J.C.; Berthoud, V.M.; Brañes, M.C.; Martínez, A.D.; Beyer, E.C. Plasma membrane channels formed by connexins: Their regulation and functions. *Physiol. Rev.* **2003**, *83*, 1359–1400.

3. Laird, D.W. Life cycle of connexins in health and disease. *Biochem. J.* **2006**, *394*, 527–543.

4. Kjenseth, A.; Fykerud, T.; Rivedal, E.; Leithe, E. Regulation of gap junction intercellular communication by the ubiquitin system. *Cell. Signal.* **2010**, *22*, 1267–1273.

5. Evans, W.H.; de Vuyst, E.; Leybaert, L. The gap junction cellular internet: Connexin hemichannels enter the signalling limelight. *Biochem. J.* **2006**, *397*, 1–14.

6. Johnstone, S.R.; Billaud, M.; Lohman, A.W.; Taddeo, E.P.; Isakson, B.E. Posttranslational modifications in connexins and pannexins. *J. Membr. Biol.* **2012**, *245*, 319–332.

7. Aasen, T. Connexins: Junctional and non-junctional modulators of proliferation. *Cell Tissue Res.* **2015**, *360*, 685–699.

8. Aasen, T.; Graham, S.V.; Edward, M.; Hodgins, M.B. Reduced expression of multiple gap junction proteins is a feature of cervical dysplasia. *Mol. Cancer* **2005**, *4*, 1–5.

9. Jamieson, S.; Going, J.J.; D'Arcy, R.; George, W.D. Expression of gap junction proteins connexin 26 and connexin 43 in normal human breast and in breast tumours. *J. Pathol.* **1998**, *184*, 37–43.

10. Bodily, J.; Laimins, L.A. Persistence of human papillomavirus infection: Keys to malignant progression. *Trends Microbiol.* **2011**, *19*, 33–39.

11. Zur Hausen, H. Papillomaviruses in the causation of human cancers—A brief historical account. *Virology* **2009**, *384*, 260–265.

12. Goon, P.K.; Stanley, M.A.; Ebmeyer, J.; Steinstrasser, L.; Upile, T.; Jerjes, W.; Bernal-Sprekelsen, M.; Gorner, M.; Sudhoff, H.H. HPV and head and neck cancer: A descriptive update. *Head Neck Oncol.* **2009**, *1*, 36.

13. Scheffner, M.; Werness, B.A.; Huibregtse, J.M.; Levine, A.J.; Howley, P.M. The E6 oncoprotein encoded by human papillomavirus types 16 and 18 promotes the degradation of p53. *Cell* **1990**, *63*, 1129–1136.

14. Kiyono, T.; Hiraiwa, A.; Fujita, M.; Hayashi, Y.; Akiyama, T.; Ishibashi, M. Binding of high-risk human papillomavirus E6 oncoproteins to the human homologue of the *Drosophila* discs large tumor suppressor protein. *Proc. Natl. Acad. Sci. USA* **1997**, *94*, 11612–11616.

15. Lee, S.S.; Weiss, R.S.; Javier, R.T. Binding of human virus oncoproteins to hDlg/SAP97, a mammalian homolog of the *Drosophila* discs large tumor suppressor protein. *Proc. Natl. Acad. Sci. USA* **1997**, *94*, 6670–6675.

16. Subbaiah, V.K.; Kranjec, C.; Thomas, M.; Banks, L. PDZ domains: The building blocks regulating tumorigenesis. *Biochem. J.* **2011**, *439*, 195–205.

17. Roberts, S.; Delury, C.; Marsh, E. The PDZ protein discs-large (DLG): The "Jekyll and Hyde" of the epithelial polarity proteins. *FEBS J.* **2012**, *279*, 3549–3558.

18. Nicolaides, L.; Davy, C.; Raj, K.; Kranjec, C.; Banks, L.; Doorbar, J. Stabilization of HPV16 E6 protein by PDZ proteins, and potential implications for genome maintenance. *Virology* **2011**, *414*, 137–145.

19. Dimitratos, S.D.; Woods, D,F.; Stathakis, D.; Bryant, P.J. Signalling pathways are focused at specialised regions of the plasma membrane by scaffolding proteins of the MAGUK family. *Bioessays* **1999**, *21*, 912–921.

20. Kim, E.; Sheng, M. PDZ domain proteins of synapses. *Nat. Rev. Neurosci.* **2004**, *5*, 771–781.

21. Woods, D.F.; Bryant, P.J. The disc-large tumor supressor gene of drosophila encodes a guanylate kinase homolog localized at septate junctions. *Cell* **1991**, *66*, 451–464.

22. Bilder, D.; Perrimon, N. Localisation of apical determinants by the basolateral PDZ protein Scribbled. *Nature* **2000**, *403*, 676–680.

23. MacDonald, A.I.; Sun, P.; Hernandez-Lopez, H.; Aasen, T.; Hodgins, M.B.; Edward, M.; Roberts, S.; Massimi, P.; Thomas, M.; Banks, L.; *et al*. A functional interaction between the MAGUK protein hDlg and the gap junction protein connexin 43 in cervical tumour cells. *Biochem. J.* **2012**, *446*, 9–21.

24. Aasen, T.; Hodgins, M.B.; Edward, M.; Graham, S.V. The relationship between connexins, gap junctions, tissue architecture and tumour invasion, as studied in a novel *in vitro* model of HPV-16-associated cervical cancer progression. *Oncogene* **2003**, *22*, 7969–7980.

25. Jeon, S.; Allen-Hoffman, B.L.; Lambert, P.F. Integration of human papillomavirus type 16 into the human genome correlates with a selective growth advantage of cells. *J. Virol.* **1995**, *69*, 2989–2997.

26. Johnstone, S.R.; Kroncke, B.M.; Straub, A.C.; Best, A.K.; Dunn, C.A.; Mitchell, L.A.; Peskova, Y.; Nakamoto, R.K.; Koval, M.; Lo, C.W.; *et al*. MAPK phosphorylation of connexin 43 promotes binding of cyclin E and smooth muscle cell proliferation. *Circ. Res.* **2012**, *111*, 201–211.

27. Gardiol, D.; Kühne, C.; Glausinger, B.; Lee, S.; Javier, R.; Banks, L. Oncogenic human papillomavirus E6 proteins target the discs large tumour suppressor for proteasome-mediated degradation. *Oncogene* **1999**, *18*, 5487–5496.

28. King, T.J.; Fukushima, L.H.; Donlon, T.A.; Hieber, A.D.; Shimabukuro, K.A.; Bertram, J.S. Correlation between growth control, neoplastic potential and endogenous connexin43 expression in HeLa cell lines: Implications for tumour progression. *Carcinogenesis* **2000**, *21*, 311–315.

29. King, T.J.; Fukushima, L.H.; Hieber, A.D.; Shimabukuro, K.A.; Sakr, W.A.; Bertram, J.S. Reduced levels of connexin 43 in cervical dysplasia: Inducible expression in a cervical carcinoma line decreases neoplastic potential with implications for tumour progression. *Carcinogenesis* **2000**, *21*, 1097–1109.

30. Jackson, R.; Togtema, M.; Zehbe, I. Subcellular localization and quantitation of the human papillomavirus type 16 E6 oncoprotein through immunocytochemistry detection. *Virology* **2013**, *435*, 425–432.

31. Jeon, S.; Lambert, P.F. Integration of human papillomavirus type 16 DNA into the human genome leads to increased stability of E6 and E7 mRNAs: Implications for cervical carcinogenesis. *Proc. Natl. Acad. Sci. USA* **1995**, *92*, 1654–1658.

32. McFarlane, M.; MacDonald, A.I.; Stevenson, A.; Graham, S.V. Human Papillomavirus 16 Oncoprotein expression is controlled by the cellular splicing factor SRSF2 (SC35). *J. Virol.* **2015**, *89*, 5276–5287.

33. Thomas, M.; Glaunsinger, B.; Pim, D.; Javier, R.; Banks, L. HPV E6 and MAGUK protein interactions: Determination of the molecular basis for specific protein recognition and degradation. *Oncogene* **2001**, *20*, 5431–5439.

34. Giepmans, B.N.G.; Moolenar, W.H. The gap junction protein connexin43 interacts with the second PDZ domain of the zona occulens-1 protein. *Curr. Biol.* **1998**, *8*, 931–934.

35. Toyofuku, T.; Yabuki, M.; Otsu, K.; Kuzuya, T.; Hori, M.; Tada, M.H. Direct association of the gap junction protein connexin-43 with ZO-1 in cardiac myocytes. *J. Biol. Chem.* **1998**, *273*, 12725–12731.

36. Hunter, A.W.; Barker, R.; Zhu, C.; Gourdie, R. Zonal Occulens-1 alters connexin43 gap junction size and organisation by influencing channel accretion. *Mol. Biol. Cell* **2005**, *16*, 5686–5698.

37. Barker, R.J.; Price, R.L.; Gourdie, R.G. Increased association of ZO-1 with connexin43 during remodeling of cardiac gap junctions. *Circ. Res.* **2002**, *90*, 317–324.

38. Segretain, D.; Decrouy, X.; Dompierre, J.; Escalier, D.; Rahman, N.; Fiorini, C.; Mograbi, B.; Siffroi, J.-P.; Huhtaniemi, I.; Fenichel, O.; *et al.* Sequestration of connexin43 in the early endosomes: An early event of Leydig cell tumor progression. *Mol. Carcinogen.* **2003**, *38*, 179–187.

39. Laing, J.G.; Chou, B.C.; Steinberg, T.H. ZO-1 alters the plasma membrane localization and function of Cx43 in osteoblastic cells. *J. Cell. Sci.* **2005**, *118*, 2167–2176.

40. Thomas, M.; Massimi, P.; Navarro, C.; Borg, J.-P.; Banks, L. The hScrib/Dlg apico-basal control complex is differentially targeted by HPV-16 and HPV-18 E6 proteins. *Oncogene* **2005**, *24*, 6222–6230.

41. Contreras-Paredes, A.; de la Cruz-Hernandez, E.; Martinez-Ramirez, I.; Duenas-Gonzalez, A.; Lizano, M. E6 variants of human papillomavirus 18 differentially modulate the protein kinase B/phosphatidylinositol 3-kinase (akt/PI3K) signaling pathway. *Virology* **2009**, *383*, 78–85.

42. Spangle, J.M.; Munger, K. The Human Papillomavirus Type 16 E6 Oncoprotein activates mTORC1 signaling and increases protein synthesis. *J. Virol.* **2010**, *84*, 9398–9407.

43. Spangle, J.M.; Munger, K. The HPV16 E6 oncoprotein causes prolonged receptor protein tyrosine kinase signaling and enhances internalization of phosphorylated receptor species. *PLoS Pathog.* **2013**, *9*, e1003237.

44. Laird, D.W.; Castillo, M.; Kasprzak, L. Gap junction turnover, intracellular trafficking, and phosphorylation of connexin43 in brefeldin A-treated rat mammary tumor cells. *J. Cell Biol.* **1995**, *131*, 1193–1203.

45. Solan, J.L.; Lampe, P.D. Connexin43 phosphorylation: Structural changes and biological effects. *Biochem. J.* **2009**, *419*, 261–272.

46. Beardslee, M.A.; Laing, J.G.; Beyer, E.C.; Saffitz, J.E. Rapid turnover of connexin43 in the adult rat heart. *Circ. Res.* **1998**, *83*, 629–635.

47. Hung, A.Y.; Sheng, M. PDZ domains: Structural modules for protein complex assembly. *J. Biol. Chem.* **2002**, *277*, 5699–5702.

48. Steinhoff, I.; Leykauf, K.; Bleyl, U.; Dürst, M.; Alonso, A. Phosphorylation of the gap junction protein Connexin43 in CIN III lesions and cervical carcinomas. *Cancer Lett.* **2006**, *235*, 291–297.

Recent Progress in Therapeutic Treatments and Screening Strategies for the Prevention and Treatment of HPV-Associated Head and Neck Cancer

Sonia N. Whang, Maria Filippova and Penelope Duerksen-Hughes

Abstract: The rise in human papillomavirus (HPV)-associated head and neck squamous cell carcinoma (HNSCC) has elicited significant interest in the role of high-risk HPV in tumorigenesis. Because patients with HPV-positive HNSCC have better prognoses than do their HPV-negative counterparts, current therapeutic strategies for HPV+ HNSCC are increasingly considered to be overly aggressive, highlighting a need for customized treatment guidelines for this cohort. Additional issues include the unmet need for a reliable screening strategy for HNSCC, as well as the ongoing assessment of the efficacy of prophylactic vaccines for the prevention of HPV infections in the head and neck regions. This review also outlines a number of emerging prospects for therapeutic vaccines, as well as for targeted, molecular-based therapies for HPV-associated head and neck cancers. Overall, the future for developing novel and effective therapeutic agents for HPV-associated head and neck tumors is promising; continued progress is critical in order to meet the challenges posed by the growing epidemic.

Reprinted from *Viruses*. Cite as: Whang, S.N.; Filippova, M.; Duerksen-Hughes, P. Recent Progress in Therapeutic Treatments and Screening Strategies for the Prevention and Treatment of HPV-Associated Head and Neck Cancer. *Viruses* **2015**, *7*(9), 5040-5065.

1. Introduction and Background

Head and neck squamous cell carcinoma (HNSCC) is the sixth most common cancer in the world, with an incidence of over half a million new cases annually [1–5]. The most common tumor sites of HNSCC include the oral cavity, nasal cavity, larynx, hypopharynx, and the oropharynx [2–4,6–8]. A few decades ago, a decline in HNSCC in relation to the carcinomas of the hypopharynx and larynx was indicated [2,5,9–11]. This was attributed to the rise in public awareness [8,12–14] and the consequential decline in excessive tobacco and alcohol consumption, factors traditionally associated with this carcinoma [2,5,15–17]. In contrast to the encouraging trend, certain types of HNSCC have risen over the past couple of decades due to an increase in the incidence of oropharynx squamous cell carcinoma (OPSCC) [2,5,10,11], which includes cancers that form in the tonsils and at the base of the tongue [4,9,10,15,17–19]. This became particularly evident in patients with no history of tobacco smoking or alcohol abuse [5,16,20], arguing for the presence of an additional etiological agent [3,5,15]. The striking increase in these cancers has been attributed to the rising prevalence of human papillomavirus (HPV)-associated tumors [10,15,16,21–24].

The link between HPV and oropharyngeal carcinoma was initially suggested four decades ago, when it was still considered a risk factor [11,15]. However, it was not until the past decade that the prevalence of HPV in the head and neck has elicited considerable attention [25], and the International Agency for Research against Cancer (IARC) has now acknowledged HPV as an emergent etiological factor in the development of OPSCC [4,11,15,19,20,26]. With up to 80% of OPSCC now related to HPV [26], research reveals that the virus has undoubtedly altered the epidemiology and survival outcome landscape of head and neck carcinoma [3,21,27]. In fact, the incidence of HPV-negative head and neck squamous cell carcinomas has statistically decreased by 50%, in step with the gradual reduction of tobacco and alcohol use since the 1980s [11,17,21,25,28,29]. In contrast, HPV-positive oropharyngeal carcinomas have escalated by a dramatic 225% in the US [11,13,17,21,25,28,29], and they will represent a large fraction of the HNSCC population in the country within the next 20 years [9,21]. In fact, at the current rate of increase, OPSCC is predicted to surpass the incidence of HPV-positive cervical cancer, the archetypal HPV malignancy, in the US by the year 2020 [2,13–15,21,23,28,30–32]. Not only does this carcinoma affect the US, but it also confers a growing public health concern internationally [2,21]. Thus, the increasing epidemic of HPV-derived HNSCC is becoming a major health care issue with significant clinical ramifications [2,21].

2. High-Risk HPV as an Etiological Factor

HPV infection has been extensively studied in the context of its association with cervical cancer [2,9,23], the second leading cancer in women in less developed countries [15,33], and multiple studies have clearly established that HPV infection in the genitals is transmitted by sexual contact [20,34]. The factors responsible for the surge in HPV-derived HNSCC were once nebulous [29,34], but accumulating evidence now indicates that HPV-initiated OPSCC may be a result of changing sexual behaviors in the population [2,10,16,17,20,21,26,28,29,35,36]. For example, it has been demonstrated that HPV is eight times more likely to be isolated from the oral cavity of sexually experienced individuals than from the oral cavity of those who are sexually inexperienced [32,37–39]. Similarly, oral infection is highly correlated with multiple lifetime sexual partners, early coital debut, oral–genital sex, as well as French kissing [11,15,26,29,31,35,37,39,40]. Osazuwa et al., surmised that within the US, a sexually active individual is likely to encounter an HPV infection at one or more points during their lifetime [11,41]. However, not every HPV infection develops into a carcinoma. In fact, a large majority of infections are transient and clear without any clinical manifestations [9,18,23,41,42], with 66% of infections clearing within 12 months and 90% within 24 months [9,42]. Despite the high level of clearance, the presence of high-risk HPV infection in the oral cavity has been associated with a five to fifty-fold increased risk of HNSCC development, depending on the HPV type [20,28,29,38]. Consequently, the chances of developing head and neck cancer (HNC) increase when interacting with more than 25 lifetime vaginal sex partners and/or more than five lifetime oral sex partners, according to a study conducted by D'Souza et al. [2,20,43,44]. Interestingly, it has been shown that an HPV infection in the head and neck is correlated with an infection in the anogenital area [10,29] as cervical cancer patients have a five-fold higher risk of head and neck cancer [32,34,45]. In addition, an increased risk for tongue and tonsil carcinomas are observed in male partners of women with cervical carcinoma [2,10,32,46], and these results have been corroborated by a match on the HPV type in those couples [29,34,47,48]. Therefore, significant accumulated evidence supports the idea that the likely transmission of this infection is primarily through oral–genital and oral–oral routes [26,34].

Since HPV-positive oropharyngeal cancers display a different etiology than do HPV-negative cancers [14,21,49], HPV-derived OPSCCs are found in a subpopulation of patients that is epidemiologically, genetically, and demographically distinct from patients presenting with the more traditional HPV-negative OPSCCs [2,9,11,22]. Unlike HPV-negative OPSCCs, which are typically found in individuals older than 60 years of age with a strong history of tobacco and alcohol consumption [11,50], HPV-related OPSCC typically appears in younger populations, between the ages of 40 and 55, with generally low levels of substance abuse [9,12,29,37,51]. This cohort of patients tends to be high functioning [28], and demonstrates a better general condition [29] as well as health [2,3,36,39,52–55]. Moreover, a recent study reported an 80% higher incidence in males than in females [2,11,19,25,32,56,57] and a lower incidence in blacks than in Caucasians (4% in blacks *vs.* 34% in their Caucasian counterparts) [2,21,32,58,59]. In addition, this patient cohort possesses higher economic status and more education [2,13]. Therefore, subjects with HPV-related HNSCC are likely to be middle-aged Caucasian males who are non-smokers and non-drinkers with a higher socioeconomic status and educational level [9,28,32].

3. Current Treatments and Therapies

Current therapeutic interventions for HNSCC patients include surgery, chemotherapy, and radiotherapy [6,15,52,60]. Each of these treatments have been employed at different clinics in the US [31], but currently no clinical guidelines differentiating treatment strategies between HPV-derived and tobacco-derived HNSCC exist [23,61,62]. Moreover, only a few clinical trials have made such a distinction [1,2,31,60,63–66], even though these two subsets represent separate disease entities pathologically and etiologically [24,26,31,49,57,63]. Presently, the standard therapy for head and neck cancer is determined by the tumor stage [2,4,15,64], the site of the tumor [4,15,64] and the expected functional outcomes [4], as well as by the preference of the practitioner and the patient, which include considerations of the level of organ preservation and the patient's quality of life [2]. Head and neck cancer is classified into the following categories: early-stage or stage I/II, locally advanced or stage III/IV, and recurrent or metastatic phase [67]. Early stages of head and neck cancer are usually treated with a single-modality treatment, such as radiotherapy or surgical resection [4,12,13,15,68]. A combination of multiple therapies for superior oncologic results are required for the management of advanced stages III/IV [4,61,67]; for example, surgery with adjuvant radiation or chemoradiation with chemotherapy being added for high risk pathologic features found from the surgical specimen [2,14,35,69,70], or radiotherapy with concomitant chemotherapy [14,64,71–73]. Therefore, patients with advanced stages of head and neck cancer are treated through a multidisciplinary and multimodal treatment approach [50,67,68,74].

3.1. Surgery

Surgery is one of the standard treatments for early stage I/II HNSCC. In the past, surgical procedures sometimes consisted of extensive open transmandibular, and open pharyngotomy procedures [2,12,62,64,75] that resulted in severe morbidities including facial deformity, dysarthria, and dysphagia [15,52,53,62], especially in more locally advanced cases. Over the past 30 years, advances in radiotherapy and chemotherapy yielding favorable oncologic outcomes shifted treatment choices away from open surgery [52,55,62], until new minimally invasive trans-oral surgery (TOS) came into prominence as a viable surgical tool for early phase OPSCC [9,54,62,66,75] within the last decade, promising to reduce morbidity and mortality while improving organ preservation [9,24,53]. This new surgical approach enables resection of a tumor through the opening of the mouth without the damage to normal tissue and musculature seen in transcervical or transmandibular approaches [62,76]. Because of these advancements in technology, HPV-associated OPSCC patients may be the most appropriate subgroup to undergo a minimally invasive TOS regimen since they tend to be younger, non-smokers, and have good odds for long-term survival [9,62]. Moreover, the restoration of surgical resection as a safe treatment modality reinstituted the advantage of acquiring surgical specimens for definitive pathological staging to guide in the determination of adjuvant therapy needed. Transoral laser microsurgery (TLM) and transoral robotic surgery (TORS) are currently the principal TOS techniques utilized for head and neck carcinoma [9,28,62].

TLM is one of the procedures available for early head and neck cancer [28]. This procedure utilizes surgical apparatus already present in many medical centers, such as a laryngoscope, operating microscope, and a CO_2 laser [28,77]. TLM is capable of conserving normal tissue by resecting the tumors *via* a direct transoral approach using transtumor cuts to assess tumor depth and microscopic magnification to aid in margin control [28,54,77]; as a result, the TLM treatment of locally advanced head and neck cancer can attain excellent cosmetic and functional outcomes [28,53].

In 2009, TORS, an alternative method for transoral surgery, became approved for small primary tumors of the head and neck region [9] and is quickly becoming a popular technique [53,54,62,75]. TORS's magnified and angled stereoscopic visualization and articulated robotic arms aid in complex resections [2,52–54] as well as the performance of oncologic extirpations *en bloc* in the oral cavity [28,62,77]. In addition, TORS offers tremor filtration and high-precision motion scaling although at a significantly higher cost [28,52,54,62]. The price for a Da Vinci robotic system surpasses a million dollars, and the additional expenses for services and expendable supplies can be a limiting factor for many clinical centers [53,54,62]. However, some of the advantages of the TORS over open surgery include low rates of complications and mortality with shorter postoperative recovery time, as well as satisfactory oncological results and improved swallowing outcomes [2,9,52,53]. There is some evidence to suggest that TORS resection may allow reduced doses of adjuvant radiation with similar oncologic control and reduced treatment morbidity [53,75]. To help clarify this, the ECOG 3311 clinical trial is evaluating the de-intensification of postoperative radiation after surgical resection of HPV-associated OPSCC [61,62,76]. Therefore, these new trans-oral surgical techniques are decreasing cosmetic disfigurement while improving function and quality of life [62,78].

3.2. Chemotherapy

Cisplatin is the most widely used chemotherapeutic agent with the best prognostic outcome, achieving about a 90% 3-year survival rate [15,67]. Cisplatin, also known as cis-Diammineplatinum (II) dichloride or CDDP, is a DNA intercalator targeting cells that replicate at a high rate [74]. This intercalator binds to guanine residues causing crosslinks between the DNA strands, and eventually leading to cell death [74]. Studies indicate that HPV-associated patients have a higher response rate to platinum-based chemotherapy than do their HPV-negative counterparts [74]. However, the benefits of this therapy come at a price due to comorbidities such as, but not limited to, xerostomia, dysphagia, neurotoxicity and renal failure [15,52,55]. This platinum-based regimen continues to be a standard treatment for organ preservation protocols [15,72,79] as well as advanced and unresectable head and neck cancers [15,80]. Other commonly used chemotherapeutic agents consist of platinum compounds such as carboplatin; taxanes such as docetaxel and paclitaxel; methotrexate; and 5-fluorouracil [67,81,82]. These chemotherapeutic drugs are showing some promise in the treatment of HNSCC patients, however, additional agents that can target the tumor cells more specifically are under investigation. Targeted chemotherapeutic agents such as cetuximab are discussed below.

3.3. Radiotherapy

Historically, radiotherapy has been thought of as a conventional treatment for HNSCC and is usually a component of a multi-modal therapy plan [8,55]. Radiotherapy induces double strand breaks of the tumor cells, reducing cell viability and increasing cell cycle arrest and death [83]. Radiation treatment delivery has evolved through the decades, and advances in radiotherapy have led to the development of intensity-modulated radiotherapy (IMRT) [84,85]. IMRT delivers radiation to tumor tissues while simultaneously reducing the dosage to non-carcinogenic cells [62,86]. In this manner, IMRT can more efficiently spare healthy tissues, enhance tumor coverage, and achieve a steady dose distribution [85]. Even though IMRT has improved survival outcomes, the toxicities concomitant to irradiation continue to deteriorate a patient's quality of life [28,86]. For instance, HNSCC treated patients have a higher likelihood of experiencing occlusive carotid artery disease and stroke [12,28]. Moreover, a considerable amount of radiotherapy-induced malignancies become apparent in HNC survivors [28]. Notwithstanding, the major cause of death in HNC survivors unrelated to cancer is cardiovascular disease associated with radiotherapy [28]. Since the HPV-dependent OPSCC population is typically younger and exhibits a favorable prognosis, the value of reducing chronic morbidities such as xerostomia [12,53], dysphagia, mucositis, lymphedema, and fibrosis is considerable [3,53,62]. Therefore, radiation protocols are actively being researched in attempts to decrease both the dosage and duration of therapy [77].

Research has shown that disease control is attainable in both HPV-related and HPV-unrelated subsets when TORS is employed as an initial surgical approach followed by chemoradiation [9,35]. Unfortunately, these patients are subject to the side effects of surgical procedures as well as those of nonsurgical interventions [9,31]. Despite the improvements in therapeutic techniques toward reducing morbidity and increasing survival, the 5-year survival rate of HNSCC patients remains at around 50% [4,6,35,57,67,75,82,87–89].

4. Management of HPV-Associated Tumors: The Debate

Clinicians are becoming increasingly aware of the need for differential therapeutic regimens between HPV-positive and HPV-negative patients [31] due to their distinct disease etiologies [14,15,22,63].

Evidence that differences in the biological aspect of these subgroups may affect their prognosis and optimal treatment is increasing [1,15,90]. For example, data collected over the past several years makes a compelling case that patients with HPV-derived OPSCC have a more favorable survival than do their matched controls, regardless of treatment strategy [1,3,20–22,28,31,35,37,57,60,63,65,91]. Research suggests that HPV expression corresponds with increased response rates to conventional chemotherapy [2,17,28,29,52,57,63,91], radiotherapy [1,2,16,17,22,29,57,63], and radiochemotherapy (RCT) [1,28,31,52,63,65,91,92]. Moreover, the 3-year overall survival of patients with HPV-associated OPSCC is about 75% as opposed to 50% for those with HPV-unassociated malignancies [10,24,31,37,57,63]. Additionally, studies of HPV-positive HNSCC revealed a drop of approximately 50% in recurrences, a 40% decrease in the risk of death [17,25,39] and a lower incidence of metastases than seen with their HPV-negative counterparts [2,23,37,65,93]. As impressive as these statistics look, recurrence and metastasis are still responsible for the leading cause of death in HPV-derived OPSCC [31,49,94]. In summary, patients with HPV-induced tumor report improved therapeutic responses to interventions and better survival rates due to increased sensitivity to chemotherapy and radiotherapy [1,15,20,28,31,35,95].

The reason(s) HPV-related HNSCC are associated with an improved survival outcome as compared to HPV-unrelated cancers remains speculative [9,14,60], but this difference could be ascribed to a variety of factors [17,63]. One set of explanations focuses on the patient population, indicating that the favorable prognosis of patients with HPV-associated cancers may be attributable to their younger age at diagnosis [1,2,9,17,74], their high functioning and superior performance status [2,9,17], as well as the presence of minimal tobacco and alcohol related co-morbidities [1,2,17,28,74].

An alternate or possibly complementary explanation focuses on differences in biological mechanisms. That is, even though the biologic mechanisms leading to divergent prognoses in HPV-dependent and independent oropharyngeal cancer have been elusive [14,57], the survival benefit enjoyed by HPV-associated patients could be connected to the molecular differences arising from virus-mediated activities as opposed to events that occur as a consequence of the carcinogens or mutations present in non-HPV cancer patients [43,53,63]. For example, in most tobacco-related tumors, the tumor suppressor gene *TP53* is mutated and inactive, while the *TP53* gene in HPV-infected tumors is wild-type and functionally intact, with the protein being degraded by the HPV oncoprotein E6 [2,5,30,35]. Research indicates that persistent treatment with certain therapeutic agents can suppress *E6* oncogenes, allowing the *TP53* gene to carry out its normal function [53,74]. Therefore, the presence of the wild-type *TP53* gene and the lower mutation rate [37] observed in HPV-derived SCC may enable these tumor cells to undergo an intact apoptotic response when treated with radiotherapy and/or chemotherapy, resulting in a high response rate [2,3,9,20,53].

Another possibility is that HPV-positive cancer cells express viral proteins that induce and enhance the immune response, which becomes involved in clearing cancer cells during treatment [2,8,74,96]. This theory was proposed after a cancer cell line treated with chemoradiotherapy *in vitro* demonstrated increased survival [3] and resistance to treatment [1,28] as compared to the same therapy applied *in vivo*, where the cells are surrounded by an immunologic microenvironment. Likewise, an apparent higher response in immunocompetent *vs.*

immunodeficient mice further supports this finding [3]. In addition, studies indicate that the majority of HPV-infected tumor patients manifest a higher titer of T cells infiltrating the tumor [1] and a high percentage of cytotoxic $CD8^+$ T cells that are specific to HPV [1,3,37] compared to non-HPV tumor patients.

Lastly, the difference in the degree of intratumor heterogeneity between HPV-dependent and HPV-independent OPSCC could contribute to their divergent prognoses. Intratumor heterogeneity refers to a tumor population comprised of subpopulations that display differing genetic makeups [28]. Assuming that certain subpopulations are more susceptible to treatment therapies than others, tumors with high intratumor heterogeneity are progressively identified as having poor therapeutic response and recurrence or metastasis [28]. HPV-driven tumors are considered to represent a homogeneous, one-agent-induced population, and are thus less intratumorally heterogeneous, possibly leading to the better therapeutic response.

To date, an effective mono-dimensional therapy approach suitable for head and neck carcinoma is not available [31]. Moreover, the classical therapies generate substantial side effects [77,96]. Traditionally, therapeutic strategies have consisted of open surgery with the option of radiochemotherapy [55,77]. The adverse effects of these therapeutic interventions have not improved in recent decades, and severe consequences associated with swallowing [15,55,77], talking [15,55,77], breathing [77], hearing [15], and even one's countenance [15,55,77] are prevalent. The current contention lies in whether the intensity level of the therapy is too high for the cohort of HPV-positive patients that exhibit better outcomes [20,23,31,55,76]. The different therapeutic strategies all have comparable oncological effects, yet the functional complications can have a particularly long lasting effect on the rising cohort of young patients with HPV-associated head and neck cancer [2,28]. In making their decisions, clinicians are dealing with a subset of patients that will most likely reach full recovery and surpass their cancer by a few decades, and hence will be severely affected by the late sequelae of cancer treatment [2,28,52,54,93]. Consequently, an intensive multidisciplinary regimen resulting in considerable morbidity might be inappropriate for the HPV-initiated HNSCC subgroup [2,9]. Accordingly, the favorable prognosis in HPV-driven oropharyngeal cancer has prompted the progression to organ preservation strategies [23,28,55] that treat the tumor with minimal cosmetic and functional complications [19]. Therefore, evaluating the options for therapeutic de-escalation to reduce toxicity and determining treatment strategy with high efficacy to optimize quality of life is of utmost importance for this HPV-associated subpopulation [9,28,31,55,64,87,97].

Some researchers contend that concurrent radiochemotherapy may confer excess treatment [9]. Moreover, evidence has surfaced denoting the overtreatment of adjuvant chemotherapy after surgical resection in locally advanced HNSCC patients [77], accruing proponents for the de-escalation regimens. Yet the establishment of a de-intensification regimen can be challenging since nearly 10% of patients with HPV-derived tumors have a poorer prognosis and a higher likelihood of developing metastases or recurrence [9,14,31,63], demanding a more potent therapy. Some advise not to change treatment decisions or management strategy on the basis of HPV, as conclusive evidence is lacking [4,18,20,24,37,97]. Others argue that the treatment of patients with HPV-associated OPSCC should depend on the tumor phase [24], the general condition and performance status of the patient, and the expected functional outcomes [9]. Their aim is to increase the opportunities to tackle early phase carcinomas with a mono-dimensional regimen [9]. Further investigation is necessary to determine whether an alternative treatment strategy is required for HPV-associated HNC patients.

5. De-Intensification Trials

Clinical trials testing various de-intensification strategies for HPV-positive head and neck carcinoma patients are under examination [23,28]. The de-escalation of therapy intensity may be achieved through several different approaches [36,52]. An initial proposal was to decrease the standard dose of definitive radiotherapy or chemoradiotherapy, since radiation is considered the most toxic component of a therapeutic regimen [23,28]. An Eastern Cooperative Oncology Group (ECOG1308) phase II trial evaluated the response to chemotherapy with paclitaxel, carboplatin, and cetuximab, and based on their complete response, determined which patients could safely undergo radiation dose reduction [23,31,53,80,93]. In 2014, the investigators revealed positive initial results in patients that underwent the dose reduction [3].

Another strategy is to employ the new minimally invasive TOS technique as a primary surgical therapy [28,52]. A randomized trial, ECOG3311, evaluating whether initial transoral surgery (TORS) can allow for decreased adjuvant dose radiotherapy for patients with HPV-positive HNC is currently in progress in the US [28,37,76,93].

Another possibility is the administration of a less toxic alternate agent, such as cetuximab, an anti-epidermal growth factor receptor (EGFR) antibody [52]. The Radiation Therapy Oncology Group study (RTOG 1016) and De-ESCALaTE phase III trials are comparing conventional cisplatin concurrently with radiotherapy to the new cetuximab with concomitant radiation in HPV-driven locally advanced oropharyngeal squamous cell carcinoma (SCC) [15,23,28,31,36,37,93].

6. Molecular Mechanisms

Ever since the presence of HPV was demonstrated in tissues of HNSCC patients in 1983, the study of molecular mechanisms in HPV-associated HNSCC has garnered significant attention [3,15,20,98]. Insight accumulated on the molecular progression of HPV derives from the extensive research performed on cervical tumorigenesis [2,9,23,74]; consequently, cervical cancer has become the standard model for HPV studies [15,18]. With an epidemic on the horizon, it will be vital to adjust our understanding of the properties of HPV in cervical carcinoma to be applicable to head and neck carcinoma [15].

Approaches already developed for the treatment and prevention of cervical cancer may be of great help in combating HPV-derived HNSCC [15]. Nonetheless, the different anatomical and molecular aspects between cervical and oropharyngeal carcinoma must be delineated to adapt the current knowledge to the oral context [15]. For example, estrogen signaling plays a significant role in cervical cancer, while hormonal dependence is not discernible in head and neck carcinomas [15,99]. Furthermore, the cervix is not as frequently exposed to elevated amounts of cytotoxic agents and chemical carcinogens as the oropharynx [9]. The distribution of specific HPV types detected in the two cancers varies as well, revealing a broad spectrum of high-risk HPV types accounting for cervical cancer in comparison to the more limited variety observed in head and neck carcinomas [15]. Another difference observed is that, contrary to the integrated HPV form predominant in cervical cancers [100,101], the HPV genome in HNSCC samples is frequently found in both episomal and integrated forms [20,32,34,102–104], indicating that integration is not essential for progression of tumorigenesis in this location [15,34]. Additionally, the presence of HPV in different cancers engenders divergent prognoses [57]. That is, while HPV-driven HNSCC have better treatment outcomes, the presence of HPV in cervical cancer is associated with poor prognosis [57,105], and HPV-associated cervical cancers are considered more chemoresistant than are other gynecological tumors [106]. These differential prognoses may be due to the distinctive properties and elements characteristic of the host cancer that come into play with the virus, and might contribute substantially to the pathogenesis of HPV malignancy [57]. Despite these considerations, the molecular virology of infection is not anticipated to be significantly different in HNSCC as compared to that present in cervical cancer. The prevailing understanding of the molecular details of HPV has therefore shed light on HPV-positive head and neck cancer.

HPV is transmitted through the mucosal and non-mucosal skin epithelia [15,37]. About 200 HPV types categorized based on the HPV L1 sequence have been detected, some of which have the ability to induce carcinogenesis [15,23,33,37,107]. Nearly 40 of these HPV types affect the mucosal tissues [107] and can be stratified into low-risk (HPV 6,11) and high-risk (*i.e.*, HPV-16, 18) categories, based on their ability to develop precancerous lesions and their potential to cause malignant transformation [1,15,28,33,37,108,109]. The oncogenic high-risk subtypes are expected to give rise to 5.2% [14,15,18,28,110] of cancers globally, being responsible for up to 70% of oropharyngeal [14,33], 99% of cervical [14], 88% of anal [14], and 70% of vaginal [2,14,33] lesions. Of the 20 identified carcinogenic high-risk HPV types [37,107], HPV-16 is the most rampant [25,39], accounting for more than 90% of HPV-positive oropharyngeal cancers [1,14–16], followed by HPV-18 [11].

The HPV is a non-enveloped, double-stranded DNA virus that displays a predilection for squamous cell epithelium [15,28,33,37,111]. The stratified squamous epithelium is composed of progenitor cells in the lower *stratum*, and as they move up the suprabasal layer [20,37], they become differentiating keratinocytes [15,74]. HPV infection occurs when small lesions or tears at the surface of the epithelium are present, granting the virus entry to the progenitor cells in the basal layer of the stratified epithelium [15,20,37,74]. Following an infection, the virus will seize the host cellular machinery to synthesize viral nucleic acids and transcribe proteins, though usually at low levels [9,15,42]. HPV then takes advantage of the differentiation process in these keratinocytes to complete its life cycle [15,42,112]. When the differentiating cells reach the top *stratum* of the epithelium, HPV will proceed with protein coat formation, assembly of the new viral components, and eventual viral release [15]. Though the process described does not normally lead to cancer, certain events can trigger HPV to transform the differentiating keratinocytes into SCC [9].

The HPV genome is composed of approximately 8,000 base pairs [109] with dual promoters that encode two separate groups of viral proteins [1,107,111,113]. The non-structural or early genes *E1*, *E2*, *E4*, *E5*, *E6*, and *E7* are involved in viral replication, and the structural or late genes *L1* and *L2* control the viral packaging [15,33,107,111]. E1 manages the replication and transcription of the virus by acting as a DNA helicase [15], and is the only viral protein with enzymatic activity [33]. E2 can regulate the HPV genome and down-regulate the expression of E6 and E7 oncoproteins by binding to their promoters [15,111]. The activity of E4 is less well understood, but findings suggest that its interactions with the intermediate filaments of the keratin cytoskeleton may assist with viral release [15,114].

The immortalizing qualities of the virus are attributable primarily to the oncoproteins E6 and E7 [1,2,15,112] with additional contributions from E5 [37]. The cooperation between these three oncoproteins and with their interacting cellular partners promotes the transformation of the host's epithelium and maintenance of the phenotype that leads to tumorigenesis [1,15,23,37,42,112]. As currently understood, the function of E5 is to subvert immune surveillance by repressing the major histocompatibility complex (MHC) class I molecules in the host cells [42,115]. Moreover, the E5 oncoprotein, particularly E5 from HPV-16, is involved with trafficking and signaling through the EGFR pathway [42,115].

The oncoproteins E6 and E7 are constitutively expressed throughout the progression of the carcinoma [90], making them attractive targets for antiviral therapy [112,114,116–119]. In the case of cervical cancer, the elevated expression of the E6 and E7 oncoproteins is attributed to the integration of HPV into the genome of the host, in such a way as to deregulate expression of the negative regulator E2 [15,19,102,111]. However, integration seems to be less necessary for the development of HNSCC, indicating that the enhanced expression of viral oncogenes in this context can be independent of viral integration [23,103]. We can speculate that the reason for the expression of oncoproteins in episomal HPV oral cancer may be exposure to exogenously derived factors, which can synergistically work in conjunction with the virus to elicit tumorigenesis.

The central role of the oncogenic protein E6 is to inhibit apoptosis of the infected cells by accelerating the degradation of apoptotic mediators, including the well-known tumor suppressor protein p53 [109,120–122], thereby removing these proteins from functioning in the intrinsic apoptotic pathway [115]. The HPV E6 oncoprotein induces ubiquitination of p53 by complexing with E6AP, an E3 ubiquitin ligase [20,123]. The resulting annihilation of p53 leads to the prevention of cell cycle arrest and/or apoptosis [20,30,112,123]. E6 proteins from high risk and low risk HPV types are both able to bind to p53, however, only the high-risk types are able to carry it through to proteasomal degradation [112,124]. In addition to blocking the intrinsic apoptotic pathway through p53 degradation, E6 is able to protect host cells from extrinsic apoptosis, which is triggered by the binding of tumor necrosis factors (TNF)-family ligands to their corresponding receptors [115]. For example, E6 has been shown to bind to major players of the extrinsic apoptotic pathway such as the initiator of the caspase cascade, procaspase 8 [125,126], as well as the adaptor molecule Fas-associated Death Domain (FADD) [127,128]. E6 binding to these substrates leads to their accelerated degradation, thereby inhibiting the transmission of apoptotic signals to effector caspases such as caspases 3 and 7. As a result, E6 prevents cells from undergoing apoptosis initiated through both the intrinsic and extrinsic pathways [129].

Another oncogene, *E7*, enhances cellular proliferation by inactivating the retinoblastoma protein (pRb) and other proteins involved in the control of cell division [2,25,109,120,130,131]. The HPV E7 protein binds to the pRb-E2F complex and removes pRb from the complex, leading to the disruption of cell cycle controls [20,123,132]. Hence, a therapeutic strategy that targets these oncogenes would target the cells that have been infected and transformed by reactivating their intrinsic and extrinsic apoptotic pathways and regaining cell cycle control. Such promising avenues could potentially augment the effectiveness of current modalities while reducing toxicity and morbidities.

7. HPV Detection and Screening Tools

The majority of head and neck carcinomas are discovered at late stages of tumor progression, arguing for the need of a reliable detection tool that is clinically relevant to facilitate early detection of HNSCC [15,27]. Considering factors of age, stage of disease, and tobacco smoking status in these cancer patients, the most significant prognostic indicator of survival found to date is HPV status [2–4,8,19,20,63,64,74]. It is estimated that HPV affects approximately 70% of all carcinomas in the oropharynx and the oral cavity [2,10,21,31,34,35,39,73,85]. Moreover, since HPV-related OPSCC has a remarkably more favorable prognosis than does HPV-unrelated cancer [35], establishing HPV status through an effective screening tool will offer significant advantages.

In contrast to the case with cervical cancer, there are no reliable screening methods or routine check-ups equivalent to the Pap smear to detect early HPV neoplasia in the oral cavity [13,15,29,35]. Moreover, since the infected tissue in the oral cavity normally arises in an inaccessible location, devising and implementing such a tool for regular diagnosis becomes challenging [15,32,133], leaving it up to the patients to consistently monitor for symptoms such as continual sore throats, oral lesions, or swollen masses or glands [13,15]. Unfortunately, these relatively mild and non-alarming manifestations tend to go unnoticed quite frequently, compounding the issue that most head and neck carcinomas are identified at later tumor stages by the time of diagnosis [6,7,15,82,83]. Consequently, finding accurate and practical methods to assess the presence of HPV in the oral cavity is a high priority [2].

At this point, the technique(s) to be employed for determining the HPV status of head and neck cancers is controversial, due to variations in available methods in terms of cost, sensitivity, technicality, specificity, and reliability [2,18,20,27–29,134–136]. Three common methods of detection are currently used: Polymerase Chain Reaction (PCR), in situ hybridization (ISH), and p16 immunohistochemistry (IHC) [2,54]. In particular, the detection of the viral DNA, such as *E6* or *E7* sequences [137,138] through PCR or ISH has been a very common practice [9,139,140]. PCR is highly sensitive, detecting as little viral DNA as 0.001 copy per genome from tumor samples, plasma or salivary collections [28,141]. It can also assess the viral load [135] and identify the viral subtype by probing for the *L1* region of the HPV genome [9,28,135,138,142]. A disadvantage of focusing on the *L1* region is that this region can be compromised or deleted following integration into the host genome [138,143], thereby leading to underestimates of the presence or the viral load of HPV [22]. Furthermore, since PCR detects a region of the viral genome indiscriminately of whether it is in the integrated or episomal form, this method does not have the ability to determine the physical status of the virus nor its activity, which are essential in assessing tumor development [28,138,139]. Additionally, this method is rather expensive and is therefore only utilized in select laboratory centers [28,144]. On the other hand, ISH is highly specific in detecting viral integration status and transcriptional activity [9,28,139]. It utilizes a fluorescent-labeled probe to localize and visualize the HPV DNA in the host genome of the tumor dissection [135,138]. Diffuse signals indicate the presence of episomal HPV, while punctate signals represent the integrated forms [145]. Nevertheless, since ISH does not amplify the viral genome, this method is not as sensitive [138] or as fast as PCR. However, the procedure can be automated and has become available in certain clinical laboratories [28,135].

The detection of HPV E6/E7 mRNA is the "gold standard" validation of active HPV oncoprotein transcription, and is considered clinically applicable in the evaluation of carcinogenesis [9,27,139,146]. Since mRNA is very fragile and easily degraded, fresh or rapidly frozen samples are required for this approach [9,139]. While the detection of mRNA through reverse-transcriptase PCR or RT-PCR is technically challenging and perceived as inappropriate for routine screening [9], the novel ISH assay, RNAscope, has been met with great interest and found to be perhaps the most promising of available methods [139,146].

Another major alternative for detecting the virus is the IHC of the CDK inhibitor p16, a transcript encoded by the *CDKN2A* gene [9,54,138]. This technique has become popular due to its high sensitivity [28], technical ease, swiftness, practicality [28,37,139,144], inexpensiveness [28,37,139,144], and adequate consistency with PCR and ISH [28]. p16 is considered a suitable surrogate marker of HPV infection [9,20], and is biologically relevant because its overexpression corresponds closely to the transformation of infected cells [15,138]. p16 becomes up-regulated when E2F is released from the E2F-pRb complex after pRb is degraded by E7 [9,15,20,37,96]. This method of detection is the most widespread across multiple clinical centers [37,139]. It should, however, be noted that not all tumors that test positive for p16 contain HPV [37]. Across various tests, HPV infection has not been identified in approximately 10%–20% of p16[+] head and neck carcinomas [37,139]. Since the practice, interpretation, and reporting of p16 IHC differ, in some cases its prognostic diagnosis can be misinformative and hence unreliable as a stand-alone method [2,9,28,139].

Many investigators propose that using RT-PCR to detect the presence of E6/E7 mRNA may be suitable as a gold standard for fresh samples, since the expression of these two oncogenes is characteristic of a functional HPV infection and cell transformation [9,17,19,27]. However, this method requires further examination [139]. According to one study, the employment of HPV-PCR or p16 IHC alone is not very reliable or clinically adequate [147]; notwithstanding, Dalianis *et al.* reported that a HPV DNA test such as PCR in addition to an evaluation of p16 overexpression through IHC is regarded as "specific and sensitive as utilizing a gold standard" [9,17,19,145,148]. Yet another panel of experts has suggested a "cost-efficient" stepwise algorithm to reliably determine HPV infections, which includes an initial testing of p16 through IHC followed by an HPV ISH to confirm the IHC results [28,139]. If the tests provide conflicting results, a PCR or an ISH probe for specific HPV types can be utilized [28]. This sequence of methods is thought to provide the highest specificity for determining HPV status [20,139]. Others have suggested variations of these detection methods and proposed a variety of combinations [9,136,139]. In order to standardize the detection methods in clinical settings and to design reliable clinical research, a unanimous agreement on the most reliable detection tool(s) for HPV status is required and requisite [27,28].

8. Prophylactic Vaccines

A steep upward shift in the incidence of HPV-derived HNSCC demands a search for a vaccine that can avert the infection of oral HPV before an opportunity to develop a malignant lesion arises, especially considering the lack of a reliable routine screening tool for those at risk of oropharyngeal SCC [15,21]. Past vaccines have been effective at immunizing against viruses such as influenza and varicella, and such prototypes should help in the development of prophylactics against oral HPV infection [15].

Preventive vaccines against HPV in the cervix have been developed and have become available to the public within the past decade [9]. The first prophylactic vaccine to be approved was Gardasil, a quadrivalent vaccine that prevents infection from high-risk HPV types 16 and 18 as well as the low-risk HPV types 6 and 11 [15,42]. Cervarix has been developed as a bivalent vaccine that immunizes against HPV types 16 and 18 [15,42]. Both prophylaxes encompass the predominant high-risk HPV types that are found in cervical malignancy, whereas the quadrivalent vaccine also targets genital warts and contains in addition the two most prevalent non-oncogenic viral types [15,28]. Despite the fact that Cervarix excludes the low-risk HPV types, a study that compares both prophylaxes indicated that Cervarix is able to produce a stronger antibody response than Gardasil against the two oncogenic HPV types [42]. Phase III trials of these vaccines established efficacy and safety in the protection against anogenital HPV infections, lesions, and warts, but these prophylaxes have not been certified for the immunization of HPV infection in the head and neck region [9,15]. Notwithstanding, there is great potential that the current HPV vaccination will prevent oral HPV infection [9,19]. A trial that was originally intended to examine the efficacy of the HPV vaccine in cervical infections has collected oral rinses that showed encouraging results of the vaccine's effectiveness in obviating HPV infection from the oral cavity [11,13,32,133].

In contrast to the large diversity of high-risk HPV types observed in cervical carcinoma [15], HPV types 16 and 18 constitute over 95% of HPV-positive tonsillar and oropharyngeal cancers [11,19,35]. Hence, the current prophylactic vaccines can be highly effective at preventing HPV-derived HNSCC, since they encompass the primary HPV types that are causal of OPSCC [15]. Moreover, although clinical evidence supporting their efficacy in the prevention of head and neck cancers is not yet documented [9,35], these vaccines have demonstrated that they can induce a systemic robust humoral response against the oncogenic HPV types 16 and 18, and hence should in principle be efficacious against oral infections [9,15,20]. Ongoing clinical trials are currently assessing the effectiveness of the quadrivalent HPV vaccine against HPV infection in the oral cavity [34]. The effect of these prophylactic HPV vaccines on oropharyngeal HPV infection and HPV-derived head and neck cancer will be clearer once further results are obtained [32,35,42].

9. Therapeutic Vaccines

Therapeutic vaccines for HPV-driven malignancies are currently undergoing clinical investigations [20,23]. Unlike the previously described prophylactic vaccines, which offer no protection against individuals already infected with HPV [2,35,112,116,120], therapeutic vaccines are intended to treat the individual by eliciting a cell-mediated response that can recognize and attack an established dysplasia or persistent infection [23,34,107]. Moreover, in contrast to prophylactic vaccines, which incite an antibody-mediated humoral response to clear the virus and to prevent access to the squamous epithelium, therapeutic vaccines must activate the T cell-mediated immune system to destroy the existing HPV-infected cells and prevent them from developing into carcinomas [42,111,118]. This can be challenging for immunocompromised patients because of their weakened immune system; hence, these vaccines are anticipated to be most effective in immunocompetent individuals.

In the design and development of therapeutic vaccines, HPV-16 E6 and E7 oncoproteins have become popular viral targets since they are consistently expressed in HPV malignancies and are critical for transformation [23,90,107,116–119]. Moreover, in contrast to tumorigenic antigens derived from mutated or overexpressed self-proteins, viral E6 and E7 are entirely foreign proteins, which express numerous antigenic epitopes and thus contribute toward an enhanced immune response [23,116,119]. More importantly, only the infected cells will express these viral proteins, making them ideal targets for therapy of HPV-derived cancers [23,118]. A majority of clinical trials for therapeutic vaccines are in their early phase and have focused on feasibility, immunogenicity, and safety [20,114]. Multiple vaccines are currently being explored as potential therapeutic strategies including DNA vaccines, peptide and protein vaccines, cell-based vaccines, as well as bacterial and viral live vector vaccines [20,23,107,116–118].

Due to their safety, ease of production, purity and stability, DNA vaccines have become attractive therapeutic candidates for HPV-associated HNSCC [23,107,111,116,118,119]. DNA vaccines introduce plasmid DNA into the host and promote its transcription and immune presentation of the encoded HPV proteins by the transfected cells [107,118,119]. This MHC presentation elicits T cell-mediated and/or antibody-mediated responses that attack the encoded antigen [107,118,119]. However, DNA vaccines can have low immunogenicity because they lack the ability to spread the DNA from the transfected cells and amplify it in the neighboring cells [111,119]. Despite such limitations, significant results from the therapeutic HPV DNA vaccine studies have progressed to various clinical investigations [119]. For example, a phase I trial at Johns Hopkins University is evaluating a DNA vaccine targeting HPV-16 E7 antigens in patients with advanced HPV-16-positive OPSCC [23,119,149]. This vaccine encodes for HPV-16 E7 fused to the immuno-modulatory agent calreticulin, a protein that can stimulate natural killer T cells and enhance MHC class I antigen presentation [23,117,119,149].

In contrast, peptide vaccines are taken up by antigen presenting cells (APC) directly without the need for encoding and are loaded onto MHC molecules for antigenic presentation [23,107]. This leads to activation of an antigen specific T cell response and putative elimination of infected cells [107]. Peptide vaccines are safe, stable, and easily prepared, but have poor immunogenicity [107,111,119]. Some adjuvants used to circumvent the low immunogenicity include costimulatory molecules, cytokines, chemokines, and Toll-like receptor (TLR) ligands [111,119]. Specific examples include calreticulin, Montanide ISA-51, and GM-CSF, [2,23,111,117]. Another disadvantage with respect to peptide vaccines is that they are MHC restricted, which limits their widespread use [111,119]. However, this restriction can be overcome by the use of overlapping long peptides that harbor several epitopes of the antigen [111]. One study has devised an HPV peptide vaccine composed of synthetic long overlapping peptides that encompass the E6 and E7 oncoproteins of HPV type 16 [42,90,111]. Additionally, a phase II clinical trial of this peptide vaccine with the adjuvant Montanide ISA-51 resulted in the mounting of a complete vaccine-induced immunologic response [42,90,111].

Protein vaccines are similar to peptide vaccines in many ways, but they can bypass MHC restriction since the protein contains a variety of antigenic epitopes [111,118]. Additionally, protein vaccines are loaded onto MHC class II molecules, creating primarily a humoral response instead of a cell-mediated response [111,118]. A phase II trial of the HspE7 protein-based vaccine, which is a chimeric protein composed of HPV-16 E7 and a Bacille Calmette-Guerin (BCG) heat shock protein (Hsp65), yielded modest results [107,118]. TA-CIN, a fusion protein composed of HPV-16 E6, E7, and L2, represents advancement in the field of HPV vaccination because it combines therapeutic as well as prophylactic vaccines. This protein-based vaccine has progressed to clinical trial [90,111].

The cell-based vaccine technique entails the pulsing of dendritic cells (DC) with an antigen [107,119], allowing for the presentation of epitopes, such as those derived from HPV E7, in association with MHC molecules, and is capable of eliciting a high immunologic response [107,111]. A phase I study has shown the approach to be safe and immunogenic, and a phase II trial is underway [107]. However, the production of this vaccine is lengthy, taxing, and expensive [111,119] due to the need to isolate immature dendritic cells from the patient, transfect or pulse the autologous DCs with the specific antigen, allow the DCs to mature, and expand the DCs *ex vivo* before injecting them back into the patient [111,118].

A live vector, consisting of either a bacteria or a virus, can be employed to deliver antigens such as those found in the E6 and E7 oncoproteins to the host APCs in order to enhance antigen presentation and the induction of a cell-mediated response [107,111,118]. These vectors generate a strong immune response by facilitating the spread and expansion of oncoproteins [107,111,118]. However, the disadvantage is that these live vectors could incite an immune response against the vector itself since it is intrinsically pathogenic and foreign to the host [107]. A bacterial vector-based vaccine composed of a genetically modified strain of *Listeria monocytogenes* fused to E7 has shown the ability to cause regression of solid tumors and has progressed to phase I clinical studies in oropharyngeal cancer patients [107,111,118,149]. Another group designed a vector vaccine using an integrase defective lentiviral vector (IDLV) to deliver a HPV-16 E7 protein fused to calreticulin [2,111,117]. A preclinical study revealed that a single intramuscular injection eradicated 90% of early stage tumors [2,117]. These encouraging outcomes along with emerging therapeutic vaccine trials may imply that an immunotherapeutic vaccine for immunocompetent patients shows a promising future [2,117].

10. Targeted Therapies Directed against Growth Factor Receptors

Current treatment for HNSCC patients is confined to standard therapies, such as irradiation, surgery, and chemotherapy [60,67]; and despite continued advances in these classic clinical modalities, survival rates remains comparable and many patients experience long-term side effects [15,60,82,150]. Consequently, advancements in molecular research have made the identification of targeted therapies an attractive therapeutic approach due to its purported reduced toxicity and improved efficacy [15,150].

We have come a long way in understanding the molecular biology of head and neck cancer over the past few decades [68]. Interestingly, the EGFR has been shown to be frequently elevated in over 90% of HNSCC patients [2,4,67,71,88,150]. EGFR contributes to the pathogenesis of HNSCC such that its overexpression is closely related to low survival, distant metastases, and radioresistance [4,36,67,71,88,150]. Studies have indicated that low EGFR levels in HPV-positive tumors were correlated with favorable therapeutic outcomes, while high EGFR levels were associated with poor survival [34,60,88,150,151].

The role of EGFR is to transmit signals to intracellular pathways that regulate a host of cellular activities including proliferation, cell cycle progression, apoptosis, migration, metastasis, differentiation and angiogenesis [36,60,80,151]. Among the mechanisms attributed to overexpression of EGFR are deregulation of *TP53* and amplification of *EGFR* [67]. Thus, this extracellular domain has been an attractive and prominent therapeutic target for treatment intervention [60]. Several agents directed against EGFR have been produced, of which monoclonal antibodies (mAb) and small tyrosine kinase inhibitors (TKIs) have been shown to be the most effective [68,150]. The mAbs bind to the extracellular binding domain of this receptor, while TKI's bind to the cytoplasmic side of EGFR and influence downstream molecular pathways [2,57,68,80].

Cetuximab is a recombinant chimeric immunoglobulin (Ig)G mAb, specifically targeting the extracellular domain of EGFR [2,60,80,150]. This mAb has been the most extensively studied of the anti-EGFR antibodies [150] and is the first and only targeted therapy approved for head and neck carcinoma [14,28,68,71,80]. Food and Drug Administration (FDA) approval of cetuximab (Erbitux, Merck; Darmstadt, Germany) was established in 2006 after a phase III randomized study yielded remarkable results in the overall survival of HNSCC patients when cetuximab was used in conjunction with radiotherapy (a survival of 45.6% *vs.* 36.4% for radiotherapy alone) [4,15,28,60,67,80,152]. Therefore, cetuximab is recommended for the treatment of locally advanced HNSCC in combination with radiation and in recurrent/metastatic disease either as a monotherapy or in conjunction with platinum-based chemotherapy and 5-fluorouracil [15,23,67,71,80,150]. Several clinical trials are active including the Radiation Therapy Oncology Group (RTOG1016) trial, which compares cetuximab to cisplatin along with radiation in locally advanced disease [15,23,28,36,37,80,93]. This study will determine whether the less toxic cetuximab can replace cisplatin as part of a de-intensification protocol in HPV-derived HNSCC [37,80].

Other fully humanized IgG anti-EGFR antibodies under consideration include zalutumumab (HuMax-EGFr, Genmab, Copenhagen, Denmark) and panitumumab (Vectibix, Amgen; Thousand Oaks, CA, USA), and these are being investigated in phase II and III studies [2,23,37,67,80]. A phase II trial on nimotuzumab (YM Biosciences; Ontario, Canada), a recombinant humanized mAb, has demonstrated remarkable outcomes [67,68]. These antibodies could potentially be used as substitutes for cetuximab [2].

EGFR TKIs have also demonstrated some clinical activity in HNSCC but without as much success as seen with the mAbs [57,80]. The small molecule TKIs gefitinib and erlotinib showed no efficacy in recurring and metastasizing tumors [68,80]. A phase II trial of gefitinib on recurrent or metastatic head and neck cancer produced a low response rate [80], and ECOG-E1302, a phase III randomized study, evaluated gefitinib in addition to docetaxel in recurrent or metastatic head and neck cancer but was terminated before its completion [80]. Despite these disappointments, other EGFR targets have yielded some early encouraging results [24]. Lapatinib, a dual reversible tyrosine kinase inhibitor of EGFR/HER2, is in a phase III trial assessing its efficacy in the maintenance of treatment [60,71]. Afatinib, also known as BIBW2992, is an irreversible dual tyrosine kinase inhibitor of EGFR/HER2 [71,80]. A randomized phase II trial is comparing cetuximab to afatinib in patients with recurrent or metastatic HNSCC where cisplatin has been unsuccessful [71,80].

EGFR is involved in downstream intracellular pathways such as the PI3K/Akt/mTOR pathway. Alterations in the phosphoinositide 3-kinase (PI3K) pathway have been found in patients with head and neck cancers, and appear even more predominately in patients with HPV-derived tumors [24,57,153]. These alterations may contribute to tumor resistance to anti-EGFR therapy [24]. Hence, targeting PI3K is a reasonable strategy for OPSCC treatment, and trials in phases I and II are in progress [24]. Research on the mammalian target of rapamycin (mTOR) inhibitors rapamycin, everolimus, and temsirolimus have shown mTOR suppression and delayed tumor advancement [87,150,154]. Additionally, rapamycin has been revealed to synergize with platinum-based chemotherapy in the eradication of OPSCC [87]. There are numerous trials in progress of mTOR inhibitors concomitant with different therapeutic modalities for head and neck carcinoma [87].

Vascular endothelial growth factor (VEGF) is another type of growth factor and is considered one of the most critical angiogenic cytokines in tumor vasculogenesis [83,150]. Target agents have been developed to block its receptor, VEGFR. Bevacizumab is a monoclonal antibody against VEGFR that is being explored in conjunction to other anti-EGFR therapies [83,150,155]. Sorafenib and sunitinib are tyrosine kinase inhibitors directed against VEGFR that have revealed notable therapeutic results in different human cancer cells with tolerable toxicity, and are showing encouraging results in OPSCC [154,156].

11. Targeted Therapies Directed against HPV Oncoproteins

Determining the molecular differences between HPV-dependent and HPV-independent head and neck cancers will be crucial in the discovery of therapeutic targets specific for HPV-dependent malignancies [15]. Various investigations have indicated that the HPV oncogenes E6 and E7 or their substrates may be efficacious anti-cancer targets [31,157]. However, approaches targeting the oncogenes have only reached very early phases of development, in contrast to the late-phase developments attained by agents targeting growth factor receptors [158]. Therapeutic agents targeting the viral oncoproteins include synthetic peptides [159], RNA aptamers [33,109], ribozymes [33,159], transcription factors [160], intrabodies [160], anti-sense oligonucleotides [33,160], small interfering RNA (siRNA) [30,33,159], and small molecule inhibitors [159]. Because small molecule inhibitors can be easily delivered and absorbed by tumor cells [159] and since they are flexible for medical use [161], they have gradually surfaced as a treatment option with notable efficacy and low toxicity (Figure 1) [30].

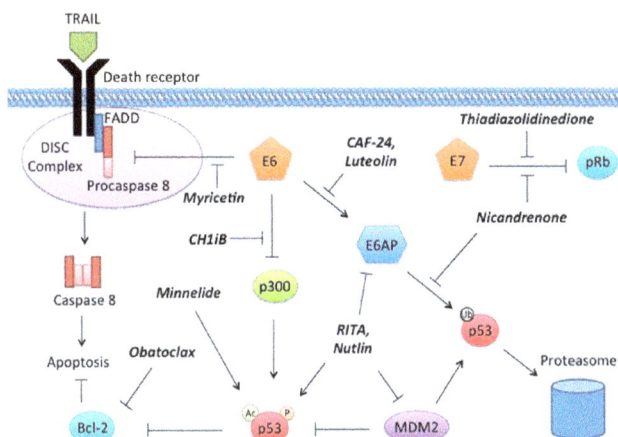

Figure 1. Involvement of small molecule inhibitors on cellular pathways affected by the E6 and E7 HPV oncoproteins.

The interaction between E6 and E6AP represents an attractive antiviral target, as agents that target this interaction may be able to inhibit the degradation of p53 and sensitize cells to agents that induce apoptosis [30,33,90,112]. One study has identified small molecules that bind to the oncoprotein E6 with great affinity [120]. In this study, the novel flavone CAF-24 and the naturally occurring flavonoid luteolin were shown to inhibit the E6-E6AP interaction by binding to the hydrophobic site between these two proteins [120]. This strategy inhibits the oncoproteins from binding to their cellular partners, thus inhibiting their oncogenic activities [120,160]. Preventing the binding of E6AP and thus the degradation of p53 can reactivate the apoptotic pathways, enhancing the outcome of available therapies [30,120,162].

A small molecule that has been widely studied in multiple types of cancer is the p53 protector, RITA (Reactivation of p53 and Induction of Tumor cell Apoptosis) [159,163]. This molecule targets p53 by changing its conformation and protecting it from binding to molecules such as E6AP and E6 that facilitate ubiquitination [33,112,159]. In this way, p53 is rescued and the apoptotic pathway reactivated, leading to the loss of tumor cells [159].

A similar approach is taken by the non-peptide small molecule compound Nutlin-3A, an imidazoline analog and potent MDM2 antagonist. Nutlin causes substantial cell death in a variety of wild-type p53 expressing cell lines [30]; however, its activity appears to be moderate as compared to RITA [159]. Another promising molecule that reactivates the wild-type p53 is Minnelide, a triptolide analog, which has shown to induce apoptosis in HPV-positive HNSCC tumors *in vitro* as well as *in vivo* [30]. CH1iB is a novel small molecule that also reactivates p53 function by inhibiting E6 from binding to p300 and thereby allowing p300 to acetylate p53 [164]. This acetylation increases p53 stability and transcriptional activity, prompting the active p53 tumor suppressor pathway to induce apoptosis when cells are treated with chemotherapeutic agents [164]. A preclinical study of Obatoclax, a small molecule antagonist of the Bcl-2 family (B-cell lymphoma 2 (Bcl-2) is a downstream substrate of E6 that is associated with resistance to treatment [151]), indicates some therapeutic value in the treatment of oropharyngeal carcinoma [165].

Another attractive target is the interaction of E6 and caspase 8, a protein involved in the extrinsic apoptotic pathway [166]. The extrinsic apoptotic pathway can be activated by several TNF-family ligands including TNF-related apoptosis-inducing ligand (TRAIL). TRAIL can initiate apoptosis in tumor cells with expression of TRAIL-specific receptors, namely DR4 and DR5 [167], and TRAIL-therapy is considered a promising anti-tumor approach. Binding of ligands to the receptor activates the apoptotic cascade, which starts with the formation of the death-inducing signaling complex (DISC) complex composed, in many instances, of the receptor, FADD and the initiator caspase, procaspase 8. The assemblage of this complex results in cleavage and activation of procaspase 8. E6 interferes with this process by binding to procaspase 8 and FADD, accelerating their degradation and preventing the successful completion of the apoptotic cascade [125–128]. If therapeutic agents such as small molecules could inhibit E6 from binding to procaspase 8 and FADD, it would restore the normal functioning of the apoptosis pathway. Proof of principle for this approach was demonstrated by the flavonol myricetin, which was able to prevent the binding of E6 to caspase 8, showing potential for reactivating the extrinsic apoptotic pathway [166]. Further studies on the identification, optimization and evaluation of small molecules of E6 inhibitors are currently underway.

Another strategy is to inhibit the interaction between E7 and pRb, thereby preventing E7 from inhibiting pRB's ability to inhibit cell division. The small compound thiadiazolidinedione inhibits HPV-E7 from disrupting the pRb-E2F complex by blocking the E7-pRb interaction [168]. Lastly, a small compound, namely nicandrenone, has demonstrated the ability to target the sites of both the E6-p53 and E7-pRb1 interactions, thereby blocking the transformative activities of both viral oncoproteins [169]. All the above strategies can lead to the development of efficient therapies against HPV-driven OPSCC and could be used in combination with current therapies to induce tumor cell death and reduce the undesirable side effects of current treatments.

Research into small molecules useful for the treatment of HPV-dependent cancers is ongoing and encouraging. However, concepts developed during studies conducted on cervical cancer will have to be assimilated and translated to oropharyngeal carcinoma. Further developments in our understanding of the molecular biology underlying the development of HNSCC will be necessary to refine the efficacy of these early phase agents.

12. Conclusions and Future Directions

The current epidemic of HNSCC has sparked significant interest in the role of HPV in oncogenesis, and the emergence of HPV-positive head and neck cancer has shifted the demographic of HNSCC from an older population to a younger generation. Current treatments, which consist of transoral surgery, platinum-based chemotherapy, and intensity-modulated radiotherapy, are increasingly recognized as requiring improvements. While advances in standard therapies have improved outcomes, the new group of younger patients is at high risk of morbidity and consequently a compromised quality of life. Therefore, the demand for major progress in the therapy and diagnosis of HPV-associated carcinoma remains current and compelling [35]. The better prognosis of

HPV-related OPSCC has broached topics of de-escalation strategies [77], leading to the emergence of various de-intensification trials for HNSCC. With this concern in mind, standardizing a screening method for HPV status would help in diagnosing and delivering appropriate treatments to this subpopulation. The commercially available HPV prophylactic vaccines have had a profound effect in the prevention of HPV infection in the context of cervical cancer, but their efficacy has not yet been proven in the context of HPV-dependent head and neck carcinomas. Ongoing trials are anticipated to address this issue. A preventive vaccine would mitigate the epidemic long-term, but will not address the more urgent issue of treating patients with existing HPV infections. Hence, the development of therapeutic vaccines has the potential to meet a pressing need for better treatments of HPV-associated tumors in immunocompetent OPSCC patients. Additionally, targeted therapies of growth factors potentially have a more widespread use, and they have progressed in clinical trials, though with mixed results and varying success.

Several advances in biotherapy have led to the identification of a number of small molecular compounds with the potential for contributing to the development of less toxic treatments. The field of small molecular targeted therapy is in its infancy, but current findings are encouraging, advocating for the rapid progression of the field. The studies presented above reveal the urgency of the burden and the impetus to identify better targets and antiviral therapies effective in attenuating the incidence of HPV infection and counteracting the growing epidemic of HPV-associated head and neck cancers [112].

Acknowledgments

We thank Vasiliy A. Loskutov for helpful discussions and editorial suggestions. We also extend our appreciation to Dr. Steve C. Lee, MD PhD, and Dr. Pedro A. Andrade Filho, MD for their helpful discussions, contributions, and suggestions to the sections addressing "Treatments and Therapies" and "Surgery".

Conflicts of Interest

The authors declare no conflict of interest.

References

1. Bol, V.; Gregoire, V. Biological basis for increased sensitivity to radiation therapy in HPV-positive head and neck cancers. *BioMed Res. Int.* **2014**, *2014*, 696028.
2. Elrefaey, S.; Massaro, M.A.; Chiocca, S.; Chiesa, F.; Ansarin, M. HPV in oropharyngeal cancer: The basics to know in clinical practice. *Acta Otorhinolaryngol. Ital.* **2014**, *34*, 299–309.
3. Friedman, J.M.; Stavas, M.J.; Cmelak, A.J. Clinical and scientific impact of human papillomavirus on head and neck cancer. *World J. Clin. Oncol.* **2014**, *5*, 781–791.
4. Machiels, J.P.; Lambrecht, M.; Hanin, F.X.; Duprez, T.; Gregoire, V.; Schmitz, S.; Hamoir, M. Advances in the management of squamous cell carcinoma of the head and neck. *F1000prime Rep.* **2014**, *6*, 44, doi:10.12703/P6-44.

5.   Van Kempen, P.M.; Noorlag, R.; Braunius, W.W.; Stegeman, I.; Willems, S.M.; Grolman, W. Differences in methylation profiles between HPV-positive and HPV-negative oropharynx squamous cell carcinoma: A systematic review. *Epigenetics* **2014**, *9*, 194–203.

6.   Chai, R.C.; Lambie, D.; Verma, M.; Punyadeera, C. Current trends in the etiology and diagnosis of HPV-related head and neck cancers. *Cancer Med.* **2015**, 4, 596–607.

7.   John, K.; Wu, J.; Lee, B.W.; Farah, C.S. MicroRNAs in head and neck cancer. *Int. J. Dent.* **2013**, *2013*, 650218.

8.   Vermeer, D.W.; Spanos, W.C.; Vermeer, P.D.; Bruns, A.M.; Lee, K.M.; Lee, J.H. Radiation-induced loss of cell surface cd47 enhances immune-mediated clearance of human papillomavirus-positive cancer. *Int. J. Cancer* **2013**, *133*, 120–129.

9.   Boscolo-Rizzo, P.; Del Mistro, A.; Bussu, F.; Lupato, V.; Baboci, L.; Almadori, G.; MC, D.A.M.; Paludetti, G. New insights into human papillomavirus-associated head and neck squamous cell carcinoma. *Acta Otorhinolaryngol. Ital.* **2013**, *33*, 77–87.

10. D'Souza, G.; Dempsey, A. The role of HPV in head and neck cancer and review of the HPV vaccine. *Prev. Med.* **2011**, *53* (Suppl. 1), S5–S11.

11. Osazuwa-Peters, N.; Wang, D.D.; Namin, A.; Vivek, J.; O'Neill, M.; Patel, P.V.; Varvares, M.A. Sexual behavior, HPV knowledge, and association with head and neck cancer among a high-risk group. *Oral Oncol.* **2015**, *51*, 452–456.

12. Monnier, Y.; Simon, C. Surgery *versus* radiotherapy for early oropharyngeal tumors: A never-ending debate. *Curr. Treat. Options Oncol.* **2015**, *16*, 362, doi:10.1007/s11864-015-0362-4.

13. Moore, K.A., 2nd; Mehta, V. The growing epidemic of HPV-positive oropharyngeal carcinoma: A clinical review for primary care providers. *J. Am. Board Fam. Med.* **2015**, *28*, 498–503.

14. Sepiashvili, L.; Bruce, J.P.; Huang, S.H.; O'Sullivan, B.; Liu, F.F.; Kislinger, T. Novel insights into head and neck cancer using next-generation "omic" technologies. *Cancer Res.* **2015**, *75*, 480–486.

15. Adams, A.K.; Wise-Draper, T.M.; Wells, S.I. Human papillomavirus induced transformation in cervical and head and neck cancers. *Cancers* **2014**, *6*, 1793–1820.

16. Antonsson, A.; Neale, R.E.; Boros, S.; Lampe, G.; Coman, W.B.; Pryor, D.I.; Porceddu, S.V.; Whiteman, D.C. Human papillomavirus status and p16 expression in patients with mucosal squamous cell carcinoma of the head and neck in queensland, australia. *Cancer Epidemiol.* **2015**, *39*, 174–181.

17. Arbyn, M.; de Sanjose, S.; Saraiya, M.; Sideri, M.; Palefsky, J.; Lacey, C.; Gillison, M.; Bruni, L.; Ronco, G.; Wentzensen, N.*; et al.* Eurogin 2011 roadmap on prevention and treatment of HPV-related disease. *Int. J. Cancer.* **2012**, *131*, 1969–1982.

18. Bosch, F.X.; Broker, T.R.; Forman, D.; Moscicki, A.B.; Gillison, M.L.; Doorbar, J.; Stern, P.L.; Stanley, M.; Arbyn, M.; Poljak, M.*; et al.* Comprehensive control of human papillomavirus infections and related diseases. *Vaccine* **2013**, *31* (Suppl. 8), I1–31.

19. Dalianis, T. Human papillomavirus (HPV) and oropharyngeal squamous cell carcinoma. *Presse Med. (Paris, Fr. 1983)* **2014**, *43*, e429–e434.

20. Ramshankar, V.; Krishnamurthy, A. Human papilloma virus in head and neck cancers-role and relevance in clinical management. *Indian J. Surg. Oncol.* **2013**, *4*, 59–66.

21. Chaturvedi, A.K.; Engels, E.A.; Pfeiffer, R.M.; Hernandez, B.Y.; Xiao, W.; Kim, E.; Jiang, B.; Goodman, M.T.; Sibug-Saber, M.; Cozen, W.*; et al.* Human papillomavirus and rising oropharyngeal cancer incidence in the united states. *J. Clin. Oncol.* **2011**, *29*, 4294–4301.

22. Gillison, M.L.; Koch, W.M.; Capone, R.B.; Spafford, M.; Westra, W.H.; Wu, L.; Zahurak, M.L.; Daniel, R.W.; Viglione, M.; Symer, D.E.*; et al.* Evidence for a causal association between human papillomavirus and a subset of head and neck cancers. *J. Natl. Cancer Inst.* **2000**, *92*, 709–720.

23. Psyrri, A.; Sasaki, C.; Vassilakopoulou, M.; Dimitriadis, G.; Rampias, T. Future directions in research, treatment and prevention of HPV-related squamous cell carcinoma of the head and neck. *Head Neck Pathol.* **2012**, *6 (Suppl. 1)*, S121–S128.

24. Urban, D.; Corry, J.; Rischin, D. What is the best treatment for patients with human papillomavirus-positive and -negative oropharyngeal cancer? *Cancer* **2014**, *120*, 1462–1470.

25. Garbuglia, A.R. Human papillomavirus in head and neck cancer. *Cancers* **2014**, *6*, 1705–1726.

26. Dalianis, T. Human papillomavirus and oropharyngeal cancer, the epidemics, and significance of additional clinical biomarkers for prediction of response to therapy (review). *Int. J. Oncol.* **2014**, *44*, 1799–1805.

27. Dreyer, J.H.; Hauck, F.; Oliveira-Silva, M.; Barros, M.H.; Niedobitek, G. Detection of HPV infection in head and neck squamous cell carcinoma: A practical proposal. *Virchows. Archiv.* **2013**, *462*, 381–389.

28. Bonilla-Velez, J.; Mroz, E.A.; Hammon, R.J.; Rocco, J.W. Impact of human papillomavirus on oropharyngeal cancer biology and response to therapy: Implications for treatment. *Otolaryngol. Clin. N. Am.* **2013**, *46*, 521–543.

29. Wierzbicka, M.; Jozefiak, A.; Szydlowski, J.; Marszalek, A.; Stankiewicz, C.; Hassman-Poznanska, E.; Osuch-Wojcikiewicz, E.; Skladzien, J.; Klatka, J.; Pietruszewska, W.; *et al.* Recommendations for the diagnosis of human papilloma virus (HPV) high and low risk in the prevention and treatment of diseases of the oral cavity, pharynx and larynx. Guide of experts ptorl and kidl. *Otolaryngol. Pol.* **2013**, *67*, 113–134.

30. Caicedo-Granados, E.; Lin, R.; Fujisawa, C.; Yueh, B.; Sangwan, V.; Saluja, A. Wild-type p53 reactivation by small-molecule minnelide in human papillomavirus (HPV)-positive head and neck squamous cell carcinoma. *Oral Oncol.* **2014**, *50*, 1149–1156.

31. Lui, V.W.; Grandis, J.R. Primary chemotherapy and radiation as a treatment strategy for HPV-positive oropharyngeal cancer. *Head Neck Pathol.* **2012**, *6* (Suppl. 1), S91–S97.

32. Rettig, E.; Kiess, A.P.; Fakhry, C. The role of sexual behavior in head and neck cancer: Implications for prevention and therapy. *Expert Rev. Anticancer ther.* **2015**, *15*, 35–49.

33. Stanley, M.A. Genital human papillomavirus infections: Current and prospective therapies. *J. Gen. Virol.* **2012**, *93*, 681–691.

34. Mannarini, L.; Kratochvil, V.; Calabrese, L.; Gomes Silva, L.; Morbini, P.; Betka, J.; Benazzo, M. Human papilloma virus (HPV) in head and neck region: Review of literature. *Acta Otorhinolaryngol. Ital.* **2009**, *29*, 119–126.

35. Haedicke, J.; Iftner, T. Human papillomaviruses and cancer. *Radiother. Oncol.* **2013**, *108*, 397–402.

36. Langer, C.J. Exploring biomarkers in head and neck cancer. *Cancer* **2012**, *118*, 3882–3892.

37. Blitzer, G.C.; Smith, M.A.; Harris, S.L.; Kimple, R.J. Review of the clinical and biologic aspects of human papillomavirus-positive squamous cell carcinomas of the head and neck. *Int. J. Radiat. Oncol. Biol. Phys.* **2014**, *88*, 761–770.

38. Gillison, M.L.; Broutian, T.; Pickard, R.K.; Tong, Z.Y.; Xiao, W.; Kahle, L.; Graubard, B.I.; Chaturvedi, A.K. Prevalence of oral HPV infection in the united states, 2009–2010. *Jama* **2012**, *307*, 693–703.

39. Nelke, K.H.; Lysenko, L.; Leszczyszyn, J.; Gerber, H. Human papillomavirus and its influence on head and neck cancer predisposition. *Postepy higieny i medycyny doswiadczalnej (Online)* **2013**, *67*, 610–616.

40. D'Souza, G.; Agrawal, Y.; Halpern, J.; Bodison, S.; Gillison, M.L. Oral sexual behaviors associated with prevalent oral human papillomavirus infection. *J. Infect. Dis.* **2009**, *199*, 1263–1269.

41. Moscicki, A.B. Impact of HPV infection in adolescent populations. *J. Adolesc. Health* **2005**, *37*, S3–S9.

42. Best, S.R.; Niparko, K.J.; Pai, S.I. Biology of human papillomavirus infection and immune therapy for HPV-related head and neck cancers. *Otolaryngol. Clin. N. Am.* **2012**, *45*, 807–822.

43. D'Souza, G.; Kreimer, A.R.; Viscidi, R.; Pawlita, M.; Fakhry, C.; Koch, W.M.; Westra, W.H.; Gillison, M.L. Case-control study of human papillomavirus and oropharyngeal cancer. *N. Engl. J. Med.* **2007**, *356*, 1944–1956.

44. Pickard, R.K.; Xiao, W.; Broutian, T.R.; He, X.; Gillison, M.L. The prevalence and incidence of oral human papillomavirus infection among young men and women, aged 18–30 years. *Sex. Transm. Dis.* **2012**, *39*, 559–566.

45. Steinau, M.; Hariri, S.; Gillison, M.L.; Broutian, T.R.; Dunne, E.F.; Tong, Z.Y.; Markowitz, L.E.; Unger, E.R. Prevalence of cervical and oral human papillomavirus infections among us women. *J. Infect. Dis.* **2014**, *209*, 1739–1743.

46. Hemminki, K.; Dong, C.; Frisch, M. Tonsillar and other upper aerodigestive tract cancers among cervical cancer patients and their husbands. *Eur. J. Cancer Prev.* **2000**, *9*, 433–437.

47. Vogt, S.L.; Gravitt, P.E.; Martinson, N.A.; Hoffmann, J.; D'Souza, G. Concordant oral-genital HPV infection in south africa couples: Evidence for transmission. *Front. Oncol.* **2013**, *3*, 303, doi:10.3389/fonc.2013.00303.

48. Widdice, L.E.; Breland, D.J.; Jonte, J.; Farhat, S.; Ma, Y.; Leonard, A.C.; Moscicki, A.B. Human papillomavirus concordance in heterosexual couples. *J. Adolesc. Health* **2010**, *47*, 151–159.

49. Trosman, S.J.; Koyfman, S.A.; Ward, M.C.; Al-Khudari, S.; Nwizu, T.; Greskovich, J.F.; Lamarre, E.D.; Scharpf, J.; Khan, M.J.; Lorenz, R.R.; *et al*. Effect of human papillomavirus on patterns of distant metastatic failure in oropharyngeal squamous cell carcinoma treated with chemoradiotherapy. *JAMA Otolaryngol. Head Neck Surg.* **2015**, *141*, 457–462.

50. Cooper, T.; Biron, V.L.; Fast, D.; Tam, R.; Carey, T.; Shmulevitz, M.; Seikaly, H. Oncolytic activity of reovirus in HPV positive and negative head and neck squamous cell carcinoma. *J. Otolaryngol. Head Neck Surg.* **2015**, *44*, doi:10.1186/s40463-015-0062-x.

51. Smith, E.M.; Ritchie, J.M.; Summersgill, K.F.; Klussmann, J.P.; Lee, J.H.; Wang, D.; Haugen, T.H.; Turek, L.P. Age, sexual behavior and human papillomavirus infection in oral cavity and oropharyngeal cancers. *Int. J. Cancer.* **2004**, *108*, 766–772.

52. Duek, I.; Billan, S.; Amit, M.; Gil, Z. Transoral robotic surgery in the HPV era. *Rambam Maimonides Med. J.* **2014**, *5*, e0010.

53. Genden, E.M. The role for surgical management of HPV-related oropharyngeal carcinoma. *Head Neck Pathol.* **2012**, *6* (Suppl. 1), S98–S103.

54. Lorincz, B.B.; Jowett, N.; Knecht, R. Decision management in transoral robotic surgery (tors): Indications, individual patient selection, and role in the multidisciplinary treatment of head and neck cancer from a european perspective. *Head Neck* **2015**, doi:10.1002/hed.24059.

55. Nichols, A.C.; Yoo, J.; Hammond, J.A.; Fung, K.; Winquist, E.; Read, N.; Venkatesan, V.; MacNeil, S.D.; Ernst, D.S.; Kuruvilla, S.; *et al*. Early-stage squamous cell carcinoma of the oropharynx: Radiotherapy *vs.* Trans-oral robotic surgery (orator)—Study protocol for a randomized phase ii trial. *BMC Cancer* **2013**, *13*, doi:10.1186/1471-2407-13-133.

56. Sanders, A.E.; Slade, G.D.; Patton, L.L. National prevalence of oral HPV infection and related risk factors in the u.S. Adult population. *Oral Dis.* **2012**, *18*, 430–441.

57. Zhang, P.; Mirani, N.; Baisre, A.; Fernandes, H. Molecular heterogeneity of head and neck squamous cell carcinoma defined by next-generation sequencing. *Am. J. Pathol.* **2014**, *184*, 1323–1330.

58. Settle, K.; Posner, M.R.; Schumaker, L.M.; Tan, M.; Suntharalingam, M.; Goloubeva, O.; Strome, S.E.; Haddad, R.I.; Patel, S.S.; Cambell, E.V., 3rd; *et al*. Racial survival disparity in head and neck cancer results from low prevalence of human papillomavirus infection in black oropharyngeal cancer patients. *Cancer Prev. Res. (Phil., PA.)* **2009**, *2*, 776–781.

59. Weinberger, P.M.; Merkley, M.A.; Khichi, S.S.; Lee, J.R.; Psyrri, A.; Jackson, L.L.; Dynan, W.S. Human papillomavirus-active head and neck cancer and ethnic health disparities. *Laryngoscope* **2010**, *120*, 1531–1537.

60. Fumagalli, I.; Dugue, D.; Bibault, J.E.; Clemenson, C.; Vozenin, M.C.; Mondini, M.; Deutsch, E. Cytotoxic effect of lapatinib is restricted to human papillomavirus-positive head and neck squamous cell carcinoma cell lines. *OncoTargets Ther.* **2015**, *8*, 335–345.

61. Kaczmar, J.M.; Tan, K.S.; Heitjan, D.F.; Lin, A.; Ahn, P.H.; Newman, J.G.; Rassekh, C.H.; Chalian, A.A.; O'Malley, B.W., Jr.; Cohen, R.B.; *et al*. HPV-related oropharyngeal cancer: Risk factors for treatment failure in patients managed with primary transoral robotic surgery. *Head & Neck* **2014**.

62. Mydlarz, W.K.; Chan, J.Y.; Richmon, J.D. The role of surgery for HPV-associated head and neck cancer. *Oral Oncol.* **2015**, *51*, 305–313.

63. Ang, K.K.; Harris, J.; Wheeler, R.; Weber, R.; Rosenthal, D.I.; Nguyen-Tan, P.F.; Westra, W.H.; Chung, C.H.; Jordan, R.C.; Lu, C.; *et al.* Human papillomavirus and survival of patients with oropharyngeal cancer. *N. Engl. J. Med.* **2010**, *363*, 24–35.

64. George, M. Should patients with HPV-positive or negative tumors be treated differently? *Curr. Oncol. Rep.* **2014**, *16*, 384, doi:10.1007/s11912-014-0384-2.

65. Guo, T.; Qualliotine, J.R.; Ha, P.K.; Califano, J.A.; Kim, Y.; Saunders, J.R.; Blanco, R.G.; D'Souza, G.; Zhang, Z.; Chung, C.H.; *et al.* Surgical salvage improves overall survival for patients with HPV-positive and HPV-negative recurrent locoregional and distant metastatic oropharyngeal cancer. *Cancer* **2015**, *121*, 1977–1984.

66. Maxwell, J.H.; Mehta, V.; Wang, H.; Cunningham, D.; Duvvuri, U.; Kim, S.; Johnson, J.T.; Ferris, R.L. Quality of life in head and neck cancer patients: Impact of HPV and primary treatment modality. *Laryngoscope* **2014**, *124*, 1592–1597.

67. Martinez-Useros, J.; Garcia-Foncillas, J. The challenge of blocking a wider family members of egfr against head and neck squamous cell carcinomas. *Oral Oncol.* **2015**, *51*, 423–430.

68. Purohit, S.; Bhise, R.; Lokanatha, D.; Govindbabu, K. Systemic therapy in head and neck cancer: Changing paradigm. *Indian J. Surg. Oncol.* **2013**, *4*, 19–26.

69. Bernier, J.; Domenge, C.; Ozsahin, M.; Matuszewska, K.; Lefebvre, J.L.; Greiner, R.H.; Giralt, J.; Maingon, P.; Rolland, F.; Bolla, M.; *et al.* Postoperative irradiation with or without concomitant chemotherapy for locally advanced head and neck cancer. *N. Engl. J. Med.* **2004**, *350*, 1945–1952.

70. Cooper, J.S.; Pajak, T.F.; Forastiere, A.A.; Jacobs, J.; Campbell, B.H.; Saxman, S.B.; Kish, J.A.; Kim, H.E.; Cmelak, A.J.; Rotman, M.; *et al.* Postoperative concurrent radiotherapy and chemotherapy for high-risk squamous-cell carcinoma of the head and neck. *N. Engl. J. Med.* **2004**, *350*, 1937–1944.

71. Burtness, B.; Bourhis, J.P.; Vermorken, J.B.; Harrington, K.J.; Cohen, E.E. Afatinib *versus* placebo as adjuvant therapy after chemoradiation in a double-blind, phase iii study (lux-head & neck 2) in patients with primary unresected, clinically intermediate-to-high-risk head and neck cancer: Study protocol for a randomized controlled trial. *Trials* **2014**, *15*, 469, doi:10.1186/1745-6215-15-469.

72. Hamoir, M.; Ferlito, A.; Schmitz, S.; Hanin, F.X.; Thariat, J.; Weynand, B.; Machiels, J.P.; Gregoire, V.; Robbins, K.T.; Silver, C.E.; *et al.* The role of neck dissection in the setting of chemoradiation therapy for head and neck squamous cell carcinoma with advanced neck disease. *Oral Oncol.* **2012**, *48*, 203–210.

73. Patel, S.N.; Cohen, M.A.; Givi, B.; Dixon, B.J.; Gilbert, R.W.; Gullane, P.J.; Brown, D.H.; Irish, J.C.; de Almeida, J.R.; Higgins, K.M.; *et al.* Salvage surgery of locally recurrent oropharyngeal cancer. *Head & Neck* **2015**.

74. Modur, V.; Thomas-Robbins, K.; Rao, K. HPV and csc in hnscc cisplatin resistance. *Front. Biosci. (Elite ed.)* **2015**, *7*, 58–66.

75. Ford, S.E.; Brandwein-Gensler, M.; Carroll, W.R.; Rosenthal, E.L.; Magnuson, J.S. Transoral robotic *versus* open surgical approaches to oropharyngeal squamous cell carcinoma by human papillomavirus status. *Otolaryngol. Head Neck Surg.* **2014**, *151*, 606–611.

76. Ridge, J.A. Surgery in the HPV era: The role of robotics and microsurgical techniques. *Am. Soc. Clin. Oncol. Educ. Book* **2014**, 154–159, doi:10.14694/EdBook_AM.2014.34.154.

77. Hinni, M.L.; Nagel, T.; Howard, B. Oropharyngeal cancer treatment: The role of transoral surgery. *Curr. Opin. Otolaryngol. Head Neck Surg.* **2015**, *23*, 132–138.

78. O'Leary, P.; Kjaergaard, T. Transoral robotic surgery and oropharyngeal cancer: A literature review. *Ear Nose Throat J.* **2014**, *93*, E14–E21.

79. Forastiere, A.A.; Goepfert, H.; Maor, M.; Pajak, T.F.; Weber, R.; Morrison, W.; Glisson, B.; Trotti, A.; Ridge, J.A.; Chao, C.; *et al.* Concurrent chemotherapy and radiotherapy for organ preservation in advanced laryngeal cancer. *N. Engl. J. Med.* **2003**, *349*, 2091–2098.

80. Psyrri, A.; Seiwert, T.Y.; Jimeno, A. Molecular pathways in head and neck cancer: Egfr, pi3k, and more. *Am. Soc. Clin. Oncol. Educ. Book* **2013**, 246–255, doi:10.1200/EdBook_AM.2013.33.246.

81. Malhotra, B.; Bellile, E.L.; Nguyen, N.P.; Fung, V.K.; Slack, M.; Bilich, R.; Papagerakis, S.; Worden, F. Carboplatin-pemetrexed in treatment of patients with recurrent/metastatic cancers of the head and neck; superior outcomes in oropharyngeal primaries. *Front. Oncol.* **2014**, *4*, 362, doi:10.3389/fonc.2014.00362.

82. Inhestern, J.; Oertel, K.; Stemmann, V.; Schmalenberg, H.; Dietz, A.; Rotter, N.; Veit, J.; Gorner, M.; Sudhoff, H.; Junghanss, C.; *et al.* Prognostic role of circulating tumor cells during induction chemotherapy followed by curative surgery combined with postoperative radiotherapy in patients with locally advanced oral and oropharyngeal squamous cell cancer. *PLoS ONE* **2015**, *10*, e0132901.

83. Hsu, H.W.; Wall, N.R.; Hsueh, C.T.; Kim, S.; Ferris, R.L.; Chen, C.S.; Mirshahidi, S. Combination antiangiogenic therapy and radiation in head and neck cancers. *Oral Oncol.* **2014**, *50*, 19–26.

84. McBride, S.M.; Parambi, R.J.; Jang, J.W.; Goldsmith, T.; Busse, P.M.; Chan, A.W. Intensity-modulated *versus* conventional radiation therapy for oropharyngeal carcinoma: Long-term dysphagia and tumor control outcomes. *Head Neck* **2014**, *36*, 492–498.

85. Surucu, M.; Shah, K.K.; Mescioglu, I.; Roeske, J.C.; Small, W., Jr.; Choi, M.; Emami, B. Decision trees predicting tumor shrinkage for head and neck cancer: Implications for adaptive radiotherapy. *Technol. Cancer Res. Treat.* **2015**, (Epub ahead of print)

86. Broglie, M.A.; Soltermann, A.; Rohrbach, D.; Haile, S.R.; Pawlita, M.; Studer, G.; Huber, G.F.; Moch, H.; Stoeckli, S.J. Impact of p16, p53, smoking, and alcohol on survival in patients with oropharyngeal squamous cell carcinoma treated with primary intensity-modulated chemoradiation. *Head Neck* **2013**, *35*, 1698–1706.

87. Coppock, J.D.; Wieking, B.G.; Molinolo, A.A.; Gutkind, J.S.; Miskimins, W.K.; Lee, J.H. Improved clearance during treatment of HPV-positive head and neck cancer through mtor inhibition. *Neoplasia (New York, N.Y.)* **2013**, *15*, 620–630.

88. Kumar, B.; Cordell, K.G.; Lee, J.S.; Prince, M.E.; Tran, H.H.; Wolf, G.T.; Urba, S.G.; Worden, F.P.; Chepeha, D.B.; Teknos, T.N.; *et al.* Response to therapy and outcomes in oropharyngeal cancer are associated with biomarkers including human papillomavirus, epidermal growth factor receptor, gender, and smoking. *Int. J. Radiat. Oncol. Biol. Phys.* **2007**, *69*, S109–S111.

89. Lohaus, F.; Linge, A.; Tinhofer, I.; Budach, V.; Gkika, E.; Stuschke, M.; Balermpas, P.; Rodel, C.; Avlar, M.; Grosu, A.L.; *et al.* HPV16 DNA status is a strong prognosticator of loco-regional control after postoperative radiochemotherapy of locally advanced oropharyngeal carcinoma: Results from a multicentre explorative study of the german cancer consortium radiation oncology group (dktk-rog). *Radiother. Oncol.* **2014**, *113*, 317–323.

90. Stern, P.L.; van der Burg, S.H.; Hampson, I.N.; Broker, T.R.; Fiander, A.; Lacey, C.J.; Kitchener, H.C.; Einstein, M.H. Therapy of human papillomavirus-related disease. *Vaccine* **2012**, *30* (Suppl 5), F71–F82.

91. Fakhry, C.; Westra, W.H.; Li, S.; Cmelak, A.; Ridge, J.A.; Pinto, H.; Forastiere, A.; Gillison, M.L. Improved survival of patients with human papillomavirus-positive head and neck squamous cell carcinoma in a prospective clinical trial. *J. Natl. Cancer Instit.***2008**, *100*, 261–269.

92. Rischin, D.; Young, R.J.; Fisher, R.; Fox, S.B.; Le, Q.T.; Peters, L.J.; Solomon, B.; Choi, J.; O'Sullivan, B.; Kenny, L.M.; *et al.* Prognostic significance of p16ink4a and human papillomavirus in patients with oropharyngeal cancer treated on trog 02.02 phase iii trial. *J. Clin. Oncol.* **2010**, *28*, 4142–4148.

93. Laskar, S.G.; Swain, M. HPV positive oropharyngeal cancer and treatment deintensification: How pertinent is it? *J. Cancer Res. Ther.* **2015**, *11*, 6–9.

94. Amine, A.; Rivera, S.; Opolon, P.; Dekkal, M.; Biard, D.S.; Bouamar, H.; Louache, F.; McKay, M.J.; Bourhis, J.; Deutsch, E.; *et al.* Novel anti-metastatic action of cidofovir mediated by inhibition of E6/E7, CXCR4 and rho/ROCK signaling in HPV tumor cells. *PLoS ONE* **2009**, *4*, e5018.

95. Turner, D.O.; Williams-Cocks, S.J.; Bullen, R.; Catmull, J.; Falk, J.; Martin, D.; Mauer, J.; Barber, A.E.; Wang, R.C.; Gerstenberger, S.L.; *et al.* High-risk human papillomavirus (HPV) screening and detection in healthy patient saliva samples: A pilot study. *BMC Oral Health* **2011**, *11*, 28, doi:10.1186/1472-6831-11-28.

96. Fertig, E.J.; Markovic, A.; Danilova, L.V.; Gaykalova, D.A.; Cope, L.; Chung, C.H.; Ochs, M.F.; Califano, J.A. Preferential activation of the hedgehog pathway by epigenetic modulations in HPV negative hnscc identified with meta-pathway analysis. *PLoS ONE* **2013**, *8*, e78127.

97. Masterson, L.; Moualed, D.; Liu, Z.W.; Howard, J.E.; Dwivedi, R.C.; Tysome, J.R.; Benson, R.; Sterling, J.C.; Sudhoff, H.; Jani, P.; *et al.* De-escalation treatment protocols for human papillomavirus-associated oropharyngeal squamous cell carcinoma: A systematic review and meta-analysis of current clinical trials. *Eur. J. Cancer (Oxf., Engl. 1990)* **2014**, *50*, 2636–2648.

98. Syrjanen, K.J.; Pyrhonen, S.; Syrjanen, S.M.; Lamberg, M.A. Immunohistochemical demonstration of human papilloma virus (HPV) antigens in oral squamous cell lesions. *Br. J. Oral Surg.* **1983**, *21*, 147–153.

99. Langevin, S.M.; Grandis, J.R.; Taioli, E. Female hormonal and reproductive factors and head and neck squamous cell carcinoma risk. *Cancer Lett.* **2011**, *310*, 216–221.

100. Jeon, S.; Allen-Hoffmann, B.L.; Lambert, P.F. Integration of human papillomavirus type 16 into the human genome correlates with a selective growth advantage of cells. *J. Virol.* **1995**, *69*, 2989–2997.

101. Yu, T.; Ferber, M.J.; Cheung, T.H.; Chung, T.K.; Wong, Y.F.; Smith, D.I. The role of viral integration in the development of cervical cancer. *Cancer Genet. Cytogenet.* **2005**, *158*, 27–34.

102. Gao, G.; Johnson, S.H.; Kasperbauer, J.L.; Eckloff, B.W.; Tombers, N.M.; Vasmatzis, G.; Smith, D.I. Mate pair sequencing of oropharyngeal squamous cell carcinomas reveals that HPV integration occurs much less frequently than in cervical cancer. *J. Clin. Virol.* **2014**, *59*, 195–200.

103. Olthof, N.C.; Huebbers, C.U.; Kolligs, J.; Henfling, M.; Ramaekers, F.C.; Cornet, I.; van Lent-Albrechts, J.A.; Stegmann, A.P.; Silling, S.; Wieland, U.; *et al.* Viral load, gene expression and mapping of viral integration sites in HPV16-associated hnscc cell lines. *Int. J. Cancer* **2015**, *136*, E207–E218.

104. Koskinen, W.J.; Chen, R.W.; Leivo, I.; Makitie, A.; Back, L.; Kontio, R.; Suuronen, R.; Lindqvist, C.; Auvinen, E.; Molijn, A.; *et al.* Prevalence and physical status of human papillomavirus in squamous cell carcinomas of the head and neck. *Int. J. Cancer* **2003**, *107*, 401–406.

105. Albers, A.; Abe, K.; Hunt, J.; Wang, J.; Lopez-Albaitero, A.; Schaefer, C.; Gooding, W.; Whiteside, T.L.; Ferrone, S.; DeLeo, A.; *et al.* Antitumor activity of human papillomavirus type 16 E7-specific T cells against virally infected squamous cell carcinoma of the head and neck. *Cancer Res.* **2005**, *65*, 11146–11155.

106. Rein, D.T.; Kurbacher, C.M. The role of chemotherapy in invasive cancer of the cervix uteri: Current standards and future prospects. *Anti-Cancer Drugs* **2001**, *12*, 787–795.

107. Kumar, S.; Biswas, M.; Jose, T. HPV vaccine: Current status and future directions. *Med. J. Armed Forces India* **2015**, *71*, 171–177.

108. Cubie, H.A.; Cuschieri, K. Understanding HPV tests and their appropriate applications. *Cytopathology* **2013**, *24*, 289–308.

109. Nicol, C.; Cesur, O.; Forrest, S.; Belyaeva, T.A.; Bunka, D.H.; Blair, G.E.; Stonehouse, N.J. An RNA aptamer provides a novel approach for the induction of apoptosis by targeting the HPV16 E7 oncoprotein. *PLoS ONE* **2013**, *8*, e64781.

110. Parkin, D.M. The global health burden of infection-associated cancers in the year 2002. *Int. J. Cancer* **2006**, *118*, 3030–3044.

111. Lin, K.; Doolan, K.; Hung, C.F.; Wu, T.C. Perspectives for preventive and therapeutic HPV vaccines. *J. Formos. Med. Assoc. = Taiwan yi zhi* **2010**, *109*, 4–24.

112. D'Abramo, C.M.; Archambault, J. Small molecule inhibitors of human papillomavirus protein—Protein interactions. *Open Virol. J.* **2011**, *5*, 80–95.

113. Blioumi, E.; Chatzidimitriou, D.; Pazartzi, C.; Katopodi, T.; Tzimagiorgis, G.; Emmanouil-Nikoloussi, E.N.; Markopoulos, A.; Kalekou, C.; Lazaridis, N.; Diza, E.; *et al.* Detection and typing of human papillomaviruses (HPV) in malignant, dysplastic, nondysplastic and normal oral epithelium by nested polymerase chain reaction, immunohistochemistry and transitional electron microscopy in patients of northern greece. *Oral Oncol.* **2014**, *50*, 840–847.

114. Almajhdi, F.N.; Senger, T.; Amer, H.M.; Gissmann, L.; Ohlschlager, P. Design of a highly effective therapeutic HPV16 E6/E7-specific DNA vaccine: Optimization by different ways of sequence rearrangements (shuffling). *PLoS ONE* **2014**, *9*, e113461.

115. Yuan, C.H.; Filippova, M.; Duerksen-Hughes, P. Modulation of apoptotic pathways by human papillomaviruses (HPV): Mechanisms and implications for therapy. *Viruses* **2012**, *4*, 3831–3850.

116. Devaraj, K.; Gillison, M.L.; Wu, T.C. Development of HPV vaccines for HPV-associated head and neck squamous cell carcinoma. *Crit. Rev. Oral Biol. Med.* **2003**, *14*, 345–362.

117. Grasso, F.; Negri, D.R.; Mochi, S.; Rossi, A.; Cesolini, A.; Giovannelli, A.; Chiantore, M.V.; Leone, P.; Giorgi, C.; Cara, A. Successful therapeutic vaccination with integrase defective lentiviral vector expressing nononcogenic human papillomavirus E7 protein. *Int. J. Cancer* **2013**, *132*, 335–344.

118. Ma, B.; Maraj, B.; Tran, N.P.; Knoff, J.; Chen, A.; Alvarez, R.D.; Hung, C.F.; Wu, T.C. Emerging human papillomavirus vaccines. *Expert Opin. Emerg. Drugs* **2012**, *17*, 469–492.

119. Monie, A.; Tsen, S.W.; Hung, C.F.; Wu, T.C. Therapeutic HPV DNA vaccines. *Expert Rev. Vaccines* **2009**, *8*, 1221–1235.

120. Cherry, J.J.; Rietz, A.; Malinkevich, A.; Liu, Y.; Xie, M.; Bartolowits, M.; Davisson, V.J.; Baleja, J.D.; Androphy, E.J. Structure based identification and characterization of flavonoids that disrupt human papillomavirus-16 E6 function. *PLoS ONE* **2013**, *8*, e84506.

121. Scheffner, M.; Werness, B.A.; Huibregtse, J.M.; Levine, A.J.; Howley, P.M. The E6 oncoprotein encoded by human papillomavirus types 16 and 18 promotes the degradation of p53. *Cell* **1990**, *63*, 1129–1136.

122. Lechner, M.S.; Laimins, L.A. Inhibition of p53 DNA binding by human papillomavirus E6 proteins. *J. Virol.* **1994**, *68*, 4262–4273.

123. Kennedy, E.M.; Kornepati, A.V.; Goldstein, M.; Bogerd, H.P.; Poling, B.C.; Whisnant, A.W.; Kastan, M.B.; Cullen, B.R. Inactivation of the human papillomavirus *E6* or *E7* gene in cervical carcinoma cells by using a bacterial CRISPR-Cas RNA-guided endonuclease. *J. Virol.* **2014**, *88*, 11965–11972.

124. Hietanen, S.; Lain, S.; Krausz, E.; Blattner, C.; Lane, D.P. Activation of p53 in cervical carcinoma cells by small molecules. *Proc. Natl. Acad. Sci. USA* **2000**, *97*, 8501–8506.

125. Filippova, M.; Johnson, M.M.; Bautista, M.; Filippov, V.; Fodor, N.; Tungteakkhun, S.S.; Williams, K.; Duerksen-Hughes, P.J. The large and small isoforms of human papillomavirus type 16 E6 bind to and differentially affect procaspase 8 stability and activity. *J. Virol.* **2007**, *81*, 4116–4129.

126. Tungteakkhun, S.S.; Filippova, M.; Fodor, N.; Duerksen-Hughes, P.J. The full-length isoform of human papillomavirus 16 E6 and its splice variant E6* bind to different sites on the procaspase 8 death effector domain. *J. Virol.* **2010**, *84*, 1453–1463.

127. Filippova, M.; Parkhurst, L.; Duerksen-Hughes, P.J. The human papillomavirus 16 E6 protein binds to fas-associated death domain and protects cells from fas-triggered apoptosis. *J. Biol. Chem.* **2004**, *279*, 25729–25744.

128. Tungteakkhun, S.S.; Filippova, M.; Neidigh, J.W.; Fodor, N.; Duerksen-Hughes, P.J. The interaction between human papillomavirus type 16 and fadd is mediated by a novel E6 binding domain. *J. Virol.* **2008**, *82*, 9600–9614.

129. Tungteakkhun, S.S.; Duerksen-Hughes, P.J. Cellular binding partners of the human papillomavirus E6 protein. *Arch. Virol.* **2008**, *153*, 397–408.

130. Dyson, N.; Howley, P.M.; Munger, K.; Harlow, E. The human papilloma virus-16 E7 oncoprotein is able to bind to the retinoblastoma gene product. *Science (New York, N.Y.)* **1989**, *243*, 934–937.

131. Boyer, S.N.; Wazer, D.E.; Band, V. E7 protein of human papilloma virus-16 induces degradation of retinoblastoma protein through the ubiquitin-proteasome pathway. *Cancer Res.* **1996**, *56*, 4620–4624.

132. Liu, X.; Clements, A.; Zhao, K.; Marmorstein, R. Structure of the human papillomavirus E7 oncoprotein and its mechanism for inactivation of the retinoblastoma tumor suppressor. *J. Biol. Chem.* **2006**, *281*, 578–586.

133. Kreimer, A.R. Prospects for prevention of HPV-driven oropharynx cancer. *Oral Oncol.* **2014**, *50*, 555–559.

134. Bishop, J.A.; Lewis, J.S., Jr.; Rocco, J.W.; Faquin, W.C. HPV-related squamous cell carcinoma of the head and neck: An update on testing in routine pathology practice. *Semin. Diagn. Pathol.* **2015**, doi:10.1053/j.semdp.2015.02.013.

135. Jarboe, E.A.; Hunt, J.P.; Layfield, L.J. Cytomorphologic diagnosis and HPV testing of metastatic and primary oropharyngeal squamous cell carcinomas: A review and summary of the literature. *Diagn. Cytopathol.* **2012**, *40*, 491–497.

136. Linxweiler, M.; Bochen, F.; Wemmert, S.; Lerner, C.; Hasenfus, A.; Bohle, R.M.; Al-Kadah, B.; Takacs, Z.F.; Smola, S.; Schick, B. Combination of p16(ink4a) /ki67 immunocytology and HPV polymerase chain reaction for the noninvasive analysis of HPV involvement in head and neck cancer. *Cancer Cytopathol.* **2015**, *123*, 219–229.

137. Dictor, M.; Warenholt, J. Single-tube multiplex pcr using type-specific E6/E7 primers and capillary electrophoresis genotypes 21 human papillomaviruses in neoplasia. *Infect. Agents Cancer* **2011**, *6*, 1, doi:10.1186/1750-9378-6-1.

138. Weiss, D.; Heinkele, T.; Rudack, C. Reliable detection of human papillomavirus in recurrent laryngeal papillomatosis and associated carcinoma of archival tissue. *J. Med. Virol.* **2015**, *87*, 860–870.

139. Melkane, A.E.; Mirghani, H.; Auperin, A.; Saulnier, P.; Lacroix, L.; Vielh, P.; Casiraghi, O.; Griscelli, F.; Temam, S. HPV-related oropharyngeal squamous cell carcinomas: A comparison between three diagnostic approaches. *Am. J. Otolaryngol.* **2014**, *35*, 25–32.

140. Wittekindt, C.; Wagner, S.; Klussmann, J.P. [HPV-associated head and neck cancer. The basics of molecular and translational research]. *Hno* **2011**, *59*, 885–892.

141. Ahn, S.M.; Chan, J.Y.; Zhang, Z.; Wang, H.; Khan, Z.; Bishop, J.A.; Westra, W.; Koch, W.M.; Califano, J.A. Saliva and plasma quantitative polymerase chain reaction-based detection and surveillance of human papillomavirus-related head and neck cancer. *JAMA Otolaryngol. Head Neck Surg.* **2014**, *140*, 846–854.

142. Kelesidis, T.; Aish, L.; Steller, M.A.; Aish, I.S.; Shen, J.; Foukas, P.; Panayiotides, J.; Petrikkos, G.; Karakitsos, P.; Tsiodras, S. Human papillomavirus (HPV) detection using *in situ* hybridization in histologic samples: Correlations with cytologic changes and polymerase chain reaction HPV detection. *Am. J. Clin. Pathol.* **2011**, *136*, 119–127.

143. Kimple, A.J.; Torres, A.D.; Yang, R.Z.; Kimple, R.J. HPV-associated head and neck cancer: Molecular and nano-scale markers for prognosis and therapeutic stratification. *Sensors (Basel, Switz.)* **2012**, *12*, 5159–5169.

144. Duncan, L.D.; Winkler, M.; Carlson, E.R.; Heidel, R.E.; Kang, E.; Webb, D. P16 immunohistochemistry can be used to detect human papillomavirus in oral cavity squamous cell carcinoma. *J. Oral Maxillofac. Surg.* **2013**, *71*, 1367–1375.

145. Smeets, S.J.; Hesselink, A.T.; Speel, E.J.; Haesevoets, A.; Snijders, P.J.; Pawlita, M.; Meijer, C.J.; Braakhuis, B.J.; Leemans, C.R.; Brakenhoff, R.H. A novel algorithm for reliable detection of human papillomavirus in paraffin embedded head and neck cancer specimen. *Int. j. Cancer* **2007**, *121*, 2465–2472.

146. Bishop, J.A.; Ma, X.J.; Wang, H.; Luo, Y.; Illei, P.B.; Begum, S.; Taube, J.M.; Koch, W.M.; Westra, W.H. Detection of transcriptionally active high-risk HPV in patients with head and neck squamous cell carcinoma as visualized by a novel E6/E7 mRNA *in situ* hybridization method. *Am. J. Surg. Pathol.* **2012**, *36*, 1874–1882.

147. Fonmarty, D.; Cherriere, S.; Fleury, H.; Eimer, S.; Majoufre-Lefebvre, C.; Castetbon, V.; de Mones, E. Study of the concordance between p16 immunohistochemistry and HPV-pcr genotyping for the viral diagnosis of oropharyngeal squamous cell carcinoma. *Eur. Ann. Otorhinolaryngol. Head Neck Dis.* **2015**, *132*, 135–139.

148. Shi, W.; Kato, H.; Perez-Ordonez, B.; Pintilie, M.; Huang, S.; Hui, A.; O'Sullivan, B.; Waldron, J.; Cummings, B.; Kim, J.; *et al.* Comparative prognostic value of HPV16 E6 mRNA compared with *in situ* hybridization for human oropharyngeal squamous carcinoma. *J. Clin. Oncol.* **2009**, *27*, 6213–6221.

149. Mirghani, H.; Amen, F.; Blanchard, P.; Moreau, F.; Guigay, J.; Hartl, D.M.; Lacau St Guily, J. Treatment de-escalation in HPV-positive oropharyngeal carcinoma: Ongoing trials, critical issues and perspectives. *Int. J. Cancer.* **2015**, *136*, 1494–1503.

150. Dorsey, K.; Agulnik, M. Promising new molecular targeted therapies in head and neck cancer. *Drugs* **2013**, *73*, 315–325.

151. Kumar, B.; Cordell, K.G.; Lee, J.S.; Worden, F.P.; Prince, M.E.; Tran, H.H.; Wolf, G.T.; Urba, S.G.; Chepeha, D.B.; Teknos, T.N.; *et al.* Egfr, p16, HPV titer, bcl-xl and p53, sex, and smoking as indicators of response to therapy and survival in oropharyngeal cancer. *J. Clin. Oncol.* **2008**, *26*, 3128–3137.

152. Bonner, J.A.; Harari, P.M.; Giralt, J.; Azarnia, N.; Shin, D.M.; Cohen, R.B.; Jones, C.U.; Sur, R.; Raben, D.; Jassem, J.; *et al.* Radiotherapy plus cetuximab for squamous-cell carcinoma of the head and neck. *N. Engl. J. Med.* **2006**, *354*, 567–578.

153. Lechner, M.; Frampton, G.M.; Fenton, T.; Feber, A.; Palmer, G.; Jay, A.; Pillay, N.; Forster, M.; Cronin, M.T.; Lipson, D.; *et al.* Targeted next-generation sequencing of head and neck squamous cell carcinoma identifies novel genetic alterations in HPV+ and HPV- tumors. *Genome Med.* **2013**, *5*, doi:10.1186/gm453.

154. Aderhold, C.; Faber, A.; Umbreit, C.; Chakraborty, A.; Bockmayer, A.; Birk, R.; Sommer, J.U.; Hormann, K.; Schultz, J.D. Small molecules alter vegfr and pten expression in HPV-positive and -negative scc: New hope for targeted-therapy. *Anticancer Res.* **2015**, *35*, 1389–1399.

155. Argiris, A.; Kotsakis, A.P.; Hoang, T.; Worden, F.P.; Savvides, P.; Gibson, M.K.; Gyanchandani, R.; Blumenschein, G.R., Jr.; Chen, H.X.; Grandis, J.R.; *et al.* Cetuximab and bevacizumab: Preclinical data and phase ii trial in recurrent or metastatic squamous cell carcinoma of the head and neck. *Ann. Oncol.* **2013**, *24*, 220–225.

156. Aderhold, C.; Faber, A.; Grobschmidt, G.M.; Chakraborty, A.; Bockmayer, A.; Umbreit, C.; Birk, R.; Stern-Straeter, J.; Hormann, K.; Schultz, J.D. Small molecule-based chemotherapeutic approach in p16-positive and -negative hnscc *in vitro*. *Anticancer Res.* **2013**, *33*, 5385–5393.

157. Griffin, H.; Elston, R.; Jackson, D.; Ansell, K.; Coleman, M.; Winter, G.; Doorbar, J. Inhibition of papillomavirus protein function in cervical cancer cells by intrabody targeting. *J. Mol. Biol.* **2006**, *355*, 360–378.

158. Duenas-Gonzalez, A.; Cetina, L.; Coronel, J.; Cervantes-Madrid, D. Emerging drugs for cervical cancer. *Expert Opin. Emerg. Drugs* **2012**, *17*, 203–218.

159. Zhao, C.Y.; Szekely, L.; Bao, W.; Selivanova, G. Rescue of p53 function by small-molecule rita in cervical carcinoma by blocking E6-mediated degradation. *Cancer Res.* **2010**, *70*, 3372–3381.

160. Malecka, K.A.; Fera, D.; Schultz, D.C.; Hodawadekar, S.; Reichman, M.; Donover, P.S.; Murphy, M.E.; Marmorstein, R. Identification and characterization of small molecule human papillomavirus E6 inhibitors. *ACS Chem. Biol.* **2014**, *9*, 1603–1612.

161. Smukste, I.; Bhalala, O.; Persico, M.; Stockwell, B.R. Using small molecules to overcome drug resistance induced by a viral oncogene. *Cancer Cell* **2006**, *9*, 133–146.

162. Cho, Y.; Cho, C.; Joung, O.; Lee, K.; Park, S.; Yoon, D. Development of screening systems for drugs against human papillomavirus-associated cervical cancer: Based on E6-E6AP binding. *Antivir. Res.* **2000**, *47*, 199–206.

163. Issaeva, N.; Bozko, P.; Enge, M.; Protopopova, M.; Verhoef, L.G.; Masucci, M.; Pramanik, A.; Selivanova, G. Small molecule rita binds to p53, blocks p53-hdm-2 interaction and activates p53 function in tumors. *Nat. Med.* **2004**, *10*, 1321–1328.

164. Xie, X.; Piao, L.; Bullock, B.N.; Smith, A.; Su, T.; Zhang, M.; Teknos, T.N.; Arora, P.S.; Pan, Q. Targeting HPV16 E6-p300 interaction reactivates p53 and inhibits the tumorigenicity of HPV-positive head and neck squamous cell carcinoma. *Oncogene* **2014**, *33*, 1037–1046.

165. Yazbeck, V.Y.; Li, C.; Grandis, J.R.; Zang, Y.; Johnson, D.E. Single-agent obatoclax (gx15–070) potently induces apoptosis and pro-survival autophagy in head and neck squamous cell carcinoma cells. *Oral Oncol.* **2014**, *50*, 120–127.

166. Yuan, C.H.; Filippova, M.; Tungteakkhun, S.S.; Duerksen-Hughes, P.J.; Krstenansky, J.L. Small molecule inhibitors of the HPV16-E6 interaction with caspase 8. *Bioorganic Med. Chem. Lett.* **2012**, *22*, 2125–2129.

167. Garnett, T.O.; Filippova, M.; Duerksen-Hughes, P.J. Bid is cleaved upstream of caspase-8 activation during trail-mediated apoptosis in human osteosarcoma cells. *Apoptosis* **2007**, *12*, 1299–1315.

168. Fera, D.; Schultz, D.C.; Hodawadekar, S.; Reichman, M.; Donover, P.S.; Melvin, J.; Troutman, S.; Kissil, J.L.; Huryn, D.M.; Marmorstein, R. Identification and characterization of small molecule antagonists of prb inactivation by viral oncoproteins. *Chem. Biol.* **2012**, *19*, 518–528.

169. Shaikh, F.; Sanehi, P.; Rawal, R. Molecular screening of compounds to the predicted protein-protein interaction site of rb1-E7 with p53- E6 in HPV. *Bioinformation* **2012**, *8*, 607–612.

The Subcellular Localisation of the Human Papillomavirus (HPV) 16 E7 Protein in Cervical Cancer Cells and Its Perturbation by RNA Aptamers

Özlem Cesur, Clare Nicol, Helen Groves, Jamel Mankouri, George Eric Blair and Nicola J. Stonehouse

Abstract: Human papillomavirus (HPV) is the most common viral infection of the reproductive tract, affecting both men and women. High-risk oncogenic types are responsible for almost 90% of anogenital and oropharyngeal cancers including cervical cancer. Some of the HPV "early" genes, particularly E6 and E7, are known to act as oncogenes that promote tumour growth and malignant transformation. Most notably, HPV-16 E7 interacts with the tumour suppressor protein pRb, promoting its degradation, leading to cell cycle dysregulation in infected cells. We have previously shown that an RNA aptamer (termed A2) selectively binds to HPV16 E7 and is able to induce apoptosis in HPV16-transformed cervical carcinoma cell lines (SiHa) through reduction of E7 levels. In this study, we investigated the effects of the A2 aptamer on E7 localisation in order to define its effects on E7 activity. We demonstrate for the first time that E7 localised to the plasma membrane. In addition, we show that A2 enhanced E7 localisation in the ER and that the A2-mediated reduction of E7 was not associated with proteasomal degradation. These data suggest that A2 perturbs normal E7 trafficking through promoting E7 ER retention.

Reprinted from *Viruses*. Cite as: Cesur, Ö.; Nicol, C.; Groves, H.; Mankouri, J.; Blair, G.E.; Stonehouse, N.J. The Subcellular Localisation of the Human Papillomavirus (HPV) 16 E7 Protein in Cervical Cancer Cells and Its Perturbation by RNA Aptamers. *Viruses* **2015**, 7, 3443–3461.

1. Introduction

Papillomaviruses have been discovered in many vertebrates including humans, cattle, dogs, birds and reptiles. Within the papillomavirus family, around 170 human and 130 animal genotypes have been sequenced, identified and subsequently classified into 37 genera to date (see Papillomavirus Episteme (PaVE) [1]. Human papillomaviruses (HPVs) are classified into five genera (Alpha, Beta, Gamma, Mu and Nu) [2]. The Alpha genus contains high-risk mucosal HPVs that are the main cause of cervical cancer [3]. Amongst high-risk genotypes, HPV16 and 18 are most frequently found in cervical biopsies [4]. The majority (70%–90%) of HPV infections with both high- and low-risk genotypes are asymptomatic and can be cleared spontaneously within one to two years. Only a small percentage of persistent infections (around ∼5%–10%) with high-risk types result in precancerous lesions. If untreated, these lesions may lead to squamous cell carcinoma (SCC) [5].

HPV-related anogenital and oropharyngeal cancers are a global health burden. Cancer Research UK reported around ∼3000 new cases of cervical cancer in 2011 alone in the UK; the twelfth most common female cancer and third most common gynaecological cancer after uterine and ovarian cancers. In a worldwide perspective, cervical cancer is the second most prevalent cancer in women with an estimated 530,000 new cases in 2012 and around 270,000 deaths mostly in the developing world, according to the World Health Organisation (WHO) [6]. There is no HPV-specific cervical cancer treatment. However, protection can be achieved by the use of barrier methods and vaccination. Currently, there are two prophylactic vaccines in use based on virus-like particles (VLPs): the bivalent Cervarix (HPV16 and 18 VLPs) and the quadrivalent Gardasil (HPV6, 11, 16 and 18 VLPs). A nine-valent vaccine covering five additional high-risk HPVs is undergoing a Phase 3 clinical trial (trial number NCT00943722).

HPV encodes oncoproteins termed E5, E6 and E7. E7 is a highly phosphorylated, acidic polypeptide of approximately 100 amino acid residues. The E7 phosphoprotein has two conserved regions: CR1, CR2 and a carboxy-terminal region containing two zinc finger domains. CR2 contains a conserved LXCXE (L, leucine; C, cysteine; E, glutamate; X, any amino acid) motif, which is required for the interaction with pRb. E7 shares sequence similarities with the adenovirus (Ad) E1A protein and the simian vacuolating virus 40 (SV40) large tumour (large T) antigen [7]. Structural characterisation of HPV45 E7 [8,9] indicates that the N-terminus is intrinsically disordered while the C-terminus is highly structured with the zinc-binding region possibly involved in dimerisation. E7 has been shown to be present in dimeric form when expressed in *Escherichia coli* by gel filtration and non-denaturing acrylamide gel electrophoresis [10]. It has also been reported to form tetramers [11] and higher order oligomers [12] by analytical ultracentrifugation sedimentation equilibrium experiments.

The intracellular localisation of E7 has been investigated in detail. Subcellular fractionation experiments in CaSki cells demonstrated the presence of E7 in soluble cytoplasmic fractions [13]. Nuclear [14] and nucleolar [15] distribution in HPV16+ CaSki cells has also been proposed. Using antibodies with high discrimination capacity against monomeric, dimeric or oligomeric forms of E7, E7 dimers were shown to distribute to the nucleus, whilst oligomeric E7 displayed cytoplasmic distribution [16]. The presence of nuclear localisation and export sequences led to the hypothesis that E7 shuttles between the cytoplasm and nucleus [17]. Consistent with this, leptomycin B treatment has been shown to lead to E7 accumulation in the nucleus [17,18]. Cell confluency has also been proposed to dictate E7 localisation, being predominantly cytoplasmic in confluent cells but locating to both the nucleus and cytoplasm in sub-confluent cells [18], suggesting that location may be cell cycle-dependent.

Aptamers are short (15–100 nucleotides), single-stranded RNA or DNA molecules generated by SELEX (Systematic Evolution of Ligands by Exponential Enrichment), reviewed in [19–21]. Aptamers fold into specific complex structures that bind target proteins in a conformation-dependent manner and can interfere with function. Aptamers have been identified which recognise a number of viral proteins including HPV16 E6 and E7 [22–24], the RNA-dependent RNA polymerase of foot-and-mouth disease virus [25,26] and hepatitis C virus non-structural protein 5B [27,28]. An aptamer, targeted to VEGF termed pegaptanib (Macugen) was approved by the Food and Drug Administration (FDA) for the treatment of age-related macular degeneration in 2004 and examples of aptamers with anti-proliferative effects in cancer cells are currently undergoing clinical trials, including a guanosine-rich DNA oligonucleotide, AS1411 [29] and a L-RNA aptamer (Spiegelmer), NOX-12 [30].

We previously described an HPV16 E7 aptamer (termed A2) that resulted in a loss of E7 expression after transfection into HPV16+ cells [24]. We postulated that E7 was being targeted for degradation. Here, we show that aptamers can endocytose into early/late endosomes and that A2 redistributes E7 to the ER from the plasma membrane. We therefore propose that A2 interferes with normal E7 trafficking and cellular localisation.

2. Materials and Methods

2.1. Cell Culture

The SiHa cell line (ATCC No. HTB-35) was derived from a human squamous cell carcinoma of the cervix and contained 1–2 copies of the integrated HPV16 genome. CaSki cells (ATCC No. CRL-1550) were derived from a human epidermoid carcinoma of the cervix, and contain approximately 600 integrated copies. The HeLa cell line (ATCC No. CCL-2) was derived from a human adenocarcinoma of the cervix and contains approximately 10–50 copies of HPV18 genome. HaCaT cells are spontaneously-immortalised human keratinocytes (HPV negative). SaOS-2 (ATCC No. HTB-85) is an osteosarcoma cell line and negative for HPV DNA. SaOS-2, CaSki, SiHa, HeLa and HaCaT cells were maintained in DMEM containing 1% L-glutamine (GE Healthcare, Little Chalfont, Buckinghamshire, UK) supplemented with 10% foetal bovine serum (FBS) (PAA, Pasching, Austria), 100 units/mL penicillin (Lonza, Slough, UK), 0.1 mg/mL streptomycin (Lonza) in T-25 flasks. Flasks were maintained in a horizontal position in a humidified incubator (37 °C; 5% CO_2). Cells were plated in 6-well (for protein extraction) or 12-well (for immunostaining) dishes for the experiments. In order to analyse the pathway of E7 degradation, CaSki cells were treated with a proteasome inhibitor, MG132 (Cayman Chemicals, Ann Arbor, MI, USA) at 100 µM for up to 6 h in the presence or absence of 100 nM aptamer.

2.2. Generation of 2'-Fluoro-Modified RNA Molecules

We have previously generated a library for templates of RNA aptamers selected against E7 [23]. A2 and SF1 aptamer templates were amplified by PCR and *in vitro* transcription was performed as previously described, incorporating 2'F U and C [22–24]. Aptamer 21-2 (5'-Cy5 and 3'-Cy3-labelled) and aptamer 47tr (5'-Cy3-labelled) both contained 2'F C and aptamer 21-2 also included 2'F U. 21-2 was purchased from Abgene (UK) and 47tr was synthesised by phosphoramidite chemistry in house [26,31].

2.3. Transfection of Cells with Aptamers Using Oligofectamine

Cells were transfected with aptamers at a final RNA concentrations of up to 100 nM using Oligofectamine (Invitrogen, Life Technologies, Waltham, MA, USA) according to the manufacturer's instructions.

2.4. Collection of Cells and Protein Extraction

Cells were lysed in radio-immunoprecipitation (RIPA) buffer [50 mM Tris-HCl (pH 8.0), 150 mM NaCl, 1% (*v/v*) Nonidet P-40, 0.5% (*w/v*) sodium deoxycholate, 0.1% (*w/v*) SDS] containing EDTA-free protease inhibitors (Roche, Penzberg, Germany) (1 tablet per 10 mL buffer), DNase I (5 µg/ mL) and 10 mM MgCl$_2$ on ice for 30 min. Protein samples were mixed with 2 × Laemmli buffer and boiled to denature at 95 °C for 5 min.

2.5. SDS-PAGE and Western Blotting

This was performed as previously described [23]. Membranes were probed with rabbit anti-GAPDH (Sigma-Aldrich, St. Louis, MO, USA, used at 1:6000), mouse monoclonal HPV16 anti-E7; clone NM2 (Santa Cruz Biotechnology, Inc., Dallas, TX, USA, used at 1:200) overnight at 4 °C. Membranes were labelled with goat anti-mouse IgG peroxidase conjugate (Sigma, USA used at 1:2000) and goat anti-rabbit IgG peroxidase (Sigma, St. Louis, MO, USA, used at 1:1000) in 5% (*w/v*) milk for 1 h at room temperature.

2.6. Immunostaining of Cells and Fluorescence Microscopy

SiHa, SaOS-2, HeLa and HaCaT cells were grown on coverslips, washed with PBS and fixed with 4% (*v/v*) formaldehyde in PBS for 10 min at room temperature. Permeabilisation of cells was performed by incubation in 0.1% (*v/v*) TritonX-100/PBS for 10 min. Cells were incubated in blocking solution [1% BSA/Triton X-100 in 1 x PBS (*w/v*)] for 1 h at room temperature. Mouse monoclonal HPV16 anti-E7; clone 289–17013 (Abcam, UK used at 1:2000), mouse monoclonal HPV16 anti-E7; cloneNM2 (Santa Cruz Biotechnology, Inc., Dallas, TX, USA, used at 1:200), mouse monoclonal HPV18 anti-E7; clone 8E2 (Abcam, Cambridge, UK, used at 1:200), rabbit polyclonal anti-EEA1 (Millipore, Billerica, MA, USA, used at 1:500), rabbit polyclonal anti-LAMP-1 (CD107a) (Millipore, Billerica, MA USA used at 1:500) or rabbit polyclonal anti-LC3 (Abcam, Cambridge, UK used at 1:2000) were added to cells and incubated overnight at 4 °C on a rotating platform. To analyse cell-surface staining, cells were incubated with anti-E7 at 4 °C prior to fixation for 1 h.

Cells were washed with PBS and incubated with fluorochrome-conjugated secondary antibodies AlexaFluor 568 F goat anti-mouse IgG (H+L), AlexaFluor 488 chicken anti-mouse IgG (H+L), AlexaFluor 488 goat anti-rabbit IgG (H+L). Secondary antibodies (all from Life Technologies, Waltham, MA, USA) were used at 1:500 in 1% BSA/Triton X-100 for 2 h at room temperature and kept in the dark. Glass coverslips were mounted on microscope slides in Vectashield-DAPI stain (Vector laboratories, Burlingame, CA, USA) and viewed on Zeiss LSM 510 upright or LSM 700 inverted confocal microscopes.

2.7. Quantitation of Co-Localisation by Imeris Software

Confocal images were analysed by Bitplane:Imeris image analysis software in order to calculate the Pearson's co-localisation coefficient values. Results are presented as % of co-localisation between two different channels and n refers to the number of cells analysed.

3. Results

3.1. A2-Mediated Degradation of E7 Is Not Mediated via Proteasomal Pathways

We have previously described a loss of E7 in A2-transfected HPV16+ CaSki cells. The effect was not observed with a control aptamer SF1, selected to an unrelated protein [24]. Here, the peptide-aldehyde proteasome inhibitor, MG132 (carbobenzoxyl-L-leucyl-L-leucyl-L-leucinal) was used to study E7 degradation in the presence or absence of A2. CaSki cells transfected with A2 or a control aptamer (SF1) for 14–16 h were treated with 100 µM MG132 for up to 6 h. Figure 1 demonstrates that transfection of the A2 aptamer appeared to result in a lower level of E7, as evidenced by the level at t = 0 (corresponding to 14–16 h post-transfection), in agreement with previous data [24]. Furthermore, in A2-transfected cells, E7 levels remained low. MG132 treatment clearly had little effect, indicating that newly-synthesised E7 was still being degraded. In contrast, the level of MG132-mediated inhibition appeared to be highly variable in the mock- or SF1-treated cells and E7 did appear to accumulate, suggesting that the protein normally undergoes proteasomal degradation. The clear differences between the E7 levels in the mock- or SF1-treated cells compared to A2-treated cells led to the suggestion that A2 acts to enhance E7 degradation by a non-proteasomal mechanism.

3.2. E7 Localises to the Plasma Membrane

We next sought to investigate the intracellular distribution of E7. Using HPV16 E7-specific antibodies in HPV16+ SiHa cells, we observed that E7 displayed diffuse cytoplasmic staining, as expected, however this appeared to extend to the cell periphery, characteristic of localisation to the plasma membrane or plasma membrane-associated sub-membrane cisternae (Figure 2A). It is interesting to note that HeLa cells (HPV18+) show a similar distribution of E7, Figure 2B. We aimed to confirm this observation by detecting E7 staining in unpermeabilised SiHa cells. These experiments (with two different anti-HPV16 antibodies) demonstrated E7 localisation at the cell surface (Figure 2A, i and ii). No such pattern of staining was evident in SaOS-2 cells or HaCaT cells (HPV-negative cell lines), confirming the staining to be E7 specific (Figure 2A, iii and iv). We also employed a plasma membrane-specific stain (CellMask Orange). This stain is an amphipathic molecule, consisting of a lipophilic region and a negatively charged hydrophilic dye for membrane loading and the attachment of the probe in the plasma membrane, respectively [32,33]. Use of CellMask also demonstrated the presence of E7 at the plasma membrane, Figure 3A.

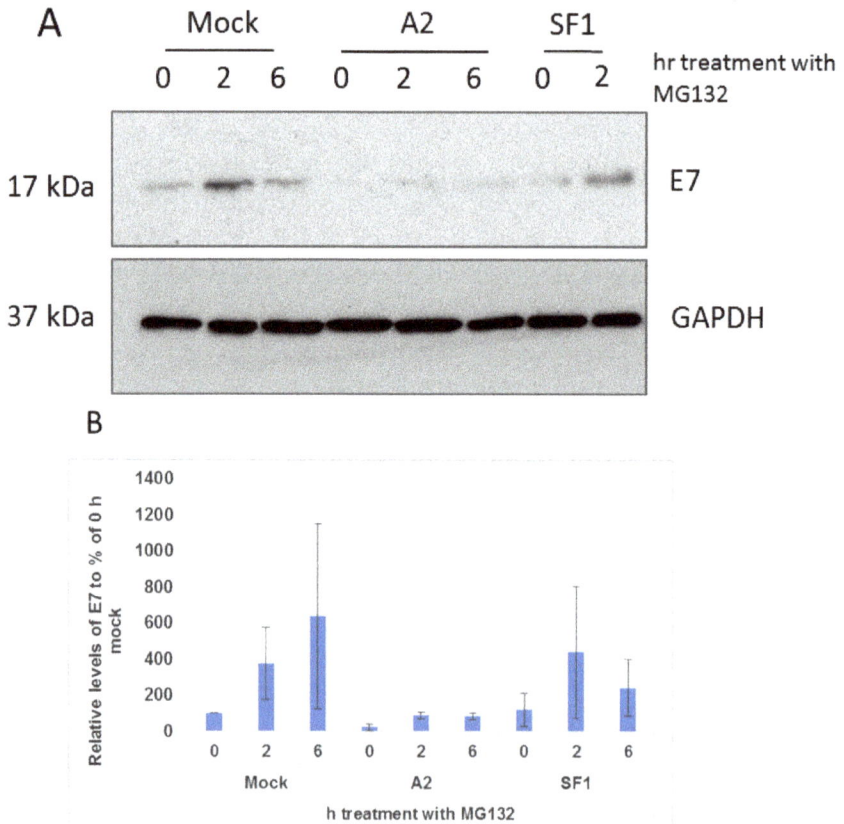

Figure 1. Proteasomal degradation of E7 was not affected by A2 treatment. CaSki cells were mock-transfected or transfected with 100 nM A2 or SF1 for 16 h. Cells were treated with MG132 (100 μM) for the indicated time points and immunoblot analysis of cell lysates was performed to detect E7 and GAPDH (as a housekeeping control) (**A**); Mean data from three independent experiments is shown, together with standard errors (**B**).

It should be noted that we were unable to produce membrane fractions that were completely free of cytoplasmic contamination, therefore were unable to confirm E7 localisation by cell fractionation (data not shown). However, we investigated the ability of E7 to undergo internalisation and removal from the plasma membrane, using protocols documented for the study of other cell-surface proteins [34]. Cells were maintained at 4 °C to allow antibody attachment followed by incubation at 37 °C for 30 min to follow the internalisation of cell-surface E7. Redistribution of E7, from a peripheral cell surface localisation (observed at $t = 0$) to the peri-nuclear region of the cells (at $t = 30$ min) was observed. This strongly suggested that E7 is endocytosed from the plasma membrane to an intracellular compartment, indicative of rapid internalisation from the cell surface (Figure 3B).

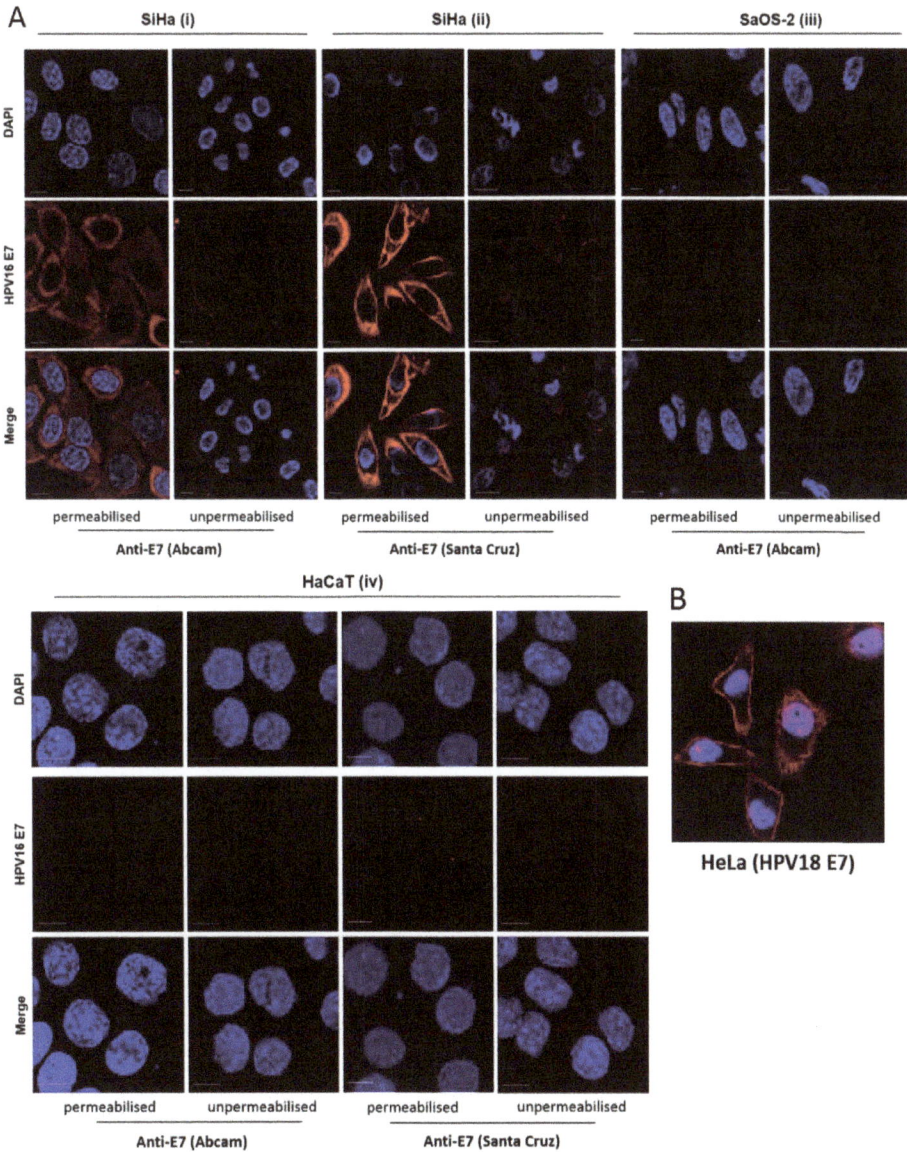

Figure 2. E7 distributes to the plasma membrane of HPV-transformed cells. (**A**) HPV16 E7+ SiHa cells (i and ii) and SaOS-2 cells (iii) HaCaT cells (iv) (as negative controls, containing no E7) were fixed, permeabilised and incubated with antibodies recognising full-length HPV16 E7 (i) and (iii) and residues 35–56 (ii). A comparable group of live cells was incubated with these antibodies in parallel. Cells were fixed and stained with AlexaFluor568 anti-mouse antibodies. Fluorescent imaging was performed using either a Zeiss LSM700 or LSM510 inverted microscope. The scale bar = 10 μm. Representative images are shown; (**B**) HeLa cells were fixed, permeabilised and labelled with anti-HPV18 E7 antibodies, which recognise residues 36–70, followed by staining with AlexaFluor594 goat anti-mouse IgG.

A (unpermeabilised)

B

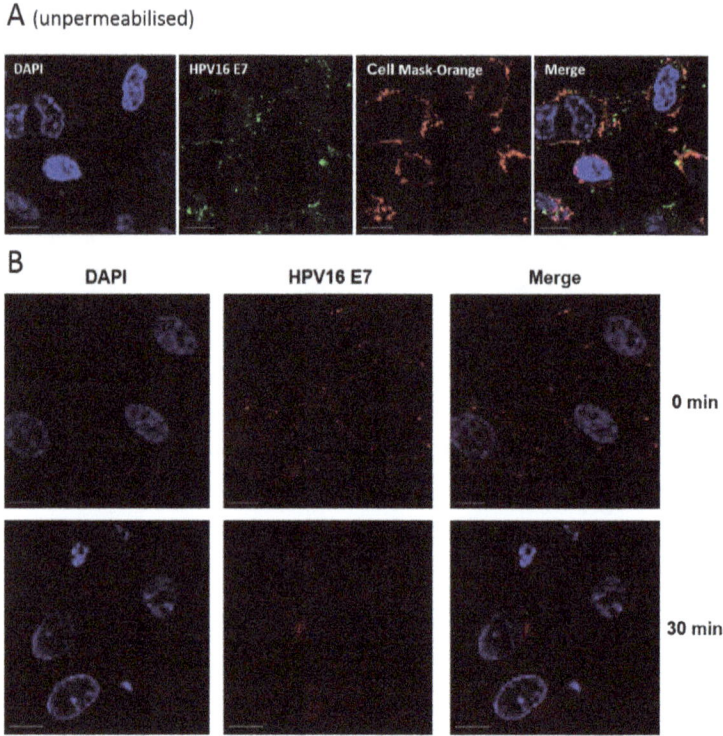

Figure 3. (**A**) Co-distribution of E7 with the cell membrane marker CellMask Orange (red) in unpermeabilised SiHa cells. Live cells were dual-stained with anti-E7 (Abcam) and CellMask Orange; (**B**) Time and temperature-dependent internalisation of cell surface E7, chased by anti-E7 antibody (Abcam). SiHa cells were incubated with the anti-E7 antibody for 1 h at 4 °C and incubated at 37 °C for 30 min to allow E7 internalisation. Cells were fixed, permeabilised and stained with AlexaFluor568 anti-mouse antibodies. Fluorescent imaging was performed using either a Zeiss LSM700 or LSM510 inverted microscope. The scale bar = 10 μm.

3.3. Aptamers Localise to Early/Late Endosomes upon Transfection

The internalisation of ligands, extracellular molecules, plasma membrane proteins and lipids commonly occur by endocytosis. Typically, endocytosed cargo is delivered to early endosomal compartments, followed by transit to the late endosomes/lysosomes for degradation, to the trans-Golgi network (TGN) or recycling endosomes in order to return the cargo back to the plasma membrane, reviewed in [35]. However, little is known about the sub-cellular localisation of RNA molecules after uptake.

To investigate uptake of RNA aptamers in E7-expressing cells, we utilised chemically-synthesised, model aptamers (*i.e.*, not selected to bind to E7), labelled with Cy3 [26,31] and assessed their transit into cellular compartments at defined time points post-incubation.

Counting the number of aptamer-postive cells revealed a transfection efficiency of 87% ($n = 90$). At 3 h post-transfection, co-localisation with EEA1 demonstrated that the cy3-labelled aptamer 21-2 predominantly localised to early endosomes (63.3%, Figure 4A,C). However, in cells transfected for 6 h, this co-localisation was reduced and the aptamer was localised to both early and late endosomes/lysosomes (45.5% and 33.2% respectively, Figure 4B,C). A second aptamer, Cy3-labelled 47tr, displayed a similar pattern of distribution to Cy3 21-2 in SiHa cells (data not shown). Taken together these data suggest that RNA aptamers are targeted to early/late endosomal/lysosomal pathways following cell entry. However, the fate of these aptamers at later time points is currently unknown.

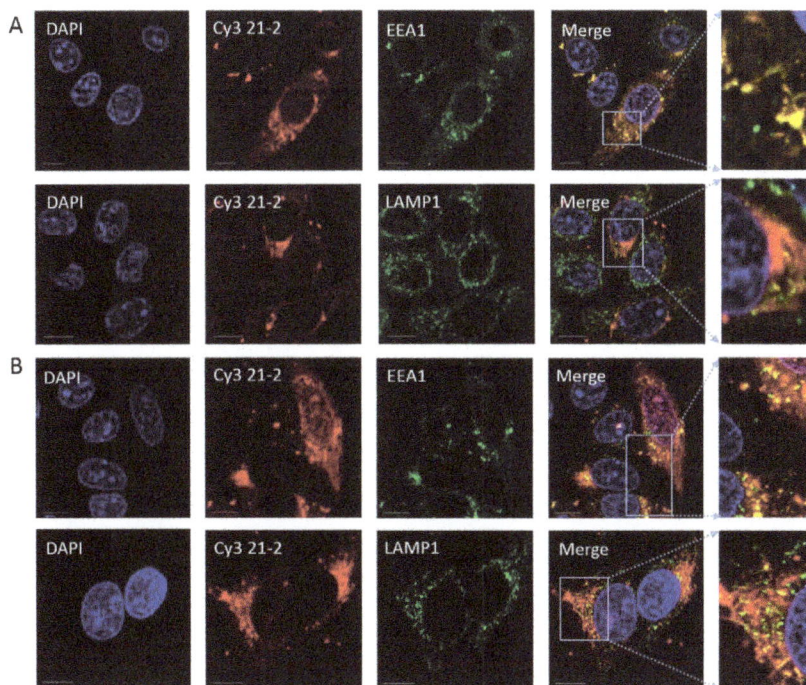

Figure 4. *Cont.*

C

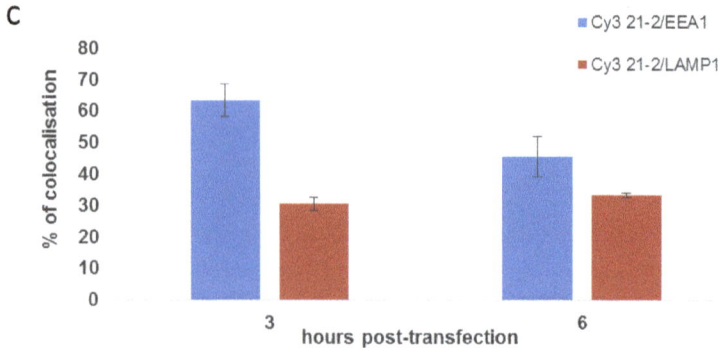

Figure 4. A model aptamer (aptamer 21-2) localises predominantly in early and late endosomes in SiHa cells following Oligofectamine transfection. Cells were transfected with 80 nM Cy3-labelled 21-2 using Oligofectamine. At 3 (Panel (**A**)) or 6 h (Panel (**B**)) post-transfection, cells were fixed, permeabilised and incubated with primary antibodies anti-EEA1 or anti-LAMP1 (staining early and late endosomes, respectively) prior to incubation with AlexaFluor488 goat anti-rabbit secondary antibody. The right hand side panel is a zoom of merged images. The scale bar = 10 μm. The level of co-localisation was analysed by Bitplane:Imeris image analysis software (Panel (**C**)). For EEA1 and LAMP1 colocalisation respectively, $n = 8$ and 29 at 3 h and $n = 30$ and 19 at 6 h. Red, green and blue are Cy3 (21-2), FITC (early or late endosomes) and DAPI (nucleus), respectively.

3.4. Localisation of E7 in Cellular Compartments upon Aptamer Transfection

We have previously shown that aptamer A2 can reduce E7 levels [24] but the mechanism of this effect was not addressed. Given our finding that E7 can distribute to the plasma membrane (Figures 2 and 3) we hypothesised that the A2 aptamer may bind cell surface E7, and mediate E7 degradation via the endosomal/lysosomal pathways. As labelling of A2 may have effects on conformation and therefore function, indirect methods were used to evaluate the effects of A2 on E7. Co-localisation studies with E7 and endosomal markers were performed to assess the cellular distribution of E7 in the presence or absence of A2 and the control aptamer, SF1. Cells were fixed 16 h post-transfection, in order to allow time for any A2-mediated effects to become apparent [24]. It should be noted that we were unable to quantitate transfection efficiency with A2 or SF1. These aptamers are similar in size, but somewhat larger than 21-2. It is therefore likely that the transfection will be less efficient.

We observed only minimal co-localisation of E7 with the endosomal marker EEA1 (Figure 5A,C) with values of 12.6%, 13.0% and 16.0% for mock-, A2- and SF1-treated SiHa cells, respectively. There were no statistical significant differences between the three treatments. It should be noted that these data were collected at a later timepoint than the data shown in Figure 3, as we aimed to allow time for A2-mediated effects on E7 to become apparent. It is therefore possible that co-localisation between E7 and endosomal markers may have been evident at earlier times post-transfection. Co-localisation with LAMP1 (Figure 5B,C) was low and very similar for the three treatments (ranged between 2.0% and 2.5%, not significant. The level of colocalisation between the autophagosomal marker LC3 (Figure 6A,C) and E7 also showed no difference between the three treatments (12.7%, 10.2%, 11.6%). In contrast, there was an increased co-localisation of E7 with the ER marker, calreticulin, upon transfection with A2 (Figure 6B,C). Co-localisation in mock- and SF1-treated cells was 15.1% and 15.1%, but significantly higher at 32.3% in A2-transfected cells ($p \leq 0.01$). Taken together, these data suggest that A2 perturbs normal E7 trafficking and cellular distribution through enhancing E7 ER retention following E7 binding.

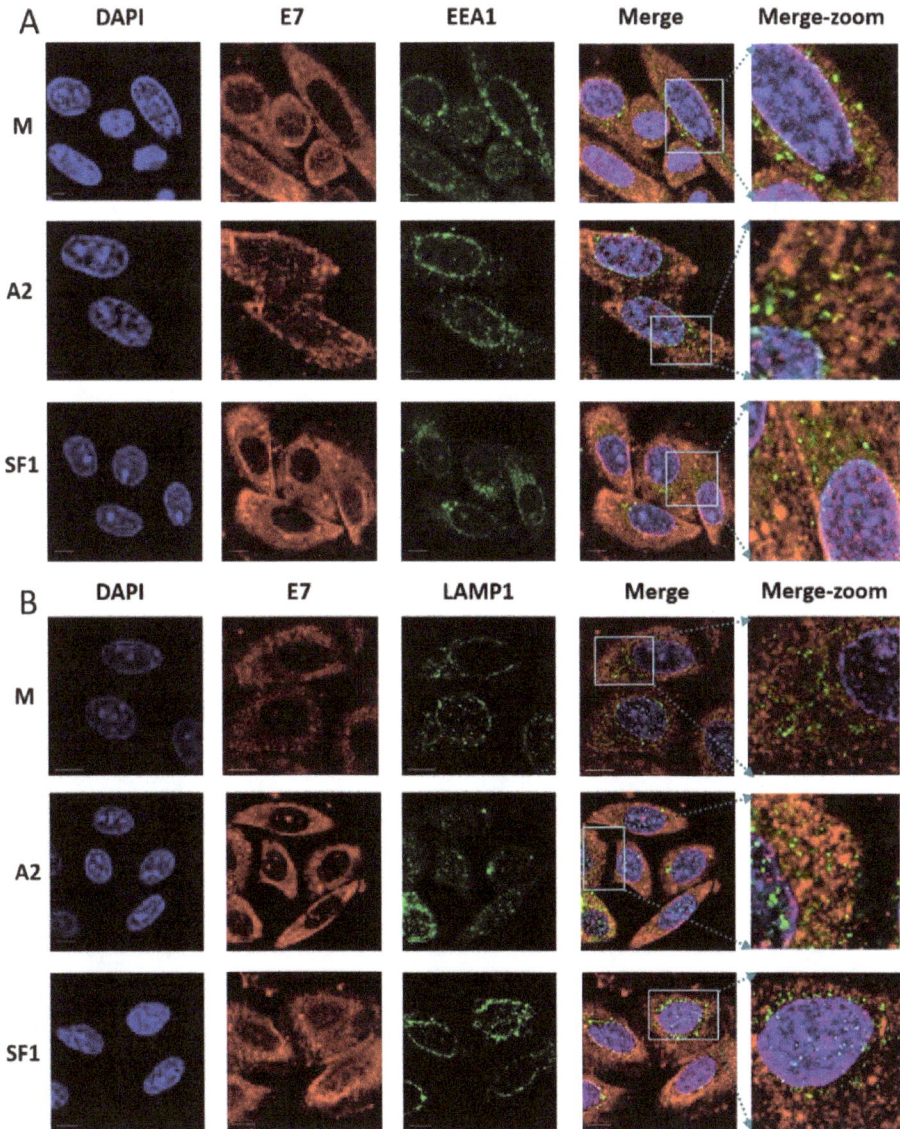

Figure 5. *Cont.*

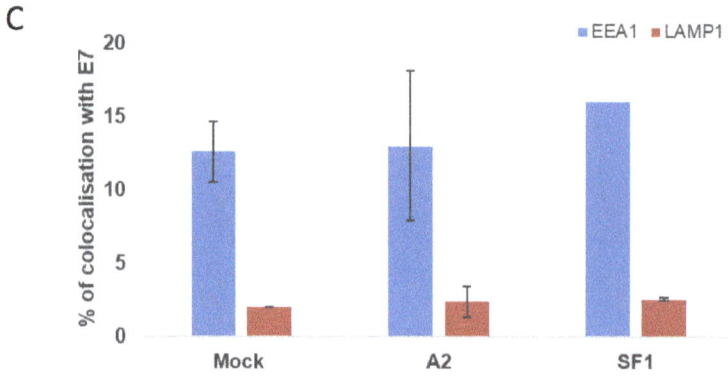

Figure 5. E7 does not co-localise with the early endosomal marker, EEA1 (**A**) or the late endosomal (lysosomal) marker, LAMP1 (**B**). SiHa cells were mock-transfected (M) or transfected with 100 nM A2 or SF1 using Oligofectamine. At 14 h post-transfection, cells were co-stained with either anti-EEA1 or anti-LAMP1 (staining early and late endosomes, respectively) prior to incubation with AlexaFluor488 goat anti-rabbit secondary antibody. Cells were then incubated with anti-E7 antibody (Abcam) followed by AlexaFluor568 rabbit anti-mouse secondary antibody. Red, green and blue are E7, FITC (early or late endosomes) and DAPI (nucleus), respectively. The last panel is a zoom of merged images. The scale bar = 10 μm. E7/EEA1 and E7/LAMP1 colocalisation were analysed using Bitplane:Imeris image analysis software based on the calculation of Pearson's correlation coefficient values and presented as percentage co-localisation (**C**). $n = 27$, 23 and 4 for E7/EEA1 and $n = 12$, 24 and 9 for E7/LAMP-1 in mock-, A2- and SFI-treated SiHa cells, respectively.

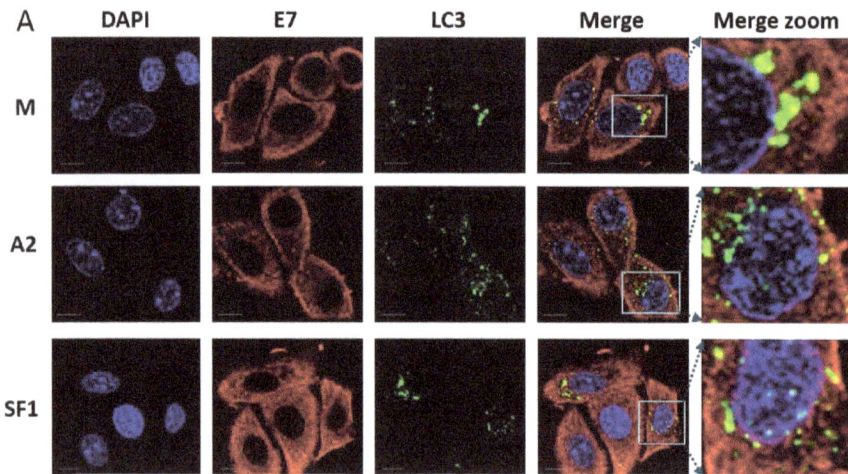

Figure 6. *Cont.*

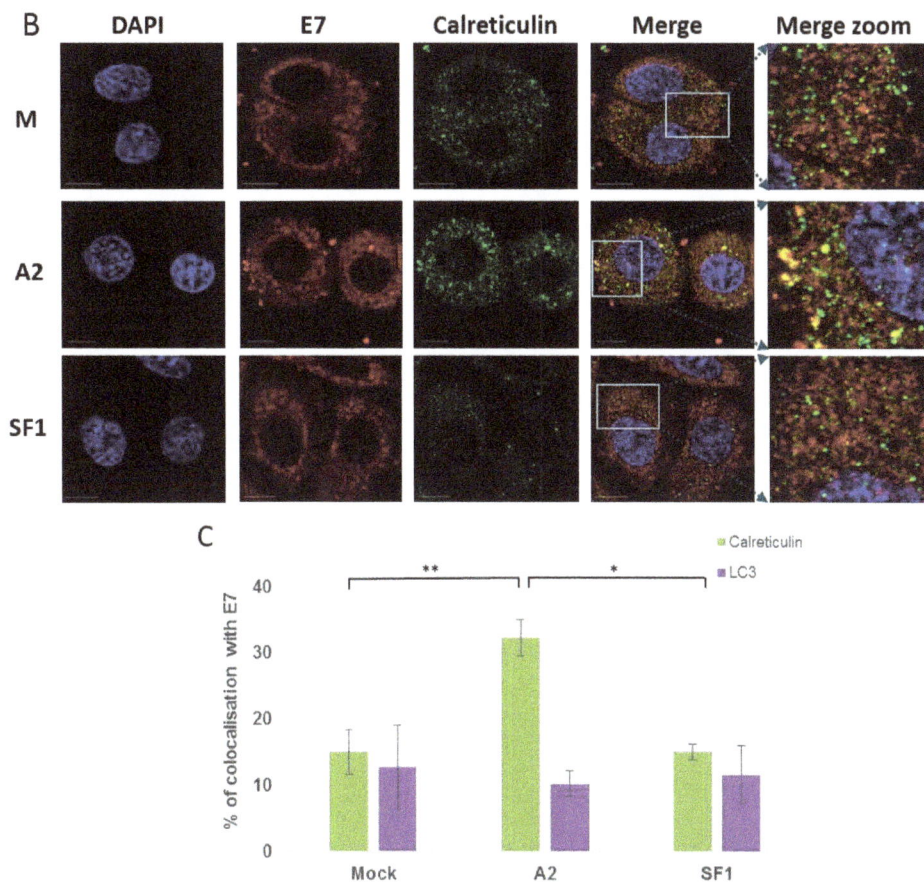

Figure 6. E7 does not co-localise with an autophagosome marker, LC3 (**A**) but appears to co-localise with endoplasmic reticulum marker calreticulin (**B**) in the presence of A2. SiHa cells were mock-transfected (M) or transfected with 100 nM A2 or SF1 using Oligofectamine. At 14 h post-transfection, cells were co-stained with anti-E7 (Abcam) and either anti-LC3 (an autophagosomal marker) or anti-calreticulin (an ER marker) prior to incubation with either AlexaFluor568 rabbit anti-mouse or AlexaFluor488 goat anti-rabbit secondary antibodies. Red, Green and Blue are E7, FITC (LC3, autophagosome) and DAPI (nucleus), respectively. The last panel is a zoom of merged images. The scale bar = 10 μm. E7/LC3 and E7/calreticulin colocalisation was analysed using Bitplane:Imeris image analysis software based on the calculation of Pearson's correlation coefficient values and presented as percentage co-localisation (**C**). $n = 21$, 23 and 17 for E7/LC3 and $n = 19$, 39 and 20 for E7/calreticulin with mock-, A2- and SFI-treatment, respectively. * $p \leq 0.0005$, ** $p \leq 0.002$.

4. Discussion

This work is the first to describe the localisation of E7 to the plasma membrane and to propose a mechanism for E7 degradation in the presence of the E7-specific aptamer, A2. Three independent antibodies recognising full-length HPV16 E7, HPV16 E7 residues 35–56 and HPV18 E7 residues 36–70 were employed to show that E7 distributed to the plasma membrane. It is therefore likely that the N-terminal region of E7 is accessible and extracellularly-exposed since we could detect E7 in unpermeabilised SiHa cells. We were unable to produce membrane fractions that were completely free of cytoplasmic contamination, therefore were unable to confirm E7 localisation by cell fractionation. However, the data above, together with the ability of E7 to undergo internalisation and removal from the plasma membrane provide compelling evidence for membrane localisation. It should be noted that the amount of membrane-associated E7 is relatively low and this could explain why this has not been detected previously [13,36]. The level of cell confluency could also be a factor [18]. Although the data with HPV18 is preliminary, it is possible that this effect may be common across other HPV types. Further work would be necessary in order to establish whether this is the case.

We have previously reported that transfection of CaSki cells with the aptamer A2 resulted in a decrease of E7 levels with an accompanying increase in pRb, which was suggested as the reason behind the apoptotic effects of A2 observed in HPV-transformed cells [24]. The loss of E7 in A2-transfected cells was previously suggested to be due to the inability of A2-bound E7 to interact with its major cellular partners including pRb, thus remaining unfolded which could lead to destabilisation and degradation [24]. The proteasome/ubiquitin system is primarily responsible for clearance of cellular misfolded/unfolded proteins. E7 is a short-lived protein with a half-life of 30–40 min [13], allowing detection of protein accumulation within a few hours of addition of MG132. Consistent with previous studies [37], we showed that degradation of E7 can be mediated by 26S proteasomes as MG132 treatment resulted in the accumulation of E7. However, in the presence A2, E7 accumulation was much less evident. Given this data, we propose an alternative non-proteasomal mechanism for aptamer-mediated E7 degradation.

Studies with a control, labelled aptamer revealed that this molecule localised to both early and late endosomes. Localisation to early endosomes was particularly striking at early timepoints post-transfection (3 h). It is interesting to note that the reduction in co-localisation seen after 6 h did not correlate with an increase in co-localisation with the late endosomal marker. Further time points might be useful, however, this could be indicative of escape of aptamers from the endocytic pathway. Whilst there is a paucity of data on the sub-cellular localisation of aptamers, transfection of siRNAs using cationic lipids, nanoparticles or cell-type-specific delivery reagents have shown that siRNAs are trafficked to the endosomal pathway (early and then late endosomes), reviewed in [38]. It is clear that such siRNA molecules must escape the endosomal pathway to function, and the observation led to a hypothesis for A2-mediated activity. E7 on the cell surface could be internalised along with the bound aptamer during transfection, resulting in its accumulation in endosomes and then late endosomes/lysosomes for degradation. In order to investigate this hypothesis, HPV16+ SiHa cells were co-stained with E7 and several endosomal markers in the presence and absence of A2. We found no evidence of E7 accumulation in lysosomes or autophagosomes in aptamer-transfected or mock-transfected SiHa cells after 14 h. However, E7 localisation in the ER (using calreticulin as a marker) increased in cells transfected with A2. Interestingly, E7 localisation to the ER has been demonstrated previously in CaSki cells (using calnexin as a marker) [39]. These authors also noted localisation to the Golgi, we have not investigated whether A2 has any effect here. Our data suggest that a possible conformational change of E7, mediated by A2 binding, leads to an accumulation in the ER. This would be predicted to perturb normal biosynthetic delivery of E7 to the plasma membrane or other organelles (Figure 7). As E7 does not possess common peptide ER retention motifs such as [K/H]DEL, KKXXX, KXKXXX and PL[Y/F][F/Y]XXN, it is likely that A2 binding prevents correct E7 folding/oligomerisationleading to its accumulation in an unfolded/aberrant state. This accumulation of unfolded E7 and subsequent stimulation of related stress responses may contribute to the stimulation of apoptosis previously observed in HPV+ cells [24,40–42]. Future work (including the use of specific inhibitors) would be necessary to validate this model.

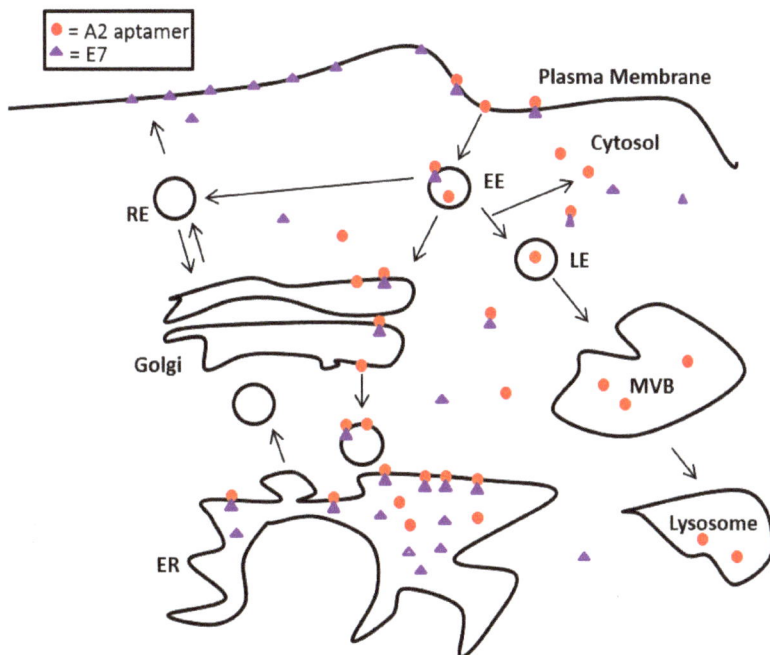

Figure 7. A2 can enter the cells alone or via the endosomal system. Upon contact with E7; either at the plasma membrane or intracellularly, A2 enhances E7 retention in the ER. This would be predicted to reduce E7 delivery to the plasma membrane via its normal biosynthetic route. ER: endoplasmic reticulum, EE: early endosome, LE: late endosome, MVB: multivesicular bodies, RE: recycling endosome.

5. Conclusions

In summary, the data reported here demonstrate the utility of RNA aptamers as molecular tools to aid our understanding of the cellular function of the HPV E7 oncoprotein.

Acknowledgments

O.C. would like to thank to the Turkish Ministry of National Education for funding. We also thank all members of the Stonehouse, Blair and Mankouri groups for helpful discussions.

Author Contributions

All authors contributed to manuscript preparation. O.C., C.N. and H.G. generated experimental data. J.M., G.E.B. and N.J.S. contributed to experimental design and data analysis.

Conflicts of Interest

The authors declare no conflict of interest.

References

1. PaVE: Papilloma Virus Genome Database. Available online: http://pave.niaid.nih.gov/#home (accessed on 23 June 2015).
2. Bernard, H.U.; Burk, R.D.; Chen, Z.; van Doorslaer, K.; zur Hausen, H.; de Villiers, E.M. Classification of papillomaviruses (PVs) based on 189 PV types and proposal of taxonomic amendments. *Virology* **2010**, *401*, 70–79. [CrossRef] [PubMed]
3. Zur Hausen, H. Papillomaviruses in the causation of human cancers—A brief historical account. *Virology* **2009**, *384*, 260–265. [CrossRef] [PubMed]
4. Naucler, P.; Mabota da Costa, F.; da Costa, J.L.; Ljungberg, O.; Bugalho, A.; Dillner, J. Human papillomavirus type-specific risk of cervical cancer in a population with high human immunodeficiency virus prevalence: Case-control study. *J. Gen. Virol.* **2011**, *92*, 2784–2791. [CrossRef] [PubMed]
5. Bosch, F.X.; Lorincz, A.; Muñoz, N.; Meijer, C.J.; Shah, K.V. The causal relation between human papillomavirus and cervical cancer. *J. Clin. Pathol.* **2002**, *55*, 244–265. [CrossRef] [PubMed]
6. Human papillomavirus vaccines: WHO position paper, October 2014-Recommendations. *Vaccine.* [CrossRef]
7. McLaughlin-Drubin, M.E.; Munger, K. The human papillomavirus E7 oncoprotein. *Virology* **2009**, *384*, 335–344. [CrossRef] [PubMed]
8. Liu, X.; Clements, A.; Zhao, K.; Marmorstein, R. Structure of the human Papillomavirus E7 oncoprotein and its mechanism for inactivation of the retinoblastoma tumor suppressor. *J. Biol. Chem.* **2006**, *281*, 578–586. [CrossRef] [PubMed]
9. Ohlenschlager, O.; Seiboth, T.; Zengerling, H.; Briese, L.; Marchanka, A.; Ramachandran, R.; Baum, M.; Korbas, M.; Meyer-Klaucke, W.; Dürst, M.; *et al.* Solution structure of the partially

folded high-risk human papilloma virus 45 oncoprotein E7. *Oncogene* **2006**, *25*, 5953–5959. [CrossRef]

10. McIntyre, M.C.; Frattini, M.G.; Grossman, S.R.; Laimins, L.A. Human papillomavirus type 18 E7 protein requires intact Cys-X-X-Cys motifs for zinc binding, dimerization, and transformation but not for Rb binding. *J. Virol.* **1993**, *67*, 3142–3150. [PubMed]

11. Clements, A.; Johnston, K.; Mazzarelli, J.M.; Ricciardi, R.P.; Marmorstein, R. Oligomerization properties of the viral oncoproteins adenovirus E1A and human papillomavirus E7 and their complexes with the retinoblastoma protein. *Biochemistry* **2000**, *39*, 16033–16045. [CrossRef] [PubMed]

12. Alonso, L.G.; García-Alai, M.M.; Smal, C.; Centeno, J.M.; Iacono, R.; Castaño, E.; Gualfetti, P.; de Prat-Gay, G. The HPV16 E7 viral oncoprotein self-assembles into defined spherical oligomers. *Biochemistry* **2004**, *43*, 3310–3317. [CrossRef] [PubMed]

13. Smotkin, D.; Wettstein, F.O. The major human papillomavirus protein in cervical cancers is a cytoplasmic phosphoprotein. *J. Virol.* **1987**, *61*, 1686–1689. [PubMed]

14. Smith-McCune, K.; Kalman, D.; Robbins, C.; Shivakumar, S.; Yuschenkoff, L.; Bishop, J.M. Intranuclear localization of human papillomavirus 16 E7 during transformation and preferential binding of E7 to the Rb family member p130. *Proc. Natl. Acad. Sci. USA* **1999**, *96*, 6999–7004. [CrossRef] [PubMed]

15. Zatsepina, O.; Braspenning, J.; Robberson, D.; Hajibagheri, M.A.; Blight, K.J.; Ely, S.; Hibma, M.; Spitkovsky, D.; Trendelenburg, M.; Crawford, L.; *et al.* The human papillomavirus type 16 E7 protein is associated with the nucleolus in mammalian and yeast cells. *Oncogene* **1997**, *14*, 1137–1145. [CrossRef] [PubMed]

16. Dantur, K.; Alonso, L.; Castaño, E.; Morelli, L.; Centeno-Crowley, J.M.; Vighi, S.; de Prat-Gay, G. Cytosolic accumulation of HPV16 E7 oligomers supports different transformation routes for the prototypic viral oncoprotein: The amyloid-cancer connection. *Int. J. Cancer* **2009**, *125*, 1902–1911. [CrossRef] [PubMed]

17. Knapp, A.A.; McManus, P.M.; Bockstall, K.; Moroianu, J. Identification of the nuclear localization and export signals of high risk HPV16 E7 oncoprotein. *Virology* **2009**, *383*, 60–68. [CrossRef] [PubMed]

18. Laurson, J.; Raj, K. Localisation of human papillomavirus 16 E7 oncoprotein changes with cell confluence. *PLoS ONE* **2011**, *6*, e21501. [CrossRef] [PubMed]

19. Tuerk, C.; Gold, L. Systematic evolution of ligands by exponential enrichment: RNA ligands to bacteriophage T4 DNA polymerase. *Science* **1990**, *249*, 505–510. [CrossRef] [PubMed]

20. Keefe, A.D.; Cload, S.T. SELEX with modified nucleotides. *Curr. Opin. Chem. Biol.* **2008**, *12*, 448–456. [CrossRef] [PubMed]

21. Blind, M.; Blank, M. Aptamer Selection Technology and Recent Advances. *Mol. Ther. Nucleic Acids* **2015**, *4*, e223. [CrossRef]

22. Belyaeva, T.A.; Nicol, C.; Cesur, O.; Travé, G.; Blair, G.E.; Stonehouse, N.J. An RNA Aptamer Targets the PDZ-Binding Motif of the HPV16 E6 Oncoprotein. *Cancers* **2014**, *6*, 1553–1569. [CrossRef] [PubMed]

23. Nicol, C.; Bunka, D.H.; Blair, G.E.; Stonehouse, N.J. Effects of single nucleotide changes on the binding and activity of RNA aptamers to human papillomavirus 16 E7 oncoprotein. *Biochem. Biophys. Res. Commun.* **2011**, *405*, 417–421. [CrossRef] [PubMed]

24. Nicol, C.; Cesur, Ö.; Forrest, S.; Belyaeva, T.A.; Bunka, D.H.; Blair, G.E.; Stonehouse, N.J. An RNA aptamer provides a novel approach for the induction of apoptosis by targeting the HPV16 E7 oncoprotein. *PLoS ONE* **2013**, *8*, e64781. [CrossRef] [PubMed]

25. Ellingham, M.; Bunka, D.H.; Rowlands, D.J.; Stonehouse, N.J. Selection and characterization of RNA aptamers to the RNA-dependent RNA polymerase from foot-and-mouth disease virus. *RNA* **2006**, *12*, 1970–1979. [CrossRef] [PubMed]

26. Forrest, S.; Lear, Z.; Herod, M.R.; Ryan, M.; Rowlands, D.J.; Stonehouse, N.J. Inhibition of the foot-and-mouth disease virus subgenomic replicon by RNA aptamers. *J. Gen. Virol.* **2014**, *95*, 2649–2657. [CrossRef] [PubMed]

27. Bellecave, P.; Cazenave, C.; Rumi, J.; Staedel, C.; Cosnefroy, O.; Andreola, M.L.; Ventura, M.; Tarrago-Litvak, L.; Astier-Gin, T. Inhibition of hepatitis C virus (HCV) RNA polymerase by DNA aptamers: Mechanism of inhibition of in vitro RNA synthesis and effect on HCV-infected cells. *Antimicrob. Agents Chemother.* **2008**, *52*, 2097–2110. [CrossRef] [PubMed]

28. Biroccio, A.; Hamm, J.; Incitti, I.; De Francesco, R.; Tomei, L. Selection of RNA aptamers that are specific and high-affinity ligands of the hepatitis C virus RNA-dependent RNA polymerase. *J. Virol.* **2002**, *76*, 3688–3696. [CrossRef] [PubMed]

29. Bates, P.J.; Kahlon, J.B.; Thomas, S.D.; Trent, J.O.; Miller, D.M. Antiproliferative activity of G-rich oligonucleotides correlates with protein binding. *J. Biol. Chem.* **1999**, *274*, 26369–26377. [CrossRef] [PubMed]

30. Sayyed, S.G.; Hägele, H.; Kulkarni, O.P.; Endlich, K.; Segerer, S.; Eulberg, D.; Klussmann, S.; Anders, H.J. Podocytes produce homeostatic chemokine stromal cell-derived factor-1/CXCL12, which contributes to glomerulosclerosis, podocyte loss and albuminuria in a mouse model of type 2 diabetes. *Diabetologia* **2009**, *52*, 2445–2454. [CrossRef] [PubMed]

31. Doble, R.; McDermott, M.F.; Cesur, O.; Stonehouse, N.J.; Wittmann, M. IL-17A RNA aptamer: Possible therapeutic potential in some cells, more than we bargained for in others? *J. Investig. Dermatol.* **2014**, *134*, 852–855. [CrossRef] [PubMed]

32. Nafissi, N.; Alqawlaq, S.; Lee, E.A.; Foldvari, M.; Spagnuolo, P.A.; Slavcev, R.A. DNA ministrings: Highly safe and effective gene delivery vectors. *Mol. Ther. Nucleic Acids* **2014**, *3*, e165. [CrossRef] [PubMed]

33. Rossignol, J.F.; La Frazia, S.; Chiappa, L.; Ciucci, A.; Santoro, M.G. Thiazolides, a new class of anti-influenza molecules targeting viral hemagglutinin at the post-translational level. *J. Biol. Chem.* **2009**, *284*, 29798–29808. [CrossRef] [PubMed]

34. Mankouri, J.; Taneja, T.K.; Smith, A.J.; Ponnambalam, S.; Sivaprasadarao, A. Kir6.2 mutations causing neonatal diabetes prevent endocytosis of ATP-sensitive potassium channels. *EMBO J.* **2006**, *25*, 4142–4151. [CrossRef] [PubMed]

35. Grant, B.D.; Donaldson, J.G. Pathways and mechanisms of endocytic recycling. *Nat. Rev. Mol. Cell Biol.* **2009**, *10*, 597–608. [CrossRef] [PubMed]

36. Greenfield, I.; Nickerson, J.; Penman, S.; Stanley, M. Human papillomavirus 16 E7 protein is associated with the nuclear matrix. *Proc. Natl. Acad. Sci. USA* **1991**, *88*, 11217–11221. [CrossRef] [PubMed]

37. Reinstein, E.; Scheffner, M.; Oren, M.; Ciechanover, A.; Schwartz, A. Degradation of the E7 human papillomavirus oncoprotein by the ubiquitin-proteasome system: Targeting via ubiquitination of the N-terminal residue. *Oncogene* **2000**, *19*, 5944–5950. [CrossRef] [PubMed]

38. Dominska, M.; Dykxhoorn, D.M. Breaking down the barriers: siRNA delivery and endosome escape. *J. Cell Sci.* **2010**, *123*, 1183–1189. [CrossRef] [PubMed]

39. Valdovinos-Torres, H.; Orozco-Morales, M.; Pedroza-Saavedra, A.; Padilla-Noriega, L.; Esquivel-Guadarrama, F.; Gutierrez-Xicotencatl, L. Different Isoforms of HPV-16 E7 Protein are Present in Cytoplasm and Nucleus. *Open Virol. J.* **2008**, *2*, 15–23. [CrossRef] [PubMed]

40. Casagrande, R.; Stern, P.; Diehn, M.; Shamu, C.; Osario, M.; Zúñiga, M.; Brown, P.O.; Ploegh, H. Degradation of proteins from the ER of S. cerevisiae requires an intact unfolded protein response pathway. *Mol. Cell* **2000**, *5*, 729–735. [CrossRef]

41. Liu, C.Y.; Kaufman, R.J. The unfolded protein response. *J. Cell Sci.* **2003**, *116*, 1861–1862. [CrossRef] [PubMed]

42. Travers, K.J.; Patil, C.K.; Wodicka, L.; Lockhart, D.J.; Weissman, J.S.; Walter, P. Functional and genomic analyses reveal an essential coordination between the unfolded protein response and ER-associated degradation. *Cell* **2000**, *101*, 249–258. [CrossRef]

Mechanisms of Cancer Cell Killing by the Adenovirus E4orf4 Protein

Tamar Kleinberger

Abstract: During adenovirus (Ad) replication the Ad E4orf4 protein regulates progression from the early to the late phase of infection. However, when E4orf4 is expressed alone outside the context of the virus it induces a non-canonical mode of programmed cell death, which feeds into known cell death pathways such as apoptosis or necrosis, depending on the cell line tested. E4orf4-induced cell death has many interesting and unique features including a higher susceptibility of cancer cells to E4orf4-induced cell killing compared with normal cells, caspase-independence, a high degree of evolutionary conservation of the signaling pathways, a link to perturbations of the cell cycle, and involvement of two distinct cell death programs, in the nucleus and in the cytoplasm. Several E4orf4-interacting proteins including its major partners, protein phosphatase 2A (PP2A) and Src family kinases, contribute to induction of cell death. The various features of E4orf4-induced cell killing as well as studies to decipher the underlying mechanisms are described here. Many explanations for the cancer specificity of E4orf4-induced cell death have been proposed, but a full understanding of the reasons for the different susceptibility of cancer and normal cells to killing by E4orf4 will require a more detailed analysis of the complex E4orf4 signaling network. An improved understanding of the mechanisms involved in this unique mode of programmed cell death may aid in design of novel E4orf4-based cancer therapeutics.

Reprinted from *Viruses*. Cite as: Tamar Kleinberger. Mechanisms of Cancer Cell Killing by the Adenovirus E4orf4 Protein. *Viruses* **2015**, *7*, 2334-2357.

1. Introduction

The adenovirus (Ad) Early region 4 open-reading-frame 4 (E4orf4) protein is a regulator of the progression of Ad infection from the early to the late phase. Its activities include down-regulation of early viral gene expression and of cellular genes affecting Ad replication [1–4], control of alternative splicing of viral mRNAs [5,6], and regulation of protein translation [4,7]. E4orf4 effects on viral gene expression also influence viral DNA replication [8,9]. Deletion of the E4orf4 sequence from the Ad genome was shown to moderately inhibit Ad replication in cells in tissue culture by only a few fold [7,10]. However, considerable homology exists between E4orf4 proteins belonging to several classes of human Ads [11], suggesting that E4orf4 function is important for the virus. The effect of E4orf4 on the efficiency of Ad infection in a whole organism is currently unknown.

The E4orf4 protein is a 14 kDa polypeptide without known non-Ad homologs. *Ab initio* modeling of the structure of E4orf4 predicted that it consists of three α-helices, as well as N- and C-terminal loops [12]. The E4orf4 protein contains a highly basic stretch of amino acids (residues 66–75), which may provide a nuclear and nucleolar targeting function [13], as well as a docking site for one of the E4orf4 partners, Src kinase [14] (Figure 1).

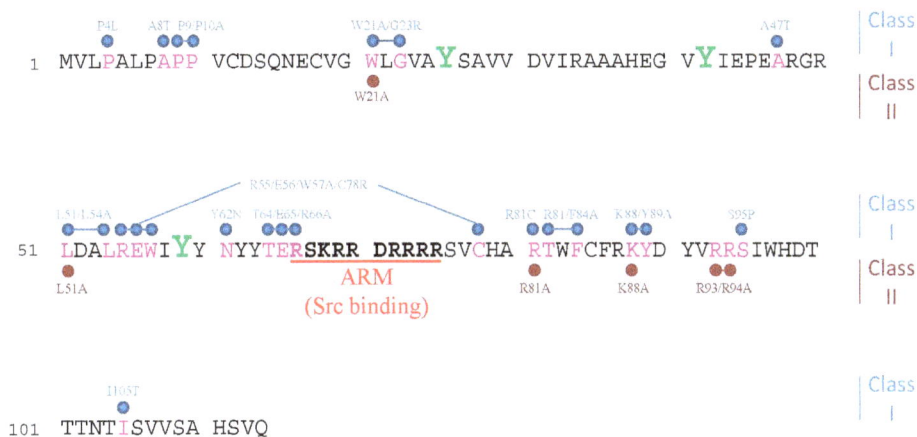

Class I
Class II

```
        P4L   A8T P9/P10A       W21A/G23R                              A47T
1   MVLPALPAPP VCDSQNECVG WLGVAYSAVV DVIRAAAHEG VYIEPEARGR
                          W21A
```

```
                 R55/E56/W57A/C78R
     L51/L54A  Y62N T64/I65/R66A        R81C R81/E84A K88/Y89A   S95P
51  LDALREWIYY NYYTERSKRR DRRRRSVCHA RTWFCFRKYD YVRRSIWHDT
    L51A              ARM          R81A      K88A    R93/R94A
                 (Src binding)
```

```
     I105T
101 TTNTISVVSA HSVQ
```

Figure 1. Mutation analysis of PP2A and Src binding sites in E4orf4. The Ad5 E4orf4 protein sequence is shown. The basic E4ARM domain is displayed in bold and underlined in red. The tyrosines that are phosphorylated by Src kinases are in bold green and a larger font. Other residues involved in E4orf4-induced cell death are marked in pink. Mutations in the E4orf4 sequence that reduced association of E4orf4 with a PP2A phosphatase activity by at least two-fold and impaired E4orf4-induced cell death ('class I' mutants [11]) are shown above the sequence in light blue. Mutations that did not reduce the E4orf4-PP2A interaction more than two-fold but were deficient in induction of cell death ('class II mutants') are shown below the sequence in dark red. Three more mutations were found to reduce PP2A binding: V19A/T102I, A25T/ΔD52/R87C and ΔV29/R81C, which are not shown for simplicity sake. The basic E4orf4 ARM domain is required for Src kinase binding, but has only a low effect on PP2A binding [14,19].

Protein phosphatase 2A is a major E4orf4 partner [15]. Phosphatases of the PP2A group are Ser/Thr phosphatases, which are involved in most cellular processes. These enzymes contain three subunits: one of two isoforms of a catalytic C subunit encoded by *PPP2CA* and *PPP2CB*, one of two isoforms of a scaffolding A subunit encoded by *PPP2R1A* and *PPP2R1B*, and one of twenty-three regulatory B subunits belonging to four unrelated gene families, each containing several isoforms (B/B55: *PPP2R2A-PPP2R2D*, B'/B56: *PPP2R5A-PPP2R5E*, B'': *PPP2R3A-PPP2R3C*, and B''': *STRN*, *STRN3*, *STRN4*). The combinatorial assembly of the various PP2A subunits yields 92 combinations of heterotrimeric enzyme complexes (ABC) and four combinations of heterodimeric complexes (AC), each possessing its own catalytic properties, substrate specificities, tissue or cell-specific expression, and subcellular localization [16]. A functional heterodimeric BC complex has also been described [17]. It has been reported that in both mammalian cells and in yeast E4orf4 associates with PP2A through an interaction with regulatory B subunits, including the mammalian B55/yeast Cdc55 and mammalian B56 subunits [11,15,18–21]. The E4orf4-PP2A complex contains all three types of PP2A subunits (ABC) and has substantial phosphatase activity [15,18,21]. All

E4orf4 activities during virus infection that are known to date require an interaction between PP2A and the viral protein [2,3,5,7,11,15].

2. The Unique Mode of E4orf4-Induced Cell Death

When E4orf4 is expressed alone, outside the context of virus infection, it induces cell death in transformed cells, which is dose-dependent [20,22], p53-independent [23–25] and possesses several interesting features described below. These include a higher susceptibility of cancer cells to E4orf4-induced cell killing compared with normal cells, participation of caspase-independent signaling pathways that are linked downstream to known cell death programs, a high degree of evolutionary conservation of the relevant signaling pathways, a link to perturbations of the cell cycle, and the involvement of at least two distinct cell death programs, in the nucleus and in the cytoplasm. In contrast, during virus infection E4orf4 does not contribute significantly to cell death for several reasons: (1) E4orf4 levels during virus infection are regulated by negative feedback [2] and may not be sufficient for induction of cell death; (2) Viral cell death-inhibitory genes may reduce premature cell killing by E4orf4; (3) During infection, E4orf4 is detected mostly in the nucleus [26], suggesting that at least the cytoplasmic cell death program induced by E4orf4 may be diminished; (4) In nature Ad infects normal, untransformed cells that may be less susceptible to E4orf4 toxicity. Moreover, E4orf4 was reported to possess a protective effect during Ad infection of non-transformed CREF cells [4].

Two major E4orf4 partners, protein phosphatase 2A (PP2A) and Src kinases, together with additional E4orf4-associating proteins, play an important role in E4orf4-induced cell death, and studies of their contribution are described in detail below (Section 3).

2.1. Cancer Specificity of E4orf4-Induced Cell Death

Experiments in primary rat embryo fibroblasts revealed that E4orf4 induced high levels of cell death when co-transfected with various oncogene combinations but not when co-transfected with an empty vector. Thus oncogene expression sensitized primary cells grown in tissue culture to killing by E4orf4 [19]. A mutant E4orf4 protein that lost the ability to bind an active PP2A also lost the ability to kill oncogene-transformed cells [19]. Consistent with these results, a later review paper described studies showing that E4orf4 induced cell death in 40 human cancer cell lines, but not in several primary human cell types derived from various tissues [27]. Unpublished results from our laboratory indicate that E4orf4 also eliminates cancer clones in the *Drosophila* eye disc more efficiently than normal clones, suggesting that E4orf4-induced cell death is cancer specific not only in tissue culture cells but also in a multicellular organism.

The basis for the differential response of normal and cancer cells to E4orf4 is not clear yet but several possible explanations have been proposed based on the nature of transformed cells and on features of E4orf4-induced cell death described below: (1) Activation of the oncogenic state leads to induction of latent apoptotic signals that are uncoupled from the basic apoptotic machinery and provide a lower threshold for activation of cell death by various signals [28]; (2) It was reported that cancer cells become addicted to crucial oncogenic pathways [29] and it may be possible that E4orf4 inhibits these pathways leading to cell death of the oncogene-addicted cells but not of normal cells; (3) E4orf4 may exploit activated oncogenes in cancer cells, such as Src, for induction of cell death (Section 3.5); (4) Cell cycle checkpoints in cancer cells are defective to some extent [30] and these cells would be more susceptible to E4orf4, which disrupts mitotic checkpoints (Section 2.5); (5) We showed in the *Drosophila* model system that E4orf4 can inhibit classical apoptosis in normal fly tissues (Section 4), and it can be hypothesized that this E4orf4 function is lost in cancer cells, leading to a more effective cell killing [31]; (6) E4orf4-induced structural changes observed in mitochondria (Section 3.5) could affect metabolic reprogramming, which may influence cancer and normal cells differentially [32]. Deciphering the mechanisms underlying E4orf4-induced cell death will facilitate a better understanding of the different susceptibility of normal and cancer cells to E4orf4 toxicity.

As E4orf4 induces a p53-independent, non-canonical programmed cell death [23–25] and a large percent of human tumors are p53-deficient [33], investigation of the unique mode of E4orf4-induced cancer cell killing may have exciting implications for cancer therapy. Although understanding of the differential sensitivity of normal and cancer cells to E4orf4-induced cell death is still minimal, several researchers began to explore the feasibility of using E4orf4-based approaches for cancer therapy [34–38]. These initial attempts to use E4orf4 to treat cancer cells *in vivo* are still very preliminary but they provide further motivation to develop basic research aimed at understanding the E4orf4 cell death network both in tissue culture cells and in animal models.

2.2. Caspase-Independent Cell Death Signaling that Feeds into Known Cell Death Pathways

Several reports have indicated that E4orf4-induced cell death is caspase-independent although crosstalk with caspase-dependent pathways also occurs [23,39–41]. Numerous lines of evidence led to these conclusions. First, addition of various broad-range caspase inhibitors did not prevent E4orf4-induced cell killing in several cell lines and in some cases no caspase activation was observed upon E4orf4 expression [23,41]. However, caspase activation was observed when E4orf4 was expressed in other types of cells [40,41]. Furthermore, in some cell lines caspase inhibition eliminated certain morphologies associated with E4orf4-induced cell death, such as accumulation of sub-G1 cells containing fragmented DNA [40,41], and in other cases caspase inhibition decreased nuclear condensation induced by E4orf4 and even increased cell survival measured by a clonogenic assay [40]. Variability in the type of caspase-dependent pathways that were induced by E4orf4 also emerged from these studies. One report described a contribution of the extrinsic apoptotic pathway including FADD/MORT1 and caspase-8 to E4orf4-induced cell death in 293T cells, but no involvement of caspase 9 [40], whereas another report described a contribution of the mitochondria-apoptosome intrinsic apoptotic pathway to E4orf4-induced DNA fragmentation in C33A cells [41].

The activation of classical caspase-dependent pathways in addition to caspase-independent cell killing may facilitate amplification of E4orf4-induced cell death. Furthermore, in addition to utilizing classical apoptosis to assist E4orf4-induced cell killing in some cell lines, it was reported that in the p53-deficient H1299 cells E4orf4 could execute cell death with necrotic characteristics through induction of mitotic catastrophe [42]. Taken together, the results suggest that E4orf4 initiates caspase-independent signaling that can be linked to various cell death pathways, such as caspase-dependent apoptosis or mitotic catastrophe-mediated necrosis. A crosstalk between caspase-dependent and independent pathways was also observed during E4orf4-induced cell death in *Drosophila melanogaster*, and both these types of cell death contributed to E4orf4-induced toxicity in the fly [22]. The mechanisms underlying the decision dictating the direction in which caspase-independent E4orf4 cell death signaling will proceed are not known, but they may depend on the physiological state of the cells and on the cell genetic or proteomic content.

2.3. Evolutionary Conservation of E4orf4 Cell Death Signaling

The major E4orf4 partners described to date are protein phosphatase 2A (PP2A) and Src kinases, and their contribution to E4orf4-induced cell death signaling is described below (Section 3). PP2A is highly conserved in evolution from yeast to mammalian cells [43], and Src kinase, although absent from yeast, is conserved from the unicellular choanoflagellate *Monosiga ovata* through the primitive multicellular sponge *Ephydatia fluviatilis* to mammals [44]. It appears that the high evolutionary conservation of the E4orf4 partners results in the ability of E4orf4 to induce its unique mode of cell death signaling in various organisms from yeast [18,45] through *Drosophila* [22] to mammalian cells (Figure 2), facilitating the use of model organisms for research on E4orf4 functions.

2.4. Morphological Hallmarks of E4orf4-Induced Cell Death and Assays for Measuring E4orf4-Induced Cell Killing

Since upstream caspase-independent events in the E4orf4-induced cell death pathway appear to be linked downstream to various types of cell death, only few morphological hallmarks of E4orf4-induced cell killing can be consistently used to assay this process. It has been shown that the most typical morphologies associated with E4orf4-induced cell death include membrane blebbing, nuclear condensation and cell detachment, whereas morphologies associated with classical apoptosis such as DNA fragmentation, caspase activation, phosphatidylserine externalization, or changes in mitochondrial membrane potential do not always accompany E4orf4-induced cell killing [23,40,42]. Therefore, the assays commonly used to quantify E4orf4-induced cell death include determination of the percentage of cells with membrane blebbing or with nuclear condensation and clonogenic cell survival assays.

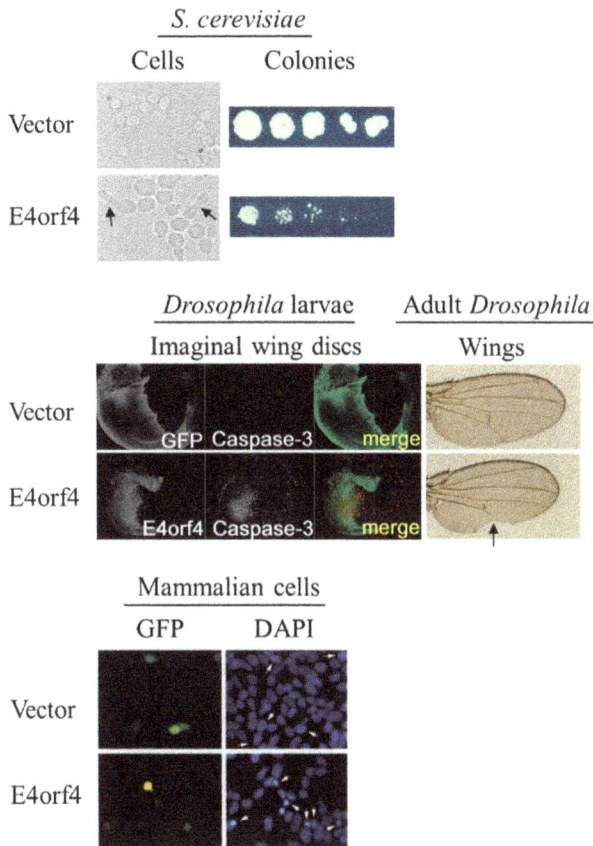

Figure 2. E4orf4-induced cell death is highly conserved in evolution. E4orf4-induced toxicity was studied in the yeast *S. cerevisiae*, in *Drosophila melanogaster* and in mammalian cells in tissue culture. The pictures shown here demonstrate E4orf4 effects in the various organisms, including budding defects in yeast (marked with arrows), reduced size of yeast colonies, caspase activation in a *Drosophila* wing disc, a wing defect in adult flies (marked with an arrow) and nuclear condensation seen by DAPI staining with cell rounding and blebbing seen by GFP staining in mammalian cells (arrows mark nuclei of transfected cells). E4orf4-induced toxicity required an interaction with PP2A in all these organisms and an interaction with Src kinases contributed to this process in flies and mammals. Studies of E4orf4 in all these model systems contributed different insights into E4orf4 biology. See the text for more details. Parts of this figure were adapted from the following sources with permissions: Top panel: [45]. Middle panel: [22]. Bottom panel: [25].

2.5. Perturbations of the Cell Cycle Precede E4orf4-Induced Cell Death

Many adenoviral genes cause alterations in the cell cycle as a consequence of their contribution to virus infection [46]. When cellular effects of E4orf4 were examined, it was noticed that induction of E4orf4 expression in HEK293-derived cell lines resulted in accumulation of cells with 4N DNA content 24 h after induction, suggesting a cell cycle arrest at G2/M. This arrest was followed at later times by an accumulation of apoptotic sub-G1 cells [25,45]. Later studies of E4orf4 in the yeast *S. cerevisiae* revealed that E4orf4 induced an irreversible growth arrest in this organism, which was also associated with G2/M arrest [18,45]. Visualization of spindles with GFP-marked tubulin demonstrated that most E4orf4-expressing yeast cells were arrested either with short pre-anaphase spindles or with extended telophase spindles [45]. At least part of the inhibitory effect of E4orf4 on cell cycle progression in yeast could be attributed to its interaction with the spindle checkpoint containing the anaphase-promoting complex/cyclosome (APC/C) and recruitment of PP2A to this complex [45]. APC/C is an E3 ubiquitin ligase, which marks cell cycle regulators for degradation. It associates with various activating subunits at specific stages of the cell cycle, including Cdc20 early in mitosis and Cdh1 in later mitosis and G1 [47]. Ase1 and Pds1 are substrates of APC/C^{Cdh1} and APC/C^{Cdc20} respectively. It was shown that E4orf4 stabilized Ase1 and increased Pds1 levels [45], suggesting that it inhibited both APC/C complexes. Furthermore, benomyl, a microtubule depolymerizing drug that transiently triggers the spindle assembly checkpoint by inhibiting APC/C^{Cdc20} sensitized yeast cells to E4orf4 toxicity [45]. This observation is consistent with inhibition of APC/C^{Cdc20} by E4orf4, as benomyl may exacerbate cell cycle arrest when APC/C is already partially inhibited by E4orf4. Moreover, using the power of yeast genetics it was shown that mutations in APC/C and its co-activators Cdc20 and Cdh1 displayed synthetic lethality with E4orf4 expression [45], further demonstrating that APC/C inhibition increased E4orf4 toxicity.

Additional cell cycle mutants that manifested synthetic lethality with E4orf4 expression in yeast contained mutations that reduced activity of Cdc28, the yeast cell cycle-dependent kinase 1 (Cdk1) [45]. Furthermore, E4orf4 increased Histone H1 phosphorylation by a Cdc28 complex containing the mitotic Clb2 cyclin and this effect was dependent on the yeast B55 subunit of PP2A, Cdc55, and on the Cdc28 phosphatase Mih-1 [18,45]. It was suggested that E4orf4 may partially counteract its own cell cycle inhibitory effect by stimulating Cdc28, an activator of APC/C [48,49], as part of a multifaceted mode of regulation [45].

In contrast to the results indicating inhibition of APC/C^{Cdc20}, it was reported that when yeast cells were arrested in S-phase by hydroxyurea treatment, APC/C^{Cdc20} was prematurely activated by E4orf4 and Pds1 levels were reduced, as were the levels of Scc1 whose stability depends on the presence of Pds1. The activity of APC/C^{Cdh1} was not altered under these conditions. Premature activation of APC/C^{Cdc20} required the E4orf4 interaction with PP2A [50]. How may these seemingly contradictory results describing both inhibition and activation of APC/C^{Cdc20} be reconciled? A possible explanation could involve the existence of different potential targets of PP2A in the APC/C complex and in its modulators that may be targeted differentially by the E4orf4-PP2A complex in S-phase and mitosis. For example, the APC/C inhibitor Emi1, which is itself inhibited by Cdk1 phosphorylation [51] could be such a target. Although Emi1 is not normally phosphorylated during S-phase, a Cdc55-dependent E4orf4-induced Cdk1 activation [18,45,50] may lead to Emi1 phosphorylation in S-phase followed by premature APC/C activation. On the other hand, during mitosis, Cdc55-dependent inhibition of APC/C may be more prominent [45], while Emi1 is degraded at mitotic entry [52,53]. However, whereas inhibition of APC/C complexes may account at least in part for the E4orf4-induced G2/M arrest, it is not clear how premature activation of APC/C^{Cdc20} in S-phase assists in this process. Thus a better understanding of the interaction between E4orf4 and the cell cycle requires more study.

The delay in the cell cycle and M-phase perturbations caused by E4orf4 may result in cells with improper chromosome numbers returning to the cell cycle, leading to cell death. It was indeed shown recently that E4orf4 expression in H1299 cells led to accumulation of cells with 4N DNA content which contained high levels of Cyclin E but not of mitotic markers such as Cyclin B1 or pSer10 Histone H3. It was suggested that these cells were in the G1 phase of the cell cycle rather than in G2/M and that they progressed from mitosis to G1 in the absence of cytokinesis [54]. These results were further confirmed by time-lapse microscopy experiments demonstrating that E4orf4 slowed down dramatically the transit of H1299 cells through mitosis and caused a delay or failure of cytokinesis [55]. E4orf4-induced perturbation of cytokinesis was also observed in yeast cells, which displayed abnormal budding [17,18,45] (Figure 2). Both the tetraploid H1299 cells and cells with 2N DNA content found in the E4orf4-expressing cell population proceeded to undergo cell death, as measured by uptake of propidium iodide. When arrested at G0/G1 and then released into the cell cycle, E4orf4-expressing cells also manifested a deficiency in initiation of DNA replication [54]. DNA replication may be affected by E4orf4 through inhibition of APC/C^{Cdh1}, as this complex is involved in degradation of proteins that control DNA replication, or through targeting of PP2A to other substrates involved in regulation of DNA replication [56,57].

The findings portrayed above, demonstrating that the activity of Cdc28 in complex with mitotic cyclins was increased in E4orf4-expressing yeast cells whereas E4orf4-expressing H1299 cells contained high levels of Cyclin E but not of the mitotic Cyclin B1 [18,45,54] indicate that these events may depend on the type of cell expressing E4orf4. However, taken together, the results indicate that E4orf4 targets various cell cycle regulators and that perturbation of the normal cell cycle may lead to cell death. A link between cell cycle disruption and induction of cell death was shown previously, using drugs or other viral genes that induced both cell cycle arrest and cell death (for example, [58–61]), but the exact mechanisms underlying this link have not been fully defined yet. Mitotic catastrophe was suggested as one mechanism that can account for the link between cell cycle perturbations and cell death [62,63]. Furthermore, mitotic catastrophe can lead to various modes of cell death, such as apoptosis and necrosis, as well as to senescence, depending on the genetic identity of the cells and the physiological conditions [64]. Besides mitotic catastrophe, other distinct mechanisms can remove cells that failed at later mitotic stages and progressed to G1, such as outcompeting cells with an abnormal number of chromosomes, a condition that leads to reduced fitness [65]. Since E4orf4 interacts with other nuclear proteins which contribute to induction of cell death, besides direct cell cycle regulators [57], and because it also induces a cytoplasmic cell death program (Section 2.6), additional cell death pathways are also likely to be important for E4orf4 toxicity.

2.6. E4orf4 Induces Nuclear and Cytoplasmic Cell Death Programs

During virus infection, GFP-tagged E4orf4 was described to concentrate in the nucleus but its localization did not overlap viral replication centers. Only very little E4orf4 staining was found in the cytoplasm [26]. The localization of native E4orf4 during virus replication was not published. When E4orf4 expression was induced outside the context of virus infection, several studies in various cell lines revealed that it first accumulated in the nucleus but was later detected in the cytoplasm, cytoskeleton and membrane regions, with a tendency to concentrate in membrane blebs [1,13,41,66]. Nuclear localization of E4orf4 required a highly basic arginine-rich motif (residues 66–75) named E4ARM [13] (Figure 1). This motif was reported to target GFP-tagged E4orf4 to nucleoli as well [13], although native E4orf4 was not observed to concentrate there [67].

When expressed alone, the shift of E4orf4 accumulation from nuclear to non-nuclear cellular locations was shown to depend on the interaction of E4orf4 with Src family kinases and on Tyr-phosphorylation of E4orf4 by these kinases [66,68]. Co-expression with constitutively active Src kinase increased E4orf4 accumulation in the cytoplasm, in membrane blebs and in a triton-insoluble cytoskeleton protein fraction [66]. Substitution of E4orf4 Tyr residues, which are normally phosphorylated by Src kinases, to phenylalanine led to E4orf4 nuclear localization, whereas mimicking phosphorylation by replacing Tyr residues with glutamic acid enhanced E4orf4 accumulation outside the nucleus [68]. Based on these results, it was suggested that E4orf4 phosphorylation by Src may result in changes to E4orf4 conformation in such a way that would promote its retention outside the nucleus. It was reported that leptomycin B, an inhibitor of CRM1-mediated export from the nucleus did not affect E4orf4 cytoplasmic accumulation, but inhibition of the myosin II motor by blebbistatin induced E4orf4 nuclear retention [41]. These results imply that the myosin II motor, but not CRM1-dependent nuclear export, is required for the accumulation of E4orf4 outside the nucleus.

Because E4orf4 shifted with time from a nuclear to a cytoplasmic-membranal localization, experiments were performed to determine the relative contribution of nuclear, cytoplasmic, and membranal E4orf4 to cell death signaling [41]. Targeting GFP-tagged E4orf4 to the cytoplasm (using a nuclear export sequence) or to membranes (using a myristoylation signal or a CAAX box) reconstructed efficiently the contribution of Src to E4orf4-induced cell death, inducing membrane blebbing and nuclear condensation at least as competently as the WT GFP-E4orf4 protein. Nuclear targeting of GFP-E4orf4 (using a nuclear localization sequence) prevented the early appearance of membrane blebbing and delayed the appearance of nuclear condensation. It was further shown, using E4orf4 Tyr substitutions, that E4orf4 phosphorylation by Src was required for the cytoplasmic morphologies associated with E4orf4-induced cell death, such as membrane blebbing, but had a much smaller effect on nuclear condensation [68]. Analysis of additional E4orf4 mutants revealed that E4orf4 binding to its major partners, PP2A and Src kinases, affected differentially the nuclear and cytoplasmic forms of E4orf4-induced cell death [14]. Based on these results it was suggested that E4orf4 induced two distinct cell death programs, a nuclear program supported by the interaction with PP2A and a cytoplasmic program supported by the interaction with Src kinases. Caspase activation occurred, but was dispensable for the execution of both cell death programs, whereas calpains were required for the Src-mediated cytoplasmic death signal. It was further shown that the relative contribution of the two branches of E4orf4-induced cell death differed in various cell lines [41]. However, it is still not absolutely clear whether E4orf4 activates two independent cell death programs or whether initial E4orf4 functions in the nucleus contribute to activation of both pathways. Thus, artificial targeting of E4orf4 to various cellular sites was not absolute and the residual presence of E4orf4 in the nucleus may have been sufficient to perform the required preliminary nuclear functions that may potentially be needed to trigger both branches of the death program. Likewise, although mutants that did not bind PP2A were still able to induce cytoplasmic cell death signaling [14], it is possible that residual PP2A binding was sufficient to initiate it. Although E4orf4-induced cell death was shown to be dose-dependent [20,22], it was not determined yet whether both PP2A- and Src-dependent death programs required high E4orf4 levels or whether even low levels of interaction with PP2A were enough to initiate cell death signaling.

3. Mechanisms Underlying the Contribution of E4orf4-Associating Proteins to E4orf4-Induced Cell Death

3.1. PP2A

Early studies of E4orf4-induced cell death addressed the question whether the interaction with PP2A was required for this process. Mutation analyses of E4orf4 (Figure 1) revealed that alterations in several residues significantly reduced PP2A binding and also extensively decreased E4orf4 toxicity [11,19]. These mutants were classified as class I mutants. On the other hand, a second class of E4orf4 mutants, class II, was also described, which contained mutants that bound PP2A at more than 50% WT efficiency but their ability to induce cell death was significantly impaired [11]. It was suggested that the E4orf4-PP2A interaction of class II mutants was different than the WT interaction and could not exert a biological function. It is also possible that these mutants failed to bind E4orf4-PP2A substrates or that they failed to bind another protein that might also contribute to induction of cell death. A more direct support for a role of the PP2A-B55 subunit, which mediates the PP2A-E4orf4 interaction, in E4orf4-induced cell death came from the findings that a B55 antisense construct that reduced B55 expression also inhibited E4orf4- but not p53-induced cell death [19] and that B55 overexpression increased E4orf4 toxicity [12,21,45]. Similarly, in the yeast *S. cerevisiae*, deletion of *cdc55*, encoding the yeast PP2A B55 subunit, or of *tpd3*, encoding the PP2A-A subunit, reduced dramatically E4orf4 toxicity [18,45]. Although E4orf4 was reported to bind not just the B55 subunit of PP2A but also the B56 subunit [21], only PP2A complexes containing the B55 subunit contributed to induction of cell death [18,21,45]. It appears that the complete PP2A holoenzyme plays a role in E4orf4 toxicity as it was reported that an E4orf4 mutant (S95P) that bound the B55 subunit alone, possibly displacing the A and C subunits from the complex, acted as a dominant negative mutant, reducing cell death induced by WT E4orf4 [21]. On the other hand, based on work in yeast, it was suggested that although the PP2A-C subunit contributes to E4orf4-mediated toxicity, the Cdc55 subunit may provide an additional contribution to this process that is independent of stable complex formation with the PP2A-C subunit [69]. However, Cdc55 did not appear to mediate E4orf4 toxicity through a Tpd3-independent pathway [69]. The nature of the PP2A-C-subunit-independent role of the Cdc55 subunit is not clear, although it was suggested that Cdc55 may form a heterotrimeric phosphatase with Tpd3 and with another PP2A-like catalytic subunit, Sit4, a yeast PP6 orthologue. Remarkably, deletion of Sit4 in yeast and knockdown of PP6 in mammalian cells increased E4orf4-mediated cell death. However, a physical interaction between E4orf4 and Sit4 could not be detected [69].

It was previously shown that the interaction of E4orf4 with an active PP2A was required for various E4orf4 functions during virus infection including down-regulation of gene expression and regulation of alternative splicing and that the PP2A inhibitor okadaic acid inhibited these E4orf4 functions [6,15]. However, the nature of the molecular consequences of the E4orf4-PP2A interaction that lead to cell death is controversial at present and two possibilities have been suggested. One suggestion indicates that upon overexpression, E4orf4 sequesters PP2A complexes containing the B55 subunit and prevents dephosphorylation of substrates that are required for cell survival [70]. Another hypothesis proposes that E4orf4 targets PP2A to novel substrates thus disrupting normal cellular regulation, creating optimal conditions for virus replication but leading to cell death in the absence of virus functions that may prevent premature cell killing.

The first hypothesis was based on the following findings: (1) Some, but not all, non-physiological substrates of the E4orf4-PP2A complex that were tested have been dephosphorylated *in vitro* less efficiently by an E4orf4-PP2A complex immunoprecipitated with an E4orf4-specific antibody than by a PP2A complex immunoprecipitated with HA-B55-specific antibodies in the absence of E4orf4 [20]. However, as acknowledged by the authors, E4orf4 associates with many PP2A complexes containing different B55 or B56 isoforms [20,21] and therefore this indirect comparison could reflect altered substrate specificity of the various holoenzymes associating with E4orf4; (2) Although E4orf4 expression in the context of the virus was associated with hypophosphorylation of some viral and cellular proteins [4,6], when E4orf4 was expressed alone hyperphosphorylation of some PP2A candidate substrates was observed [7,20]. However, this effect was not shown to be direct and may occur through activation of kinases, such as mTOR [7]; (3) Treatment of E4orf4-expressing cells with the PP2A inhibitors okadaic acid or I_1^{PP2A} enhanced E4orf4-induced cell death [20]. However, okadaic acid can inhibit other PP2A-like phosphatases at least as efficiently as PP2A itself [71], and the I_1^{PP2A} inhibitor was shown to associate with PP1 and modify its substrate specificity [72], indicating that these inhibitors may have off-target effects that are not dependent on PP2A. Indeed, as described above, reduced expression of PP6, a PP2A-like phosphatase, was shown to increase E4orf4-induced cell death [69], and PP6 is inhibited by okadaic acid even more efficiently than PP2A [73]. These observations suggest that the increase in E4orf4 toxicity that was reported to be caused by okadaic acid treatment may stem from PP6 inhibition; (4) The E4orf4-binding site in the PP2A-B55 subunit was mapped by bioinformatics predictions and mutation analyses [12,70]. These studies revealed that residues located on two sides of a putative substrate binding groove described in the PP2A-B55 structure [74] contributed to E4orf4 binding. Consequently it was suggested that E4orf4 competes with PP2A-B55 substrates for binding to the substrate binding site and prevents their dephosphorylation [70]. In support of this conclusion it was demonstrated that the presence of E4orf4 led to inhibition of p107 binding to B55 and to its hyperphosphorylation. On the other hand, only part of the B55 residues reported to be important for dephosphorylation of the PP2A-B55 substrate, Tau, were shown to be necessary for E4orf4 binding. This suggests that the binding sites for E4orf4 and PP2A substrates only partially overlap. Furthermore, during virus infection E4orf4 was shown to recruit PP2A-B55 to new substrates, such as a subgroup of the SR splicing factors [5,70], indicating that while E4orf4 may prevent accessibility of some substrates to PP2A, it also recruits novel substrates to this enzyme. In addition, the hypothesis that upon overexpression E4orf4 sequesters PP2A holoenzymes containing B55 and prevents dephosphorylation of important substrates, thus, leading to cell death, appears to contradict additional findings reported previously. First, overexpression of B55 increased E4orf4-induced cell death [12,21], while the sequestration hypothesis would predict that B55 overexpression would reduce cell killing. Second, knockdown of mammalian B55 and deletion of yeast Cdc55 inhibited E4orf4-induced cell killing instead of increasing it [18,21,45]. Therefore, a second hypothesis was proposed to explain PP2A-dependent E4orf4-induced cell death. This hypothesis postulates that E4orf4 recruits PP2A to novel substrates or enhances the affinity of substrate binding to PP2A, leading to an imbalance in normal cellular pathways and to cell death. In support of this hypothesis, overexpressed E4orf4, outside the context of virus infection, was reported to recruit PP2A to substrates such as the ACF chromatin remodeling factor in mammalian cells ([57], Section 3.2) and the mitotic regulator APC/C in yeast ([45], Section 3.3), both of which contributed to E4orf4 toxicity. Nevertheless, the possibility of a combined scenario may also be considered, in which the dephosphorylation of diverse proteins by PP2A is influenced in various ways by E4orf4 and the integrated effects of altered phosphorylation of these proteins lead to cell death. To further examine the proposed mechanisms underlying PP2A-dependent E4orf4 toxicity, direct E4orf4-PP2A targets must be identified and their contribution to the death process will have to be determined.

3.2. The ATP-Dependent Chromatin Remodeling Factor, ACF

ACF is an ATP-dependent chromatin remodeling factor consisting of the SNF2h ATPase and the Acf1 regulatory subunit [56]. Both subunits were found to co-immunoprecipitate with E4orf4, and the PP2A C subunit associated with ACF in the presence of WT E4orf4 but not in its absence or in the presence of a mutant E4orf4 protein that could not bind PP2A [57]. Moreover, when chromatin proteins were fractionated by extraction in increasing salt concentrations, the PP2A-B55 subunit was enriched in higher salt fractions in the presence of E4orf4 compared to its distribution in the absence of E4orf4 or in the presence of the mutant E4orf4 protein. These results suggest that E4orf4 alters the chromatin-binding properties of PP2A, possibly by recruiting it to ACF. It was also demonstrated that the ACF complex was involved in E4orf4-induced cell death. Reduced SNF2h expression or expression of a dominant-negative catalytically-inactive SNF2h mutant inhibited E4orf4-induced cell death whereas, surprisingly, knockdown of Acf1 increased cell death. Knockdown of an Acf1 homolog, WSTF, which also associates with SNF2h, inhibited E4orf4-induced toxicity. The hypothesis proposed to explain these results inferred that the E4orf4-PP2A complex inhibited ACF and facilitated enhanced chromatin remodeling activities of other SNF2h-containing complexes, such as WSTF-SNF2h. An E4orf4-induced switch in chromatin remodeling may determine life *versus* death decisions [57] (Figure 3). This hypothesis predicts phosphorylation changes in ACF or ACF-associating proteins that could be important to E4orf4 function, but such alterations have not been reported yet. It is not likely, however, that E4orf4 recruits PP2A to a new substrate like ACF, with the sole purpose of inhibiting PP2A activity towards it.

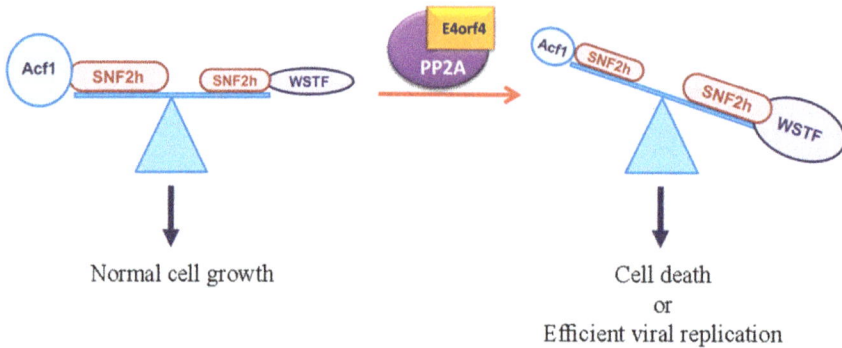

Figure 3. A working model of E4orf4 function in chromatin. Various SNF2h-containing complexes participate in chromatin remodeling and affect transcription, DNA replication, DNA repair, *etc.* The model shown in this figure proposes that E4orf4 together with PP2A inhibits the ACF chromatin remodeler. This inhibition leads to a shift in the balance between various SNF2h-containing remodeling complexes, allowing, for example, more activity of a WSTF-SNF2h complex. The variation in chromatin remodeling activity alters chromatin structure and induces changes in transcription, DNA replication, DNA repair, or other processes that require remodeling. These events contribute to E4orf4 functions during virus infection and lead to cell death when E4orf4 is expressed alone. This figure is adapted from [57] with permission of Oxford University Press.

3.3. The APC/C Ubiquitin Ligase Complex

A second potential E4orf4-PP2A target is the mitotic regulator APC/C. As described above (Section 2.5), work in yeast revealed that E4orf4 recruited PP2A to the APC/C complex and affected degradation of APC/C substrates [45,50]. As a result of the physical and functional interactions between E4orf4 and the APC/C, E4orf4 caused perturbations in the cell cycle which progressed to cell death. Genetic experiments revealed a functional interaction between the yeast B55 subunit, Cdc55, and the APC/C activating subunits Cdc20 and Cdh1 even in the absence of E4orf4 [45,75,76]. It was also shown that a PP2A complex containing the Cdc55 subunit dephosphorylated APC/C subunits [77]. It is possible therefore that E4orf4 may stabilize an existing transient interaction between PP2A and the APC/C [45]. It was reported that E4orf4 increased nuclear accumulation of Cdc55 [50], thus possibly enhancing the targeting of nuclear substrates, such as the APC/C by PP2A holoenzymes containing the B55/Cdc55 subunit. However, phosphorylation sites in the APC/C complex or its activating subunits, which may be affected by the E4orf4-PP2A complex, have not been identified to date. Interestingly, APC/C was found to be a target of many viral proteins and some of these proteins cause cytotoxicity specifically in tumor cells, providing evidence that targeting the APC/C could be exploited to selectively eliminate cancer cells [78].

3.4. The Golgi UDPase, Ynd1

Inducible expression of E4orf4 in the yeast *S. cerevisiae* was shown to result in PP2A-dependent irreversible growth arrest [18,45], suggesting that at least part of the E4orf4 signaling network responsible for induction of cell death is conserved from yeast to mammals. A further confirmation of the high degree of evolutionary conservation of E4orf4-induced cell death came from a genetic screen in yeast for E4orf4 mutants that lost their toxicity. These mutants were shown to have impaired abilities to bind PP2A and to induce cell death in mammalian cells [79]. Thus the results suggest that E4orf4-induced toxicity in yeast reflects accurately E4orf4-induced cell death in mammalian cells and therefore yeast may serve as a powerful genetic model system for the study of E4orf4-induced cell killing. Studies in this model organism revealed that not all of E4orf4-induced toxicity was generated via a B55/Cdc55-dependent pathway. Although much of the E4orf4 toxicity was abrogated in yeast cells lacking Cdc55, some residual growth inhibitory effect of E4orf4 could be detected [18,76]. Moreover, E4orf4 mutants that were deficient in PP2A binding could elicit low levels of toxicity [18]. A yeast screen for mediators of E4orf4 toxicity indeed demonstrated that a second E4orf4 partner, the Golgi apyrase Ynd1, contributed to E4orf4-induced irreversible growth arrest, and the contributions of Cdc55 and Ynd1 to this process were additive, indicating that they were not part of the same pathway [76].

Apyrases are nucleoside triphosphate diphosphohydrolases that hydrolyze a variety of nucleoside 5'-triphosphates and 5' diphosphates with different nucleotide preferences. Most members of this family are integral membrane glycoproteins, and their catalytic sites face the extracellular medium or the lumen of intracellular organelles [80]. Ynd1 is a Golgi apyrase whose enzymatic activity is required for regulation of nucleotide-sugar import into the Golgi lumen. This protein is inserted in the Golgi membrane, its 500 N-terminal amino acids, including its catalytic domain, are located in the Golgi lumen, whereas its 113 C-terminal residues are found at the cytoplasmic face of the Golgi membrane [81,82].

Although Cdc55 and Ynd1 had an additive effect on E4orf4 toxicity, there was also a functional interaction between them, since deletion of *YND1* sensitized the cells to killing by E4orf4 in the presence of over-expressed *CDC55*. It was further shown that the Ynd1 and Cdc55 proteins interacted physically [76]. The physical and functional interactions between Ynd1 and Cdc55 together with their additive effects on E4orf4 toxicity are consistent with the hypothesis that the Ynd1-mediated E4orf4 function has PP2A-dependent as well as PP2A-independent aspects. Investigation of the role of mammalian Ynd1, Golgi UDPase, in E4orf4-induced cell death supported this possibility. On the one hand, E4orf4 was shown to stabilize the UDPase protein in a PP2A-dependent manner and the rise in UDPase levels increased E4orf4-induced cell death, whereas reduced UDPase expression inhibited E4orf4 toxicity. On the other hand, E4orf4 disrupted large molecular weight protein complexes containing the Golgi UDPase in a PP2A-independent manner [83]. A functional interaction between *YND1* and the APC/C activators, *CDC20* and *CDH1* was also observed, although the nature of this link is not currently understood [76].

Surprisingly, the Ynd1 cytosolic tail was found to mediate E4orf4-induced toxicity whereas the apyrase activity was dispensable for this process [76,84]. It was suggested that the Ynd1 cytosolic tail could potentially mediate E4orf4 toxicity by acting as a scaffold for a multi-protein complex which is targeted and disrupted by E4orf4, releasing some of its protein components [84]. The *Saccharomyces* Genome Database cites at least 10 membrane proteins, which physically associate with Ynd1, probably through its cytosolic tail. Six of these proteins associate with each other and may all be part of a large protein complex. Some of them participate in both early and late secretory pathways in yeast, and it is conceivable that E4orf4 targets a secretory protein complex anchored to the Ynd1 cytosolic tail. It was shown that E4orf4 affected protein trafficking in mammalian cells through its association with Src kinases ([85], Section 3.5). It is possible that in yeast cells, which lack Src family members, E4orf4 may utilize a backup mechanism which allows it to interact directly with components of the secretory pathway and to influence protein trafficking, thus further transducing its toxic signal [84]. This mechanism is likely used by E4orf4 in mammalian cells as well.

3.5. Src Family Kinases (SFKs)

When expressed alone, E4orf4 was found to associate with SFKs [14,66] but this interaction was not reported during virus infection. Absence or low levels of E4orf4-Src interaction during virus infection may stem from the inaccessibility of E4orf4 to Src, as E4orf4 is mostly nuclear in virus-infected cells [26]. E4orf4 interacts with less than 1-5% of endogenous Src proteins, depending on the cell type. This interaction is mediated by the kinase domain of Src and the E4orf4 ARM domain [14]. The Src-binding domain of E4orf4 slightly overlaps with the PP2A binding domain, but E4orf4 complexes containing both PP2A and Src were detected, demonstrating that binding of these proteins to E4orf4 was not mutually exclusive [14]. Src binding is required for the cytoplasmic cell death program induced by E4orf4 [14]. The interaction of E4orf4 with SFKs leads to alterations in both protein partners. E4orf4 is phosphorylated on tyrosine residues within motifs that are typical of SFK substrates, including Tyr-26, -42, and -59 and these phosphorylation events affect E4orf4 cellular localization and cell death signaling [68]. Conversely, WT E4orf4 protein, but not E4orf4 mutants that are unable to bind SFKs, modulates SFK signaling. Thus, E4orf4 inhibits phosphorylation of substrates that are involved in SFK signaling to survival pathways, such as focal adhesion kinase (FAK) and ERK [14,66], and on the other hand it promotes SFK phosphorylation of substrates that are involved in regulation of actin dynamics, such as the F-actin-binding protein cortactin, JNK, and myosin light chain [14,66,86]. As a result, E4orf4 appears to recruit SFK signaling to stimulate actin-dependent membrane remodeling, leading to cell death through multiple mechanisms [87].

Following phosphorylation by SFKs, E4orf4 accumulates at cytoplasmic sites and induces the assembly of a juxtanuclear actin-myosin network. The juxtanuclear actin-myosin structures are connected to the cell cortex by actin cables, which converge on enlarged focal adhesions that reflect an increase in cell tension, [86,87]. Activation of the myosin II motor and the juxtanuclear increase in actin-myosin-based contraction drives E4orf4-induced blebbing, an early characteristic morphology that accompanies the E4orf4 cell death process [66].

The dramatic actin remodeling induced by E4orf4 requires two major pathways involving the family of Rho GTPases. The first is a Src-Rho-Rho kinase (ROCK) signaling pathway that activates JNK to phosphorylate Paxillin. Paxillin phosphorylation leads to deregulation of adhesion dynamics and disrupts tension homeostasis in the cell [88]. The second pathway consists of Cdc42, N-Wasp, and the Arp2/3 complex, and promotes actin polymerization on internal membranes and alterations in recycling endosome (RE) dynamics [86]. REs are a heterogeneous population of tubulovesicular endosomes usually concentrated in the pericentriolar region, which form the endocytic recycling compartment (ERC). RE trafficking is regulated by the small GTPase Rab11 and is involved in retrieval of internalized membranes and signaling molecules to the plasma membrane or to the trans-Golgi via retrograde membrane transport. Both pathways involving the Rho GTPases cooperate to transduce the cytoplasmic death-promoting activity of E4orf4, which acts via actin remodeling [86]. Indeed, inhibition of actin polymerization by low doses of drugs such as cytochalasin D reduced E4orf4-induced cell death, indicating a causal role for actin dynamics in this process [66,86].

A more detailed investigation of the role of actin remodeling and RE trafficking in E4orf4 death signaling [85] revealed that in the early stages of E4orf4-induced cell death, some SFKs, Cdc42, and actin interfered with the organization of the endocytic recycling compartment and enhanced RE transport to the Golgi apparatus while decreasing the recycling of protein cargos back to the plasma membrane. The resulting changes in Golgi membrane dynamics required actin-regulated Rab11a membrane trafficking and triggered scattering of Golgi membranes. This process functionally contributed to the progression of cell death as shown by the findings that Rab11a knockdown inhibited cell death as did knockdown of syntaxin-6, a TGN trafficking factor involved in fusion of RE with Golgi membranes, or overexpression of golgin-160, a Golgi matrix protein associated with Golgi dynamic changes during apoptosis. Thus, E4orf4 acts by recruiting SFK signaling to promote transport of REs to the Golgi where they may facilitate the dynamic rearrangement of membranes. It was suggested that E4orf4-induced assembly of the perinuclear actin network may be the outcome of polarized membrane traffic, which may facilitate delivery of actin-remodeling factors [85]. It is currently not known how fission of Golgi membranes brings about cell death, although an appealing suggestion was made whereby factors released from the Golgi may contribute to this process, similarly to factors released from the mitochondria [85].

In addition to its effects in the Golgi apparatus, E4orf4 was reported to cause dramatic changes in the morphology of mitochondria and to stimulate their mobilization to the polarized actin network using SFK- and Rab11a-dependent mechanisms. Based on these findings it was proposed that Rab11a may be responsible for regulation of both polarized membrane trafficking and mitochondrial dynamics during rearrangement of the cell to coordinate organelle functions with cytoskeletal dynamics [32]. It was recently realized that oxidative phosphorylation in the mitochondria plays a role in ATP production in cancer cells, as well as in proliferating cells and not just in quiescent cells [89]. The E4orf4-induced structural changes in mitochondria may lead to metabolic reprogramming which could contribute to the E4orf4-induced cell death process.

4. E4orf4-Induced Cell Death in an Animal Model

Because the mechanisms underlying PP2A-dependent E4orf4-induced toxicity appeared to be highly conserved in evolution from yeast to mammals, the effect of E4orf4 in a multicellular organism was investigated in another model organism, *Drosophila melanogaster*, which provides many genetic tools that facilitate the efficient investigation of regulatory pathways. The study of E4orf4 in *Drosophila* revealed that the characteristics of E4orf4-induced cell death in the fly were very similar to those in mammalian cells [22] (Figure 2). Thus in normal *Drosophila* tissues, E4orf4 induced low levels of cell killing, caused by both caspase-dependent and –independent mechanisms. *Drosophila* PP2A-B55 (*twins/abnormal anaphase resolution*) and Src64B contributed additively to this form of cell-death. However, this study, which addressed for the first time the consequences of E4orf4 expression in a whole multicellular organism, revealed a surprising finding: E4orf4 not only induced cell death but also inhibited classical apoptosis induced by the fly proapoptotic genes *reaper (rpr)*, *head involution defective (hid)*, and *grim*. E4orf4 also inhibited cell death induced by JNK signaling. However, whereas inhibition of *rpr*, *hid* and *grim* partially reduced cell killing, JNK inhibition did not diminish E4orf4-induced toxicity and even enhanced it. These results indicate that E4orf4-induced cell killing is a distinctive form of cell death that differs from classical cell death pathways induced by *rpr/hid/grim* or JNK signaling. Although E4orf4 appeared to inhibit JNK signaling in *Drosophila* [22], it was reported to activate JNK in certain transformed mammalian tissue culture cells [14,88]. This apparent contradiction can be explained by findings described previously, demonstrating that JNK signaling can be highly dependent on cellular context and on the nature of the stimulus [90–92]. Alterations of these conditions may impact the interaction between E4orf4 and the JNK pathway [22]. Future studies will have to determine whether JNK inhibition or activation by E4orf4 depend on the tumorigenic state of the cells, on the cell environment (monolayer or tissue) or on the type of organism studied.

The combination of both induction and inhibition of cell death by E4orf4 which was observed in normal *Drosophila* resulted in minor effects, leading to minimal tissue damage [22]. However, unpublished results from our laboratory indicated that expression of E4orf4 in cancer clones that were induced in larval eye discs by a combination of an activated oncogene and deletion of a tumor suppressor led to a dramatic reduction in cancer clone size and to significantly enhanced survival of adult flies that would not have emerged in the absence of E4orf4. These findings suggest one possible explanation for the differential effect of E4orf4 in normal and cancer cells. It can be hypothesized that E4orf4 is unable to inhibit apoptosis in cancer cells, thus inducing higher cell death levels [31]. This hypothesis is one of several possible explanations for enhanced cell killing by E4orf4 in cancer cells, which will have to be tested in the future (Section 2.1). As described above, E4orf4-mediated protection from toxicity was also observed in untransformed mammalian cells, albeit in the context of virus infection [4].

5. Perspectives

E4orf4 research has progressed significantly in the past several years as summarized in Figure 4. Several mechanisms appear to contribute additively to E4orf4-induced cell death although their relative importance in this process may vary in different cells. However, many questions remain unanswered. Two major E4orf4 partners, PP2A and Src kinases were identified and appear to provide much of the contribution to E4orf4-induced cell death. However, several other E4orf4 partners, such as the Ynd1 Golgi UDPase, may make additional contributions. Downstream effectors of both PP2A and Src have been identified but a detailed analysis of the connectivity of the E4orf4 network is still lacking. Thus, for example, what are the direct targets of the E4orf4-PP2A complex in chromatin and in the APC/C and how exactly does their altered phosphorylation influence E4orf4-induced cell death? What is the molecular nature of the contribution of changes in protein trafficking and Golgi membrane disruption to nuclear condensation and cell death? Is there a crosstalk between PP2A and Src kinases during the E4orf4 cell death process? What is the contribution of Ynd1 and of other as yet undescribed E4orf4-associating proteins to induction of cell death? These are just a few examples of questions that remain open. Furthermore, our as yet unpublished results demonstrate that E4orf4 contributes to inhibition of the cell DNA damage response by Ad, and this function may also aid in cancer cell killing by E4orf4, requiring further study. Once the molecular signaling involved in E4orf4-induced cell death is further elucidated, it will be easier to compare the underlying mechanisms in normal and cancer cells and identify the reasons for the different susceptibility of these cells to E4orf4. Moreover, the E4orf4 cell death network should be investigated not only in tissue culture cells but also in animal models to provide additional insights. The knowledge obtained in such experiments will further aid in design of novel E4orf4-based cancer therapeutics.

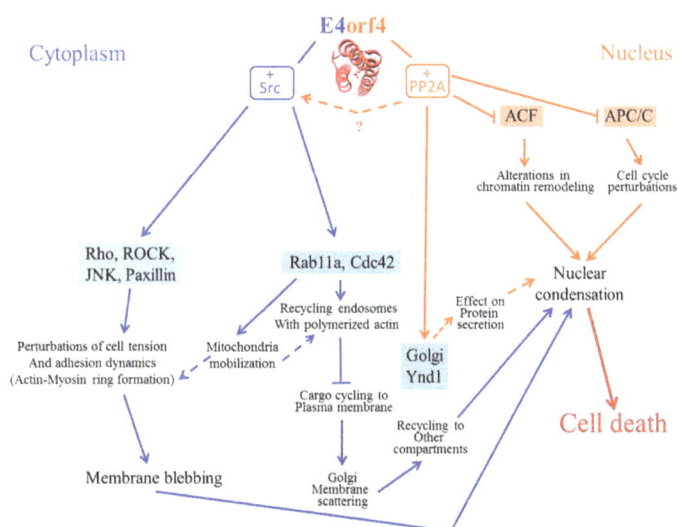

Figure 4. An integrated model of the mechanisms underlying E4orf4-induced cell death. E4orf4, in collaboration with its partners, PP2A and Src, induces alterations in the nucleus (chromatin remodeling, perturbations in cell cycle regulation) and in the cytoplasm (actin remodeling, changes in protein and membrane trafficking, changes in mitochondria morphology and mobilization), which result in blebbing, nuclear condensation and cell death: see the text for details. Arrows representing PP2A-dependent signaling are marked in orange and arrows showing Src-dependent signaling are marked in blue. Nuclear effectors are highlighted in orange and non-nuclear effectors are highlighted in blue. Connections in the E4orf4 network that were suggested but not proven yet are shown by discontinuous arrows. E4orf4 is represented by a structural model containing three alpha-helices as suggested by *ab initio* modeling [12]. This model was originally published in J. Biol. Chem. by Ben Horowitz, Rakefet Sharf, Meirav Avital-Shacham, Antonina Pechkovsky, and Tamar Kleinberger. 2013. 288: 13718-13727. © The American Society for Biochemistry and Molecular Biology.

414

Acknowledgments

The author thanks R. Sharf and H. Nabel-Rosen for helpful comments on the manuscript and K. Nebenzahl for comments on figure composition.

Work in our laboratory described in this review was supported in part by the Israel Science Foundation, by the Deutsche Forschungsgemeinschaft (DFG) within the framework of the German-Israeli Project Cooperation (DIP), by the Israel Cancer Association, by the Israel Cancer Research Fund, and by the Rappaport Faculty of Medicine and Research Institute, Technion—Israel Institute of Technology.

Conflicts of Interest

The author declares no conflict of interest.

References

1. Ben-Israel, H.; Sharf, R.; Rechavi, G.; Kleinberger, T. Adenovirus E4orf4 protein downregulates MYC expression through interaction with the PP2A-B55 subunit. *J. Virol.* **2008**, *82*, 9381–9388.

2. Bondesson, M.; Ohman, K.; Mannervik, M.; Fan, S.; Akusjarvi, G. Adenovirus E4 open reading 4 protein autoregulates E4 transcription by inhibiting E1A transactivation of the E4 promoter. *J. Virol.* **1996**, *70*, 3844–3851.

3. Mannervik, M.; Fan, S.; Strom, A.C.; Helin, K.; Akusjarvi, G. Adenovirus E4 open reading frame 4-induced dephosphorylation inhibits E1A activation of the E2 promoter and E2F-1-mediated transactivation independently of the retinoblastoma tumor suppressor protein. *Virology* **1999**, *256*, 313–321.

4. Muller, U.; Kleinberger, T.; Shenk, T. Adenovirus E4orf4 protein reduces phosphorylation of c-fos and E1A proteins while simultaneously reducing the level of AP-1. *J. Virol.* **1992**, *66*, 5867–5878.

5. Estmer Nilsson, C.; Petersen-Mahrt, S.; Durot, C.; Shtrichman, R.; Krainer, A.R.; Kleinberger, T.; Akusjarvi, G. The adenovirus E4-orf4 splicing enhancer protein interacts with a subset of phosphorylated SR proteins. *EMBO J.* **2001**, *20*, 864–871.

6. Kanopka, A.; Muhlemann, O.; Petersen-Mahrt, S.; Estmer, C.; Ohrmalm, C.; Akusjarvi, G. Regulation of adenovirus alternative RNA splicing by dephosphorylation of SR proteins. *Nature* **1998**, *393*, 185–187.

7. O'Shea, C.; Klupsch, K.; Choi, S.; Bagus, B.; Soria, C.; Shen, J.; McCormick, F.; Stokoe, D. Adenoviral proteins mimic nutrient/growth signals to activate the mTOR pathway for viral replication. *EMBO J.* **2005**, *24*, 1211–1221.

8. Bridge, E.; Medghalchi, S.; Ubol, S.; Leesong, M.; Ketner, G. Adenovirus early region 4 and viral DNA synthesis. *Virology* **1993**, *193*, 794–801.

9. Medghalchi, S.; Padmanabhan, R.; Ketner, G. Early region 4 modulates adenovirus DNA replication by two genetically separable mechanisms. *Virology* **1997**, *236*, 8–17.

10. Huang, M.M.; Hearing, P. Adenovirus early region 4 encodes two gene products with redundant effects in lytic infection. *J. Virol.* **1989**, *63*, 2605–2615.

11. Marcellus, R.C.; Chan, H.; Paquette, D.; Thirlwell, S.; Boivin, D.; Branton, P.E. Induction of p53-independent apoptosis by the adenovirus E4orf4 protein requires binding to the Balpha subunit of protein phosphatase 2A. *J. Virol.* **2000**, *74*, 7869–7877.

12. Horowitz, B.; Sharf, R.; Avital-Shacham, M.; Pechkovsky, A.; Kleinberger, T. Structure- and modeling-based identification of the adenovirus E4orf4 binding site in the protein phosphatase 2A B55alpha subunit. *J. Biol. Chem.* **2013**, *288*, 13718–13727.

13. Miron, M.J.; Gallouzi, I.E.; Lavoie, J.N.; Branton, P.E. Nuclear localization of the adenovirus E4orf4 protein is mediated through an arginine-rich motif and correlates with cell death. *Oncogene* **2004**, *23*, 7458–7468.

14. Champagne, C.; Landry, M.C.; Gingras, M.C.; Lavoie, J.N. Activation of adenovirus type 2 early region 4 ORF4 cytoplasmic death function by direct binding to Src kinase domain. *J. Biol. Chem.* **2004**, *279*, 25905–25915.

15. Kleinberger, T.; Shenk, T. Adenovirus E4orf4 protein binds to protein phosphatase 2A, and the complex down regulates E1A-enhanced junB transcription. *J. Virol.* **1993**, *67*, 7556–7560.

16. Haesen, D.; Sents, W.; Lemaire, K.; Hoorne, Y.; Janssens, V. The basic biology of PP2A in hematologic cells and malignancies. *Front. Oncol.* **2014**, *4*, e347.

17. Koren, R.; Rainis, L.; Kleinberger, T. The scaffolding A/Tpd3 subunit and high phosphatase activity are dispensable for Cdc55 function in the *Saccharomyces cerevisiae* spindle checkpoint and in cytokinesis. *J. Biol. Chem.* **2004**, *279*, 48598–48606.

18. Roopchand, D.E.; Lee, J.M.; Shahinian, S.; Paquette, D.; Bussey, H.; Branton, P.E. Toxicity of human adenovirus E4orf4 protein in *Saccharomyces cerevisiae* results from interactions with the Cdc55 regulatory B subunit of PP2A. *Oncogene* **2001**, *20*, 5279–5290.

19. Shtrichman, R.; Sharf, R.; Barr, H.; Dobner, T.; Kleinberger, T. Induction of apoptosis by adenovirus E4orf4 protein is specific to transformed cells and requires an interaction with protein phosphatase 2A. *Proc. Natl. Acad. Sci. USA* **1999**, *96*, 10080–10085.

20. Li, S.; Brignole, C.; Marcellus, R.; Thirlwell, S.; Binda, O.; McQuoid, M.J.; Ashby, D.; Chan, H.; Zhang, Z.; Miron, M.J.; *et al.* The adenovirus E4orf4 protein induces G2/M arrest and cell death by blocking PP2A activity regulated by the B55 subunit. *J. Virol.* **2009**, *83*, 8340–8352.

21. Shtrichman, R.; Sharf, R.; Kleinberger, T. Adenovirus E4orf4 protein interacts with both Bα and B' subunits of protein phosphatase 2A, but E4orf4-induced apoptosis is mediated only by the interaction with Bα. *Oncogene* **2000**, *19*, 3757–3765.

22. Pechkovsky, A.; Lahav, M.; Bitman, E.; Salzberg, A.; Kleinberger, T. E4orf4 induces PP2A- and Src-dependent cell death in *Drosophila melanogaster* and at the same time inhibits classic apoptosis pathways. *Proc. Natl. Acad. Sci. USA* **2013**, *110*, E1724–E1733.

23. Lavoie, J.N.; Nguyen, M.; Marcellus, R.C.; Branton, P.E.; Shore, G.C. E4orf4, a novel adenovirus death factor that induces p53-independent apoptosis by a pathway that is not inhibited by zVAD-fmk. *J. Cell Biol.* **1998**, *140*, 637–645.

24. Marcellus, R.C.; Lavoie, J.N.; Boivin, D.; Shore, G.C.; Ketner, G.; Branton, P.E. The early region 4 orf4 protein of human adenovirus type 5 induces p53-independent cell death by apoptosis. *J. Virol.* **1998**, *72*, 7144–7153.

25. Shtrichman, R.; Kleinberger, T. Adenovirus type 5 E4 open reading frame 4 protein induces apoptosis in transformed cells. *J. Virol.* **1998**, *72*, 2975–2982.

26. Miron, M.J.; Blanchette, P.; Groitl, P.; Dallaire, F.; Teodoro, J.G.; Li, S.; Dobner, T.; Branton, P.E. Localization and importance of the adenovirus E4orf4 protein during lytic infection. *J. Virol.* **2009**, *83*, 1689–1699.

27. Branton, P.E.; Roopchand, D.E. The role of adenovirus E4orf4 protein in viral replication and cell killing. *Oncogene* **2001**, *20*, 7855–7865.

28. Fearnhead, H.O.; Rodriguez, J.; Govek, E.E.; Guo, W.; Kobayashi, R.; Hannon, G.; Lazebnik, Y.A. Oncogene-dependent apoptosis is mediated by caspase-9. *Proc. Natl. Acad. Sci. USA* **1998**, *95*, 13664–13669.

29. Weinstein, I.B.; Joe, A. Oncogene addiction. *Cancer Res.* **2008**, *68*, 3077–3080.

30. Williams, G.H.; Stoeber, K. The cell cycle and cancer. *J. Pathol.* **2012**, *226*, 352–364.

31. Pechkovsky, A.; Salzberg, A.; Kleinberger, T. The adenovirus E4orf4 protein induces a unique mode of cell death while inhibiting classical apoptosis. *Cell Cycle* **2013**, *12*, 2343–2344.

32. Landry, M.C.; Champagne, C.; Boulanger, M.C.; Jette, A.; Fuchs, M.; Dziengelewski, C.; Lavoie, J.N. A functional interplay between the small GTPase Rab11a and mitochondria-shaping proteins regulates mitochondrial positioning and polarization of the actin cytoskeleton downstream of Src family kinases. *J. Biol. Chem.* **2014**, *289*, 2230–2249.

33. Fuster, J.J.; Sanz-Gonzalez, S.M.; Moll, U.M.; Andres, V. Classic and novel roles of p53: Prospects for anticancer therapy. *Trends Mol. Med.* **2007**, *13*, 192–199.

34. Mitrus, I.; Missol-Kolka, E.; Plucienniczak, A.; Szala, S. Tumour therapy with genes encoding apoptin and E4orf4. *Anticancer Res.* **2005**, *25*, 1087–1090.

35. Wang, D.M.; Zhou, Y.; Xie, H.J.; Ma, X.L.; Wang, X.; Chen, H.; Huang, B.R. Cytotoxicity of a recombinant fusion protein of adenovirus early region 4 open reading frame 4 (E4orf4) and human epidermal growth factor on p53-deficient tumor cells. *Anticancer Drugs* **2006**, *17*, 527–537.

36. Zhou, Y.; Chen, H.; Ma, X.L.; Xie, H.J.; Wang, C.L.; Zhang, S.H.; Wang, X.; Huang, B.R. Fusion protein of adenovirus E4orf4 and human epidermal growth factor inhibits tumor cell growth. *Int. J. Cancer* **2009**, *125*, 1186–1192.

37. Galioot, A.; Godet, A.N.; Maire, V.; Falanga, P.B.; Cayla, X.; Baron, B.; England, P.; Garcia, A. Transducing properties of a pre-structured alpha-helical DPT-peptide containing a short canine adenovirus type 2 E4orf4 PP2A1-binding sequence. *Biochim. et Biophys. Acta* **2013**, *1830*, 3578–3583.

38. Wolkersdorfer, G.W.; Morris, J.C.; Ehninger, G.; Ramsey, W.J. Trans-complementing adenoviral vectors for oncolytic therapy of malignant melanoma. *J. Gene Med.* **2004**, *6*, 652–662.

39. Marcellus, R.C.; Teodoro, J.G.; Wu, T.; Brough, D.E.; Ketner, G.; Shore, G.; Branton, P.E. Adenovirus type 5 early region 4 is responsible for E1A-induced p53-independent apoptosis. *J. Virol.* **1996**, *70*, 6207–6215.

40. Livne, A.; Shtrichman, R.; Kleinberger, T. Caspase activation by adenovirus E4orf4 protein is cell line-specific and is mediated by the death receptor pathway. *J. Virol.* **2001**, *75*, 789–798.

41. Robert, A.; Miron, M.J.; Champagne, C.; Gingras, M.C.; Branton, P.E.; Lavoie, J.N. Distinct cell death pathways triggered by the adenovirus early region 4 orf 4 protein. *J. Cell Biol.* **2002**, *158*, 519–528.

42. Li, S.; Szymborski, A.; Miron, M.J.; Marcellus, R.; Binda, O.; Lavoie, J.N.; Branton, P.E. The adenovirus E4orf4 protein induces growth arrest and mitotic catastrophe in H1299 human lung carcinoma cells. *Oncogene* **2009**, *28*, 390–400.

43. Janssens, V.; Goris, J. Protein phosphatase 2a: A highly regulated family of serine/threonine phosphatases implicated in cell growth and signalling. *Biochem. J.* **2001**, *353*, 417–439.

44. Segawa, Y.; Suga, H.; Iwabe, N.; Oneyama, C.; Akagi, T.; Miyata, T.; Okada, M. Functional development of Src tyrosine kinases during evolution from a unicellular ancestor to multicellular animals. *Proc. Natl. Acad. Sci. USA* **2006**, *103*, 12021–12026.

45. Kornitzer, D.; Sharf, R.; Kleinberger, T. Adenovirus E4orf4 protein induces PP2A-dependent growth arrest in *S. Cerevisiae* and interacts with the anaphase promoting complex/cyclosome. *J. Cell Biol.* **2001**, *154*, 331–344.

46. Ben-Israel, H.; Kleinberger, T. Adenovirus and cell cycle control. *Front. Biosci.* **2002**, *7*, d1369–d1395.

47. Peters, J.M. The anaphase promoting complex/cyclosome: A machine designed to destroy. *Nat. Rev. Mol. Cell Biol.* **2006**, *7*, 644–656.

48. Rudner, A.D.; Murray, A.W. Phosphorylation by Cdc28 activates the Cdc20-dependent activity of the anaphase-promoting complex. *J. Cell Biol.* **2000**, *149*, 1377–1390.

49. Shteinberg, M.; Protopopov, Y.; Listovsky, T.; Brandeis, M.; Hershko, A. Phosphorylation of the cyclosome is required for its stimulation by Fizzy/cdc20. *Biochem. Biophys. Res. Commun.* **1999**, *260*, 193–198.

50. Mui, M.Z.; Roopchand, D.E.; Gentry, M.S.; Hallberg, R.L.; Vogel, J.; Branton, P.E. Adenovirus protein E4orf4 induces premature APCCdc20 activation in *Saccharomyces cerevisiae* by a protein phosphatase 2A-dependent mechanism. *J. Virol.* **2010**, *84*, 4798–4809.

51. Moshe, Y.; Bar-On, O.; Ganoth, D.; Hershko, A. Regulation of the action of early mitotic inhibitor 1 on the anaphase-promoting complex/cyclosome by cyclin-dependent kinases. *J. Biol. Chem.* **2011**, *286*, 16647–16657.

52. Guardavaccaro, D.; Kudo, Y.; Boulaire, J.; Barchi, M.; Busino, L.; Donzelli, M.; Margottin-Goguet, F.; Jackson, P.K.; Yamasaki, L.; Pagano, M. Control of meiotic and mitotic progression by the F box protein beta-Trcp1 *in vivo*. *Dev. Cell* **2003**, *4*, 799–812.

53. Margottin-Goguet, F.; Hsu, J.Y.; Loktev, A.; Hsieh, H.M.; Reimann, J.D.; Jackson, P.K. Prophase destruction of EMI1 by the SCF(betaTrcp/Slimb) ubiquitin ligase activates the anaphase promoting complex to allow progression beyond prometaphase. *Dev. Cell* **2003**, *4*, 813–826.

54. Cabon, L.; Sriskandarajah, N.; Mui, M.Z.; Teodoro, J.G.; Blanchette, P.; Branton, P.E. Adenovirus E4orf4 protein-induced death of p53-/- H1299 human cancer cells follows a G1 arrest of both tetraploid and diploid cells due to a failure to initiate DNA synthesis. *J. Virol.* **2013**, *87*, 13168–13178.

55. Sriskandarajah, N.; Blanchette, P.; Kucharski, T.J.; Teodoro, J.; Branton, P.E. Analysis by live imaging of effects of the adenovirus E4orf4 protein on passage through mitosis of H1299 tumor cells. *J. Virol.* **2015**, *89*, 4685–4689.

56. Collins, N.; Poot, R.A.; Kukimoto, I.; Garcia-Jimenez, C.; Dellaire, G.; Varga-Weisz, P.D. An ACF1-ISWI chromatin-remodeling complex is required for DNA replication through heterochromatin. *Nat. Genet.* **2002**, *32*, 627–632.

57. Brestovitsky, A.; Sharf, R.; Mittelman, K.; Kleinberger, T. The adenovirus E4orf4 protein targets PP2A to the ACF chromatin-remodeling factor and induces cell death through regulation of SNF2h-containing complexes. *Nucleic Acids Res.* **2011**, *39*, 6414–6427.

58. Zhang, D.; Yang, R.; Wang, S.; Dong, Z. Paclitaxel: New uses for an old drug. *Drug Des. Dev. Ther.* **2014**, *8*, 279–284.

59. Sun, B.; Cai, Y.; Li, Y.; Li, J.; Liu, K.; Yang, Y. The nonstructural protein NP1 of human bocavirus 1 induces cell cycle arrest and apoptosis in Hela cells. *Virology* **2013**, *440*, 75–83.

60. Begemann, M.; Kashimawo, S.A.; Lunn, R.M.; Delohery, T.; Choi, Y.J.; Kim, S.; Heitjan, D.F.; Santella, R.M.; Schiff, P.B.; Bruce, J.N.; *et al.* Growth inhibition induced by Ro 31–8220 and calphostin C in human glioblastoma cell lines is associated with apoptosis and inhibition of CDC2 kinase. *Anticancer Res.* **1998**, *18*, 3139–3152.

61. Zhou, B.-B.; Li, H.; Yuan, J.; Kirschner, M.W. Caspase-dependent activation of Cyclin-dependent kinases during Fas-induced apoptosis in Jurkat cells. *Proc. Natl. Acad. Sci. USA* **1998**, *95*, 6785–6790.

62. Castedo, M.; Perfettini, J.L.; Roumier, T.; Kroemer, G. Cyclin-dependent kinase-1: Linking apoptosis to cell cycle and mitotic catastrophe. *Cell Death Differ.* **2002**, *9*, 1287–1293.

63. Castedo, M.; Perfettini, J.L.; Roumier, T.; Andreau, K.; Medema, R.; Kroemer, G. Cell death by mitotic catastrophe: A molecular definition. *Oncogene* **2004**, *23*, 2825–2837.

64. Blagosklonny, M.V. Mitotic arrest and cell fate: Why and how mitotic inhibition of transcription drives mutually exclusive events. *Cell Cycle* **2007**, *6*, 70–74.

65. Hayashi, M.T.; Karlseder, J. DNA damage associated with mitosis and cytokinesis failure. *Oncogene* **2013**, *32*, 4593–4601.

66. Lavoie, J.N.; Champagne, C.; Gingras, M.-C.; Robert, A. Adenovirus E4 open reading frame 4-induced apoptosis involves dysregulation of Src family kinases. *J. Cell Biol.* **2000**, *150*, 1037–1055.

67. Kleinberger, T. Induction of cancer-specific cell death by the adenovirus E4orf4 protein. *Adv. Exp. Med. Biol.* **2014**, *818*, 61–97.

68. Gingras, M.C.; Champagne, C.; Roy, M.; Lavoie, J.N. Cytoplasmic death signal triggered by Src-mediated phosphorylation of the adenovirus E4orf4 protein. *Mol. Cell. Biol.* **2002**, *22*, 41–56.

69. Li, Y.; Wei, H.; Hsieh, T.C.; Pallas, D.C. Cdc55p-mediated E4orf4 growth-inhibition in *S. cerevisiae* is mediated only in part via the catalytic subunit of PP2A. *J. Virol.* **2008**, *82*, 3612–3623.

70. Mui, M.Z.; Kucharski, M.; Miron, M.J.; Hur, W.S.; Berghuis, A.M.; Blanchette, P.; Branton, P.E. Identification of the adenovirus E4orf4 protein binding site on the B55alpha and Cdc55 regulatory subunits of PP2A: Implications for PP2A function, tumor cell killing and viral replication. *PLOS Pathog.* **2013**, *9*, e1003742.

71. Swingle, M.; Ni, L.; Honkanen, R.E. Small-molecule inhibitors of ser/thr protein phosphatases: Specificity, use and common forms of abuse. *Methods Mol. Biol.* **2007**, *365*, 23–38.

72. Katayose, Y.; Li, M.; Al-Murrani, S.W.; Shenolikar, S.; Damuni, Z. Protein phosphatase 2A inhibitors, I(1)(PP2A) and I(2)(PP2A), associate with and modify the substrate specificity of protein phosphatase 1. *J. Biol. Chem.* **2000**, *275*, 9209–9214.

73. Prickett, T.D.; Brautigan, D.L. The alpha4 regulatory subunit exerts opposing allosteric effects on protein phosphatases PP6 and PP2A. *J. Biol. Chem.* **2006**, *281*, 30503–30511.

74. Xu, Y.; Chen, Y.; Zhang, P.; Jeffrey, P.D.; Shi, Y. Structure of a protein phosphatase 2A holoenzyme: Insights into B55-mediated Tau dephosphorylation. *Mol. Cell* **2008**, *31*, 873–885.

75. Wang, Y.; Burke, D.J. Cdc55p, the B-type regulatory subunit of protein phosphatase 2A, has multiple functions in mitosis and is required for the kinetochore/spindle checkpoint in *Saccharomyces cerevisiae*. *Mol. Cell. Biol.* **1997**, *17*, 620–626.

76. Maoz, T.; Koren, R.; Ben-Ari, I.; Kleinberger, T. YND1 interacts with CDC55 and is a novel mediator of E4orf4-induced toxicity. *J. Biol. Chem.* **2005**, *280*, 41270–41277.

77. Vernieri, C.; Chiroli, E.; Francia, V.; Gross, F.; Ciliberto, A. Adaptation to the spindle checkpoint is regulated by the interplay between Cdc28/Clbs and PP2ACdc55. *J. Cell Biol.* **2013**, *202*, 765–778.

78. Smolders, L.; Teodoro, J.G. Targeting the anaphase promoting complex: Common pathways for viral infection and cancer therapy. *Expert Opin. Ther. Targets* **2011**, *15*, 767–780.

79. Afifi, R.; Sharf, R.; Shtrichman, R.; Kleinberger, T. Selection of apoptosis-deficient adenovirus E4orf4 mutants in *S. cerevisiae*. *J. Virol.* **2001**, *75*, 4444–4447.

80. Robson, S.C.; Sevigny, J.; Zimmermann, H. The E-NTPDase family of ectonucleotidases: Structure function relationships and pathophysiological significance. *Purinerg. Signal.* **2006**, *2*, 409–430.

81. Zhong, X.; Guidotti, G. A yeast Golgi E-type ATPase with an unusual membrane topology. *J. Biol. Chem.* **1999**, *274*, 32704–32711.

82. Zimmermann, H. Ectonucleotidases: Some recent developments and a note on nomenclature. *Drug Dev. Res.* **2001**, *52*, 44–56.

83. Avital-Shacham, M.; Sharf, R.; Kleinberger, T. NTPDASE4 gene products cooperate with the adenovirus E4orf4 protein through PP2A-dependent and -independent mechanisms and contribute to induction of cell death. *J. Virol.* **2014**, *88*, 6318–6328.

84. Mittelman, K.; Ziv, K.; Maoz, T.; Kleinberger, T. The cytosolic tail of the Golgi apyrase Ynd1 mediates E4orf4-induced toxicity in *Saccharomyces cerevisiae*. *PLOS ONE* **2010**, *5*, e15539.

85. Landry, M.C.; Sicotte, A.; Champagne, C.; Lavoie, J.N. Regulation of cell death by recycling endosomes and golgi membrane dynamics via a pathway involving Src-family kinases, Cdc42 and Rab11a. *Mol. Biol. Cell* **2009**, *20*, 4091–4106.

86. Robert, A.; Smadja-Lamere, N.; Landry, M.C.; Champagne, C.; Petrie, R.; Lamarche-Vane, N.; Hosoya, H.; Lavoie, J.N. Adenovirus E4orf4 hijacks rho GTPase-dependent actin dynamics to kill cells: A role for endosome-associated actin assembly. *Mol. Biol. Cell* **2006**, *17*, 3329–3344.

87. Lavoie, J.N.; Landry, M.C.; Faure, R.L.; Champagne, C. Src-family kinase signaling, actin-mediated membrane trafficking and organellar dynamics in the control of cell fate: Lessons to be learned from the adenovirus E4orf4 death factor. *Cell Signal* **2010**, *22*, 1604–1614.

88. Smadja-Lamere, N.; Boulanger, M.C.; Champagne, C.; Branton, P.E.; Lavoie, J.N. Jnk-mediated phosphorylation of paxillin in adhesion assembly and tension-induced cell death by the adenovirus death factor E4orf4. *J. Biol. Chem.* **2008**, *283*, 34352–34364.

89. Ward, P.S.; Thompson, C.B. Metabolic reprogramming: A cancer hallmark even warburg did not anticipate. *Cancer Cell* **2012**, *21*, 297–308.

90. Chen, F. Jnk-induced apoptosis, compensatory growth, and cancer stem cells. *Cancer Res.* **2012**, *72*, 379–386.

91. Eferl, R.; Wagner, E.F. AP-1: A double-edged sword in tumorigenesis. *Nat. Rev.* **2003**, *3*, 859–868.

92. Lin, A. Activation of the JNK signaling pathway: Breaking the brake on apoptosis. *Bioessays* **2003**, *25*, 17–24.

Impact of Adenovirus E4-ORF3 Oligomerization and Protein Localization on Cellular Gene Expression

Elizabeth I. Vink, Yueting Zheng, Rukhsana Yeasmin, Thomas Stamminger, Laurie T. Krug and Patrick Hearing

Abstract: The Adenovirus E4-ORF3 protein facilitates virus replication through the relocalization of cellular proteins into nuclear inclusions termed tracks. This sequestration event disrupts antiviral properties associated with target proteins. Relocalization of Mre11-Rad50-Nbs1 proteins prevents the DNA damage response from inhibiting Ad replication. Relocalization of PML and Daxx impedes the interferon-mediated antiviral response. Several E4-ORF3 targets regulate gene expression, linking E4-ORF3 to transcriptional control. Furthermore, E4-ORF3 was shown to promote the formation of heterochromatin, down-regulating p53-dependent gene expression. Here, we characterize how E4-ORF3 alters cellular gene expression. Using an inducible, E4-ORF3-expressing cell line, we performed microarray experiments to highlight cellular gene expression changes influenced by E4-ORF3 expression, identifying over four hundred target genes. Enrichment analysis of these genes suggests that E4-ORF3 influences factors involved in signal transduction and cellular defense, among others. The expression of mutant E4-ORF3 proteins revealed that nuclear track formation is necessary to induce these expression changes. Through the generation of knockdown cells, we demonstrate that the observed expression changes may be independent of Daxx and TRIM33 suggesting that an additional factor(s) may be responsible. The ability of E4-ORF3 to manipulate cellular gene expression through the sequestration of cellular proteins implicates a novel role for E4-ORF3 in transcriptional regulation.

Reprinted from *Viruses*. Cite as: Vink, E.I.; Zheng, Y.; Yeasmin, R.; Stamminger, T.; Krug, L.T.; Hearing, P. Impact of Adenovirus E4-ORF3 Oligomerization and Protein Localization on Cellular Gene Expression. *Viruses* **2015**, *7*, 2428-2449.

1. Introduction

The outcome of adenovirus (Ad) infection is determined by the interplay between the ability of the host cell to mount an effective antiviral response and the ability of the virus to restrict host cell defenses. Successful Ad replication relies on functions provided by the early region four (E4). This region encodes seven known proteins required for counteracting the host cell antiviral response, effective shutoff of host-cell protein synthesis, late viral mRNA accumulation, and late viral protein synthesis [1–3].

The E4-ORF3 and E4-ORF6 proteins have functionally redundant properties sufficient to facilitate virus DNA replication and infectious particle production [1,2]. Together with Ad E1B-55K, E4-ORF6 primarily functions as an adaptor molecule in an E3 cullin-RING ligase complex [4,5], promoting the ubiquitination and proteasome-dependent degradation of substrates, such as p53 [6,7], Mre11-Rad50-Nbs1 (MRN complex proteins) [8], DNA ligase IV [9], integrin α3 [10], and bloom helicase [11]. Rather than targeting proteins for degradation, the 14 kDa

E4-ORF3 protein promotes productive Ad infection by oligomerizing into filamentous nuclear inclusions termed tracks [12]. E4-ORF3 nuclear track assembly creates protein binding interfaces and results in the sequestration and inhibition of a variety of cellular proteins [13,14]. This inhibits cellular antiviral properties [15] and serves as a hub for post-translational modifications [16]. Like E4-ORF6, E4-ORF3 targets the MRN DNA repair complex and p53 for inactivation [8,17,18]. Cellular targets unique to E4-ORF3 include PML, TRIM24, and TRIM33 [12,19,20]. Sequestration of cellular proteins into E4-ORF3 nuclear tracks results in inhibition of the DNA damage response, altered p53-mediated signaling, disruption of the interferon-mediated antiviral response, and may influence transcriptional regulation [8,17,18,20–22].

Relocalization of MRN complex components Mre11, Rad50, and Nbs1 by E4-ORF3 disrupts activation of the double-strand break repair pathway and may influence cell cycle checkpoint signaling [15]. In the absence of E4 protein products, the MRN complex detects the linear, double-strand Ad genome. The resulting activation of the non-homologous end-joining pathway generates viral genome concatamers, too large to be packaged into the viral capsid [8,17,23,24]. Interestingly, activation of the DNA damage response and concatamer formation is not sufficient to inhibit viral genome replication [25,26]. Associating directly with viral DNA, the MRN complex inhibits genome synthesis by blocking access to the origin of replication [27]. As early as six hours post-infection (hpi), Ad serotype 5 (Ad5) E4-ORF3 sequesters the MRN complex away from virus replication centers and promotes the sumoylation of Mre11 and Nbs1 [8,16,17,25]. The physical removal of MRN proteins is sufficient to allow genome replication and inhibits double-strand break repair [8,17]. Supplementary to E4-ORF3, Ad inhibits the MRN complex as well as downstream non-homologous end-joining repair protein DNA ligase IV through E1B-55K/E4-ORF6-dependent degradation [9].

E4-ORF3 sequesters PML into tracks [12]. As a multifunctional protein, PML has been linked to many different processes through its ability to form punctate structures termed nuclear bodies (PML-NB) [28]. E4-ORF3-dependent relocalization of PML disrupts PML-NB and sequesters PML-NB components Daxx and Sp100. This interferes with the ability of the cell to mount an effective interferon-mediated antiviral response [21,22]. PML-NB protein composition varies between tissue types and cellular conditions; the vast PML interactome implicates PML-NB in a wide variety of cellular processes [28]. PML-NB-associated functions include cell cycle regulation, post-translational modification, DNA damage response, apoptosis, and transcriptional regulation [29]. It remains unclear if relocalization by E4-ORF3 impacts each of these PML-NB-associated functions.

Although E4-ORF3 does not alter p53 localization, E4-ORF3 down-modulates p53-mediated signaling [18]. E4-ORF3 track formation facilitates the colocalization of histone methyltransferases SUV39H1 and SUVH2 at dense cellular DNA; this is thought correlate with the upregulation of heterochromatin production. Forming a continuous scaffold with H3K9me3 heterochromatin at p53 promoter elements, E4-ORF3 indirectly antagonizes the ability of p53 to associate with DNA [18].

The E4-ORF3 protein relocalizes several members of the TRIM protein family into nuclear tracks including TRIM19 (PML), TRIM24 (TIF1α) and TRIM33 (TIF1γ) [12,19,20], although the functional significance of TRIM24 and TRIM33 relocalization remains unclear. TRIM (TRIpartite Motif) family members share conserved N-terminal Ring-B box-Coiled Coil (RBCC) motifs;

TRIM24 and TRIM33 contain C-terminal PHD/Bromo domains [30–32]. TRIM24 and TRIM33 are multifunctional proteins involved in cellular processes including transcriptional regulation, growth control, regulation of development, and protein post-translational modification [30–32]. It is interesting to hypothesize that relocalization of these proteins by E4-ORF3 alters one or more of these properties. The RBCC domains of both TRIM24 and TRIM33 have ubiquitin ligase activity that targets substrates including p53 and Smad4 for ubiquitin-dependent proteasome degradation [33,34]. TRIM33 also directs sumoylation of cellular substrates [35]. TRIM24 acts as a ligand-dependent coregulator of gene expression from nuclear receptor-bound promoter elements, influencing retinoic acid and estrogen-mediated signaling pathways [36–38]. Like TRIM24, TRIM33 serves as a multifunctional coregulator of gene expression, influencing TGFβ signaling [38,39]. Functioning as a reader of heterochromatin modifications, the PHD/Bromo domain positions TRIM33 on promoter elements upstream of target genes. This activates the ubiquitin ligase activity of TRIM33 which then alters the composition of regulatory Smad proteins at promoter elements [40,41].

Here, we characterized how E4-ORF3-dependent relocalization of transcriptional regulators alters cellular gene expression. A series of microarray time course experiments spanning 24 hours highlighted over 400 E4-ORF3-regulated genes. By sequentially disrupting E4-ORF3 functions or knocking-down cellular effectors, we determined that these differential expression events rely on nuclear track formation, but that changes in cellular gene expression by E4-ORF3 appear to be independent of MRN and resistant to Daxx and TRIM33 knockdown. These results suggest that an additional factor(s) may be responsible for E4-ORF3 activity. The ability of E4-ORF3 to manipulate cellular gene expression through the sequestration of cellular proteins implicates a novel role for E4-ORF3 in transcriptional regulation.

2. Materials and Methods

2.1. Cell Culture and Virus Infection

Experiments were carried out in U2OS-Tet (Clontech, Mountain View, CA, USA), U2OS (ATCC, Manassas, VA, USA), and 293FT (Life Technologies, Grand Island, NY, USA) cells. Cells were grown in Dulbecco's modified eagle medium (DMEM) supplemented with 10% Fetal Clone III (Hyclone, Logan, UT, USA; U2OS-Tet and U2OS cells) or 10% Fetal Bovine Serum (Hyclone; 293FT cells), 2 mM L-glutamine, 100 µM MEM non-essential amino acids, and 1 mM sodium pyruvate. Cells were transfected using Lipofectamine (Life Technologies, Grand Island, NY, USA) or Fugene 6 (Roche, Indianapolis, IN, USA) using the manufacturer's instruction. Cells were infected with Ad5 E1-replacement viruses that express HA-tagged wild-type or mutant E4-ORF3 proteins for 1 h using 500 particles per cell followed by the addition of fresh medium. E1A replacement viruses that express E4-ORF3 fused to an HA epitope under the control of a CMV promoter include Ad-CMV-HA-ORF3-WT, Ad-CMV-HA-ORF3-N82A, Ad-CMV-HA-ORF3$_{L103A}$, and Ad-CMV-HA-ORF3$_{D105A/L106A}$ [17,26].

2.2. Immunofluoresence and Western Blot Analysis

Cells were grown on glass coverslips, transfected, and then infected under the conditions described above. Between 16 and 18 hours post-infection (hpi), cells were washed with PBS, fixed with −20 °C methanol, and blocked for 1 h at room temperature with 10% goat serum diluted into PBS. Primary antibodies were diluted into 10% goat serum block and applied to coverslips for 1 h at room temperature. Coverslips were then washed with PBS and incubated with secondary antibody consisting of tetramethyl rhodamine isothiocyanate (TRITC) labeled anti-rat (Invitrogen) or fluorescein isothiocyanate (FITC) labeled anti-rabbit (Invitrogen) antibody for 30 min at room temperature in the dark. Coverslips were mounted on slides using ImmunoMount (Thermo Shandon, Pittsburgh, PA, USA).

Cell lysates prepared for western blot analysis were resolved on SDS polyacrylamide gels. Proteins were transferred to polyvinylidene fluoride (PVDF) (Hybond-P, GE Healthcare, Pittsburgh, PA, USA) membranes or nitrocellulose membranes (Protran, GE Healthcare, Pittsburgh, PA, USA) overnight at 40 mA. Membranes were blocked for 1 h with 3% BSA in PBS, and then incubated with primary antibody overnight at 4 °C. The membranes were washed and incubated in anti-mouse HRP, anti-rabbit HRP, or anti-rat HRP secondary antibodies for 30 min. Immobilon chemiluminescent HRP substrate (Millipore, Billerica, MA, USA) was used for detection.

Primary antibodies used for immunofluorescence and Western blot analysis include: tubulin (T5192, Sigma Aldrich), PML (PG-M3, sc-966, Santa Cruz Biotechnology, Dallas, TX, USA), Daxx (25C12, Cell Signaling Technology, Danvers, MA, USA), HA peptide (600-401-384, Rockland Immunochemcials, Limerick, PA, USA), rat monoclonal E4-ORF3 (mAb 6A11, [42]), and rabbit polyclonal anti-TRIM33 (generated at Lampire Biological Laboratories, Pipersville, PA, USA).

2.3. Creation of an E4-ORF3 Inducible Cell Line

Clontech's Tet-On system was used to generate a stable E4-ORF3-inducible cell line. A PCR-generated amplicon consisting of the Ad2 E4-ORF3 coding region was cloned into the BamHI and EcoRI restriction enzyme sites of the pUHD10-3 vector, fused at the N-terminus with an HA epitope and placing it under the control of a doxycycline (dox)-inducible promoter element (pTet-HA-E4-ORF3). The resulting plasmid construct was cotransfected into U2OS-Tet-On osteosarcoma cells with pTK-Hyg (Clontech) using Fugene 6. Two days post transfection, cells were treated with 200 μg/mL G418 and 100 μg/mL Hygromycin B to select for cells with pTet-HA-E4-ORF3. Individual, drug-resistant colonies were isolated, amplified, and treated with 1 μg/mL dox for 48 h to screen for HA-E4-ORF3 expression. One cell subclone that exhibited no E4-ORF3 expression minus the addition of dox and levels of E4-ORF3 equivalent to that observed following Ad5 infection was chosen for subsequent use (termed Tet-E4-ORF3 cells).

2.4. Microarray Analysis

Three independent aliquots of dox-induced Tet-E4-ORF3 cells were collected at 6, 12, 18, and 24 h time points. Two aliquots of dox-induced parental U2OS-Tet-On cells were collected at 6 and 24 h post induction. Aliquots of untreated Tet-E4-ORF3 and untreated U2OS-Tet-On cells also were collected. Total cellular RNA was isolated from cells using the RNeasy kit (Qiagen, Valencia, CA, USA) according to the manufacturer's instruction. RNA quality was measured on an Agilent 2100 Bioanalyzer with the RNA 6000 Nano LabChip®. Transcripts were prepared with Agilent's Quick Amp Labeling Kit, two-color, then hybridized to an Agilent 4 × 44k whole human genome expression array. Initial processing of the raw data was performed using Agilent's Feature Extraction (FE) software v9.5 (Santa Clara, CA, USA). Data were then normalized using the Limma Bioconductor Package (http://www.bioconductor.org/packages/release/bioc/html/limma.html). Results were subject to loess normalization within each array and Aquantile normalization between arrays and then $\log_2$ transformed. [43,44]. A hierarchical clustering algorithm with centroid linkage implemented by Cluster 3.0 software was used to group expression data from each array replicate and the results were visualized with Java Treeview [45,46].

The data were subject to enrichment analysis via the Database for Annotation, Visualization, and Integrated Discovery (DAVID) [47,48]. Agilent probe names corresponding to genes of interest were uploaded to the database along with a list of probe names corresponding to all genes represented on the microarray as background. Gene Ontology (GO) terms associated with P-values less than 0.0001 were considered.

2.5. RT-qPCR

Total cellular RNA was isolated from cells using the RNeasy system (Qiagen). 2 µg of RNA were subject to reverse transcription PCR using SSII RT (Invitrogen) according to the manufacturer's instructions. cDNA pools were diluted 1:10 in dH_2O. 1 µL of this dilution served as a template for qRT-PCR along with 0.5 µM each forward and reverse primer and 10 µL 2× DyNAmo HS sybr green master mix (Thermo Scientific, Waltham, MA, USA) in a total volume of 20 µL per reaction. Primer pairs were designed with the aid of qPrimerdepot [49] and RTPrimerDB [50] (Supplemental Table S4). Each RT-qPCR run contained two technical replicates of each sample. RT-qPCR was carried out and analyzed on an Applied Biosystem 7500 Real Time PCR System, according to the program: 95 °C for 10 min hot start followed by 40 cycles of 95 °C for 15 s, 60 °C for 1 min. The Pfaffl method of relative quantification was used to convert the resulting threshold cycle data for each sample to relative fold change information [51]. Primer efficiencies were calculated by performing serial dilutions of template cDNA and generating a standard curve. The slope of the standard curve was applied to the equation Efficiency $= 10^{(-1/slope)}$. Efficiencies listed in Supplemental Table 4, column E represent and average value of at least two results.

2.6. Generation of Knockdown Cell Lines

Oligonucleotides to express short hairpins RNAs directed against TRIM33 [34] were cloned into the pSIREN-RetroQ vector (Clontech). Constructs containing non-functional shRNA (shControl) or shRNA directed against Daxx (shDaxx) were previously described [52]. 4 µg Retrovirus construct was cotransfected in to 293FT cells along with 4 µg pVSV-G and 4 µg helper virus vector using Lipofectamine 2000. Aliquots of the resulting retrovirus-containing media from transfected cells were collected at 48 h and 72 h post-transfection, clarified by centrifugation, filtered through a 0.45-µm filter, and stored at −80 °C. Undiluted retrovirus stock was added to Tet-E4-ORF3 cells directly or diluted 1:500 into general media before infection in the presence of 5 µg/mL polybrene. Media was replaced after overnight incubation, and the virus was allowed to recombine with host cell genomes. At 48 hpi, cells were treated with 2 µg/mL puromycin to select for stably transfected cells. Pools of cells were screened for knockdown by Western blot analysis for the corresponding protein of interest.

3. Results

3.1. Creation and Characterization of a Tet-Inducible E4-ORF3 Cell Line

To characterize changes in gene expression that result from E4-ORF3 expression, we generated a stable U2OS-derived cell line that expresses HA-tagged E4-ORF3 in a dox-inducible manner (termed Tet-E4-ORF3 cells). Human U2OS osteosarcoma cells were selected for their ability to express wild-type p53. This approach allows for consistent E4-ORF3 expression while minimizing background transcriptional changes that may have resulted from the use of a viral expression vector. HA-E4-ORF3 was cloned into the pUHD10-3 vector under the control of a Tet response element (TRE). This construct was transfected into U2OS-Tet cells, a cell line that constitutively expresses the reverse tetracycline transactivator protein (rtTA). The introduction of dox allows rtTA to associate with the TRE and induce expression of HA-E4-ORF3. After drug selection, the resulting colonies were screened for their ability to express HA-E4-ORF3 in a dox-inducible manner.

HA-E4-ORF3 expression in Tet-E4-ORF3 cells was characterized over a 24 h period following dox treatment. Cell lysates were collected at 6, 12, 18, and 24 h post induction (1 µg/mL dox) and analyzed for HA-E4-ORF3 expression by Western blot analysis (Figure 1A). Minimal HA-E4-ORF3 expression was observed in untreated cells. HA-E4-ORF3 accumulation could be detected as early as 6 h post induction, with peak expression at 24 h post-induction. Dox treatment for periods longer than 24 h failed to yield further increases in HA-E4-ORF3 expression. Induced Tet-E4-ORF3 cells express HA-E4-ORF3 to a similar degree and over a similar time frame as E4-ORF3 expression during wild-type Ad infection (Supplementary Figure S1A). To assay for potential cytotoxicity of HA-E4-ORF3 expression or dox treatment, cellular proliferation of dox-treated Tet-E4-ORF3 and parental U2OS-Tet cells was monitored over a period of 96 h. HA-E4-ORF3-expressing cells and untreated Tet-E4-ORF3 cells proliferated to a similar extent for the first 48 h of dox induction (Supplementary Figure 1B). The presence of HA-E4-ORF3 corresponded with impaired proliferation after 48 h. This suggests that a time course experiment spanning 24 h of dox treatment should be free of cytotoxic effects of dox treatment or E4-ORF3 expression.

To confirm the formation of functional E4-ORF3 nuclear tracks, Tet-E4-ORF3 cells were grown on coverslips, stimulated with dox, and analyzed for HA-E4-ORF3 localization and PML reorganization by immunofluoresence (Figure 1). In accordance with the Western blot results, HA-E4-ORF3 tracks could be detected in cells as early as 6-h post induction (G). Over the ensuing 18 h, HA-E4-ORF3 tracks became more elongated and densely packed the nuclei in the majority of cells (H–J). Costaining with an antibody directed against PML revealed that HA-E4-ORF3 efficiently disrupted PML-NB and relocalized PML into nuclear tracks (C–F, merged images K–N). We conclude that Tet-E4-ORF3 cells express functional E4-ORF3 protein in an inducible manner.

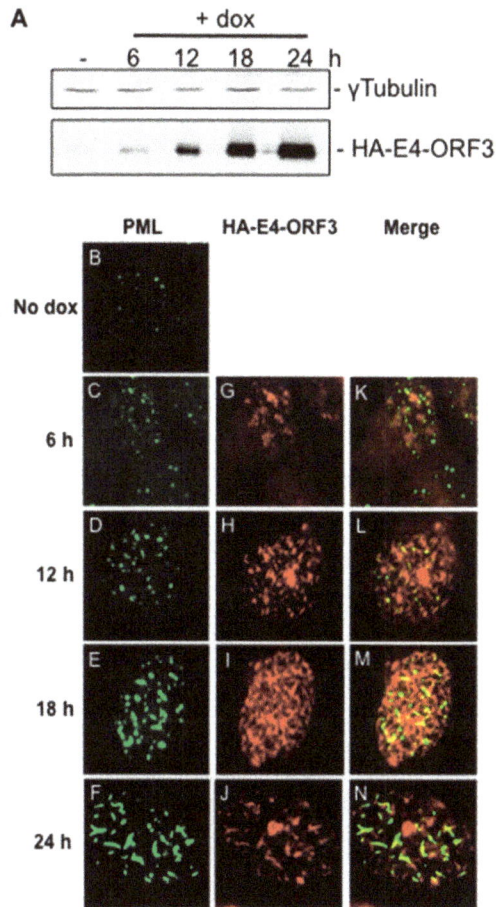

Figure 1. Doxycycline-treated Tet-E4-ORF3 cells produce HA-E4-ORF3 capable of recruiting endogenous PML into nuclear tracks. Tet-E4-ORF3 cells in tissue culture dishes (**A**) or grown on coverslips (**B–N**) were mock-treated or supplemented with 1 μg/mL dox for 6, 12, 18, or 24 h. Lysates were subject to Western blot analysis using α-HA antibody or α-γtubulin antibody (A). Cells on coverslips were fixed and stained with an antibody directed against E4-ORF3 or PML followed by TRITC and FITC labeled secondary antibodies. PML localization is shown in (**B–F**). HA-E4-ORF3 localization is shown in (**G–J**). Merged images are shown in (**K–N**).

3.2. Microarray Experiments and Quality Control

Dox-induced HA-E4-ORF3 expression increased over time with maximum detectable protein levels at 24 h post induction (Figure 1). We carried out a time course experiment to observe changes in cellular gene expression over a 24 h period of E4-ORF3 expression. Tet-E4-ORF3 cells were left untreated or were treated with 1 µg/mL dox for 6, 12, 18, and 24 h. Aliquots of cells were taken from each time point and subject to Western blot and immunofluoresence analyses to verify E4-ORF3 expression and track formation. Total cellular RNA was purified from the remaining cells and analyzed by microarray. Transcripts generated from each dox-induced time point were cohybridized to an Agilent 4 × 44k whole human genome expression array along with transcripts generated from an untreated, reference population of Tet-E4-ORF3 cells. This time course experiment was repeated to yield a total of three biological replicates.

To minimize the consideration of genes displaying E4-ORF3-independent changes in expression, we carried out a control microarray experiment. Duplicate batches of parental U2OS-Tet cells were treated with 1 µg/mL dox for 6 or 24 h before RNA purification. The four pools of transcripts were cohybridized to an Agilent 4 × 44k array against the same reference pool of transcripts used in the HA-E4-ORF3 experiment set. Expression changes highlighted in this experiment may represent dox-dependent changes in gene expression, variation resulting from the additional passage of U2OS-Tet cells beyond the creation of the E4-ORF3 expressing cell line, as well as random background expression changes.

Results from both E4-ORF3 time course and control microarrays were normalized using the limma Bioconductor package and were subject to Loess normalization within each array and Aquantile normalization between arrays [43,44]. Gene expression data were determined by comparing transcript levels in the presence of E4-ORF3 or dox relative to transcript levels in the absence of treatment. This was calculated by determining the log$_2$ ratio of green to red fluorescent intensity levels. We performed quality control analyses on the resulting data. Fluorescent intensity data from each of the arrays were subject to hierarchical cluster analysis to assess the interrelatedness of each microarray time course replicate (Supplementary Figure 2A). The four control arrays formed a distinct cluster, independent of the E4-ORF3 expression arrays, indicating a high degree of interrelatedness. Arrays from replicates 2 and 3 segregated into early (6 h, 12 h) and late (18 h, 24 h) clusters. Time points produced in replicate 1 formed a distinct cluster in the dendrogram, indicating that a high degree of differential expression in this replicate may be due to technical variation. The distribution of fluorescent intensity across each array was graphed on a density plot (Supplementary Figure S2B). A wider distribution of fluorescent intensities was seen associated with genes on the control arrays than with the test arrays. This suggests that a greater number of genes were differentially expressed under control conditions than in the presence of E4-ORF3. This could indicate that many of the observed changes in transcript levels result from technical, rather than biological variation. To account for any technical variation, we used two methods to select target genes of interest (see below). Identification of a target gene by both methods provides strong evidence of E4-ORF3-dependent expression changes. For one of these methods—selecting genes of interest that display a high degree of differential expression at 24-hours post induction—we only considered data from replicates 2 and 3.

3.3. Data Analysis and Target Gene Selection

Two approaches were used to analyze the resulting microarray data cluster analysis and identify genes that show E4-ORF3-dependent differential expression at 24-h post induction relative to 6-h post induction. First, we identified genes with a significant two-fold or greater change in response to

E4-ORF3 expression and then subjected that list to gene ontology enrichment analysis. Second, we identified gene with unique E4-ORF3-dependent gene expression profiles by hierarchical clustering. These two approaches selected for genes using different criteria, yielding overlapping, but non-identical pools of target genes.

Determining differential expression at 24-h post induction relative to 6-h post induction selects for genes that share a similar expression pattern as HA-E4-ORF3 (Figure 1). Transcript levels in dox-induced Tet-E4-ORF3 cells were determined relative to transcript levels in dox-treated, parental U2OS-Tet cells to eliminate the inclusion of dox-dependent expression changes. Dox-independent expression changes at 24-h post induction were calculated relative to 6-h post induction to yield a value representing E4-ORF3-dependent expression changes at 24-h post induction relative to 6-h post induction. To search for E4-ORF3-regulated targets, we considered genes that show a two-fold or greater expression change and possess a p-value less than 0.01. 409 genes met these criteria; 289 were up-regulated and 120 were down-regulated at 24-h post induction (Supplementary Table S1).

To look for potential cellular processes influenced by E4-ORF3, we subjected the list of 409 genes to enrichment analysis. The Database for Annotation, Visualization, and Integrated Discovery (DAVID) was used to identify Gene Ontology (GO) terms over-represented in the list of E4-ORF3-regulated targets compared to a background list encompassing all genes represented on the microarray. GO terms associated with p-values less than 0.0001 were considered. The 409 E4-ORF3 targets are associated with a variety of terms including cell signaling and cellular defense. This gene list is also enriched in terms implicating growth factor activity and suggests many targets localize to the plasma membrane and extracellular space (Table 1).

As a complementary approach to E4-ORF3 target selection, expression change data from each array were subject to hierarchical cluster analysis with centroid linkage using Cluster3.0 [45] then visualized with Java TreeView [46]. Genes that cluster together in the presence of E4-ORF3 but not under control conditions may be coregulated by a common promoter element and/or share functional relevance. Two cluster analyses were carried out. Expression data from 6, 12, 18, and 24 h post-induction time points from Tet-E4-ORF3 replicates 1, 2, and 3 were subject to cluster analysis. Expression data from 6 and 24 h time points from the control, dox-only array were used in a second cluster analysis. Potential clusters of interest were considered on the basis of forming a distinct node in the associated dendrogram and contained genes that display similar expression tendencies within each time point. Finally, genes must not cluster in the absence of E4-ORF3 expression. As most genes did not display differential expression in the presence of E4-ORF3, a high proportion of genes did not fall into distinct, easily discernible clusters. Two clusters were initially considered: cluster 1 contains 31 genes (Figure 2A, Supplementary Table S2) and cluster 2 contains 79 genes (Figure 2B, Supplementary Table S3). Both clusters consist of genes that were either down-regulated or not differentially expressed at 6 h post-induction but were up-regulated after 24 h of E4-ORF3 expression. Gene expression under control conditions failed to match expression patterns seen in the presence of E4-ORF3. Of the 110 genes highlighted by the cluster analysis, 27 also fall into the group of 289 genes up-regulated at 24-h post induction relative to 6-h post induction.

Table 1. Enrichment analysis of E4-ORF3 target genes.

Category	Term	Count	%	p value
Biological process	Cell surface receptor-linked signal transduction	59	18.97	1.01E-09
	Response to wounding	29	9.32	5.77E-08
	Inflammatory response	19	6.11	8.57E-06
	Defense response	25	8.04	4.95E-05
	Cell-cell signaling	25	8.04	6.65E-05
	Immune response	25	8.04	1.94E-04
	G protein coupled receptor signaling pathway	28	9.00	3.44E-04
	Cell migration	14	4.50	8.35E-04
Cellular component	Extracellular region	74	23.79	2.86E-11
	Extracellular space	36	11.58	1.42E-08
	Integral to plasma membrane	50	16.08	2.03E-08
	Intrinsic to plasma membrane	50	16.08	4.20E-08
	Extracellular region part	42	13.50	1.10E-07
	Plasma membrane part	62	19.93	1.49E-04
Molecular function	Growth factor activity	14	4.50	3.75E-06
	Polysaccharide binding	13	4.18	1.36E-05
	Pattern binding	13	4.18	1.36E-05
	Cytokine activity	14	4.50	2.47E-05
	Glycosoaminoglycan binding	12	3.86	2.86E-05
	Ligand-gated ion channel activity	11	3.54	6.42E-05
	Ligand-gated channel activity	11	3.54	6.42E-05
	Heparin binding	10	3.22	6.47E-05
	Carbohydrate binding	18	5.79	1.00E-04
	Gated channel activity 14	16	5.14	2.71E-04

A list of 409 E4-ORF3-regulated genes were subject to enrichment analysis using DAVID. Category indicates the ontology under consideration, whereas the term refers to the specific Gene Ontology (GO) term. Values in the count column specify the number of E4-ORF3-regulated genes that associate with each GO term. The % column indicates the percentage of genes associated with each GO term present in the list of 409 E4-ORF3-regulated genes. The P Value column lists modified Fischer Exact P-values associated with each enrichment event. Terms with p-values less than 0.0001 were considered.

Expression data from E4-ORF3 or control microarray analyses were subject to hierarchical cluster analysis with centroid linkage. Clusters were selected for further analysis on the basis of forming a distinct node unique to the E4-ORF3 dendrogram and maintaining consistent expression patterns within each time point. Blue signal represents down-regulation, whereas yellow indicates up-regulation. Results of E4-ORF3 and control cluster analysis were combined in each heat map for ease of visualization.

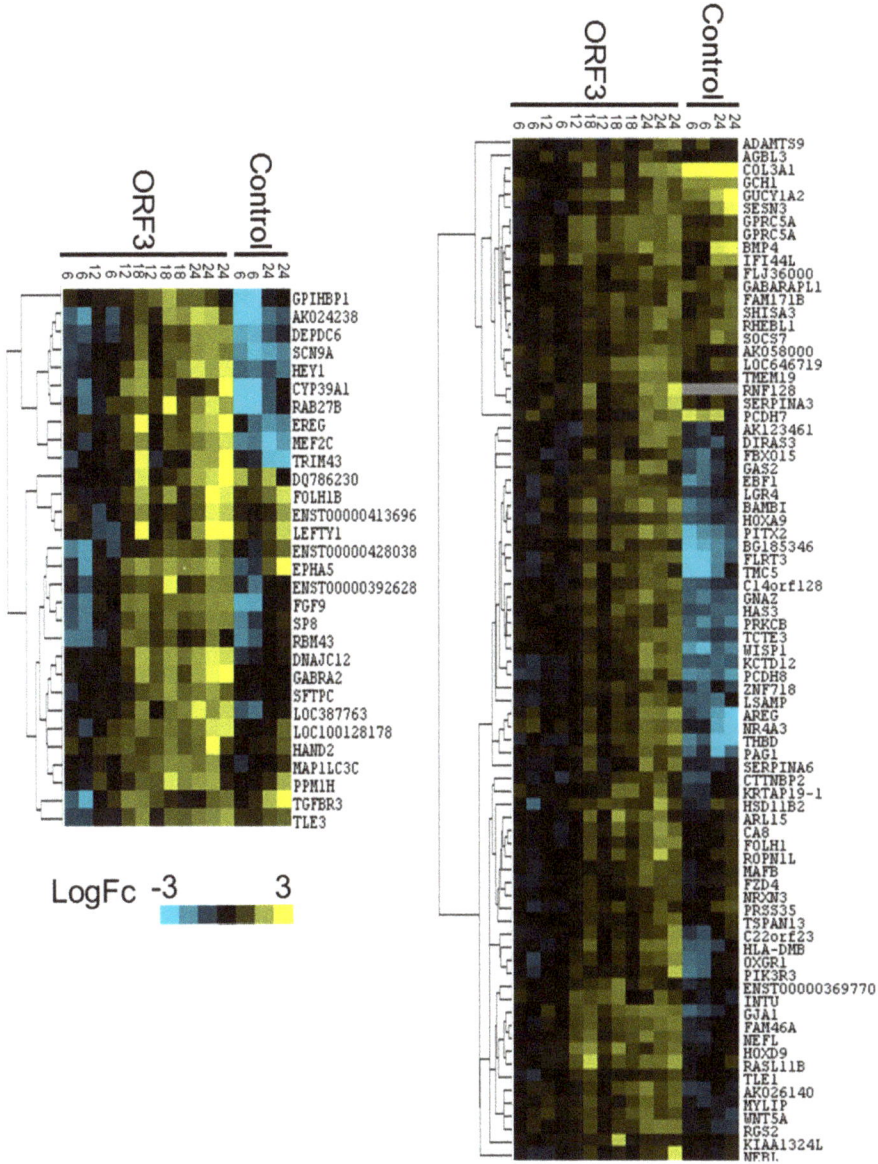

Figure 2. Heat maps and dendrograms corresponding to cluster 1 (**A**) and cluster 2 (**B**).

3.4. Target Gene Validation

E4-ORF3 targets were selected for validation on the basis of displaying a consistently large E4-ORF3-specific change in expression in each biological replicate. Four genes were selected from each cluster of interest: HEY1, TLE3, SP8, and FGF9 from cluster 1 (Figure 2A, Supplementary Table S2) and AREG, BAMBI, PITX2, and RGS2 from cluster 2 (Figure 2B, Supplementary Table S3). Five of these genes—AREG, BAMBI, RGS2, HEY1, and TLE3—were indicated as potential E4-ORF3 targets in both analyses.

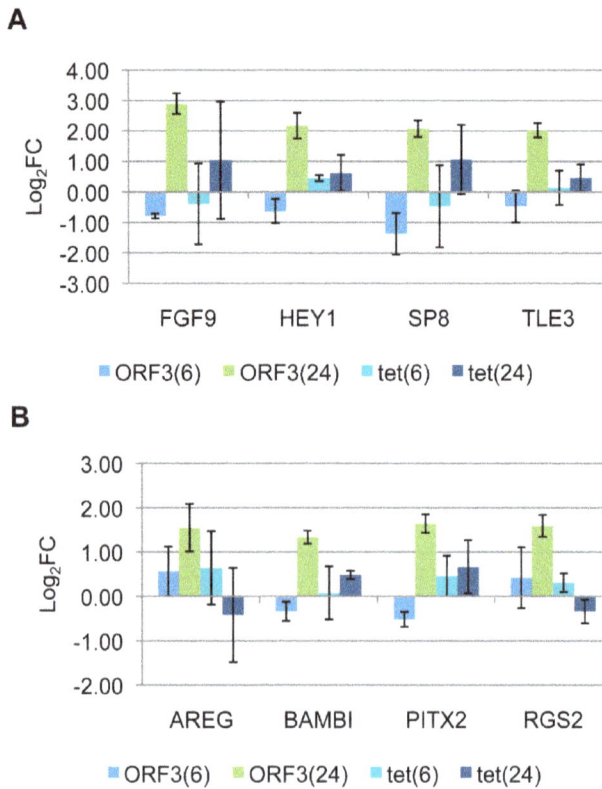

A

B

Figure 3. Validation of select E4-ORF3 targets from (**A**) cluster 1 or (**B**) cluster 2 by RT-qPCR with relative quantification. Bars represent Log_2 fold expression change in dox-treated Tet-E4-ORF3 or U2OS-Tet cells relative to mock treated Tet-E4-ORF3 or U2OS-Tet cells, normalized to GAPDH at 6 and 24 hours post-induction. Error bars = S.D., n = 3.

The ability of E4-ORF3 to influence these eight genes was validated by RT-qPCR (Figure 3). We performed reverse transcription PCR to generate cDNA from mock, 6 h and 24 h RNA pools used in the microarray experiment or from a pool of RNA generated from newly induced Tet-E4-ORF3 cells for a total of three biological replicates. To verify that the expression changes are specific to E4-ORF3, we also generated three sets of cDNA from mock-treated parental U2OS-Tet cells or U2OS-Tet cells treated with 1 μg/mL dox for 6 or 24 h. These batches of cDNA served as templates for qPCR using primers directed against the eight genes of interest. The Pfaffl method of relative quantification was used to calculate changes in gene expression in the presence of dox relative to untreated conditions and normalized to GAPDH [51]. FGF9, HEY1, SP8, and TLE3 demonstrated greater than four-fold E4-ORF3-dependent expression changes by 24-h post induction, although FGF9 and SP8 yielded a two-fold increase in response to dox alone (Figure 3A). AREG, BAMBI,

PITX2, and RGS2 displayed greater than a two-fold increase in expression change specifically in the presence of HA-E4-ORF3 by 24-h post induction (Figure 3B).

3.5. Linking Differential Gene Expression to E4-ORF3 Functions

The wide array of transcriptional changes induced by E4-ORF3 has the potential to influence many cellular processes. Linking the differential expression of the validated genes to previously characterized E4-ORF3 functions may help to suggest a functional significance for these transcriptional changes in the context of Ad infection. To link differential expression events to E4-ORF3 function, we utilized a series of Ad expression vectors that express wild-type and mutant versions of E4-ORF3. The mutant proteins relocalize a subset of cellular proteins normally targeted by wild-type E4-ORF3 or fail to form tracks entirely.

The formation of nuclear tracks serves as a necessary prerequisite for the relocalization of cellular proteins by E4-ORF3. The ability of E4-ORF3 to induce the differential expression of target genes independent of track formation would suggest that E4-ORF3 does not rely on the sequestration of a cellular protein to influence gene expression. This could imply that E4-ORF3 regulates target genes directly, or that HA-E4-ORF3 over-expression induces a cellular stress response independent of E4-ORF3 function. E4-ORF3$_{L103A}$ contains a point mutation in a region necessary for protein oligomerization [13,14]. Although this protein folds properly, it fails to form nuclear tracks. We normalized infection conditions to express mutant or wild-type E4-ORF3 at a level similar to that produced in Tet-E4-ORF3 cells after 24 h of dox treatment (Figure 4A). Tet-E4-ORF3 cells were infected with HA-E4-ORF3$_{WT}$ or HA-E4-ORF3$_{L103A}$ expression viruses for 24 h. As a positive control, Tet-E4-ORF3 cells were treated with 1 µg/mL dox for 24 h. Total cellular RNA was isolated from cells and subject to RT-qPCR. The resulting pool of cDNA served as template with primer pairs directed against the eight validated genes of interest. qPCR results were subject to the Pfaffl method of relative quantification [51]. The experiment was repeated a total of three times (Figure 4B). Whereas wild-type E4-ORF3, either expressed in Tet-E4-ORF3 cells by dox induction or by an Ad expression vector, facilitated the induction of the eight target genes, mutant E4-ORF3$_{L103A}$ failed to induce expression. These results demonstrate that nuclear track formation is necessary for the up-regulation of the eight tested E4-ORF3 target genes and supports a hypothesis that suggests that a cellular protein sequestration event, rather than another E4-ORF3 activity, facilitates differential cellular gene expression. Failure of HA-ORF3$_{L103A}$ to up-regulate the eight genes of interest also suggests that the differential expression events highlighted by the microarray study do not result from a general stress response induced by dox addition and the expression of an ectopic protein.

E4-ORF3 relocalizes proteins of the MRN complex, interfering with the ability of the host cell to detect and repair double-strand DNA breaks. E4-ORF3$_{D105A/L106A}$ contains two point mutations in a region that forms the MRN complex binding interface. This version of E4-ORF3 forms tracks, sequesters PML, TRIM24, and TRIM33, but fails to relocalize the MRN complex [17,26]. To determine if the recruitment of the MRN complex influences the differential expression of the eight validated genes, we mock-infected Tet-E4-ORF3 cells or expressed HA-E4-ORF3$_{WT}$ and HA-E4-ORF3$_{D105A/L106A}$ by virus infection for 24 h (Figure 4A). These cells were used to generate total cellular RNA which served as a template for RT-qPCR, as described above. Both

HA-E4-ORF3$_{WT}$ and HA-E4-ORF3$_{D105/L106}$ clearly induced an increase in all target genes (Figure 4B). The increase in the level of induction with the HA-E4-ORF3$_{D105A/L106A}$ mutant protein compared with HA-E4-ORF3$_{WT}$ correlated with increased E4-ORF3 mutant protein expression (Figure 4A). These results suggest that the sequestration of the MRN complex is not necessary for the observed E4-ORF3-dependent up-regulation of cellular gene expression.

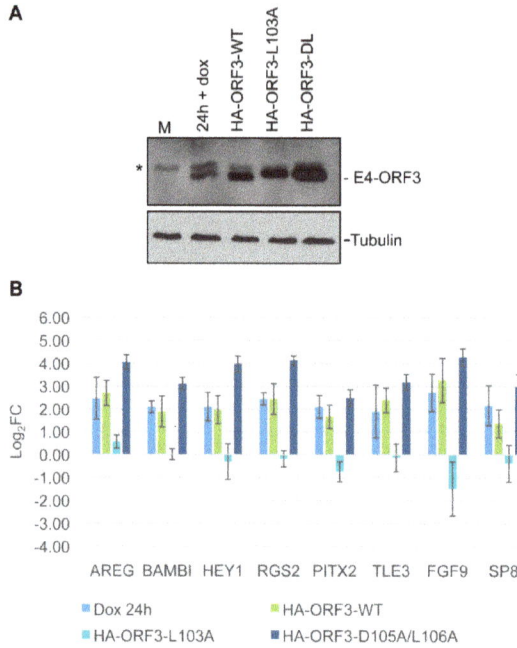

Figure 4. Nuclear track formation, but not sequestration of the MRN complex, is necessary for the up-regulation of select E4-ORF3 target genes. Cells were induced with doxycycline or infected with wild-type or mutant E4-ORF3-expressing Ad vectors to yield an equivalent amount of protein product. E4-ORF3$_{L103A}$ folds properly but fails to form nuclear tracks. E4-ORF3$_{D105A/L106A}$ forms tracks and relocalizes all known targets except the MRN complex. (**A**) Protein levels were verified by Western blot analysis. * Indicates a background band; (**B**) Cellular transcript levels were recorded by relative quantification RT-qPCR. Bars represent Log2 fold change in expression at 24hpi + E4-ORF3, relative to mock, normalized to GAPDH. Error bars = S.D., n = 3. Gene expression differences in the presence of HA-ORF3-WT *vs*. HA-ORF3-L103A are statistically significant ($p < 0.05$) for every target except SP8 as determined by one-way ANOVA analysis with multiple comparisons.

Several cellular proteins sequestered into E4-ORF3 nuclear tracks regulate cellular gene expression. Daxx generally represses gene regulation through chromatin remodeling although is may serve as a transcriptional activator in certain cases [53]. TRIM33 serves as a multifunctional coregulator of gene expression, influencing TGFβ signaling and other pathways [38,39]. Daxx and

TRIM33 serve as transcriptional effectors that respond to multiple signaling pathways. To examine a potential link between the sequestration of Daxx and TRIM33 with the regulation of cellular gene expression by E4-ORF3, we generated knock-down cell lines in the Tet-E4-ORF3 cell background. shRNAs directed against Daxx or TRIM33, or a non-specific control, were introduced into the inducible E4-ORF3 cell line using retroviruses and reductions in corresponding protein levels was analyzed by Western blot (Figures 5A and 6A). shRNAs against Daxx and TRIM33 knocked-down target protein expression ~5-fold for each protein relative to shControl cells (Figures 5A and 6A).

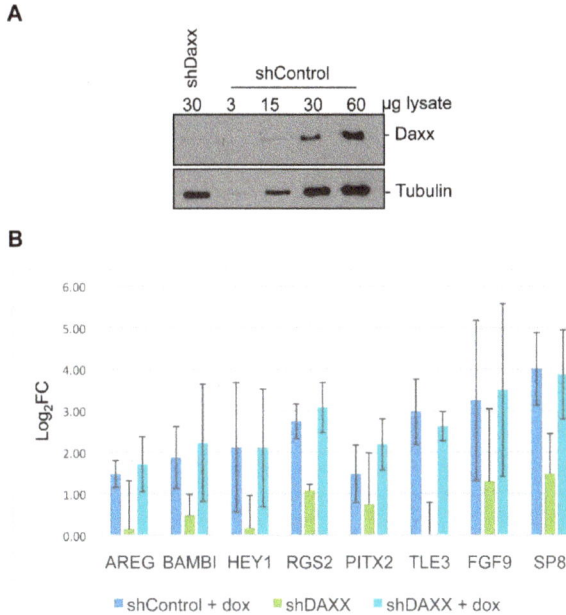

Figure 5. Loss of Daxx does not impede the up-regulation of E4-ORF3 target genes. (**A**) Pools of Tet-E4-ORF3 cells were stably infected with retrovirus that express short hairpin RNA directed against Daxx or a nonspecific control. 60, 30, 15, and 3 μg of shControl lysate were loaded on a polyacrylamide gel with 30 μg of shDaxx and subject to Western blot analysis; (**B**) shDaxx and shControl cells were mock-treated or supplemented with dox to express E4-ORF3 for 24 hours. Transcript levels were recorded by relative quantification RT-qPCR. Bars represent $\log_2$ fold change in expression in mock treated shDaxx cells, dox-induced shDaxx cells, or dox-induced shControl cells, relative to mock-treated shControl cells, normalized to GAPDH. Error bars = S.D., n = 3. No statistically significant difference in transcript levels could be detected between dox-induced control cells and dox-induced shDaxx cells as determined by one-way ANOVA analysis with multiple comparisons.

436

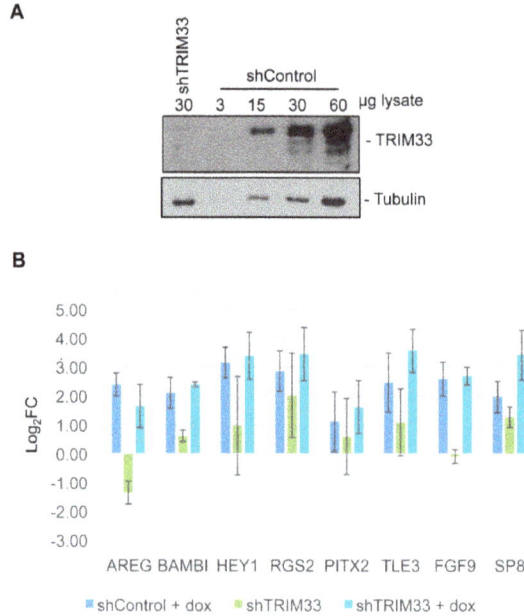

Figure 6. Loss of TRIM33 does not prevent the up-regulation of E4-ORF3 target genes. (A) Pools of Tet-E4-ORF3 cells were stably infected with retrovirus that express short hairpin RNA directed against TRIM33 or a nonspecific control. 60, 30, 15, and 3 μg of shControl lysate were loaded on a polyacrylamide gel with 30 μg of shTRIM33 and subject to Western blot analysis; (B) shTRIM33 and shControl cells were mock-treated or supplemented with dox to express E4-ORF3 for 24 h. Transcript levels were recorded by relative quantification RT-qPCR. Bars represent $\log_2$ fold change in expression in mock-treated shTRIM33 cells, dox-induced shTRIM33 cells, or dox-induced shControl cells, relative to mock-treated shControl cells, normalized to GAPDH. Error bars = S.D., n = 3. No statistically significant difference in transcript levels could be detected between dox-induced control cells and dox-induced shTRIM33 cells as determined by one-way ANOVA analysis with multiple comparisons.

Control or knock-down cells were left untreated or treated with dox to induce HA-E4-ORF3 expression. At 24-h post induction, total cellular RNA was isolated and used to generate cDNA to be used as a qPCR template. The Pfaffl method of relative quantification was used to determine the transcript levels of the eight genes of interest in knockdown cells relative to uninduced shControl cells (Figures 5B and 6B). Each experimental set was repeated a total of three times. Knockdown of Daxx or TRIM33 did not prevent the E4-ORF3-dependent up-regulation of any of the eight genes analyzed. Knockdown of Daxx alone modestly induced the expression of RGS2, FGF9, and SP8 (Figure 5B). Knockdown of TRIM33 alone reduced AREG expression and induced SP8 and RGS2 expression 2- to 4-fold, respectively (Figure 6B).

4. Discussion

Previous studies link E4-ORF3-induced nuclear track formation to disruption of the interferon-mediated antiviral response, inhibition of the DNA damage response, and altered p53-mediated signaling [8,17,18,21,22,25]. Here, we implicate a novel role for E4-ORF3 track formation in the regulation of cellular gene expression. We have identified over four hundred genes that are differentially expressed in response to E4-ORF3 expression, and provide evidence that track formation is necessary for these changes to occur with eight genes that were analyzed further. These E4-ORF3-dependent changes in cellular gene expression appear to be independent of MRN activity and resistant to Daxx and TRIM33 knockdown.

To allow for a more robust examination of potential E4-ORF3 target genes, we utilized two methods of microarray data analysis. First, we considered genes that displayed a two-fold or greater expression change at 24-h post induction relative to 6-h post induction in the presence of E4-ORF3 relative to dox alone (Supplementary Table S1). This method highlights genes that display a high degree of differential expression at times that correspond to high levels of E4-ORF3 expression. As an alternate approach, we performed a cluster analysis (Figure 2, Supplementary Tables S2 and S3). Highlighting genes with similar expression patterns over the time course study, this analysis generates clusters of potentially coregulated genes. These two approaches select for genes using different criteria, yielding overlapping, but non-identical pools of target genes. We considered gene expression trends between 6 and 24 h post induction, as well as the absolute magnitude of differential expression at individual time points. We investigated all genes that met the minimal standards of showing a significant two-fold or greater E4-ORF3-dependent expression change. In addition, we considered genes displaying smaller expression changes that cluster with other potentially relevant E4-ORF3 targets. We believe that using two methods of data analysis allows for a more comprehensive identification of potential genes of interest.

To gain a better understanding of how the E4-ORF3 target genes highlighted by the microarray experiment relate to Ad infection, we subjected the group of 409 E4-ORF3 target genes to gene ontology enrichment analysis (Table 1). Results implicate E4-ORF3 as influencing a wide variety of cellular processes including the regulation of cellular growth properties. Many of the E4-ORF3-influenced genes localize in the extracellular space or in the plasma membrane. It is interesting to speculate that E4-ORF3 influences genes that alter gene expression in neighboring cells, perhaps to prepare them to accept incoming Ad. E4-ORF3-regulated genes also fell into categories related to antiviral effects including inflammatory response, defense response, immune response, and cytokine activity. There was no obvious common transcriptional regulatory mechanism or pathway that relates to the regulation of these diverse genes. The fact that E4-ORF3 up-regulated ~70% of the target genes and down-regulated the remaining ~30% of the target genes (Supplementary Table S1) indicates that E4-ORF3 may influence the activities of different cellular effectors of gene regulation. This idea is consistent with the observation that E4-ORF3 regulates gene expression of some genes early (6-h post induction) and other genes later (by 24-h post induction; Supplementary Tables S2 and S3). The latter observation also is consistent with the possibility that

E4-ORF3 influences primary effectors early that result in secondary events at later times after E4-ORF3 expression.

Interestingly, genes that encode E4-ORF3 track-associated proteins were not highlighted by the cluster analysis or through the selection of highly differentially expressed genes at 24-h post induction. PML, MRN components, TRIM24, TRIM33, Daxx, and Sp100 transcript levels remained consistent in the presence or absence of E4-ORF3 expression. This suggests that the cell does not compensate for a decrease in free protein by increasing transcription of these target genes.

E4-ORF3 has been previously shown to inhibit p53-mediated signaling. However, we were unable to detect an enrichment of p53 targets in the population of E4-ORF3-influenced genes. Furthermore, E4-ORF3 did not alter p53 stability or phosphorylation status in Tet-E4-ORF3 cells (data not shown). Stressors including DNA damage, oncogenes, and the Ad E1A protein trigger the activation of p53-mediated signaling [54,55]. P53 likely remains repressed in Tet-E4-ORF3 cells in the absence of these stressors. This may prevent E4-ORF3 from further down-regulating p53-mediated signaling.

Eight genes were validated by RT-qPCR: HEY1, TLE3, SP8, and FGF9 were selected from cluster 1, and AREG, BAMBI, RGS2, and PITX2 were selected from cluster 2 (Figure 2, Supplementary Tables S2 and S3). These targets were selected on the basis of showing a large E4-ORF3-dependent change in expression. Furthermore, these genes of interest correspond to high quality probes on the microarray; displaying high levels of fluorescence relative to background, as well as pixel correlation values approaching one. These genes serve as readout of E4-ORF3-influenced transcriptional regulation. Identifying factors that impact the up-regulation of these genes helps to elucidate how E4-ORF3 may be influencing transcription.

Comparison of transcript levels after the introduction of wild-type or mutant E4-ORF3-expressing Ad vectors revealed that differential expression of the eight validated genes relied on E4-ORF3 nuclear track formation but was independent of the MRN complex. Mutant E4-ORF3$_{L103A}$ folds properly but fails to form nuclear tracks [14]. Failure of this mutant protein to up-regulate target gene expression suggests that the differential expression changes highlighted by the microarray experiment do not result from a stress response induced by an ectopic protein, but likely relies on the sequestration of a cellular protein. Mutant E4-ORF3$_{D105A/L106A}$ targets PML, TRIM24, and TRIM33 for track localization, but fails to sequester the MRN complex [17,26]. Like wild-type E4-ORF3, this mutant protein induced the up-regulation of select target genes. These results indicate that sequestration of the MRN complex does not influence expression of the tested genes, suggesting they function independent of the DNA damage response. To explore the requirement for the track localization of Daxx and TRIM33 on E4-ORF3 transcriptional regulation, we knocked down these proteins in Tet-E4-ORF3 cells. Neither reduction in Daxx nor TRIM33 protein levels prevented the E4-ORF3-dependent up-regulation of any of the eight validated genes. Interestingly, knockdown of Daxx alone modestly induced the expression of RGS2, FGF9, and SP8 (Figure 5B). Knockdown of TRIM33 alone reduced AREG expression and induced SP8 and RGS2 expression 2- to 4-fold, respectively (Figure 6B). TRIM33 and Daxx both have transcriptional repressor properties. It is interesting to speculate that Daxx and TRIM33 contribute to the down-regulation of RGS2, FGF9 and SP8 under normal cellular conditions. Sequestration into E4-ORF3 tracks could inhibit this transcriptional repression. Multiple E4-ORF3 track proteins may influence an overlapping pool of target genes. The simultaneous knockdown of multiple cellular effectors may have a greater influence on transcription than what was observed through knocking down of individual factors.

5. Conclusions

We conclude that the Ad E4-ORF3 protein influences the expression of a wide range of cellular genes involved in different cellular functions. E4-ORF3-induced nuclear track formation is necessary to induce these expression changes. The observed expression changes may be independent of Daxx and TRIM33 suggesting that an additional factor(s) may be responsible. The ability of E4-ORF3 to manipulate cellular gene expression through the sequestration of cellular proteins implicates a novel role for E4-ORF3 in transcriptional regulation.

Acknowledgments

We thank Thomas Dobner (Heinrich-Pette Institute) for the anti-E4-ORF3 monoclonal antibody (6A-11) and members of our laboratories for informed discussions. This work was supported by NIH grant CA122677 to P.H. E.V. was supported by NIH training grants T32CA009176 and T32GM007964. L.T.K. was supported by an American Cancer Research Scholar Grant, RSG-11-160-01-MPC, and NIH grant AI097875. T.S. was supported by the DFG (SFB796) and the IZKF Erlangen (project A62).

Author Contributions

E.V, Y.Z., and P.H. conceived and designed the experiments. E.V. performed the experiments. E.V., R.Y., L.K, and P.H. analyzed the data. Y.Z. and T.S. contributed reagents/materials/analysis tools. E.V. and P.H. wrote the manuscript. All authors read and approved the manuscript.

Conflicts of Interest

The authors declare no conflict of interest.

References

1. Bridge, E.; Ketner, G. Redundant control of adenovirus late gene expression by early region 4. *J. Virol.* **1989**, *63*, 631–638.
2. Huang, M.M.; Hearing, P. Adenovirus early region 4 encodes two gene products with redundant effects in lytic infection. *J. Virol.* **1989**, *63*, 2605–2615.
3. Sandler, A.B.; Ketner, G. Adenovirus early region 4 is essential for normal stability of late nuclear RNAs. *J. Virol.* **1989**, *63*, 624–630.
4. Cheng, C.Y.; Blanchette, P.; Branton, P.E. The adenovirus E4orf6 E3 ubiquitin ligase complex assembles in a novel fashion. *Virology* **2007**, *364*, 36–44.
5. Cheng, C.Y.; Gilson, T.; Dallaire, F.; Ketner, G.; Branton, P.E.; Blanchette, P. The E4orf6/E1B55K E3 ubiquitin ligase complexes of human adenoviruses exhibit heterogeneity in composition and substrate specificity. *J. Virol.* **2011**, *85*, 765–775.

6. Harada, J.N.; Shevchenko, A.; Pallas, D.C.; Berk, A.J. Analysis of the adenovirus E1B-55K-anchored proteome reveals its link to ubiquitination machinery. *J. Virol.* **2002**, *76*, 9194–9206.

7. Querido, E.; Blanchette, P.; Yan, Q.; Kamura, T.; Morrison, M.; Boivin, D.; Kaelin, W.G.; Conaway, R.C.; Conaway, J.W.; Branton, P.E. Degradation of p53 by adenovirus E4orf6 and E1B55K proteins occurs via a novel mechanism involving a Cullin-containing complex. *Genes Dev.* **2001**, *15*, 3104–3117.

8. Stracker, T.H.; Carson, C.T.; Weitzman, M.D. Adenovirus oncoproteins inactivate the Mre11-Rad50-NBS1 DNA repair complex. *Nature* **2002**, *418*, 348–352.

9. Baker, A.; Rohleder, K.J.; Hanakahi, L.A.; Ketner, G. Adenovirus E4 34k and E1b 55k oncoproteins target host DNA ligase IV for proteasomal degradation. *J. Virol.* **2007**, *81*, 7034–7040.

10. Dallaire, F.; Blanchette, P.; Groitl, P.; Dobner, T.; Branton, P.E. Identification of integrin alpha3 as a new substrate of the adenovirus E4orf6/E1B 55-kilodalton E3 ubiquitin ligase complex. *J. Virol.* **2009**, *83*, 5329–5338.

11. Orazio, N.I.; Naeger, C.M.; Karlseder, J.; Weitzman, M.D. The adenovirus E1b55K/E4orf6 complex induces degradation of the Bloom helicase during infection. *J. Virol.* **2011**, *85*, 1887–1892.

12. Doucas, V.; Ishov, A.M.; Romo, A.; Juguilon, H.; Weitzman, M.D.; Evans, R.M.; Maul, G.G. Adenovirus replication is coupled with the dynamic properties of the PML nuclear structure. *Genes Dev.* **1996**, *10*, 196–207.

13. Ou, H.D.; Kwiatkowski, W.; Deerinck, T.J.; Noske, A.; Blain, K.Y.; Land, H.S.; Soria, C.; Powers, C.J.; May, A.P.; Shu, X.; *et al.* A structural basis for the assembly and functions of a viral polymer that inactivates multiple tumor suppressors. *Cell* **2012**, *151*, 304–319.

14. Patsalo, V.; Yondola, M.A.; Luan, B.; Shoshani, I.; Kisker, C.; Green, D.F.; Raleigh, D.P.; Hearing, P. Biophysical and functional analyses suggest that adenovirus E4-ORF3 protein requires higher-order multimerization to function against promyelocytic leukemia protein nuclear bodies. *J. Biol. Chem.* **2012**, *287*, 22573–22583.

15. Weitzman, M.D.; Ornelles, D.A. Inactivating intracellular antiviral responses during adenovirus infection. *Oncogene* **2005**, *24*, 7686–7696.

16. Sohn, S.Y.; Hearing, P. Adenovirus regulates sumoylation of Mre11-Rad50-Nbs1 components through a paralog-specific mechanism. *J. Virol.* **2012**, *86*, 9656–9665.

17. Evans, J.D.; Hearing, P. Relocalization of the Mre11-Rad50-Nbs1 complex by the adenovirus E4 ORF3 protein is required for viral replication. *J. Virol.* **2005**, *79*, 6207–6215.

18. Soria, C.; Estermann, F.E.; Espantman, K.C.; O'Shea, C.C. Heterochromatin silencing of p53 target genes by a small viral protein. *Nature* **2010**, *466*, 1076–1081.

19. Vink, E.I.; Yondola, M.A.; Wu, K.; Hearing, P. Adenovirus E4-ORF3-dependent relocalization of TIF1alpha and TIF1gamma relies on access to the Coiled-Coil motif. *Virology* **2012**, *422*, 317–325.

20. Yondola, M.A.; Hearing, P. The adenovirus E4 ORF3 protein binds and reorganizes the TRIM family member transcriptional intermediary factor 1 alpha. *J. Virol.* **2007**, *81*, 4264–4271.

21. Ullman, A.J.; Hearing, P. Cellular proteins PML and Daxx mediate an innate antiviral defense antagonized by the adenovirus E4 ORF3 protein. *J. Virol.* **2008**, *82*, 7325–7335.

22. Ullman, A.J.; Reich, N.C.; Hearing, P. Adenovirus E4 ORF3 protein inhibits the interferon-mediated antiviral response. *J. Virol.* **2007**, *81*, 4744–4752.

23. Karen, K.A.; Hoey, P.J.; Young, C.S.; Hearing, P. Temporal regulation of the Mre11-Rad50-Nbs1 complex during adenovirus infection. *J. Virol.* **2009**, *83*, 4565–4573.

24. Weiden, M.D.; Ginsberg, H.S. Deletion of the E4 region of the genome produces adenovirus DNA concatemers. *Proc. Natl. Acad. Sci. USA* **1994**, *91*, 153–157.

25. Lakdawala, S.S.; Schwartz, R.A.; Ferenchak, K.; Carson, C.T.; McSharry, B.P.; Wilkinson, G.W.; Weitzman, M.D. Differential requirements of the C terminus of Nbs1 in suppressing adenovirus DNA replication and promoting concatemer formation. *J. Virol.* **2008**, *82*, 8362–8372.

26. Evans, J.D.; Hearing, P. Distinct roles of the Adenovirus E4 ORF3 protein in viral DNA replication and inhibition of genome concatenation. *J. Virol.* **2003**, *77*, 5295–5304.

27. Mathew, S.S.; Bridge, E. Nbs1-dependent binding of Mre11 to adenovirus E4 mutant viral DNA is important for inhibiting DNA replication. *Virology* **2008**, *374*, 11–22.

28. Geoffroy, M.C.; Chelbi-Alix, M.K. Role of promyelocytic leukemia protein in host antiviral defense. *J. Interferon Cytokine Res.* **2011**, *31*, 145–158.

29. Van Damme, E.; Laukens, K.; Dang, T.H.; van Ostade, X. A manually curated network of the PML nuclear body interactome reveals an important role for PML-NBs in SUMOylation dynamics. *Int. J. Biol. Sci.* **2010**, *6*, 51–67.

30. Cammas, F.; Khetchoumian, K.; Chambon, P.; Losson, R. TRIM involvement in transcriptional regulation. *Adv. Exp. Med. Biol.* **2012**, *770*, 59–76.

31. Ikeda, K.; Inoue, S. TRIM proteins as RING finger E3 ubiquitin ligases. *Adv. Exp. Med. Biol.* **2012**, *770*, 27–37.

32. Napolitano, L.M.; Meroni, G. TRIM family: Pleiotropy and diversification through homomultimer and heteromultimer formation. *IUBMB Life* **2012**, *64*, 64–71.

33. Allton, K.; Jain, A.K.; Herz, H.M.; Tsai, W.W.; Jung, S.Y.; Qin, J.; Bergmann, A.; Johnson, R.L.; Barton, M.C. Trim24 targets endogenous p53 for degradation. *Proc. Natl. Acad. Sci. USA* **2009**, *106*, 11612–11616.

34. Dupont, S.; Zacchigna, L.; Cordenonsi, M.; Soligo, S.; Adorno, M.; Rugge, M.; Piccolo, S. Germ-layer specification and control of cell growth by Ectodermin, a Smad4 ubiquitin ligase. *Cell* **2005**, *121*, 87–99.

35. Ikeuchi, Y.; Dadakhujaev, S.; Chandhoke, A.S.; Huynh, M.A.; Oldenborg, A.; Ikeuchi, M.; Deng, L.; Bennett, E.J.; Harper, J.W.; Bonni, A.; Bonni, S. TIF1gamma protein regulates epithelial-mesenchymal transition by operating as a small ubiquitin-like modifier (SUMO) E3 ligase for the transcriptional regulator SnoN1. *J. Biol. Chem.* **2014**, *289*, 25067–25078.

36. Khetchoumian, K.; Teletin, M.; Tisserand, J.; Mark, M.; Herquel, B.; Ignat, M.; Zucman-Rossi, J.; Cammas, F.; Lerouge, T.; Thibault, C.; Metzger, D.; Chambon, P.; Losson, R. Loss of Trim24 (Tif1alpha) gene function confers oncogenic activity to retinoic acid receptor alpha. *Nat. Genet.* **2007**, *39*, 1500–1506.

37. Tsai, W.W.; Wang, Z.; Yiu, T.T.; Akdemir, K.C.; Xia, W.; Winter, S.; Tsai, C.Y.; Shi, X.; Schwarzer, D.; Plunkett, W.; *et al.* TRIM24 links a non-canonical histone signature to breast cancer. *Nature* **2010**, *468*, 927–932.

38. Zhong, S.; Delva, L.; Rachez, C.; Cenciarelli, C.; Gandini, D.; Zhang, H.; Kalantry, S.; Freedman, L.P.; Pandolfi, P.P. A RA-dependent, tumour-growth suppressive transcription complex is the target of the PML-RARalpha and T18 oncoproteins. *Nat. Genet.* **1999**, *23*, 287–295.

39. He, W.; Dorn, D.C.; Erdjument-Bromage, H.; Tempst, P.; Moore, M.A.; Massague, J. Hematopoiesis controlled by distinct TIF1gamma and Smad4 branches of the TGFbeta pathway. *Cell* **2006**, *125*, 929–941.

40. Agricola, E.; Randall, R.A.; Gaarenstroom, T.; Dupont, S.; Hill, C.S. Recruitment of TIF1gamma to chromatin via its PHD finger-bromodomain activates its ubiquitin ligase and transcriptional repressor activities. *Mol. Cell* **2011**, *43*, 85–96.

41. Xi, Q.; Wang, Z.; Zaromytidou, A.I.; Zhang, X.H.; Chow-Tsang, L.F.; Liu, J.X.; Kim, H.; Barlas, A.; Manova-Todorova, K.; Kaartinen, V.; *et al.* A poised chromatin platform for TGF-beta access to master regulators. *Cell* **2011**, *147*, 1511–1524.

42. Nevels, M.; Tauber, B.; Kremmer, E.; Spruss, T.; Wolf, H.; Dobner, T. Transforming potential of the adenovirus type 5 E4orf3 protein. *J. Virol.* **1999**, *73*, 1591–1600.

43. Smyth, G.K. Linear models and empirical bayes methods for assessing differential expression in microarray experiments. *Stat. Appl. Genet. Mol. Biol.* **2004**, *3*, 1–25.

44. Smyth, G.K.; Speed, T. Normalization of cDNA microarray data. *Methods* **2003**, *31*, 265–273.

45. de Hoon, M.J.; Imoto, S.; Nolan, J.; Miyano, S. Open source clustering software. *Bioinformatics* **2004**, *20*, 1453–1454.

46. Saldanha, A.J. Java Treeview—Extensible visualization of microarray data. *Bioinformatics* **2004**, *20*, 3246–3248.

47. Huang da, W.; Sherman, B.T.; Lempicki, R.A. Bioinformatics enrichment tools: Paths toward the comprehensive functional analysis of large gene lists. *Nucl. Acids Res.* **2009**, *37*, 1–13.

48. Huang da, W.; Sherman, B.T.; Lempicki, R.A. Systematic and integrative analysis of large gene lists using DAVID bioinformatics resources. *Nat. Protoc.* **2009**, *4*, 44–57.

49. Cui, W.; Taub, D.D.; Gardner, K. qPrimerDepot: A primer database for quantitative real time PCR. *Nucl. Acids Res.* **2007**, *35*, D805–D809.

50. Lefever, S.; Vandesompele, J.; Speleman, F.; Pattyn, F. RTPrimerDB: The portal for real-time PCR primers and probes. *Nucl. Acids Res.* **2009**, *37 (Suppl. 1)*, D942–D945.

51. Pfaffl, M.W. A new mathematical model for relative quantification in real-time RT-PCR. *Nucl. Acids Res.* **2001**, *29*, e45.

52. Tavalai, N.; Papior, P.; Rechter, S.; Leis, M.; Stamminger, T. Evidence for a role of the cellular ND10 protein PML in mediating intrinsic immunity against human cytomegalovirus infections. *J. Virol.* **2006**, *80*, 8006–8018.

53. Salomoni, P. The PML-Interacting Protein DAXX: Histone Loading Gets into the Picture. *Front. Oncol.* **2013**, *3*, e152.

54. Lowe, S.W.; Ruley, H.E. Stabilization of the p53 tumor suppressor is induced by adenovirus 5 E1A and accompanies apoptosis. *Genes Dev.* **1993**, *7*, 535–545.

55. Debbas, M.; White, E. Wild-type p53 mediates apoptosis by E1A, which is inhibited by E1B. *Genes Dev.* **1993**, *7*, 546–554.

Role of Host MicroRNAs in Kaposi's Sarcoma-Associated Herpesvirus Pathogenesis

Zhiqiang Qin, Francesca Peruzzi, Krzysztof Reiss and Lu Dai

Abstract: MicroRNAs (miRNAs) are small non-coding RNA species that can bind to both untranslated and coding regions of target mRNAs, causing their degradation or post-transcriptional modification. Currently, over 2500 miRNAs have been identified in the human genome. Burgeoning evidence suggests that dysregulation of human miRNAs can play a role in the pathogenesis of a variety of diseases, including cancer. In contrast, only a small subset of human miRNAs has been functionally validated in the pathogenesis of oncogenic viruses, in particular, Kaposi's sarcoma-associated herpesvirus (KSHV). KSHV is the etiologic agent of several human cancers, such as primary effusion lymphoma (PEL) and Kaposi's sarcoma (KS), which are mostly seen in acquired immune deficiency syndrome (AIDS) patients or other immuno-suppressed subpopulation. This review summarizes recent literature outlining mechanisms for KSHV/viral proteins regulation of cellular miRNAs contributing to viral pathogenesis, as well as recent findings about the unique signature of miRNAs induced by KSHV infection or KSHV-related malignancies.

Reprinted from *Viruses*. Cite as: Qin, Z.; Peruzzi, F.; Reiss, K.; Dai, L. Role of Host MicroRNAs in Kaposi's Sarcoma-Associated Herpesvirus Pathogenesis. *Viruses* **2014**, *6*, 4571-4580.

1. Introduction

MicroRNAs (miRNAs) are small (~19–24 nucleotides in length), non-coding RNAs that bind to both, untranslated and coding regions of target mRNAs, causing their degradation or post-transcriptional modification. The biogenesis of miRNAs begins in the nucleus where RNA polymerase II generates primary miRNA (pri-miRNAs) transcripts, which are subsequently processed by the RNase III enzyme Drosha, generating precursor miRNAs (pre-miRNAs). Pre-miRNAs are transported from the nucleus to the cytoplasm where they are cleaved by the cytoplasmic RNase III enzyme, Dicer, generating mature miRNAs, which are incorporated into the RNA-induced silencing complex (RISC) [1,2]. Published literature has demonstrated that miRNAs regulate a variety of physiological and pathological processes in the cell, such as cell proliferation, apoptosis, differentiation, development, mobility, invasiveness, and angiogenesis [3].

miRNAs are encoded by many different organisms including viruses, in which miRNA sequences and functions are often different from human miRNAs [4]. For example, several oncogenic herpesviruses have been found encoding multiple miRNAs in their genomes, such as Kaposi's sarcoma-associated herpesvirus (KSHV), which is the etiologic agent of human cancers including multicentric Castleman's disease (MCD), primary effusion lymphoma (PEL), and Kaposi's sarcoma (KS) [5–7]. Thus far, 12 KSHV pre-miRNAs (miR-K12-1~miR-K12-12), encoding 18 mature miRNAs, have been identified within the viral genome [8,9]. These miRNAs are located in the KSHV latency-associated region (KLAR), together with several KSHV-encoded latent proteins, which are critical for maintenance of the viral episome and for KSHV-mediated oncogenesis. In fact, most of these KSHV-miRNAs are expressed in different KSHV-infected host cells and/or KSHV-related tumor tissues, and play important roles in viral pathogenesis and tumorigenesis, which have been comprehensively reviewed by us and others before [10–13]. Interestingly, several KSHV-miRNAs, including miR-K12-11, miR-K12-10a and miR-K12-3, can act as viral analogs of the human cellular miRNAs miR-155, miR-142-3p and miR-23, respectively. As a result of such homology in the seeding sequence, cellular and viral miRNAs can share the repertoire of targeted genes [14–17]. In contrast to the well-defined KSHV-miRNAs, we know little about how human miRNAs can be regulated by KSHV/viral proteins and functionally involved in viral pathogenesis and tumorigenesis. The current review will summarize recent findings regarding the regulatory function of human miRNAs contributing to the pathogenesis of KSHV-infected host cells and KSHV-related malignancies.

2. Human miRNAs and KSHV-Induced Cell Mobility and Angiogenesis

Acquisition of a migratory or invasive phenotype represents one of the hallmarks of KSHV-infected endothelial cells, with implications for both, viral dissemination and angiogenesis within KS lesions [18,19]. Moreover, KS is characterized by the proliferation of infected spindle cells of vascular and lymphatic endothelial origin, accompanied by intense angiogenesis with erythrocyte extravasation and inflammatory infiltration [20,21]. Recent reports provide the solid evidence that cellular miRNAs are involved in KSHV-induced cell mobility and angiogenesis. Tsai and colleagues reported that KSHV-encoded K15 protein, minor form (K15M), can induce cell migration and invasion, potentially through upregulation of cellular miR-21 and miR-31 via its conserved Src-Homology 2 (SH2)-binding motif [22]. In contrast, knocking down both miR-21 and miR-31 inhibited K15M-mediated cell motility, which indicated that targeting K15 or its downstream-regulated microRNAs may represent novel therapies for treatment of KSHV-associated neoplasia [22]. Upregulation of miR-31 by KSHV was further confirmed in virally infected lymphatic endothelial cells (LECs), in which depletion of miR-31 reduced cell mobility [23]. One of the mechanisms identified for the miR-31 mediated increasing cell motility was through direct repression of a novel tumor suppressor and inhibitor of migration, FAT4; moreover, a reduction of FAT4 enhanced EC mobility [23]. Through the analysis of miRNAs microarray data, the miR-221/miR-222 cluster was found significantly downregulated in KSHV-infected LECs and resulted in an increase of EC migration, potentially through KSHV-encoded latency-associated nuclear antigen (LANA) and Kaposin B proteins [23]. Further experimental data confirmed that the transcription factors, ETS2 and ETS1, were the downstream targets of miR-221 and miR-222, respectively, and overexpression of ETS1 or ETS2 alone was sufficient to induce EC migration [23]. In addition to these factors, KSHV can also downregulate miR-30b and miR-30c, whereas increasing the expression of their direct target, Delta-like 4 (DLL4), a functional protein in vascular development and angiogenesis [24], can induce KSHV-mediated LECs angiogenesis [25]. Interestingly, these miRNAs (miR-21, miR-31, miR-221/222, miR-30) can act as either "oncogenes" or "tumor-suppressor genes" in a variety of cancers in which they can regulate tumor cell proliferation, apoptosis, invasion, angiogenesis, metastasis and other important cellular functions [26–30], indicating functional relevance of these regulatory miRNAs in virus-related malignancies.

3. Human miRNAs and KSHV Lifecycle/Replication

KSHV lifecycle involves two distinct phases: latent and lytic. During latent infection, which represents the predominant phase in the majority of infected cells, only a limited number of viral genes are expressed. Exposure to a variety of stimuli induces lytic replication, resulting in virion assembly and release of infectious viral particles [31]. Maintenance of latent KSHV infection, coordinated with lytic reactivation within a small subset of infected cells, is critical for promotion of KSHV persistence and dissemination. Several published studies have demonstrated a role for specific KSHV-miRNAs in maintaining viral latency through either direct targeting of the viral lytic reactivation activator, *Rta* (ORF50) [32,33], or via indirect mechanisms including targeting host factors such as IκBα, nuclear factor I/B (NFIB), Rbl2, BCLAF1, and IKKε [34–38]. In contrast, little is known about the role of human miRNAs regulating "latent to lytic" switch in KSHV lifecycle. Recent miRNA profiling studies indicated that ectopic expression of HIV-encoded Nef protein can suppress the expression of KSHV lytic proteins and the production of infectious viral particles, potentially through regulation of cellular miRNAs [39]. Indeed, at least five of the 99 miRNAs upregulated by Nef (miR-557, miR-766, miR-1227, miR-1258, and miR-1301) had putative binding sites in the 3' UTR of viral lytic reactivation activator, *Rta* [39]. Further data confirmed that ectopic expression of miR-1258 impaired RTA synthesis and enhanced Nef-mediated inhibition of KSHV replication [39]. Based on the complex mechanisms (direct and indirect) for KSHV lytic reactivation mentioned above, it is reasonable to conceive that there should be even more cellular miRNAs involved in the regulation of KSHV lifecycle and replication.

4. Human miRNAs and KSHV-Induced Cytokine Response and Immune Recognition

KSHV infection can induce a variety of pro-migratory, pro-angiogenic and pro-inflammatory cytokines and chemokines to promote viral pathogenesis and survival of the infected cells [21]. Moreover, to establish a life-long persistent infection, KSHV has evolved a complex mechanism by which unique viral proteins antagonize host innate and adaptive immunity [40]. For example, KHSV infection and ectopic expression of KSHV-encoded viral FLICE inhibitory protein (vFLIP) suppressed the expression of one chemokine receptor, CXCR4 [41]. Suppression of CXCR4 by KSHV and vFLIP was associated with the upregulation of cellular miR-146a expression, a miRNA that is known to bind to the 3'UTR of CXCR4 mRNA. Further data confirmed that upregulation of miR-146a required vFLIP-induced NF-κB activities, because vFLIP NF-κB-defective mutant lost such ability. For clinical relevance, downregulation of CXCR4 accompanied by increased expression of miR-146a has been found in the KS tissues derived from patients, and it could contribute to KS development by promoting premature release of KSHV-infected endothelial progenitors into the circulation [41].

KSHV encodes a viral interleukin 6 (vIL-6) that mimics many functions of human IL-6 (hIL-6), since they can both stimulate the proliferation of tumors caused by KSHV and play a role in the inflammatory cytokine syndrome associated with HIV and KSHV co-infection [42–45]. Kang and colleagues recently identified a direct repression of vIL6 by cellular miR-1293 and hIL6 by miR-608 through binding sites in their ORF sequences [46]. More importantly, miR-1293 is primarily expressed in the germinal center, but is not present in the mantle zone of human lymph nodes where the expression of vIL6 is often found in patients with KSHV-associated MCD [46]. Interestingly, the KSHV-encoded ORF57 protein appeared to compete with miR-1293 and/or miR-608 for the same binding site in the vIL-6 and/or hIL-6 RNAs, thereby preventing vIL-6 and/or hIL-6 RNA degradation from association with the miR-1293/miR-608-specified RISC [47]. These data have demonstrated that KSHV has evolved a "smart strategy" by using viral proteins against host miRNAs regulation to its own advantage, including mechanisms that would allow survival of infected cells and promote virus-associated malignancies.

In addition, Lagos *et al.* reported two groups of cellular miRNAs induced during primary KSHV infection of LECs: the "early" group reached its peak of expression at six hours post-infection, and included miR-146a, miR-31 and miR-132; the "late" group, which included miR-193a and Let-7i, steadily increased its expression during the next 72 hours [48]. One of the lately expressed miRNAs, the highly upregulated miR-132, has been shown to negatively affect the expression of interferon-stimulated genes through suppression of the p300 transcriptional co-activator, facilitating viral gene expression and replication [48]. These data clearly indicate that this oncogenic virus can use host miRNAs to regulate antiviral innate immunity to promote survival of the infected cells. Interestingly, a similar induction of functional miR-132 was also observed during infection of monocytes with herpes simplex virus-1 (HSV-1) and cytomegalovirus (CMV) [48], which indicated that this kind of miRNA-mediated antiviral immune response could be triggered by several viruses.

5. Human miRNAs Profile in KSHV-Related Malignancies

miRNA microarray combined with bioinformatics analysis represents a powerful tool to understand miRNA global profile unique to certain cancer cells/tissues when compared with normal controls. Recent studies using different miRNA microarray have gained insight into the cellular miRNAs profile altered in KSHV-related malignancies including KS and PEL. Wu *et al.* performed a miRNA profiling by analyzing six paired KS and matched adjacent healthy tissues using the miRCURY™ LNA Array (v.18.0) (Exiqon, Woburn, MA, USA) which contains 3100 capture probes, covering all human, mouse, and rat miRNAs annotated in miRBase 18.0, as well as all viral microRNAs related to these species [49]. They identified 170 differentially expressed miRNAs (69 upregulated and 101 downregulated) in KS *versus* adjacent healthy tissues. Among them, the most significantly upregulated human miRNAs included miR-126-3p, miR-199a-3p, and miR-16-5p, while the most significantly downregulated miRNAs included miR-125b-1-3p and miR-1183. Of those, miR-125b-1-3p and miR-16-5p had statistically significant associations with KSHV and HIV infections in KS. Catrina Ene and colleagues performed miRNA microarray analysis of 17 KS specimens and three normal skin specimens as controls, using the miRNA Microarray Kit V2 platform which contains 723 human and 76 human viral miRNAs from the Sanger database v.10.1 (Agilent Technologies, Santa Clara, CA, USA) [50]. They detected 185 differentially expressed miRNAs in KS *versus* normal skin; of those, 76 were upregulated and 109 were downregulated. The most significantly upregulated human miRNAs were miR-513a-3p, miR-298, and miR-206;

whereas miR-99a, miR-200 family, miR-199b-5p, miR-100, and miR-335 were the most significantly downregulated miRNAs. We assume that the differential signature of miRNAs found in KS samples from these two studies is probably caused by the different patient race, geography area (the first study was conducted in Asia and the second was in Europe), miRNA microarray platforms and data analysis methods. Although informative, perhaps larger and more systematic studies should be designed to find a definitive miRNA signature that could distinguish those different KS subtypes, including classic KS, endemic KS, iatrogenic KS, and epidemic KS (AIDS-KS) [21].

In another KSHV-caused malignancy, PEL, O'Hara *et al.* identified 68 PEL-specific miRNAs by using a TaqMan-based miRNA array [51]. Interestingly, some tumor suppressor miRNAs, including miR-221/222 and let-7 family members, were found significantly downregulated in KSHV-related malignancies, such as PEL and KS [52]. Therefore, downregulation of these tumor suppressor miRNAs may represent an alternative mechanism of KSHV-mediated transformation.

In addition to intracellular miRNAs, circulating miRNAs such as those found in exosomes, have emerged as powerful diagnostic tools and may act as minimally invasive, stable biomarkers. Transfer of tumor-derived exosomal miRNAs to surrounding cells may represent an important form of cellular communication [53–55]. Recently, Chugh and colleagues measured the host circulating miRNAs in plasma, pleural fluid or serum from patients with KS or PEL and from two mouse models of KS [56]. They found that many host miRNAs in particular the members of miR-17-92 cluster, were detectable within patient exosomes and circulating miRNA profiles from KSHV mouse models. Moreover, a subset of miRNAs including miR-19a, miR-21, miR-27a, miR-130, and miR-146a, seemed to be preferentially incorporated into exosomes, suggesting their potential use as biomarkers for KSHV-associated diseases [56]. In addition to circulating miRNAs and exosomes, more than 100 cellular miRNAs and abundant U2 small nuclear RNAs (snRNA)-derived unusual small RNAs (usRNAs) were detected in KSHV virions using the deep-sequencing technology [57]. Similarly, some miRNAs including miR-143, miR-23a, miR-130b, miR-451, and miR-185, were found preferentially packaged into KSHV virions when compared with intracellular miRNAs profile in KSHV-infected cells [57]. Taken together, these recent findings indicate that KSHV-related malignancies have a unique signature, not only for intracellular miRNAs, but also extracellular miRNAs (circulating and virion-packed miRNAs), although the underlying mechanisms and individual miRNA-mediated functions remain largely unknown within these malignancies.

6. Concluding Remarks

Until now, over 2500 human miRNAs have been identified and the list is still growing based on the miRBase database [58]. However, only a very small subset of human miRNAs as "tip of the iceberg" has been functionally validated in KSHV-infected cells or KSHV-related malignancies (summarized in Table 1). In fact, recent array-based data have indicated that either KSHV infection or KSHV-related malignancies can induce the unique signature of human intracellular and extracellular miRNAs [49–52,56,57], suggesting that more host miRNAs are likely involved in the regulation of KSHV pathogenesis. Functional validation of these human miRNAs and their respective target genes in future investigations will eventually help to better understand this virus-host interaction network and provide the framework for the development of more effective strategies targeting host miRNAs-mediated regulatory network against KSHV-related diseases.

Table 1. Overview of human miRNAs regulated by KSHV/viral proteins.

Human miRNAs	Validated Targets	Regulated by Viral Proteins	Functions	References
miR-21	-	K15M	Cell mobility	[22]
miR-31	FAT4	K15M	Cell mobility	[22,23]
miR-221/222	ETS2/ETS1	LANA and Kaposin B	Cell migration	[23]
miR-30b/c	DLL4	-	Angiogenesis	[25]
miR-557/766/1227/1258/1301	RTA	-	Viral replication	[39]
miR-146a	CXCR4	vFLIP	Immune response	[41]
miR-1293	vIL-6	ORF57	Immune response	[46,47]
miR-608	hIL-6	ORF57	Immune response	[46,47]
miR-132	p300	-	Immune escape	[48]

Acknowledgments

This work was supported by grants from a Center for Biomedical Research Excellence P20-GM103501 subaward (RR021970), the Ladies Leukemia League Grant (2014-2015), and the National Natural Science Foundation (NNSF) of China (81101791, 81272191, 81472547 and 81400164).

Author Contributions

LD and ZQ: wrote the paper. FP and KR: scientific discussion and editorial corrections.

Conflicts of Interest

The authors declare no conflict of interest.

References and Notes

1. Moens, U. Silencing viral microRNA as a novel antiviral therapy? *J. Biomed. Biotechnol.* **2009**, *2009*, 419539. doi:10.1155/2009/419539.
2. Bartel, D.P. MicroRNAs: Genomics, biogenesis, mechanism, and function. *Cell* **2004**, *116*, 281–297.
3. Ambros, V. The functions of animal microRNAs. *Nature* **2004**, *431*, 350–355.
4. Cullen, B.R. Viral and cellular messenger RNA targets of viral microRNAs. *Nature* **2009**, *457*, 421–425.
5. Soulier, J.; Grollet, L.; Oksenhendler, E.; Cacoub, P.; Cazals-Hatem, D.; Babinet, P.; d'Agay, M.F.; Clauvel, J.P.; Raphael, M.; Degos, L.; *et al.* Kaposi's sarcoma-associated herpesvirus-like DNA sequences in multicentric Castleman's disease. *Blood* **1995**, *86*, 1276–1280.

6. Cesarman, E.; Chang, Y.; Moore, P.S.; Said, J.W.; Knowles, D.M. Kaposi's sarcoma-associated herpesvirus-like DNA sequences in AIDS-related body-cavity-based lymphomas. *N. Engl. J. Med.* **1995**, *332*, 1186–1191.

7. Chang, Y.; Cesarman, E.; Pessin, M.S.; Lee, F.; Culpepper, J.; Knowles, D.M.; Moore, P.S. Identification of herpesvirus-like DNA sequences in AIDS-associated Kaposi's sarcoma. *Science* **1994**, *266*, 1865–1869.

8. Cai, X.; Lu, S.; Zhang, Z.; Gonzalez, C.M.; Damania, B.; Cullen, B.R. Kaposi's sarcoma-associated herpesvirus expresses an array of viral microRNAs in latently infected cells. *Proc. Natl. Acad. Sci. USA* **2005**, *102*, 5570–5575.

9. Samols, M.A.; Hu, J.; Skalsky, R.L.; Renne, R. Cloning and identification of a microRNA cluster within the latency-associated region of Kaposi's sarcoma-associated herpesvirus. *J. Virol.* **2005**, *79*, 9301–9305.

10. Zhu, Y.; Haecker, I.; Yang, Y.; Gao, S.J.; Renne, R. gamma-Herpesvirus-encoded miRNAs and their roles in viral biology and pathogenesis. *Curr. Opin. Virol.* **2013**, *3*, 266–275.

11. Qin, Z.; Jakymiw, A.; Findlay, V.; Parsons, C. KSHV-Encoded MicroRNAs: Lessons for Viral Cancer Pathogenesis and Emerging Concepts. *Int. J. Cell Biol.* **2012**, *2012*, 603961.

12. Ziegelbauer, J.M. Functions of Kaposi's sarcoma-associated herpesvirus microRNAs. *Biochim Biophys Acta* **2011**, *1809*, 623–630.

13. Ganem, D.; Ziegelbauer, J. MicroRNAs of Kaposi's sarcoma-associated herpes virus. *Semin. Cancer Biol.* **2008**, *18*, 437–440.

14. Gottwein, E.; Corcoran, D.L.; Mukherjee, N.; Skalsky, R.L.; Hafner, M.; Nusbaum, J.D.; Shamulailatpam, P.; Love, C.L.; Dave, S.S.; Tuschl, T.; *et al.* Viral microRNA targetome of KSHV-infected primary effusion lymphoma cell lines. *Cell Host Microbe* **2011**, *10*, 515–526.

15. Gottwein, E.; Mukherjee, N.; Sachse, C.; Frenzel, C.; Majoros, W.H.; Chi, J.T.; Braich, R.; Manoharan, M.; Soutschek, J.; Ohler, U.; *et al.* A viral microRNA functions as an orthologue of cellular miR-155. *Nature* **2007**, *450*, 1096–1099.

16. Skalsky, R.L.; Samols, M.A.; Plaisance, K.B.; Boss, I.W.; Riva, A.; Lopez, M.C.; Baker, H.V.; Renne, R. Kaposi's sarcoma-associated herpesvirus encodes an ortholog of miR-155. *J. Virol.* **2007**, *81*, 12836–12845.

17. Manzano, M.; Shamulailatpam, P.; Raja, A.N.; Gottwein, E. Kaposi's sarcoma-associated herpesvirus encodes a mimic of cellular miR-23. *J. Virol.* **2013**, *87*, 11821–11830.

18. Qin, Z.; Dai, L.; Slomiany, M.G.; Toole, B.P.; Parsons, C. Direct activation of emmprin and associated pathogenesis by an oncogenic herpesvirus. *Cancer Res.* **2010**, *70*, 3884–3889.

19. Qin, Z.; Dai, L.; Defee, M.; Findlay, V.J.; Watson, D.K.; Toole, B.P.; Cameron, J.; Peruzzi, F.; Kirkwood, K.; Parsons, C. Kaposi's Sarcoma-Associated Herpesvirus Suppression of DUSP1 Facilitates Cellular Pathogenesis following De Novo Infection. *J. Virol.* **2013**, *87*, 621–635.

20. Ganem, D. KSHV and the pathogenesis of Kaposi sarcoma: Listening to human biology and medicine. *J. Clin. Investig.* **2010**, *120*, 939–949.

21. Mesri, E.A.; Cesarman, E.; Boshoff, C. Kaposi's sarcoma and its associated herpesvirus. *Nat. Rev. Cancer* **2010**, *10*, 707–719.

22. Tsai, Y.H.; Wu, M.F.; Wu, Y.H.; Chang, S.J.; Lin, S.F.; Sharp, T.V.; Wang, H.W. The M type K15 protein of Kaposi's sarcoma-associated herpesvirus regulates microRNA expression via its SH2-binding motif to induce cell migration and invasion. *J. Virol.* **2009**, *83*, 622–632.

23. Wu, Y.H.; Hu, T.F.; Chen, Y.C.; Tsai, Y.N.; Tsai, Y.H.; Cheng, C.C.; Wang, H.W. The manipulation of miRNA-gene regulatory networks by KSHV induces endothelial cell motility. *Blood* **2011**, *118*, 2896–2905.

24. Phng, L.K.; Potente, M.; Leslie, J.D.; Babbage, J.; Nyqvist, D.; Lobov, I.; Ondr, J.K.; Rao, S.; Lang, R.A.; Thurston, G.; *et al.* Nrarp coordinates endothelial Notch and Wnt signaling to control vessel density in angiogenesis. *Dev. Cell* **2009**, *16*, 70–82.

25. Bridge, G.; Monteiro, R.; Henderson, S.; Emuss, V.; Lagos, D.; Georgopoulou, D.; Patient, R.; Boshoff, C. The microRNA-30 family targets DLL4 to modulate endothelial cell behavior during angiogenesis. *Blood* **2012**, *120*, 5063–5072.

26. Sekar, D.; Hairul Islam, V.I.; Thirugnanasambantham, K.; Saravanan, S. Relevance of miR-21 in HIV and non-HIV-related lymphomas. *Tumour Biol.* **2014**, *35*, 8387–8393.

27. Laurila, E.M.; Kallioniemi, A. The diverse role of miR-31 in regulating cancer associated phenotypes. *Genes Chromosomes Cancer* **2013**, *52*, 1103–1113.

28. Chen, W.X.; Hu, Q.; Qiu, M.T.; Zhong, S.L.; Xu, J.J.; Tang, J.H.; Zhao, J.H. miR-221/222: Promising biomarkers for breast cancer. *Tumour Biol.* **2013**, *34*, 1361–1370.

29. Garofalo, M.; Quintavalle, C.; Romano, G.; Croce, C.M.; Condorelli, G. miR221/222 in cancer: Their role in tumor progression and response to therapy. *Curr. Mol. Med.* **2012**, *12*, 27–33.

30. Fuziwara, C.S.; Kimura, E.T. MicroRNA Deregulation in Anaplastic Thyroid Cancer Biology. *Int. J. Endocrinol.* **2014**, *2014*, 743450.

31. Schulz, T.F. The pleiotropic effects of Kaposi's sarcoma herpesvirus. *J. Pathol.* **2006**, *208*, 187–198.

32. Lin, X.; Liang, D.; He, Z.; Deng, Q.; Robertson, E.S.; Lan, K. miR-K12–7-5p encoded by Kaposi's sarcoma-associated herpesvirus stabilizes the latent state by targeting viral ORF50/RTA. *PLoS One* **2011**, *6*, e16224.

33. Bellare, P.; Ganem, D. Regulation of KSHV lytic switch protein expression by a virus-encoded microRNA: An evolutionary adaptation that fine-tunes lytic reactivation. *Cell Host Microbe* **2009**, *6*, 570–575.

34. Lei, X.; Bai, Z.; Ye, F.; Xie, J.; Kim, C.G.; Huang, Y.; Gao, S.J. Regulation of NF-kappaB inhibitor IkappaBalpha and viral replication by a KSHV microRNA. *Nat. Cell. Biol.* **2010**, *12*, 193–199.

35. Lu, C.C.; Li, Z.; Chu, C.Y.; Feng, J.; Sun, R.; Rana, T.M. MicroRNAs encoded by Kaposi's sarcoma-associated herpesvirus regulate viral life cycle. *EMBO Rep.* **2010**, *11*, 784–790.

36. Lu, F.; Stedman, W.; Yousef, M.; Renne, R.; Lieberman, P.M. Epigenetic regulation of Kaposi's sarcoma-associated herpesvirus latency by virus-encoded microRNAs that target Rta and the cellular Rbl2-DNMT pathway. *J. Virol.* **2010**, *84*, 2697–2706.

37. Liang, D.; Gao, Y.; Lin, X.; He, Z.; Zhao, Q.; Deng, Q.; Lan, K. A human herpesvirus miRNA attenuates interferon signaling and contributes to maintenance of viral latency by targeting IKKepsilon. *Cell Res.* **2011**, *21*, 793–806.

38. Ziegelbauer, J.M.; Sullivan, C.S.; Ganem, D. Tandem array-based expression screens identify host mRNA targets of virus-encoded microRNAs. *Nat. Genet.* **2009**, *41*, 130–134.

39. Yan, Q.; Ma, X.; Shen, C.; Cao, X.; Feng, N.; Qin, D.; Zeng, Y.; Zhu, J.; Gao, S.J.; Lu, C. Inhibition of Kaposi's sarcoma-associated herpesvirus lytic replication by HIV-1 Nef and cellular microRNA hsa-miR-1258. *J. Virol.* **2014**, *88*, 4987–5000.

40. Lee, H.R.; Lee, S.; Chaudhary, P.M.; Gill, P.; Jung, J.U. Immune evasion by Kaposi's sarcoma-associated herpesvirus. *Future Microbiol.* **2010**, *5*, 1349–1365.

41. Punj, V.; Matta, H.; Schamus, S.; Tamewitz, A.; Anyang, B.; Chaudhary, P.M. Kaposi's sarcoma-associated herpesvirus-encoded viral FLICE inhibitory protein (vFLIP) K13 suppresses CXCR4 expression by upregulating miR-146a. *Oncogene* **2010**, *29*, 1835–1844.

42. Aoki, Y.; Yarchoan, R.; Wyvill, K.; Okamoto, S.; Little, R.F.; Tosato, G. Detection of viral interleukin-6 in Kaposi sarcoma-associated herpesvirus-linked disorders. *Blood* **2001**, *97*, 2173–2176.

43. Jones, K.D.; Aoki, Y.; Chang, Y.; Moore, P.S.; Yarchoan, R.; Tosato, G. Involvement of interleukin-10 (IL-10) and viral IL-6 in the spontaneous growth of Kaposi's sarcoma herpesvirus-associated infected primary effusion lymphoma cells. *Blood* **1999**, *94*, 2871–2879.

44. Nicholas, J.; Ruvolo, V.R.; Burns, W.H.; Sandford, G.; Wan, X.; Ciufo, D.; Hendrickson, S.B.; Guo, H.G.; Hayward, G.S.; Reitz, M.S. Kaposi's sarcoma-associated human herpesvirus-8 encodes homologues of macrophage inflammatory protein-1 and interleukin-6. *Nat. Med.* **1997**, *3*, 287–292.

45. Uldrick, T.S.; Wang, V.; O'Mahony, D.; Aleman, K.; Wyvill, K.M.; Marshall, V.; Steinberg, S.M.; Pittaluga, S.; Maric, I.; Whitby, D.; *et al.* An interleukin-6-related systemic inflammatory syndrome in patients co-infected with Kaposi sarcoma-associated herpesvirus and HIV but without Multicentric Castleman disease. *Clin. Infect. Dis.* **2010**, *51*, 350–358.

46. Kang, J.G.; Majerciak, V.; Uldrick, T.S.; Wang, X.; Kruhlak, M.; Yarchoan, R.; Zheng, Z.M. Kaposi's sarcoma-associated herpesviral IL-6 and human IL-6 open reading frames contain miRNA binding sites and are subject to cellular miRNA regulation. *J. Pathol.* **2011**, *225*, 378–389.

47. Kang, J.G.; Pripuzova, N.; Majerciak, V.; Kruhlak, M.; Le, S.Y.; Zheng, Z.M. Kaposi's sarcoma-associated herpesvirus ORF57 promotes escape of viral and human interleukin-6 from microRNA-mediated suppression. *J. Virol.* **2011**, *85*, 2620–2630.

48. Lagos, D.; Pollara, G.; Henderson, S.; Gratrix, F.; Fabani, M.; Milne, R.S.; Gotch, F.; Boshoff, C. miR-132 regulates antiviral innate immunity through suppression of the p300 transcriptional co-activator. *Nat. Cell Biol.* **2010**, *12*, 513–519.

49. Wu, X.J.; Pu, X.M.; Zhao, Z.F.; Zhao, Y.N.; Kang, X.J.; Wu, W.D.; Zou, Y.M.; Wu, C.Y.; Qu, Y.Y.; Zhang, D.Z.; *et al.* The expression profiles of microRNAs in Kaposi's sarcoma. *Tumour Biol.* **2014**, doi:10.1007/s13277-014-2626-1.

50. Catrina Ene, A.M.; Borze, I.; Guled, M.; Costache, M.; Leen, G.; Sajin, M.; Ionica, E.; Chitu, A.; Knuutila, S. MicroRNA expression profiles in Kaposi's sarcoma. *Pathol. Oncol. Res.* **2014**, *20*, 153–159.

51. O'Hara, A.J.; Vahrson, W.; Dittmer, D.P. Gene alteration and precursor and mature microRNA transcription changes contribute to the miRNA signature of primary effusion lymphoma. *Blood* **2008**, *111*, 2347–2353.

52. O'Hara, A.J.; Wang, L.; Dezube, B.J.; Harrington, W.J., Jr.; Damania, B.; Dittmer, D.P. Tumor suppressor microRNAs are underrepresented in primary effusion lymphoma and Kaposi sarcoma. *Blood* **2009**, *113*, 5938–5941.

53. Moltzahn, F.; Olshen, A.B.; Baehner, L.; Peek, A.; Fong, L.; Stoppler, H.; Simko, J.; Hilton, J.F.; Carroll, P.; Blelloch, R. Microfluidic-based multiplex qRT-PCR identifies diagnostic and prognostic microRNA signatures in the sera of prostate cancer patients. *Cancer Res.* **2011**, *71*, 550–560.

54. Huang, Z.; Huang, D.; Ni, S.; Peng, Z.; Sheng, W.; Du, X. Plasma microRNAs are promising novel biomarkers for early detection of colorectal cancer. *Int. J. Cancer* **2010**, *127*, 118–126.

55. Mitchell, P.S.; Parkin, R.K.; Kroh, E.M.; Fritz, B.R.; Wyman, S.K.; Pogosova-Agadjanyan, E.L.; Peterson, A.; Noteboom, J.; O'Briant, K.C.; Allen, A.; *et al.* Circulating microRNAs as stable blood-based markers for cancer detection. *Proc. Natl. Acad. Sci. USA* **2008**, *105*, 10513–10518.

56. Chugh, P.E.; Sin, S.H.; Ozgur, S.; Henry, D.H.; Menezes, P.; Griffith, J.; Eron, J.J.; Damania, B.; Dittmer, D.P. Systemically circulating viral and tumor-derived microRNAs in KSHV-associated malignancies. *PLoS Pathog.* **2013**, *9*, e1003484.

57. Lin, X.; Li, X.; Liang, D.; Lan, K. MicroRNAs and unusual small RNAs discovered in Kaposi's sarcoma-associated herpesvirus virions. *J. Virol.* **2012**, *86*, 12717–12730.

58. Acunzo, M.; Romano, G.; Wernicke, D.; Croce, C.M. MicroRNA and cancer—A brief overview. *Adv. Biol. Regul.* **2014**, doi:10.1016/j.jbior.2014.09.013.

MDPI AG
Klybeckstrasse 64
4057 Basel, Switzerland
Tel. +41 61 683 77 34
Fax +41 61 302 89 18
http://www.mdpi.com/

Viruses Editorial Office
E-mail: viruses@mdpi.com
http://www.mdpi.com/journal/viruses

www.ingramcontent.com/pod-product-compliance
Lightning Source LLC
Chambersburg PA
CBHW051926190326

41458CB00026B/6420